Nuclear Magnetic Resonance in Agriculture

Editors

Philip E. Pfeffer, Ph.D.
Walter V. Gerasimowicz, Ph.D.

Lead Scientists
Plant and Soil Biophysics Laboratory
Agricultural Research Service
Eastern Regional Research Center
United States Department of Agriculture
Philadelphia, Pennsylvania

CRC Press, Inc.
Boca Raton, Florida

Library of Congress Cataloging-in-Publication Data

Nuclear magnetic resonance in agriculture / editors, Philip E. Pfeffer, Walter V. Gerasimowicz.
p. cm.
Includes index.
ISBN 0-8493-6864-2
1. Agriculture—Research—Technique. Nuclear magnetic resonance spectroscopy. I. Pfeffer, Philip E., 1941- . II. Gerasimowicz, Walter V., 1952- .
S540.N82N83 1989
631′.072—dc19

88-27454
CIP

Direct all inquiries to CRC Press, Inc., 2000 Corporate Blvd., N.W., Boca Raton, Florida, 33431.

International Standard Book Number 0-8493-6864-2

Library of Congress Number 88-27454
Printed in the United States

PREFACE

This book has been assembled for those scientists working in the area of agricultural research. It is intended to define the most important roles NMR spectroscopy is playing in pioneering new avenues of investigation in agriculture.

We have purposely not focused our attention on any one particular area of agricultural research or any specific NMR methodology to emphasize the enormous range of experiments, be it high-resolution solution spectroscopy, 2D spectroscopy, solid-state spectroscopy, *in vivo,* or imaging that can be tailored to solve a specific problem.

Some chapters also provide the reader with a general introduction into the NMR methodology used. Others concentrate on specific applications of these experimental techniques for investigating and answering a variety of complex questions. Some of the NMR techniques which are described are applicable to the study of biological systems and may not be considered unique to the study of agricultural systems. However, some methods, such as solid-state NMR, are particularly well suited for evaluating the kind of complex, ill-defined, amorphous matrices one encounters when examining soils, intact plant cell wall polymers, plant structural polymer composites, as well as protein mixtures and foods.

The methodology of *in vivo* NMR, both solution and solid state, is beginning to have a major impact on the area of plant tissues and cultured plant cell metabolism associated with mineral stress, ion transport, protein turnover, pesticide metabolism, and hypoxia. These investigations are especially important because of the mechanistic implications associated with their findings. Thus, it is now possible to equate specific molecular level responses directly with controlled perturbations and thereby observe how the metabolic machinery of a system responds. This approach can be powerful, for example, in support of the discipline of genetic engineering in which new functioning systems need to be analyzed in terms of the expression of a functional response.

High-resolution NMR still serves an important role in the area of structure identification. For this purpose, 2D methods have become particularly powerful and sophisticated in unraveling the structure of unknown compounds as well as defining their molecular mobilities and interactions in an active biological state. Broad-banded instruments now allow us to examine nuclei such as ^{1}H, ^{17}O, ^{31}P, ^{15}N, ^{35}Cl, ^{2}H, ^{19}F, ^{13}C, and ^{11}B making most molecules accessible to a wide range of investigations.

The area of magnetic resonance imaging (MRI) is covered in a single last chapter. Although there are not many agricultural applications that presently make use of this methodology, many are on the horizon. MRI coupled with spectroscopy promises to be the most powerful magnetic resonance tool of all. It is not the purpose of this book to provide an exhaustive compilation of all the literature available concerning the application of NMR in agriculture, but to selectively highlight the most important developments of today. We hope that this collection can provide the reader with current, basic information, and updated bibliographic references which can serve as the ultimate source of the state-of-the-art.

The editors would like to acknowledge the support of their respective families during the production of this volume. Dr. Pfeffer would especially like to dedicate his contribution to his wife Judy for a year of understanding, support, patience, and many lost weekends.

THE EDITORS

Philip E. Pfeffer, Ph.D., is a lead scientist in the Plant and Soil Biophysics Research Unit at the Eastern Regional Research Center of the U.S. Department of Agriculture in Philadelphia.

Dr. Pfeffer received his A.B. degree from Hunter College of the City University of New York in 1962 and his M.S. and Ph.D. degrees in 1964 and 1966, respectively, from the Department of Chemistry, Rutgers University, New Brunswick, NJ. After 2 years of postdoctoral training in NMR spectroscopy at the University of Chicago in 1968, he joined the Eastern Regional Research Center in Philadelphia.

Dr. Pfeffer is a member of the American Chemical Society, American Association for the Advancement of Science, Society for Applied Spectroscopy, International Society of Magnetic Resonance, and the honorary Society Sigma Xi.

Among other awards, he has received the Philadelphia Section Award (1982) from the American Chemical Society, U.S. Department of Agriculture Science and Education Award (1982), the Eastern Regional Research Center Distinguished Research Scientist Award (1983), and was the USDA North Atlantic Area Scientist of the Year (1986). He was also the recipient of a 6-month visiting scientist grant from the Centre d'Etudes Nucleaaires de Grenoble in 1986.

Dr. Pfeffer has presented over 30 invited lectures at national and international meetings, universities, and institutes. He has published more than 100 research papers and is the author of 8 book chapters and co-editor of a monograph on solid-state NMR spectroscpy. He is a member of the editorial board of the *Journal of Carbohydrate Chemistry*. His current major research interests include *in vivo* NMR studies of ion transport and symbiosis in plant tissues and cells.

Walter V. Gerasimowicz is a Supervisory Lead Scientist in the Plant and Soil Biophysics Laboratory at the Eastern Regional Research Center, U.S. Department of Agriculture. Dr. Gerasimowicz received his B.S. in chemistry from Ursinus College in 1974, and his M.S. and Ph.D. degrees in physical chemistry from Villanova University in 1977 and 1981, respectively. After 2 years as an ARS postdoctoral research associate (1981 to 1983), he joined the staff of the USDA as a research chemist.

Currently, Dr. Gerasimowicz also holds an appointment as an adjunct faculty member and teaches at Villanova University. He serves on the Editorial Advisory board of the *Microchemical Journal*. He is a member of the American Chemical Society, American Association for the Advancement of Science, Society for Applied Spectroscopy, and the honorary societies Phi Kappa Phi and Sigma Xi. During 1987 and 1988 he was a visiting scientist at the Naval Research Laboratory in Washington, D.C. Dr. Gerasimowicz has authored or coauthored over 50 publications and abstracts, and has presented numerous invited lectures at national and international meetings, and at universities. His current major research efforts are focused upon multiple-quantum and solid-state NMR studies of complex materials and polymeric systems.

CONTRIBUTORS

Ion C. Baianu, Ph.D., Dr. Physics
Associate Professor
Department of Food Science
Physical Chemistry and NMR Laboratory
College of Agriculture
University of Illinois at Urbana
Urbana, Illinois

Richard Bligny, Ph.D.
Research Director
Laboratoire de Physiologie Cellulaire
Végétale, CENG
Grenoble, France

George Coker III, Ph.D.
Department of Anatomy
Washington University Medical School
St. Louis, Missouri

Roland Douce, Ph.D.
Professor
Department of Biology
Grenoble University
Grenoble, France

James S. Frye, Ph.D.
Department of Chemistry
Colorado State University
Fort Collins, Colorado

George Gassner, Ph.D.
Research Leader
Insect Biochemistry and Molecular Genetics
USDA/ARS/Bioscience
Research Laboratory
Fargo, North Dakota

Walter V. Gerasimowicz, Ph.D.
Lead Scientist
Plant and Soil Biophysics Laboratory
Eastern Regional Research Center
Agricultural Research Service
U.S. Department of Agriculture
Philadelphia, Pennsylvania

James F. Haw, Ph.D.
Department of Chemistry
Texas A&M University
College Station, Texas

Fumitaka Horii, Ph.D.
Professor
Institute for Chemical Research
Kyoto University
Uji, Kyoto, Japan

Peter L. Irwin, Ph.D.
Research Plant Physiologist
Eastern Regional Research Center
U.S. Department of Agriculture
Philadelphia, Pennsylvania

Thomas F. Kumosinski, Ph.D.
Lead Scientist
Macromolecular and Cell Structure
Eastern Regional Research Center
U.S. Department of Agriculture
Philadelphia, Pennsylvania

Gary E. Maciel, Ph.D.
Professor
Department of Chemistry
Colorado State University
Fort Collins, Colorado

Helmut Pessen, Ph.D.
Research Scientist
Eastern Regional Research Center
Agricultural Research Service
U.S. Department of Agriculture
Philadelphia, Pennsylvania

Philip E. Pfeffer, Ph.D.
Lead Scientist
Plant and Soil Biophysics Laboratory
Eastern Regional Research Center
Agricultural Research Service
U.S. Department of Agriculture
Philadelphia, Pennsylvania

Justin K. M. Roberts, Ph.D.
Department of Biochemistry
University of California
Riverside, California

Claude Roby, Ph.D.
RMBM/SPH/DRF
Centre de'Etudes Nucleaires
Grenoble, France

Venceslav Rutar, Ph.D.
Department of Chemistry
Iowa State University
Ames, Iowa

Jacob Schaefer, Ph.D.
Professor
Department of Chemistry
Washington University
St. Louis, Missouri

David Weisleder, Ph.D.
USDA/ARS/NRRC
Peoria, Illinois

TABLE OF CONTENTS

In Vivo NMR

Chapter 1

INTRODUCTION TO HIGH-RESOLUTION NMR SPECTROSCOPY AND ITS APPLICATION TO *IN VIVO* STUDIES OF AGRICULTURAL SYSTEMS

Philip E. Pfeffer and Walter V. Gerasimowicz

TABLE OF CONTENTS

I. INTRODUCTION

Nuclear magnetic resonance (NMR) spectroscopy is a most powerful and versatile technique that continues to be applied to ever-increasing disciplines including physics, chemistry, biochemistry, physiology, and medicine. The development of modern high-speed computers coupled with the introduction of Fourier transform (FT) NMR methodology and high-field superconducting magnets has significantly increased the speed, sensitivity, and flexibility of this instrumentation. In addition, because NMR is nondestructive and noninvasive by design, it has developed into the method of choice for examining the molecular level chemistry of functioning biological systems.

Beginning with the early studies of Moon and Richards[1] in 1973, which demonstrated the correlation between intracellular pH in red blood cells and ^{31}P NMR chemical shifts, *in vivo* NMR has rapidly expanded its application to the elucidation of biochemical pathways and bioenergetics of microorganisms,[2] cells,[3] plant tissues,[4] perfused organs,[5] and organs in intact animals.[6] Although these NMR experiments can measure the rates at which various reactions, e.g., those of the glycolytic pathway, take place *in vivo,* we would also like to understand how these rates are being controlled by the cells to provide the required fuel on demand. For this reason it is important to establish the level of metabolic activity in fully functioning systems by assessing simultaneously the *in vivo* concentration of metabolites such as ATP, ADP, etc. that play a key role in regulating the enzyme pools. Consequently, if we are to know the roles of these metabolites it is particularly important to relate their change in concentration to change in metabolic activity.

Little is known about the relationship between metabolism and physiological activity largely because of the experimental difficulties of examining a living system. NMR can give us a window for measuring metabolite levels, fluxes through metabolic pathways, rates of transport, or reactions under steady-state conditions and at the same time allow us to assess intracellular pH and compartmentation. Such an integrated approach can potentially create a bridge between how physiological and metabolic processes are controlled and integrated in a fully functioning system.

Although most of the early work has concentrated primarily on establishing the characteristic criteria associated with metabolic stability in living systems or verifying well-known biochemical pathways, a recent scrutiny of the literature indicates that the emerging work has begun to uncover novel findings that are making important contributions to our understanding of diverse biological systems in metabolically stressed and diseased states.[7-9]

In this chapter we will give a short, general description of the NMR methodology with emphasis on the basic resonance phenomena, design of the instrumentation, and instrumental parameters that influence the experimental results. This nonmathematical, descriptive approach is designed to acquaint the neophyte with some of the elementary principals of NMR. For a thorough and comprehensive treatment of modern high-resolution NMR spectroscopy, the reader is advised to consult three excellent texts by Becker,[10] Shaw,[11] and Sanders and Hunter.[12]

The subsequent section of this chapter will give an overview of the application of *in vivo* NMR and a description of the kind of information that is obtained from such experiments utilizing a variety of nuclei and methodologies. The areas of investigation will be largely restricted to studies of microorganisms as well as plant and some animal tissues. Examples will be taken from work on agricultural systems whenever possible. Adaptation of the methods presented for characterizing nonagricultural systems should be obvious to the reader. ^{31}P will be the nucleus of discussion for the majority of this portion although ^{13}C, ^{14}N, ^{15}N, ^{1}H, ^{19}F, ^{23}Na, and ^{39}K will also be considered to a minor extent.

II. GENERAL DESCRIPTION OF NMR METHODOLOGY

In general, spectroscopy deals with the interaction of electromagnetic energy (radiation) with

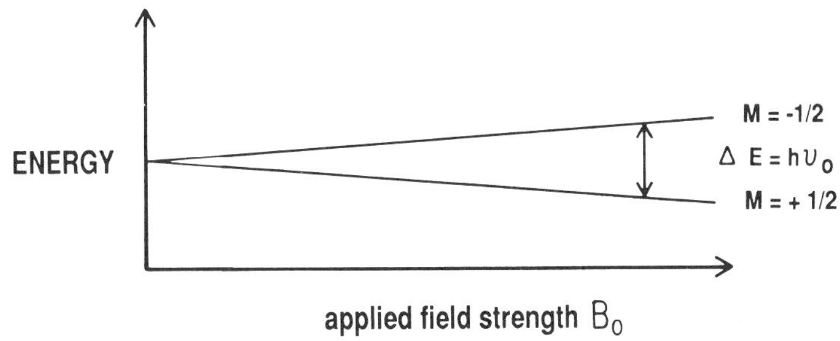

FIGURE 1. Energy of two orientations of a nucleus of spin one half as a function of magnetic field strength B_0.

matter. The radiation that is absorbed or emitted by the material is then detected as a function of wavelength or frequency and is defined as the spectrum. This spectrum is then interpreted with regard to the atomic or molecular structure of the material under study. The frequencies associated with NMR are typically in the range of 100 to 600 MHz and are therefore associated with transitions between energy levels that are relatively closely spaced. These levels correspond to different magnetic states of the nuclei and are directly influenced by the strength of the applied magnetic field according to the equation (see Figure 1)

$$\Delta E = \hbar \gamma B_0 \tag{1}$$

where: ΔE = the difference in energy between the two states, B_0 = applied magnetic field, γ = magnetogyric ratio characteristic of each nucleus, and $\hbar$ = Plank's constant.

In the NMR experiment the observation of a signal relies on the fact that the atomic nucleus with an odd atomic mass (i.e., sum to its neutrons and protons) has a net magnetic moment or spin, when placed in a magnetic field. This moment essentially turns the nucleus into a magnetic dipole or bar magnet. Unfortunately, these nuclear magnets interact very weakly with the applied field B_0 and separate into the two energy levels according to the Boltzman distribution (population excess in the higher state of <1 part in 10^4). This small magnetic interaction accounts for the small energy separation ΔE and low frequency range of NMR (see Figure 1). In addition, the small population excess is also responsible for the inherent low signal intensity obtained which is characteristic of this methodology. Common nuclei that exhibit these magnetic properties are the highly abundant (100%) isotopes of ^{1}H, ^{31}P, and ^{19}F or low natural abundance isotopes ^{13}C (1.1%), ^{15}N (.37%), and ^{113}Cd (12.3%). These are the most widely utilized nuclei since they have only two quantized orientations (nuclear spin = 1/2, quantum number I = 1/2) in the magnetic field which yield narrow line spectra. In contrast, those nuclei with higher quantum numbers such as the quadrupolar nuclei ^{23}Na, ^{14}N, ^{2}H with I = 3/2, 1, and 1, respectively, have many more orientations and correspondingly more complex and broadened spectral lineshapes and multiple transitions. Those nuclei with no net spin, i.e., the sum of their neutrons and protons = even number such as ^{16}O or ^{12}C, do not produce any resonance signal.

Two fields need to be used in order to observe an NMR spectrum. First, a strong static magnetic field designated B_0 is applied to the sample. This causes the nuclear dipoles to orient themselves with and against the magnetic field, i.e., into lower and higher energy states, respectively, as mentioned above (Figure 1). A second field, B_1, is then generated in a coil perpendicular to the B_1 field via electromagnetic radiation in the radio frequency region of the spectrum. At every magnetic field B_0, there is a specific frequency of electromagnetic radiation (Larmor frequency) that can induce an applied field B_1. This B_1 field when applied can align or energetically flip the particular nucleus by way of the absorption of a discreet quantum of energy to a higher energy state. Thus, if a homogeneous B_0 field (superconducting magnetic field) is maintained, we can vary the radio-frequency rf-generated field B_1 until the characteristic Larmor frequency of the nucleus under study is matched. The characteristic Larmor frequencies of each nucleus is determined by its individual shielding properties (magnetogyric ratio). For example, a static B_0 field of 9.3Tesla (T) requires a B_1-generated radio frequency or Larmor frequency, for example, of approximately 400 MHz to produce a resonance spectrum for ^1H, 100 MHz for ^{13}C, and 162 MHz for ^{31}P.

In addition to the static B_0 and applied B_1 field, there is a third small localized field, $B_0\sigma$, present around the nuclei which is produced by the electronic currents of the atoms and molecules. If it were not for $B_0\sigma$, where σ is designated as a shielding constant, all nuclei associated with a particular Larmor frequency, e.g., ^{13}C, would absorb energy at exactly the same frequency. However, because of σ, each nucleus within a molecule feels a slightly different effective magnetic field due to its chemical environment and absorbs energy at a slightly different frequency or resonance position. The separation of these resonance frequencies from an arbitrarily chosen reference is called the chemical shift (δ), a unitless number expressed in parts per million which is independent of field. In the spectrum each absorption is given a chemical shift value based on the relationship:

$$\delta_{ppm} = \frac{\upsilon_S - \upsilon_R}{\upsilon_R} \times 10^6 \tag{2}$$

where υ_S = resonance frequency of the sample peak and υ_R = resonance frequency of the reference peak.

For example, a chemical shift of 1 ppm observed for an ^1H $C\underline{H}_3$ resonance relative to some internal reference peak obtained with an NMR spectrometer operating at 9.3T would also resonate at a 1-ppm chemical shift position in a spectrum taken at 2.3T. However, the difference in the two spectra would be that at 9.3T the separation of the $C\underline{H}_3$ resonance from the reference would be 400 Hz, which corresponds to 1 ppm at this field while at a 2.3T field the separation would be only 100 Hz. Thus, higher static magnetic fields (B_0) yield greater chemical shift dispersion (resolution) in addition to increased sensitivity.

Tetramethylsilane (TMS) is commonly used as a reference compound (chemical shift $\delta = 0.0$) for ^{13}C and ^1H spectra. Phosphoric acid (85%) is the most widely used reference for ^{31}P spectra, while others such as methylene diphosphonate (MDP),[13] hexamethylphosphoramide (HMPA),[14] or hexachlorocyclotriphosphazene (HCCTP)[15] have also been used as either primary[14] or secondary[16] external references. Figure 2 shows the ^{31}P spectrum obtained from a perchloric acid extract of metabolites from soybean nodules. The chemical shift positions of the scale have been measured relative to an external reference peak of HMPA, whose shift has been established relative to 85% H_3PO_4 a primary reference whose ^{31}P resonance is assigned a value of 0.0 ppm. For most *in vitro* studies it is convenient to dissolve the reference compound in the solution except for those cases in which the reference compound may react with those compounds under study. However, for *in vivo* work, one must use an external standard such as a capillary or spherical cell containing the reference compound placed in the NMR tube. This kind of referencing sometimes presents problems in comparing chemical shifts because of the differences in susceptibility or chemical environment of the compartments, i.e., capillary vs. cells.[17]

The observed intensities (areas) of the resonances in the spectra are proportional to the number of nuclei they represent. However, as we will describe later, certain factors such as spin-lattice relaxation times, T_1, and nuclear Overhauser enhancements can distort these intensities. In addition each resonance may be split or coupled into multiple lines as a result of its interaction with neighboring nuclei. This interaction, which is transmitted through bonds, is independent of magnetic field strength and is known as spin-spin scalar coupling. The splittings are often useful for identifying resonances of spin 1/2 nuclei because the n + 1 multiplicity of the splitting

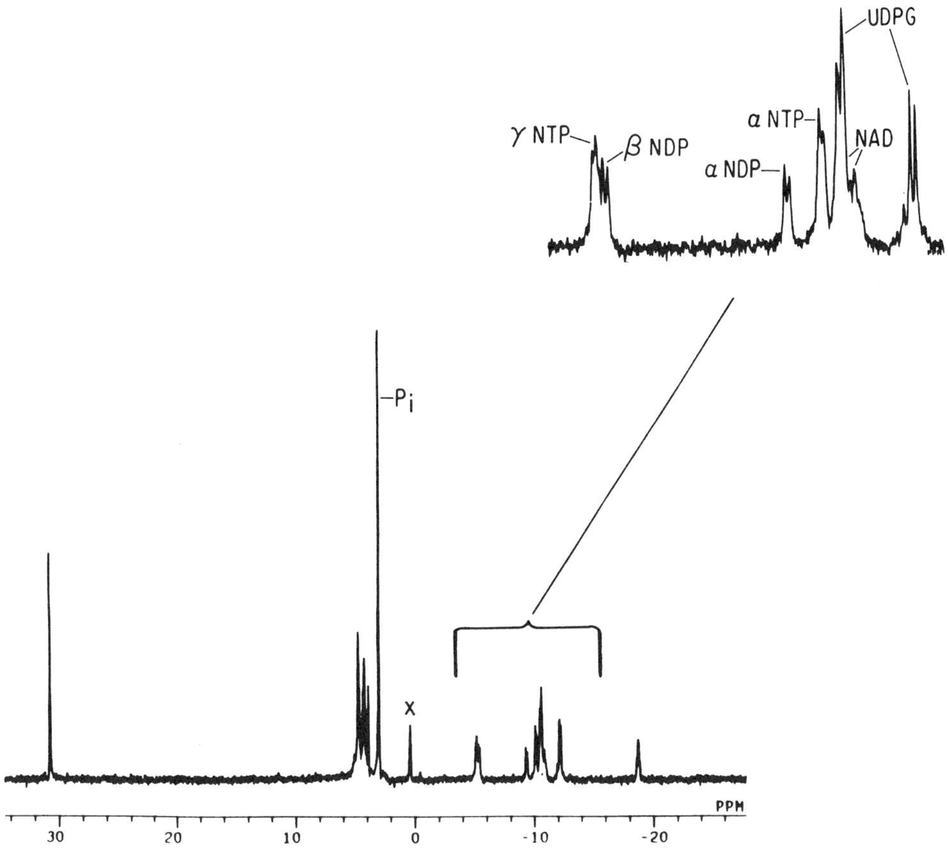

FIGURE 2. 161.7-MHz ^{31}P spectrum of phosphorus metabolites extracted with perchloric acid from soybean nodules. The HMPA reference peak at 30.73 ppm is assigned relative to 85% H_3PO_4 which is the primary reference peak assigned a value of 0.00 ppm.

pattern gives information about the number of adjacent bonded nuclei, and the magnitude of the coupling can be correlated with the dihedral angle between them. The insert in Figure 2 shows the mutual 16-Hz ^{31}P-^{31}P doublet, homonuclear splitting in each of γ- and α-nucleotide triphosphate (NTP) and α- and β-nucleotide diphosphate (NDP) ^{31}P resonances due to the two (+1/2, −1/2) ^{31}P orientations of each nucleus in the magnetic field. The area of each resonance represents the molar concentration of the corresponding compound. The shift positions are reported relative to the external standard (HMPA) at 30.73 ppm (0.0 ppm being assigned to the H_3PO_4 resonance).

The line shapes of the NMR signals are largely determined by the spin-spin relaxation time T_2, since the natural line width $\delta_{1/2}$ (width at half height) is given by:

$$1/T_2 = \pi\Delta\delta_{1/2} \quad (3)$$

In the case of large or immobilized molecules, broader line widths are observed and in principle these line widths can be used to assess molecular motion. However, one must be cautious in this interpretation since other factors such as chemical exchange and field inhomogeneity can also contribute to line widths. For the most part, high-resolution NMR is only used to study soluble molecules which are relatively mobile and whose correlation times, T_c, (a characteristic time scale of molecular motion or tumbling) are in the range of 10^{-8} to 10^{-12} s.

Early continuous wave NMR spectrometers used constant, continuously applied radio frequency power in the presence of a slowly swept (because of spin population saturation) variable magnetic field. Because single slow field sweep was utilized, (although computer time averaging was later applied) only very highly concentrated samples could be examined. Each resonance signal was generated independently as the field was incremented. Modern spectrometers utilize high-power rf pulses and Fourier transformation to give much greater sensitivity. In the Fourier transform instruments, very strong and homogeneous B_0 fields (4.6 to 13.8T, 100 to 600 MHz) are produced by superconducting magnets and each spectrum is obtained by applying a short, high-power rf pulse to produce a B_1 radio frequency field. This pulse, which may only last for 15 to 25 μs, can simultaneously excite the full spectrum of frequencies for each nucleus being examined (Figure 3A and B). The response obtained from this pulse, referred to as a time domain spectrum or free induction decay (FID), is stored in the computer memory and subsequently mathematically transformed into the conventional frequency domain (frequency vs. intensity) spectrum (Figure 4). In order to improve the signal/noise (S/N) ratio, the accumulated time domain spectra of many of these pulsed experiments are added together in the computer, and following the completion of the experiment they are retrieved and treated by Fourier transformation. The improvement in S/N for the resulting spectrum is proportional to the $\sqrt{\#\text{scans}}$ because the signal increases by a factor of the number of scans while the noise being random increases by the $\sqrt{\#\text{scans}}$. Thus, if a 10-fold increase in S/N is needed to give an acceptable spectrum, 100-fold more scans (pulsed spectra) should be required.

The sample is placed in the instrument probe, situated in the strong B_0 field (Figure 3A), and the transmitter sends radiation of the appropriate frequency through this compartment. Following the transmission of a specific amount of rf radiation corresponding to a 90° pulse applied in the xy plane, the magnetization is tipped 90° from its vertical Z axis alignment in the B_0 field to the xy plane (Figure 3B). The net magnetization (M_{xy}) then rotates coherently about, in the xy plane at the Larmor frequency ω_0, and induces an electromagnetic force (emf) in the coil surrounding the sample. The rf-generated B_1 field is turned off and the nuclei undergo relaxation in the xy plane by T_2 ($e^{-t/T}2$), and relaxation in the Z direction by T_1 ($e^{-t/T}1$), (Figure 3C). The FID which results from the rf pulse corresponds to the T_2 decay of the induced signal, generated from the free precession of the nuclei in the B_0 field. This response is amplified and processed in the receiver of the spectrometer and stored in the computer. The same coil is normally used for transmitting the B_1 field as well as for detecting the resulting signal. Figure 5 shows a simple schematic representation of the basic FT NMR spectrometer.

In the NMR experiment, the value of the required interval between successive rf pulses is dependent on the magnitude of the spin-lattice relaxation time, T_1, the time constant associated with the reestablishment of the equilibrium magnetization state in the Z direction (Figure 3A) according to the equation,

$$\frac{dM_z}{dt} = \frac{M_0 - M_z}{T_1} \tag{4}$$

where M_0 = the component of magnetization along the Z axis of equilibrium and M_z = the component of magnetization produced after a 90° rf pulse. To reestablish >99% of the initial magnetization along the Z axis, prior to the next successive pulse, requires an interval equal to $5 \times T_1$.

Spin-lattice relaxation is facilitated by dipole-dipole interactions of neighboring nuclei randomly tumbling in solution. This process is most efficient when the reciprocal of the molecular correlation time, T_c (see above), is equal to the resonance or Larmor frequency. At the same time magnetization in also decaying exponentially or spreading out in the xy plane (Figure 3C) with a time constant, T_2, according to the relationship:

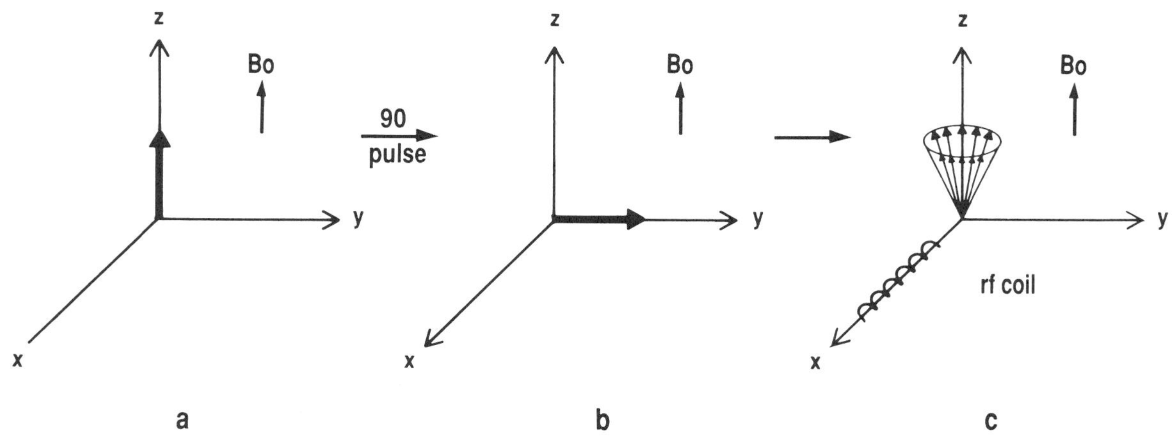

FIGURE 3. (a) Alignment of excess nuclei (Boltzman distribution) in the magnetic field prior to the rf pulse; (b) following a specific pulse of rf power the nuclei are typed 90° in the xy plane; and (c) the B_1 field is turned off and the nuclei reestablish their orientation in the B_0 field via T_1 relaxation and dephase in the xy plane by the T_2 mechanism.

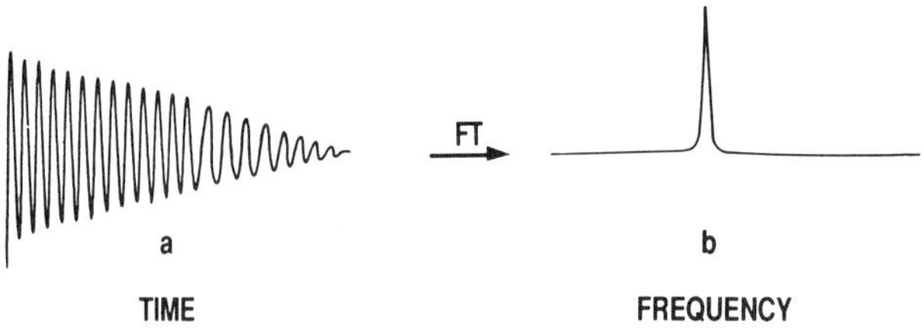

FIGURE 4. (a) Time domain response following a 90° rf pulse and (b) frequency vs. intensity spectrum resulting from Fourier transformation of a.

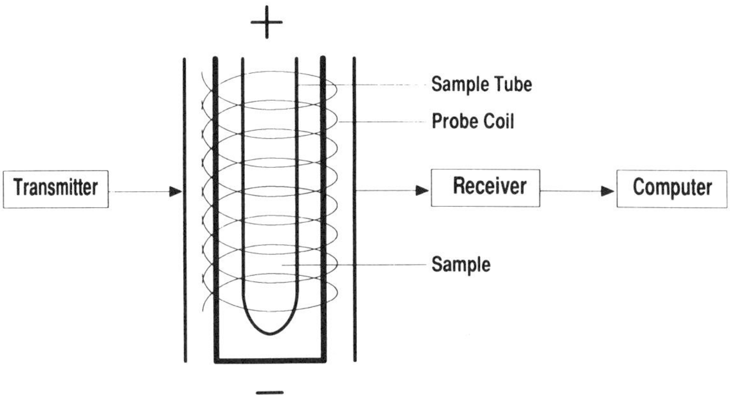

FIGURE 5. A simple schematic representation of an NMR spectrometer.

$$\frac{dM_{xy}}{dt} - \frac{M_{xy}}{T_2} \qquad (5)$$

where M_{xy} is the net magnetization in the xy plane.* As this dephasing process takes over, with a time constant of T_2, the observed resonance lines experience broadening according to Equation 3. Both spin-lattice T_1, and spin-spin T_2 relaxation processes are governed by fluctuating magnetic fields associated with molecular motion. However, the time scales for these two relaxation processes differ. Slow motions in the kilohertz range contribute only to spin-spin relaxation, whereas components of motion at the resonance frequency level (MHz) contribute to both spin-spin and spin-lattice relaxation. Figure 6 shows how molecular correlation times in the resonance frequency range give rise to the shortest T_1 values, while those at 10^{-6} s promote rapid T_2 relaxation. Note that those molecules having T_c values less than 10^{-9} s are generally high molecular weight macromolecules, viscous liquids, or solids, whose wide line widths are a result of the rapid loss of phase coherence in the xy plane. For small freely tumbling molecules with $T_c < 10^{-9}$, $T_2 = T_1$. Over the whole range of correlation times T_2 can never be longer than T_1. With NMR frequencies in the range of 100 to 300 MHz, the minimum in the T_1 curve will occur in

* Actually, the line broadening of the NMR signal is due to T_2*, which is a time constant that takes into account other relaxation processes, such as B_0 field inhomogeneities. Thus, the decay of the M_{xy} magnetization is always more rapid than would be anticipated, based on a value of T_2 resulting from spin-spin interactions alone. For this reason, it is essential to maintain a relatively homogeneous B_0 field for high-resolution NMR. Unfortunately, because of the inherent inhomogeneity of tissue used in *in vivo* studies, the control of line narrowing by careful attention to B_0 homogeneity is minimal.

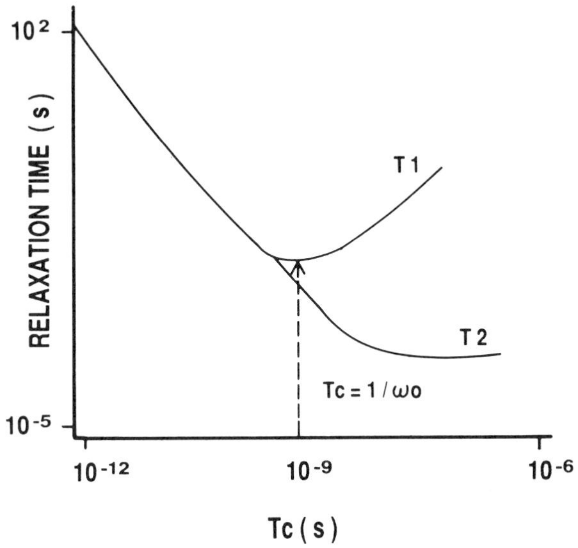

FIGURE 6. Efficiency of T_1 and T_2 relaxation processes as a function of molecular correlation times, T_c.

the vicinity of 5×10^{-10} to 2×10^{-7} s. In non-viscous solution, small molecules with molecular weights of less than 200 almost always have correlation times and relaxation times to the left of the minimum (Figure 6). Polymers such as small proteins of molecular weights in the range of 10,000 to 25,000 Da may have relaxation times on either side of the minimum depending on the Larmor frequency of the experiment.

A number of methods for making relaxation time measurements, i.e., inversion recovery, saturation recovery, and progressive saturation, have been amply discussed in the literature.[8,10,12] Suffice it to say that one must pay careful attention to relaxation parameters in any NMR experiment if a quantitative assessment of the resulting spectrum is to be made. Failure to implement the proper recycle time following each pulsed spectrum can lead to incorrect conclusions regarding the ratio of various resonance intensities. Examples of these problems and their solutions will be mentioned in the following section.

In the final analysis there are three essential features in an NMR experiment that can be utilized to give information about a living system. These considerations are (1) chemical shifts, which ultimately allow us to identify and quantitate the concentration of compound(s) in question, as well to probe (in some instances) the intracellular pH of the cells, (2) the resonance splitting patterns which are diagnostic for determining the geometric relationships between chemically bonded species, and (3) the two different relaxation processes, T_1 and T_2, that can be used to simplify spectra and probe molecular motion and interaction in two distinct frequency ranges. In the chapters to follow we will elaborate on the exploration of such parameters to help define the chemical state of living systems.

Clearly, modern NMR methodology can solve problems in ways that could barely be imagined 10 years ago, but many of these new methods have only been successfully employed by a relatively small number of agricultural scientists, i.e., animal and plant physiologists, food scientists, and nutritionists. How do these scientists decide whether *in vivo* NMR or any other NMR approach is appropriate for solving their problems? We hope after digesting the material in this and the other chapters in this book the readers will have better insights into the selection of those NMR experiments that would benefit their research

Ultimately, one must gain access to an NMR spectrometer or spectroscopist with whom to collaborate.

Table 1
RELATIVE SENSITIVITIES AND CONCENTRATION THRESHOLDS OF COMMON NUCLEI REQUIRED FOR COMPARABLE SECTRA

Nucleus	Sensitivity	Concentration threshold[a] (m*M*)
^{1}H	1.0	<<1
^{23}Na	9.25×10^{-2}	5
^{31}P	6.64×10^{-2}	1
^{14}N	1.01×10^{-3}	10[b]
^{39}K	4.7×10^{-4}	20
^{13}C	1.76×10^{-4}	>>20
^{15}N	3.85×10^{-6}	>>20

[a] This concentration threshold takes into account the difference in relaxation properties. Concentrations are averaged over the total volume of the sample in the detecting coil, corresponding to 1 g fresh weight of sample and a spectral accumulation time of 20 min (see Reference 18).

[b] This threshold is considered for molecules with local fields that are tetrahedrally symmetrical around the nitrogen atom, e.g., NH_4^+, NO_3^-.

From Ratcliffe, R. G., *Inorg. Biochem.*, 28, 347, 1986. With permission.

III. PARAMETERS OF THE *IN VIVO* NMR EXPERIMENT

A. Selecting a Nucleus (An Overview)

The choosing of an appropriate nucleus to study a particular biological system must be based on a number of considerations. First and foremost, one must consider the quality and quantity of information that is required. For example, what is the concentration of the active metabolites in the cells, and at what level of abundance is the magnetically active nucleus present? Can one use isotopically enriched compounds where the natural level is too low to detect these molecules? What is the spin state of the naturally abundant isotopes vs. the enriched nonabundant one? In some instances it may be useful to enrich a tissue with an isotopically labeled nucleus (spin 1/2), e.g., ^{15}N (natural abundance 2.24×10^{-2}) in favor of examining the more highly abundance (5.7%) nucleus ^{14}N which is quadrupolar and would generally, except in certain circumstances, give poorly interpretable, broadened spectra. Some examples demonstrating the advantages, disadvantages, and methods of circumventing these problems will be discussed in subsequent sections.

Ratcliffe[18] has compiled a list of nuclei and their inherent sensitivities as they relate to the S/N ratio of resulting spectra. In addition, he has compared this to the threshold concentration of each nucleus needed to produce comparable spectra in a period of 15 to 20 min. Table 1 contains a listing of this information for quick reference. To date, the most commonly used nucleus for *in vivo* NMR work is ^{31}P because of its spin 1/2, relatively simple intrepretable spectra, and the key metabolic significance of the compounds it represents. Although ^{23}Na and ^{1}H have lower concentration thresholds, sodium is quadrupolar and has a smaller chemical shift range (~10 ppm) while *in vivo* proton NMR requires special spectrometer techniques because of the dominance of the water resonance in the spectra. ^{13}C has a very meager level of natural abundance (1.1%), but is useful in studies of tissues that accumulate carbon-containing storage compounds; however, most *in vivo* investigations require ^{13}C-labeled compound precursors to examine metabolic sequences in functioning tissues and cells. The latter methodology is an excellent alternative to radio tracer experiments since ^{13}C NMR allows the direct observation of specific metabolic pathways and distributions of label within metabolites without the need for complex degradation procedures. On a one-to-one concentration basis ^{13}C is actually two to three times more sensitive than ^{31}P. This greater sensitivity is primarily a result of its narrow line widths and signal enhancement gained from broad-band proton decoupling (nuclear Overhauser enhancement, NOE). As you will see, the extremely large chemical shift range of ^{13}C makes this nucleus ideal to utilize when resolving a large assortment of cellular metabolites. As with ^{13}C, the natural abundance of ^{15}N is very low; in fact, it is only 1/50th as sensitive as ^{13}C. Therefore, ^{15}N NMR can only be examined with labeled substrates *in vivo*. However, the ^{15}N chemical shift range is 50% larger than ^{13}C making ^{15}N more sensitive to structural changes associated with the nitrogen environment. Potassium has similar problems to sodium in that it is quadrupolar in nature in addition to being a less sensitive nucleus than sodium. Consequently, ^{39}K is rarely used for *in vivo* investigations even though it is a physiologically important nucleus. The ^{19}F nucleus is close to ^{1}H in sensitivity and its chemical shift range far outstretches that of ^{13}C. Thus, one would expect this nucleus to be the perfect probe for biological studies. Unfortunately, naturally occurring ^{19}F-containing compounds are rarely found in biological systems. To take advantage of the excellent spectroscopic properties of fluorine, many researchers have incorporated this nucleus into compounds that could be diffused into cells. These spy molecules can then be examined to probe intracellular pH and the binding of other NMR invisible ions. The various NMR nuclei used to detect inorganic ions are summarized in Table 2, adapted from Ratcliffe.[18] These studies will be discussed under the heading of each nucleus that follows.

B. Maintaining Viability

In vivo applications put additional experimental demands on NMR spectroscopy. One must pay careful attention to the condition and maintenance of the living system being examined. How well is it being perfused? What are the optimum conditions necessary for sustaining it for extended periods of time? How does one prevent bacterial contamination? In addition, we must also consider the conditions of the spectrometer. We must have a leakproof perfusion system to prevent damage to the probe and shim coils. Figure 7 shows the configuration of the perfusion apparatus commonly used in *in vivo* tissue studies. Often multiple connected reservoirs are used so perfusates with different reagents may be perfused and switched without having to stop ongoing experiments.[14] NMR tubes from 10 to 75 mm in diameter have typically been used for studies of small excised plant tissues,[19] cells,[20] organs,[8] and intact anesthetized small animals;[8] however, a recent study reported a more sophisticated system that can use a 5-mm tube.[21] In a circulating system it is convenient to bubble the required gas such as O_2, air, N_2, He, etc. into the reservoir containing the perfusing liquid. This technique prevents problems associated with the loss of field homogeneity when gas bubbles are passed directly into the NMR tube containing the sample. In some instances microbubbling techniques or enzymatic generation of oxygen have been employed to circumvent possible homogeneity problems.[22] Figure 8 shows a typical NMR assembly used for the perfusion of plant tissue.[14] Note the two failsafe features: pressure-sensitive connections positioned close to the peristaltic pump and the spinner housing overflow vacuum line. Observe also that the flow in the tube is downward, i.e., the liquid is drawn out from the bottom of the tube. This arrangement is utilized to prevent any trapped gas in the lines from accumulating in the interstitial space between the tissue.[23] This assembly is quite convenient for examining most small or excised tissues and cultured cells (plant and mammalian). The ^{31}P spectra shown to the right in Figure 8 demonstrate the maintenance of a consistently healthy physiological state of excised maize root tips during 34 h of perfusion. To study small cells with continual perfusion requires a somewhat more elaborate design since these small organisms must be kept from flowing along with the perfusing liquid. To circumvent this problem, researchers have devised a dialysis microfiber arrangement that allows liquid to flow through and evenly bathe the cells.[23] Figure 9 shows three ^{31}P NMR spectra of maize root tips at different levels of viability as a function of perfusion condition.

In Figure 9A, we see a spectrum representing a highly energized state, i.e., high levels of NTP

Table 2
NMR DETECTION OF INORGANIC IONS IN PLANT TISSUES

Ion	NMR nucleus	Comments
H^+	^{1}H $^{13}C(*)$ $^{15}N(*)$ ^{31}P	pH-dependent chemical shifts and line widths (^{15}N) for metabolites containing ionizable groups can be used to estimate intracellular pH; ^{31}P NMR is of only limited use for measuring the pH of the acidic vacuolar compartment ($[H^+] \geq 1$ μM)
Carbonate	$^{13}C(*)$	Detectable in tissues labeled with $^{13}CO_2$
NH_4^+	^{14}N $^{15}N(*)$	The naturally abundant ^{14}N is quadrupolar (1 = 1) and most ^{14}N resonances are unacceptably broad; NH_4^+ and NO_3^- give narrow lines, but to follow their metabolic transformation by NMR it is necessary to use ^{15}N NMR in isotopically labeled tissues
NO_3^-	^{14}N $^{15}N(*)$	
Na^+	^{23}N	Potentially complicated, but informative, line shapes for this nucleus (1 = 3/2)
Mg^{2+}	^{31}P	The ^{13}P resonances of ATP are Mg^{2+} dependent
Ortho-phosphate (P_i)	^{31}P	The most commonly observed nucleus *in vivo*
SO_4^{2-}	$^{33}S(*)$	^{33}S is a difficult NMR nucleus, and even in a fully labeled system it is unlikely to be of any use for the detection of intracellular SO_4^{2-}
Cl^-	^{35}Cl	Potentially complicated, but informative line shapes for the nucleus (1 = 3/2)
K^+	^{39}K	Similar to ^{23}Na, but the sensitivity is much less (1 = 3/2)
Ca^{2+}	^{19}F	An indirect method requiring infiltration of a ^{19}F labeled probe; problems in infiltrating the related fluorescent probes into plant cells suggest that this NMR method is inapplicable to plant tissues
Mn^{2+}	^{1}H	Extensive literature on chloroplast suspensions (with confusing differences in interpretation between the two groups involved)
Mn^{2+}	^{31}P	Mn-induced line-broadening effects in ^{31}P spectra can be interpreted in terms of the intracellular distribution of Mn^{2+}
Al^{3+}	^{31}P	Estimates the effect of this ion on energetics of the cells, ATP levels, intracellular pH, and Pi depletion

Adapted from Ratcliffe, R. G., *Inorg. Biochem.*, 28, 347, 1986.

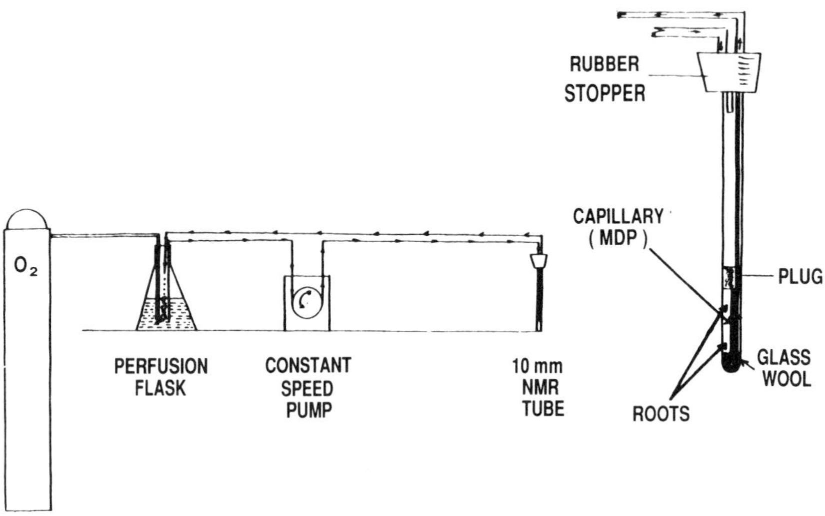

FIGURE 7. Schematic representation of tissue perfusion system.

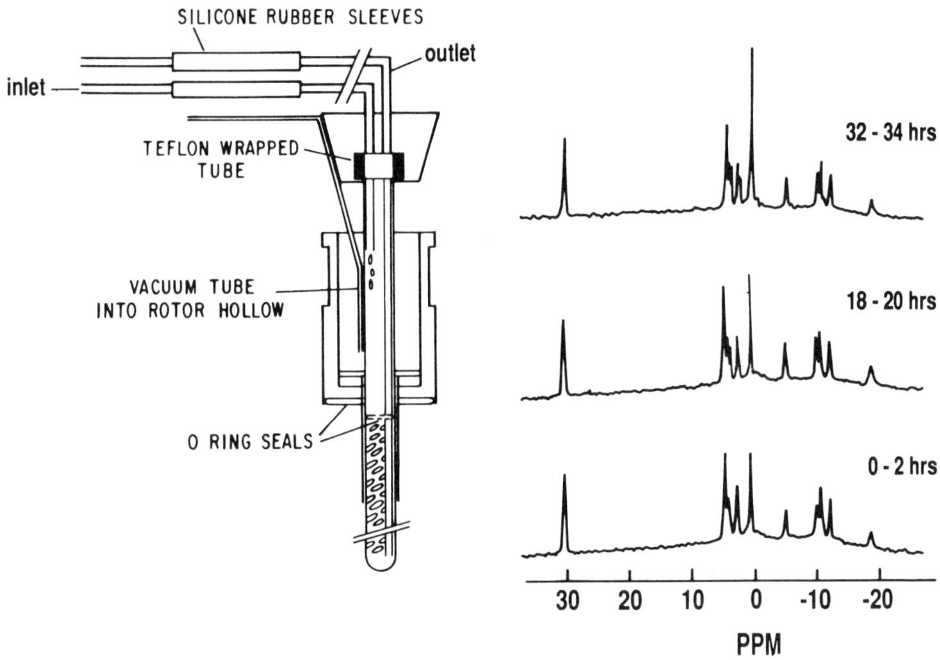

FIGURE 8. NMR tube perfusion device with overflow prevention tube and breakaway sleeves and 161.7-MHz ^{31}P spectra of perfused maize root tips after 34 h.

and glucose-6-phosphate. In Figure 9B, these levels have dropped significantly due to only trace amounts of oxygen (hypoxia) perfusing through the tissue; however, the tissue still remains viable. With no perfusion, Figure 9C, anoxia is complete within 20 min following a concomitant buildup of localized CO_2 and membrane destruction. More details concerning these experiments will follow in this chapter and Chapter 3. Whole organs such as hearts, kidneys, and livers are generally perfused with a cannula and often the pumping of the fluid is gated or timed to the pulsing of the spectrometer so as not to obtain spectral artifacts from the movement of the flowing liquid.[24] Studies of organs of small intact animals only require that the animal be properly anesthetized.[25] Whole limb, NMR spectroscopy, usually of a leg or arm muscle, is performed with a wide-bore (25 cm or greater) horizontal magnet. Spectra from such experiments are used for clinical examinations for diagnosing diseased muscle function or exercising physiology studies.[26]

C. Qualitative Aspects

Identification of the resonances in an *in vivo* NMR spectrum is not always a simple matter. Often the observed resonances are broadened and overlap badly in the spectrum, making it difficult to resolve and quantify each peak. This resonance overlap is a particular problem for ^{31}P and ^{1}H NMR. For ^{31}P NMR a simple comparison of shifts of known compounds in solution whose ionic strength and pH are close to those of the cell can help eliminate some compounds as contributors to the *in vivo* spectrum. Following this procedure, careful cold perchloric acid extraction[20] of the tissue, lyophilization, and subsequent inspection of the extract in a solution containing CDTA or other paramagnetic metal ion sequestrant will yield a relatively narrow line spectrum of metabolites. Figure 10 shows the downfield portion of the ^{31}P NMR spectrum of a perchloric acid extract of soybean nodules with small amounts of suspected nucleotides added to verify the identity of each resonance. To the left of the extract spectra (Figure 10) is the *in vivo* ^{31}P spectrum of the intact nodules from which the extract was prepared. Note that the two resonances labeled Pi cyt and Pi vac in the *in vivo* spectrum have become one Pi resonance at

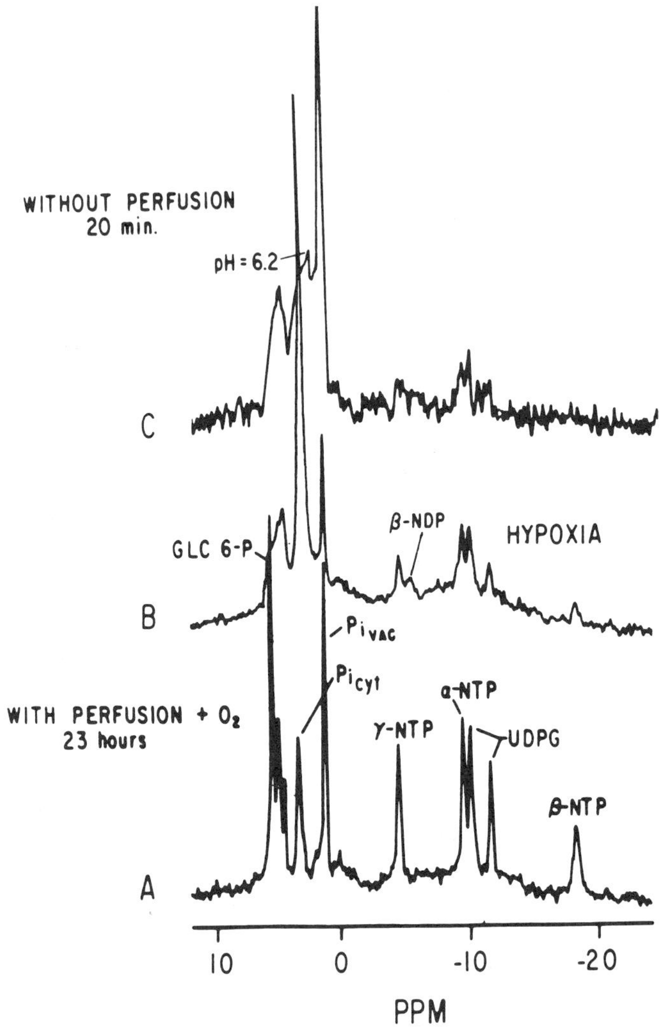

FIGURE 9. 161.7-MHz ^{31}P spectra of maize root tips, 10,000 scans each (27-min accumulation), (A) after 23 h of perfusion with an oxygen satured perfusate (B) following 1 1/2 h of perfusion with a nitrogen-saturated perfusate and (C) without perfusion for 20 min.

23.1 ppm in the extract spectra (pH = 7.8). Thus, the extract spectra helps to identify the resonances representing Pi in two different chemical environments (membrane separated) within the cell and distinguishes them from other nucleotides whose resonances may be in close proximity. If one suspects that there is more than one Pi peak present in the *in vivo* spectrum due to different compartmentation, the extract spectra will clarify this point, since only one Pi resonance will appear. Another check on the identity of each resonance can be made by titration of the extracts of phosphorus metabolites since most of these ^{31}P compounds exhibit characteristic titration curves that correlate with chemical shift in the vicinity of their respective pKas. This topic will be discussed in more detail in the section on ^{31}P studies. Additionally, extracts can be treated with various specific enzymes, e.g., a phosphodiesterase or mutase, to establish the identity of individual compounds.[27]

The identification of ^{13}C resonances is somewhat more complex because of the large number of observable metabolites. Also with some exceptions that we will mention later, ^{13}C resonances do not always exhibit the broad titration/chemical shift range of ^{31}P, especially at physiological pH. For this reason one must often rely primarily on resonance peak matching. This method is

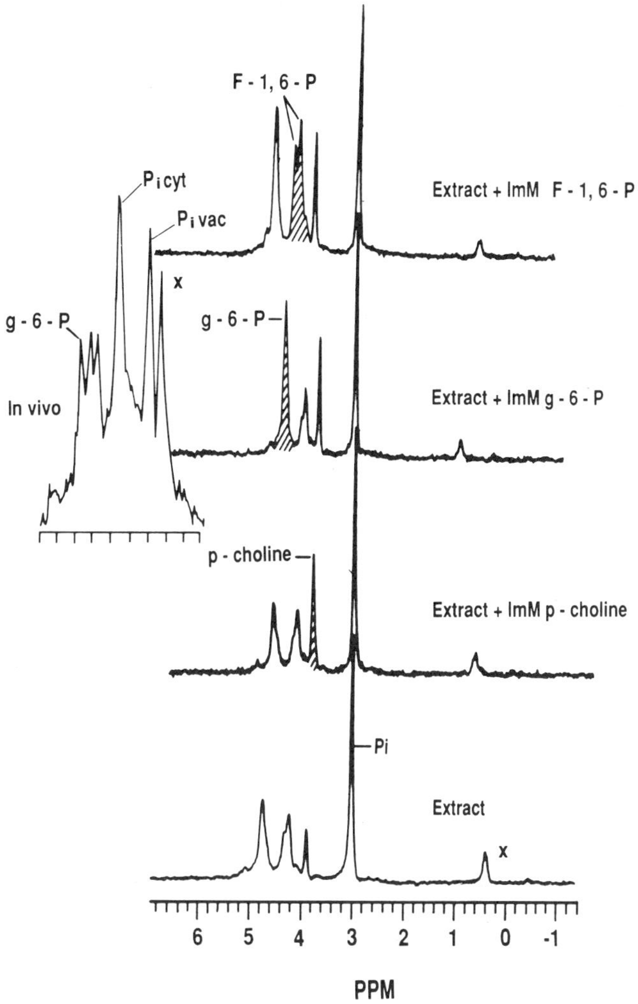

FIGURE 10. Low-field section of 161.7-MHz ^{31}P spectra of the *in vivo* and perchloric acid extracts of soybean nodules. Each resonance is identified by the successive addition of a known metabolite. Each spectrum required 1000 scans with a 3-s pulse delay.

usually quite reliable since resonance patterns for ^{13}C spectra are well established and reproducible.[28] Additionally, the identity of a carbon resonance, i.e., a CH, CH_2, CH_3 or quarterary carbon, can be verified by examining the proton-coupled ^{13}C spectrum of an extract. In such spectra the C̲H carbon resonance are doublets, CH_2s are triplets, CH_3s quartets, and quaternary Cs are seen as singlets (n + 1 lines), due to the coupling to n magnetically equivalent protons.[29] This analysis may also be accomplished using recently devised NMR pulse sequences (INEPT, DEPT, etc.) to obtain the multiplicity of each resonance.[30]

^{13}C, which is a spin 1/2 nucleus, shows ^{13}C-^{13}C n + 1 line splitting in labeled metabolites. This splitting can be diagnostic for evaluating biosynthetic and metabolic labeling patterns. The splitting attributed to ^{13}C-^{13}C coupling between neighboring carbons gives direct insight into pathways for the incorporation of ^{13}C-labeled precursors.[31] When ^{13}C is next to an unlabeled natural abundance ^{12}C (~99%), the former gives rise to a singlet resonance, indicating that only one carbon of the ^{13}C-labeled precursor was incorporated at that site in the metabolite.[32] This method is useful for establishing intermediate steps in biosynthetic pathways.[31] A new double quantum coherent 2D NMR pulse sequence[30] (INADEQUATE) has recently become useful for detecting ^{13}C-^{13}C bonds in biosynthetic intermediates.[33] However, this method can only be used if relatively large amounts of the labeled metabolites are available.[30]

Identification of ^{15}N resonances generally offers no problem because each represents a unique isolated nucleus associated with a specific compound. For example, when enriched ^{15}N-labeled NH_4Cl was used as a nutrient source for fungi, the resulting ^{15}N spectra of the labeled amino acids gave a clear profile of the sequence of nitrogen incorporation.[34] ^{14}N on the other hand, because of its quadrupolor nature, is not a practical nucleus for evaluating compounds other than those that are tetrahedrally symmetrical such as NO_3 or NH_4.[35] All other natural abundance ^{14}N compounds within a tissue are not readily visible by NMR because of their broad lines (100 to 1000 Hz), whereas ^{14}N line widths of cellular $^{14}NH_4$ and $^{14}NO_3$ are only 10 Hz.[36]

Proton, with the most sensitive nucleus natural abundance levels, offers the most complex spectra to analyze largely because of its limited chemical shift range (10 ppm) and complex homonuclear spin-spin coupling pattern and strong dipole-dipole interactions. Typically, the ^1H signals derived from large macromolecules such as proteins *in vivo* are very broad due to short T_2 (slow motion), whereas, the signal from the small rapidly tumbling metabolites are sharp (longer T_2). Taking advantage of this difference in T_2 values, Brown et al.[37] were able to "edit" the ^1H spectrum of human erythrocytes with a spin echo pulse sequence and suppress the broad overlapping resonances due to the relatively immobile proteins. Identification of the remaining metabolite resonances could then be made by spectral comparison with authentic compounds or titration of selected residues associated with specific chemical shifts.[37] Fan and co-workers[21] have utilized homonuclear correlated two-dimensional NMR both *in vivo* and *in vitro* in addition to gas chromatography-mass spectrometry to identify sugars, organic acids, amino acids, and ethanol in carrot roots, maize roots, and rice shoots. Figure 11 and the corresponding Table 3 show an impressive number of metabolites that may be identified using this methodology.[30]

Other nuclei such as ^{23}Na and ^{39}K do not present any resonance identification problems since these peaks are not associated with any particular compound *in vivo*, but can reflect the chemical environment characteristic of a cell compartment. The ^{19}F nucleus is generally used as a label on compounds to probe pH or metal ion concentrations and as such does not represent a problem in terms of chemical shift assignment.

D. Quantitative Aspects

As mentioned earlier, the spin-lattice relaxation time (T_1) is the most important factor contributing to quantitation of the NMR spectral responses for spin 1/2 nuclei, such as ^{31}P, ^{13}C, ^{15}N, and ^1H. To obtain full resonance responses, one should initiate each pulse after a waiting period of five times that required for the resonance having the longest T_1. However, this waiting is not always convenient since T_1 can often be >5 s. Also, because of the constraints of time resolution, the metabolic process one follows may be past history by the time the final spectrum has been acquired. Furthermore, it is sometimes advantageous to recycle the scans faster and excite the frequencies with a smaller than 90° pulse angle to emphasize certain features of the spectrum. Figure 12 illustrates ^{31}P spectra of maize root tips which were acquired under (A) nonquantitative and (B) quantitative conditions.[14,38] Clearly, one observes an obvious difference in the ratio of the resonances in each of the spectra due to the differences in the acquisition conditions used. If we look at Table 4 we can immediately see what has occurred. Those resonances whose T_1 values are very long, i.e., Pi vac, Pi cyt, and sugar phosphates are seriously suppressed during rapid recycling and those with short T_1s, e.g., NTP and uridine diphosphoglucose (UDPG), are more strongly emphasized under these conditions. Although spectrum A is distorted in terms of the true ratios of resonances representing each component, it offers the advantage that one can readily quantify changes taking place at the concentration level of the nucleotides over shorter periods of time.[14] However, if one wants to compare the ratios, for example, of nucleotides to phosphomonoesters and Pi under the fast acquisition conditions, one would necessarily have to first perform the long acquisition time experiments, to establish the true relative intensities. After performing this experiment the fast acquisition data can be

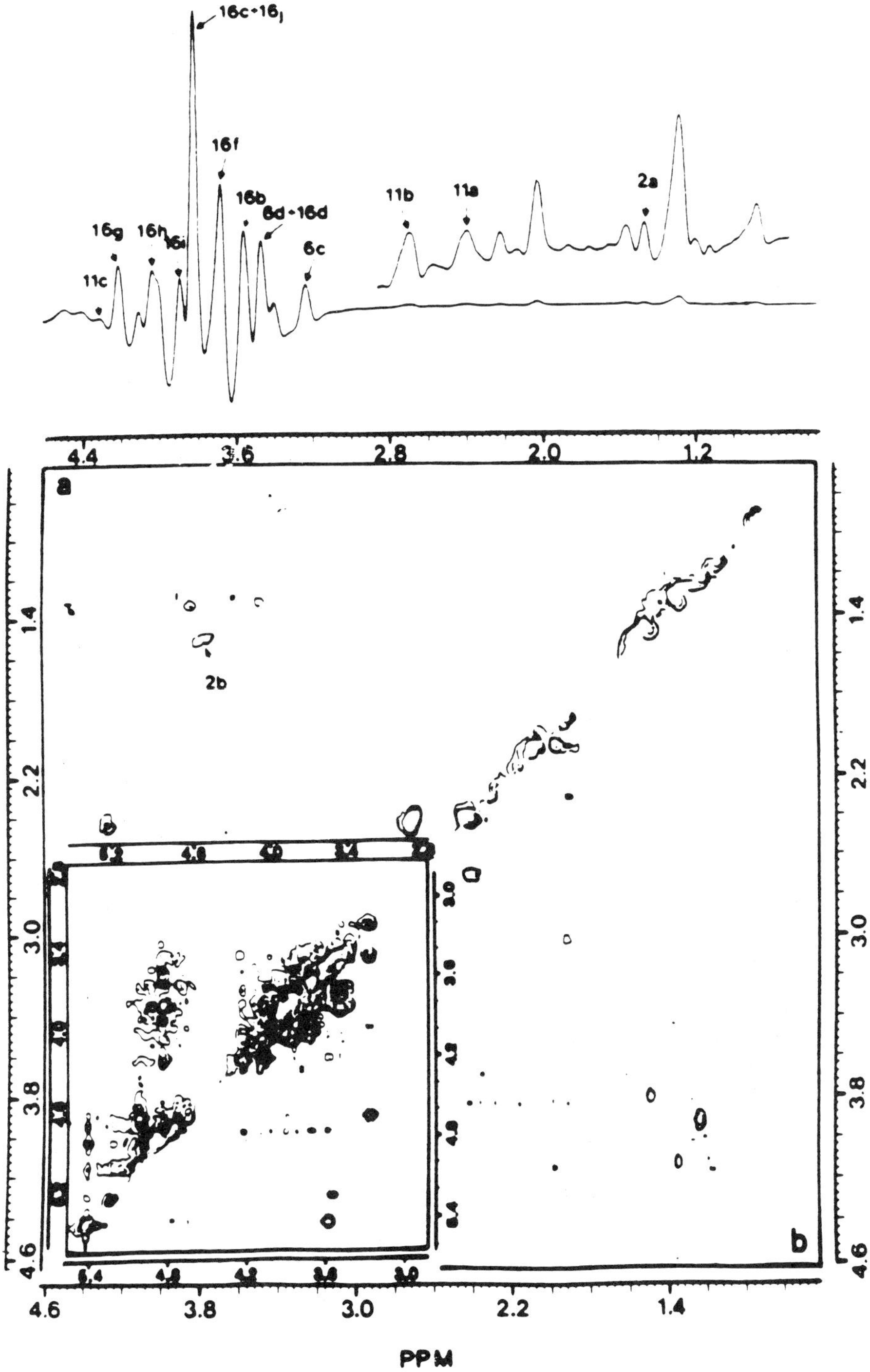

FIGURE 11. *In vivo* COSY spectrum of a carrot root under superfusion. The *in vivo* COSY part is displayed in the upper triangle (a) as contour plots along with the 1-D spectrum while the COSY spectrum of the PCA extract of the same tissue is shown in the lower triangle (b). The inset represents the sugar region and was plotted at higher contour levels than the 0.6- to 4.6-ppm region. These two regions were presented separately because of the much greater peak intensities in the sugar region than in the 0.6- to 4.6-ppm region, which would obscure the cross-peaks in the latter if they were plotted at the same contour levels. The peaks are numbered according to Table 3. (From Fan, W.-M., Higashi, R. M., and Lane, A., *Arch. Biochem. Biophys.*, 251, 674, 1986. With permission.)

Table 3
IN VIVO **^{1}H NMR ASSIGNMENT OF RESONANCES IN CARROT ROOT, MAIZE ROOT, AND RICE SHOOT TISSUES**

Peak number	Assignment	δ *in vivo*	δ *in vivo*	Tissues in which *in vivo* resonances were observed
1	*t*-Acon-αCH	6.80	6.58	Maize root
2a	Ala-βCH_3	1.48	1.47	Carrot, maize root, rice shoot
2b	Ala-αCH	3.77	3.76	Carrot
3a	Asn-βCH	2.86	2.84	Maize root, rice shoot
3b	Asn-βCH	2.95	2.94	Carrot, maize root, rice shoot
3c	Asn-αCH	4.03	4.00	Maize root
4	ETOCH-βCH_3	1.18	1.16	Maize root, rice shoot
5a	GAB-βCH_2	1.89	1.89	Maize root, rice shoot
5b	GAB-αCH_2	2.29	2.29	Carrot, maize root, rice shoot
5c	GAB-γCH_2	3.00	3.01	Carrot, maize root, rice shoot
6a	Glc-C_1β	5.24	5.23	Carrot, maize root, rice shoot
6b	Glc-C_1α	4.65	4.64	Carrot
6c	Glc-C_2β	3.24	3.24	Carrot, maize root, rice shoot
6d	Glc-C_3β	3.47	3.47	Carrot
7a	Gln-βCH_2	2.13	2.13	Carrot, maize root, rice shoot
7b	Gln-γCH_2	2.45	2.44	Carrot, maize root, rice shoot
8a	Glu-βCH_2	2.07	2.06	Maize root, rice shoot
8b	Gln-γCH_2	2.36	2.34	Carrot, maize root, rice shoot
9	His-C_2H	8.66	7.86	Maize root
10	Ile-δCH_3	0.96	0.93	Carrot, maize root
11a	Malate-βCH	2.43	2.40	Carrot
11b	Malate-β′CH	2.72,[a,c] 2.68	2.67	Carrot, maize root, rice shoot
11c	Malate-αCH	4.32,[a] 4.30	4.30	Carrot, maize root
12a	Lactate-βCH_3	1.32	1.32	Carrot, maize root, rice shoot
12b	Lactate-αCH	4.11	4.11	Carrot, maize root
13	Leu-δCH_3	0.96	0.94	Carrot, maize root, rice shoot
	Leu-δCH_3	0.96	0.96	Carrot, maize root, rice shoot
14	OAc-αCH_3	2.01	1.91	Maize root, rice shoot
15	Succ-α and -βCH	2.38	2.39	Maize root, rice shoot
16a	Sucrose-G_1	5.42	5.41	Carrot, maize root, rice shoot
16b	Sucrose-G_2	3.56	3.56	Carrot
16c	Sucrose-G_3	3.81	3.75	Carrot
16d	Sucrose-G_4	3.47	3.47	Carrot
16e	Sucrose-G_5	3.81	3.83	Carrot
16f	Sucrose-F_1	3.70	3.68	Carrot
16g	Sucrose-F_3	4.20	4.20	Carrot, maize root
16h	Sucrose-F_4	4.03	4.04	Carrot, maize root
16i	Sucrose-F_5	3.89	3.88	Carrot
16j	Sucrose-F_6	3.81	3.88	Carrot
17	Thr-γCH_3	1.32	1.32	Carrot, maize root, rice shoot
18	Val-γCH_3	0.96	0.98	Carrot, maize root, rice shoot

Adapted from Fan, W.-M. T., Higashi, R. M., and Lane, A., *Arch. Biochem. Biophys.*, 251, 674, 1986.

modified by the appropriate response factors to yield reliable ratios.[16,39] It should be emphasized that interconversion of metabolites can be continually occurring during these experiments, which means that a resonance representing a particular component with a specific relaxation time may be converted into another component whose resonance relaxation time is different. For example, γ-NTP → Pi represents a relaxation change of the corresponding resonances of 0.35 to 2.78 s, respectively. To account for this difference, under fast acquisition conditions one must compensate for the disparity in magnetization of the two resonances, because after the γ-NTP phosphate is hydrolyzed to Pi *in vivo* the γ-NTP contributes less of a response (area) to the Pi

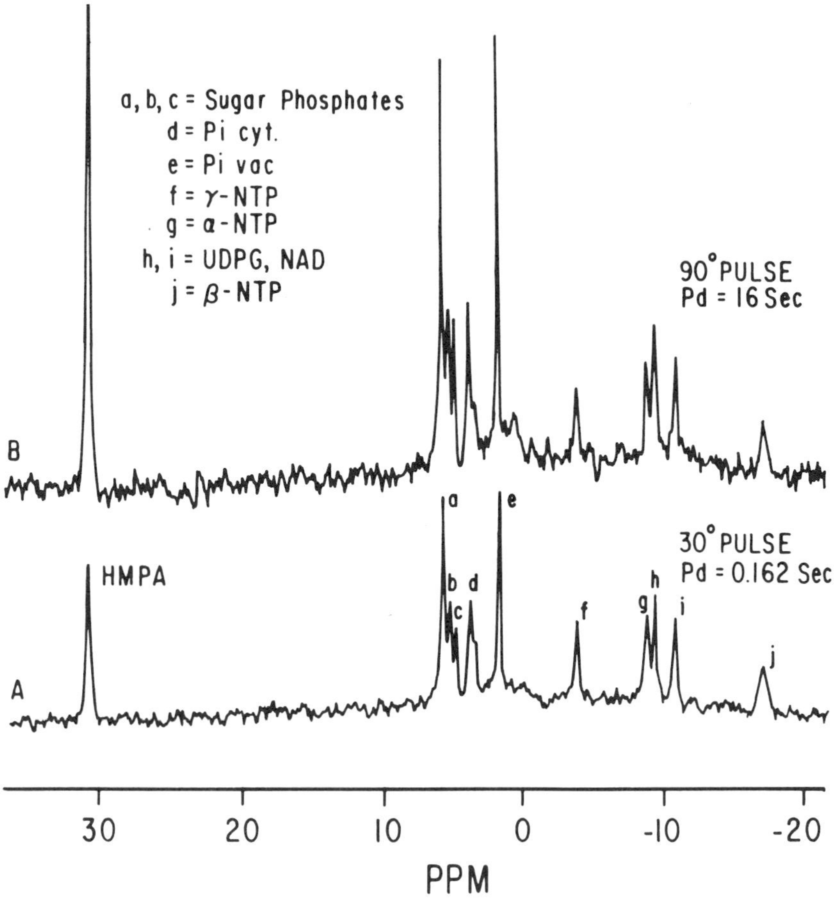

FIGURE 12. Intensity distortion due to rapid acquisition. (A) 161.7-MHz ^{31}P spectrum of approximately 900 excised root tips (5 to 7 mm) taken under fast acquisition accumulation parameters, 30° pulse (12 μs) 2000 data points zero filled to 16,000 recycling time of 0.162 s, 10,000 transients, frequency of 16 KHz, 15-Hz line broadening, and total time for acquisition of spectrum 27 min. (B) Slow recycling time full 90° pulse-generated spectrum obtained with a 16-s recycling time, 1000 scans, 16,000-Hz frequency, 15-Hz line broadening, and total time for acquisition of spectrum 3.5 h. (See Reference 14.)

resonance than it did when it was observed as a γ-NTP resonance.[16] This kind of treatment allows one to assess the material balance of mobile phosphorus metabolites in tissues and thereby account for any generation or leakage of material during the study.[16]

Under quantitative acquisition conditions, the ^{31}P line intensities correspond to the concentration of each particular compound contained in the collection of cells or tissue residing in the volume of the detection coil of the probe. Each of these resonances can be measured against the reference peak area which has previously been calibrated with standard solutions of the various compounds being examined. In essence, the spectrum displays changes in the overall concentration of components within the enclosed coil volume. However, a knowledge of the cell volume as well as the cytoplasmic and vaculolar compartment volumes is necessary to provide a meaningful estimate of intracellular concentrations.[40,41]

^{13}C and ^{15}N T_1 relaxation is largely dependent on molecular motion as it modulates the dipolar interaction of these nuclei with their directly bonded protons. In addition to the spin-lattice relaxation phenomenon discussed above, the resonance intensity of ^{13}C and ^{15}N is also subject to another phenomenon associated with relaxation known as the nuclear Overhauser effect.[11] When we observe the spectra of both ^{13}C or ^{15}N the full frequency range of the corresponding

Table 4
^{31}P NMR SPIN-LATTICE RELAXATION TIMES (T_1) OF INTRACELLULAR COMPONENTS OF MAIZE ROOT TIPS

Compound	T_1 (s)
G-6-P	2.89 ± 0.24
Pi vacuolar	3.58 ± 0.5
Pi cytoplasmic	2.78 ± 0.31
NTP	
γ	0.35 ± 0.06
α	0.39 ± 0.05
β	0.21 ± 0.02
UDPG	0.63 ± 0.06

Adapted from Pfeffer, P. E., Tu, S. I., Gerasimowicz, W. V., and Cavanaugh, J. R., *Plant Physiol.*, 80, 77, 1986.

proton spectrum is routinely broad-band noise decoupled to remove spin-spin scalar splitting.[11] During the saturation of all the proton lines, a secondary effect produces a rearrangement of energy level populations resulting in a modification of the ^{13}C or ^{15}N line intensities. ^{13}C resonances always experience an enhancement of signal from 0 to 3 depending on the number of directly bonded protons, and molecular motion. The greater the number of bonded protons, the greater is the enhancement for ^{13}C, while the efficiency of the NOE as molecular motion slows, becomes impaired.[11] Under the conditions where $T_c \cong 1/\omega_o$, the NOE shows a magnetic field dependence which can be used to calculate molecular motion.[42] Because of its negative magnetogyric ratio, the NOEs for the ^{15}N resonance have a range of 0 to –5 and in some instances may not be visible or are observed as negative peaks in the spectrum under complete proton decoupling conditions.[43] Typically NOE for ^{31}P is much smaller because the magnetogyric ratio for ^{31}P is larger relative to that of protons and also because it is rare to find ^{31}P directly bonded to protons (NOE is dependent on internuclear distance and proportional to $1/r^6$ where r is the internuclear distance between ^{1}H and ^{31}P). Also the ^{31}P-^{1}H couplings are usually very small. Consequently, proton decoupling yields little change in the ^{31}P spectrum but may generate some rf heating that can be detrimental to biological samples. To achieve quantitative responses for both ^{13}C or ^{15}N resonances, one must not only pay attention to T_1- dependent recycling times but Overhauser effects as well. The former is easily accomplished as described above, while the latter can be corrected by decoupling the proton spectrum only during the time the rf pulse is being applied.[11] This method of gated decoupling removes the scalar ^{1}H splitting from the ^{13}C or ^{15}N spectra without transmitting the proton polarization effects responsible for the distorted intensities. Of course, one must recognize that in these experiments it can take up to three times longer to obtain ^{13}C spectra and up to five times longer to obtain ^{15}N spectra* under these conditions.

Quadrupolar nuclei such as ^{23}Na, ^{39}K, or ^{14}N present no problem as far as spin-lattice relaxation is concerned, that is, because their relaxation mechanism is dominated by efficient quadrupolar interactions with neighboring nuclei (T_1 ~ millisecond range). In spite of the relatively low sensitivity of ^{14}N, one can pulse its spectrum and recycle rapidly without being concerned with saturation. Rapid scanning enables the accumulation of a large number of transients in a relatively short period of time.

* ^{15}N spectra are often shown with peaks in a negative direction because of its negative NOE.

^{23}Na and ^{39}K have problems which make them difficult nuclei to quantify. In particular spin 3/2 quadupolar nuclear interactions within living cells and tissues can give rise to two different T_2 values. One has a T_2 so short that its corresponding resonance is not detectable.[44,45] This translates into a resonance "visibility" of only 40% because the transition corresponding to the longer T_2 represents two fifths of the total signal.[45,46] Recently, a study has revealed that these "invisible" populations may be "unmasked" using a sophisticated two-dimensional double-quantum NMR method.[47]

IV. *IN VIVO* NMR APPLICATIONS

A. ^{31}P Studies

^{31}P NMR spectroscopy is the most widely used technique for examining biological systems. The most recent review describing some NMR studies of plant tissue including the use of ^{31}P has been given by Ratcliffe.[48] This nucleus combines the advantages of 100% natural abundance, relatively high sensitivity, and a chemical shift range for metabolic compounds of 30+ ppm to make it most attractive for the examinations of microorganisms, plants, and animal tissues. This spectrum contains only 8 to 12 resonances representing, for example, 3 to 4 different phosphomonoesters, orthophosphate (Pi), NTP, in some instances NDP, uridine diphosphoglucose UDPG and nicotinamide adenine dinucleotide (NAD), as seen in Figure 12. However, the spectrum provides valuable information concerning many critical cell functions.*. For example, the Pi resonance can provide valuable information concerning Pi utilization which modulates glycolysis,[52] respiration,[53] photosynthesis,[54] and starch breakdown.[55] The bioenergetics of the cell can also be examined by the assessment of the NTP/NDP ratio,[8] the intracellular pH,[8] and intracellular metabolic fluxes of Pi.[55] While spectral analysis is often relatively straightforward, one must bear in mind that we can only examine mobile, low molecular weight compounds with standard high-resolution instrumentation. Thus, a large fraction of the larger phosphorus-containing materials such as membrane phospholipids, nucleic acids, and phosphoproteins are NMR invisible. In some rare situations, this NMR invisibility can present a problem if, for example, one wants to follow the pathway of the production of polymeric materials in which fluxes are rapid through low steady-state concentrations of mobile metabolities.[56] Methods such as saturation transfer, which shall be discussed in this section, can give quantitative information concerning the interconversion of key metabolites under various cell-stressing conditions. Obviously, high resolution *in vivo* ^{31}P NMR cannot give the complete picture of the functioning cell. However, in conjunction with other methods, *in vivo* NMR can strengthen conclusions concerning various mechanisms affecting metabolic processes.

1. pH and Compartmentation

^{31}P NMR has gained the reputation of being an "expensive" pH meter. However, it may be worth the price,** since the NMR can perform this task continuously and simultaneously on different intracellular compartments as it monitors other metabolic functions. In addition, one can also correlate the levels of high- and low-energy phosphates, membrane integrity, and intracellular pH in a single experiment.

The measurement of intracellular pH is most often made utilizing the Pi resonance as a spy or probe molecular within the cellular compartments, although the resonances of other phosphorus compounds such as glucose-6-phosphate (G-6-P),[38] 2-deoxy-glucose-6-phosphate

* In animal tissues phosphocreatin and phosphodiester resonances are also observed,[8] while fungi and yeast exhibit resonances associated with stored polyphosphates[49] and phosphodiester[49] such as phosphorylserine (PS), phosphorylethanolamine (PE), and phosphorylinositol (PI); in algae,[50] glycerol 1,2-cyclic phosphate has been observed and fast-growing *Rhizobium* exhibit glycerophosphorylated cyclic β-1,2-glucans.[51]

** The price of a basic broad-banded high-resolution NMR instrument is generally $1,000/MHz, so a 400-MHz instrument would cost approximately $400,000.

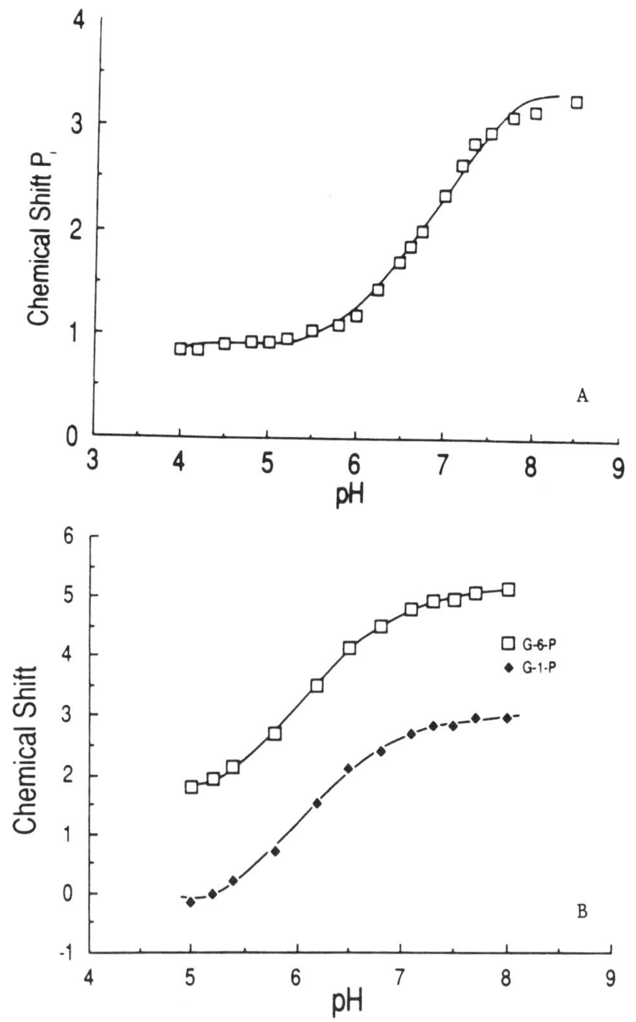

FIGURE 13. pH vs. ^{31}P chemical shift at 22°C. Solutions contained 5.0 m*M* K_2HPO_4, 2.0 m*M* $MgCl_2$, and 100 m*M* KCl for the Pi titration curve and 2.5 m*M* G-6-P and 2.5 G-1-P plus 2.0 m*M* $MgCl_2$ and 100 m*M* KCl for the sugar phosphate titration curve. The shift positions are given relative to HMPA which is assigned a value of 30.73 ppm (relative to 85% H_3PO_4 = 0.00).

(2DG-6-P),[13] methylphosphonate (MP),[57] and inosine monophosphate (IMP)[58] have also been used. The application relies on the fact that Pi exists as an equilibrium mixture of HPO_4^{-2} and $H_2PO_4^-$ under physiological conditions (~pH 7). Since the equilibrium between these two ions is rapid on the NMR time scale, we only see a single resonance representing the concentration of Pi at any particular pH. The observed δ_o or chemical shift of this Pi resonance is the weighted average shift of $^{\delta}HPO_4^{-2}$ and $^{\delta}H_2PO_4^-$ given by the relationship:

$$pH = pk + \log \frac{^{\delta}HPO_4^{-2} - \delta_o}{\delta_o - {}^{\delta}H_2PO_4^-} \tag{6}$$

By plotting the change in chemical shift δ_o of Pi vs. the known pH of standard solutions we obtain a titration curve as shown in Figure 13. The resonance position of two other metabolites within a cell can also be used to establish if they are in the same pH environment. For example, the chemical shift position of G-6-P or G-1-P should correspond to the same pH as reported by the Pi resonance if they are contained in the same cellular compartment (Figure 13). Since the pKas of these compounds are in the range of 6.2 to 6.8, they can report intracellular pH in the range of 5.5 to 7.8 ± 0.1 most accurately. Unfortunately, one cannot monitor pH changes for compartments with pHs below 5.5 or above 7.8 with any accuracy using the compounds shown in Figure 13 because of the flattened slope of the pH curve in these regions. However, in certain tissues, it is possible to introduce a nontoxic compound such as MP (pKa = 7.5) to report pH changes in the range of 8.[57] When constructing the pH vs. chemical shift plots, one must take into account the ionic strength, metal ion concentration, and interaction of other metabolites. Roberts[59] has thoroughly investigated the effects of a number of factors on the titration curves and pKas of Pi and G-6-P.

Once the ability to evaluate cellular pH was realized,[1] the method was exploited to evaluate subcellular compartmentation. The question was whether NMR would be able to provide information concerning the distribution of metabolites between the cell cytoplasm and mitochondria. Ogawa et al.[60] used the resonances of ATP, ADP, and Pi to monitor two environments (intra- and extramitochondrial) in suspensions of rat liver mitochondria. The two sets of observed Pi resonances were attributed to the difference in pH between the external and internal environments. The downfield shift of resonances corresponding to internal ATP and ADP was ascribed to their binding to divalent metal ions within the mitochondria. Metal ion binding, particularly that of magnesium, causes predictable chemical shift changes on the resonances of nucleotides such as ATP. These shifts give a direct measure of the ratio of ATP/Mg^{2+} *in vivo*.[38,61,62] As seen in Figure 12A, the γ-, α-, and β-NTP resonances show no ^{31}P-^{31}P coupling, but rather are broad and featureless. This is due primarily to the intracellular Mg^{2+} ion exchange broadening which provides an efficient mechanism for rapid spin-spin relaxation ($T_2 \sim 50$ ms).

Cohen et al.[63] studied suspensions of hepatocytes in the presence of valinomycin and observed a split in the Pi resonance corresponding to cytoplasmic and mitochondrial pHs. The cytoplasmic Pi resonance assignment was confirmed when fructose was added to the cells and produced a fructose-6-phosphate resonance whose chemical shift corresponded to the pH given by the upfield Pi resonance.

Roberts [13] and Kime et al.[38] examined the compartmentation of excised maize root tips. The resonances reported in Figure 12 corresponding to vacuolar Pi and cytoplasmic Pi have been assigned based on three criteria: (1) morphology, (2) pH, and (3) phosphorylation of glucose. In one of the first plant tissue NMR experiments, Roberts[13] demonstrated that upper root sections containing large vacuoles exhibited ^{31}P spectra in which the upfield Pi peak dominated, whereas, 1 to 2 mm root tip sections that are known to contain cells with small vacuoles gave spectra in which the lower field Pi resonance was very large. Kime et al.[38] showed that G-6-P generated from perfusion of maize root tips with 50 m*M* glucose exhibited a peak whose shift corresponded to a pH of 7.5, the same as the low-field Pi resonance. The upfield Pi had a shift corresponding to a pH of approximately 5.5, characteristic of an acidic vacuolar compartment.

The addition of 2-deoxyglucose to maize root tips can be useful as a pH marker for intracellular and metabolite compartmentation. Because of the large buildup of 2DG-6-P (see Figure 14), we can also observe the appearance of 2-deoxyglucose-1-phosphate which is not normally visible because of the high cellular ratio of G-6-P/G-1-P of approximately 19.[64] We also note that although 2DG-6-P cannot be metabolized through the normal glycolytic pathway, this compound can be incorporated via 2-deoxyglucose-1-phosphate into a new uridine derivative carrying the deoxysugars, whose chemical shifts are slightly downfield from UDPG.[65]

Kime et al.[38] examined the excised portions of potato tissue, i.e., the pith, parenchyma, cortex, radial pith, and a collection of potato buds. The pith tissue from the center of the potato gave a ^{31}P spectrum corresponding to phytic acid, whereas the parenchyma and cortex only showed a Pi resonance corresponding to an acidic vacuolar compartment. The radial pith tissue which links the central pith tissue to the potato bud gave a spectrum of phytic acid. The bud, however,

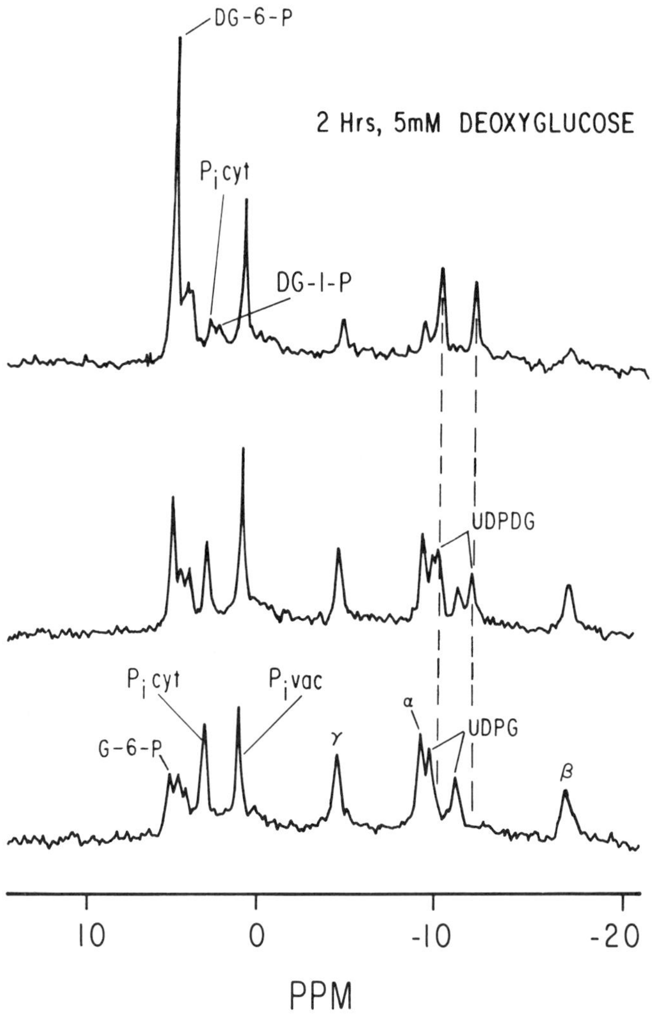

FIGURE 14. Incubation with 2-deoxyglucose 161.7-MHz ^{31}P spectra of maize root tips; lower spectrum before addition of 5 m*M* 2-deoxyglucose; middle spectrum, following 30-min treatment with 5 m*M* 2-deoxyglucose; and upper spectrum, after 2-h treatment with 5 m*M* 2-deoxyglucose. (See Reference 65.)

displayed a typical spectrum characteristic of meristematic tissue, exhibiting a cytoplasmic and vacuolar Pi as well as monophosphate esters. Earlier reports had suggested that phytic acid was distributed uniformly throughout potato tubers.[66] This study emphasizes the fact that certain key metabolites may be compartmented within a diverse differentiated tissue. However, to establish such differentiation, one must physically separate the parts of a tissue before examining them by ^{31}P NMR. If the whole tissue including the bud were examined intact, no spectral resolution or interpretation of the spectra could have been achieved. These authors[38] have also demonstrated in a similar experiment the differentiation and compartmentation of immobilized phytic acid in a seed, mobile, vacuolar Pi in a proximal root section, and mobile monophosphate esters and vacuolar and cytoplasmic Pi in a root tip from the same maize seedlings. This tissue differentiation is clearly illustrated in the ^{31}P NMR spectra shown in Figure 15. Others have also examined the ^{31}P spectra of developing turnip,[67] lettuce,[68] and cotton seeds.[69] In the spectra of the turnip and lettuce seeds, compartmented metabolites were visualized following approximately 20 h of imbibition.

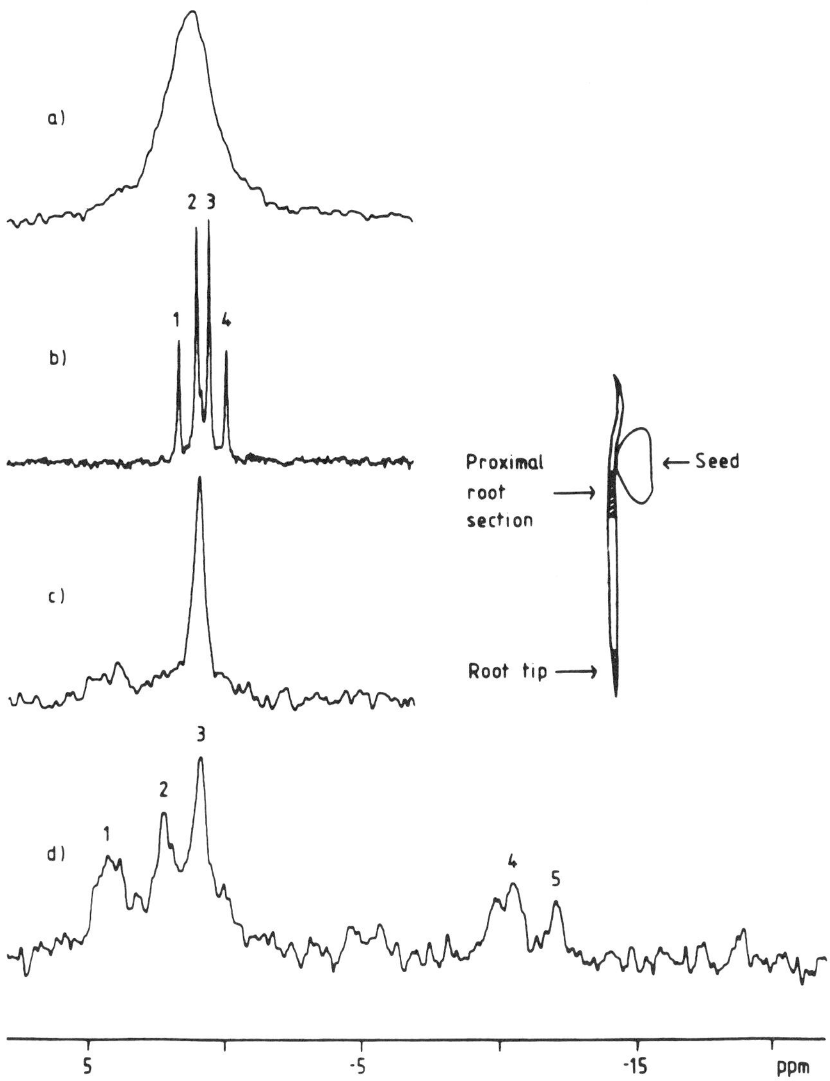

FIGURE 15. 121.49-MHz ^{31}P spectra of sections of 2-d-old maize seedlings (*Zea mays*, L. var. Aurelia): (a) 3 half seeds; (b) an aqueous extract from the same seeds at pH 5.2; (c) 100, 5.0-mm proximal section 2 s from the 4.0-cm primary roots of the Aurelia plants; and (d) 100, 3.0-mm root tips from the 4.0-cm primary roots of the Aurelia plants. The schematic diagram in the inset defines the different regions of the tissue. All the spectra are the result of 4000 scans with the slow acquisition cycle, using the coaxial 8 in 10 tube system. In the extract spectrum (b), the phytic acid resonances are labeled 1 to 4 and the small P_i resonance is visible as a shoulder to peak 2. In the root tip spectrum (d), the assignments are peak 1, monophosphate esters; peak 2, cytoplasmic P_i; peak 3, vacuolar P_i; and peaks 4 and 5, UDPG. (From Kime, M. J., Ratcliffe, R. G., and Williams, R. J. P., *J. Exp. Bot.*, 33, 656, 1982. With permission.)

^{31}P studies of dormant *Artemia* embryo spores showed that, as these spores develop following perfusion with O_2, their interior changed from pH 6.3 to 7.9 as reported by their Pi shift.[70] Moreover, nucleotide resonances appeared and an additional Pi resonance was seen and tentatively identified as a nonviable fraction of the cysts (~35%). This rather large and reversible shift in interior pH is very remarkable; however, the source of the proton equivalents generated during the induction of anaerobic dormancy of these spores is unknown.

Many workers have attempted to evaluate the effect of external environment on intracellular pH in different plant tissues and cells. An early study[71] of the temperature dependence of the cytoplasm of maize root tips indicated that the pH of the cytoplasm dropped approximately 0.5 pH units when the temperature was raised from 4 to 28°C. The poor quality of these spectra due possibly to the bubbling of air without perfusion make it unclear whether the upfield shift (acidification) of the cytoplasmic Pi resonance was due to mild hypoxia rather than a temperature-dependent pH change associated with an alteration in metabolism. Agents such as K_2SO_4 which induce the extrusion of four to five microequivalents of H^+ per gram per hour from maize root tips do not produce any measurable change in the cytoplasmic or vacuolar pH.[72] However, hypoxia, a subject that will be covered in Chapter 2, does cause a significant acidification of the cytoplasm. Dilute NH_4OH, on the other hand, induced a pH rise in the vacuolar compartment more rapidly and to a greater extent than the cytoplasm. This difference is presumed to be due to tighter pH regulation in the cytoplasm than in the vacuole, or possibly may be attributed to the fact that the vacuole is a better, more acidic trap for NH_4OH. That the cytoplasm was found to return more rapidly to the initial pH after treatment suggests that the former explanation is correct. This study points to the fact that cytoplasmic and vacuolar pHs are regulated for the most part by separate mechanisms in meristematic cells. Reid et al.[73] measured the cytoplasmic pH change caused by sodium azide treatment and anaerobiosis of maize root tips. Their findings suggest that when respiration was inhibited by azide, H^+ influx was the dominant effect in the acidification of the cytoplasm, whereas during anoxia, lactate formation was the main source of protons. Butyrate at a concentration of 1 m*M* had no effect but fusicoccin caused a 0.1 to 0.2 pH rise in barley root tips.[74] The effects of external acidity have been examined in both maize root tips[14] and barley shoot tissue[75] to simulate the effects of acid soil and acid rain, respectively, on intracellular pH, metabolism, and growth. While excised maize root tips show almost no change in cytoplasmic or vacuolar pH, as revealed by the pH-dependent Pi signal after perfusion at pH 4.0 for 20 h, barley shoot tissue exhibited a small measurable pH lowering of vacuolar pH following 2 d of spraying with acidic (pH 3 to 4) solutions. Although no growth reduction was noted in the barley tissue, the changes in vacuolar pH and plastidic sulfate content may have an energetic cost to the plant that could translate into reductions in growth.[75]

At high concentrations (5×10^{-4} *M*) and pH 6.0, 2,4-dinitrophenol (DNP) caused a loss of the pH gradient across the tonoplast of maize root tips.[38] Membrane integrity was restored, however, following continuous washing of the tissue. In contrast, at 1×10^{-5} *M* and pH 5 or 7, no ATP uncoupling or loss of tonoplast gradient was observed in barley root tips.[76] These workers[76] indicate that the collapse of the pH tonoplast gradient and loss of ATP previously reported[38] for maize root tips was caused by a large increase in permeability resulting from the high concentration of protonated DNP (5 μ*M* at pH 6) and was not a result of uncoupling per se. Treatment of maize root tips with uncoupler carbonyl cyanide p-trifluoro-methoxy-phenhydrazone (FCCP) at 10^{-5} *M* in 2% $DMSO_4$ (used for solubilization) caused a full collapse of the tonoplast gradient as well as the plasmalemma within 3 $^1/_2$ h (see Figure 16). This effect could not be reversed by tissue washing (unpublished results). Similar results have been reported for *Acer pseudoplatanus* cells.[27] Intracellular pH and K^+ ion concentration have been evaluated in mung bean roots following treatment with Ca^{2+} ionophores.[77] These experiments demonstrated that the Ca^{2+} ionophore A23187 increased [K^+] and decreased cytoplasmic pH from 7.2 to 6.6 (pH 6.6 is near the optimum pH for K^+-ATPase). This activation of K^+ uptake by Ca^{2+} makes a convincing argument that Ca^{2+} enhances the K^+-ATPase activity through acidification of the cytoplasm.

Martin and co-workers[27] have made a thorough evaluation of different factors (e.g., O_2, hypoxia, anoxia, Mn^{2+}, PO_4^{2-} content, temperature, and FCCP) affecting the viability and ^{31}P NMR spectra of three varieties of cultured plant cells (*A. pseudoplatanus*, *Catharanthus roseus*, and *Glycine max*), in an unperfused state. This paper is an excellent introduction into the use of *in vivo* ^{31}P NMR with cultured plant cells. Following this work Hughes et al.[78] reported an

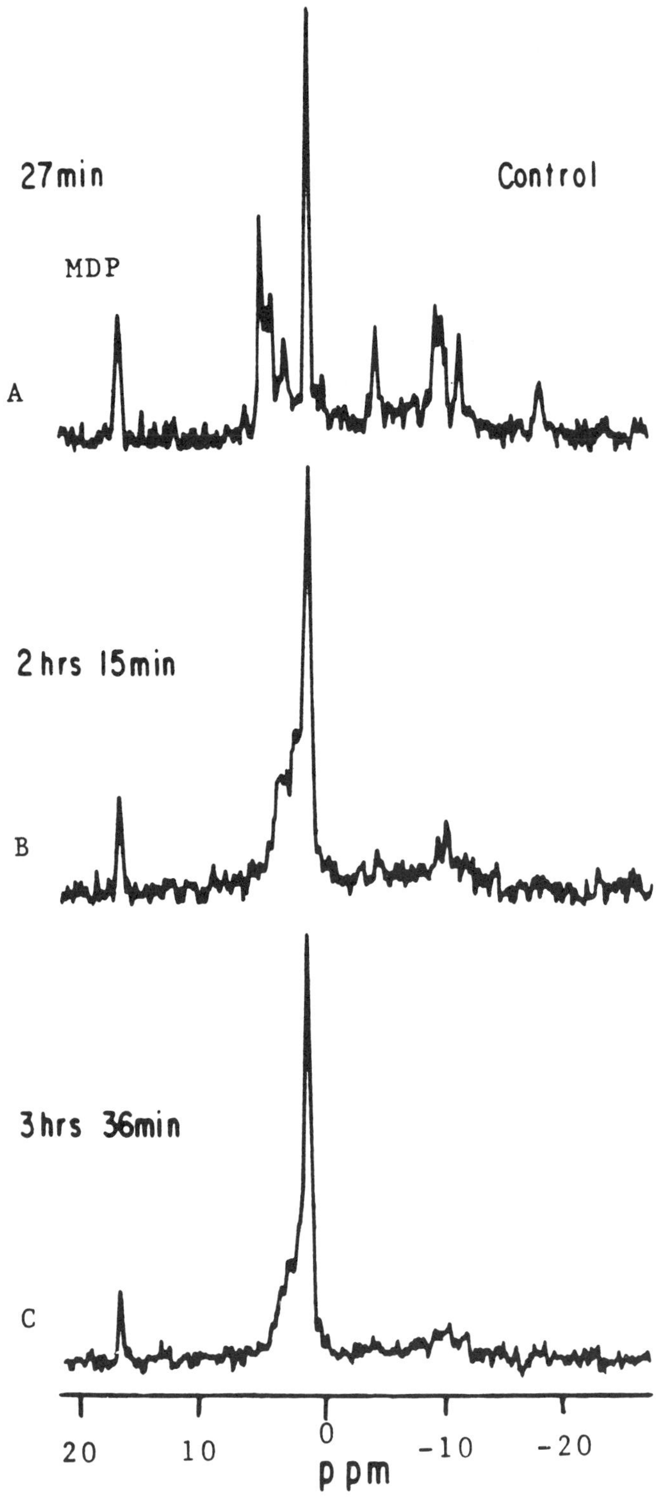

FIGURE 16. Uncoupling and loss of pH gradient. 161.7-MHz ^{31}P spectra of maize root tips (A) prior to treatment and (B) after 2 h, 15 min of treatment with 10^{-5} *M* FCCP and 2% DMSO in the circulating perfusion system saturated with O_2. (C) same as B after 3 h, 36 min.

improved method for obtaining extraordinarily good ^{31}P spectra of cell suspensions of *Elaeis guineensis* during perfusion. This paper demonstrates that the spectra can be utilized to probe the action of the uncoupling behavior of the auxin analogue, 2,4-dichlorophenoxy acetic acid (2,4-D). The effects of acid loading and cytoplasmic pH regulation in *A. pseudoplatanus* have been evaluated by Guern et al.[79] Acid loading of cultured plant cells stimulates the plasmalemma proton pump and can be used to study the relationship between cytoplasmic pH and growth. Their findings show that cultured cells can regulate their cytoplasmic pH during acid loading and, on the alkaline side, after removal of acid. That there is a strong pH regulatory mechanism in plant cells is indicated by the rapid recovery of the cytoplasmic pH.[79] ^{31}P studies of *Phaseolus vulgaris* cells under nonperfusion conditions (oxygen bubbling only) have demonstrated four changes following the addition of the elicitor phaseollin.[80] These are decreases in the vacuolar pH, cytoplasmic pH, ATP, and cytoplasmic Pi. Although the authors claim to observe a drop in vacuolar pH from 5.3 to 4.8, a change possibly associated with a metabolic process leading to phytoalexin formation, it seems doubtful that such a small pH drop could be accurately noted from the Pi shift in this acidic environment.[80] In addition, the loss of ATP, which was not found to be equated with any increase in Pi, was attributed to its utilization by the cells in phosphorylation reactions required for the synthesis of mRNA, which is NMR invisible. The slow recovery of the cell suspension to its original metabolic state took 48 h, indicating that enzyme degradation of the elicitor was complete after this time period.

Other kinds of plant materials that have been examined by ^{31}P NMR include an intact stem segment from sunflower[81] and pollen grains from flower buds of tobacco.[81] Spectra of these tissues exhibit evidence of cytoplasmic and vacuolar compartmentation. In addition, the spectrum of the pollen grain extract showed a large quantity of a phosphodiester, tentatively identified as glycerophosphorylcholine (GPC). GPC as well as glycerophosphorylethanolamine has also been recently reported in rice shoots.[82] These metabolites are both presumed to be present in the cytoplasm but only the GPC is modulated by the environmental oxygen level, i.e., GPC is produced in large quantities over long periods of anoxia. The authors hypothesized that GPC is produced to maintain cell membrane functionally under environmental stress, since GPC is a known inhibitor of lysolecithinase.

2. *Saturation Transfer*

By simply measuring the equilibrium levels of metabolites, e.g., NTP from the ^{31}P NMR spectrum of different tissues under a variety of stressing conditions, we cannot assess whether the rate of synthesis of NTP is increasing, decreasing, or remaining unchanged. We cannot assess the real change in these metabolite concentrations because the concentration of nucleotides and products of nucleotide breakdown do not bear a simple relationship to variations in the rate of nucleotide turnover.[83] To gain insight into true rates of production or consumption of a nucleotide such as ATP, one must be able to evaluate the rate constant for ATP interconversion. Saturation transfer is an NMR technique whereby low-power rf field saturation of a spin on a nucleus will transmit that saturation to any other nucleus with which it is exchanging. The rate of appearance of the nonequilibrium spin magnetization of the second resonance is monitored to determine the reaction rate. In order to measure these unidirectional reaction rates, the system must be in a steady state; the exchanging partners, e.g., γ-ATP and Pi must have separate and detectable NMR signals and the rate constants for both the forward and reverse reactions must be of the magnitude of the spin-lattice relaxation rate of each of the exchanging nuclei. In general, processes that are occurring with rate constants of roughly $10\ s^{-2}$ to $10\ s^{-1}$ may be probed by saturation transfer. A review covering the application of saturation transfer to the *in vivo* study of animal tissues and microorganisms has recently been published.[84] Using oxygen tension measurements in conjunction with saturation transfer experiments, Roberts and co-workers[85] have demonstrated that the P/O ratio, which is 3 in normal metabolizing root tips drops to 2 when succinate is substituted for glucose. This observation indicates that a secondary pathway in the electron transport system has been activated in which $FADH_2$ is the principal electron donor. Addition of cyanide with succinate suppresses the rate of ATP synthesis completely while cyanide and glucose still gave a P/O ratio of 1. These results indicated that this plant tissue contains a cyanide-resistant respiration pathway in which electrons travel to a terminal oxidase from a point between the first and second coupling sites of the mitochondrial electron transfer chain and, therefore, only the first coupling site of the electron transfer chain is used.

Although comparable experiments have not been reported in agricultural systems, in vivo unidirectional fluxes between creatin phosphate and γ-ATP have been determined by an alternate two-dimensional exchange experiment (2D-NOESY) in the head and leg of an anesthetized rat.[25] The 2D-NOESY experiment monitored an exchange process by taking advantage of a transient frequency label that is imposed on each of the exchanging species, e.g., ATP and creatin phosphate during some specific mixing time. The resulting two-dimensional spectrum is then plotted along two frequency axes, F_1 and F_2. F_1 is the precessional frequency of the nucleus before mixing with the second nucleus and F_2 is the frequency obtained following the mixing time. Nuclei that have not undergone exchange during the mixing time should show resonance signals only along the diagonal since they have the same precessional frequency in F_1 and F_2. Those nuclei that have undergone exchange will have two different frequencies, one on the diagonal corresponding to F_1 and a new position in F_2 corresponding to a change in chemical shift due to the dipolar interaction with the second nucleus during the mixing time. These second peaks representing the interaction frequencies are visualized as cross or off-diagonal peaks as illustrated in Figure 17. The unidirectional flux of creatin phosphate to γ-ATP was found to be 13 μmol/s/gwt (gram-weight) in the leg and 2 μmol/s/gwt in the head of the rat. These values are in good agreement with the standard saturation transfer measurements.[86] No unidirectional flux between γ-ATP and Pi was noted because it was assumed to be too slow (<0.1 μmol/s/gwt) to measure.[25] Some discrepancy exists between these results and those reported for the saturation transfer experiment,[86] since magnetization transfer was observed between γ-ATP and Pi in the latter. The authors suggest that the disparity in the results could come from the fact that some of the "bound" substrate pools which exchange with the observed free pools may have been saturated in the saturation transfer experiment.

3. Phosphate Uptake and Compartmentation

As mentioned earlier, phosphate is a key nutrient that is involved in a variety of life functions, which include muscle contraction in mammalian tissue,[8] regulation of glycogen[87] and starch breakdown in animal and plant cells,[54] glycolysis,[51] and mitochondrial function.[88] Thus, in order to understand these regulatory functions we must understand (1) how phosphate is distributed in the cell, (2) in what form is it stored, and (3) how Pi uptake and distribution in the cell is affected by different stresses.

Pi uptake has been extensively studied by ^{31}P NMR in various plant tissues, cells, and microorganisms. Kime et al.[38] carried out the first ^{31}P NMR experiments on aged potato parenchyma tissue that demonstrated uptake of Pi into the vacuole compartment. Pea root tips, under aerobic conditions show the same selective accumulation of Pi in the vacuole.[40] By measuring the relative areas of the vacuole and cytoplasmic Pi resonances under quantitative conditions, these workers were able to establish the ratio of Pi in the two compartments. Following the wet chemical determination of total Pi per unit fresh weight and an estimation of the size of the cytoplasmic and vacuolar volumes, the actual concentrations of vacuolar and cytoplasmic Pi were calculated. Vacuolar Pi concentration varied significantly as a function of the concentration of Pi supplied, yet the cytoplasmic Pi concentration remained remarkably constant over the entire range of Pi nutrition. The cytoplasmic Pi concentration was estimated to be approximately 18 m*M* while the vacuolar Pi concentration varied from 3 to 14 m*M*.[40] These studies assumed any Pi which could be taken up in the mitochondria would only represent approximately 6% of the total Pi in the cytoplasm and, therefore, not contribute to any serious

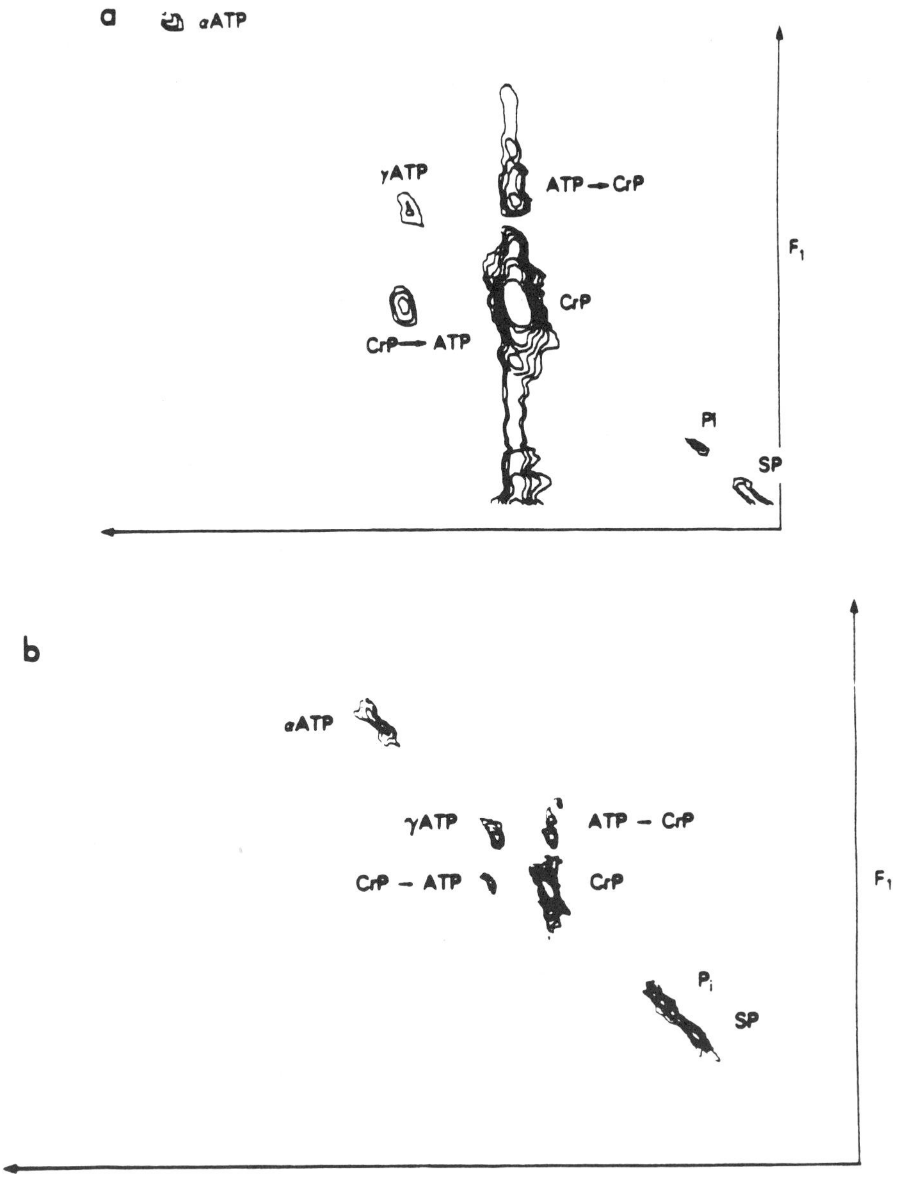

FIGURE 17. Two-dimensional contour plots of the (a) rat leg and (b) head. Diagonal peaks represent chemical species which have not undergone exchange. The lack of symmetry in the cross-peaks is caused by the added noise in the F_1 axis produced by pulse sychronization distortion. Contour *a* was collected with a t_{mix} of 0.6 s without quadrature detection with 16 scans per spectrum, and a sweep width of 2000 Hz. Contour *b* was with a t_{mix} of 0.4 s with 32 scans per spectrum with a 4000-Hz sweep width. (From Balaban, R. S., Kantor, H. L., and Ferretti, J. A., *Biol. Chem.*, 21, 12787, 1983. With permission.)

error in these calculations.[40] Rebeillé et al.[55] carried out a similar study of Pi uptake in *A. pseudoplatanus* cells and obtained comparable results as far as relative uptake of Pi in the vacuole and cytoplasm was concerned. When the cells were starved of Pi, the intracellular Pi concentration fell from 3 to 0.1 m*M* and the concentration of phosphate esters dropped from 8 to 2 m*M*. However, following starvation, uptake of Pi was stimulated, indicating that Pi uptake depends on the intracellular Pi and/or cytoplasmic ester-phosphate pool. These data suggested that Pi and/or ester-phosphates might act as modifiers of the transport system to alter its affinity or as a competitive inhibitor. Low extracellular pH fostered the uptake of Pi which the authors attribute to an H^+ symport mechanism. Further studies[89] of this system under sucrose starvation showed a slow down in metabolism, as evidenced by lowered NTP levels and a replacement of sucrose breakdown with starch hydrolysis. During this time the intracellular Pi concentration increased and ^{31}P NMR indicated that the majority of Pi was accumulated in the vacuole. Introduction of sucrose, following starvation led to the disappearance of the cytoplasmic Pi resonance and a significant increase in the G-6-P resonance.[90] However, the vacuolar Pi concentration was not depleted to replace the utilized cytoplasmic Pi. During sucrose deprivation a resonance corresponding to the choline phosphate (P-choline) was noted. These authors suggest that this compound is a good marker for membrane breakdown under stressing conditions.[90,91] Subsequently, Brodelius and Vogel[92] studied the Pi uptake and storage of *C. roseus* and *Daucus carota* cells in a perfused state. These workers noted that in *C. roseus* Pi was the first stored in the vacuoles. One part of this pool was subsequently used to maintain a constant cytoplasmic Pi level, while loss of additional Pi signal was attributed to the accumulation of Pi in an NMR-invisible state (immobile or bound state) in another cell organelle. *D. carota* did not accumulate Pi in its vacuoles and took up Pi from the medium only as it was required during growth.[91] Yeast *(Candida utilis)* vacuolar-stored Pi as well as polyphosphates (PPs) were mobilized to replenish cytoplasmic Pi when glucose was metabolized.[93] Following glucose depletion, resynthesis of vacuolar PP takes place under aerobic conditions. However, under anaerobic conditions these yeast continue to hydrolyze PP.[93] This hydrolysis is probably due to the lower level of ATP present during this time. Figure 18 shows the sequence of events following glucose addition relating to PP breakdown, movement of vacuolar Pi to the cytoplasm, synthesis of sugar phosphates, and resynthesis of vacuolar PP following depletion of glucose under aerobic conditions. When ethanol was used as a substitute, only PP synthesis was noted. These results suggest that PP may have a dual role as a store of Pi as well as an energy source.[93] Nitrogen-starved yeast cells (*Saccharomyces cerevisiae*) respond to treatment with ammonium salts at pH 8.0 with a rapid hydrolysis of intravacuolar PP to tripolyphosphate (TPP) and Pi, followed by movement of the Pi to the cytoplasm.[94] This sequential process could be examined by ^{31}P NMR since the intermediate PP breakdown products, i.e., TPPs and tetrapolyphosphates, and the decrease in the ratio of the central to terminal phosphate resonances could be observed throughout the time course of treatment. The authors[94] hypothesize that the ammonium-initiated PP breakdown provides another mechanism of maintaining pH homeostasis in the vacuoles of energy-depleted cells when they are stressed with a highly basic metabolite. A fungal pathogen *C. albicans* was examined in its transition from yeast-form cells to germ tubes and hyphae by ^{31}P NMR.[95] In the exponentially growing yeast-form cells, the PP resonance was high but progressively declined in the stationary phase and became very low and finally undetectable in the germ tube and hyphae stages, respectively. At the same time, cytoplasmic Pi grew in intensity. Thus, PP appears to accumulate early during growth and is disposed of when the organism shows its growth due to limitations of its carbon source. When the organism was grown with a surplus of phosphate, both the PP and vacuolar Pi signals increased in intensity relative to the cytoplasmic Pi signal, indicating that there is probably a metabolic interchange between PP and free Pi in the vacuole,[95] similar to that mentioned above.[93] Maize root tips exposed to high concentrations of NaCl (100 m*M*) took up twice as much Pi from a 10-m*M* KH_2PO_4 solution.[96] This stimulated uptake of Pi was associated with an increase in the concentration of cytoplasmic Pi. Following the Pi uptake, Pi was transferred to the vacuole. The NaCl appeared to increase the permeability of the plasmalemma to Pi, which resulted in a large influx of Pi from the external solution. In contrast, the movement of Pi from the cytoplasm to the vacuole depended on the ratio of Pi to total phosphate (Pi + G-6-P + NTP + UDPG) in the cytoplasm.

Ben-Hayyim and Navon[97] examined the uptake of Pi under salt stress for wild-type and NaCl-tolerant citrus cultured cells. The wild type was found to accumulate higher amounts of Pi than the NaCl-tolerant variety when exposed to equal external concentrations of Pi. As observed in

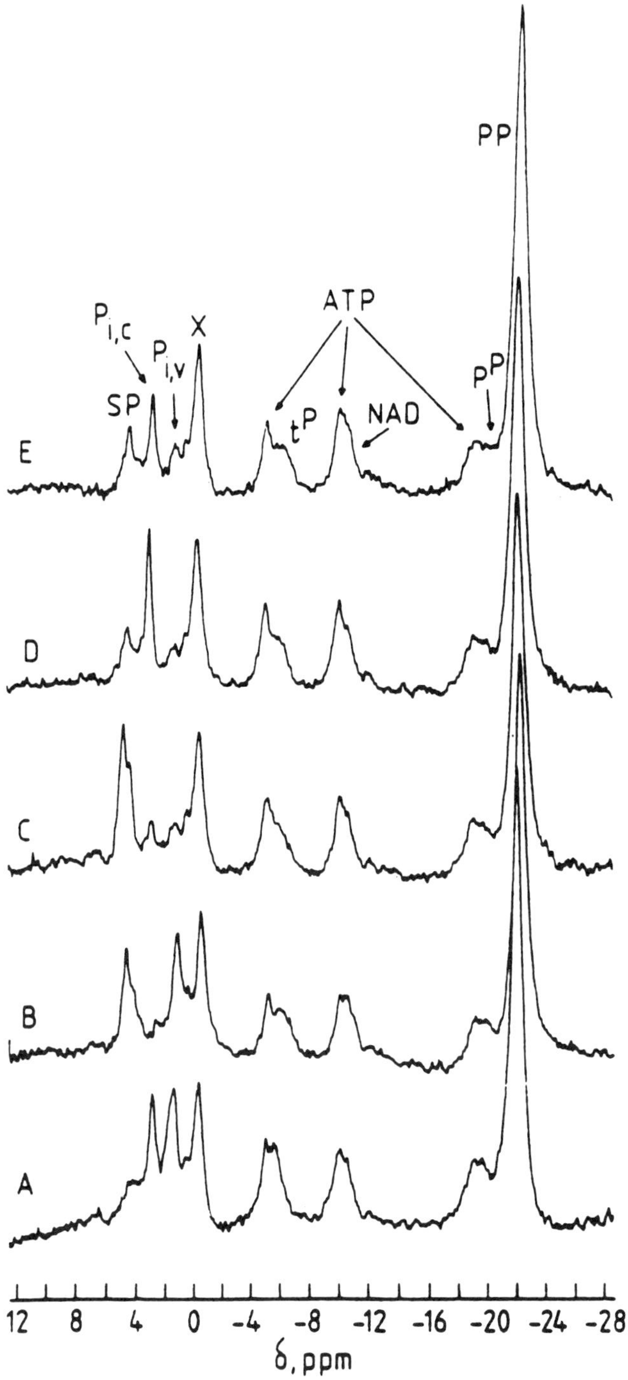

FIGURE 18. A to E. ^{31}P NMR spectra of an aerobic suspension of *C. utilis*. The temperature of the sample was maintained at 8°C. Each spectrum represents the sum of 2 consecutive blocks of 250 scans. A spectrum accumulated immediately before introduction of 100 m*M* glucose into the cell suspension: B, C, D, and E spectra taken 1.4, 22.7, 68.0, and 99.2 min after glucose addition, respectively. Assignments are SP, sugar phosphates; $P_{i,c}$, cytoplasmic P_i; $P_{i,v}$, vacuolar P_i; X, unassigned; ATP, adenosine triphosphate; NAD, nicotinamide adenine dinuleotides (reduced plus oxidized); and t^P, P^P, and PP, terminal, penultimate, and middle phosphate groups of polyphosphates, respectively. Resonance t^P includes inorganic pyrophosphate. (From Nicolay, K., Scheffers, W. A., Bruinenberg, P. M., and Kaptein, R., *Arch. Microbiol.*, 134, 270, 1983. With permission.)

the previous study,[96] the additional Pi finally moved to the vacuole. The cytoplasmic Pi concentration reached saturation at a relatively low external Pi concentration whereas the vacuolar Pi showed an increase many times its original.

4. Metal Ion Transport and Compartmentation

Both soil acidity and its effects on the liberation of toxic metals such as Al, Mn, and Cd can produce adverse conditions for the growth and production of plants.[98] Although much work has gone into the development of new strains of metal-tolerant plants to combat such problems, little is known concerning the physiological mechanisms responsible for stress tolerance or susceptibility in plants.[99]

Low pH (<4.5) conditions in the soil can perturb the overall functioning of a plant and its ability to assimilate adequate levels of organic and inorganic nutrients. However, the exact effect of hydrogen ion concentrations on this phenomenon is often difficult to determine because of the occurrence of solubilized forms of toxic metals that are present under these conditions.[98] With regard to the plant root system, it is thought that acid conditions are responsible for greater membrane permeability,[98] impaired cation transport,[100] metal displacement, and inhibition of metal uptake.[101] A paper reviewing general ion absorption studies in maize and potato tissue with ^{31}P NMR has recently appeared.[102]

Al^{3+} is among the most studied metals in terms of its toxic effects on plant function.[98] The specific action of this metal has been attributed to interactions with nucleotides, nucleic acids, and membrane lipids, as well as important calcium-regulating ATPase and phosphorylating enzymes.[103]

Another metal which poses a problem for plant development and production is divalent manganese. In general, Mn^{2+} is not as toxic to such a wide variety of plants as Al^{3+}, yet its ease of transport across the cell membrane is well known.[98] In some instances healthy roots have been shown to reduce Mn^{2+} toxicity by precipitating oxidized Mn^{2+} as MNO_2 on root surfaces.[98] Plants such as maize, however, may avoid Mn^{2+} toxicity by entrapment of relatively high concentrations of the metal in nonmetabolic centers such as the vacuole.[104]

Although many of the biologically important metals have isotopes that are directly detectable by NMR, often they have too low a natural abundance level, are paramagnetic, quadrupolar, and/or are too low in concentration in the biological system to be observed. A way around this problem is to examine the indirect consequences of a particular invading metal ion on the metabolism, bioenergetics, intracellular pH, and, when the metal ion is paramagnetic, the line broadening of metabolite resonances associated with the metal ion following its entry into the cell.

Prior to characterizing the effects of mobile metal ions in acid soils on the function of maize root tissue, Pfeffer et al.[14,16] studied the role of the acid environment on the viability of these cells. The presence of an acid medium on excised maize root tips after 20 h of perfusion fostered the production of significant amounts of excess (>75% over the initial amount) mobile Pi residing in the vacuole (see Figure 19). Based on a quantitative treatment of the ^{31}P NMR data and calculated resonance response factors for each hydrolyzable metabolite, we concluded that the excess phosphate was not generated from mobile metabolites.[16] The significant increase of the choline phosphate resonance at approximately 4.0 ppm gave a strong indication that membrane breakdown was responsible for the additional phosphorus signal[90,91] (see Figure 19B). In contrast, treatment of the same tissue at pH 4.0 with 0.5 m*M* Al^{3+} for 20 h resulted in a loss of 50% of Pi content of the cells (Figure 20) with little effect on the nucleotide levels. This Pi loss was principally attributed to cell wall precipitation of $Al_2(PO_4)_3$ and acid phosphatase inhibition by Al^{3+}.[16] The presence of Ca^{2+} did not suppress the effects of Al^{3+} and when used alone Ca^{2+} also showed a similar but less intense modulation of Pi generation. Over a growing period these effects could be manifested in phosphate deficiency resulting in diminished plant production.

Under conditions of lowered metabolism induced by the deprivation of a carbohydrate source

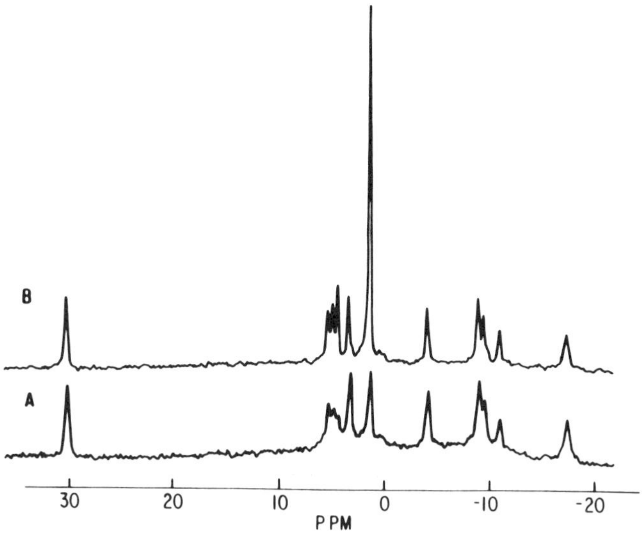

FIGURE 19. (A) 161.7-MHz ^{31}P spectrum of approximately 900 excised root tips (5 to 7 min) taken under the fast acquisition conditions following 2 h in the NMR perfusion system, pH = 4.0, 0.1 m*M* $CaSO_4$, 50 m*M* glucose, 4000 transients. (B) Same as (A) after 20 h of perfusion. (See Reference 14.)

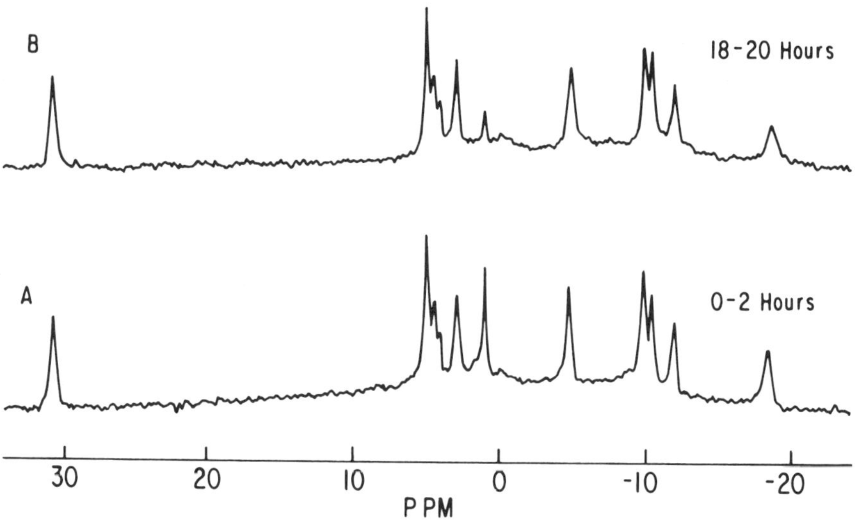

FIGURE 20. 161.7-MHz ^{31}P spectra of 3-d-old, 3 to 5 mm corn root tips perfused with a solution containing 50 m*M* glucose, 0.1 m*M* $CaSO_4$, and 2.5 m*M* $Al_2(SO_4)_3$, pH 4.0 for (A) 0 to 2 h and (B) 18 to 20 h. (See Reference 16.)

(glucose), the maize root tips underwent a drop in respiration and a drop in nucleotide levels when challenged with Al^{3+}. A similar loss of nucleotides and membrane integrity was noted in the ^{31}P spectrum of aluminum-treated tips under hypoxia.[16] Thus, cells already under a metabolic stress such as carbohydrate deficiency or hypoxia are much less able to maintain their energetic status to exclude toxic metal ions such as Al^{3+}.

Mn^{2+} is a useful paramagnetic metal ion for exploring metal interactions with NMR-active nuclei.[105] The presence of manganese in a cellular compartment is noted by the occurrence of broadened resonances (resulting from shortened T_1 and T_2 relaxation times) associated with the ites of metal binding. Several groups[14,65,105] have examined the sequential movement of Mn^{2+} into maize root tips by ^{31}P NMR to assess its metabolic fate. Under aerobic conditions, a slow entry of the metal ion into the cytoplasm was observed as evidenced by the gradual broadening of the cytoplasmic nucleotides and Pi resonances (Figure 21). Both the UDPG and phosphate monoesters were only broadened slightly because of their lower binding constant for Mn^{2+}, whereas NTP, which has a strong binding constant for Mn^{2+}, underwent extreme resonance broadening within 81 min. Final entry of Mn^{2+} into the vacuole compartment was verified by complete broadening of the vacuolar Pi resonance after 135 min. Following a thorough washing of the tissue with Mn^{2+}-free buffer, we observed that the vacuolar Pi resonance could not be regenerated,[14,65,106] whereas the cytoplasmic resonances were restored. Thus, Mn^{2+} was irreversibly trapped in the vacuolar compartment. Omission of glucose from the perfusion medium slowed the migration of Mn^{2+} into the vacuole, while rapid movement into the cytoplasm was still evident. A complete inhibition of the glycolytic pathway by incubation of the tissue with 5 m*M* 2-deoxyglucose depleted the cells of cytoplasmic Pi and nucleotides[65] (Figure 14). Subsequent treatment with Mn^{2+} followed by washing of the tissue showed the full restoration of the vacuole Pi resonance, indicating that Mn^{2+} had not traversed the tonoplast.[65]

During hypoxia, maize root tips can be sustained by an anaerobic metabolism with much reduced levels of NTP.[107] In this state, Mn^{2+} was slowly taken up in the cytoplasm in trace quantities; however, following a washing of the tissue with Mn^{2+}-free buffer, no sequestering of the metal was noted in the vacuole[14,65] (see Figure 22). A full description of the proposed mechanism responsible for the metal ion entrapment under aerobic conditions and nontransport under hypoxia are given in this report.[65] Ca^{2+} is known to protect plants from the invasion of toxic metals; however, whether this is a direct consequence of Ca^{2+} competing more favorably than other metal ions for movement across the plasmalemma is unknown. In competitive experiments with 10:1 Ca^{2+}:Mn^{2+}, it was demonstrated that a much-diminished level of Mn^{2+} buildup in the cytoplasm was observed while only minimal vacuole sequestering was noted.[65] Cd^{2+} toxicity, as observed by interference in the metabolic functioning of intact heart cells, by ^{31}P NMR has been reported by Kopp et al.[108] A similar effect has also been observed in plant tissues.[65] Ca^{2+} can effectively compete with toxic Cd^{2+} entry into these cells. Figure 23A and B show the ^{31}P spectra depicting the effect of 1 m*M* Cd^{2+}/0.1 m*M* Ca^{2+} on the metabolism of corn root tips following $3^1/_2$ h of perfusion. Note the rapid loss of cytoplasmic Pi and drop of the cytoplasmic pH from 7.6 to 7.0. In contrast, treatment of the tissue with a perfusate containing 10 m*M* Ca^{2+}/ 1 m*M* Cd^{2+} showed only marginal cytoplasmic "poisoning" after 3 1/2 h of perfusion (see Figure 23C and D).

The data obtained in these studies demonstrated that the uptake of metal ions (e.g., Mn^{2+}) in maize root tips is regulated by the energetic status (O_2 + exogenous glucose > O_2 > O_2 + 2-deoxyglucose > N_2 + glucose). Furthermore, different divalent cations may compete for the same root tip membrane transport apparatus. Based on the accepted general concepts[109] of ion transport developed mainly from research in nonplant systems, a detailed mechanism was proposed by the authors to account for the observations of these studies.[65]

Subsequent work describing a comparative study of facilitated transport of Mn^{2+} in *A. pseudoplatanus* (sycamore cells) and excised maize root tips by ^{31}P NMR has uncovered an important difference between the dicotyledon and monocotyledon cells, respectively.[110] Although passage of Mn^{2+} across the tonoplast of both sycamore cells and maize root tip cells was comparable, the entry through the plasmalemma was found to be much faster in the root tips. The slower movement across the sycamore cell plasmalemma was attributed to the occurrence of a high content of acidic polysaccharides in the cell walls of sycamore cells relative to the cell walls of root tissue. In the former, the high protonation of a plasma membrane carrier resulted in a lower affinity for metal ion binding and consequently a slower rate of Mn^{2+} movement. In addition, the greater ion exchange properties of these cell walls lowered the effective free metal ion concentrations in proximity to the membrane surface.[110]

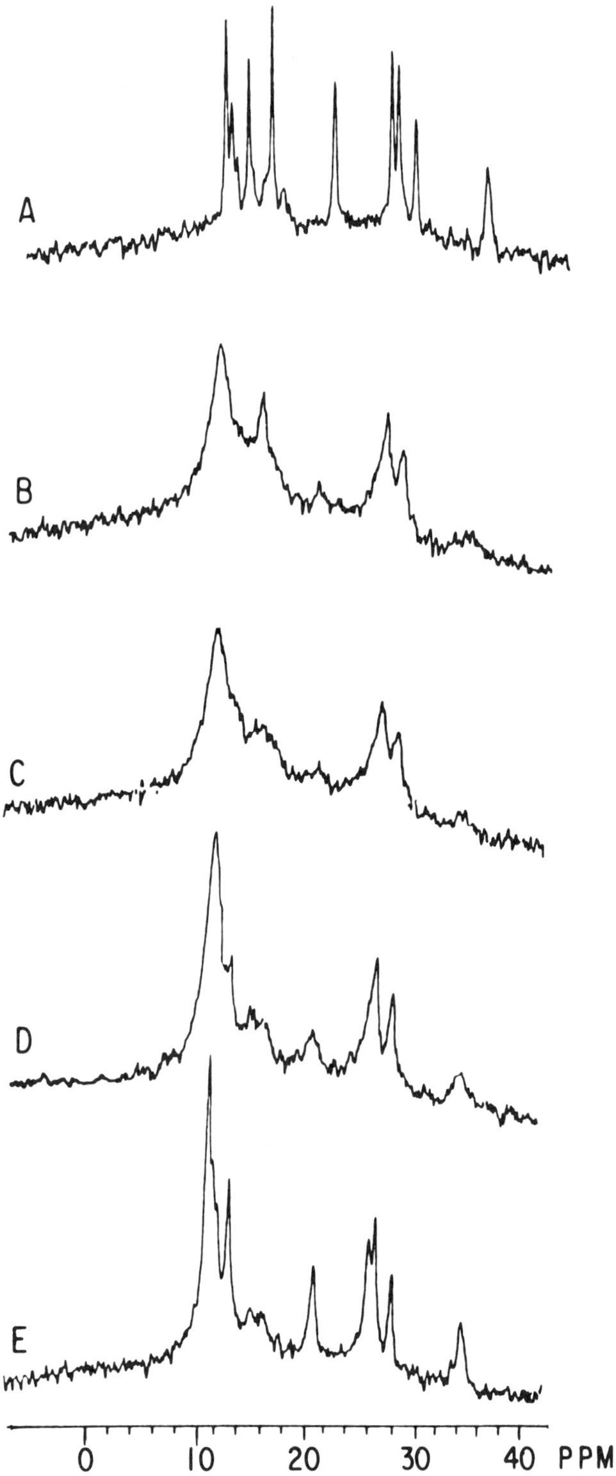

FIGURE 21. (A)^{31}P spectrum (10,000 transients) taken in Figure 20 after 3 h of perfusion with a solution of 0.1 m*M* $CaSO_4$, 50 m*M* glucose, 10 m*M* Mes buffer (pH 6.0) O_2. (B) Same as above except for the addition of 1.0 m*M* $MnCl_2$ and exposure for 54 to 81 min. (C) Same as (B), exposure 108 to 135 min. (D) Sample perfused with buffer as in (A) for 0 to 27 min. (E) Same as (D) for 81 to 108 min. (See Reference 14.)

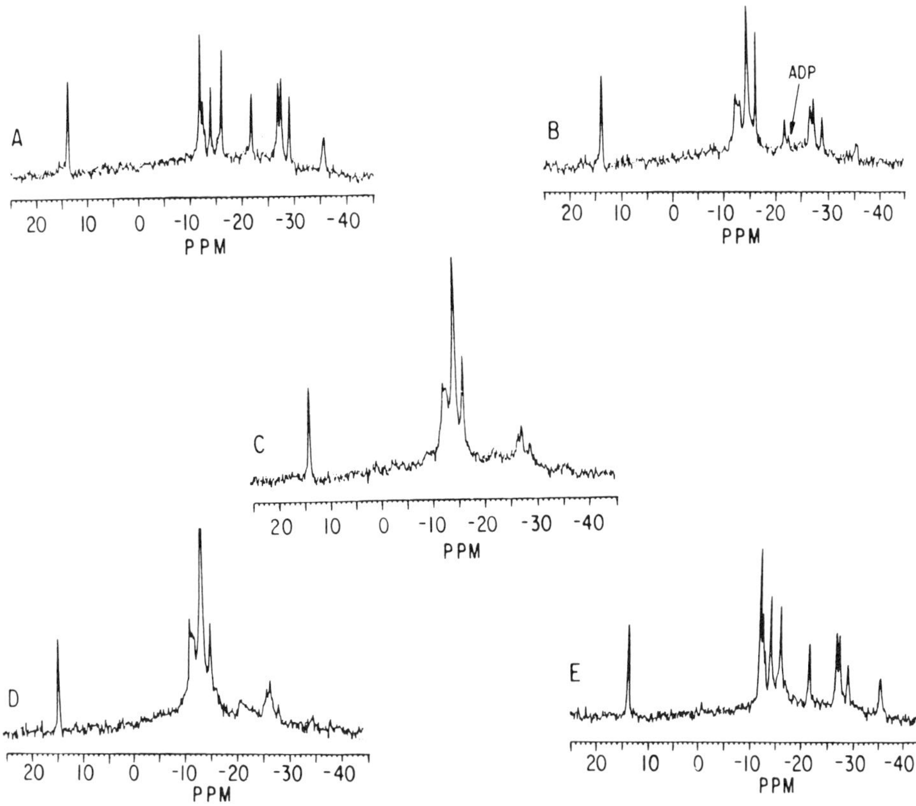

FIGURE 22. (A) ^{31}P spectrum (100,000 transients) taken as in Figure 21 above and perfused with a solution containing 0.1 m*M* $CaSO_4$, 50.0 m*M* glucose, 10.0 m*M* MES buffer (pH 6.0) O_2. (B) Same as (A) except perfusate is saturated with N_2 54 to 81 min. (C) Hypoxic state 189 to 216 min, treatment with perfusate containing 1.0 m*M* $MnCl_2$ for 108 to 135 min. (D) Hypoxic state 297 to 324 min after being washed out with non-Mn-containing buffer for 81 to 108 min. (E) Washout as in (D) with O_2 after 81 to 108 min. (See Reference 14.)

5. *Photosynthetic and Symbiotic Systems*

Kallas and Dahlquist[111] demonstrated the first application of ^{31}P NMR to an active photosynthesizing system. They examined the ^{31}P spectra of actively photosynthesizing and darkened suspensions of *Cyanobacterium synechococcus*. These photosynthesizing microorganisms exhibited a single intracellular Pi resonance corresponding to the cytoplasmic compartment, but the second photosynthetic intrathylakoidal compartment was not resolved. The failure to resolve this compartment was presumably partly a result of the low S/N of the low-field instrument used and the small volume of the intrathylakoidal space (7 to 10%) in the bacteria.[111] In general, the pH of the cytoplasm was maintained at a higher level than the external pH when the external pH was in the range of 6.3 to 6.7. Exposure to light caused an increase in intracellular pH of 0.4 of a pH unit. The additon of CCCP destroyed the internal pH control in both light and darkness. Following this work, Foyer et al.[112] studied compartmentation from ^{31}P NMR spectra of intact cells, protoplasts, and chloroplasts from photosynthetic tissues. These workers observed individual pools (vacuolar and extravacuolar compartments) of Pi in the dark from asparagus cells and spinach protoplasts. The spectra from whole leaves of spinach, wheat, and rye grass were found to be difficult to intepret because of the broadening of the ^{31}P signals due to the large air spaces that occur between the misophyll cells and the presence of intracellular paramagnetic metal ion.[112] However, the intact cells, protoplasts, and chloroplasts gave better defined spectra. Unlike previously examined bacteria,[111] the intracellular pH was not tightly controlled in these cells, but determined by the pH of the suspending medium. Waterton et al.[113] have been able to

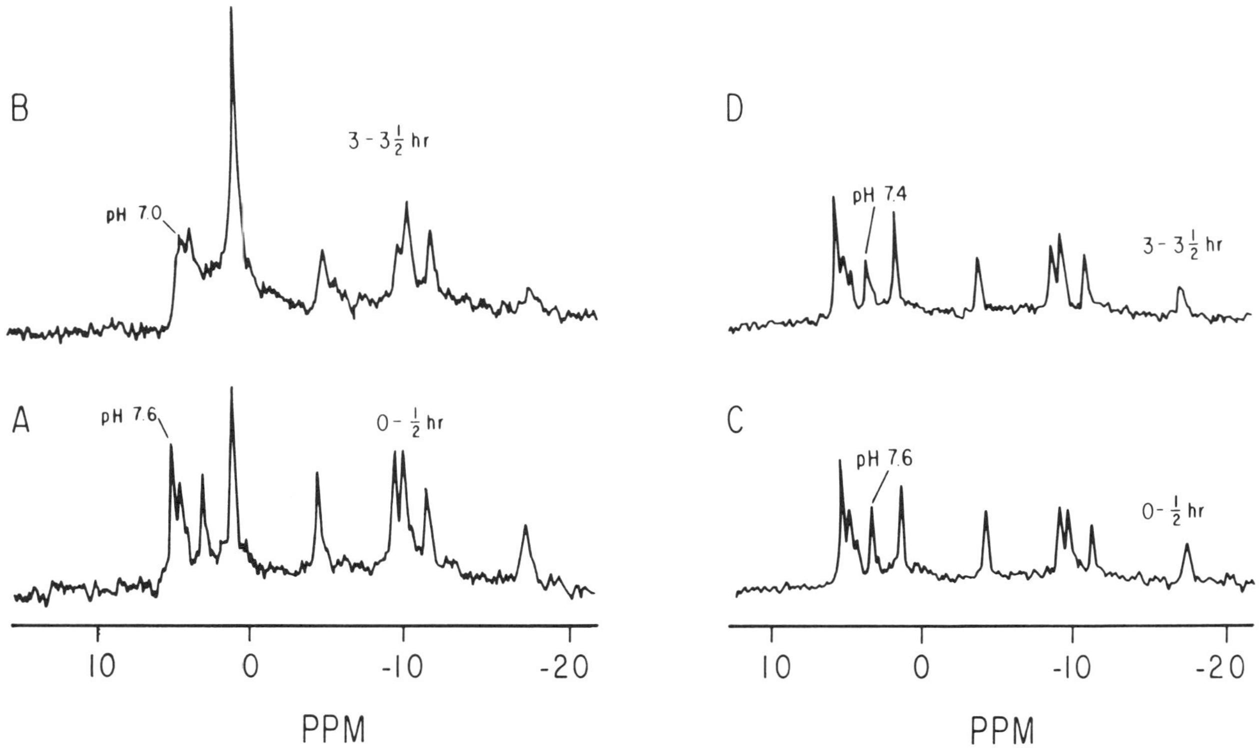

FIGURE 23. 161.7-MHz spectra of maize root tips (A) perfused with 50 m*M* glucose, 10 m*M* MES buffer, pH 6.0, 0.1 m*M* $CaSO_4$ for 1/2 h; (B) 3 1/2 h after the addition of 1 m*M* $CdCl_2$; (C) same as A except 10 m*M* $CaSO_4$ was used for 1/2 h; and (D) 3 1/2 h after the addition 1 m*M* $CdCl_2$.

establish the nature of the line broadening in the ^{31}P spectra of intact leaf tissue. Using wheat leaf disks, they measured the line width of the ^{31}P (in hertz) and showed that they were linearly dependent on the size of the external magnetic field. Thus, the line broadening observed in the previous study[112] was found to be a function of the leaf tissue air spaces. This effect arises because of the difference in the magnetic susceptibilities of the tissue water and intracellular air. Furthermore, these workers verified the phenomenon by elimination of the broadening with high-speed magic angle spinning (see Chapter 9 for details of this technique). Vacuum infiltration with buffered sorbitol was also found to be useful for improving the line width and resolution.[113,114] Two peaks, an extravacuolar Pi resonance corresponding to pH ~7.0 and vacuolar Pi (below pH 6.0) were detected in the dark with the ratio of 1:4, respectively.[113] Under illumination the extravacuolar Pi signal moved to lower field, indicating a more alkaline environment consistent with the extrusion of protons from the chloroplast stoma. Incubation with mannose produced mannose-6-phosphate outside of the vacuole, while the vacuolar Pi was depleted to maintain a constant level of Pi in the cytoplasm.

To obtain useful spectra from pea or spinach leaves, Foyer and Spencer[115] utilized ATP in the infiltrating medium as a chelating agent to distinquish between the acidic and neutral Pi pools. Barley tissue did not require the ATP treatment to distinguish the vacuolar and cytoplasmic Pi pools. Barley, soybean, and spinach leaf tissue showed a uniform response to Pi deficiency by increasing starch synthesis relative to sucrose, but the limitation under these conditions on photosynthetic capacity showed great variation among these plants.[115]

Unicellular algae such as *Chlorella* are ideal models for studying plant cell biology since they are simple systems (one cell and one chloroplast) and their extracellular growth conditions can be easily controlled. Also, because *Chlorella* can be treated as homogeneous cell suspensions, they are very amenable to examination by *in vivo* NMR methodology. Mitsumori and Osamu[116] were the first to simultaneously observe two compartments corresponding to the stroma of chloroplasts and cytoplasm in *Chlorella* by ^{31}P NMR. The two Pi signals from these compartments were assigned based on their responses to light and dark cycles and treatment with 3(3,4-dichlorophenyl)-1,1 dimethylurea. In the light, the chloroplastic pH became more alkaline, while the cytoplasmic pH became nonacidic. An increase in ATP was also observed upon illumination (see Figure 24). However, a more recent investigation of this organism using 2DG-6-P as a pH marker has reassigned the light-dependent resonance to the cytoplasmic Pi and not the stromal fraction.[116] Additional studies of *C. vulgaris* 11-h cells showed that no acidic compartment was detectable by ^{31}P NMR when the cells were treated with NH_4OH.[117] The vacuolar pH was also lowered by anaerobic treatment in the presence of glucose, but was not affected by external pH during preincubation in medium with pHs ranging from 3 to 10. These results suggest that acid fermentation takes place in *Chlorella* cells in the presence of glucose under anaerobic conditions, and therefore vacuoles may play a role as lysosomes as well as "metabolic buffer" in the cytoplasm.[117] A ^{31}P NMR study of unicellular alga acidophilic *Cyanidium caldarium* revealed a change in intracellular pH (acidification in darkness), but did not detect the vacuolar compartment in this species.[118] Illumination conditions maintained the intracellular pH from 6.8 to 7.0 even in environments with pH ranges from 1.2 to 8.4. These authors[118] contend that *Cyanidium* cells must have light to maintain their intracellular pH at a constant physiological level against passive H^+ leakage. The halotolerant alga *Dunaliella parva* studied by Ginzburg et al.[118] showed only a single pool of intracellular Pi. The intracellular pH was also found to be sensitive to changes in illumination as well as oxygenation. Spectra of the extract from these alga demonstrated the presence of a relatively rare metabolite, glycerophosphorylglycerol, whose function is not understood.

For the study of mycorrhizal systems, ^{31}P NMR affords the opportunity to examine an intact symbiotic assembly and evaluate the nature and distribution of phosphate compounds between the fungi and plant cell partners. Some preliminary studies of mycorrhizal root tips of beech *(Fagus sylvatica)* have shown the presence of a PP resonance (fungal) in addition to the normal

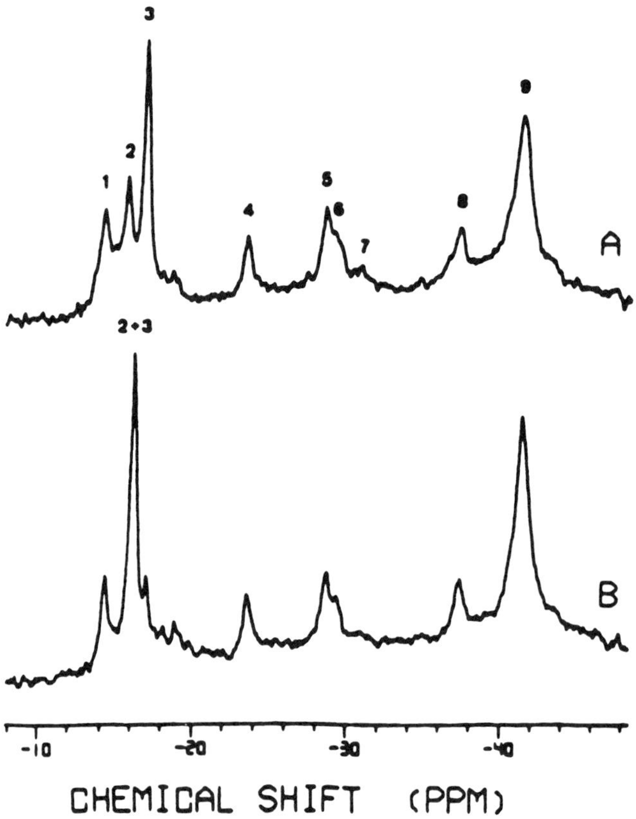

FIGURE 24. 161.8-MHz ^{31}P NMR spectra of intact autotrophically grown *Chlorella* cells measured in the dark (A) 27 min and (B) 137 min after light was turned off. Spectral assignments are 1, sugar phosphates; 2, chloroplastic P_i; 3, cytoplasmic P_i; 4, ATPγ; 5 ATPα; 6, NAD + UDP-glucose; 7, UDP-glucose; 8, ATPβ; and 9, polyphosphate. Chemical shift values are in ppm from external MDP. Packed *Chlorella* cells (1.0 ml) were suspended in the equal volume of medium containing KCl (100 m*M*), several mineral nutrients, Hepes (20 m*M*), and D_2O (10%). The initial pH of the cell suspension was adjusted to 7.0. External P_i had been removed by washing cells with P_i-free medium three times. The 2500 transients were accumulated with the recycle time of 0.22 s at 25°C. O_2 bubbling(10 ml/min) was performed to maintain aerobic conditions during NMR measurements. (From Mitsumori, F. and Osamu, I., *FEBS Lett.*, 174, 248, 1984. With permission.)

peaks of Pi.[4] Rapid hydrolysis of the PP following homogenation of the tissue suggested to the authors that the PP and polyphosphatase were located in different cellular compartments in the intact matrix.[4] Martin[119] has analyzed intact ectomycorrhizal fungi grown in pure and batch culture by ^{31}P NMR and has found them to contain a surprisingly low level (17% of total P) of PPs. However, the bulk of this polymer is observable by high-resolution ^{31}P NMR. Estimates of the chain lengths (average = 12) of soluble PP were made by evaluating the fungal extract spectra in terms of the ratio of the ^{31}P resonances associated with the inner and penultimate positions of the polymer. The observed relatively low concentration of short-chain PP is consistent with the fact that PPs are not normal metabolites of microoganisms, but are only rapidly accumulated by them under conditions of nutritional imbalance in older cultures.[119] Further studies of the two strains of ectomcycorrhizal fungi, *Cenococcum graniforme (CG)* and *Hebeloma cylindroporum (HC),* verified that the PP content was highest in the older and nitrogen-starved cultures and lowest in the rapidly growing mycelia.[120] The two varieties showed a difference in their

Table 5
LONGITUDINAL RELAXATION TIMES (T_1) AND LINE WIDTHS ($\Delta\upsilon_{1/2}$) OF ^{31}P NUCLEI OF INTRACELLULAR PI AND POLYP IN ECTOMYCORRHIZAL FUNGI COMPARED TO THOSE OF THE COMPOUNDS IN MODEL SYSTEMS

Phosphorylated compounds	$\Delta\upsilon^{1/2}$ (Hz)	T_1 (ms)
Pi		
Growth medium	40 ± 0.3	650
Intracellular		
H. crustuliniforme	14 ± 4	150 ± 44
C. graniforme	22 ± 11	141 ± 27
PolyP		
Growth medium	21 ± 2	790 ± 85
Growth medium + glycerol	—	960 ± 100
Intracellular		
H. crustuliniforme	109 ± 21	290 ± 3.0
C. graniforme	187 ± 33	4.4 ± 2.2

Note: All measurements were obtained at 36.4 MHz with a sample temperature of 4°C. The pH is 5.6 in all model systems as well in the growth medium. The line widths given represent the average of 3 measurements for model systems and at least 10 measurements for living mycelia. T_1 values were the average of three measurements (±SD).

Adapted from Martin, F., Marchal, J.-P., Timinska, A., and Canet, D., *New Phytol.*, 101, 275, 1985.

respective rates of PP degradation which appeared to be related to difference in the intracellular state of the PP. ^{31}P line width and spin-lattice relaxation measurements for Pi and PP in both intact strains and in culture medium (*in vitro* solution) demonstrated that the intracellular relaxation times were significantly shorter (see Table 5). This is attributed to a combination of factors such as partial immobilization of the polymer by cellular componenents and/or possible interaction with paramagnetic metal ions such as Mn^{2+}. The greater mobility (longer T_1) of the PP in *HC* compared with *CG* seems to explain the more rapid degradation of PP in the former.

Symbiotic nodules produced from the infection of soybean seeds *(Glycine max)* with rhizobium can be readily examined in a detached state with ^{31}P NMR.[81,121] In this condition the nodules fix little nitrogen. However, the respiration of the assembly is maintained. Because nodules contain a mixture of uninfected plant and cortical cells which contain vacuoles, in addition to the infected cells that contain no vacuole, one would anticipate that the ^{31}P spectra would be more complex than those obtained from a single plant or organism. In fact, even at magnetic fields as high as 161.7 MHz, the ^{31}P spectra show considerable overlap of the resonance associated with the plant and micro-organism cells. Continuous perfusion of the nodules[121] with an oxygen-saturated solution of 50 m*M* glucose, pH 7.5, does however improve the resolution of the spectrum over those obtained previously[81] in an unperfused state. Figure 25 shows the clear separation of the three phosphate monoesters, centered at approximately 4.4 δ and an unknown phosphate diester at approximately 0.40 δ that were not resolvable without perfusion. The latter resonance appears to represent a marker compound associated with the encapsulated

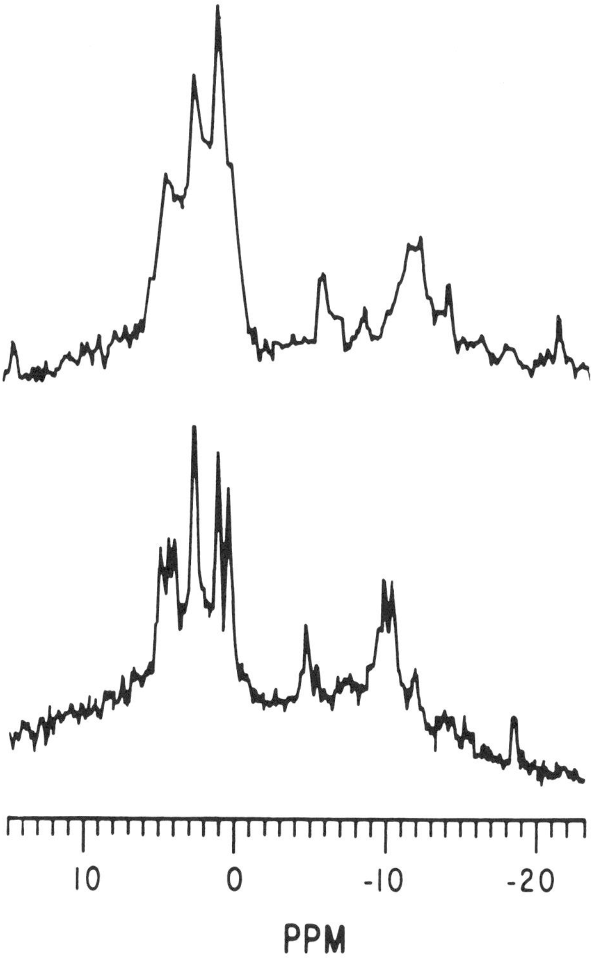

FIGURE 25. 161.7 MHz ^{31}P spectra of detached soybean root nodules; upper spectrum, 20,000 scans, no perfusion only bubbling with O_2. (From Mitsumori, F., Yoneyama, T., and Ito, O., *Plant Sci.*, 38, 87, 1985. With permission.) Lower spectrum, 20,000 scans with perfusion (unpublished results).

bacteriods, since it has also been observed in the spectra of extracts of the free-living rhizobia.[121] This compound does not correspond to any of the known phosphate diesters in any previously described *in vivo* plant or microorganism study.

^{31}P spectra of the perchloric acid extracts of the freshly detached nodules (Figure 2) showed an unusually low ratio of NTP/NDP (1.5), reflecting the anaerobic environment of the intracellular bacteroids.[121] These values are consistent with those obtained by wet chemical luciferm analysis (1 to 1.5) of the same nodule tissue.[121] With the exception of hypoxic conditions, all plant tissues and plant cells that have been examined by ^{31}P NMR in an aerobic state have shown NTP/NDP ratios approaching 25.[83]

The near coalescence of the Pi resonances at approximately 0.89 and 3.0 δ following treatment with 200 μ*M* CCCP verified the presence of two membrane- and gradient-separated compartments with pHs of approximately 7.1 and 5.9.[81] Application of 1 m*M* Mn^{2+} to the perfusion medium caused a preferential broadening of the Pi resonance at 0.89 δ representing the presumed acidic plant vacuole compartment (unpublished results) as observed previously in the spectra of maize root tips.[14] Steam girdling treatment of the stem of the host soybean plants cut off the supply of carbohydrate to the root system as demonstrated in the ^{31}P spectra of the nodules which showed diminished levels of sugar phosphates and NTP.[81]

B. ^{13}C Studies

This section will cover the application of *in vivo* ^{13}C NMR, primarily to plants and microorganisms in both natural abundance (1.1%) and isotopic enrichment studies. A more comprehensive review of the use of ^{13}C, ^{14}N, and ^{15}N NMR for monitoring plant metabolism has been published by Martin.[29] For this reason we will try to cover the more recent developments in this area and higlight only some of the older studies that have already been reviewed. The same will be true for the sections on ^{14}N and ^{15}N NMR applications.

1. ^{13}C Natural Abundance Studies

Natural abundance (1.1%) in *in vivo* ^{13}C NMR spectroscopy, because of its inherent insensivity, is only a practical technique for examining high concentrations of small, mobile, storage compounds that are metabolized over long periods of time. In principle this methodology is well suited to the study of certain microorganisms, fruits, storage roots, and seeds. The analysis of intact seeds will be covered in Chapter 4.

Martin et al.[122] have examined the natural abundance ^{13}C spectrum of intact ectomycorrhizal fungi mycelia and observed a large quantity of the storage carbohydrates, trehalose and mannitol, as well as triacylglycerols and glycogen. The concentration of the stored carbohydrates was shown to change as the fungi aged. In this way the NMR was able to examine the interchange of carbohydrates and lipids and the biochemical relationship between them.[122] ^{13}C has also been useful for studying the role of organic solutes in controlling cell osmolarity. Glucosyl glycerol, for example, was found to be produced in response to osmotic stress in cyanobacteria,[123] while glycerol was observed to be accumulating in the ^{13}C spectrum of halophilic alga (*Duwaliella salina*) under similar conditions.[124] However, only about 60% of the total glycerol in *D. salina* was NMR visible in the intact cells.[124] This may be a consequence of a slow exchange between free and cell-bound glycerol or an interaction with paramagnetic metal ions within the chloroplast.[124] ^{13}C T_1 measurements can be used to estimate the intracellular viscosity in mycelia and other intact organisms. Martin et al.[125] measured the T_1 values for fatty acid acyl chain carbon resonances in ectomycorrhizal fungi and found them to reflect a high degree of molecular motion, comparable to that found in mammalian tissue. However, a highly viscous environment characterized by a shortened T_1 of intracellular mannitol was observed in the fungi *Penicillium janczewskii*.[126]

Kainosho and Konishi[127] examined the ^{13}C spectrum of the intact pericarp of star anise and found it to contain anethole, an essential oil. Upon crushing the pericarp, the ^{13}C spectrum became broadened and highly degraded, indicating that the free motional state of the molecule was highly restricted. This change was attributed to the fact that the anethole existed as a neat, freely mobile liquid in the parenchyma of the intact pericarp prior to crushing. However, following crushing, the oil leaked from the cells and was absorbed on the woody tissue where it became motionally restricted.[127]

^{13}C spectra of intact white potatoes only showed the resonances corresponding to the mobile sugars, glucose, sucrose, and fructose, and fatty acids which represent about 1% of the potato tissue.[128] Boiling the potato, however, suppressed the spectrum of the low molecular weight species and revealed the intense spectrum of starch. These studies indicate that the large molecules of starch in the potato tend to break down into small units when exposed to boiling water. We have examined the ^{13}C spectrum of intact nonboiled potatoes to evaluate changes in glucose and sucrose storage and interconversion following different genetic manipulations and periods of storage.[129] In addition, the chemical shift of the citrate resonance was useful for evaluating the intracellular pH.[129] This methodology was used to follow the composition changes in malate, tartrate, glucose, and fructose in developing unpicked and picked grapes.[130] The changes were then correlated with known changes in berry deformability. Thomas and

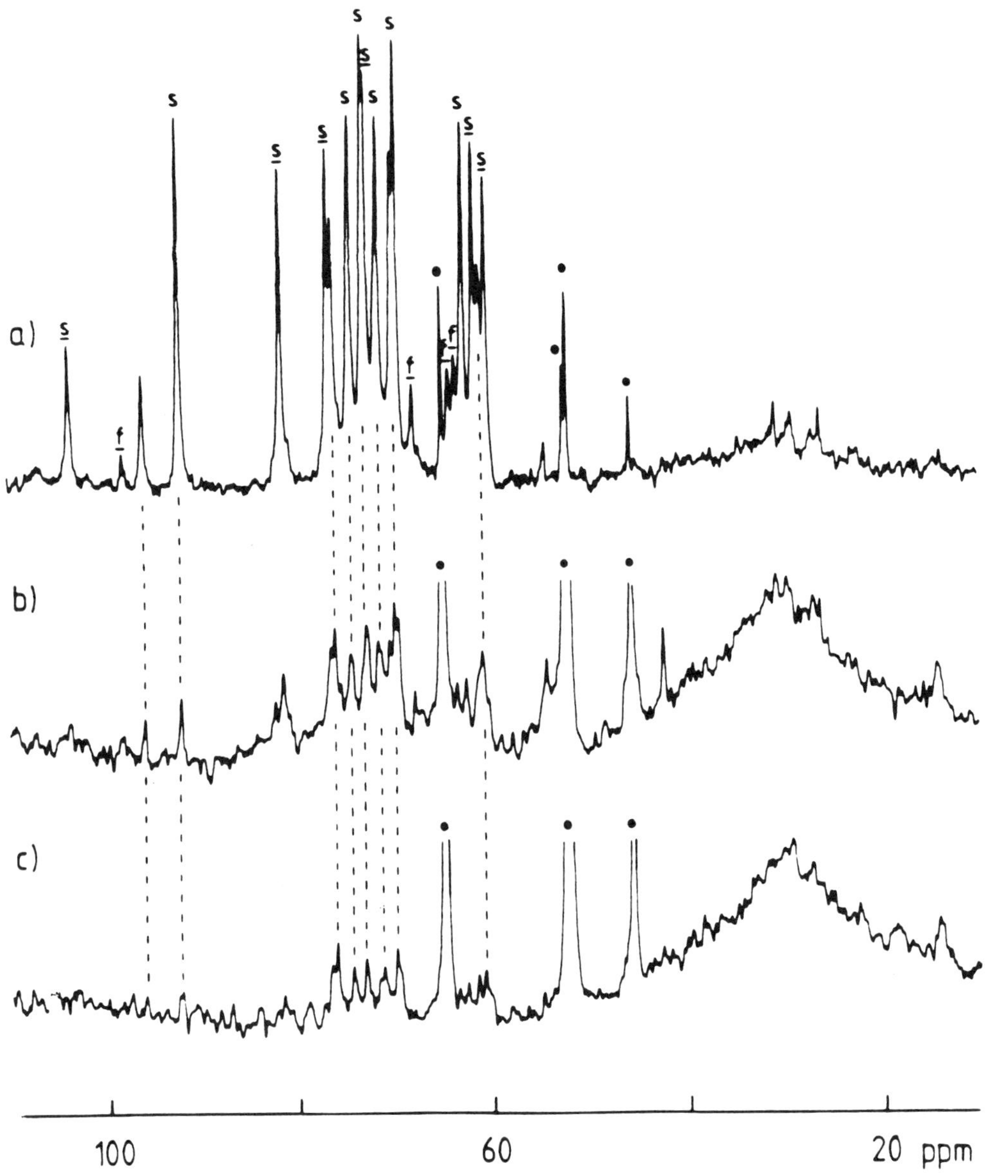

FIGURE 26. Natural abundance [1]H-decoupled 75.46 MHz ^{13}C NMR spectra of root tissues from carrot plants grown in nutrient culture: (a) a single tap root (FW = 1.7 g) from a 12-week-old plant, 4,000 transients, total acquisition time 153 min; (b) six tap root (total FW = 0.79 g) from 8-week-old plants, 16,000 transients, total acquisition time 613 min, and (c) the fibrous root system (FW = 0.65 g) from a 12 week old plant. 16,000 transients, total acquisition time 61.3 min. A line broadening of either (a) 8 Hz or (b) and (c) 16 Hz was applied before Fourier transformation. The spectra can be assigned by comparison with the characteristic spectra of sucrose, glucose, and fructose. The 12 sucrose resonances are labeled in (a) as either s or s̲, where the underlining indicates that sucrose is the only contibutor to the observed peak. The eight glucose resonances are indicated by the vertical dotted lines. Several of the fructose resonances are overlapped by the sucrose and glucose, but four clearly resolved resonances are labeled in (a) as f. The MES buffer contributed the four peaks labeled o; these resonances were truncated when plotting (b) and (c). The broad hump centered at 30 ppm was caused by the plastic tubing in the system used to circulate the aerated buffer through the NMR tube. (From Thomas, T. H. and Ratcliffe, R. G., *Physiol. Plant*, 63, 284, 1985. With permission.)

Ratcliffe[131] examined the ^{13}C spectrum of detached fibrous and storage roots of carrots (see Figure 26). The relative amounts of sucrose, glucose, and fructose in the storage roots of two ages and in fibrous roots were found to be similar to those detected by the standard, destructive enzymatic procedures. In tap roots, all three sugars were present, while in fibrous roots glucose was the major sugar. The organic acids, sucrose, citrulline, and phenolic compounds have also been analyzed in a similar nondestructive manner in excised root nodules of *Alnus glutineas* to examine their pattern of accumulation and evaluate the effect of the nutritional state on the biochemistry of symbiosis.[29]

2. ^{13}C Isotopic Enrichment Studies

As we have seen from the above, natural abundance ^{13}C NMR spectra are useful for the study of highly concentrated storage compounds. However, one cannot examine the less concentrated metabolites with this methodology. Labeled ^{13}C-enriched compounds can be used to investigate the products of various metabolic pathways and yield information regarding relatives fluxes through these pathways. Typically, a ^{13}C NMR spectrum can be obtained from a low intracellular concentration (approximately 1 μmol) of a ^{13}C-enriched metabolite (minimum 30%) within a few minutes.

While ^{31}P NMR can examine some cellular processes, it has two important limitations: (1) a large number of metabolic processes do not directly involve significant changes in the concentration of key phosphorylated species and (2) ^{31}Ps high natural abundance precludes its use to trace the fate of a particular ^{31}P metabolite. Thus, the only way to measure the turnover of phosphorylated metabolites *in vivo* is by saturation transfer and that is only possible if the rate of the process is fast enough and the relaxation time, T_1, is of the order of the exchange rates (see Section A.2 above on saturation transfer). ^{13}C NMR studies with ^{13}C-enriched compounds can directly monitor the rates of incorporation or degradation of a label, the level of its steady-state incorporation or exchange, and the specific position of the label in the metabolite. In addition, the pH-sensitive shift can be used to report the intracellular pH of compartment cells. A survey by Avison et al.[9] has reviewed the recent literature of *in vivo* ^{13}C NMR studies of animals, humans, and perfused organs and tissues which include brain, heart, kidney, and skeletal muscle.

The use of ^{13}C enrichment has dominated the voluminous numbers of reports describing carbon fluxes in microorganisms.[2,132] This methodology has not yet attained widespread applications in the area of *in vivo* plant studies although a number of reports on algae and fungi has recently appeared and been reviewed.[29] Stidham et al.[133] have successfully used the ^{13}C signal from malate in intact *Kalanchoe tubiflora* leaves (introduced by exposure to $^{13}CO_2$) to follow the decarboxylation of malate of this Crassulacean acid metabolism (CAM) plant in the light. After a period of darkness in air, a resynthesis was observed in which all four carbons of malate were found to be labeled with ^{13}C. The position of the ^{13}C resonance was also useful for determining the pH of the vacuole during light and dark periods. Since the malate was only found in the vacuole, the change in chemical shift of the four carbons was used to evaluate the change in vacuolar pH under light and dark conditions. In contrast to the Pi resonance, the four ^{13}C resonances of malate are excellent pH indicators for acidic compartments such as the vacuole since their pKas are of the order of 4.5 (see Figure 27) as opposed to 6.8 for Pi. Following a dark period, the pH of the vacuole was 4.0; however, it rose to 6.0 after 6 h of light.

In a most elegant study Holtum et al.[134] were able to directly evaluate the activity of an enzyme in the dark synthesis of malate in CAM and other plant leaves. By observing the ^{18}O-labeled malate indirectly via the isotopic shift and J coupling of its ^{13}C-carbonyl resonance (see Figure 28), the authors were able to determine if dark CO_2 fixation required the intervention of carbonic anhydrase. If malate synthesis in the dark did require high carbonic anhydrase activities, then oxygen exchange between CO_2 and cellular water (not enriched in ^{18}O) would occur more rapidly than carboxylation and the ^{13}C-malate formed would contain little more than the natural abundance level of ^{18}O. Following the ^{13}C-$^{18}O_2$ exposure, the leaves were harvested and the ^{13}C spectrum of the extracts were recorded. From the relative area of the ^{13}C-^{18}O shifted resonance of the malic acid carbonyl group, it was demonstrated that only minor ^{18}O labeling of malate had occurred, proving that CAM leaves do have a substantial carbonic anhydrase activity.

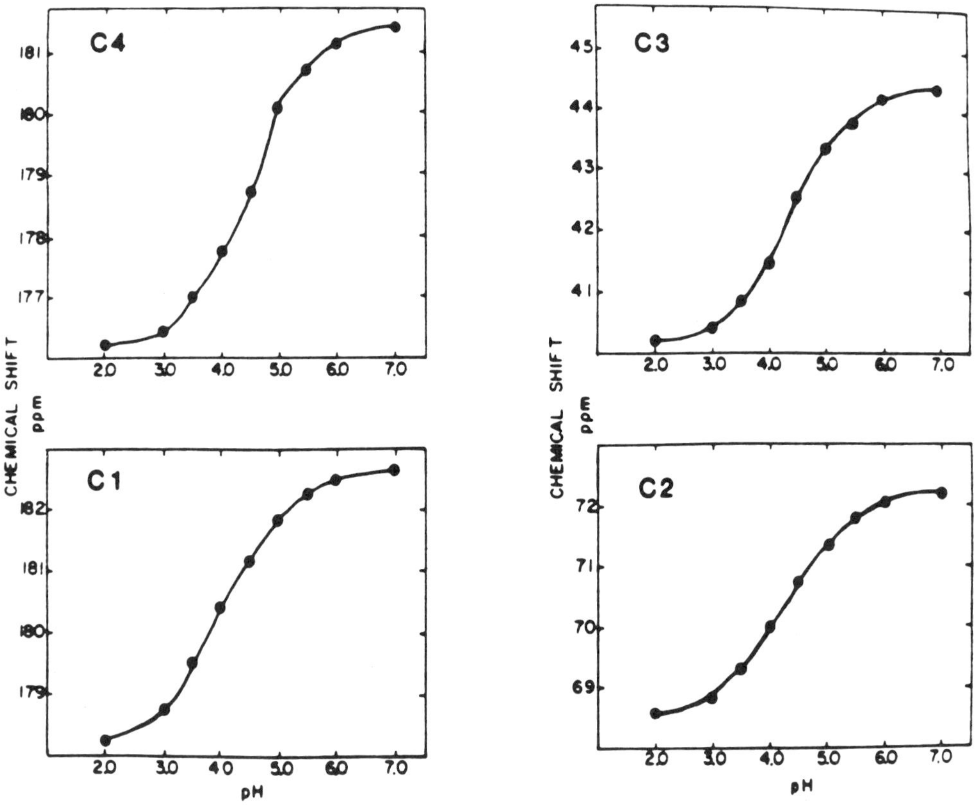

FIGURE 27. pH dependence of the chemical shifts of the ^{13}C NMR signals of malic acid. Malic acid (0.5 *M*) was prepared and adjusted to the appropriate pH with 5 *N* KOH. A 5-mm coaxial insert containing D_2O and a 0.5-mm capillary containing acetone was used in a 10-mm tube containing the malic acid. In other trials, 0.1 *M* malic acid was used with different references and lock compounds, and identical results were obtained. (From Coombe, B. G. and Jones, G. P., *Phytochemistry*, 22, 2185, 1983. With permission.)

Ashworth and Mettler[32] have carried out metabolic studies in both cultured tobacco and corn cells to establish intracellular glycine and serine pool sizes. These workers utilized 2-^{13}C-glycine and observed signals from 2-^{13}C-glycine, 2-^{13}C-serine, 3-^{13}C-serine and 2,3-^{13}C serine *in vivo*. Assignments were verified by ^{13}C homonuclear two-dimensional COSY experiments. The results of these studies suggest that serine formation from glycine is the result of a tight coupling between glycine decarboxylase and serine hydroxymethyl transferase. The metabolism of 1-^{13}C-glucose has also been followed in the elaboration of cell wall polysaccharide biosynthesis in protoplasts from cell suspensions of millet.[135] In these studies the plant cells were grown in the NMR tube with the 1-^{13}C-glucose substrate so the production of β-glucan could be measured directly in real time. Roberts et al.[136] has utilized 1-^{13}C-glucose feeding with excised maize root tips to visualize the production of lactate, ethanol, and alanine under hypoxia. The quantity of each compound was determined by measuring the area of each corresponding methyl resonance.

Martin and Canet[137] have studied the accumulation of free amino acids in the presence of ammonium ion in nitrogen-starved mycelia of the ectomycorrhizal ascomycete *Cenococcum geophilium*. ^{13}C-NMR spectroscopy of extracts from cells grown on 1-^{13}C-glucose were used to examine the biosynthesis of amino acids. Incorporation of the ^{13}C label resulted in the formation of 3-^{13}C-alanine and glutamine, glutamate, γ-aminobutyrate, and arginine labeled at the C2, C3, and C4 positions (see Figure 29). These results indicate that labeled pyruvate enters the alanine pool via alanine aminotranferase and the Krebs cycle through both pyruvate carboxylase (and/or phosphoenolpyruvate carboxykinase) and pyruvate dehydrogenase activities. The pyruvate and acetate metabolism has been studied in cell cultures of *Zea mays* by *in vivo* ^{13}C-NMR.[135] This

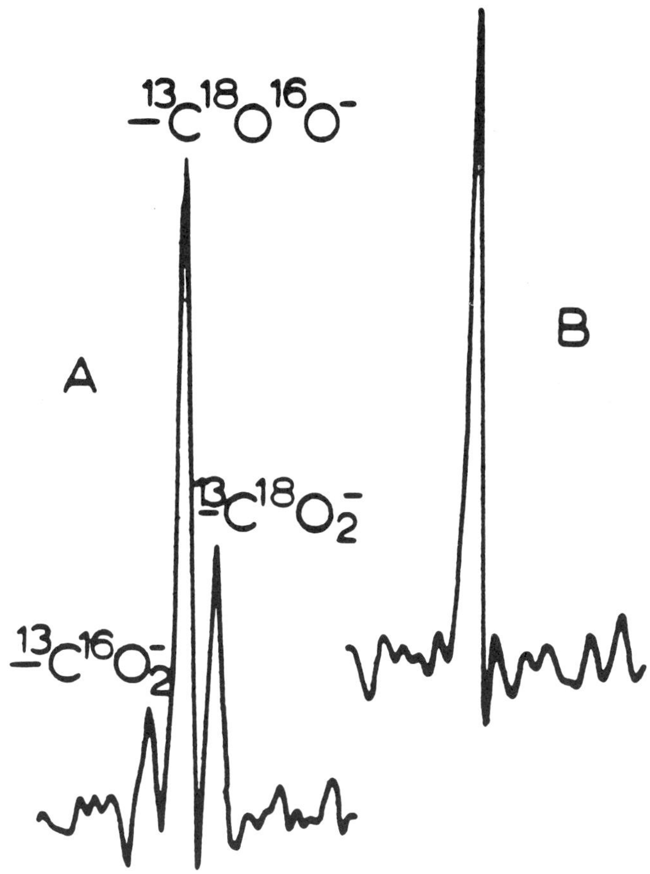

FIGURE 28. ^{13}C NMR spectra of the carboxyl region of malic acid produced by reaction of P-enolpyruvate and P-enolpyruvate carboxylase in the presence of $^{13}C^{18}O_2$. A, in the absence of carbonic anhydrase. B, in the presence of carbonic anhydrase. The single resonance seen in B represents only $^{13}C^{16}O^-_2$. (From Holtum, J.A., Summons, R., Roeska, C. A., Comins, H. N., and O'Leary, M., *J. Biol. Chem.*, 259, 6870, 1984. With permission.)

work demonstrated that growth of the corn cells in the presence of 2 m*M* [2-^{13}C] acetate produced primarily labeled glutamate and malate. Further metabolism of these intermediates resulted in labeled glutamine, aspartate, and alanine. With [1-^{13}C] acetate, little incorporation was observed due to the loss of label as $^{13}CO_2$. The uptake of [3-^{13}C] pyruvate gave similar labeling patterns to the [2-^{13}C] acetate but was considerably slower. Ito and co-workers[139] examined the environmental factors such as NO_2 and O_3 pollution on photosynthetic $^{13}CO_2$ assimilation in kidney bean primary leaves. NO_2 and, to a greater extent, O_3 inhibited the production of ^{13}C-labeled sucrose and fructose from photosynthetic pathways as evidenced by the lack of these compounds in the ^{13}C spectra of leaf tissue extracts. This result implies that carbon flow through the glycolate pathway (nonphotosynthetic) is stimulated in leaves exposed to NO_2 and/or O_3. In their studies of *Staphylococcus aureus* cells, Ezra et al.[140] have examined the intra- and extracellular ^{13}C resonances of lactate to measure the distribution and transport of this metabolite. Due to the fact that the interior of the cells contained manganese, these worker observed a downfield shift and broadening of the intracellular ^{13}C resonances of lactate relative to that of the exterior (see Figure 30). Thus, ^{13}C-NMR was used to monitor the internal and external lactate independently. The ratio of the two resonances, external (L^{ex}) and internal lactate (L^{in}), was correlated with lactate uptake in the presence of oxygen and a rapid efflux as the cells

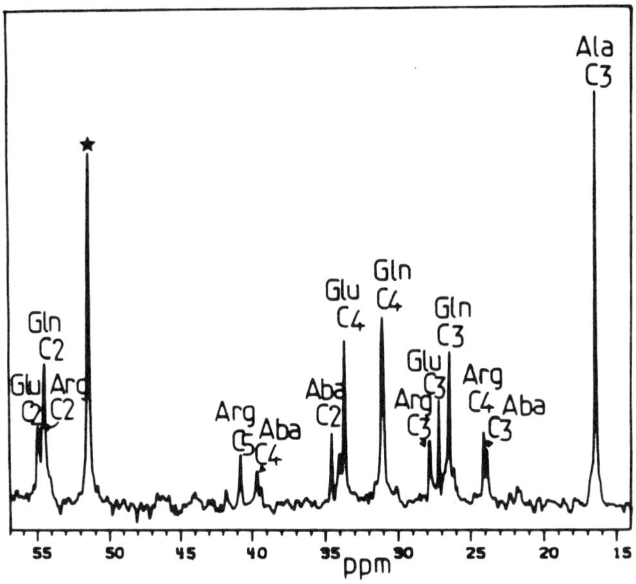

FIGURE 29. ^{13}C NMR (50.4 MHz) of an aqueous mycelial extract taken 5 h after addition of 20 m*M* [1-^{13}C] glucose to a suspension of *Cenococcum geophilium*. The region between 15 and 60 ppm, which contains most of the amino acid peaks is shown. The spectrum is a Fourier transform of the accumulation of 1000 radio frequency pulses of 90° free induction decays, with 2-s recovery time between pulses. Samples were dissolved in 50 m*M* ethylenediaminetetraacetate solution (pH 7.0) containing 20% D_2O. Identified are signals from alanine, arginine, γ–aminobutyrate (Aba), glutamate, glutamine, and ethylenediaminetetraacetate (*). (From Martin, F. and Canet, D., *Physiol. Veg.*, 24, 209, 1986. With permission.)

reverted to an anaerobic state. Treatment with fatty acid antimicrobial agents lowered the transmembrane pH gradient and showed a loss of lactate from the cells during glycolysis. In additon, the uptake of lactate under these conditions in the presence of oxygen was completely inhibited.

C. Other Nuclei

1. ^{14}N and ^{15}N

Natural abundance ^{14}N NMR has found very limited application in *in vivo* studies for the reasons outlined in Section III.A. In spite of these drawbacks, Belton and co-workers[141] examined the intracellular uptake of nitrate and ammonium ion in maize, barley, and pea root tips. Both NO_3^- and NH_4^+ gave rise to relatively sharp resonances ($\delta_{1/2} = 15$ Hz), while the amide and amino nitrogen resonances were broad and featureless ($\delta_{1/2} = 600$ Hz). Figure 31 shows the ^{14}N spectra observed for both NH_4^+- and NO_3^-- grown barley and pea roots. The assignments are as follows: 0.0 δ, NO_3^-; –200.0 δ, amide nitrogen; –335.0 δ, amino nitrogen; and 355.0 δ, NH_4^+. The spin-lattice relaxation times were very short for the ^{14}N resonances associated with the large proteins (<10 ms) while the small nitrate and ammonium molecules had T_1 values of 0.1 and 0.5 s, respectively. Since the average nitrate concentration in nitrate-grown barley roots was as high as 87 mmol dm^{-3} tissue, satisfactory spectra could be obtained in as little as 5 min. The change in internal nitrate concentration followed a single exponential and was rationalized in terms of nitrate exchange and reduction. Only one pool of nitrate ion was observed and assigned to the vacuolar compartment. In contrast, the ammonium resonance showed diphasic behavior indicative of some form of compartmentation in the tissue.

Despite the need for isotopic enrichment, ^{15}N is generally the nitrogen nucleus of choice for

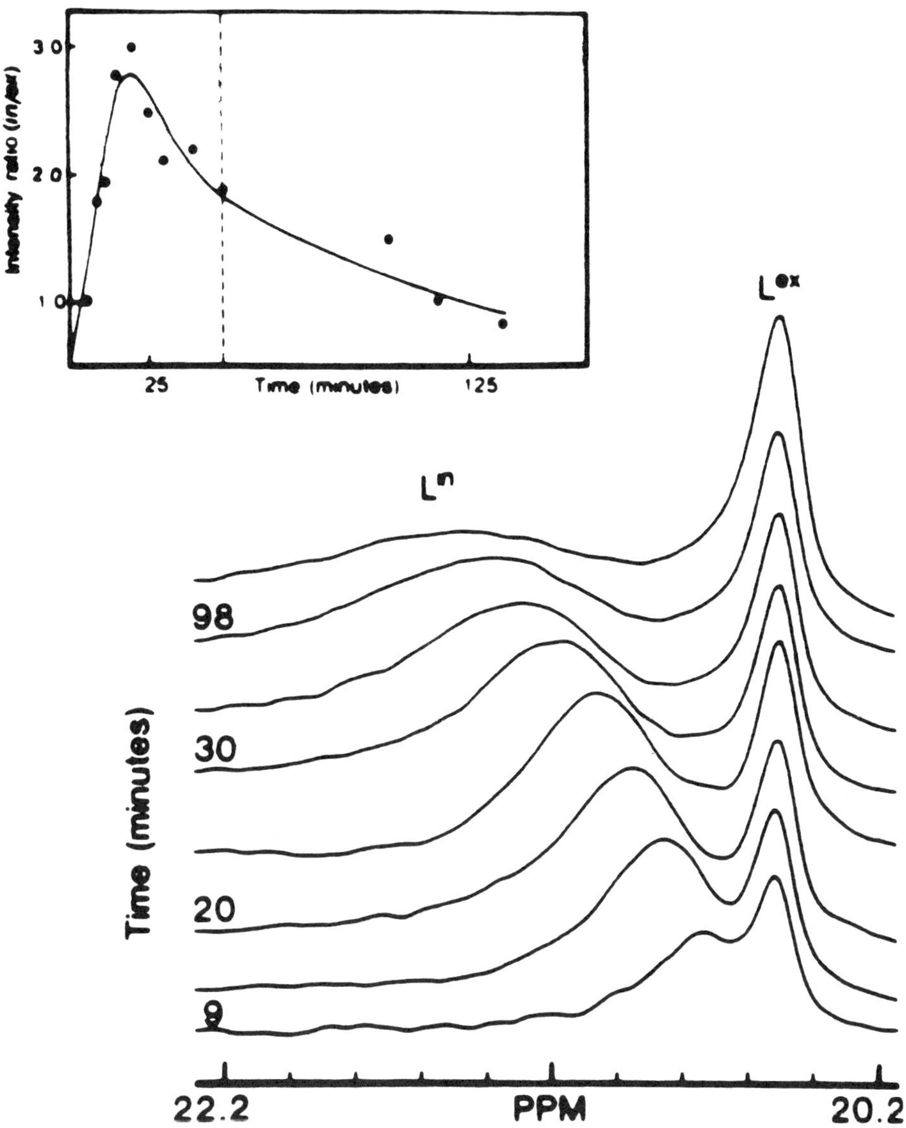

FIGURE 30. Lactate region (20.2 to 22.2 ppm) of the proton-decoupled ^{13}C NMR spectra of glycolyzing *S. aureus* cells as a function of time. The insert shows the time course of the intracellular/extracellular lactate integrated intensity ratio. Glucose consumption is completed at ~50 min (vertical dashed line). L^{in}, intracellular lactate and L^{ex}, extracellular lactate. (From Ezra, F., Lucas, D. S., Mustacich, V., and Russel, A. F., *Biochemistry*, 22, 3841, 1983. With permission.)

most *in vivo* applications for the reasons mentioned in Section III.A. Kanamori et al.[142] capitalized on the dipolar relaxation of intracellular ^{15}N-labeled glutamine, alanine, and arginine to establish that the viscosity of the cytoplasm of *Neurospora crassa* is substantially less than its vacuole due to polyanion interactions in the latter. ^{15}N has also been used as an intracellular pH probe for *N. crassa* mycelia. Line widths as well as chemical shifts correlated with the pH environment of ^{15}N-labeled amino acids, histidine, arginine, alanine, and proline used in the study. The pH dependence of the chemical shift of N_{τ} of histidine was found to be most sensitive in the pH range 5 to 7 (characteristic of the vacuole), whereas P_i whose pK is closer to 6.8 is relatively insensitive for reporting the relatively acidic compartment.

Both cytosolic amino acids such as glutamine and alanine and vacuolar amino acids such as

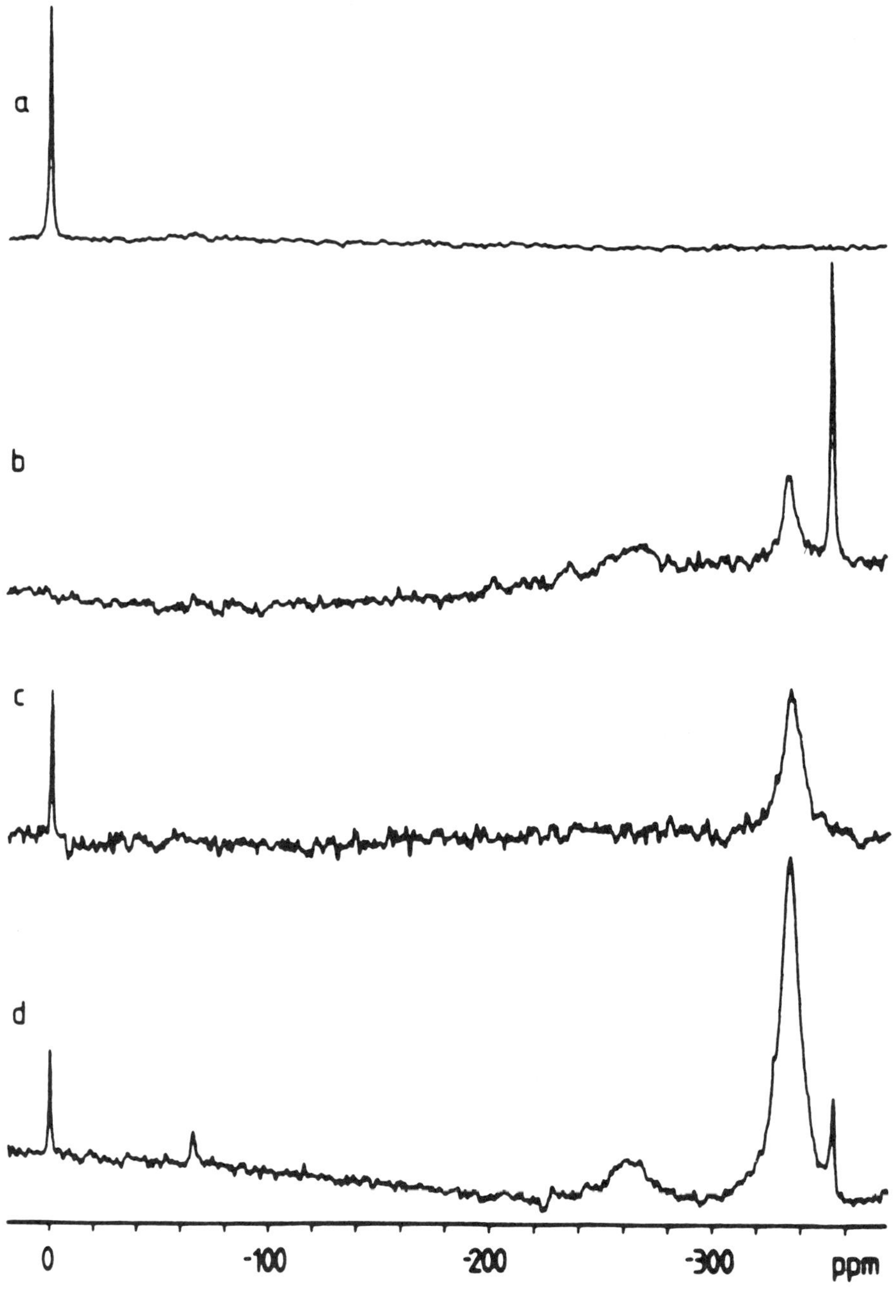

FIGURE 31. The 21.68-MHz ^{14}N spectra of (a) 11-d-old barley roots grown on 5.0 mol m^{-3} nitrate, (b) 11-d-old barley roots grown on 5.0 mol m^{-3} ammonium; (c) 5.0-mm root tips from 10-d-old pea roots grown on 5.0 mol m^{-3} nitrate; and (d) 5.0-mm root tips from 10-d-old pea roots grown on 1.0 mol m^{-3} nitrate, supplemented with 5.0 mol m^{-3} ammonium for the 12 h before excision. The spectra were accumulated (a) in 20 min with a 65° pulse and a 0.25-s recycle; (b) in 40 min with a 65° pulse and a 1-s recycle; (c) as (a); (d) in 1 h with a 90° pulse and a 0.2- s recycle. The approximate spin-lattice relaxation times for the tissue resonances are amide/amino, <10 ms; nitrate, ~ 0.1 s; and ammonium, >0.5 s. It follows that the optimun acquisition conditions, both for qualitative and quantitative spectra, differ widely for the various resonances. The resonance assignments are discussed in the text. (From Belton, P. S., Lee, R. B., and Ratcliffe, G., *J. Exp. Bot.*, 36, 190, 1985. With permission.)

arginine and lysine gave excellent ^{15}N spectra in metabolic studies of *N. crassa* mycelia and *Saccharomyces cerevisiae* cells incubated with ^{15}N-labeled substrate.[142,143] Martin[144] investigated ^{15}N- labeled metabolites in ectomycorrhizal fungi mycelia and cell-free extracts. These spectra are shown in Figure 32. A time course of incorporation of $^{15}NH_4^+$ (kinetic labeling) demonstrated glutamine synthetase activity because the glutamine amide-N was the most highly labeled component over the 1st h. The failure to observe the arginine $N_{\omega,\omega}$ resonance in the intact mycelia while it is present in the extracts (see Figure 32B) suggests that the positively charged guanidino group of arginine may be associated with negatively charged PP in the fungal vacuole. This observation is in accord with the observation of Kanamori et al.[142] that PPs decrease the T_1 of the arginine $^{15}N_{\omega,\omega'}$ resonance.

By feeding [guanidino-^{15}N] arginine to detached shoots of beech *(Fagus sylvatica)*, Martin[29] was able to demonstrate with *in vivo* ^{15}N NMR that the ammonium nitrogen produced by arginine catabolism is reassimilated via glutamine synthetase.

The pathway of NH_4^+ assimilation in *Lemna gibba*, a floating aquatic angiosperm, has recently been elucidated with labeled ^{15}N NH_4^+.[145] In this instance, the $^{15}NH_4Cl$ was taken up by the floating plant for 24 h. Following this period, ^{15}N NMR analysis of the aqueous extracts of the plant tissue revealed ^{15}N incorporation into glutamine and glutamic acid with no detectable free ammonium ion. When the tissue was exposed to methionine sulfoximine, an inhibitor of glutamine synthetase, no incorporation of ^{15}N was observed. In the presence of azaserine, a glutamate synthase inhibitor, only ^{15}N incorporation was seen in the amide group of glutamate. These results indicate glutamine synthetase and glutamate synthase are major pathways for the assimilation of ammonium ions.[145]

2. *^1H and ^{19}F*

A great number of ^1H NMR studies of heterogeneous tissues has been carried out solely through the examination of the characteristic water resonance. The reason is partly because of the complexity of the ^1H spectrum due to its high sensitivity and small chemical shift range and partly because of the sheer dominance of the water resonance over everything else in the spectrum. By examining the water resonance, one can obtain important information concerning the tissue, i.e., the water content and the relaxation properties (T_1 and T_2 and thereby the mobility and interaction with paramagnetic centers). A number of T_1 and T_2 relaxation rates of water in plant tissues and their relationship to frost hardiness have been thoroughly reviewed by Loughman and Ratcliffe.[4] These relaxation properties were also exploited to measure the exchange between the water in the cortical cells and extracellular space in maize roots in the presence of Mn-EDTA (a nontoxic form of Mn^{2+}).[146] In addition, the water-free space of the root and viscosity of the cell cytoplasm were also estimated. The proton water relaxation properties were used to monitor the translocation of the paramagnetic ion directly, since these properties measure the relaxation of the tissue water affected by the paramagnetic ions (assuming the enhancement of water relaxation is proportional to the intracellular ion concentration).[146] This same principle is presently being exploited to increase the resolution of proton magnetic resonance imaging (MRI) in plant and other agricultural materials (see Chapter 15 for full details).

High-resolution ^1H studies of heterogeneous tissues requires special spectroscopic techniques as mentioned above. First, one must suppress the dominant water resonance to begin to observe other components. Second, it is necessary to reduce the broad resonances due to proteins and relatively immobile lipids. Along with the new instrumentation, computers, and software, which have proliferated in the past years, has come a variety of pulse sequences capable of making ^1H NMR a practical tool for studying *in vivo* metabolism. The pioneering work of Campbell and co-workers[147,148] on the application of spin-echo techniques in the study of erthyrocytes laid the groundwork for most of the work that is continuing today.

The broad ^1H resonances of protein and immobile lipids can be eliminated by exploiting their

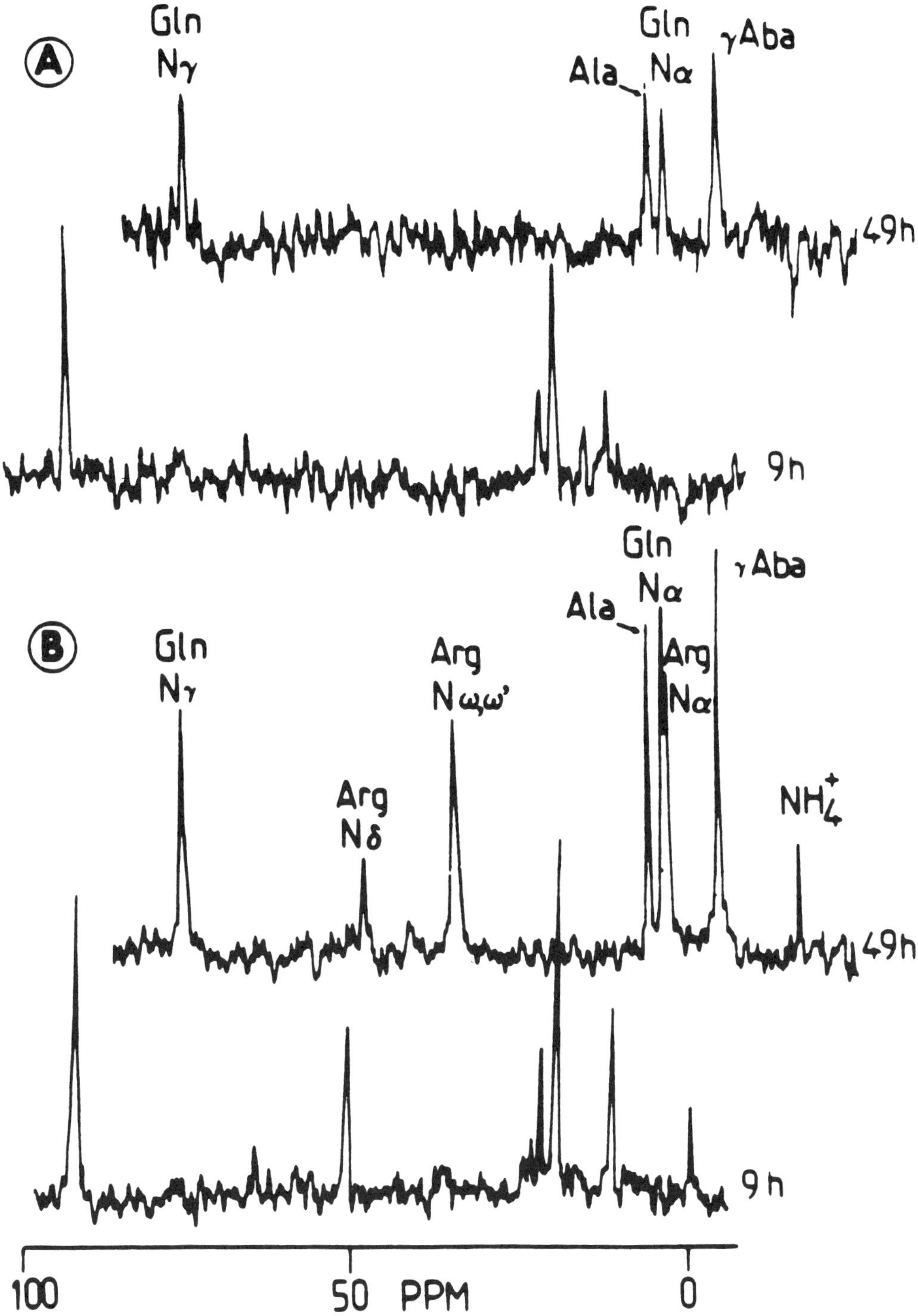

FIGURE 32. The ^{15}N spectra of ^{15}N amino acids in living mycelia (A) and cell-free extract (B) of *Cenococcum graniforme*. The fungus was fed 5 m*M* ($^{15}NH_4$)$_2$$SO_4$ (50 atom percent excess) at time zero. Each spectrum (40.5 MHz) consists of 1000 (A) or 2500 (B) scans of 2-s pulse intervals. The first peaks to appear are due to glutamine; these are followed by alanine, γ-aminobutyrate, and arginine nitrogen signals. (From Martin, F., *Physiol. Veg.*, 4, 463, 1985. With permission.)

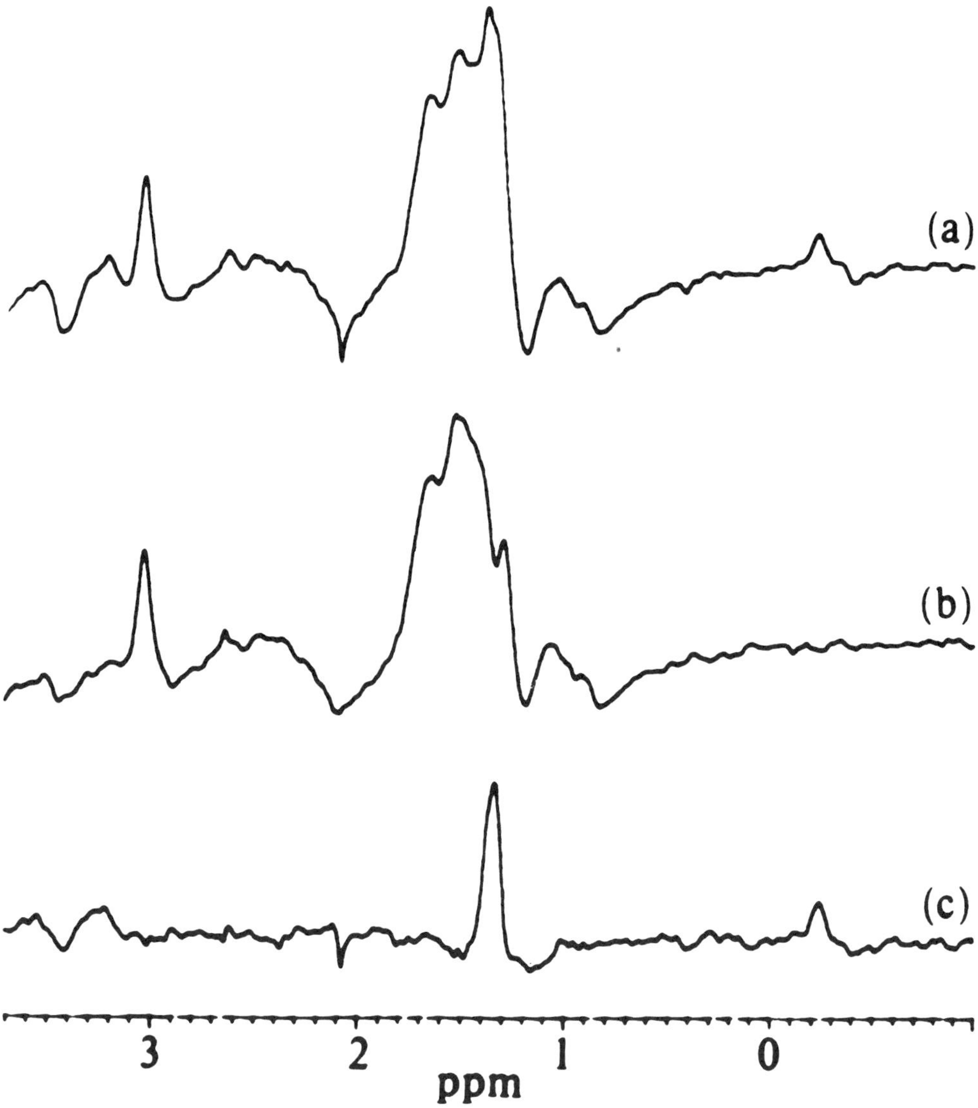

FIGURE 33. The ^{1}H NMR spectra obtained from the rat leg *in vivo* about 1 h after applying a ligation around the upper part of the leg. The spectra were obtained using the Hahm spin-echo sequence (90°-τ-180°-τ-observe), and the phase of the 180° pulse relative to the 90° pulse was shifted by +90° and −90° on alternate scans. (a) Accumulation of 32 scans at intervals of 1.9 s using a 68-ms delay between the 90 and 180° pulses. Homonuclear decoupling ($\gamma B_1 = 50$ Hz) was applied at 4.11 ppm during the 136 ms (2τ) echo delay period. The lactate signal is obscured by the large fat peak. (b) As for (a) except that the irradiation during the echo delay was switched to the opposite side of the lactate CH_3 signal, i.e., to −1.47 ppm. (c) The difference spectrum (a) to (b), showing the lactate signal at 1.32 ppm. (From Williams, S. R., Gadian, D. G., Proctor, E., Sprague, D. B., and Talbot, D. F., *J. Magn. Reson.*, 63, 406, 1985. With permission.)

relatively short T_2s, using a variety of the spin-echo sequence.[148] In addition, the large H_2O resonance can be suppressed significantly by a number of different methods including single frequency irradiation,[149] a Dante saturation pulse sequence,[150] and spin-echo sequences composed of frequency-selective Dante pulse trains.[151] Using this methodology, ^{1}H NMR has been used to evaluate lactic acid formation in various tissues under different metabolic stresses.[152,153] Figure 33 demonstrates the spectral resolution of the lactate resonance that can be achieved from the *in vivo* ^{1}H spectrum of a rat leg using a Hahn spin-echo sequence with selective proton decoupling.[154] As we have seen in Section II. A, two-dimensional (2-D) COSY methodology can be used to help resolve a complex ^{1}H spectrum of plant tissue in a superfused state[21] (see Figure 11). In this particular example, the one-dimensional (1-D) water-suppressed ^{1}H spectrum of carrot root tissue was of sufficiently good quality, i.e., line widths were relatively narrow, that molecular connectivities and dipolar coupling could be determined with the 2-D experiment *in vivo*. As we have seen in Figure 11 and Table 3, the spectrum is dominated by mobile, low molecular weight sugars, organic acids, amino acids, and ethanol.[21] In addition, T_1 values and the chemical shift position of the B-protons of malate in carrot roots suggest that malate is located in a relatively viscous and acidic compartment.[21]

Another technique has recently been developed to exploit the sensitivity of the proton resonance in combination with the greater shift dispersion of the ^{13}C spectrum to observe metabolites that could not be resolved previously.[155] Observation of the ^{13}C-coupled protons during the specific decoupling of the attached ^{13}C nuclei and subseqently subtraction of the non-^{13}C-decoupled proton spectrum yields an ^{1}H spectrum of the ^{13}C-labeled metabolite without the broad ^{1}H background resonances of the cell.[155] This method of indirect ^{1}H detection of ^{13}C NMR signals has one major drawback. Because of the overlap in resonance frequencies, one cannot single-frequency decouple unless the chemical shift difference between the ^{13}C resonances is much greater than the proton-carbon coupling constant. Figure 34 shows the ^{1}H difference spectra of *Saccharomyces cerevisiae* following incubation with ^{13}C-enriched acetate at ($\delta = 1.9$). Note that in the intial spectrum only the ^{13}C acetate ^{1}H resonance was present, but as ^{13}C acetate was consumed the ^{13}C-associated ^{1}H resonances of the subsequently labeled amino acids appeared. This technique is particularly useful for examining systems where rapid metabolic processes require that the NMR time constant for data aquisition be very short.

^{19}F NMR has seen limited application to *in vivo* studies because there are no naturally occurring fluorinated metabolites of sufficient concentration in plant or animal tissues to detect. There are, however, some fluorinated drugs in current use whose uptake has been examined by ^{19}F NMR. For example, some have been recently used for the study of malignant tumors.[156] Fluorinated anesthetics have also been studied in various tissues,[157] e.g., holothane and isoflurane, which have been utilized to differentiate normal and tumor tissue.[158]

Fluorine compounds such as mono-, di- and trifluoromethylamine whose pKas are 8.5, 7.3, and 5.9, respectively, have been found to be effective spy molecules for assessing cytoplasmic pH.[159] All three can be loaded simultaneously into cells as methyl esters which permeate the plasma membrane easily. Once inside the cell, they are enzymatically hydrolyzed to the fluoroamino acids which are stable and noncytotoxic.[160] In combination, these pH indicators can determine intracellular vs. extracellular pH profiles over a wide range (>2 pH units) of extracellular pH.[157] Dimethylfluoroalanine is very useful as an internal pH indicator since its proton-decoupled spectrum is a doublet whose splitting is pH dependent.[161] This is convenient since no other signal in the spectrum is required as a reference to measure the intracellular pH.

Symmetrically substituted difluoro derivatives of 1,2-bis-(*O*-aminophenoxy)-ethane-*N,N,N′,N′*, primes tetraacetic acid (nFBAPTA) can be used as ^{19}F NMR probes for divalent cations.[162] In particular, 4FBAPTA and 5FBAPTA show large ^{19}F NMR chemical shifts on chelating Ca^{2+} and since the two complexes show fast and slow exchange behavior, respectively, they can be used to determine the free intracellular Ca^{2+} concentrations. This measurement is unaffected by free Mg^{2+} concentrations <10 m*M*, by pH to 8, or by other contaminating ions such as Zn^{2+}, Fe^{2+}, or Mn^{2+}.[162] As in the previous studies, these chelators are loaded as their corresponding methyl esters and are hydrolyzed within the cells. Since the 5FBAPTA derivatives display two ^{19}F shifts in the presence of Ca^{2+} (one for free and one for bound Ca^{2+}), it is easy to follow the change in intracellular Ca^{2+} concentration as a function of the area of these two resonances. Figure 35 shows the change in internal Ca^{2+} concentration after treatment of mouse thymocytes with succinyl Con A and A23187, which allows extracellular Ca^{2+} to flood the cells. In addition to the Ca^{2+}–binding affinities of 5FBAPTA, this molecule also has a ^{19}F pH-sensitive

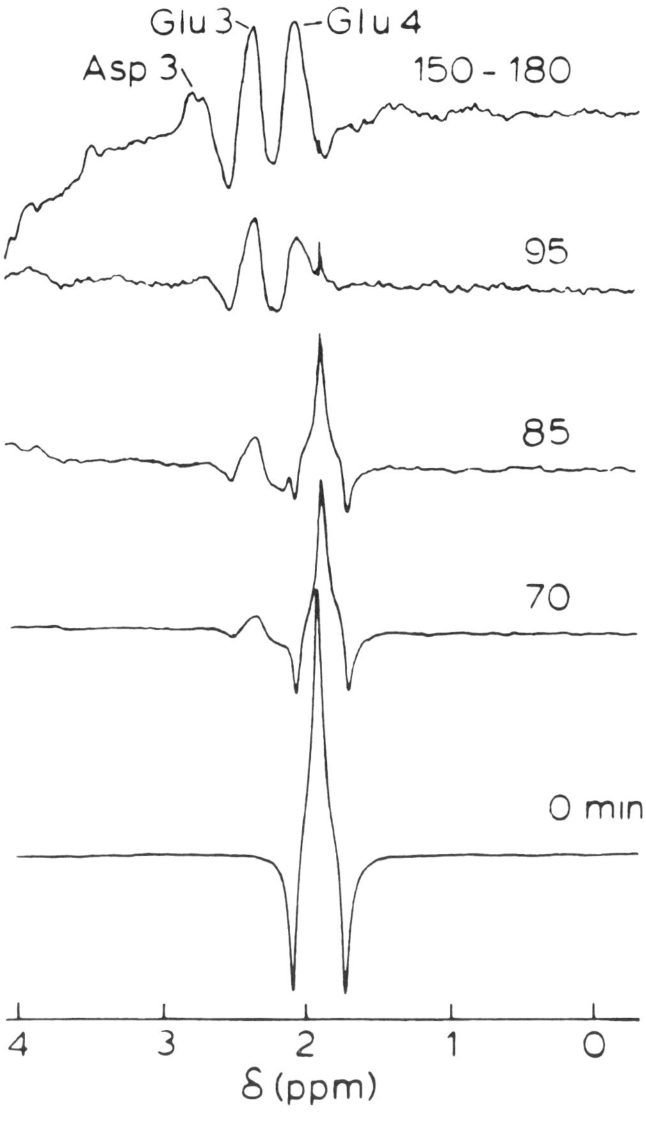

FIGURE 34. Time course of changes after oxygenation in the difference spectra obtained from the *S. cerevisiae* sample under conditions of broadband ^{13}C decoupling. Identified are signals arising from glutamate C_3, C_4 aspartate C_3, and acetate C_2 with ^{13}C coupling to the observed ^1H resonances at 2.12 ppm (Glu H_3), 2.33 ppm (Glu H_1), and 2.87 ppm (Asp H_3). (From Sillerud, L. O., Alger, J. R., and Shulman, R. G., *J. Magn. Reson.*, 45, 142, 1981. With permission.)

shift (pKa 5.7) that can be useful for determining intracellular pH. Furthermore, both 4FBAPTA and 5FBAPTA can be used for studies with opaque cell suspensions and tissues that cannot be examined by fluorescence measurements. Present applications include lymphocytes, thymocytes, tumor cells, and rat hearts.[163] Unfortunately, problems of diffusing these chelators into plant cells has made this methodology thus far inapplicable to plant tissues,[164] presumably because of the presence of the cell wall. However, studies of plant cell protoplasts may prove more successful.

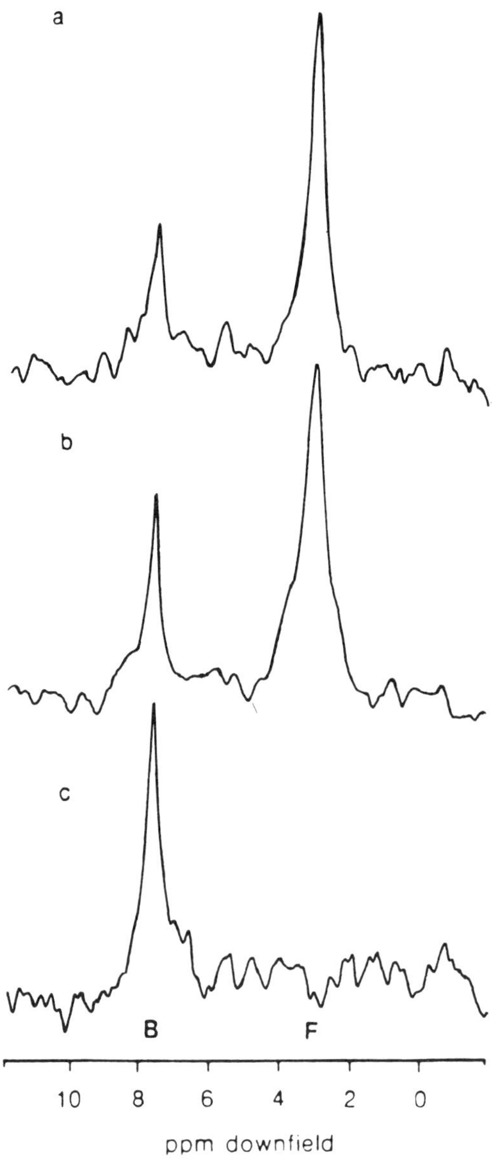

FIGURE 35. (a) ^{19}F spectrum at 37°C and 188.3 MHz of [^3H]5FBAPTA in BALB/c mouse thymocytes accumulated in 20 min at 37°C. (b) Effect of addition of succinyl Con A at 100 μg/ml. (c) Effect of addition of 50 μ*M* A23187 to the same cell preparation. The intracellular 5FBAPTA is saturated with Ca^{2+} (B) and a single resonance at the Ca-5FBAPTA position is observed. F, resonance of 5FBAPTA without Ca^{2+}. (From Smith, G. A., Hesketh, R. T., Metcalfe, J. C., Feeney, J., and Morris, P. G., *Proc. Natl. Acad. Sci. U.S.A.*, 80, 7178, 1983. With permission.)

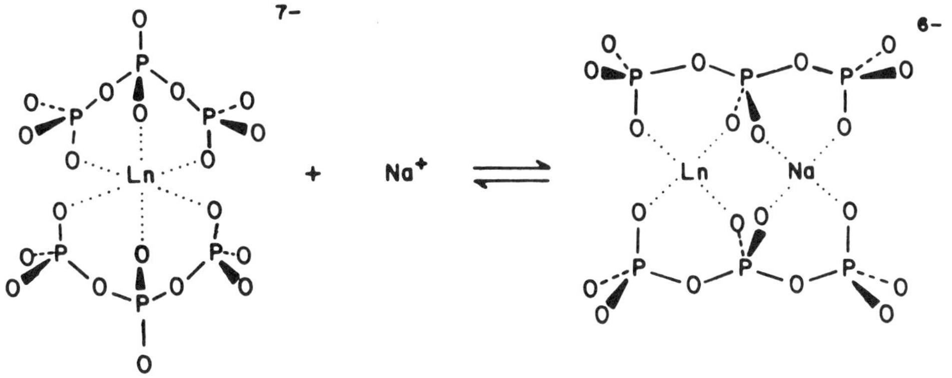

FIGURE 36. Hypothetical structures for Ln(PPP)$_2^{-7}$ and Na$^+$·Ln(PPP)$_2^{-7}$. (From Chu, S. C., Pike, N. M., Fossel, E. T., Smith, T. W., Balschi, J. A., and Springer, C. S., Jr., *J. Magn. Reson.*, 56, 33, 1984. With permission.)

3. ^{23}Na and ^{39}K

In the past, the lack of spectral resolution has limited the utility of NMR to the study of intracellular Na$^+$ ions. The relatively small internal ^{23}Na resonance and the overwhelming extracelluar ^{23}Na resonance occur at the same frequency in the spectra of intact cells, living tissues, organelles, and cell suspensions. Highly anionic hyperfine shift reagents such as bis(tripolyphosphate)dysprosium(III) Dy(PPP$_i$)$_2^{-7}$ provide a convenient means whereby this problem can be circumvented, and separate NMR signals from intra- and extracellular sodium or potassium can be detected.[165-168]

Chu et al.[169] have published a hypothetical structure (shown in Figure 36) which portrays the possible type of interaction which alkali metal ions can undergo with lanthanide shift reagents. The ability to detect these separate resonances from internal and external sodium and potassium ions depends on the fact that a hyperfine shift in frequency will only be induced on the resonances of the sodium or potassium ions which are accessible to the shift reagent. Because of its very high ionic charge and subsequent inability (under most circumstances) to penetrate a cell membrane during the time course of typical NMR measurements, the agents remains outside of the cells. Therefore, only the extracellular sodium or potassium ions are affected and undergo a resonance shift.

The cytosolic enzymes of many plants are inhibited by high intracellular concentrations of sodium ion.[170] The manner in which various plant hybrids deal with sodium, especially its exclusion, represents a key feature in their ability to survive in saline environments. Salinity can increase drastically during periods of prolonged irrigation due to the salts which accumulate in the aqueous environment. The observation of ^{23}Na resonances provides a rapid, noninvasion method for the determination of sodium ion concentrations and distributions in cells. The data can essentially be acquired within seconds to a minute. Furthermore, the living tissue is not disturbed, extracted, or disrupted. Finally, influx and efflux rates can be monitored *in vivo* in the case of plants, plant tissues, and single cells. Minimally one can measure as little as 1.0 m*M* sodium ion concentrations with typical S/N ratios on the order of 10:1.

The external and internal sodium signals in corn roots were distinguished by means of a shift reagent, 3.0 m*M* Dy(PPP$_i$)$_2^{-7}$.[166-168] We see in Figure 37 that the two signals from the internal and the external sodium are well resolved. The uptake of sodium by the corn root is also readily apparent in the same figure as the intensity of the internal resonance increases with time of exposure to the salt.[168]

The rates of Na$^+$ uptake can easily be monitored under conditions where respiration and/or metabolism are pertubed. In Figure 38 the relative rates of Na$^+$ influx are depicted for the same corn root system under a variety of metabolic states.[168] Na$^+$ ion influx is seen to decrease upon

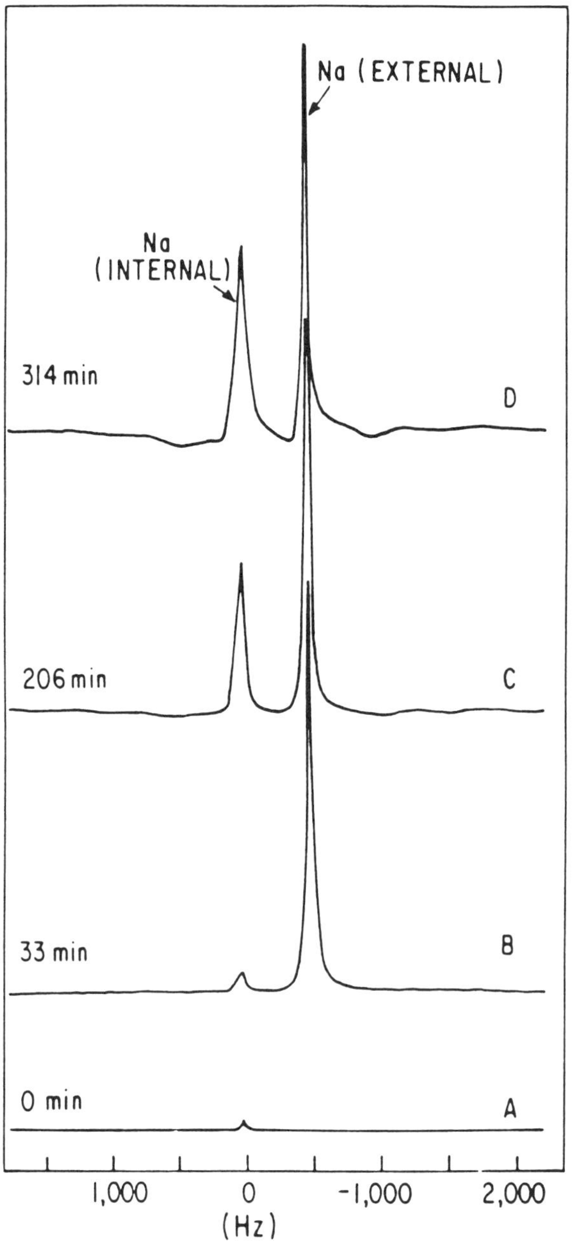

FIGURE 37. Sodium influx into aerobic root tissue. A total of 3.0 m*M* of $Dy(PPP_i)_2^{-7}$ was added along with the appropriate amount of NaCl to keep the concentration of Na^+ at 160 m*M*. Spectrum (A) represents the endogenous Na^+ concentration. Spectra (B), (C), and (D) represent the distribution of Na^+ between cellular and external spaces at the indicated time after the addition of shift reagent and NaCl. Each spectrum was time averaged for 10.83 min. Other experimental conditions used were frequency range 8 KHz; 16 K data points zero filled to 16 K; 548 transients per spectrum; repetition time 1.124 s; 28.5 ms (90° pulse); and zero broadening factor. (From Gerasimowicz, W. V., Tu, S., and Pfeffer, P. E., *Plant Physiol.*, 81, 925, 1986. With permission.)

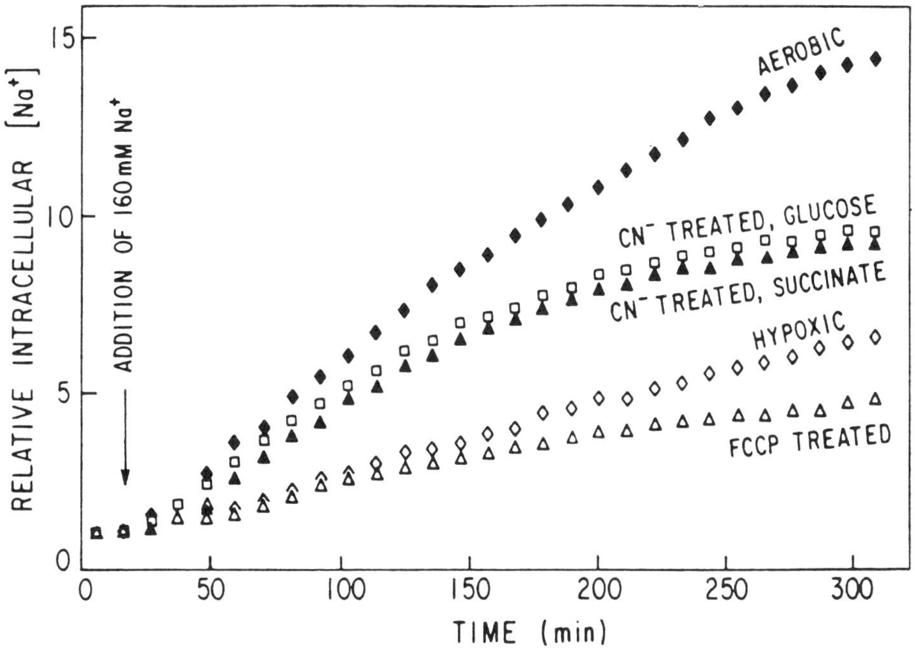

FIGURE 38. Relative rates of Na^+ influx under different conditions. The plant tissue was first treated as mentioned in Figure 37 except, in indicated cases, glucose was replaced by 25 m*M* Tris-succinate. Modifications during perfusion are as indicated in the figure. The intensities of the intracellular sodium resonances were normalized to the ^{23}Na signal observed for the roots prior to any addition of external Na^+ to the perfusion medium. The normalized intensities were plotted at the middle time of each spectral acquisition. (From Gerasimowicz, W. V., Tu, S., and Pfeffer, P. E., *Plant Physiol.*, 81, 925, 1986. With permission.)

treatment with CN^- and upon depression of ATP under the influence of hypoxia. Furthermore, when the roots are subjected to FCCP treatment, the uptake rate is even slower than that observed under anaerobic conditions.

Sillerud has studied the uptake and efflux of sodium ion in NaCl-adapted and -nonadapted millet *(Panicum miliaceum)* suspensions.[171] Figure 39 shows the time dependence of the intracellular sodium concentration in suspensions of cultured millet cells as monitored by means of natural-abundance sodium-23 NMR. The efflux curve was explained on the basis of a biexponential fit. The implication inherent in this finding is the presence of two sodium compartments in the system. One compartment equilibrates rapidly with the medium, whereas the intracellular compartment equilibrates at a slower rate.[171]

Sillerud et al.[135] have studied sodium transport in roots from intact *Distichlis spicata*. This particular plant has the ability to thrive in highly saline environments such as salt marshes. However, the ability of such species to adapt to high, constant levels of sodium ion is not well characterized. When plants grown in 130 m*M* NaCl-containing medium were subjected to conditions where no sodium was available, the efflux curve shown in Figure 40 was obtained.[135] Once again biphasic kinetics were seen. Upon completion of the efflux experiments, the authors place the roots into media containing $Dy(PPP_i)_2^{-7}$ and NaCl in order to ascertain the dynamics of ion influx. Figure 41 shows the result. Therefore, one can readily begin to correlate plant saline stress adaptability with the measured ion transport and translocation rate data.

In a very unique study, simultaneous ^{23}Na and ^{31}P NMR spectra were acquired for a number of yeast suspensions.[172] Before all of the NMR experiments were performed, the cells were preloaded with sodium. These particular yeast cells were also distinguished by the fact that no PP species were evident indicating that the cells were somewhat deficient in phosphorus. These cells were suspended in media which were low in sodium but contained inorganic phosphorus.

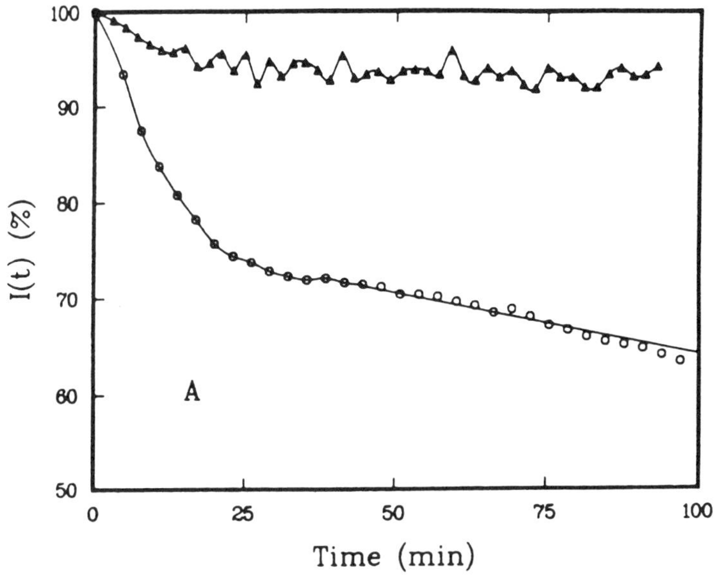

FIGURE 39. Net efflux of ^{23}Na in the presence of DyTp at pH 6.9 in the external medium. The percent change in integrals [I(t)(%)] of internal ^{23}Na peak vs. time is shown. For efflux the NaCl-adapted cells (O) were grown for 3 weeks on 130 m*M* NaCl and had resumed normal growth, while nonadapted cells (s) were preloaded for 5 d. (From Sillerud, L. O. and Heyser, J. W., *Plant Physiol.*, 75, 269, 2984. With permission.)

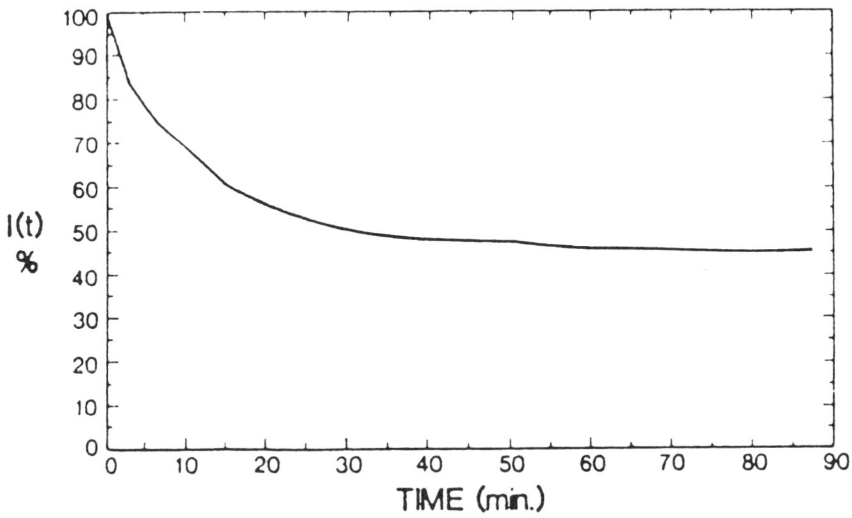

FIGURE 40. Time dependence of the intracellular sodium-23 NMR signal for *Distichlis spicata*. Seen here is sodium efflux from the roots of this whole plant after the roots were placed into a medium containing 0 m*M* NaCl and 5 m*M* DyTP. In this and the following figure, the plants were grown hydroponically on 130 m*M* NaCl and either washed extensively with sodium-free medium for the influx experiments, or placed directly into the NMR tube in sodium-free medium for the efflux runs. (From Sillerud, L. O., Heyser, J. W., Han, C. H., and Bitensky, M. W., *NMR in Living Systems*, Axenrod, T. and Ceccarelli, G., Eds., D. Reidel, New York, 1986. With permission.)

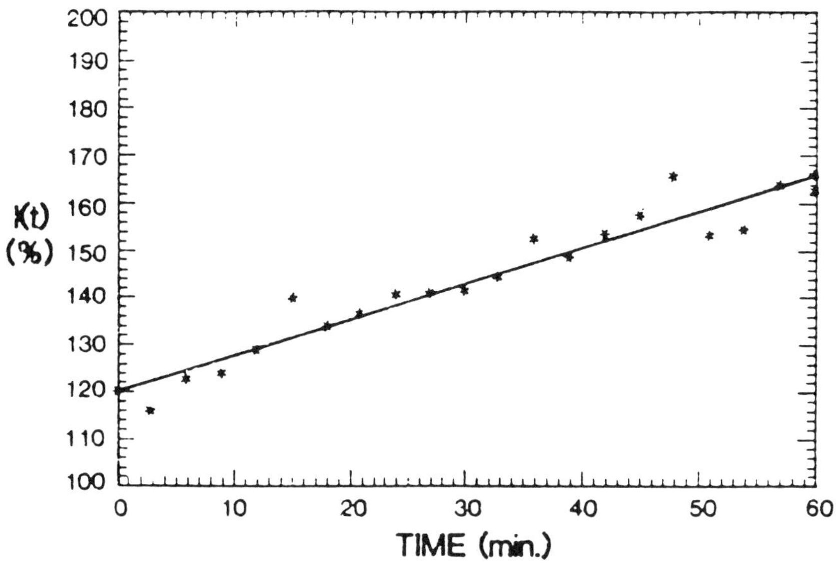

FIGURE 41. Influx of sodium into the roots of a salt-adapted whole plant, *Distichlis spicata*, as monitored with the aid of real-time sodium-23 NMR. (From Sillerud, L. O., Heyser, J. W., Han, C. H., and Bitensky, M. W., *NMR in Living Systems*, Axenrod, T. and Ceccarelli, G., Eds., D. Reidel, New York, 1986. With permission.)

The spectra obtained in these studies indicated that a net efflux of sodium occurred in all cases with a concomitant uptake of P_i. Such dual observation NMR experiments cannot be performed on standard NMR equipment typically available from vendors. The authors modified their instruments to include two entirely separate NMR transmitters, receiver channels, as well as pulse programmers and computers in order to observe and acquire data on the two nuclei simultaneously.[172]

Several recent reports on sodium transport in bacteria have appeared.[45,173,174] Shulman's group found that while extracellular sodium was completely visible in *Escherichia coli* by NMR, only 45% of the intracellular sodium could be accounted for in this fashion.[45] Internal sodium concentrations were found to be correlated with the development and maintenance of proton-driven Na^+/H^+ exchange mechanisms and/or intracellular pH condition.[45] Packer et al.[173] studied the effects of salt stress on the freshwater cyanbacterium *Synechoccus* 6311. Their findings indicated that when the bacteria were subjected to elevated levels of salt (0.5 *M* NaCl), NTP concentrations decreased while P_i and cytoplasmic pH both increased. The cell suspensions were illuminated during the spectral acquisition. The authors claim they can monitor light and dark reactions in such systems. However, an examination of the ^{23}Na spectra brings into question even qualitative interpretations of the data presented due to its relatively poor quality, i.e., extremely low intensity for the sodium and the unresolved nature of the internal and external sodium peaks in their spectra.[173]

One cautionary note should also be sounded at this point. Martin et al.[174] recently studied the intracellular sodium ion concentration in amoeba from the slime mold *Dictyostelium discordeum*. Two distinct peaks could be distinguished in these systems with the use of shift reagents. Cells which had been starved of Na^+ could easily be reloaded. However, the most interesting observation made by the authors was that $Dy(PPP_i)_2^{-7}$ was being degraded during the NMR experiments. They concluded that an acid phosphatase produced by the amoeba was responsible for this degradation. In all such experiments with living systems, one must ascertain the toxicity (if any) of the shift reagent on the system and establish the stability of the reagents being employed.

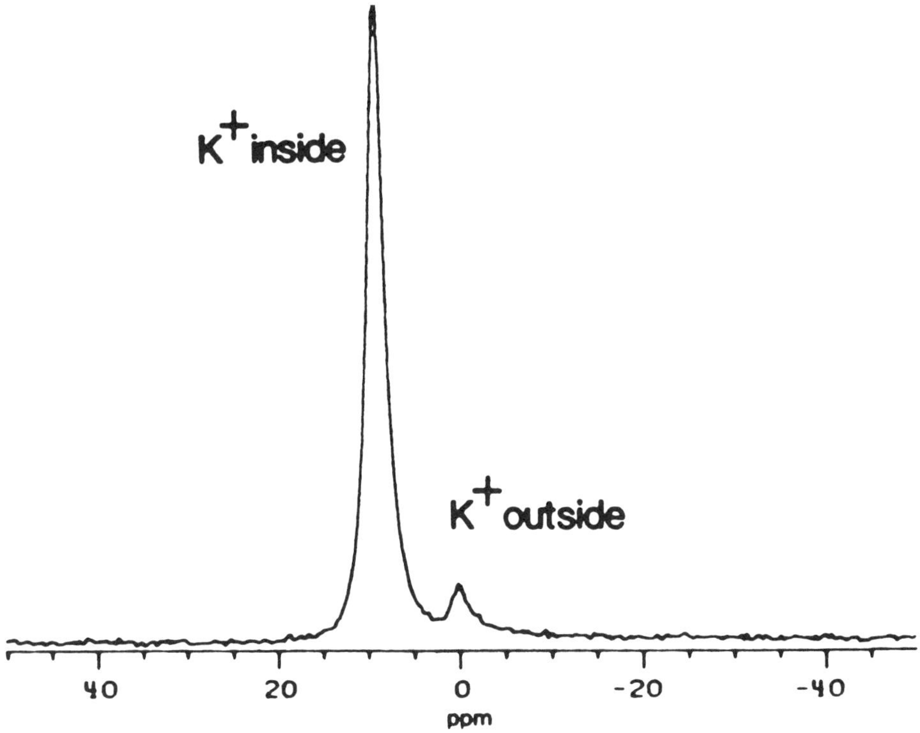

FIGURE 42. ^{39}K NMR spectrum at 16.9 MHz, 25°C in the same solution as above. A total of 35,000 pulses were collected with an acquisition time of 0.15. (Spectra courtesy of Peter Lundberg and Hans J. Vogel.) (From Braunlin, W. H., Drakenberg, T., and Forsén, S., *Current Topics in Bioenergetics*, Vol. 14, Academic Press, New York, 1985, 97. With permission.)

Intracellular ^{39}K concentrations (if visible) can also be studied by the same shift reagent methods described above. However, the ^{39}K nucleus presents more severe problems to the NMR spectroscopist than does ^{23}Na. The receptivity of potassium is lower than any other alkali metal in terms of the NMR experiment. On the other hand, the concentration of potassium in most cell types is generally an order of magnitude greater than that of sodium. Therefore, some compensation is apparent. We show an example of the separation of intracellular and extracellular ^{39}K signals in the presence of shift reagent which is possible from the spectrum of human erythrocytes in Figure 42.[44] Very little has been done to exploit ^{39}K NMR in agriculture. Some studies are being carried out in the medical areas.[175] If one can correct for losses in signal intensities, a vast potential exists for this technique in all areas of agricultural science (analogous to the work ongoing in the medical areas). At the time of this writing no literature exists on ^{39}K NMR in areas of whole plants, plant tissues, and plant metabolism.

REFERENCES

1. **Moon, R. B. and Richards, J. H.,** Determination of intracellular pH by ^{31}P nuclear magnetic resonance, *J. Biol. Chem.*, 248, 7276, 1973.
2. **Baxter, R. L.,** Microbiological applications of NMR spectroscopy. I, *Microbiol. Sci.*, 2, 203, 1985; **Baxter, R. L. and Baxter, R. L.,** Microbiological applications of NMR spectroscopy. II, *Microbiol. Sci.*, 2, 340, 1985.

3. **Shulman, R. G., Brown, T. R., Ugurbil, K., Ogawa, S., and Cohen, S. M.,** Cellular application of ^{31}P and ^{13}C nuclear magnetic resonance, *Science,* 205, 160, 1979.
4. **Loughman, B. C. and Ratcliffe, R. G.,** Nuclear magnetic resonance and the study of plants, in *Advances in Plant Nutrition,* Vol. 1, Praeger, New York, 1984; **Roberts, J. K. M.,** Study of plant metabolism *in vivo* using NMR spectroscopy, *Annu. Rev. Plant Physiol.,* 35, 375, 1984.
5. **Degane, H., Laughlin, M., Campbell, S., and Shulman, R. G.,** Kinetics of creation kinase in heart: ^{31}P NMR saturation- and inversion-transfer study, *Biochemistry,* 24, 5510, 1985.
6. **Alger, J. R., Behar, K. L., Rothman, D., and Shulman, R. G.,** *J. Magn. Reson.,* 56, 334, 1984.
7. **Roberts, J. K. M. and Jardetzky, O.,** Monitoring of cellular metabolism by NMR, *Biochim. Biophys. Acta,* 639, 53, 1981.
8. **Gadian, D. G.,** *Nuclear Magnetic Resonance and Its Applications to Living Systems,* Clarendon Press, Oxford, 1982.
9. **Avison, M. J., Hetherington, H. P., and Shulman, R. G.,** Applications of NMR to studies of tissue metabolism, *Annu. Rev. Biophys. Chem.,* 15, 337, 1986.
10. **Becker, E. D.,** High resolution NMR, *Theory and Chemical Applications,* Academic Press, New York, 1980.
11. **Shaw, D.,** *Fourier Transform NMR Spectroscopy,* Elsevier, Amsterdam, 1979.
12. **Sanders, J. K. M. and Hunter, B. K.,** *Modern NMR Spectroscopy a Guide for the Chemist,* Oxford University Press, Oxford, 1987.
13. **Roberts, J. K. M., Ray, P. M., Jardetzsky, N. W., and Jadetzky, O.,** Estimation of cytoplasmic and vacuolar pH in higher plants cells by ^{31}P NMR, *Nature,* 283, 870, 1980.
14. **Pfeffer, P. E., Tu, S.-I., Gerasimowicz, W. V., and Cavanaugh, J. R.,** *In vivo* ^{31}P NMR studies of corn root tissue and its uptake of toxic metals, *Plant Physiol.,* 80, 77, 1986.
15. **Gard, J. K. and Ackerman, J. H.,** A ^{31}P NMR external reference for intact biological systems, *J. Magn. Reson.,* 51, 124, 1983.
16. **Pfeffer, P. E., Tu, S.-I., Gerasimowicz, W. V., and Boswell, R. T.,** Effects of aluminum on the release and-or immobilization of soluble phosphate in corn root tissues, *Planta,* 172, 200, 1987.
17. **Batley, M. and Redmond, J. W.,** ^{31}P NMR reference standards for aqueous samples, *J. Magn. Reson.,* 49, 172, 1982.
18. **Ratcliffe, R. G.,** NMR and the inorganic composition of plants, *Inorg. Biochem.,* 28, 347, 1986.
19. **Lee, R. B. and Ratcliffe, R. G.,** Development of an aerobic system for use in plant tissue NMR experiments, *Exp. Bot.,* 34, 1213, 1983.
20. **Roby, C., Martin, J. B., Bligny, R., and Douce, R.,** Biochemical changes during sucrose deprivation in higher plant cells. II. Phosphorus-31 nuclear magnetic resonance studies, *J. Biol. Chem.,* 262, 5000, 1987.
21. **Fan, W.-M. T., Higashi, R. M., and Lane, A.,** Monitoring hypoxic metabolism in superfused plant tissue by *in vivo* ^{1}H NMR, *Arch. Biochem. Biophys.,* 251, 674, 1986.
22. **den Hollander, J. A., Ugurbil, R., Brown, T. R., and Shulman, R. G.,** Phosphorus-31 nuclear magnetic resonance studies of the effect of oxygen upon glycolysis in yeast, *Biochemistry,* 20, 5871, 1981.
23. **Boulanger, Y., Vinay, D., Phan Viet, M. T., Guardo, R., and Desroches, M.,** An improved perfusion system for NMR study of living cells, *Magn. Reson. Med.,* 2, 495, 1985.
24. **Geoffrion, Y., Lareau, S., Deslauriers, R., Butler, K., Pass, M., and Smith, I. C. P.,** A versatile perfusion technique for metabolic studies by NMR, *Magn. Res. Med.,* 2, 65, 1985.
25. **Balaban, R. S., Kantor, H. L., and Ferretti, J. A.,** *In vivo* flux between phosphocreatine and adenosine triphosphate determined by two-dimensional phosphorus NMR, *Biol. Chem.,* 21, 12787, 1983.
26. **Chance, B., Clark, B. J., Nioka, S., Subramanian, H., Maris, J. M., Argov, Z., and Bode, H.,** Phosphorus nuclear magnetic resonance spectroscopy *in vivo, Circulation,* 72 (4), 103, 1985.
27. **Martin, J. B., Bligny, R., Rebeille, F., Douce, R., Leguay, J. J., Mathiew, Y., and Guern, J.,** ^{31}P nuclear magnetic resonance study of intracellular pH of plant cells cultivated in liquid medium, *Plant Physiol.,* 70, 1156, 1982.
28. **Stothers, J. B.,** *Carbon-13 NMR Spectroscopy,* Academic Press, New York, 1972.
29. **Martin, F.,** Monitoring plant metabolism by ^{13}C, ^{15}N and ^{14}N nuclear magnetic resonance spectroscopy. A review of the applications to algae, fungi and higher plants, *Physiol. Veg.,* 4, 463, 1985.
30. **Reinhard, B. and Günther, H.,** Modern pulse methods in high-resolution NMR spectroscopy, *Angew. Chem. Int. Ed. Engl.,* 22, 350, 1983.
31. **Walker, T. E., Han, C. H., Kollman, V. H., London, R. E., and Matwiyoff, N. A.,** ^{13}C NMR studies of the biosyntheses by microbacterium ammoniaphilum of L-glutamate selectively enriched with carbon-13, *J. Biol. Chem.,* 257, 1189, 1982.
32. **Ashworth, D. J. and Mettler, I. J.,** Direct observation of glycine metabolism in tobacco suspension cells by carbon-13 NMR spectroscopy, *Biochemistry,* 23, 2252, 1984.
33. **Stinson, E. E., Wise, W. B., Moreau, R. A., Jurewicz, A. J., and Pfeffer, P. E.,** Alternariol: evidence for biosythesis via norlichexanthone, *Can. J. Chem.,* 64, 1590, 1986.
34. **Martin, F.,** Nitrogen-15 NMR studies of nitrogen assimilation and amino acid biosythesis in the ectomycorrhizal fungus cenoccum graniforme, *FEBS Lett.,* 182, 350, 1985.

35. **Mason, J.**, Nitrogen NMR spectroscopy in inorganic, organometallic and bio-inorganic chemistry, *Chem. Rev.*, 81, 205, 1981.
36. **Belton, P. S., Lee, R. B.,** and **Ratcliffe, R. G.**, A ^{14}N NMR study of inorganic nitrogen metabolism in barley, maize and pea roots, *J. Exp. Bot.*, 36, 190, 1985.
37. **Brown, F., Campbell, I. D., and Kuchel, P. W.**, Human erythrocyte metabolism studies by ^{1}H spin echo NMR, *FEBS Lett.*, 82, 12, 1977.
38. **Kime, M. J., Ratcliffe, R. G., and Williams, R. J. P.**, The application of ^{31}P nuclear magnetic resonance to higher plants, *J. Exp. Bot.*, 33, 656, 1982.
39. **Schleich, T., Willis, J. A., and Matson, G. B.**, Longitudinal (T_1) relaxation times and phosphorus metabolites in bovine and rabbit lens, *Exp. Eye Res.*, 39, 455, 1984.
40. **Lee, R. B. and Ratcliffe, R. G.**, Phosphorus nutrition and the intracellular distribution of inorganic phosphate in pea root tips. A quantitative study using ^{31}P-NMR, *J. Exp. Bot.*, 34, 1222, 1983.
41. **Reid, R. J., Loughman, B. C., and Ratcliffe, R. G.**, ^{31}P NMR measurements of cytoplasmic pH changes in maize root tips, *J. Exp. Bot.*, 36, 889, 1985.
42. **Kanamori, K., Legerton, T. L., Weiss, R. L., and Roberts, J. D.**, Nitrogen-15 spin-lattice relaxation times of amino acids in *Neurospora crassa* as a probe for intracellular environment, *Biochemistry*, 21, 4916, 1982.
43. **Kanamori, K. and Roberts, J. D.**, ^{15}N NMR studies of biological systems, *Acc. Chem. Res.*, 16, 35, 1983.
44. **Braunlin, W. H., Drakenberg, T., and Forsén, S.**, Metal in NMR: application to biological systems, in *Current Topics in Bioenergetics*, Vol. 14, Academic Press, New York, 1985, 97.
45. **Castle, A. M., Macnab, R. M., and Shulman, R. G.**, Measurement of intracellular sodium concentration and sodium transport in *Escherichia coli* by ^{23}Na NMR, *J. Biol. Chem.*, 261, 3288, 1986.
46. **Gupta, R. K.**, NMR spectroscopy of intracellular sodium ions in living cells, in *NMR in Living Systems*, Axenrod, T. and Ceccurelli, G., Eds., D. Reidel, New York, 1986, 291.
47. **Rooney, W. D., Barbara, T. M., and Springer, C. S., Jr.**, Two-dimensional double-quantum NMR spectroscopy of isolated spin 3/2 systems, *J. Am. Chem. Soc.*, 110, 674, 1988.
48. **Ratcliffe, R. G.**, The application of NMR methods to plant tissues, *Methods Enzymol.*, 148, 683, 1986.
49. **Navone, G., Shulman, R. G., Yamane, T., Eccleshall, T. R., Lam, K.-B., Baronsky, J. J., and Murmur, J.**, P-31 NMR studies of wild-type and glycolytic pathway mutants of *Saccharomyces cerevisiae*, *Biochemistry*, 21, 4487, 1979.
50. **Boyd, R. K., DeFreitas, A. S. W., Hoyle, J., McCulloch, A. W., McInnes, A. G., Rogerson, A., and Walter, J. A.**, Glycerol 1, 2-cyclic phosphate in centric diatoms, *J. Biol. Chem.*, 262, 12406, 1987.
51. **Batley, M., Redmond, J. W., Djordjevic, S. P., and Rolfe, B. C.**, Characterization of glycerophosphorylated cyclic β-1,2-glucans from fast-growing Rhizobium species, *Biochim. Biophys. Acta*, 901, 119, 1987.
52. **Turner, J. F. and Turner, D. H.**, The regulation of glycolysis and the pentose phosphate pathway, in *The Biochemistry of Plants*, Vol. 2, Davies, D. D., Ed., Academic Press, New York, 1980, 279.
53. **Scarpa, A.**, in *Membrane Transport in Biology*, Giebish, G., Tosteson, D. C., and Ussing, H. H., Eds., Springer-Verlag, Berlin, 1979, 263.
54. **Preiss, J. and Levi, C.**, in *Proc. 4th Int. Congr. Photosynthesis*, Hall, D. O., Coombs, J., and Goodwin, T. W., Eds., The Biochemical Society, London, 1982, 457.
55. **Rebeillé, F., Bligny, R., and Douce, R.**, Regulation of Pi uptake by *Acer pseudoplatanus* cells, *Arch. Biochem. Biophys.*, 219, 371, 1982.
56. **Jacob, G. S., Schaefer, J., Stejskal, E. O., and McKay, R. A.**, Solid state NMR determination of glyphosate metabolism in a *Pseudomonas* sp., *J. Biol. Chem.*, 260, 5899, 1985.
57. **Slonczewski, J. L., Rosen, B. P., Alger, J. R., and Macnab, R. M.**, pH homeostasis in *Escherichia coli*: measurement by ^{31}P NMR of methylphosphonate and phosphate, *Proc. Natl. Acad. Sci. U.S.A.*, 78, 6271, 1981.
58. **Kushmerick, M. J. and Meyer, R. A.**, *Am. J. Physiol.*, 248, C542, 1985.
59. **Roberts, J. K. M., Jardetzky, N. W., and Jardetzky, O.**, Intracellular pH measurements by ^{31}P NMR. Influence of factors other than pH on ^{31}P chemical shifts, *Biochemistry*, 20, 5389, 1981.
60. **Ogawa, S., Rottenberg, H., Brown, T. R., Shulman, R. G., Castillo, C. L., and Glynn, P.**, *Proc. Natl. Acad. Sci. U.S.A.*, 75, 1796, 1978.
61. **Gupta, R. K., Gupta, P., Yushok, W. D., and Rose, Z.**, On the non-invasive measurement of intracellular free magnesium by ^{31}P NMR spectroscopy. High-resolution phosphorus-31 nuclear magnetic resonance study of rat liver mitochondria, *Physiol. Chem. Phys. Med. NMR*, 15, 265, 1983.
62. **Sondheimer, G. M., Kuhn, W., and Kalbitzer, H. R.**, Observation of Mg^{2+}·ATP and uncomplexed ATP in slow exchange by ^{31}P-NMR at high magnetic fields, *Biochem. Biophys. Res. Commun.*, 134, 1379, 1986.
63. **Cohen, S. M., Ogawa, S., Rothenberg, H., Glynn, P., and Yamane, T.**, *Nature*, 273, 554, 1978.
64. **Mahler, H. R. and Cordes, E. H.**, *Biological Chemistry*, Harper & Row, New York, 1966, 420.
65. **Pfeffer, P. E., Tu, S.-I., Gerasimowicz, W. V., and Boswell, R. T.**, Role of the vacuole in metal ion trapping as studied by *in vivo* ^{31}P NMR spectroscopy, in *Plant Vacuoles*, Marin, B., Ed., Plenum Press, New York, 1987, 349.
66. **Samofus, B. and Schwimmer, S.**, Phytic acid as a phosphorus reservoir in the developing potato tuber, *Nature*, 194, 578, 1962.

67. **McCain, D. C.,** ^{31}P-NMR study of turnip seed germination and sprout growth inhibition by peroxydisulfate ion, *Biochim. Biophys. Acta,* 783, 231, 1983.
68. **Defini, M., Angeline, R., Bruno, F., Conti, F., Dicocco, M. E., Guildiani, A. M., and Manes, F.,** Phosphate metabolites in germinating lettuce seeds observed by ^{31}P NMR spectroscopy, *Cell. Mol. Biol.,* 31, 385, 1985.
69. **Hendrix, D. L., Radin, J. W., and Nieman, R. A.,** Intracellular pH of cotton embryos and seed coats during fruit development determined by ^{31}P NMR spectroscopy, *Plant Physiol.,* 85, 588, 1987.
70. **Busa, W. B.,** Cellular dormancy and the scope of pH mediated metabolic regulation, in *Intracellular pH: Measurement and Regulation,* Alan R. Liss, New York, 1982, 64.
71. **Aducci, P., Federico, R., Carpinelli, G., and Podo, F.,** Temperature dependence of intracellular pH in higher plant cells, *Planta,* 156, 579, 1982.
72. **Roberts, J. K. M., Wemmer, D., Ray, P. N., and Jardetzky, O.,** Regulation of cytoplasmic and vacuolar pH in maize root tips under different experimental conditions, *Plant Physiol.,* 69, 1344, 1982.
73. **Reid, R. J., Loughman, B. C., and Ratcliffe, R. G.,** ^{31}P NMR measurements of cytoplasmic pH changes in maize root tips, *J. Exp. Bot.,* 36, 889, 1985.
74. **Reid, R. J., Field, L. D., and Pitman, M. G.,** Effects of external pH, fussicoccin and butyrate on the cytoplasmic pH in barley root tips measured by ^{31}P-NMR spectroscopy, *Planta,* 166, 341, 1985.
75. **Wofender, J. and Wellburn, A. R.,** Cellular readjustment of barley seedlings to simulated acid rain, *New Phytol.,* 104, 97, 1986.
76. **Jackson, P. C., Pfeffer, P. E., and Gerasimowicz, W. V.,** Use of ^{31}P NMR to assess effects of DNP on ATP levels *in vivo* in barley roots, *Plant Physiol.,* 81, 1130, 1986.
77. **Torimitsu, K., Hayashi, M., Ohta, E., and Sakata, M.,** Effect of K^+ and H^+ stress and role of Ca^{2+} in the regulation of intracellular K^+ concentration in mung bean, *Physiol. Plant.,* 63, 247, 1985.
78. **Hughes, W. A., Bociek, S. M., Barrett, J. N., and Ratcliffe, R. G.,** An investigation of the growth characteristics of oil-palm *(Elaeis guineensis)* suspension cultures using ^{31}P NMR, *Biosci. Rep.,* 3, 1141, 1983.
79. **Guern, J., Mathieu, Y., Pean, M., Corinne, P., Beloeil, J.-C., and Lallemand, J.-Y.,** Cytoplasmic pH regulation in *Acer pseudoplatanus* cells, *Plant Physiol.,* 82, 840, 1986.
80. **Ojalva, J., Roken, J. S., Navon, G., and Goldberg, I.,** ^{31}P NMR study of elicitor treated *Phaseolus velganis* cell suspension cultures, *Plant Physiol.,* 85, 716, 1987.
81. **Mitsumori, F., Yoneyama, T.,** and **Ito, O.,** Phosphorus-31 NMR studies on higher plant tissue, *Plant Sci.,* 38, 87, 1985.
82. **Menegus, F. and Fronza, G.,** Modulation of glycerophosphorylcholine and glycerophosphorylethanolamine in rice shoots by the environment oxygen level, *FEBS Lett.,* 187, 151, 1985.
83. **Roberts, J. K. M., Lane, A. N., Clark, R. A., and Nieman, R. H.,** Relationship between the rate of synthesis of ATP and the concentration of reactants and products of ATP hydrolysis in maize root tips, determined by ^{31}P NMR, *Arch. Biochem. Biophys.,* 240, 712, 1985.
84. **Alger, J. R. and Shulman, R. G.,** *Q. Rev. Biophys.,* 17, 83, 1984.
85. **Roberts, J. K. M., Wemmer, D., and Jardetzky, O.,** Measurement of mitochondrial ATPase activity in maize root tips by saturation transfer ^{31}P NMR, *Plant Physiol.,* 74, 632, 1984.
86. **Shoubridge, E. A., Briggs, R. W., and Radda, G. K.,** ^{31}P NMR saturation transfer measurements of the steady state rates of creatine kinase and ATP synthetase in rat brain, *FEBS Lett.,* 140, 288, 1982.
87. **Dawson, M. J. and Gadian, D. G.,** Studies of the biochemistry of contracting and relaxing muscle by the use of P-31 NMR in conjunction with other techniques, *Philos. Trans. R. Soc. B,* 289, 445, 1980.
88. **Hanson, J. B., Bertagnolli, B. L., and Shepard, W. D.,** Phosphate-induced stimulation of acceptorless respiration in corn mitochondria, *Plant Physiol.,* 50, 347, 1972.
89. **Rebeille, F., Bligny, R., Martin, J. B., and Douce, R.,** Effect of sucrose starvation on sycamore *(Acer pseudolplatanus)* cell carbohydrates and Pi status, *Biochem. J.,* 226, 679, 1985.
90. **Roby, C., Martin, J. B., Bligny, R., and Douce, R.,** Biochemical changes during sucrose deprivation in higher plant cells. II. Phosphorus-31 NMR studies, *J. Biol. Chem.,* 262, 5000, 1987.
91. **Dorne, A. J., Bligny, R., Rebeille, F., Roby, C., and Douce, R.,** Fatty acid disappearance and phosphorylcholine accumulation in higher plant cells after a long period of sucrose deprivation, *Plant Physiol. Biochem.,* 25, 589, 1987.
92. **Brodelius, P. and Vogel, H. J.,** A phosphorus-31 NMR study of phosphate uptake and storage in cultured *Catharanthus roseus* and *Daucus carota* plant cells, *J. Biol. Chem.,* 260, 3556, 1985.
93. **Nicolay, K., Scheffers, W. A., Bruinenberg, P. M., and Kaptein, R.,** *In vivo* ^{31}P NMR studies of the role of the vacuole in phosphate metabolism of yeasts, *Arch. Microbiol.,* 134, 270, 1983.
94. **Greenfield, N. J., Hussain, M., and Lenard, J.,** Effects of growth state and amines on cytoplasmic and vacuolar pH, phosphate and polyphosphate levels in *Saccharomyces cerevisiae:* a ^{31}P-NMR study, *Biochim. Biophys. Acta,* 926, 205, 1987.
95. **Cassone, A., Carpinelli, G., Angiolella, L., Maddaluno, G., and Podo, F.,** ^{31}P NMR study of growth and dimorphic transition in *Candida albicans, J. Gen. Microbiol.,* 129, 1569, 1983.
96. **Roberts, J. K. M., Linker, C. S., Benoit, A. G., Jardetzky, O., and Nieman, R. H.,** Salt stimulation of phosphate uptake in maize root tips studied by ^{31}P NMR, *Plant Physiol.,* 74, 947, 1984.

97. **Ben-Hayyim, G. and Navon, G.**, Phosphorus-31 NMR studies of wild-type and NaCl-tolerant citrus cells, *J. Exp. Bot.*, 36, 1877, 1985.
98. **Foy, C. D., Chaney, R. L., and White, M. C.**, The physiology of metal toxicity in plants, *Annu. Rev. Plant Physiol.*, 29, 511, 1978.
99. **Foy, C. D.**, The physiology of plant adaptation to mineral stress, *Iowa State J. Res.*, 57, 355, 1983.
100. **Maas, E. V.**, Calcium uptake by excised maize roots and interactions with alkali cations, *Plant Physiol.*, 44, 985, 1969.
101. **Hussain, F., Overstreet, R., and Jacobson, L.**, The influence of hydrogen ion concentration on cation absorption by barley roots, *Plant Physiol.*, 29, 234, 1954.
102. **Loughman, B. C.**, The application of *in vivo* techniques in the study of metabolic aspects of ion absorption in crop plants, *Plant Soil*, 99, 63, 1987.
103. **Haug, A.**, Molecular aspects of aluminium toxicity, *Crit. Rev. Plant Sci.*, 1, 345, 1984.
104. **Sirkar, S. and Amin, J. V.**, Influence of anxin on respiration of manganese toxic cotton plants, *Indian J. Exp. Biol.*, 17, 618, 1979.
105. **Dwek, R. A.**, *NMR in Biochemistry*, Clarendon Press, Oxford, 1973.
106. **Kime, M. J., Loughman, B. C., and Ratcliffe, R. G.**, The application of ^{31}P NMR to higher plant tissue. II. Detection of intracellular changes, *J. Exp. Bot.*, 33, 670, 1982.
107. **Roberts, J. K. M., Callis, J., Jardetzky, O., Walbot, U., and Freeling, M.**, Cytoplasmic acidosis as a determinant of flooding intolerance in plants, *Proc. Natl. Acad. Sci. U.S.A.*, 81, 6029, 1984.
108. **Kopp, S. J., Prentice, R. C., Glonek, T., Feliksik, J. M., and Tow, J. P.**, Cadmium interaction with sarcoplasmic inorganic phosphate detected in the intact isolated heart, in *Trace Substances in Environmental Health, Part XVII*, Hemphill, D. D., Ed., University of Missouri Press, Columbia, MO, 1982, 25.
109. **Racker, E.**, Transport of ions, *Chem. Res.*, 12, 338, 1979.
110. **Roby, C., Bligny, R., Douce, R., Tu, S.-I., and Pfeffer, P. E.**, Facilitated transport of MN^{2+} in sycamore *(Acer pseudoplatanus)* cells and excised maize root tips. A comparative *in vivo* ^{31}P study, *Biochem. J.*, 252, 401, 1988.
111. **Kallas, T. and Dahlquist, F. W.**, Phosphorus-31 NMR analysis of internal pH during photosynthesis in the *Cyanobacterium synechoccus*, *Biochemistry*, 20, 5900, 1981.
112. **Foyer, C., Walker, D., Spencer, C., and Mann, B.**, Observations on the phosphate status and intracellular pH of intact cell, protoplasts and chloroplasts from photosynthetic tissue using phosphorus-31 NMR, *Biochem. J.*, 202, 429, 1982.
113. **Waterton, J. C., Bridges, I. G.**, and **Irving, M. P.**, Intracellular compartmentation detected by ^{31}P-NMR in intact photosynthetic wheat-leaf tissue, *Biochim. Biophys. Acta*, 763, 315, 1983.
114. **Foyer, C. and Spencer, C.**, Observations on the phosphorus status of photosynthetic tissue using phosphorus-31 NMR, *Agric. Res. Counc. Rep. Photosynth. Group*, 24, 1982.
115. **Foyer, C. and Spencer, C.**, The relationship between phosphate status and photosynthesis in leaves, *Planta*, 167, 369, 1986.
116. **Mitsumori, F. and Osamu, I.**, Phosphorus-31 NMR studies of photosythesizing *Chlorella*, *FEBS Lett.*, 174, 248, 1984; **Sianoudis, J., Kusel, A. C., Meyer, A., Grimme, L. H., and Leibfritz, D.**, The cytoplasmic pH in photosynthesizing cells of the green alga *Chlorella fusca*, measured by P-31 NMR spectroscopy, *Arch. Microbiol.*, 147, 25, 1987.
117. **Kuchitsu, K., Oh-hama, T., Tsuzuki, M., and Miyachi, S.**, Detection and characterization of acidic compartments (vacuoles) in *Chlorella vulgaris* 11h cells by ^{31}P-*in vivo* NMR spectroscopy and cytochemcal techniques, *Arch. Microbiol.*, 148, 83, 1987.
118. **Enami, I., Akutsu, H., and Kyogoku, Y.**, Intracellular pH regulation in an acidophilic unicellular alga, *Cyanidium caldarium:* ^{31}P-NMR determination of intracellular pH, *Plant Cell Physiol.*, 27, 1351, 1986; **Ginzburg, M., Ratcliffe, R. G., and Southon, T. E.**, Phosphorus metabolism and intracellular pH in the halotolerant alga *Dunaliella parva* studied by ^{31}P-NMR, *Biochim. Biophys. Acta*, in press.
119. **Martin, F., Canet, D., Rolin, D., Marchal, J. P., and Larher, F.**, Phosphorus-31 NMR study of polyphosphate metabolism in intact ectomycorrhizal fungi, *Plant Soil*, 71, 469, 1983.
120. **Martin, F., Marchal, J. P., Timinska, A., and Canet, D.**, The metabolism and physical state of polyphosphates in ectomycorrhizal fungi. A ^{31}P NMR study, *New Phytol.*, 101, 275, 1985.
121. **Pfeffer, P. E., Rolin, D. B., Gerasimowicz, W. V., Tu, S.-I., Boswell, R. T., and Sloger, C.**, Legume-rhizobium symbiosis as examined by *in vivo* ^{31}P NMR spectroscopy, in 14th Annu. Meet. Fed. Analytical Chemistry and Spectroscopy Societies, Detroit, October 1987, 69
122. **Martin, F., Canet, D., and Marchal, J. P.**, *In vivo* natural abundance ^{13}C NMR studies of the carbohydrate storage in ectomycorrhizal fungi, *Physiol. Veg.*, 22, 733, 1984.
123. **MacKay, M. A., Norton, R. S., and Borowitzka, L. J.**, Marine blue-green algae have a unique osmoregulatory system, *Mar. Biol.*, 73, 301, 1983.
124. **Norton, R. S., MacKay, M. A., and Borowitzka, L. J.**, The physical state of osmoregulatory solutes in unicellular algae. A natural-abundance carbon-13 NMR relaxation study, *Biochem. J.*, 202, 699, 1982.

125. **Martin, F., Marchal, J. P., and Brondeau, J.,** *In vivo* natural abundance ^{13}C NMR studies of living ectomycorrhizal fungi. Observation of fatty acid in *Cenococcum graniforme* and *Helelona crustuliniforme*, *Plant Physiol.*, 75, 151, 1984.
126. **Hocking, A. D. and Norton, R. S.,** Natural abundance ^{13}C NMR studies on the internal solutes of xerophilic fungi, *J. Gen. Microbiol.*, 129, 2915, 1983.
127. **Kainosho, M. and Konishi, H.,** ^{13}C NMR spectrum of dried star anise fruits and its histological implications, *Tetrahedron Lett.*, 51, 4757, 1976.
128. **Kainosho, M.** and **Ajisaka, K.,** Carbon-13 NMR spectra of gross plant tissues containing starch, *Tetrahedron Lett.*, 18, 1563, 1978.
129. **Pfeffer, P. E., Osman, S. F., and Vincendon, M.,** unpublished results.
130. **Coombe, B. G. and Jones, G. P.,** Measurement of the changes in the composition of developing undetached grape berries by using ^{13}C NMR techniques, *Phytochemistry*, 22, 2185, 1983.
131. **Thomas, T. H. and Ratcliffe, R. G.,** ^{13}C NMR determination of sugar levels in storage roots of carrot *(Daucus carota)*, *Physiol. Plant.*, 63, 284, 1985.
132. **den Hollander, J. A. and Shulman, R. G.,** ^{13}C studies of *in vivo* kinetic rates of metabolic processes, *Tetrahedron Lett.*, 39, 3529, 1983.
133. **Stidham, M. A., Moreland, D. E., and Siedow, J. N.,** ^{13}C NMR studies of Crassulacean acid metabolism in intact leaves of *Kalanchoe tubiflora*, *Plant Physiol.*, 73, 517, 1983.
134. **Holtum, J. A., Summons, R., Roeska, C. A., Comins, H. N.,** and **O'Leary, M.,** Oxygen-18 incorporation into malic acid during nocturnal carbon dioxide fixation in crassulacean acid metabolism plants. A new approach to estimating *in vivo* carbonic anhydrase activity, *J. Biol. Chem.*, 259, 6870, 1984.
135. **Sillerud, L. O., Heyser, J. W., Han, C. H., and Bitensky, M. W.,** Applications of carbon-13 and sodium-23 NMR in the study of plants, animals and human cells, in *NMR in Living Systems*, Axenrod, T. and Ceccarelli, G., Eds., D. Reidel, New York, 1986, 309.
136. **Roberts, J. K. M., Callis, J., Wemmer, D., Walbot, V., and Jardetzky, O.,** The mechanism of cytoplasmic pH regulation in hypoxic maize root tips and its role in survival under hypoxia, *Proc. Natl. Acad. Sci. U.S.A.*, 81, 3379, 1984.
137. **Martin, F. and Canet, D.,** Biosynthesis of amino acids during ^{13}C glucose utilization by the ectomycorrhizal ascomycete *Cenococcum geophilium* monitored by ^{13}C NMR, *Physiol. Veg.*, 24, 209, 1986.
138. **Ashworth, D. J., Lee, R. Y., and Adams, D. O.,** Characterization of acetate and pyruvate metabolism in suspension cultures of *Zea mays* by ^{13}C NMR spectroscopy, *Plant Physiol.*, 85, 463, 1987.
139. **Ito, O., Mitsumori, F., and Totsuka, T.,** Effects of NO_2 and O_3 alone or in combination on kidney bean plants. III. Photosynthetic $^{13}CO_2$ assimilation observed by ^{13}C NMR, *Res. Rep. Natl. Inst. Environ. Stud. Jpn.*, 66, 27, 1984.
140. **Ezra, F., Lucas, D. S., Mustacich, V., and Russel, A. F.,** Phosphorus-31 and carbon-13 NMR studies of anaerobic glucose metabolism and lactate transport in *Staphylococcus aureus* cells, *Biochemistry*, 22, 3841, 1983.
141. **Belton, P. S., Lee, R. B., and Ratcliffe, G.,** A ^{14}N NMR study of inorganic nitrogen metabolism in barley, maize and pea roots, *J. Exp. Bot.*, 36, 190, 1985.
142. **Kanamori, K., Legerton, T. L., Weiss, R. L., and Roberts, J. D.,** Effect of nitrogen source on glutamine and alanine biosynthesis in *Neurospora crassa*. An *in vivo* ^{15}N NMR study, *J. Biol. Chem.*, 257, 14168, 1982.
143. **Rüterjan, H. and Juretschke, H. P.,** NMR study of cells and tissues, in *Proc. 4th Int. Sym. Rapid Methods and Automation in Microbiology and Immunology*, Springer-Verlag, Berlin, 1985, 1.
144. **Martin, F.,** ^{15}N-NMR studies of nitrogen assimilation and amino acid biosynthesis in ectomycorrhizal fungus *Cenococcum graniforme*, *FEBS Lett.*, 182, 350, 1985.
145. **Monselise, E. B.-I., Kost, D., Porath, D., and Tal, M.,** ^{15}N NMR of ammonium ion assimulation by *Lemna gibba* L., *New Phytol.*, 107, 341, 1987.
146. **Bacic, G. and Ratkovic, S.,** Water exchange in plant tissue studied by proton NMR in the presence of paramagnetic centers, *Biophys. J.*, 45, 767, 1984.
147. **Campbell, I. D.** and **Dobson, C. M.,** Spin-echo double resonance: a novel method for detecting decoupling in Fourier transform NMR, *J. Chem. Soc. Commun.*, 18, 750, 1975.
148. **Brown, F. F., Campbell, I. D., Kuchel, P., and Rabenstein, D. L.,** Human, erythrocyte metabolism studies by ^{1}H spin-echo NMR, *FEBS. Lett.*, 82, 12, 1977.
149. **Rabenstein, D. L. and Isab, A. A.,** Water elimination by T_2 relaxation in ^{1}H spin-echo FT NMR studies of intact human erythrocytes and protein solutions, *J. Magn. Reson.*, 36, 281, 1979.
150. **Ugurbil, K., Petein, M., Maiden, R., Michurski, S., Cohen, J. N., and From, A. H.,** High resolution NMR studies of perfused hearts, *FEBS Lett.*, 167, 73, 1984.
151. **Jue, T., Arias-Mendoza, F., Gonnella, N. C., Shulman, G. I., and Shulman, R. G.,** A ^{1}H technique for observing metabolite signals in the spectrum of perfused liver, *Proc. Natl. Acad. Sci.*, 82, 4246, 1985.

152. **Behar, K. L., Rothman, D. L., Shulman, R. G., Petroff, O. A. C., and Prichard, J. W.,** Detection of cerebral lactate *in vivo* during hypoxemia by ^{1}H NMR at relatively low field strengths (1.9T), *Proc. Natl. Acad. Sci., U.S.A.*, 81, 2517, 1984.
153. **Schnall, M. D., Subramanian, V. H., Leigh, J. S., Gyalai, A., McLaughlin, A., and Chance, B.,** A technique for simultaneous ^{1}H and ^{31}P NMR at 2.2T *in vivo*, *J. Magn. Reson.*, 63, 401, 1985.
154. **Williams, S. R., Gadian, D. G., Proctor, E., Sprague, D. B., and Talbot, D. F.,** Proton NMR studies of muscle metabolites *in vivo*, *J. Magn. Reson.*, 63, 406, 1985.
155. **Sillerud, L. O., Alger, J. R., and Shulman, R. G.,** High resolution proton NMR studies of intracellular metabolites in yeast using ^{13}C decoupling, *J. Magn. Reson.*, 45, 142, 1981.
156. **Stevens, A. N., Morris, P. G., Iles, R. A., Sheldon, P. W., and Griffiths, J. R.,** *Br. J. Cancer*, 50, 113, 1984.
157. **Wyrwicz, A. M., Bzenny, M. H., Schofied, J. C., Tillman, P. C., Gordon, R. E., and Martin, P. A.,** *Science*, 222, 428, 1983.
158. **Burt, C. T., Moore, R. R., Roberts, M. F., and Brady, T. J.,** The fluorinated anesthetic halothane as a potential NMR biologic probe, *Biochim. Biophys. Acta*, 805, 375, 1984.
159. **Taylor, J. S. and Deutsch, C.,** Fluorinated α-methylamino acids as indicators of intracellular pH, *Biophys. J.*, 43, 261, 1983.
160. **Deutsch, C. J. and Taylor, J. S.,** Use of ^{19}F NMR in the measurement of intracellular pH, in *NMR Spectroscopy of Cells and Organisms*, Gupta, R. K., Ed., CRC Press, Boca Raton, FL, 1987, 55.
161. **Deutsch, C. J., Kashiwagura, T., Taylor, J., Wilson, D. F., and Erecinska, M.,** The effect of glucagon and adrenergic agonists on intracellular pH of isolated rat hepatocytes, *J. Biol. Chem.*, 260, 6808, 1985.
162. **Smith, G. A., Hesketh, R. T., Metcalfe, J. C., Feeney, J., and Morris, P. G.,** Intracellular calcium measurements by ^{19}F NMR of fluorine-labelled chelators, *Proc. Natl. Acad. Sci. U.S.A.*, 80, 7178, 1983.
163. **Metcalf, J. C., Hesketh, T. R., and Smith, G. A.,** Free cytosolic Ca^{2+} measurements with fluorine labelled indicators using ^{19}F NMR, *Cell Calcium*, 6, 183, 1985.
164. **Cook, R. J.,** *Plant Cell Environ.*, 9, 157, 1986.
165. **Gupta, R. K. and Gupta, P.,** Direct observation of resolved resonances from intra- and extracellular sodium-23 ions in NMR studies of intact cells and tissues using dysprosium (III) tripolyphosphate as a paramagnetic shift reagent, *J. Magn. Reson.*, 47, 344, 1982.
166. **Gupta, R. K., Kostellow, A. B., and Morrill, G. A.,** NMR studies of intracellular ^{23}Na in amphibian oocytes, ovulated eggs, and early embryos, *Biophys. J.*, 41, 128a, 1983.
167. **Balshi, J. A., Cirillo, V. P.,** and **Springer, C. S.,** Direct high resolution nuclear magnetic resonance studies of cation transport *in vivo*, *Biophys. J.*, 38, 323, 1982.
168. **Gerasimowicz, W. V., Tu, S.,** and **Pfeffer, P. E.,** Energy facilitated Na^{+} uptake in excised corn roots via ^{31}P and ^{23}Na NMR, *Plant Physiol.*, 81, 925, 1986.
169. **Chu, S. C., Pike, N. M., Fossel, E. T., Smith, T. W., Balschi, J. A., and Springer, C. S., Jr.,** Aqueous shift reagents for high-resolution cationic nuclear magnetic resonance. III. Dy(TTHA)$^{3-}$, Tm (TTHA)$^{3-}$, and Tm (PPP)$^{-7}$, *J. Magn. Reson.*, 56, 33, 1984.
170. **Greenway, H. and Munns, R.,** Mechanisms of salt tolerance in nonhalophytes, *Annu. Rev. Plant Physiol.*, 31, 149, 1980.
171. **Sillerud, L. O. and Heyser, J. W.,** Use of ^{23}Na-nuclear magnetic resonance to follow sodium uptake and efflux in NaCl-adapted and nonadapted millet *(Panicum miliaceum)* suspensions, *Plant Physiol.*, 75, 269, 1984.
172. **Höfeler, H., Jensen, D., Pike, M. M., Delayre, J. L., Cirillo, V. P., Springer, C. S., Fossel, E. T., and Balschi, J. A.,** Sodium transport and phosphorus metabolism in sodium-loaded yeast: simultaneous observation with sodium-23 and phosphorus-31 NMR spectroscopy *in vivo*, *Biochemistry*, 26, 4953, 1987.
173. **Packer, L., Spathy S., Martin, J. P., Roby, C., and Bligny, R.,** ^{23}Na and ^{31}P NMR studies of the effects of salt stress on the fresh water *Cyomobacterium synechococcus* 6311, *Arch. Biochem. Biophys.*, 256, 354, 1987.
174. **Martin, J. B., Klein, G., and Satre, M.,** ^{23}Na NMR study of intracellular sodium ions in *Dictyostelium discoideum* amoeba, *Arch. Biochem. Biophys.*, 254, 559, 1987.
175. **Adam, W. R., Koretsky, A. P., and Wiener, M. W.,** Measurement of tissue potassium *in vivo* using potassium-39 NMR, *Biophys. J.*, 51, 265, 1987.

Chapter 2

PHOSPHORUS-31 NUCLEAR MAGNETIC RESONANCE STUDIES IN HIGHER PLANT CELLS

Richard Bligny, Claude Roby, and Roland Douce

TABLE OF CONTENTS

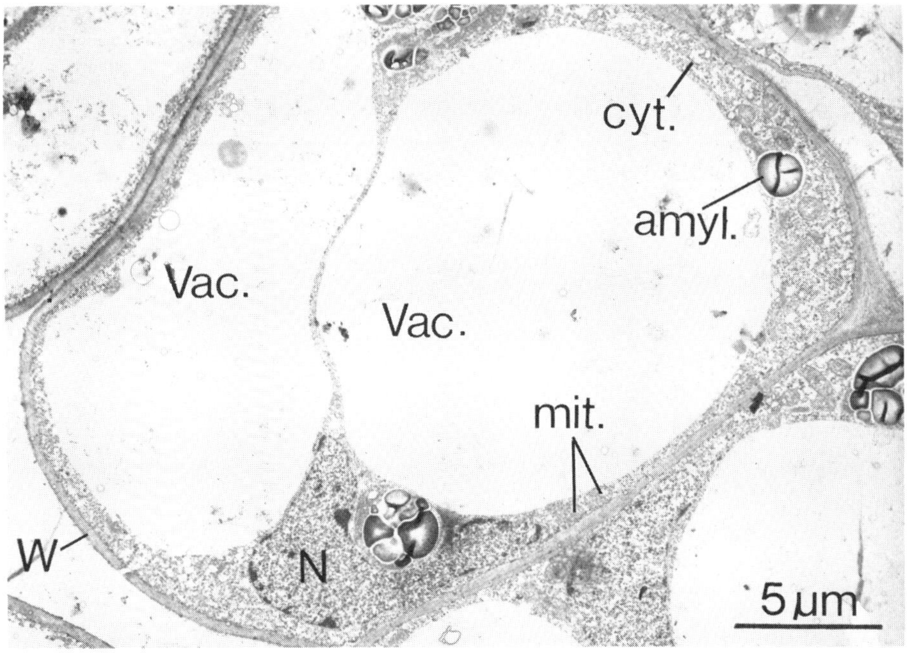

FIGURE 1. Electron micrograph of suspension cultured sycamore cells harvested at the end of the exponential phase of growth. Cells were fixed for 2 h in 4% glutaraldehyde, 20 m*M* Na-phosphate (pH 7.2), at 4°C. Samples were washed with phosphate buffer, postfixed with 1% OsO_4 in buffer for 1 h, dehydrated with a graded alcohol series, and embedded in Spurr's[41] resin (ERL 4206). Thin sections were stained with 2% (w/v) alcoholic uranyl acetate and Reynold's[42] lead citrate. Amyl., amyloplast; cyt., cytoplasm; mit., mitochondria; N, nucleus; Vac., vacuole; and W, cell wall (courtesy of Michèle Montlahuc).

I. INTRODUCTION

In plant cells the active cytoplasm occupies a peripheral shell between the central vacuole and a rigid cell wall. More than 90% of the volume of mature parenchyma cells often consists of a vacuole, which is a large watery compartment surrounded by a thick membrane (tonoplast) containing a wide variety of substances (flavonoids, various colorless phenolic compounds, etc.).[1] Consequently, the observation of freely mobile cytoplasmic organic and inorganic phosphates in mature plant cells by ^{31}P-NMR is hampered by the presence of a generally acidic vacuole which occupies much of the cell volume (Figure 1). Because of this condition,in addition to the relatively low sensitivity of NMR, it is often necessary to accumulate individual spectra over extended periods of time (e.g., See References 2 to 4). Obviously not all plant cells are so "empty"; the vacuoles of cells, e.g., the meristems of excised maize roots, can be minute. In this case the availability of tissues with different proportions of the two compartments facilitated the assignment of resonances corresponding to the vacuolar and cytoplasmic Pi pools.[5]

In the present chapter an experimental arrangement is described that enables NMR spectra of compressed plant cells to be recorded while circulating a nutrient medium through the sample. This system considerably increases the potential of NMR for the study of plant cells in three ways. First, the quality and reproducibility of the spectra is improved considerably because of the higher signal-to-noise ratio and better homogeneity of the magnetic field. Second, this system can be used to measure kinetics of metabolic interconversions and to follow phosphate (Pi) movements between the vacuole and the cytoplasm. Third, this system provides a convenient arrangement for monitoring the behavior of plant cells over a long period of time under different conditions (sucrose deprivation, temperature variation, etc).

The theory of NMR methods and relevant experimental techniques for plant tissues are largely discussed in review articles.[5,6]

II. HIGH RESOLUTION ^{31}P-NMR OF DENSELY PACKED PLANT CELLS

Suspension cultures of isolated plant cells are excellent for monitoring metabolism by ^{31}P-NMR. For this purpose it is important to be able to obtain easily large quantities of cells under rigorously defined culture conditions. Different methods have been described for cultivating higher plant cells in liquid medium.[7-9] For example, the best time to harvest large amounts of sycamore (*Acer pseudoplatanus* L.) cells in good physiological state is about 2 d before the end of the exponential phase of growth.[7] In order to obtain an intense signal, NMR experiments on isolated plant cells require a densely packed suspension (i.e., to sequester cells within a limited volume approximating that of the NMR receiver coil). However, the problems associated with such dense suspensions of cells include hypoxia, rapid consumption of nutrients, and maintenance of sterility. This has been solved by using an experimental arrangement that enables ^{31}P-NMR spectra of packed plant cells to be recorded while circulating a well-oxygenated medium through the packed suspension under controlled conditions, which simulate as closely as possible the normal physiological state. This technique has been used to study the biochemical effects of sucrose starvation of sycamore cells.[10]

A. The Apparatus

For long-term experiments (more than 3 d), all glassware and the medium are sterilized by autoclaving before use. Cells are compressed by hand between two circular perforated Teflon plates from a volume of 150 ml (2×10^6 cells per milliliter) to a volume of 20 ml (15×10^6 per milliliter)* fitting exactly into a standard 25-mm NMR tube and carrying (through the center of the plates) a narrow-bore glass tube (inlet tube), the lower end of which almost reaches the flat bottom of the NMR tube (Figure 2). A reference capillary, approximately 1.5-mm outer diameter, containing 180 m*M* methylene diphosphonate (pH 8.9 in 30 m*M* Tris) is inserted inside the inlet tube along the symmetry axis of the cell sample. A plastic screw cap adapted to the 25-mm NMR tube is used to support the inlet tube and two short glass tubes (output tube and safety output tube). A peristaltic pump is connected to these three tubes (inlet and exit tubes) and the nutrient medium is pumped through the system under slight pressure, that is, the liquid (saturated with O_2) passes slowly through the circular plates and the compressed cells (Figure 2). The flow rate used routinely (30 ml/min) provided more than ten times the O_2 consumed by the respiration of the packed sycamore cells at 25°C. Anoxia is therefore continuously avoided. Such a system prevents gas bubbles from collecting in the cell mass which would otherwise decrease the homogeneity of the magnetic field. We have verified that under these conditions cells can survive at least 7 d using a sterile sucrose-containing nutrient medium.

For kinetic experiments on respiration, the gas flow can be switched alternately from O_2 to argon or nitrogen. By modifying the composition of the circulating medium, it is possible to perturb the cell metabolism and to monitor the spectral changes simultaneously, for example, to obtain several successive spectra from the same sample.

B. The NMR Spectra

^{31}P-NMR spectra obtained from compressed sycamore cells under aerobic conditions show two distinct peaks of intracellular Pi (Cyt-Pi and Vac-Pi, Figure 3) at approximately 2.2 and 0.4 ppm [at physiological pH near neutrality, the equilibrium mixture $H_2PO_4^- \Leftrightarrow H^+ + HPO_4^{2-}$ (pK# 7.2) will be referred to as Pi]; these positions roughly correspond to pH 7.5 and 5.9, respectively. These values reflect the presence of the vacuolar Pi pool at the acidic pH and the cytoplasmic pool at the slightly alkaline pH. This is in agreement with the original observation of Roberts

* Under our conditions 30 mg (wet weight) sycamore cells correspond to 10^6 cells.[7]

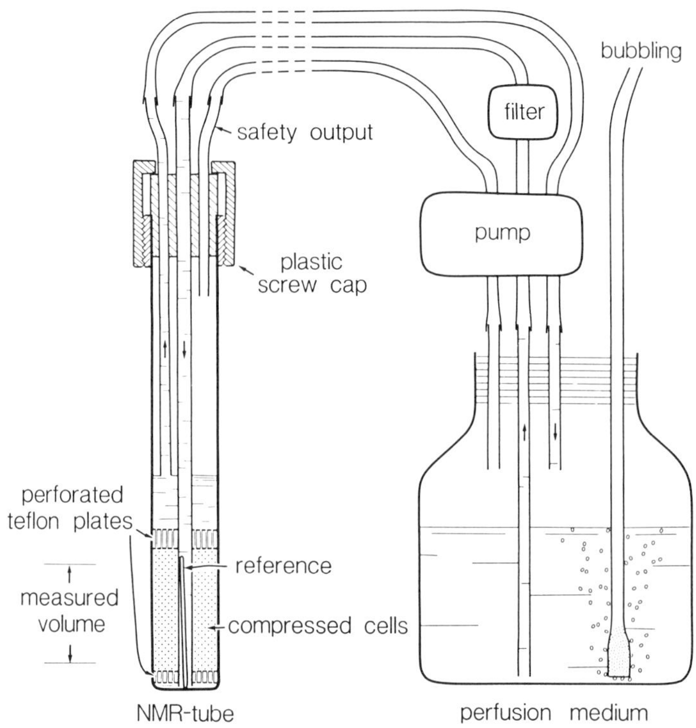

FIGURE 2. Apparatus for circulating a nutrient medium through compressed plant cells in a 25-mm (diameter) NMR tube. Cells (about 9 g) were slightly compressed between two circular perforated Teflon plates (measured volume: 16 ml). The cells comprised about 50% of the total volume.

et al.[11] on maize root tips showing that Pi resonances from the two major intracellular compartments (cytoplasm and vacuole) can be detected (see also Reference 12). The concentrations of Pi and of mobile phosphorus compounds in plant cells (glucose 6-phosphate, nucleotides, etc.) are determined by comparing the area of the signals from the methylene diphosphonate capillary reference, with its area relative to the area of the phosphorus resonances obtained from various concentrations of standard solutions of Pi, glucose 6-phosphate, nucleotides, etc.* In sycamore cells the curve thus generated and the knowledge of the vacuolar-volume/cytoplasmic-volume ratio (roughly 5) gave estimates of vacuolar Pi levels of 1.1 to 1.5 m*M* and cytoplasmic Pi levels of 0.9 to 1.2 m*M* including Pi present in cytosol and various cell organelles. Pfeffer et al.[13] have reported a value of 0.7 m*M* for cytoplasmic Pi in the case of maize root tips. However, in green cells the cytoplasmic Pi concentrations seem very large (these range from 15 to 40 m*M* for the average concentration in the chloroplast, cytosol, and other organelles in the dark).[14] Parenthetically, the vacuolar Pi magnetization has a higher T_1 value than the cytoplasmic Pi, presumably a consequence of the greater viscosity inherent in the cytoplasm.[13] In the spectra obtained from plant cells, no other Pi peaks can be identified unambiguously (see also References 2 to 5). This means that either the Pi amounts present in the mitochondrial matrix and the amyloplastid stroma of the sycamore cell are below the sensitivity of NMR and/or that the pH difference between those organelles and the cytoplasm is too small for the signals to be discriminated and/or there is a very rapid exchange between these compartments. Furthermore, the mobility of Pi in the matrix or in the stroma which are more viscous than the cytosol should be considered.

In the spectra shown in Figure 3A, there were also signals from cytoplasmic glucose 6-

* Care must be taken that sugar Pi and Pi resonances can have spin lattice relaxation (T_1) values in the 3-s range, whereas the nucleotide resonances can have T_1s of less than 1 s.[13]

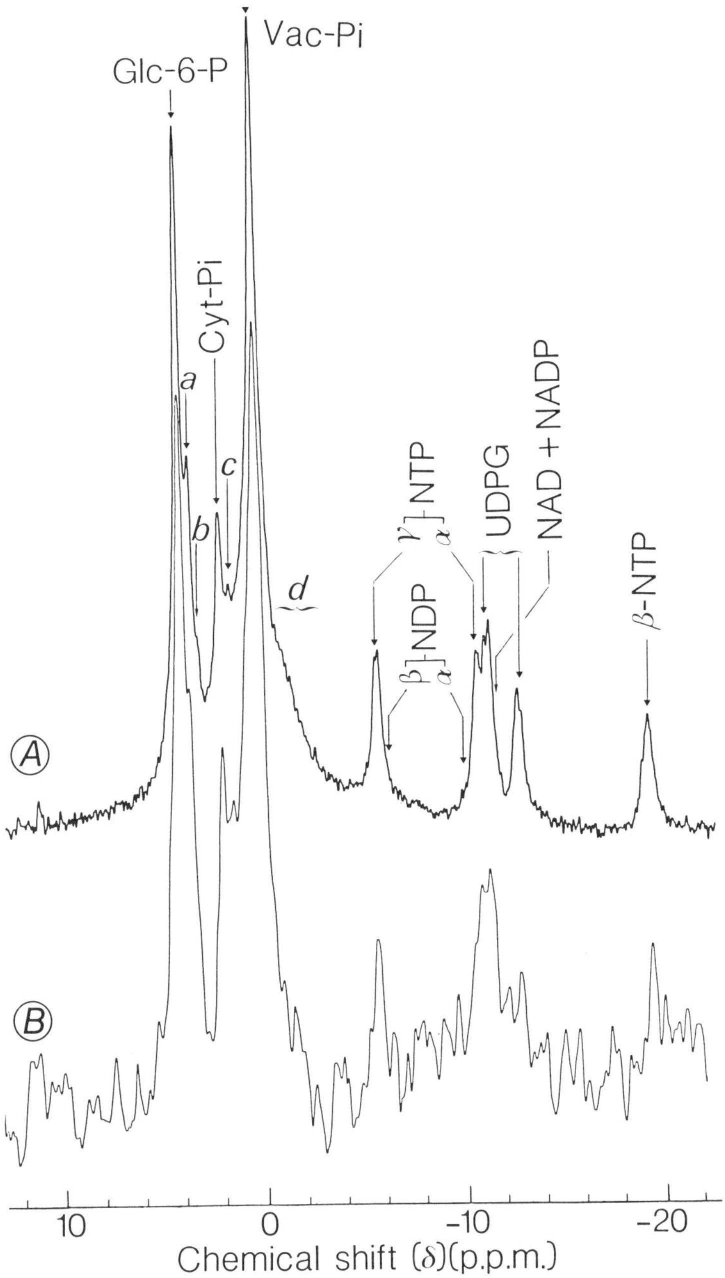

FIGURE 3. Representative ^{31}P-NMR spectra (81 MHz) of sycamore cells harvested at the end of the exponential phase of growth. Compressed cells (9 g) were perfused with a continuously aerated solution (Pi- and Mn-free culture medium). During the first 12 h cells were acclimated to the perfusion conditions. This acclimation time is necessary to obtain homogeneous spectra showing sharp vacuolar Pi peaks. The spectra recorded at 25° with a 45° pulse angle are the result of (A), 3000 transients with a repetition time of 1.2 s (1 h); and (B), 100 transients with a repetition time of 1.2 s (2 min). Peak assignments: Glc-6-P, glucose 6-P; a, position of various P-compounds including fructose 6-P, ribose 5-P, and P-ethanolamine; b, NMP (nucleotides monophosphate) and P-choline; Cyt-Pi, cytoplasmic Pi; c, position of glucose 1-P; Vac-Pi, vacuolar Pi; d, position of different phosphodiesters and myo-inositol hexakisphosphate (phytate); NTP, nucleotides triphosphate (mainly ATP and UTP); NDP, nucleotides diphosphate (mainly ADP and UDP); and UDPG, uridine-5′-diphosphate-α-D-glucose. For peak assignments, see also the spectrum obtained from a perchloric acid extract, Figure 4.

phosphate, β– and α-uridine-5′-diphosphate α-D-glucose (UDP-glucose) [with contribution of some nicotinamide adenine dinucleotide (phosphate) (NAD(P)) to the β-peak of UDP-glucose] and γ-, α-, and β-phosphorus of nucleotide triphosphate (NTP). Adenosine triphosphate (ATP) has three phosphorus signals,[15] but only one is unique to it (that of its β-phosphate); its estimated concentration in the cytoplasm was nearly 1.2 m*M*. The frequencies of the α- and β-phosphates of ADP are so close to those of the α- and γ-phosphate of ATP, respectively, as to be almost indistinguishable in NMR experiments *in vivo*. It is common practice, therefore, for ADP to be quantified by subtraction of β-peak from γ-peak. All the plant cells[10,16] or tissues[5,17-20] examined so far in contrast with animal cells contain in their cytoplasm rather high concentrations of UDP-glucose and this sugar nucleotide should be considered as a biochemical fingerprint for plant cells. As pointed out by Bieleski,[21] UDP-glucose has been frequently underestimated in plant tissues because it hydrolyzes rapidly under alkaline conditions. The presence of large amounts of UDP-glucose in plant cells is probably related to the initial breakdown of sucrose catalyzed by sucrose synthase (EC 2.4.1.13) to give UDP-glucose. UDP-glucose is subsequently metabolized by UDP-glucose pyrophosphorylase (EC 2.7.7.9) at the expense of pyrophosphate generated by pyrophosphate fructose-6-P-1-phosphotransferase (EC 2.7.1.90) operating in the reverse direction.[22–23] Sucrose synthase is lacking in animal cells. The high concentration of glucose 6-phosphate also observed indicates that in plant cells phosphofructokinase (EC 2.7.1.11) is one of the major regulatory enzymes in glycolysis. This enzyme, which catalyzes a nonequilibrium reaction, is strongly affected by a number of intracellular constituents including Pi, phosphoenolpyruvate, and ATP.[24] In aerobic cells this enzyme will normally be in an inhibited state (high level of cytoplasmic ATP, low level of cytoplasmic Pi). Such a situation could explain the high concentration of glucose 6-phosphate found in the cytosol (we appreciate that a portion of glucose 6-phosphate is located in the amyloplasts but suggest that the small volume of the stroma in these plastids means that this proportion is relatively small).

The region of the spectrum (Figure 3A) to high field of glucose 6-phosphate (peak a, 3.7 ppm) corresponds to phosphomonoesters. In other systems,[6,10,13,25] this region has been assigned to sugar phosphates including phosphoglyceric acid, fructose 6-phosphate, ribose 5-phosphate and fructose 1,6-bisphosphate. However, other peaks reside underneath it. The shoulder at 3.3 ppm (peak b) could be attributable to nucleotide monophosphate (NMP) and phosphorylcholine. As the β-NTP peak intensity is approximately equal to that of γ-NTP peak [which has some small contribution from β-nucleotide diphosphate (NDP)] we may assume that *in vivo* the NTP/NDP (free nucleotides) ratio is high (at least 10). It is clear, therefore, that sycamore cells, like all the nongreen plant cells examined so far, exhibit a very high adenylate energy charge.[26] Because mitochondria and plastids in nongreen cells contain a small fraction of the total adenine nucleotides, the high adenylate energy charge measured in whole cells reflects cytosolic adenylate energy charge. Such a result is easily interpretable if one considers that in plant mitochondria the proton motive force arises from the membrane potential (250 mv) due to the presence of a powerful K^+/H^+ antiporter (for a review see Reference 27). The transmembrane potential (negative inside) is the driving force for an electrophoretic asymmetric nucleotide exchange (ATP^{4-}/ADP^{3-}). It drives ADP inside the mitochondria and ATP outside. Accordingly at equilibrium the cytosolic ATP/ADP should be considerable and is greater than that inside, and the log of the ratios of the external (cytosolic) ATP/ADP ratio to the internal (matrix) ATP/ADP ratio is equal to $\Delta\psi$. On the contrary, in photosynthetic cells from higher plants, chloroplasts occupy much of the cytoplasmic volume[1] and these cell organelles contain, therefore, a major fraction of the total cell adenine nucleotides. Consequently, in cells derived from green leaves, the adenylate energy charge measured in whole cells does not necessarily reflect the true cytosolic situation. Unfortunately, resolution of narrowly separated ATP and ADP resonances has not been obtained so far using cells derived from mature leaves.*[14,28,29] However, utilizing

* Paramagnetic metal ions such as Mn^{2+} broaden lines by shortening the spin-spin relaxation time of diamagnetic nuclei, like ^{31}P in ATP. This problem is particularly acute when studying tissues containing fully mature chloroplasts that require Mn ions at level of photosystem II.

an ingenious arrangement of syringe, nylon net, and membrane filters, Stitt et al.[30] were able to rupture the cells (protoplasts), separate the released subcellular fractions, and quench them in $HClO_4$ in about one tenth of a second, thereby ensuring that measured adenine nucleotides levels best reflected those present *in vivo* immediately prior to protoplast rupture. The importance of the time factor is apparent from the fact that, at a rate of 200 μmol CO_2 reduction per milligram chlorophyll per hour, the choloroplastic pool of ATP turns over about six times per second. Under these conditions Stitt et al.[30] showed that in green cells even in the light (the stromal ATP/ADP ratio is higher in the light than in the dark, as is commonly observed using enzymic measurements with isolated chloroplasts and with chloroplasts in protoplasts and leaves) the stromal ATP/ADP values are not high compared to the cytosol. This showed that *in situ* plant leaf mitochondria resemble mitochondria from nongreen cells in that they are highly energized; that is, green cell mitochondria maintains, independently from the functioning chloroplasts (the innermembrane of the envelope surrounding mature chloroplasts does not possess a nucleotide translocator), a high cytosolic ATP/ADP quotient.

The spectra of compressed sycamore cells presented in this chapter (Figure 3A) are not different from those reported by Martin et al.[4] using the air-bubbling technique. However, the mean separation between the cytoplasmic and vacuolar Pi resonances was consistently better in the compressed cells than in the stationary system[4,31,32] or in the circulating system introduced by Lee and Ratcliffe[17] for maintaining the aeration of plant tissues or in the system used by Brodelius and Vogel[3] where cells were immobilized in agarose. We have verified that the average shape of the spectra described above was remarkably stable with time, even after 5 d, when the growth is stopped and when the well-oxygenated, sterile circulating Pi-free medium contained sucrose. However, under these conditions the addition of small amounts of Pi to the circulating medium leads to a marked increase in vacuolar Pi content. Conversely, if the cell growth (cell enlargement or multiplication) is not stopped, Pi required for synthesis is utilized at the expense of vacuolar Pi unless the circulating medium contains low amounts of Pi. This system, therefore, provides a convenient arrangement for monitoring the behavior of plant cells over a long period of time under different conditions such as sucrose or O_2 deprivation. On the other hand, short accumulation times of 2 min are sufficient (Figure 38) to allow the examination of the main Pi movements during rapid metabolic transitions.

However, as pointed out above, meaningful interpretation of these spectra requires the rigorous identification of resonances corresponding to specific phosphorus-containing metabolites. Numerous studies have shown that the region of the ^{31}P-NMR spectrum, generally ascribed to sugar phosphates in previous investigations of a variety of plant tissues[5] or isolated plant cells[2,4] may contain other phosphomonoesters including lipid metabolites.[33] These considerations point to the need for unambiguous determination of the origin of Pi resonances in the spectra of *in vivo* plant cells.

III. ^{31}P-NMR SPECTRA OF NEUTRALIZED PERCHLORIC ACID EXTRACTS

For perchloric acid extract, cells (9 g wet weight) were quickly frozen at liquid nitrogen temperature to avoid ATP destruction and finely pestled with 1 ml of 70% (w/v) perchloric acid. The frozen powder was then placed at –10°C and thawed. The thick suspension thus obtained was centrifuged at 10,000 × g for 10 min to remove particular matter. The supernatant was neutralized with 2 *M* $KHCO_3$ to about pH 6.0 and centrifuged at 10,000 × g for 10 min to remove the precipitated $KClO_4$. Next, 120 μmol of *trans*-1,2-diaminocyclohexane *N,N,N′,N′*-tetraacetic-acid (CDTA), a strong chelating agent,[34] was then added. CDTA prevented the precipitation of specific Pi esters as Mg salts such as myo-inositol hexakisphosphate (phytate). It is also more effective than ethylenediamine *N,N,N′,N′*-tetraacetic-acid (EDTA) to eliminate free divalent metal ions and leads to sharp lines for phosphorus resonances of metabolites such as ATP. The resulting supernatant was lyophilized and stored at liquid nitrogen temperature. For the NMR measurements this freeze-dried material containing 120 μmol CDTA was redissolved in 2 ml of 40 m*M* HEPES buffer, pH 7.8, containing 10% D_2O (perchloric acid extract).

Figure 4 shows a typical proton-decoupled ^{31}P-NMR spectrum of a perchloric acid extract of sycamore cells (9 g wet weight) at pH 7.8 in the presence of a large excess of CDTA. From literature data and by adding signals from sugar phosphates including glucose 6-P, glucose 1-P, phosphoglyceric acid, and fructose 6-P (unfortunately fructose 6-P is obscured by the resonance of P-ethanolamine and ribose 5-P); Pi; γ-, α-, and β-ATP; γ-, α-, and β-UTP; β-, and α-ADP; β- and α-UDP; β- and α-UDP-glucose; and β- and α-UDP-galactose (UDP-gal). As expected, ATP is the most abundant triphosphate, but other nucleotides are present in significant amounts. Since it was possible to detect separately the triplets of the β-phosphates of ATP, GTP, and (CTP + UTP), a typical sycamore cell had the following nucleotide composition: 65% ATP and 35% UTP (+CTP). GTP was barely detectable. All the phosphorylated compounds present at concentrations lower than 100 μ*M* were not distinguishable from the background noise. To estimate intracellular concentrations in the case of sycamore cells containing rather large vacuoles (in contrast with meristematic tissues such as maize root tips), we assumed that 6 g of wet cells contain 1 ml of cytoplasm. This means that there are many Pi metabolites that are known to be present but are not seen in the NMR spectra of the extracts. Among these are fructose 1,6 bis-P, pyrophosphate, and glyceraldehyde 3-P. The peak at 3.3 ppm close to that of NADPH (5′-P) and NMP (primarily AMP) was assigned to phosphorylcholine. The NAD and NADP peaks which yields a quadruplet centered at –10.9, –11.0, –11.2, and –11.3 ppm on the high-field side of the α-NTP peak are well identified. However, solutions of NADH and NADPH are acid labile. Consequently, these reduced coenzymes are probably destroyed during the course of perchloric acid treatment of intact cells.

The high NTP/NDP ratio as seen under *in vivo* conditions (Figure 3A) is also observed for cell extracts (Figure 4). A low level of NDP relative to NTP indicates that little if any hydrolysis of NTP had occurred during the extraction procedure. It has been suggested that in some mammalian tissues such as kidney, ADP may be bound to some protein or paramagnetic substance which renders it invisible to NMR[35] since chemical extraction and assay of ADP in many animal tissues give a value one or two orders of magnitude higher than that observed by NMR.[36] Apparently such a situation does not seem to exist in plant cells since ^{31}P-NMR spectra of intact cells and perchloric extracts give almost similar results for ADP concentrations (Figures 3 and 4). The spectrum of a cell extract in the presence of CDTA showed the β-phosphorus signal of ATP at higher field (–21.2 ppm) than did the spectrum of an intact cell suspension (–19.6 ppm). This finding indicates that in intact sycamore cells ATP is predominantly bound to a divalent metal ion, most probably Mg^{2+} (Figure 3), which was sequestered in the cell extract by the addition of CDTA. Such an observation raises the problem of enzymatic systems or carrier such as the mitochondrial nucleotide carrier that can utilize free ATP^{4-}. A specific system removing Mg^{2+} ions in the vicinity of the enzyme should also be involved. Interestingly we have observed that large amounts of phytate are present in a number of plant tissues including green leaves. For example, in the case of perchloric acid extracts from sycamore cells, phytate appears as a complex multi-peak resonance showing absorption bands at 2.3, 2.1, 1.9, and 1.6 ppm. The estimated concentration of phytate was about 1 m*M* on the basis of the whole cells indicating that it may serve as a store of Pi, cations (Mg^{2+} and perhaps free iron), or of high energy phosphoryl groups which can be metabolized by phytase and phytate-nucleotide diphosphate phosphotransferases during cell growth. Thus phytate may buffer the variations of free Mg^{2+} and Pi concentration in the cytosol. However, since this phosphorous-compound is not easily NMR-detectable *in vivo* (Mg^{2+} phytate is insoluble) it was not possible to determine in what compartment it was accumulated and what the concentration of free phytate in the cytoplasm was. These points are presently under investigation.

Figure 5 shows a typical proton-decoupled ^{31}P-NMR spectra of perchloric acid extracts of cells isolated from pea leaves maintained under dark or light conditions. This figure indicates that under illumination the ribulose 1,5-bisphosphate pool is maintained at a high value by the

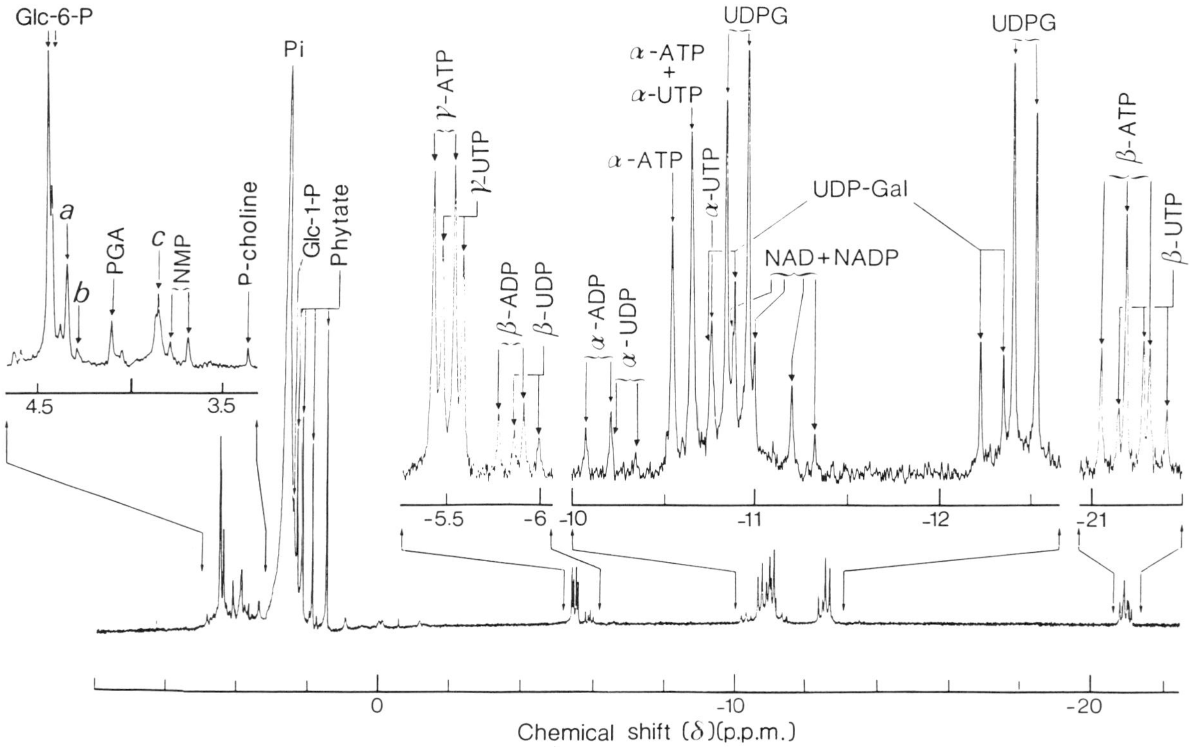

FIGURE 4. Proton-decoupled ^{31}P-NMR spectrum of perchloric acid extracts of sycamore cells. Perchloric extracts were prepared from oxygenated cells (9 g wet weight) as described in text. The samples (2.5 ml) containing 250 µl of ^2H_2O and 120 µmol CDTA were put in a 10-mm diameter NMR tube and spectra recorded at 162 MHz on a Bruker AM400 spectrometer. Spectra were acquired over a period of 2 h with a 50° pulse angle, a sweep width of 6000 Hz, and a recycling time of 4 s. Free induction decays were accumulated using 8 K data points and zero-filled to 32 K prior Fourier transformation. A 0.1-Hz line broadening was applied. Two levels of broad-band H decoupling were employed: 2.5 W during the data acquisition (0.68 s) and 0.5 W during the delay period. All the interesting regions are shown on an expanded scale. Peak assignments: Glc-6-P, glucose 6-P; a, unidentified peak; b, position of *sn* -glycerol 3-P; PGA, phosphoglyceric acid; c, position of various phosphomonoesters including fructose 6-P, ribose 5-P, and P-ethanolamine; NMP, nucleotides monophosphate (mainly AMP at 3.95 ppm and UMP at 3.8 ppm); Pi, inorganic phosphate; Glc-1-P, glucose 1-P; phytate, myo-inositol hexakisphosphate; UDPG, uridine-5′-diphosphate-α-D-glucose, and UDP-Gal, uridine -5′-diphosphate galactose.

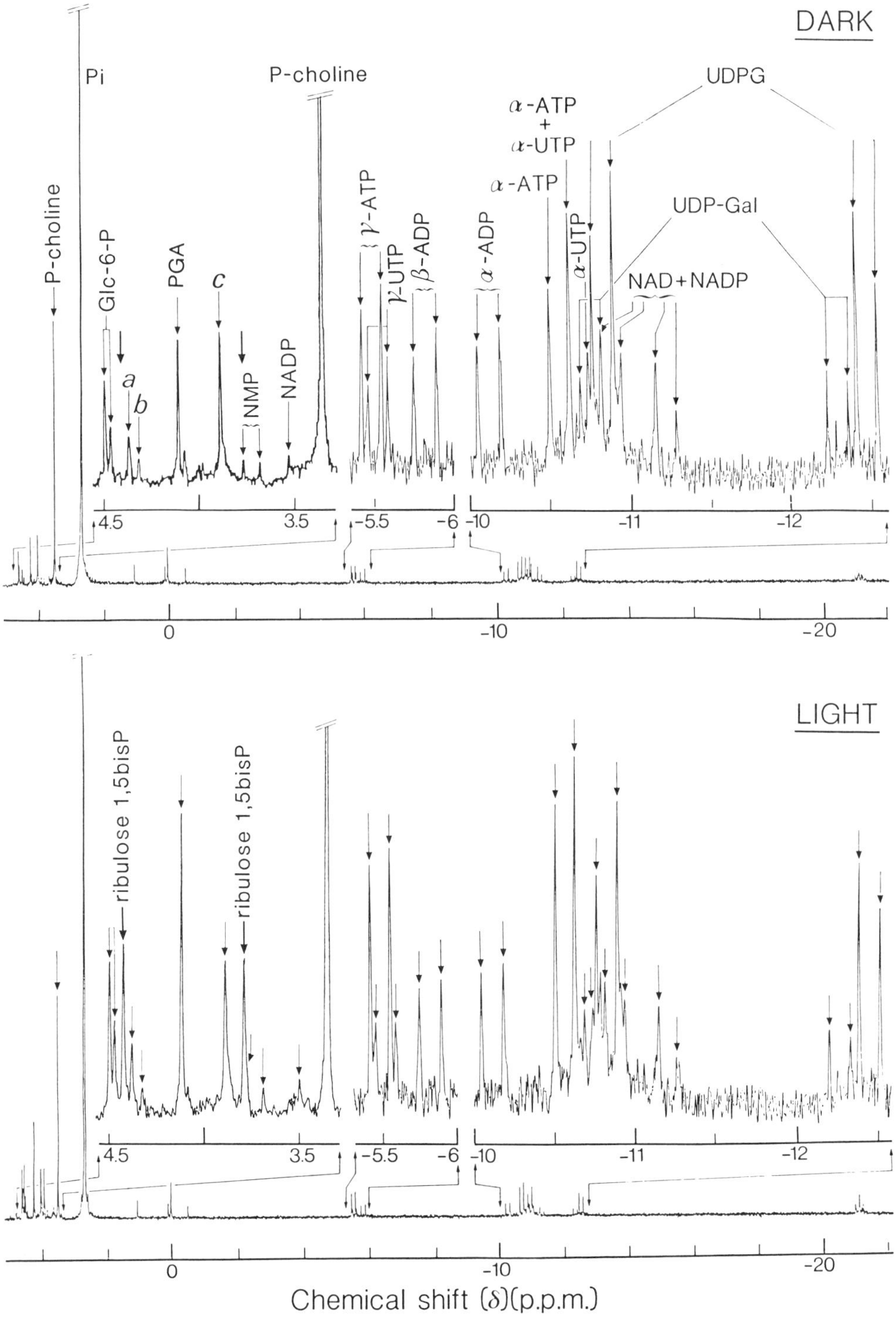

FIGURE 5. Proton-decoupled ^{31}P-NMR spectra of perchloric acid extracts of pea leaves. Perchloric extracts were prepared from 9 g (wet weight) of leaves as described in text. Dark condition: leaves were submitted for 5 min in the dark at 22°C before freezing at liquid nitrogen temperature. Light condition: leaves were illuminated under warm light from fluorescent tubes (10 to 40 μE/m^2/s of photosynthetically active radiation) at 22°C for 1 h before freezing at liquid nitrogen temperature. For NMR measurements and peak assignments, see Figure 4.

regenerative phase of the Benson-Calvin cycle which turns over rapidly (for a review see Reference 37). Upon switching to dark conditions, the concentration of ribulose 1,5-bisphosphate fell dramatically within 5 min. In fact, under these transient conditions the rate of ribulose 1,5-bisphosphate utilization by carboxylation exceeded the rate of regeneration because ribulose 1,5-bisphosphate carboxylase/oxygenase (rubisco) activity declined more slowly than did the capacity of the plant to regenerate ribulose 1,5-bisphosphate.[38,39] These results strongly emphasize the importance of NMR as a powerful tool to study photosynthetic carbon metabolism in higher plant cells. Interestingly Figure 5 also indicates that the ATP/ADP ratio observed in a perchloric extract of green cells is much lower than that previously observed in sycamore cells (Figures 3 and 4), or root tips.[5]

IV. RELATIONSHIPS BETWEEN THE CYTOPLASM AND THE VACUOLE PHOSPHATE POOL IN PLANT CELLS

Pi plays a central role in cell functioning. It is involved in most metabolic energy transductions, acts as an intermediate in the synthesis of numerous metabolites and complex molecules (phospholipids, DNA, RNA, etc.), and it regulates the activity of many enzyme reactions and metabolic sequences such as respiration, glycolysis, photosynthesis, and starch metabolism. For example, plastids and mitochondria must be considered as Pi-requiring organelles and the cytosolic fraction is the major site of Pi recirculation. Consequently, an important question concerning the role of the vacuole in general cell metabolism is whether or not Pi present in the cytosolic compartment rapidly equilibrates with Pi in the vacuole. Unfortunately, the investigation of this subject has been hindered by the lack of suitable methods for measuring the Pi concentration of these two major intracellular compartments. The most widely used method for studying these intracellular compartments is subcellular fractionation followed by the isolation of vacuolar and cytoplasmic fractions. However, this methodology could never reveal a possible coupling between both cell compartments because the major problems associated with cell fractionation is that Pi leaks out from vacuoles and many cell organelles after cell rupturing.

To solve this important problem, Rébeillé et al.[40] first described a detailed ^{31}P-NMR investigation of the time course of the flux of cytoplasmic and vacuolar Pi concentrations in sycamore cells transferred to either a phosphate-deficient or a phosphate-rich medium. They demonstrated that when Pi was omitted from the nutrient solution, the endogenous Pi pool decreased to 10% of the control within 72 h. During this time, growth slows down and the vacuolar Pi pool decreased considerably, relative to the cytoplasmic pool. These results demonstrate, therefore, that during the first hours of Pi starvation, the Pi and phosphate ester contents of the cytoplasm are largely maintained at the expense of the vacuolar Pi pool, that is, Pi slowly effuxes from vacuole to cytoplasm in order to sustain the growth requirement. If the same experiment is carried out in the presence of a growth inhibitor, the vacuolar Pi level does not decrease and Pi starvation did not settle. Rébeillé et al.[40] also demonstrated that when the Pi starved cells were transferred to a medium containing a low Pi concentration this anion entered the cells rapidly and accumulated in the cytoplasm. Once the cytoplasmic Pi pool was filled, Pi was continuously taken into the vacuole (Figure 6). In the case of sycamore cells, the vacuole does not constitute a large Pi reservoir because the cytosolic and the vacuolar Pi concentrations seem to approach each other with the passage of time. Furthermore when the non-Pi-starved cells were transferred to a Pi-rich medium (i.e., containing more than 20 mM Pi), Pi entered the cells slowly by diffusion* and accumulated in the vacuole suggesting that most of the excess Pi

* At external Pi concentrations up to 2 mM, sycamore cells incorporate Pi against a concentration gradient by a process which is energy dependent. Under these conditions, the intracellular Pi concentration is maintained constant. On the contrary, at external Pi concentrations above 15 mM, i.e., at external Pi concentrations higher than that which counterpoised the electrochemical gradient (Pi gradient + electrical membrane potential ($\Delta\psi$) which is in the opposite direction to that which would favor Pi diffusion), Pi enters the cell by slow diffusion and the intracellular Pi concentration increases proportionally to the external Pi concentration.[40]

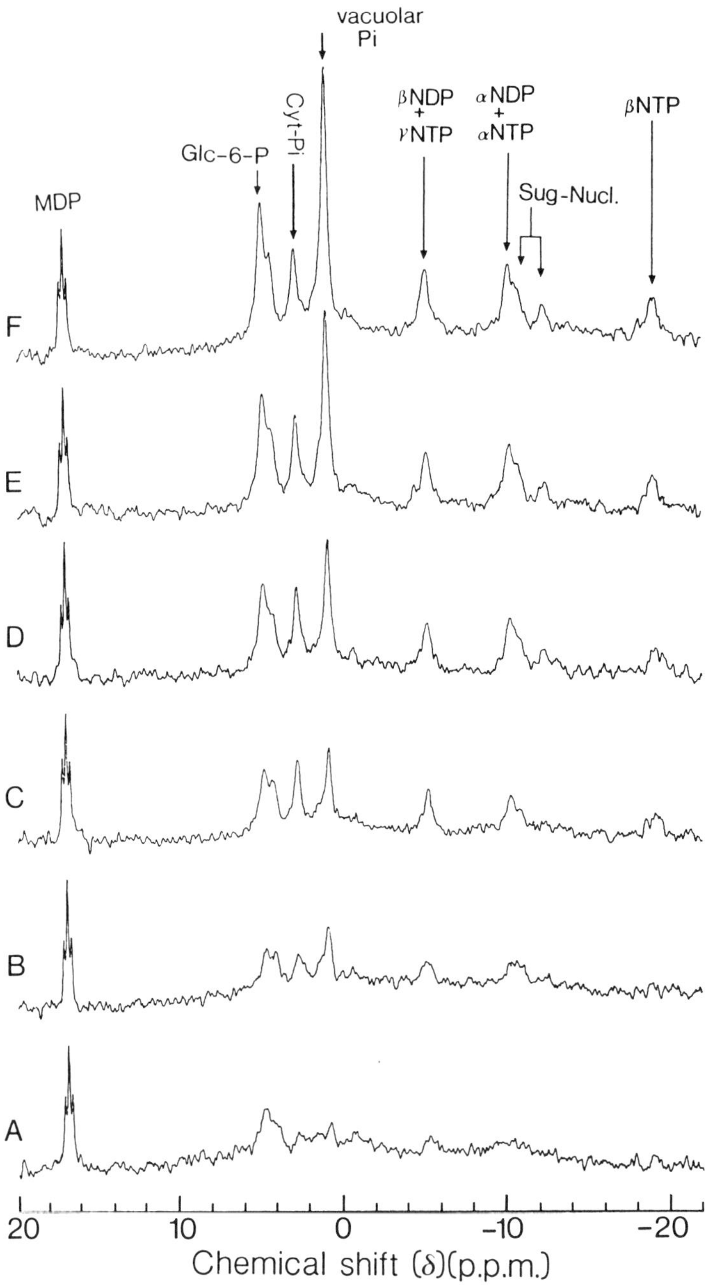

FIGURE 6. Time-course evolution of the most abundant ^{31}P-NMR detectable phosphorus compounds in Pi-starved sycamore cells after addition of 200 μ*M* Pi in the perfusion medium. Cells were first submitted to Pi starvation up to 5 d followed by 3 h recovery after addition of 200 μ*M* Pi to the circulating medium. A, spectrum from cells at 5 d following Pi starvation. B to F, spectra from cells recovering from 5 d Pi starvation 30, 60, 90, 120, and 150 min, respectively, after the addition of 200 μ*M* Pi to the circulating medium. Each spectrum was the result of 2000 transients with a repetition time of 0.6 s (20 min) and a 30° pulse angle. In spectrum A, (Pi starvation control), note that only Glc-6-P can be clearly distinguished. Phosphate (200 μ*M*) was added in the perfusion medium at the end of the accumulation time of the spectrum A. In spectrum B the peak at 0.5 ppm (position of vacuolar Pi) was attributable to the externally added Pi (pH of the perfusion medium, 5.5). In spectra C, D, and E, note that cytoplasmic Pi and NTP were first recovered, then glucose 6-P, and finally vacuolar Pi. Spectrum F shows a full recovery of NMR-detectable Pi compounds. Peak assignments as in Figure 3.

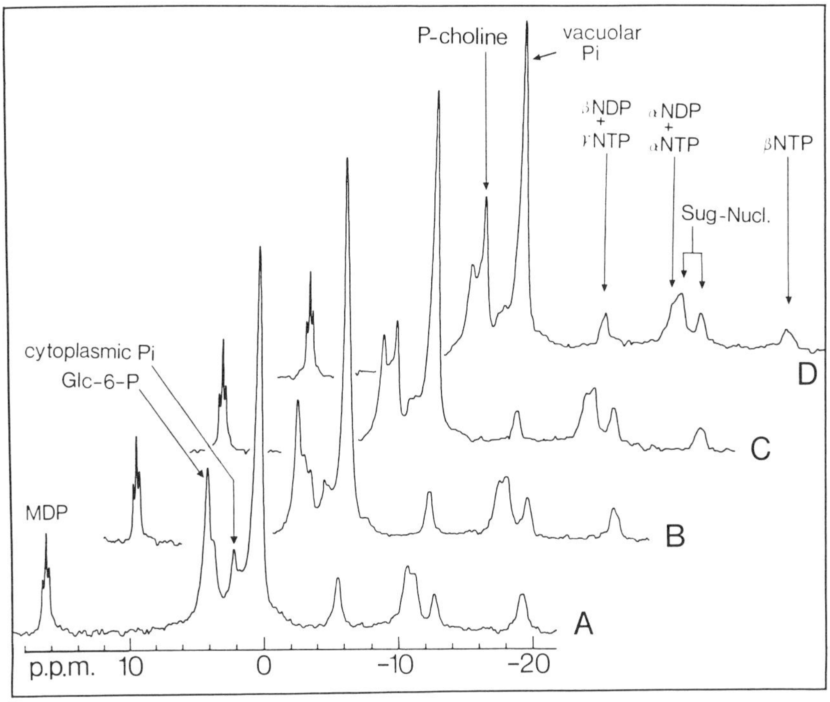

FIGURE 7. Time course accumulation of phosphorylcholine in sycamore cells perfused with a Mn^{2+}-, and Pi-free nutrient medium containing 10 μm choline. The spectra recorded A 25°C with a 30° pulse angle were the result of 12,000 transients with a repetition time of 0.6 s (2 h). A, standard spectrum and B, C, and D, spectra obtained after 2, 4, and 6 h, respectively, of perfusion of the cells with a nutrient medium containing 10 μm choline. Note the steady accumulation of phosphoryl choline (P-choline) with time. In order to maintain constant the concentration of choline in the perfusion medium a 20-l reservoir of perfusion medium was utilized. Peak assignments as in Figure 3.

molecules present in the cytoplasm were expelled in the vacuolar compartment. Brodelius and Vogel[3] also reported ^{31}P-NMR studies of Pi uptake and storage in cultured *Catharanthus roseus* plant cells. They indicated in agreement with Rébeillé et al.[40] the importance of the vacuole in the uptake, storage, and utilization of Pi (see also Reference 16). All these experiments together clearly demonstrate that the vacuole compartment is able to slowly "buffer" the fluctuations of Pi concentrations occurring in the cytoplasm and more precisely for the purpose of Pi storage or growth requirement.

However, several recent reports suggest that in fact the vacuolar apparatus is unable to buffer the rapid fluctuation of Pi concentrations occurring in the cytosol during the course of cell metabolism. For example, when mannose is supplied to spinach leaf disks,[28] the cytosolic Pi is almost totally depleted due to the rapid phosphorylation of mannose by cytosolic hexokinase. Under these conditions, on the time scale of several minutes, the Pi in the vacuole hardly changes to compensate the dramatic decrease in cytosolic Pi. Similarly, we have recently shown that the addition of 10 μ*M* choline to the perfusion medium triggered the rapid synthesis of phosphorylcholine (P-choline) in the cytoplasm of sycamore cells (Figure 7). Within 4 h about 3 m*M* P-choline accumulated in the cytoplasmic compartment. During that time, in the absence of Pi in the nutrient medium, almost all the Pi present in the cytoplasm was consumed and the glucose 6-P content was reduced from 3.5 m*M* to about 1.2 m*M* . The addition of glycerol to the nutrient medium led to a similar observation [i.e., the accumulation of glycerol phosphate (unshown result)]. All these experiments help in understanding the regulation of photosynthesis *in situ* because it shows that the vacuolar Pi pool does not fluctuate rapidly enough to buffer the rapid decrease in the cytoplasmic Pi which necessarily occurs between the dark and light for the synthesis of phosphorylated intermediates. In fact, in a leaf cell, the pool of Pi present in the thin film of cytosol is rather low; this implies as suggested by Walker[37] that the photosynthetic rate might, in some circumstances, become limited by the supply of Pi delivered to the chloroplast by the Pi translocator.

Roby et al.[10] have used ^{31}P-NMR to investigate the changes in Pi, phosphate esters, and nucleotides of sycamore cells during a prolonged period of sucrose deprivation followed by sucrose replenishment. Their data shed new light on the controlled flux of Pi to and from the vacuole. Perfusion of compressed sycamore cells with sucrose-free culture medium triggered a progressive decrease in the glucose 6-P and sugar nucleotides resonance over 30 h (Figures 8 and 9). During this period Pi molecules liberated from phosphate ester molecules were slowly expelled into the vacuole where they accumulated. Addition of 50 m*M* sucrose in the circulating system after a long period of sucrose deprivation resulted in a very rapid disappearance of the cytoplasmic Pi resonance and a marked increase in that of glucose 6-P. Again in these series of experiments the vacuolar Pi pool did not fluctuate to buffer the Pi in the cytoplasm. In order to increase further the intracellular concentration of glucose 6-P to its original level, it was necessary to add a small amount of Pi to the circulating medium. This experiment demonstrates that although the vacuole had sequestered almost all of the cell Pi liberated during the course of sucrose starvation, the Pi necessary to restart cell metabolism was utilized first from the cytoplasm and then, if present, from the external medium. It is clear, therefore, that the consumption of vacuolar Pi is a very slow process which requires several hours and occurs only in response to the depletion of Pi from the external medium.

All these results illustrate the problem of Pi movement between the vacuole and the cytoplasm (and vice versa) and raise the question of the regulation of cytoplasmic Pi concentration in plant cells. Presently we have no evidence to suggest a mechanism by which plant cells regulate their intracellular Pi concentration. However, whenever the cytosolic ATP concentration is substantial, cytosolic Pi in excess is preferentially expelled toward the vacuole. On the contrary, when the ATP concentration is low (i.e., during transition from aerobic to anaerobic conditions)* Pi deriving from the nucleotide, and the glucose 6-P pools stays in the cytoplasmic compartment (Figure 10). These observations strongly suggest that the sequestration of Pi in the vacuole is an energy-dependent process which is under the control of cytoplasmic ATP.

V. CONCLUDING REMARKS

These results demonstrate that plant cells slightly compressed between two circular Teflon plates can survive several days as long as a well-aerated nutrient medium is pumped through the system under slight pressure. Such a system which simulates as closely as possible the normal physiological state enables NMR spectra of plant cells to be continuously recorded. Under these experimental conditions, the temporal resolution of the kinetics of metabolic processes is therefore considerably improved because a high cell density can be used within a limited volume (that of the receiver coil) giving a better signal-to-noise ratio.

As shown in this chapter, our present knowledge about Pi transport processes across the tonoplast is still quite fragmentary. However, we have demonstrated that the homeostatic process of Pi transport across the tonoplast is slow and allows short-term changes in the cytosolic Pi pool size to be used as a means of regulating metabolic processes.

* Perfusion of plant cells with an O_2-free culture medium triggers the following cascade of reactions: (1) ATP declined sharply within 2 min (it was calculated the O_2 initially dissolved in plant cells was consumed within about 2 min); (2) glucose 6-P declined progressively; and (3) cytoplasmic pH decreased about 0.5 units due to lactate accumulation. Addition of an uncoupler to the medium led to the same cascade of reaction. Decreases of ATP and glucose 6-P associated to a cytosolic Pi increase are attributable to an arrest of cell respiration and to an acceleration of glycolysis.

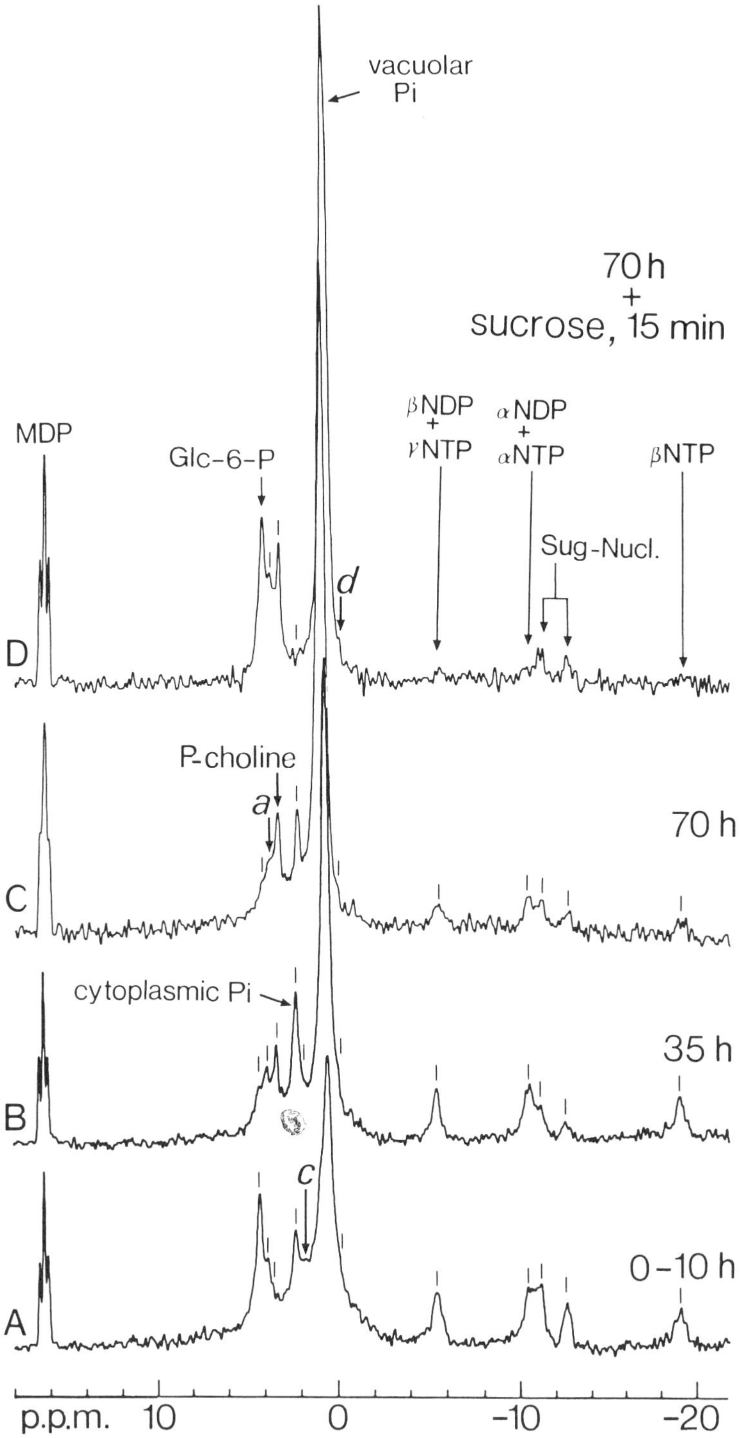

FIGURE 8. Representative ^{31}P-NMR spectra (81 MHz) of sucrose-starved sycamore cells (9 g). Each spectrum recorded at 25° represents the time average of 1500 free induction decays with a repetition time of 1.2 s (30 min) and a pulse angle of 45°. A, spectrum from cells at 10 h following sucrose starvation; B and C, spectra from cells at 35 h and 70 h following sucrose starvation; and D, recovery from 70 h sucrose starvation 15 min after the addition of 50 m*M* sucrose to the perfusion medium. Peak assignments as in Figure 3 (see also Figure 9).

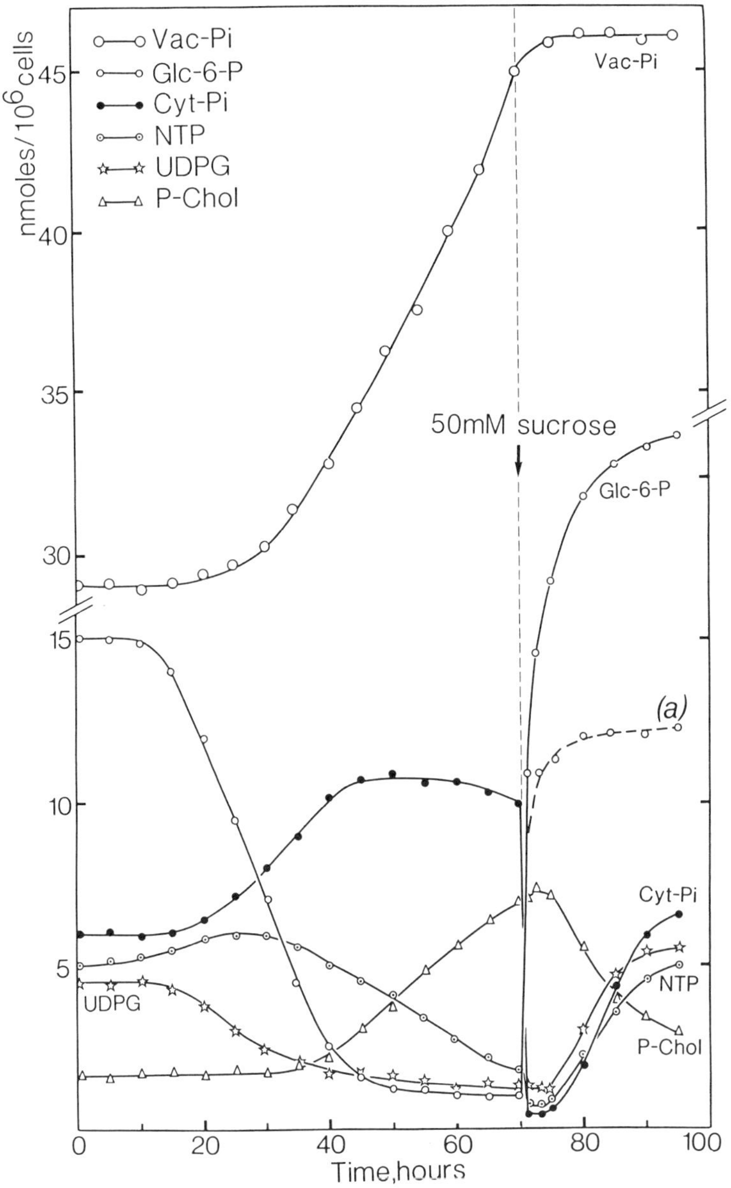

FIGURE 9. Time-course evolution of the most abundant ^{31}P-NMR-detectable phosphorus compounds in sycamore cells subjected to sucrose starvation. Sycamore cells were subjected to sucrose starvation up to 70 h followed by 24 h recovery after the addition of 50 m*M* sucrose and 50 μ*M* Pi (at arrow) to the perfusion medium. The concentrations of mobile phosphorus compounds in the cell sample were determined either biochemically[43] or by comparing the area of their signals to the area of methylene diphosphonate signal (capillary reference) and by using the same method to determine the area of the phosphorus resonances obtained from standard solutions of nucleotide triphosphate (NTP), glucose 6-P (Glc-6-P), Pi, UDP-glucose (UDPG), and P-choline (P-chol) at known concentrations. The broken line (a) shows the recovery of the glucose 6-P peak in the absence of Pi in the circulating medium. Note that in this case the vacuolar Pi pool does not fluctuate rapidly to buffer Pi in the cytoplasm. (From Roby, C., Martin, J. B., Bligny, R., and Douce, R., *J. Biol. Chem.*, 262, 5000, 1987. With permission.)

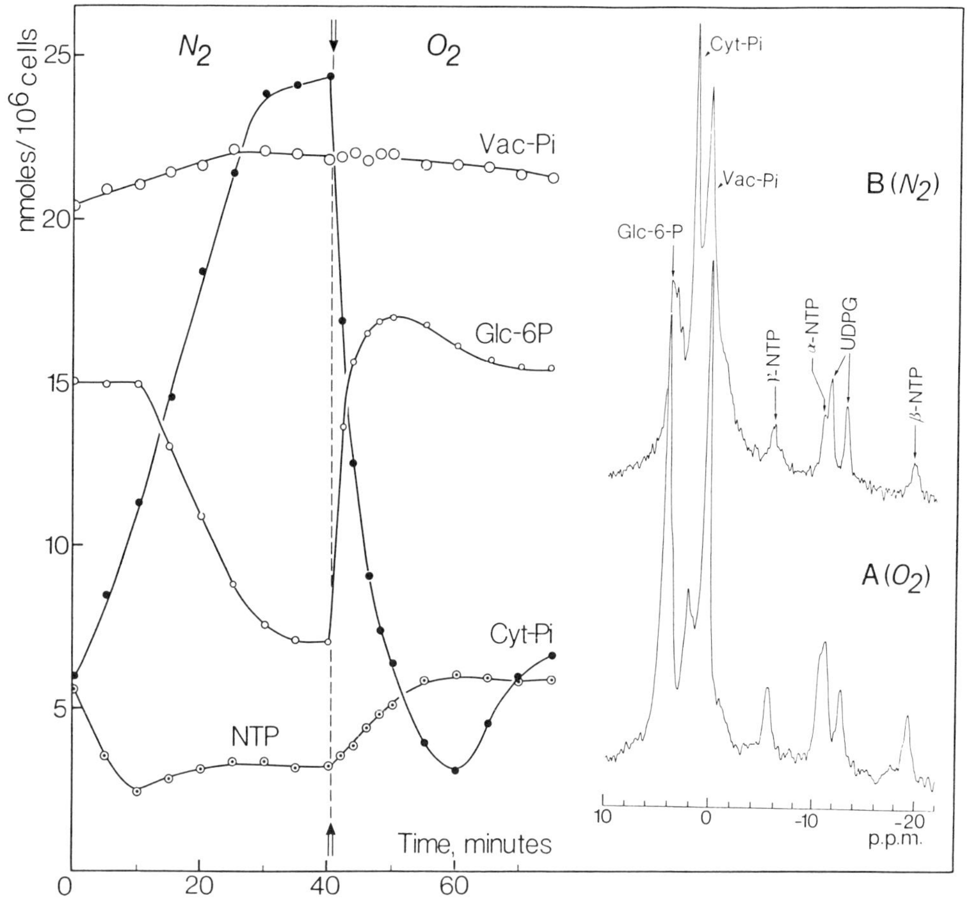

FIGURE 10. Time-course evolution of the most abundant ^{31}P-NMR-detectable phosphorus compounds in sycamore cells subjected to a transient hypoxia. At time zero the O_2 bubbling in perfusion medium was replaced by N_2 bubbling. A normal O_2 bubbling was restored after 40 min (arrow). Spectrum A, control (+O_2); and spectrum B, hypoxia (after 30 min N_2 bubbling). Each spectrum is the result of 1500 transients with a repetition time of 0.6 s (15 min) and a 30° pulse angle. Note that hypoxia induced a strong decrease (60%) of NTP and glucose 6-P, and that the released Pi accumulates within the cytoplasmic space, not in the vacuole. O_2 bubbling after hypoxia (arrow) triggered the resynthesis of NTP and glucose 6-P within less than 5 min at the expense of cytoplasmic Pi. The vacuolar Pi was not utilized. Peak assignments as in Figure 3.

REFERENCES

1. **Gunning, B. E. S. and Steer, M. W.,** *Ultrastructure and the Biology of Plant Cells,* Edward Arnold, London, 1975, 33.
2. **Ben-Hayyim, G. and Navon, G.,** Phosphorus-31 NMR studies of wild-type and NaCl-tolerant *citrus* cultured cells, *J. Exp. Bot.,* 36, 1877, 1985.
3. **Brodelius, P. and Vogel, H. J.,** A Phosphorus-31 nuclear magnetic resonance study of phosphate uptake and storage in cultured *Catharanthus roseus* and *Daucus carota* plant cells, *J. Biol. Chem.,* 260, 3556, 1985.
4. **Martin, J. B., Bligny, R., Rébeillé, F., Douce, R., Leguay, J. J., Mathieu, Y., and Guern, J.,** A ^{31}P nuclear magnetic resonance study of intracellular pH of plant cells cultivated in liquid medium, *Plant Physiol.,* 70, 1156, 1982.

5. **Roberts, J. K. M.,** Study of plant metabolism *in vivo* using NMR spectroscopy, *Annu. Rev. Plant Physiol.*, 35, 375, 1984.
6. **Roberts, J. K. M. and Jardetzky, O.,** Monitoring of cellular metabolism by NMR, *Biochim. Biophys. Acta*, 639, 53, 1981.
7. **Bligny, R.,** Growth of suspension-cultured *Acer pseudoplatanus* L. Cells in automatic culture units of large volume, *Plant Physiol.*, 59, 502, 1977.
8. **Lamport, D. T. A.,** Cell suspension cultures of higher plants: isolation and growth energetics, *Exp. Cell Res.*, 33, 195, 1964.
9. **Short, K. C., Brown, E. G., and Street, H. E.,** Studies on the growth in culture of plant cells. V. Large-scale culture of *Acer pseudoplatanus* L. cell suspensions, *J. Exp. Bot.*, 20, 572, 1969.
10. **Roby, C., Martin, J. B., Bligny, R., and Douce, R.,** Biochemical changes during sucrose deprivation in higher plant cells. II. Phosphorus-31 nuclear magnetic resonance studies, *J. Biol. Chem.*, 262, 5000, 1987.
11. **Roberts, J. K. M., Ray, P. M., Wade-Jardetzky, N., and Jardetzky, O.,** Estimation of cytoplasmic and vacuolar pH in higher plant cells by ^{31}P-NMR, *Nature*, 283, 870, 1980.
12. **Roberts, J. K. M., Wade-Jardetzky, N., and Jardetzky, O.,** Intracellular pH measurements by ^{31}P nuclear magnetic resonance. Influence of factors other than pH on ^{31}P chemical shifts, *Biochemistry*, 20, 5389, 1981.
13. **Pfeffer, P. E., Tu, S. I., Gerasimowicz, W. V., and Cavanaugh, J. R.,** *In vivo* ^{31}P-NMR studies of corn root tissue and its uptake of toxic metals, *Plant Physiol.*, 80, 77, 1986.
14. **Stitt, M., Wirtz, W., Gerhardt, R., Heldt, H. W., Spencer, C. A., Walker, D., and Foyer, C.,** A comparative study of metabolite levels in plant leaf material in the dark, *Planta*, 166, 354, 1985.
15. **Cohn, M. and Hughes, T. R., Jr.,** Nuclear magnetic resonance spectra of adenosine di- and triphosphate. II. Effect of complexing with divalent metal ions, *J. Biol. Chem.*, 237, 176, 1962.
16. **Wray, V., Schiel, O., and Berlin, J.,** High field phosphorus-31 nuclear magnetic resonance investigation of the phosphate metabolites in cell suspension cultures of *Nicotiana tabacum*, *Z. Pflanzenphysiol.*, 112, 215, 1983.
17. **Lee, R. B. and Ratcliffe, R. G.,** Development of an aeration system for use in plant tissue NMR experiments, *J. Exp. Bot.*, 34, 1213, 1983.
18. **Reid, R. J., Loughman, B. C., and Ratcliffe, R. G.,** ^{31}P-NMR measurements of cytoplasmic pH changes in maize root tips, *J. Exp. Bot.*, 36, 889, 1985.
19. **Reid, R. J., Field, L. D., and Pitman, G. G.,** Effects of external pH, fusicoccin and butyrate on the cytoplasmic pH in barley root tips measured by ^{31}P-nuclear magnetic resonance spectroscopy, *Planta*, 166, 341, 1985.
20. **Mitsumori, F., Yoneyama, T., and Ito, O.,** Phosphorus-nuclear magnetic resonance studies on higher plant tissue, *Plant Sci.*, 38, 87, 1985.
21. **Bieleski, R. L.,** Phosphate pools, phosphate transport and phosphate availability, *Annu. Rev. Plant Physiol.*, 24, 225, 1973.
22. **Edwards, J. and ap Rees, T.,** Metabolism of UDP-Glucose by developing embryos of round and wrinkled varieties of *Pisum sativum*, *Phytochemistry*, 25, 2033, 1986.
23. **Huber, S. C. and Akazawa, T.,** A novel sucrose synthase pathway for sucrose degradation in cultured sycamore cells, *Plant Physiol.*, 81, 1008, 1986.
24. **Turner, J. F. and Turner, D. H.,** The regulation of glycolysis and the pentose phosphate pathway, in *The Biochemistry of Plants. Metabolism and Respiration*, Vol. 2, Davies, D. D., Ed., Academic Press, New York, 1980, 279.
25. **Kime, J. J., Ratcliffe, R. G., Williams, R. J. P., and Loughman, B. C.,** The application of ^{31}P-nuclear magnetic resonance to higher plant tissue. I. Detection of spectra, *J. Exp. Bot.*, 33, 656, 1982.
26. **Pradet, A. and Raymond, P.,** Adenine nucleotide ratios and adenylate energy charge in energy metabolism, *Annu. Rev. Plant Physiol.*, 34, 199, 1983.
27. **Douce, R.,** *Mitochondria in Higher Plants. Structure Function and Biogenesis*, Academic Press, Orlando, FL, 1985, chap. 2.
28. **Foyer, C., Walker, D., Spencer, C., and Mann, B.,** Observations on the phosphate status and intracellular pH of intact cells protoplasts and chloroplasts from photosynthetic tissue using phosphorus-31 NMR, *Biochem. J.*, 202, 429, 1982.
29. **Waterton, J. C., Bridges, I. G., and Irving, M. P.,** Intracellular compartmentation detected by ^{31}P-NMR in intact photosynthetic wheat-leaf tissue, *Biochim. Biophys. Acta*, 763, 315, 1983.
30. **Stitt, M., Lilley, R. McC., and Heldt, H. W.,** Adenine nucleotide levels in the cytosol, chloroplasts and mitochondria of wheat leaf protoplasts, *Plant Physiol.*, 70, 971, 1982.
31. **Hughes, W. A., Bociek, S. M., Barrett, J. N., and Ratcliffe, R. G.,** An investigation of the growth characteristics of oil-palm (*Elaeis guineensis*) suspension cultures using ^{31}P-NMR, *Biosci. Rep.*, 3, 1141, 1983.
32. **Schibeci, A., Henry, R. J., Stone, B. A., and Brownlee, R. T. C.,** ^{31}P-NMR measurements of cytoplasmic and vacuolar pH in endosperm cells of ryegrass (*Lolium multiflorum*) grown in suspension culture, *Biochem. Int.*, 6, 837, 1983.

33. **Navon, G., Navon, R., Shulman, R. G., and Yamane, T.,** Phosphate metabolites in lymphoid, Friend erythroleukemia, and Hela cells observed by high-resolution ^{31}P nuclear magnetic resonance, *Proc. Natl. Acad. Sci. U.S.A.*, 75, 891, 1978.
34. **Bass, M. B. and Fromm, H. J.,** *trans*-1,2-Diaminocyclohexane *N,N,N′,N′*-tetraacetic acid is superior to ethylenediamine *N,N,N′,N′*-tetraacetic acid for sequestering Mg^{2+} in ^{31}P-NMR experiments involving ATP spectra at neutral and acidic pH, *Anal. Biochem.*, 145, 292, 1985.
35. **Stubb, M., Freeman, D., and Ross, B. D.,** Formation of n.m.r. invisible ADP during renal ischaemia in rats, *Biochem. J.*, 224, 241, 1984.
36. **Gadian, D. G.,** *Nuclear Magnetic Resonance and its Application to Living Systems*, Clarendon Press, Oxford.
37. **Walker, D.,** *The Intact Chloroplast*, Elsevier, Amsterdam.
38. **Ogren, W. L., Salvucci, M. E., and Portis, A. R., Jr.,** The regulation of rubisco activity, *Phil. Trans. R. Soc. Lond. B*, 313, 337, 1986.
39. **Mott, K. A., Jensen, R. G., O'Leary, J. W., and Berry, J. A.,** Photosynthesis and ribulose-1,5-bisphosphate concentration in intact leaves of *Xanthium strumarium* L., *Plant Physiol.*, 76, 968, 1984.
40. **Rébeillé, F., Bligny, R., Martin, J. B., and Douce, R.,** Relationship between the cytoplasm and the vacuole phosphate pool in *Acer pseudoplatanus* cells, *Arch. Biochem. Biophys.*, 225, 143, 1983.
41. **Spurr, A. R.,** A low viscosity epoxy resin embedding medium for electron microscopy, *J. Ultrastruct. Res.*, 26, 31, 1969.
42. **Reynolds, E. S.,** The use of citrate lead at high pH as an electron opaque stain in electron microscopy, *J. Cell. Biol.*, 17, 208, 1963.
43. **Rébeillé, F., Bligny, R., Martin, J. B., and Douce, R.,** Effect of sucrose starvation on sycamore (*Acer pseudoplatanus*) cell carbohydrate and Pi status, *Biochem. J.*, 226, 679, 1985.

Chapter 3

NMR STUDIES OF METABOLISM IN HYPOXIC MAIZE ROOT TIPS

Justin K. M. Roberts

TABLE OF CONTENTS

I. INTRODUCTION

In this chapter, the use of ^{31}P- and ^{13}C-NMR spectroscopy to study aspects of metabolism in hypoxic maize root tips is described. We are interested in metabolism in hypoxic plants for two reasons. First, studies of hypoxic plants can be aimed at understanding energy metabolism, because energy metabolism, and so metabolism in general, is profoundly affected at low oxygen tension.[1] Such studies are the focus here. Second, productivity of crop plants in the field is often limited because of low soil oxygen tension.[2,3] NMR can be used to look for metabolic traits associated with hypoxia-induced injury. Our studies of this second aspect have been summarized elsewhere.[4]

Primary root tips of very young maize seedlings have long been a popular material for investigations into plant metabolism,[1] and are one of the easiest tissues to work with for *in vivo* NMR studies. In addition to their high metabolic rate and reproducible performance, they are both large and dense enough to make perfusion of tightly packed samples straightforward.[5,6] Furthermore, cytoplasm occupies most of the volume of cells in the apical 2 mm of plant root tips,[7] so that the concentration of nonvacuolar metabolites per gram fresh weight is higher than in most plant tissues. These factors permit detection of cytoplasmic metabolites even at relatively low field strengths.[5]

II. ^{31}P-NMR STUDIES

^{31}P is the one naturally occurring isotope of phosphorus. ^{31}P-NMR spectra of maize root tips, as in Figure 1, contain signals from phosphorus in freely mobile metabolites.[8,9] Only the more abundant metabolites can be detected. Limits of detection range from about 1 m*M* (on a tissue fresh weight basis) for spectra obtained in a few minutes, or at low magnetic field strength, to about 0.02 m*M* for spectra obtained at high field strength over several hours.[10] In addition to different NMR signals from the different chemical groups, a unique feature of ^{31}P-NMR spectra of plants is the distinct signals from cytoplasmic and vacuolar pools of inorganic phosphate (Figure 1).[8] The positions of these phosphate signals reflect the pH in these two intracellular compartments, and can be used as naturally occurring pH indicators.[8,11] While *in vivo* inorganic phosphate resonances are primarily affected by hydrogen ion concentration, the positions and relative line widths of nucleoside triphosphate resonances primarily reflect their complexation with magnesium ion *in vivo*.[6]

As indicated by Figure 1, ^{31}P-NMR signals from cytoplasmic metabolites in oxygenated maize root tips are for the most part constant over long periods of time.[5,12,13] Only resonances from glucose-6-phosphate and uridine diphosphoglucose can change in intensity, both increasing when sugars are supplied, both decreasing when sugars are depleted.[14] In contrast, when maize root tips become hypoxic, NMR spectra show large changes in levels of many cytoplasmic phosphates (Figure 2).[8,10] These include decreases in glucose-6-phosphate and nucleoside triphosphates, and increases in cytoplasmic inorganic phosphate and nucleoside diphosphates.[10] Hence, the ratio of nucleoside tri- to diphosphates decreases from between 20 to 30 in oxygenated maize root tips to between 1 to 2 in root tips under severe hypoxia (Figure 3).[10] We have discussed and speculated on the factors which may determine nucleotide ratios in maize root tips elsewhere; the ratio [ATP]/[ADP][Pi] appears to reflect the rate of synthesis of ATP in maize root tips.[10]

In Figure 2, the rightward shift of the cytoplasmic Pi resonance indicates that cytoplasmic pH falls when maize root tips become hypoxic.[5,15] Figure 4 plots the sequential change in the position of this resonance observed in spectra obtained every minute. Within about 15 min, cytoplasmic pH decreases by about 0.5 pH unit, after which cytoplasmic pH stabilizes.[5,15] Figure 4 also shows that the period of stable cytoplasmic pH is associated with ethanolic fermentation by the hypoxic tissue. Fermentation to ethanol does not produce acid in the cytoplasm, provided

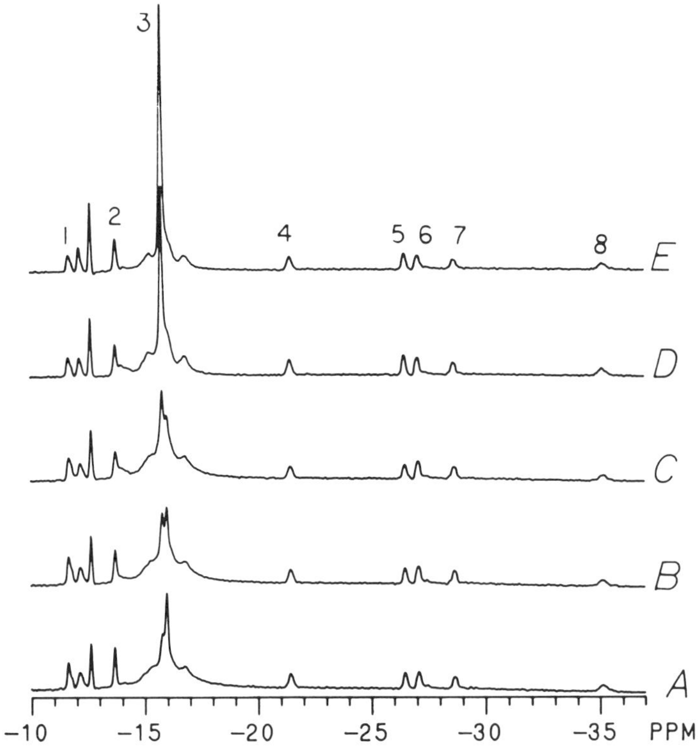

FIGURE 1. The 202 MHz ^{31}P-NMR spectra of maize root tips perfused with oxygenated 0.1 m*M* $CaSO_4$. Spectra resolution enhanced with double exponential multiplication. Each spectrum was obtained over 1 h. A, 0.5 to 1.5 h of perfusion; B, 2 to 3 h; C, 3 to 4 h; D, 10 to 11 h; and E, 22 to 23 h. Assignments: 1, glucose-6-phosphate; 2, cytoplasmic Pi; 3, vacuolar Pi; 4, γ-nucleoside triphosphate; 5, α-nucleoside triphosphate; 6 and 7, uridine diphosphoglucose and nicotinamide adenine nucelotides; and 8, β-nucleoside triphosphate.

the carbon dioxide produced comcommitantly can escape the tissue.[5,16] ^{13}C-NMR studies discussed below have shown that the initial cytoplasmic acidification in hypoxic maize root tips is associated with a burst of lactic acid synthesis.[15] The sequence of reactions just outlined is not seen in root tips of mutants lacking ADH1, the principal isozyme of alcohol dehydrogenase in maize. These mutants produce lactic acid continuously during hypoxia, resulting in severe cytoplasmic acidosis.[15,16]

The final feature of metabolism we have identified in hypoxic maize root tips using ^{31}P-NMR is a rise in vacuolar pH.[16] Only relatively large changes in vacuolar pH can be observed by ^{31}P-NMR, because vacuolar pH in maize root tips is much lower than the pK_a of inorganic phosphate,[11] so that the chemical shift of the vacuolar Pi resonance is not as sensitive to changes in pH as the cytoplasmic Pi resonance. Therefore, the quite small changes in the position of the vacuolar Pi resonance we observe, as in Figure 2, indicate increases in vacuolar pH on the order of 0.5 pH unit during hypoxia.[167] We interpret this increase in vacuolar pH to result from leakage of protons from vacuole to cytoplasm. Hence, the vacuolar sap is one source of protons that can contribute to cytoplasmic acidosis.[16]

III. ^{13}C-NMR STUDIES

^{13}C is found at about 1.1% abundance in nature. A consequence of this low natural abundance, together with the insensitivity of NMR spectrometers, is that natural abundance ^{13}C-NMR

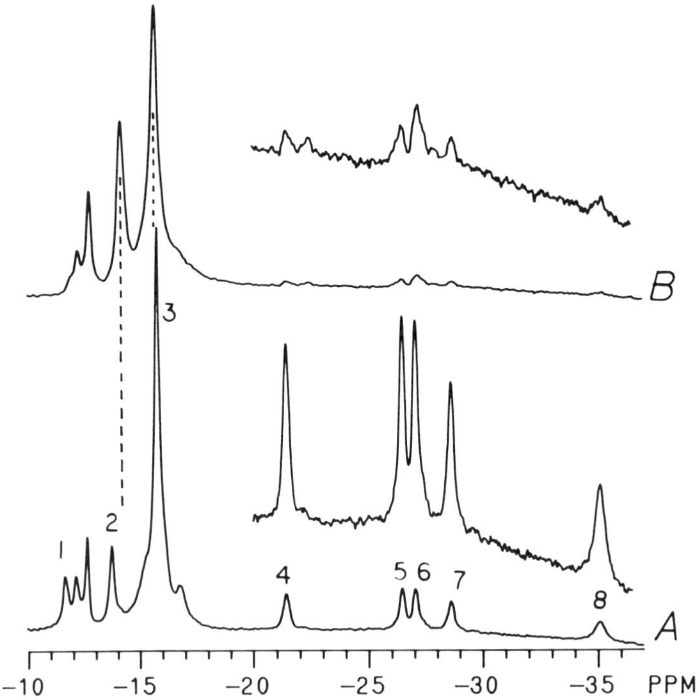

FIGURE 2. Effect of hypoxia on phosphorus NMR spectrum of maize root tips. Spectra obtained after 15-Hz line broadening. Root tips perfused with either oxygenated (A) or nitrogen-saturated (B) 0.1 m*M* $CaSO_4$. Assignments as in Figure 1.

spectra contain signals only from metabolites present at high tissue concentrations (greater than micromoles).[17] For example, such spectra of maize root tips are dominated by signals from sugars and organic acids (Figure 5A). Because of this low background, when ^{13}C-enriched chemicals are introduced to plant tissues, the ^{13}C-NMR spectra obtained subsequently contain predominantly signals from the introduced chemical and its metabolites. This is illustrated in Figure 5. When maize root tips are fed 50 m*M* [1-^{13}C]-glucose, very strong signals are obtained first from glucose (Figure 5B) and later from the glucose metabolites sucrose (Figure 5C) and amino acids (Figure 5D). Note that in the glucose-fed tissue, the β anomer of glucose is significantly less abundant than the α anomer (Figure 5B), a result essentially the reverse seen with glucose at equilibrium in free solution. A similar result was reported in studies of *Escherichia coli*,[18] and indicates low anomerase activity and preferential utilization of the β anomer.

Fermentation reactions result in conversion of the C1 of glucose into methyl groups of lactic acid, alanine, and ethanol.[1] ^{13}C-NMR spectra of hypoxic maize root tips prelabeled with glucose as in Figure 5 therefore provide information on the amounts of various fermentation end products formed.[15] Results of such an experiment are given in Figure 6. These spectra contain signals primarily from intracellular metabolites, because the tissue is continually perfused with fresh hypoxic medium. Most of the ethanol produced by the tissue leaks out (compare with Figure 4), whereas lactic acid and alanine, being charged species, are for the most part retained in the tissue. It is clear from Figure 6 that accumulation of lactic acid is restricted to the first few minutes of hypoxia, accumulation being coincident with the decrease in cytoplasmic pH at the onset of hypoxia (Figure 4). In contrast, intracellular ethanol increases principally after about 10 min of hypoxia, reaching a steady level after about 25 min of hypoxia. This behavior is similar to the time course of ethanol production inferred from the assay of ethanol in the perfusion

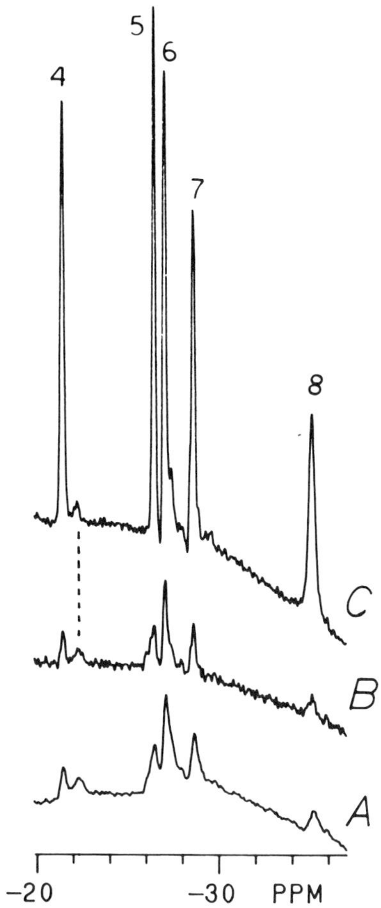

FIGURE 3. Effect of oxygen tension on nucleotide concentrations in maize root tips by ^{31}P-NMR. C, spectrum of roots perfused with oxygenated 0.1 mM $CaSO_4$, obtained over 20 h, with double exponential multiplication of the free induction decay. B, spectrum of roots perfused with nitrogen-saturated 0.1 m*M* $CaSO_4$, obtained over 3 h, with double exponential multiplication. A, same raw data as in B, but spectrum is a result of single exponential multiplication, line broadening 15 Hz. Assignments as in Figure 1; small peak to right of peak 4, b-nucleoside diphosphate.

effluent (Figure 4). In addition to lactic acid and ethanol, Figure 6 shows that large amounts of alanine are produced in hypoxic root tips.

IV. CONCLUSIONS

In vivo NMR spectroscopy can be used to monitor a number of metabolic processes in maize root tips. The sensitivity and time resolution compares favorably to other analytical biochemical techniques. Further, NMR permits some unique measurements to be made, such as simultaneous measurements of cytoplasmic and vacuolar pH. The studies described here reveal a sequence of metabolic events in maize root tips following the onset of hypoxia. Initially, glucose is fermented to lactic acid, resulting in cytoplasmic acidification. After about 10 min, production of lactic acid begins to decline as ethanol becomes the predominant fermentation end product — a product that does not lead to cytoplasmic acidification. We have shown that the initial cytoplasmic

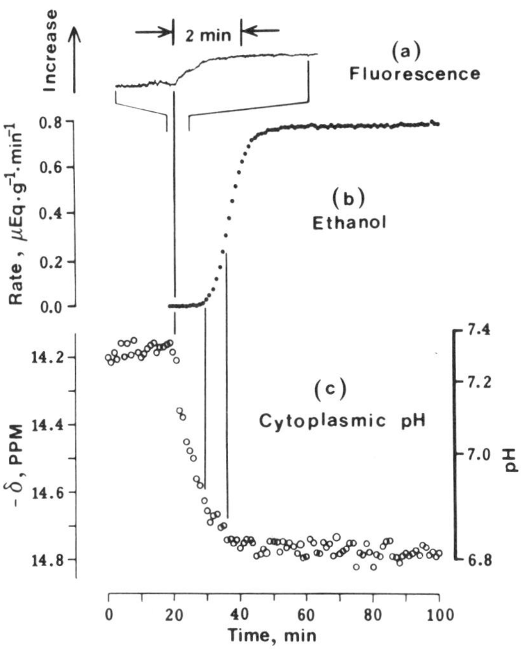

FIGURE 4. Time course of NADH fluorescence (a), rate of ethanol production (b), and cytoplasmic pH (c) in perfused maize root tips during hypoxia, determined *in vivo*. Root tips were initially perfused at 50 ml/min with oxygen-saturated 50 m*M* glucose per 0.1 m*M* $CaSO_4$. After 20 min, they were perfused at 4 ml/min with the same solution now saturated with nitrogen. (From Roberts, J. K. M., Callis, J., Wemmer, D., Walbot, V., and Jardetzky, O., *Proc. Natl. Acad. Sci. U.S.A.*, 81, 3379, 1984. With permission.)

acidification serves to turn on ethanolic fermentation, i.e., the transient lactic fermentation acts as a switch.[15] This pattern of regulation permits continued energy production for several hours while the cytoplasm is maintained near neutrality. The most striking feature of these studies that remains to be fully understood is the production of alanine. Neither the enzymes responsible for, nor the factors which control, alanine production have been properly identified, although the involvement of transaminases and glutamate dehydrogenase, and redox potential, seem likely candidates.

In cells, individual biochemical reactions are often in competition. For example, dehydrogenases compete for coenzymes which are present at concentrations lower than the "catalyst". This competition represents an important component of metabolic regulation — one that NMR can be used to understand.

ACKNOWLEDGMENTS

Spectra shown here were obtained on a General Electric GN500, purchased in part from grants from the National Science Foundation (DMB 8604091), the Department of Energy (FG03-86FR13535), and the National Institutes of Health (PHS BRSG 2 507).

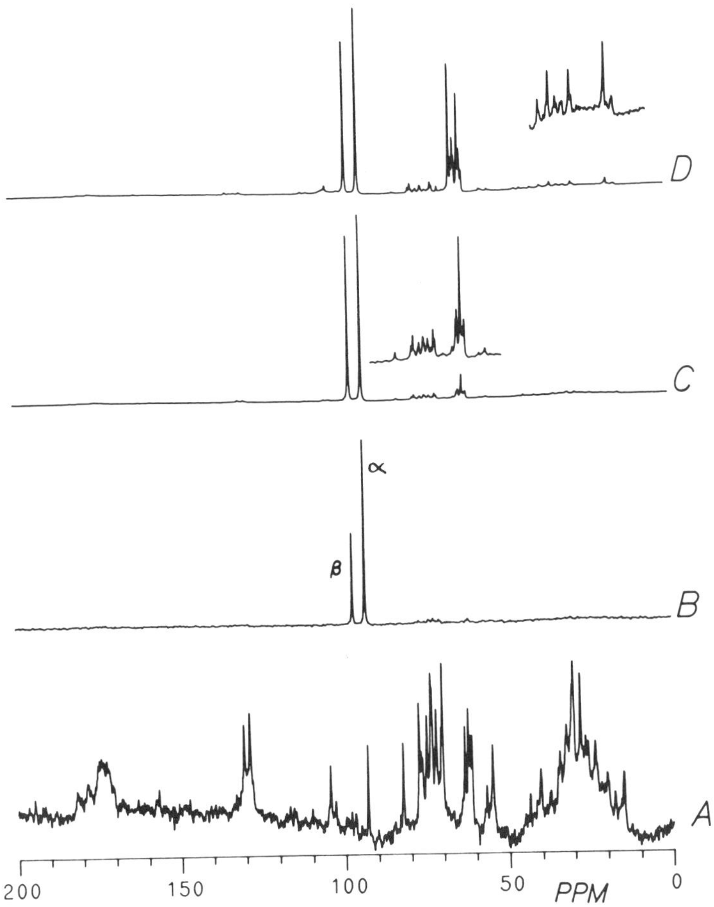

FIGURE 5. The 125 MHz ^{13}C-NMR spectra of maize root tips perfused with oxygenated 0.1 m*M* $CaSO_4$. Line broadening 15 Hz. A, natural abundance spectrum, obtained over 1 h. B, just after addition of 50 m*M* [1-^{13}C]-glucose to perfusate, spectrum obtained over 1.5 min. C, 10 to 70 min after addition of glucose, 1 h spectrum. D, 15-min spectrum after 17 of perfusion as in B. Spectral peaks from chemical groups dispersed approximately as methyls, 15 to 40 ppm; methylenes, 40 to 70 ppm; >C-O, 50 to 95 ppm; aromatics, 110 to 140; carboxylic acids, 170 to 190 ppm. The labels α and β refer to the two anomers of glucose.

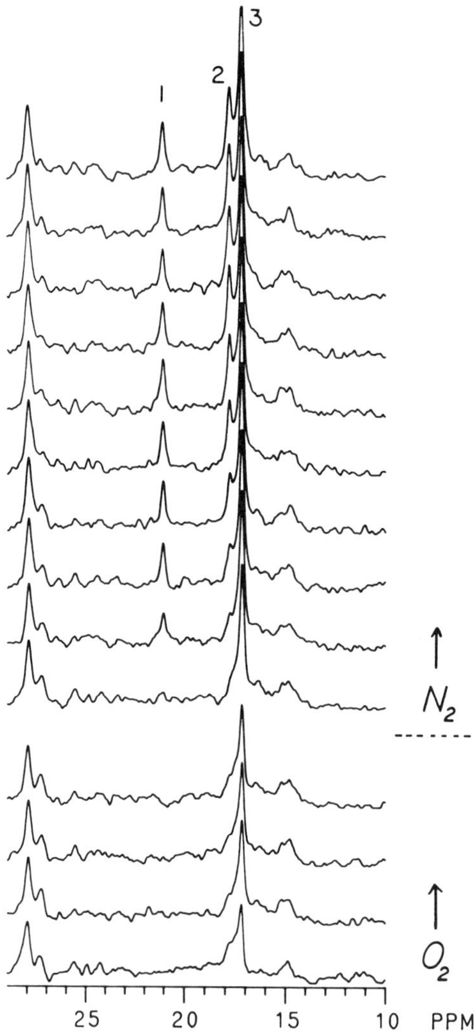

FIGURE 6. ^{13}C-NMR spectra of methyl resonances in perfused maize root tips; same sample as in Figure 5. Each spectrum was obtained over 4 min. Line broadening 15 Hz. The bottom four spectra are of oxygenated root tips, taken just after spectrum D in Figure 5. The remaining spectra were obtained sequentially, bottom to top, after switching perfusion to nitrogen-saturated 0.1 m*M* $CaSO_4$. Assignments: 1, lactic acid; 2, ethanol; and 3, alanine.

REFERENCES

1. **Beevers, H.,** *Respiratory Metabolism in Plants,* Harper & Row, New York, 1961.
2. **Boyer, J. S.,** Plant productivity and environment, *Science,* 218, 443, 1982.
3. **Kozlowski, T. T., Ed.,** *Flooding and Plant Growth,* Academic Press, New York, 1984.
4. **Roberts, J. K. M.,** Cytoplasmic acidosis and flooding tolerance in crop plants, in *Ecology and Management of Wetlands,* Vol. 1, Hook, D. D. et al., Eds., Croom Helm, Beckenham, England, 1988, 392.
5. **Roberts, J. K. M., Wemmer, D., Ray, P. M., and Jardetzky, O.,** Regulation of cytoplasmic and vacuolar pH in maize root tips under different experimental conditions, *Plant Physiol.,* 69, 1344, 1982.

6. **Roberts, J. K. M.,** Determination of the energy status of plant cells by ^{31}P-nuclear magnetic resonance spectroscopy, in *Modern Methods of Plant Analysis,* New Series, Vol. 2, Linskens, H. F. and Jackson, J. F., Eds., Springer-Verlag, New York, 1986, 43.
7. **Steer, M. W.,** *Understanding Cell Structure,* Cambridge University Press, London, 1981.
8. **Roberts, J. K. M., Ray, P. M., Wade-Jardetzky, N. G., and Jardetzky, O.,** Estimation of cytoplasmic and vacuolar pH in higher plant cells by ^{31}P-NMR, *Nature,* 283, 870, 1980.
9. **Roberts, J. K. M.,** NMR in plant biochemistry, in *The Biochemistry of Plants,* Vol. 13, Davies, D. D., Ed., Academic Press, Orlando, FL, 1987, 181.
10. **Roberts, J. K. M., Lane, A. N., Clark, R. A., and Nieman, R. H.,** Relationships between the rate of synthesis of ATP and the concentrations of reactants and products of ATP hydrolysis in maize root tips determined by ^{31}P nuclear magnetic resonance, *Arch. Biochem. Biophys.,* 240, 712, 1985.
11. **Roberts, J. K. M., Wade-Jardetzky, N. G., and Jardetzky, O.,** Intracellular pH measurements by ^{31}P-nuclear magnetic resonance. Influence of factors other than pH on ^{31}P chemical shifts, *Biochemistry,* 20, 5389, 1981.
12. **Roberts, J. K. M., Linker, C. S., Benoit, A. G., Jardetzky, O., and Nieman, R. H.,** Salt stimulation of phosphate uptake in maize root tips studied by ^{31}P nuclear magnetic resonance, *Plant. Physiol.,* 75, 632, 1984.
13. **Pfeffer, P. E., Tu, S. I., Gerasimowicz, W. V., and Cavanaugh, J. R.,** *In vivo* ^{31}P-NMR studies of corn root tissue and its uptake of toxic metals, *Plant Physiol.,* 80, 77, 1986.
14. **Roberts, J. K. M., Wemmer, D., and Jardetzky, O.,** Measurement of mitochondrial ATPase activity in maize root tips by saturation transfer ^{31}P nuclear magnetic resonance, *Plant Physiol.,* 74, 632, 1984.
15. **Roberts, J. K. M., Callis, J., Wemmer, D., Walbot, V., and Jardetzky, O.,** Mechanism of cytoplasmic pH regulation in hypoxic maize root tips and its role in survival under hypoxia, *Proc. Natl. Acad. Sci. U.S.A.,* 81, 3379, 1984.
16. **Roberts, J. K. M., Callis, J., Jardetzky, O., Walbot, V., and Freeling, M.,** Cytoplasmic acidosis as a determinant of flooding tolerance in plants, *Proc. Natl. Acad. Sci. U.S.A.,* 81, 6029, 1884.
17. **Martin, F.,** Monitoring of plants metabolism by ^{13}C, ^{15}N, and ^{14}N nuclear magnetic resonance spectroscopy, *Physiol. Veg.,* 23, 463, 1985.
18. **Gadian, D. G.,** *Nuclear Magnetic Resonance and Its Applications to Living Systems,* Oxford University Press, New York, 1982.

Chapter 4

NMR STUDIES OF INTACT SEEDS

Venceslav Rutar

TABLE OF CONTENTS

I. INTRODUCTION

Plant seeds represent obvious objects for NMR studies, since they contain several nuclei (^{1}H, ^{13}C, ^{14}N, ^{15}N,...) which may give rise to observable signals. Responses are quite complicated due to a wide variety of seeds and seed constituents. Therefore, this chapter does not intend to review all possible applications. An emphasis is given to those measurements which leave seeds essentially intact and viable, so that one can utilize NMR to its main advantage as a *nondestructive* method. Since samples can be planted upon completion of the measurements, breeding programs may be developed which fully exploit the genetic potential of a plant as revealed by NMR. New plant varieties with controlled nutritional value can be developed within short periods of time.

It seems that success in breeding high-oil hybrids lead to the impression that "additional improvements" in NMR techniques would solve other analytical problems. Determinations of oil composition and protein content are indeed possible, but routine applications do not appear very common at the present moment. While NMR represents the method of choice in measurements of total oil, characterization of other structural parameters suffers from several well-known drawbacks. The results are not always reliable, measuring times can be unacceptably long, and (what might be the most serious problem) instruments are very expensive.

II. STATE OF WATER

Hydrogen nuclei (^{1}H) are abundantly present in all biological systems, and they give rise to strong NMR signals even in low-field instruments. The main emphasis must be focused on the interpretation of results because spectral responses of different constituents overlap in studies of intact samples. Dominant contribution arising from water often obscures signals of other substances (such as oil) which might be the main object of the intended measurement.

Characterization of NMR signals implies exact determination of frequencies (chemical shifts and J coupling constants), intensities, and relaxation effects. High-resolution spectroscopy is not always possible in heterogeneous systems where differences in magnetic susceptibilities contribute to intrinsic line broadening. Therefore, the state of water had to be studied by wide-line NMR. Since featureless shapes of such resonances do not easily reveal relevant information, spin-lattice (T_1) and spin-spin (T_2) relaxation times become measurable parameters which are used for verification of theoretical models describing the state of water.

Customary NMR methodology is not very suitable in studies of heterogeneous systems. Early experiments indeed showed that the regrowth or decay of magnetization was exponential (as predicted by Bloch[1]), and theoretical descriptions were used to explain various relaxation contributions.[2] However, conditions affecting T_1 and T_2 values were well understood only in homogeneous samples. In direct contrast, studies of complex biological systems lead to ambiguous results which were dependent on the model used to explain the NMR data. The observed relaxation decay is a sum of contributions from nuclei which may reflect their different environments during the course of the experiment. The heterogeneous nature of such samples requires a new description[3] avoiding any a priori assumptions about the number of the components in the relaxation curve. Experimental measurements are utilized to determine distributions of T_1 and T_2 parameters, so that better characterization will provide deeper insight into the state of water in biological systems.

Major models currently used to describe NMR relaxation in plant seeds imply separate environments where water is immobilized to various extents. Resulting states are distributed between two extremes which are pictorially described as "tightly bound" and "free" water. Diffusion averages contributions, so that rapid exchange can give rise to a monoexponential dependence which apparently results in a single-valued T_1 or T_2 parameter. This situation is often found in dry seeds, where the residual moisture comes from water adsorbed on the solid seed matrix. Diffusion across short distances is sufficiently fast to average relaxation contributions. On the other hand, fresh seeds contain large compartments which are filled with water. The water may not come into contact with surfaces which could affect relaxation during the measuring time. Then the multicomponent nature of the NMR signal becomes clearly pronounced.

Spin-lattice relaxation measurements on intact maize seeds have shown[4] that the complete curve must be decomposed into two components, if water content exceeds approximately 5 to 7%. It is believed that the slowly relaxing contribution arises from "free" water, because it disappears in dry seeds. On the other hand, protons in contact with the solid matrix are relaxing much faster, and the corresponding T_1 could be followed as a function of water content over a wide range of hydration levels. Drying of seeds from 38% down to 10 to 15% water content reduces relaxation time, because solid surfaces become more and more efficient in their contribution to the NMR relaxation. The value of T_1 goes through a minimum at a moisture content approximately of 15%, and begins to increase upon further drying. This part of the dependence suggests that the residual moisture is immobilized within the seed matrix, and correlation times become so long[2] that increases in T_1 occur.

Spin-spin relaxation times (T_2) consistently decrease with the lower moisture content.[5] Seed drying turns out to be very efficient, if water signals must be suppressed, because a smaller number of spins and shorter T_2 values *jointly* reduce the observable response.

III. OIL CONTENT

Plant seeds contain oil which exists in a liquidlike form. From the NMR point of view, it can be treated as a viscous fluid in small droplets surrounded by a solid structure which spoils homogeneity of the magnetic field and affects relaxation. Precise definitions and measurements of relaxation times are complicated, because oil contains a large number of nonequivalent ^1H spins with different chemical shifts and intrinsic mobilities. Therefore, all reported values must be treated as illustrative parameters. In any case, T_2 relaxation times are longer than ~20 ms, clearly reflecting the liquidlike state of oil, while water has $T_2 < 0.4$ ms in dry seeds with moisture contents less than 5%. Since hydrogen nuclei in other constitutents (carbohydrates, proteins, etc.) belong to the solid part, only oil remains "NMR visible" in experiments which are designed to detect signals of liquids.

The advantage of such measurements was realized around 1960 when the first successful applications were reported.[6,7] Early experiments utilized continuous wave detection which follows response as a function of the external magnetic field when it is swept through the resonance. Signal amplitude was proportional to the amount of oil in the seed as long as interfering water was removed by drying. Further development of NMR instruments showed that pulsed Fourier-transform techniques are superior for research purposes. However, spectrometers operating in the continuous wave mode are still useful, especially for routine oil determinations in bulk samples.

Since pulsed NMR detects signals in the time domain, measurements are finished within seconds.[8] Another important consideration centers on the fact that various tailored pulse sequences of radiofrequency irradiation have been designed which eliminate certain undesired effects. Examples are spin-echo (SE)[9] or Carr-Purcell-Meiboom-Gill (CPMG)[10,11] experiments which can refocus precession arising from magnetic field inhomogeneities, so that the signal decay is governed by the intrinsic spin-spin relaxation time (T_2).

Determination of oil content can be accomplished by three techniques:

1. Free induction decay (FID) uses a single 90° pulse which excites the spin system, and the signal intensity is recorded after a delay of 100 to 500 μs which must be of sufficient length to eliminate any possible signals from tightly bound protons in the solid matrix or adsorbed moisture. On the other hand, it should be short enough that the effects of magnetic field inhomogeneity do not become pronounced.

2. SE spectroscopy uses an additional 180° pulse which refocuses precessing spins. Resulting echoes are recorded as a function of the separation τ between the pulses and the values can be extrapolated to $\tau = 0$, thus eliminating effects of individual relaxation times.
3. The CPMG pulse sequence uses a train of 180° pulses and echoes are samples between them. Extrapolation does not require separate increments of the interval τ, because a series of echoes is obtained in a single experiment.

NMR responses are amplifed electric signals which must be correlated with the amount of oil. Calibration curves are obtained by measuring artificially prepared samples of known composition or by a destructive analysis of a selected number of seeds. It is advisable that some representative samples are saved as standards for future studies, because they allow quick recalibration which may become necessary in the next set of measurements.

Several authors[12-25] have studied parameters affecting NMR determination of oil content and related topics. They also discussed possible sources of experimental errors as well as corrective actions which increase the reliability of the results. Their main conclusions are summarized in a short list with respect to possible effects which must be considered in practical applications.

A. Oil Mobility

Since tightly bound ^1H nuclei give rise to fast-decaying signals, it is possible that some lipids are not "visible" in NMR determinations. Although this problem has not been studied carefully, it is usually assumed that oil is found mostly in liquidlike droplets and NMR results are regarded as a reliable measurement.

B. Oil Composition

The NMR signal is proportional to the number of ^1H spins in the sample, and different fatty acids do not make the same contribution due to their various hydrogen/molecular weight ratios. Resulting systematic errors are partially reduced, if the results are calibrated using oil from the same variety. Differences between individual seeds cannot be eliminated in this fashion. Further complications arise when other proton-containing substances are dissolved in oil or when they exist in a liquidlike form.

C. Water Content

If seeds are not sufficiently dry, water contributes to an additional signal which may diminish the reliability of the measurement. It becomes necessary to separate contributions by following the time dependence of the signal using CPMG or SE, since moderate amounts of moisture (5 to 20%) give rise to a water signal decaying within less than 20 ms. The CPMG pulse sequence is preferred, in that it is not sensitive to diffusion effects which may decrease echoes if the interval between the 90 and 180° pulses is too long. Seeds with excessive water content are not amenable to oil content determinations with existing procedures. Signals cannot be clearly separated by wide-line NMR if T_2 values of both components are comparable.

D. Temperature

Magnetization of nuclear spins and resulting amplitudes of the NMR signal are inversely proportional to the absolute temperature of the sample. A difference of 1 K can affect the result by approximately 0.3%. Recalibration of results using a reference seed reduces fluctuations if good temperature stabilization of the NMR probe and laboratory is not available.

E. Spin-Lattice Relaxation

Accumulation of signals from repetitive experiments improves the signal-to-noise ratio and it is frequently used in NMR experiments. Fast repetition can partially saturate responses due to competition between excitation and relaxation of nuclear spins. Systematic errors are avoided only if delays of more than approximately 1 s are inserted between the signal acquisition and subsequent radiofrequency pulse. Such intervals are necessary to allow complete relaxation of the spin system, but they also increase the overall measuring time.

F. Magnetic Field Inhomogeneity

The initial 90° pulse creates spin coherence which starts to diminish due to different precessing frequencies. The rate of the decay is governed by magnetic field inhomogeneities arising mostly from intrinsic differences in magnetic susceptibility. Improved homogeneity of the external magnetic field does not help very much, because the seed itself introduces perturbation. It is therefore not surprising that an angular dependence of the signal height has been observed when corn seeds were rotated. If the magnetic field inhomogeneity affects the reliability of measurements using simple FID, SE or CPMG pulse sequences must be tried instead.

G. Pulse Imperfections

Radiofrequency pulses can become an important source of errors, if they are not adjusted correctly. The use of the FID method is not critical when results are regularly recalibrated by the reference seed providing that the probe is larger than all measured samples. Spin echoes, however, strongly depend on the accurate instrument settings, because 180° pulses must refocus precession of nuclear spins. It is possible to improve performance of the NMR hardware by software procedures, such as composite radiofrequency pulses.[26]

H. Signal Detection Method

Since the cost of instruments should be kept low, many pulsed NMR spectrometers designed for measurements of total oil content utilize "diode detection method". This cheap solution makes amplitude of the signal independent of small fluctuations or drifts of the magnetic field. On the other hand, this approach generally does not give rise to a linear calibration curve, because a substantial part of the detected voltage drops across the diode threshold and small amounts of oil cannot be measured at all. The problem can be alleviated by the "double-tube addition method" when a known amount of oil is added so that the total signal exceeds the lower detection limit.

NMR becomes a reliable, quick, and nondestructive method for determinations of oil content in single viable seeds, if the above precautions are considered in planning of experiments. Most samples can be measured within seconds and it is also possible to automate the entire procedure.[17]

IV. OIL COMPOSITION

Breeding high-oil varieties may alter oil composition thus producing an undesirable side effect.[27] A negative correlation between the total amount of oil and iodine value exists, but it is not so pronounced as to prevent development of hybrids with increased oil content and acceptable quality.

High-resolution NMR is one of the possible choices, because it has been shown by Johnson and Shoolery[28] that spectra contain enough information for calculation of iodine values. These early experiments showed well-resolved resonances of chemically nonequivalent groups in some natural fats and comparison of integrals facilitated intended measurements. The main problem lies in the fact that high-resolution NMR approaches can not be simply transferred from extracted liquids into plant seeds. Although oil has the same microscopic structure and it gives rise to resonances with the same chemical shifts and coupling constants, compartment structure of the seed creates nonuniform differences in magnetic susceptibility thus broadening responses.

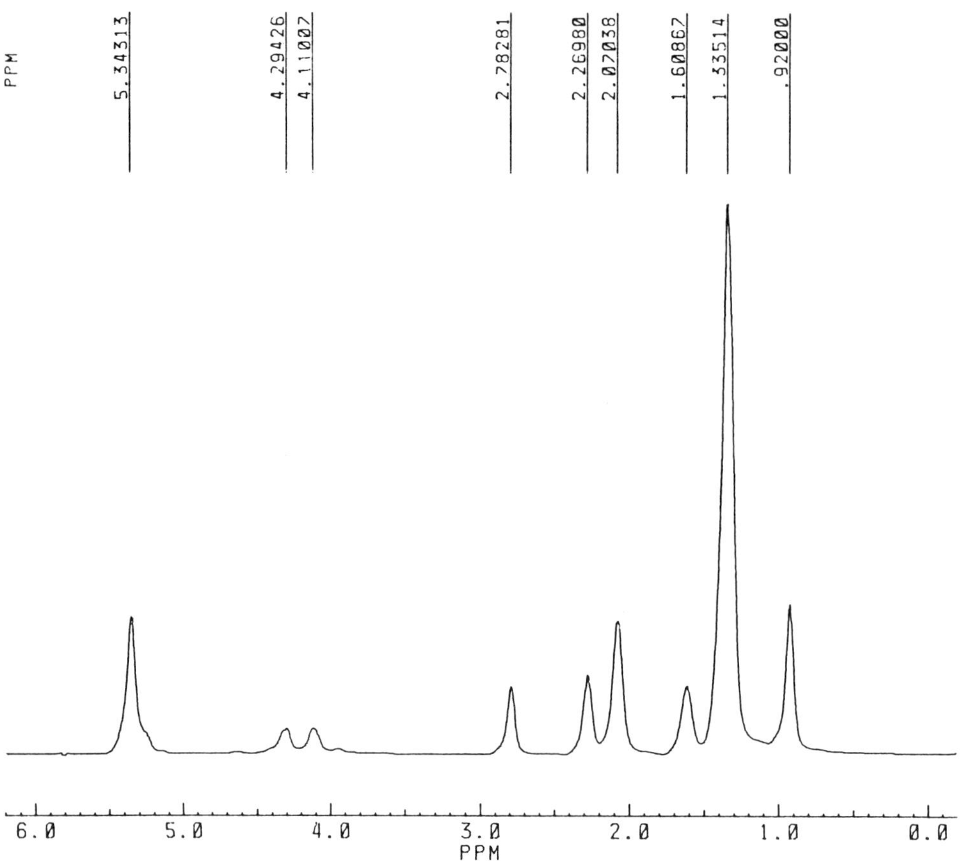

FIGURE 1. MASS proton spectrum of a single sunflower seed as obtained at 300 MHz. Since sample spinning reduced line broadening, well-resolved signals allow nondestructive determination of oil composition. This approach requires measuring times less than 1 min, and it may find successful applications in large-scale breeding programs designed to improve oil quality.

Modern attempts to characterize oil quality in intact seeds should utilize magic angle sample spinning (MASS) which reduces susceptibility effects as shown by Garroway.[29] The resulting ^1H spectrum (Figure 1) identifies signals of various functional groups and their typical chemical shifts allow nondestructive measurements of the oil composition. The signal at 0.92 ppm represents three protons in CH_3 groups, while the response at 2.78 ppm arises from underlined protons in the fragment =CH-C$\underline{H}_2$-CH= which is found in the linoleic acid. Comparison of integrals shows that this particular seed contains 76% linoleic acid.

There were also other attempts to record ^1H high-resolution spectra of viable seeds. Conway and Johnson[30] tried laboratory drying to remove water and subsequent soaking in freon to reduce susceptibility differences, so that studies of corn became possible. On the other hand, they found that sample heating did not improve resolution, because increased mobility of oil molecules cannot overcome line broadening.

Improvement of instruments has opened new possibilities, because modern Fourier transform spectrometers can detect ^{13}C signals in natural abundance and resolution problems can be solved by using ^1H and ^{13}C NMR spectroscopy together. Carbon resonances are spread over an approximately 20 times larger interval of chemical shifts and they facilitate a detailed study of oil composition[31-34] (see also Figure 2). Since ^{13}C sensitivity is only 1.8×10^{-4} as compared to ^1H, signals must be accumulated for a few minutes, even using the best instruments. The combination of long measuring times and the high cost of the equipment makes this technique somewhat less attractive.

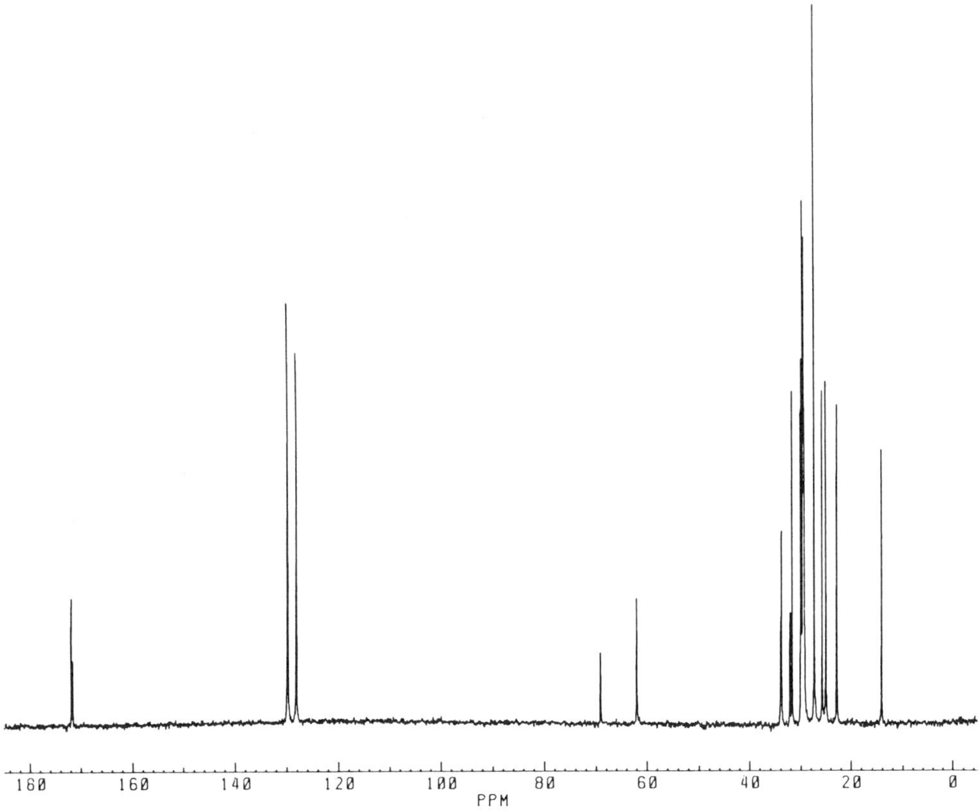

FIGURE 2. Proton-decoupled ^{13}C spectrum of a single sunflower seed. MASS simultaneously improved resolution and sensitivity thus keeping the total measuring time relatively short (15 min). Detection of the low-abundance ^{13}C nuclei represents a promising alternative to ^{1}H NMR spectroscopy.

High-resolution ^{13}C spectrum of a single sunflower seed shows only resonances of oil, since other constituents are solid and their signals are too broad to be observed in this version. Broadband ^{1}H decoupling eliminated effects of heteronuclear interactions (J_{CH}), and the carbon peaks are assigned by their characteristic chemical shifts. Terminal CH_3 groups resonate at 14 ppm, while bulk CH_2 groups contribute to peaks in the interval from 22 to 34 ppm. Glyceride carbons are shown at 62 and 69 ppm, and the olefinic ones are shifted to 127 and 129 ppm due to deshielding effects of the double bond. Carbonyl peaks are at 171 ppm.

Since olefinic resonances (=CH–) are characteristic of unsaturated fatty acids, it is possible to use the intensity of peaks at 127 to 130 ppm as indicators of their relative abundance. Oleic acid (C-18:1) contributes to the resonance at 129 ppm, while linoleic acid (C-18:2) has two double bounds and gives rise to two peaks at 127 and 129 ppm, respectively. The ratio between these two fatty acids is determined by comparing intensities (Figure 3). Presence of the linolenic acid (C-18:3) gives rise to an additional peak at approximately 131 ppm and it has been observed in studies of single soybean seeds.[32]

In quantitative evaluation of ^{13}C results, one must consider differences in nuclear Overhauser enhancements and relaxation times. Therefore, it is not possible to establish relative abundances of chemically nonequivalent groups by simply comparing signal intensities. Protonated carbons give rise to stronger responses than nonprotonated ones, because heteronuclear decoupling is essential for high-resolution spectroscopy in seeds. Additionally, segments of fatty acid chains show different mobilities and T_1 values are distributed between 0.2 and 2.3 s as shown by measurements in peanuts.[33] Long waiting times are not acceptable and results must be calibrated by comparing similar samples, so that effects of partial saturation are compensated for.

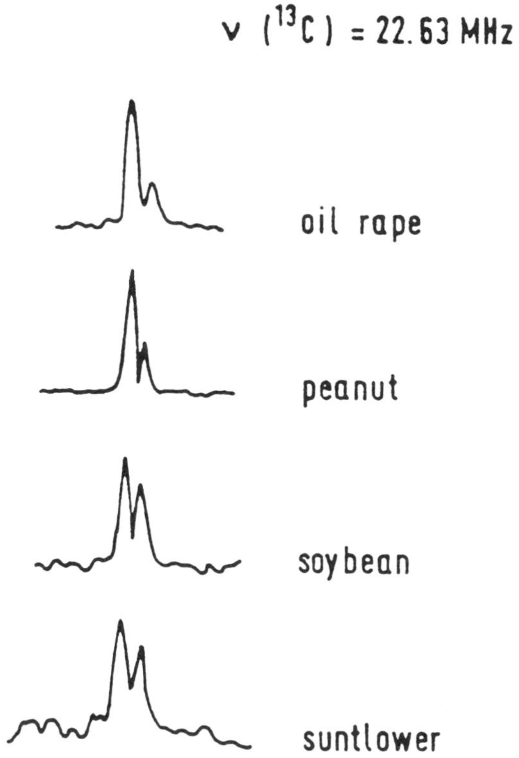

FIGURE 3. Characteristic appearance of olefinic ^{13}C signals at 129 and 127 ppm which characterize oil composition. Note that oil rape seeds contain erucic acid which gives rise to the signal at the same position as oleic acid. Replotted from Reference 33.

It was noticed also that higher magnetic fields ($B_0 > 4$ T) did not improve the quality of spectra, because susceptibility differences broadened lines thus diminishing favorable increases of the signal intensity. Shoolery[35] introduced MASS which successfully reduced line width down to about 2.5 Hz and signals of individual fatty acids became well resolved (Figure 2). This result demonstrated again that NMR visible oil is not immobilized by binding to membranes, since MASS can eliminate only spatial variations in magnetic susceptibility.[29] Effective line-narrowing of signals of oil requires only moderate spinning frequencies (typically less than 1 kHz) which can be readily accomplished even in studies of intact seeds.

The state of oil can be characterized also by two-dimensional NMR techniques which separate information about the spin system along the F_1 and F_2 dimension. Figure 4 illustrates that a special pulse sequence[36] can generate cross-sections along the F_1 dimension which reveal heteronuclear one-bond coupling constants, while responses of chemically nonequivalent groups are separated by ^{13}C chemical shifts along the F_2 dimensions. Overlap of multiplets is avoided and unimportant long-range coupling constants are suppressed in this experiment.

^{13}C NMR will find applications in nondestructive studies of oil composition, especially if mechanical stress arising from MASS does not affect the viability of samples. Liquid state versions can be used also to detect other constituents in intact seeds[37,38] or to follow the process of germination.[39]

V. PROTEIN CONTENT

Since proteins contain nitrogen, detection of ^{14}N or ^{15}N NMR signals would be extremely

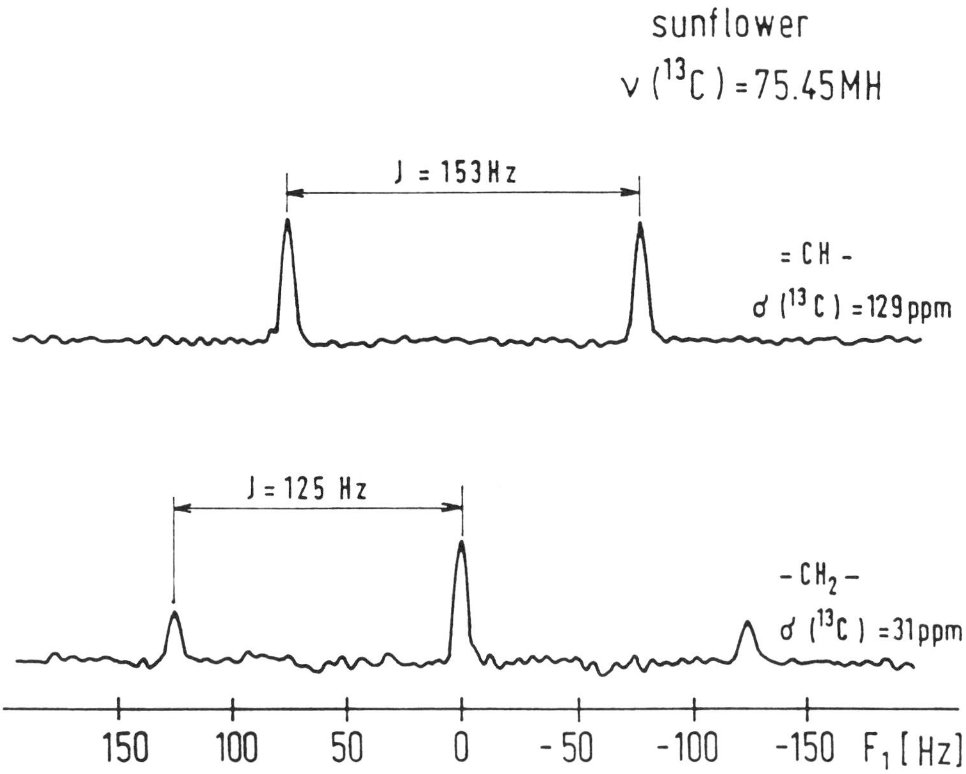

FIGURE 4. Cross-sections along the F_1 dimension as obtained by the 2-D technique which suppresses long-range heteronuclear couplings. Difference between CH and CH_2 groups is clearly shown and overlap of multiplets is avoided. Courtesy T. C. Wong, University of Missouri, Columbia.

useful for detailed studies of seed composition and subsequent breeding for higher nutritional value of new plant varieties. Unfortunately, the abundant isotope, ^{14}N, has spin I = 1 and quadrupolar interactions broaden resonances beyond present detection possibilities. ^{15}N has spin I = $^1/_2$, but its low natural abundance 0.37% imposes severe problems which can be solved only by isotopic labeling.[40] Such techniques will find successful applications in those research laboratories which are following biochemical processes in various plants.

Measurements of protein content for breeding purposes require simple approaches which can be adopted for routine use. Nitrogen resonances have inherent drawbacks and there are no obvious solutions. Therefore, ^{13}C NMR emerges as the only promising method. Since proteins are contained in the solid part of the seed, their presence can be detected only by cross polarization[41] which starts with excitation of the ^1H spin system. ^{13}C spins are brought into contact using additional irradiation fulfilling the Hartmann-Hahn condition[42] and mutual spin flips in the rotating frame relay polarization. This technique enhances ^{13}C signals and it also provides an important distinction between solids and liquids. Since seeds contain oil in a liquidlike form, any simple observation of ^{13}C signals favors detection of narrow resonances (as shown in the previous section). On the other hand, cross polarization utilizes ^{13}C-^1H dipolar interactions which are present only in solids and such signals are selectively enhanced. It is possible to suppress resonances of oil, while carbohydrates, proteins, and other solid constituents become "visible" by ^{13}C NMR.[43,44]

Although strong irradiation of ^1H spins is continued during data acquisition to decouple ^{13}C signals, resonances remain broad due to chemical shift anisotropy. This problem can be solved only by fast rotation around an axis tilted 54° 44′ with respect to the direction of the external magnetic field. Since anisotropic contributions are averaged, highly resolved spectra of intact

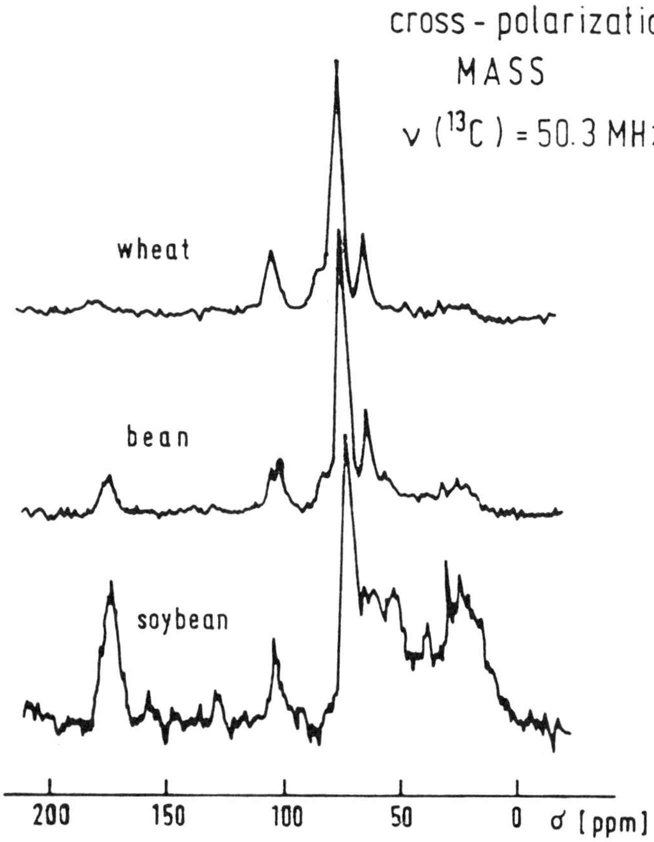

FIGURE 5. MASS spectra showing that the carbonyl peak at ~172 ppm can be used as a relevant indicator of the protein content. Measuring time was about 5 min for each spectrum. Replotted from Reference 45.

seeds can be recorded.[40,45,46] Figure 5 shows some typical results which clearly indicate that the carbonyl peak at ~172 ppm serves as a means to monitor protein content in intact seeds.

MASS spectroscopy has an undisputed potential value and it might become routinely applicable in plant breeding programs. However, reliability of results has not been tested and several problems have been only addressed and are not yet solved.

Obviously, seeds have irregular shapes and they must be rotated with frequencies more than approximately 3 kHz. Single-bearing rotor designs tolerate imbalances and fast spinning can be accomplished on most samples except, for instance, weaker sunflower seeds which may disintegrate as a result of excessive centrifugal forces. Changing of seeds is more awkward, because such probes must be taken out of the magnet. Newer versions utilizing double air bearings allow faster change of samples and they appear more suitable for routine applications. Balancing of the spinner remains one of the problems must be solved before any further progress towards nondestructive protein determinations.

ACKNOWLEDGMENTS

The author of this chapter acknowledges assistance and advice from D. E. Alexander, T. F. Conway, S. Ratkovic, J. N. Shoolery, V. T. Srinivasan, D. Skoric, and T. C. Wong.

REFERENCES

1. **Bloch, F.,** Nuclear induction, *Physiol. Rev.*, 70, 460, 1946.
2. **Bloembergen, N., Purcell, E. M., and Pound, R. V.,** Relaxation effects in nuclear magnetic resonance absorption, *Physiol. Rev.*, 73, 679, 1948.
3. **Kroeker, R. M. and Henkelman, R. M.,** Analysis of biological NMR relaxation data with continuous distributions of relaxation times, *J. Magn. Reson.*, 69, 218, 1986.
4. **Ratkovic, S.,** Proton NMR of maize seed water: the relationship between spin-lattice relaxation time and water content, *Seed Sci. Technol.*, 15, 147, 1987.
5. **Aksenov, S. I., Askochenskaya, N. A., and Golovina, E. A.,** The state of water studied in seeds of different qualitative composition and its changes upon the action of temperature, *Fiziol. Rast.*, 24, 1251, 1977.
6. **Bauman, L. F., Conway, T. F., and Watson, S.A.,** Heritability of variations in oil content of individual corn kernels, *Science*, 139, 498, 1963.
7. **Conway, T. F. and Earle, F. R.,** Nuclear magnetic resonance for determining oil content of seeds, *J. Am. Oil Chem. Soc.*, 40, 265, 1963.
8. **Zupancic, I., Vrscaj, S., Porok, J., Levstik, I., Erzen, V., Blinc, R., Pausek, S., Ehrenberg, L., and Dumanovic, J.,** Transient NMR selection method in plant breeding, *Acta Chem. Scand.*, 21, 1664, 1967.
9. **Hahn, E. L.,** Spin echoes, *Physiol. Rev.*, 80, 580, 1950.
10. **Carr, H. Y. and Purcell, E. M.,** Effects of diffusion on free precession in nuclear magnetic resonance experiments, *Physiol. Rev.*, 94, 630, 1954.
11. **Meiboom, S. and Gill, D.,** Modified spin-echo method for measuring nuclear relaxation times, *Rev. Sci. Instrum.*, 29, 688, 1958.
12. **Bauman, L. F., Conway, T. F., and Watson, S. A.,** Inheritance of variations in oil content of individual corn (*Zea mays* L.) kernels, *Crop Sci.*, 5, 137, 1965.
13. **Belikov, I. F. and Sazonenko, M. K.,** Determination of fat in soybean seeds using nuclear magnetic resonance, *Dokl. Akad. Nauk SSSR*, 167, 225 1966.
14. **Alexander, D. E., Silvela, L. S., Collins, F. I., and Rodgers, R. C.,** Analysis of oil content of maize by wide-line NMR, *J. Am. Oil Chem. Soc.*, 44, 555, 1967.
15. **Collins, F. I., Alexander, D. E., Rodgers, R. C., and Silvela, L. S.,** Analysis of oil content of soybeans by wide-line NMR, *J. Am. Oil Chem. Soc.*, 44, 708, 1967.
16. **Nikolic-Vig, V., Skoric, D., and Bedov, S.,** Variability of oil and husk percentage in sunflower seed of varietal populations of Peredovik and Vniimka 8931 and their heritability, *Savremena poljoprivreda*, 19, 23, 1971.
17. **Conway, T. F., Hickman, C. J., Kurtz, F. A., and Winkler, G. A.,** Computer controlled wide-line nuclear magnetic resonance system for automatically determining total oil content of individual maize seeds, presented at 18th Ampere Congress, Nottingham, England, 1974.
18. **Tiwari, P. N., Gambhir, P. N., and Rayan, T. S.,** Rapid and nondestructive determination of seed oil by pulsed nuclear magnetic resonance technique, *J. Am. Oil Chem. Soc.*, 51, 104, 1974.
19. **Robertson, J. A. and Morrison, W. H., III,** Analysis of oil content of sunflower seed by wide-line NMR, *J. Am. Oil Chem. Soc.*, 56, 961, 1979.
20. **Srinivasan, V. T.,** A comparison of different pulse sequences in the nondestructive estimation of seed oil by pulsed nuclear magnetic resonance technique, *J. Am. Oil Chem. Soc.*, 56, 1000, 1979.
21. **Miller, R. L., Dudley, J. W., and Alexander, D. E.,** High intensity selection for percent oil in corn, *Crop Sci.*, 21, 433, 1981.
22. **Ratkovic, S., Denic, M., Lahajnar, G., and Zupancic, I.,** Biological systems with low content: NMR approach to the state of water in plant seed, in *Biophysics of Water*, Franks, F. and Mathias, S., Eds., John Wiley & Sons, Chichester, England, 1982, 312.
23. **Ratkovic, S.,** The use of pulsed ^{1}H NMR techniques in breeding new high oil hybrids of maize, in *New Techniques and Applications of Magnetic Resonance*, Proc. 7th Ampere International Summer School, Blinc, R. and Vilfan, M., Eds., Portoroz, 1982, 325.
24. **Srinivasan, V. T., Singh, B. B., Chidambareswaran, P. K., and Sundaram, V.,** Cottonseed oil estimation by pulsed nuclear magnetic resonance technique, *J. Am. Oil Chem. Soc.*, 62, 1021, 1985.
25. **Wang, W. T., Hou, K. Q., and Mayinur, Y.,** A new method of measuring oil content by pulsed NMR, *Uljarstvo*, 22, 303, 1985.
26. **Levitt, M. H. and Freeman, R.,** NMR population inversion using a composite pulse, *J. Magn. Reson.*, 33, 473, 1979.
27. **Weber, E. J. and Alexander, D. E.,** Breeding for lipid composition in corn, *J. Am. Oil Chem. Soc.*, 52, 370, 1975.
28. **Johnson, L. F. and Shoolery, J. N.,** Determination of unsaturation and average molecular weight of natural fats by nuclear magnetic resonance, *Anal. Chem.*, 34, 1136, 1962.
29. **Garroway, A. N.,** Magic-angle sample spinning of liquids, *J. Magn. Reson.*, 49, 168, 1982.

30. **Conway, T. F. and Johnson, L. F.,** Nuclear magnetic resonance measurement of oil "unsaturation" in single viable corn kernels, *Science*, 164, 827, 1969.
31. **Shoolery, J. N.,** ^{13}C NMR studies of oil composition in viable seeds, Varian Application Note NMR-73-3, Varian, Palo Alto, CA, 1973.
32. **Schaefer, J. and Stejskal, E. O.,** Carbon-13 nuclear magnetic resonance measurement of oil composition in single viable soybeans, *J. Am. Oil Chem. Soc.*, 51, 210, 1974.
33. **Rutar, V., Burgar, M., Blinc, R., and Ehrenberg, L.,** ^{13}C NMR determination of the oil composition in individual plant seeds, *J. Magn. Reson.*, 27, 83, 1977.
34. **Shoolery, J. N.,** Applications of high resolution nuclear magnetic resonance to the study of lipids, Proc. 1981 AOCS Conference on Dietary Fats and Health, American Oil Chemists' Society, Chicago, December 6 to 11, 1981, 220.
35. **Shoolery, J. N.,** private communication.
36. **Rutar, V.,** Utilization of selective spin flips for manipulated two-dimensional NMR spectroscopy, *J. Magn. Reson.*, 56, 87, 1984.
37. **Kainosho, M.,** ^{13}C NMR studies of the intact plant tissues. Cytoplasmic aucubin and sucrose in a single seed of *Aucuba japonica*, *Tetrahedron Lett.*, 47, 4279, 1976.
38. **Kainosho, M. and Konishi, H.,** ^{13}C nuclear magnetic resonance spectrum of dried star anise fruits and its histological implications, *Tetrahedron Lett.*, 51, 4757, 1976.
39. **Colnago, L. A. and Seidl, P. R.,** Application of carbon-13 nuclear magnetic resonance to the germination of soybean seeds *in vivo*, *J. Agric. Food Chem.*, 31, 459, 1983.
40. **Schaefer, J., Stejskal, E. O., and McKay, R. A.,** Cross-polarization NMR of N-15 labeled soybeans, *Biochem. Biophys. Res. Commun.*, 88, 274, 1979.
41. **Pines, A., Gibby, M. G., and Waugh, J. S.,** Proton-enhanced NMR of dilute spins in solids, *J. Chem. Phys.*, 59, 569, 1973.
42. **Hartmann, S. R. and Hahn, E. L.,** Nuclear double resonance in the rotating frame, *Phys. Rev.*, 128, 2042, 1962.
43. **Schaefer, J. and Stejskal, E. O.,** Carbon-13 nuclear magnetic resonance analysis of intact oilseeds, *J. Am. Oil Chem. Soc.*, 52, 366, 1975.
44. **Rutar, V. and Blinc, R.,** Nondestructive determination of protein content of viable seeds by proton enhanced ^{13}C NMR, *Z. Naturforsch.*, 35c, 12, 1980.
45. **Rutar, V.,** A new possibility for nondestructive protein content determinations in viable seeds, *Appl. Spectrosc.*, 36, 259, 1982.
46. **Haw, J. F. and Maciel, G. E.,** Carbon-13 nuclear magnetic resonance spectrometry of oil seeds with cross polarization and magic-angle sample spinning, *Anal. Chem.*, 55, 1262, 1983.

In Vitro Solution NMR Studies (1D and 2D High-Resolution Methods)

Chapter 5

THE USE OF NMR IN STRUCTURE DETERMINATION OF BIOACTIVE NATURAL PRODUCTS DERIVED FROM PLANTS

David Weisleder

TABLE OF CONTENTS

I. INTRODUCTION

Until the current decade most nuclear magnetic resonance (NMR) spectrometers including the then state-of-the-art pulsed, Fourier transform machines were limited to routine proton (^{1}H) and carbon-13 (^{13}C) chemical shift determinations. Experimental measurements then available usually involved one or at most two pulses and are commonly referred to today as, one dimensional (1D) experiments. The relatively low sensitivity of older iron magnet-based spectrometers placed lower limits on the sample size necessary to obtain quality data. To determine the number of protons attached to each carbon, an off-resonance experiment was required.[1] The off-resonance experiment often took a longer time than the original experiment and sometimes resulted in a difficult to interpret overlap of resonances.

The advent of superconducting high-field spectrometers operating at ^{1}H observation frequencies greater than 200 MHz, and their widespread dissemination, has greatly expanded experimental possibilities. Experiments to determine the number of protons attached to a particular carbon can now be carried out in much less time, and with less ambiguous results. Because NMR sensitivity increases exponentially with field strength, the sample size, or lack thereof, is not as great a problem as it once was. Beside the high-field magnets, NMR spectrometers have been equipped with more sophisticated computers and software. The ever-increasing array of pulse sequences made possible by computer control has opened up many approaches to the structure determination of natural products. Some of these approaches will be the subject of this chapter.

The implementation of multipulse techniques in high-resolution NMR dates back to the mid-1970s and the independent work of Ernst[2-4] and Freeman[5,6] and their many co-workers. Several recent reviews[7-9] cover the theory of the experiments we will describe in this chapter. Many of these new pulse sequences are made possible by the increasingly sophisticated computers and software which now control every aspect of the NMR spectrometer. Each NMR manufacturer uses a unique computer and software. NMR data is not readily transferable between spectrometers of different manufacturers thereby making communication among scientists difficult. Because advances take place more rapidly in computer technology than in spectrometer design, it is advantageous to be able to efficiently transfer NMR data to an off-line computer. Standard means of communication between computers will enable NMR data to be processed more rapidly. Taking into account software bugs, ambiguities, and high fog index, it is sometimes difficult to know just what is happening to a particular sample in the spectrometer.

In this chapter experiments and strategies employed in structure determination of natural products will be illustrated with examples taken from the most recent work available. The goal is not to turn the reader into a spectroscopist but to take away some of the intimidation on the part of those who must interact with NMR spectroscopists. Frequently organic chemists need to consult with NMR spectroscopists about strategies for obtaining and interpreting data. A varied selection of experiments, some yielding plots of seemingly endless squiggles, must be considered and a strategy adopted to best utilize the spectrometer time. The recognition of a need to organize the various multipulse techniques into an analytical strategy was comprehensively described by Hall and Sanders.[10]

II. NMR STRATEGY

Table 1 is a listing of some favored NMR experiments. The list is not meant to be an exhaustive survey of those experiments currently in vogue. Table 1 should serve as a menu and starting point. The experiments represent a good beginning for an efficient approach to the structure determination problems arising in laboratories dealing with small (molecular weight under 1000), naturally occurring, active molecules.

The experimental techniques will be discussed and illustrated in subsequent sections. One may choose, as in certain restaurants, from selections under A or B. In general an overall proton

Table 1
NMR STRATEGY FOR STRUCTURE DETERMINATION

A. Proton NMR
- Standard spectrum
- 2D shift correlated (COSY)
- Standard decoupling
- NOE difference experiment

B. Carbon NMR
- Standard decoupled spectrum
- DEPT
- 2D ^{1}H-^{13}C correlation
- Selective decoupling

spectrum would be obtained first to ascertain if the peaks are sharp, well defined, and reasonably spaced. The integral would be checked for any inconsistencies which might indicate if the sample contained one or a mixture of components. After the proton spectrum, the broadband decoupled ^{13}C spectrum is usually collected. Again the spectrum is checked to note where all the resonances are located before deciding what other experiments are required to answer any structural questions. At this point some of the multipulse and two-dimensional (2D) experiments may be contemplated and employed as we will see below.

A. Two-Dimensional NMR Spectroscopy

The extensive use of 2D NMR spectroscopy in structure determination requires a brief practical explanation of this important technique. All 2D NMR experiments are made possible by pulse sequences. Pulse sequences share certain characteristics. Each pulse sequence consists of distinct time periods commonly referred to as the preparation, evolution, and detection time. Some experiments also require a mixing time period. The preparation time is the time needed to allow the nuclei to return to some reproducible state before the pulse sequence can be repeated. A pulse sequence is repeated to allow the phase of the various pulses to be altered in some predetermined manner necessary to produce the desired information. The sequence is also repeated to insure that an adequate signal-to-noise ratio is obtained. After a sufficient number of repetitions of the sequence, the data is stored in one of what will become a series of files. During the evolution period, the nuclei evolve as a function of t_1, the evolution time. The evolution time is increased in equal increments, which distinguishes each data file in the series of files that makes up the 2D experiment. The detection period (t_2) follows the last pulse in the sequence and is the acquisition time of the experiment. The acquired signal is a function of both t_1 and t_2. A double Fourier transformation of the data with respect to t_2 and t_1 converts the information into a matrix with frequencies F_2 and F_1. Depending on the pulse sequence selected, the 2D spectrum can yield structural information by correlating the chemical shifts for coupled nuclei in either a homonuclear or heteronuclear experiment as described below.

III. TECHNIQUES

A. DEPT

Of all the experiments listed in Table 1 the most generally useful one is the DEPT sequence. DEPT stands for distortionless enhancement by polarization transfer as defined by Doddrell et al.[11-14] This multipulse sequence is representative of the current pace of NMR developments mentioned in the introduction. In the standard DEPT sequence, shown in Figure 1, both protons and carbons are pulsed in order to bring about polarization transfer from protons to carbon.

The advantages of DEPT will become apparent in our discussion of specific examples below, but can be summarized as follows: when the signal-to-noise ratio is measured after the same

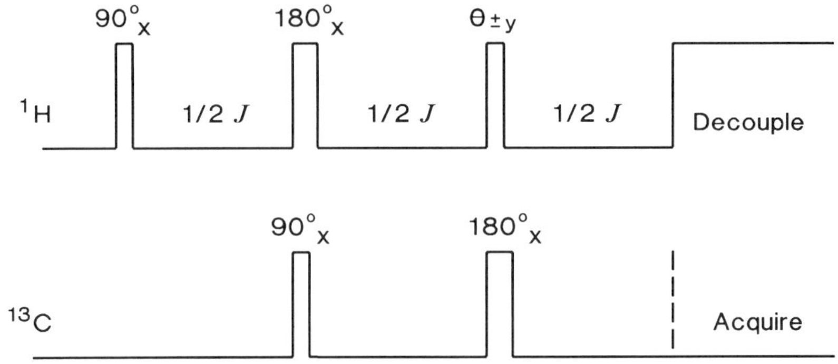

FIGURE 1. Pulse sequence for DEPT experiment. The final proton pulse is varied (45, 90, or 135°) to obtain a set of files which can be combined to yield edited spectra. If the final proton pulse is 135° a spectrum is obtained in which the methylene carbon resonances are inverted relative to the methine and methyl carbon resonances.

number of scans, the DEPT sequence results in a nearly threefold increase in sensitivity compared to the usual broadband-decoupled ^{13}C acquisition, plus the number of protons attached to each carbon is immediately available by inspection.

A simple computer program which automatically obtains, processes, and edits DEPT spectra into the individual methine, methylene, and methyl subspectra should be included in all software packages available from the various manufacturers. Figure 2 shows the edited DEPT spectra, as well as the complete ^{13}C spectrum of the mycotoxin T-2 (**1**) obtained in our laboratory with spectrometer automation.

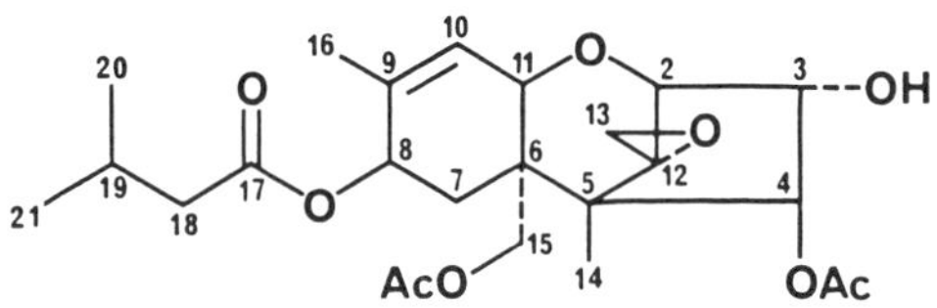

1

T-2 was originally obtained from the toxic fungus, *Fusarium tricinctum*, isolated from moldy corn in the mid-1960s.[15] The primary virtues of editing a ^{13}C spectrum into individual components are illustrated in Figure 2, that is, the low field methylene (C-15) is easily distinguished and the high field methine (C-19) which could otherwise be thought to be a methyl or methylene resonance is unambiguously assigned. The ability to recognize the various carbon multiplicities is the first important step in determining structure.

Recently DEPT has been extended to 2D NMR experiments,[16-18] the final proton flip has been changed to 30, 90, or 150° so that the data can be combined on a 1:1 basis, instead of using a set of empirically derived coefficients.

DEPT is only one of a number of sequences which take advantage of polarization transfer from protons to carbons or other less sensitive nuclei.[19] The increase in sensitivity to be gained by polarization transfer techniques makes possible results such as those illustrated in Figure 3.

The sample which gave rise to the DEPT spectrum in Figure 3 was a mixture of two anomeric forms of the monosaccharide, 4-*O*-[1-carboxyethyl]-D-galactose (**2**) and contained a total of 2.3 mg.

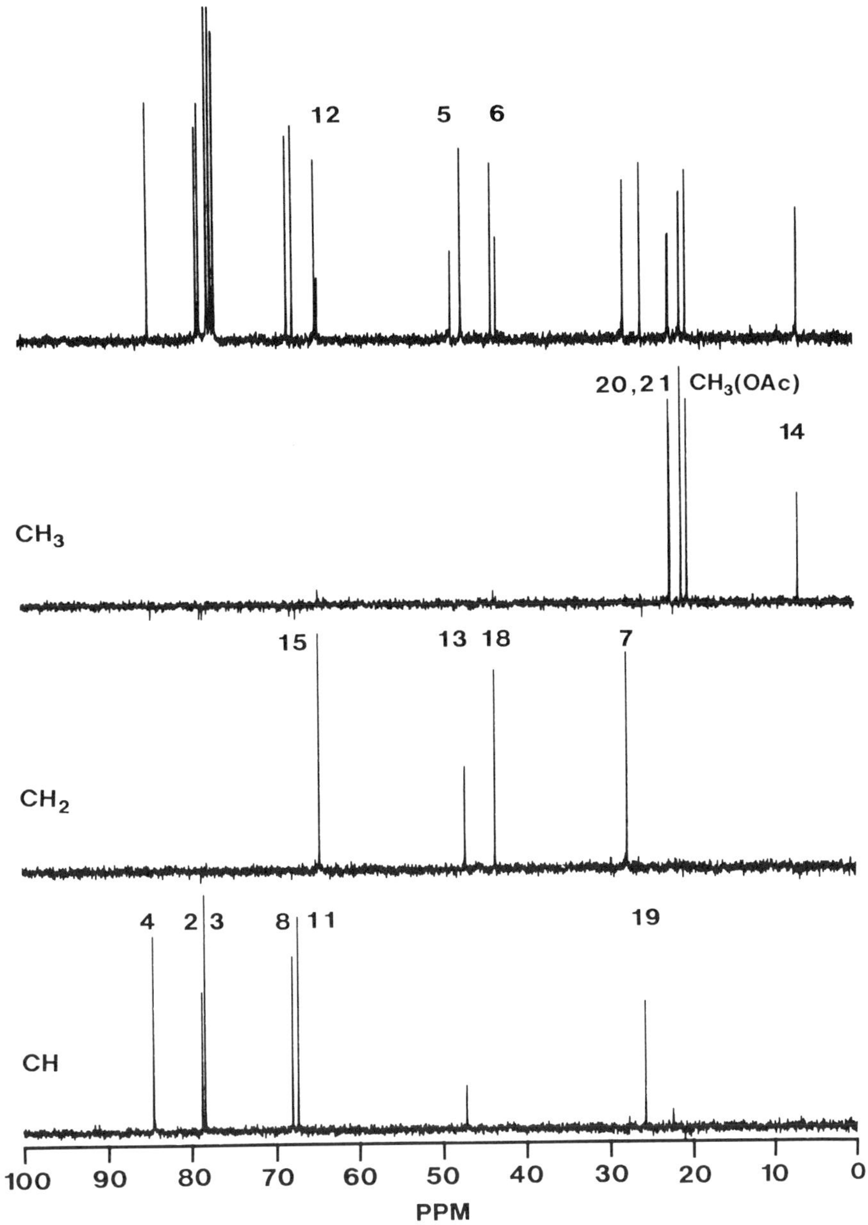

FIGURE 2. High-field portion of the 75.47 MHz-edited DEPT spectra for T-2 (Structure 1). The top spectrum is the normal ${}^{13}C$ spectrum. The subspectra are labeled according to the types of carbons exhibited.

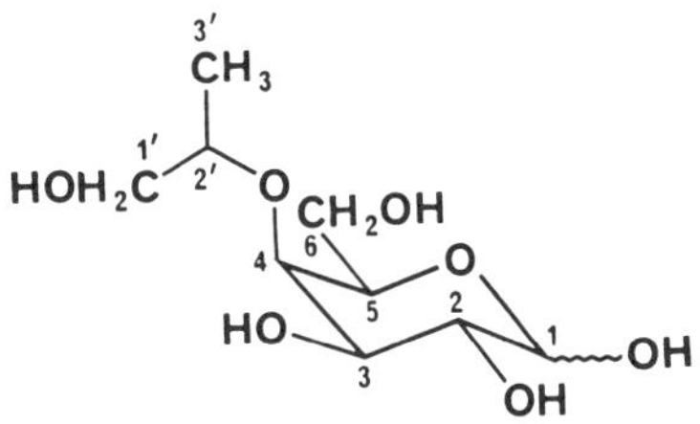

2

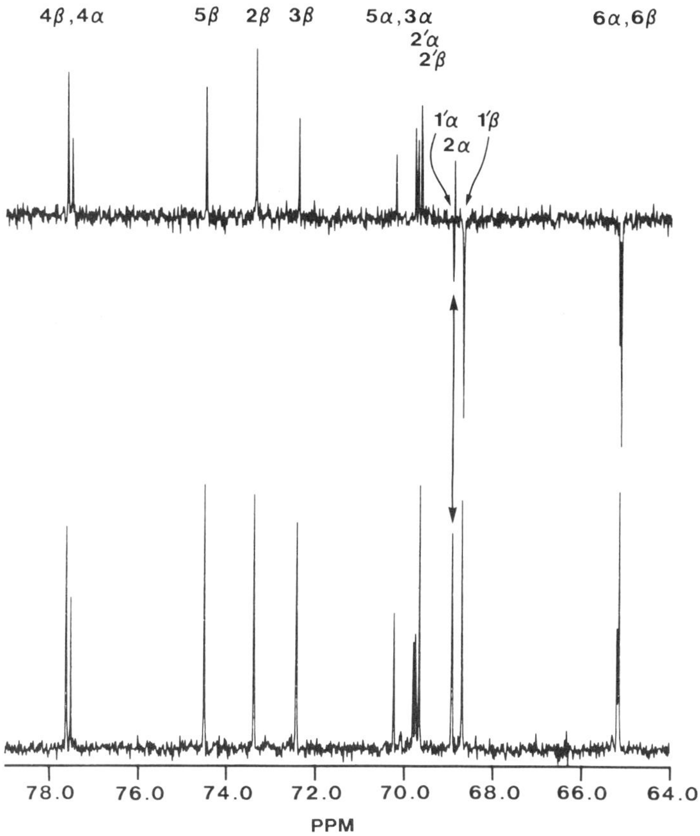

FIGURE 3. An expanded region of the ^{13}C and DEPT spectrum of a mixture of the α and β anomers of 2. Note that the peak at 68.9 ppm is made up of a methylene and methine carbon.

Table 2
[13]C CHEMICAL SHIFTS OF THE α AND β ANOMERS OF 4-*O*-[1-CARBOXYETHYL]-D-GALACTOSE

	1	2	3	4	5	6	1′	2′	3′
α	92.9	68.9	69.8	77.5	70.2	65.2	68.9	69.8	16.0
β	97.0	73.4	72.4	77.6	74.5	65.1	68.7	69.6	16.0

Compound **2** is a major acidic sugar component obtained from the extracellular polysaccharide produced by a strictly anaerobic bacterial species, *Butyrivibrio fibrisolvens,* commonly found in cecal or ruminal contents of mammals.[20] Utilizing the DEPT sequence with a final proton pulse of 135°, the spectrum (Figure 3) was obtained which enables one to determine where all the carbons, of each anomer, resonate. A single peak appears at 68.9 ppm in the carbon spectrum (Table 2). The DEPT spectrum reveals the presence of two carbons with differing multiplicities. Because the anomeric ratio is approximately 2:1 in favor of the β anomer, the two carbon resonances at 68.9 ppm must both belong to the α anomer.

B. COSY[21]

Following the examination of the usual ^{1}H, ^{13}C, and DEPT spectra, it is the usual practice in our laboratory to obtain a proton correlation spectrum. In this experiment which is commonly

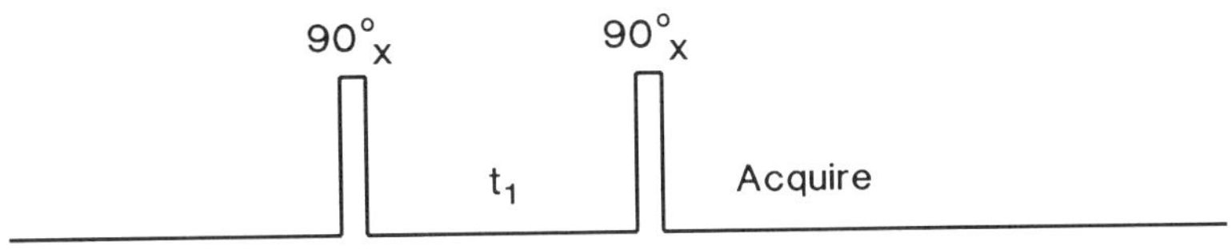

FIGURE 4. Pulse sequence for COSY experiment.

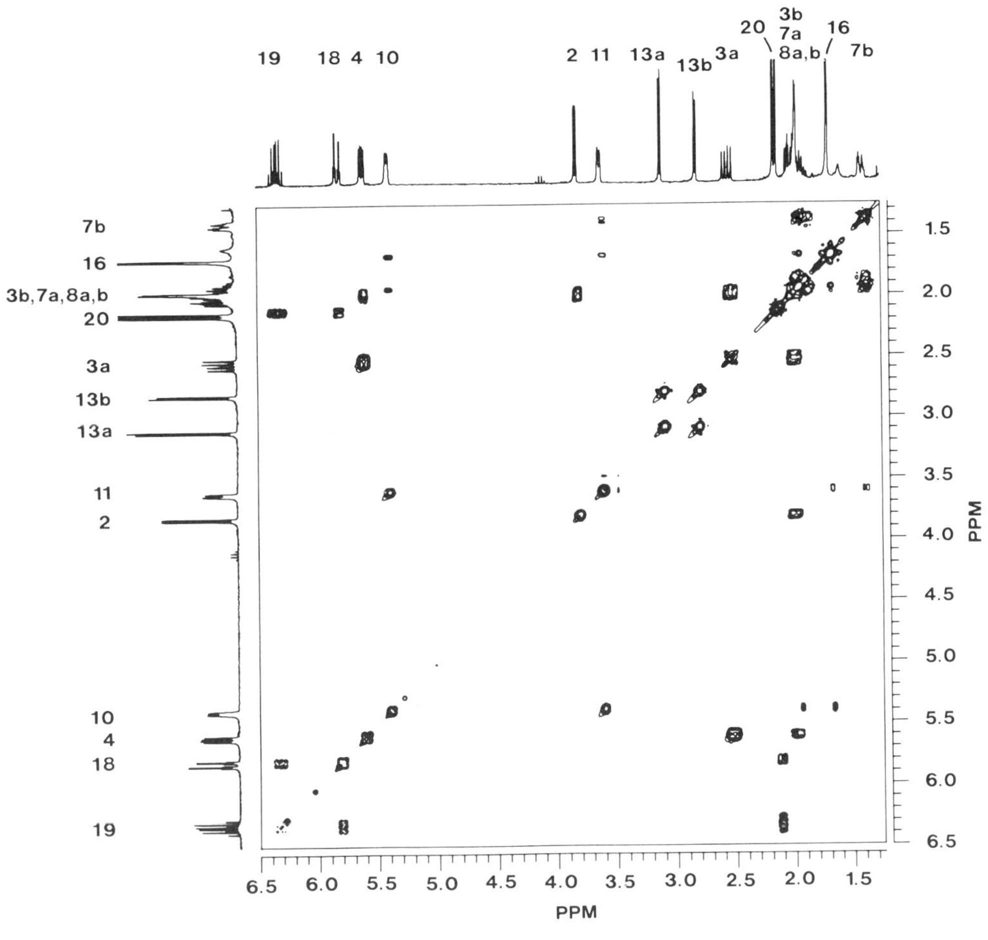

FIGURE 5. COSY spectrum of trichothec-9-en,12,13-epoxy-4-[1-oxo-2-butenyl)oxy] (Structure 3) obtained with pulse sequence.

referred to as COSY (correlation spectroscopy), the chemical shifts of all the protons can be correlated with each other.

The simplest pulse sequence for COSY is shown in Figure 4. The variable time delay, t_1, gives rise to a series of data files. Magnetization is transferred between coupled nuclei when the pulse is applied following the variable delay.

In the usual presentation of the data as a contour plot as in Figure 5, the chemical shifts appear on the diagonal and cross-peaks are symmetrically displaced from the diagonal by the chemical shift difference of the mutually coupled protons. In theory the entire coupling network of the molecule can be traced out provided enough chemical shift dispersion is present.

The COSY spectrum illustrated in Figure 5 was obtained to aid in the structure determination of a new tricothecene,[22] trichothec-9-en,12,13-epoxy-4-[(1-oxo-2-butenyl)oxy] (**3**).

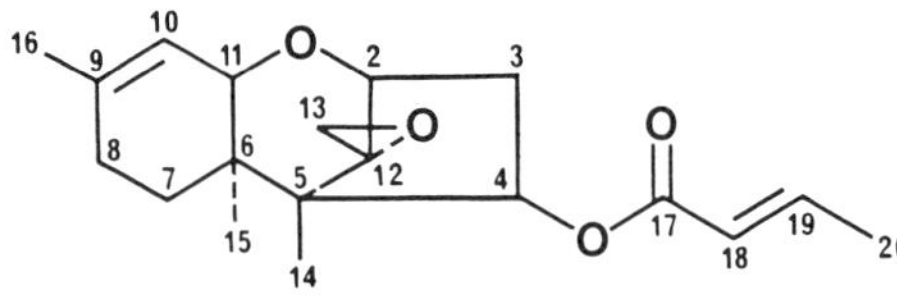

3

Trichothecenes are fungal secondary metabolites. These naturally occurring cereal crop contaminants are capable of producing toxicity in animals and humans. Compound **3**, which contains an ester at C-4 and lacks oxygenation at C-8, has been obtained from liquid culture of two isolates of *Tricothecium roseum* from soybean seeds. Proton assignments were verified with the COSY spectrum. The resonance position of H-4 is known from analogous compounds and it can be seen that H-4 is coupled to two nonequivalent C-3 protons. It can also be seen that the H-2 proton is coupled to only one of the C-3 protons. Nonequivalent protons on C-7 are found at 2.00 and 1.44 ppm.

There are many versions of the basic COSY experiment,[23-27] but for our purposes we can consider just a few of the most common sequences. The basic experiment (Figure 4) consists of a 90° pulse followed by a variable delay (t_1) and a second 90° pulse which allows for magnetization transfer between mutually coupled protons. A variation of COSY has a 45° pulse replacing the second 90° pulse. In the contour plot of a COSY 45 experiment, signals near the diagonal are diminished giving a "cleaner" and easier to interpret spectrum. Another variation of COSY introduces a small, (<0.5 s) fixed delay before and after the second pulse. The advantage of the fixed delay is that it allows the direct observation of long-range couplings which are very important in determining conformation and/or configuration in molecules. No other NMR method allows the direct observation of the structurally important long range couplings. Other variations of the COSY experiment exist,[23-27] which suppress large singlet peaks and improve cross-peak resolution. COSY variations have many general uses especially in very large molecules but we need not concern ourselves with the experimental details here. Suffice it to say that a COSY experiment can yield a wealth of structural information in one spectrum.

C. Homonuclear Decoupling

One of the oldest NMR techniques that is often neglected is the use of simple 1D proton-proton decoupling experiments, especially the difference decoupling techniques so elegantly reviewed by J. K. M. Sanders and J. D. Mersh.[28] It takes a little more time and effort to set up the decoupling experiments than to obtain a COSY spectrum and has therefore become less fashionable. For the sake of completeness, it is not a bad idea to check some COSY observations with the simple homonuclear decoupling technique.

An important variation of the difference decoupling technique is the nuclear Overhauser effect (NOE) difference decoupling experiment. In the difference NOE decoupling experiment a very weak decoupling power is applied to one and only one proton, other protons which are spatially close enough to affect the spin lattice relaxation of the irradiated nucleus become enhanced. The NOE depends on the interatomic distance of the affected protons and the correlation time. Because the NOE is inversely proportional to the interatomic distance raised to the sixth power, it is necessary to measure enhancements as small as 1%. Again the importance of computer technology has been shown in making possible measurements of very small NOE effects. An application of NOE difference spectroscopy to aid in the solution of a structure problem is illustrated in Section IV.

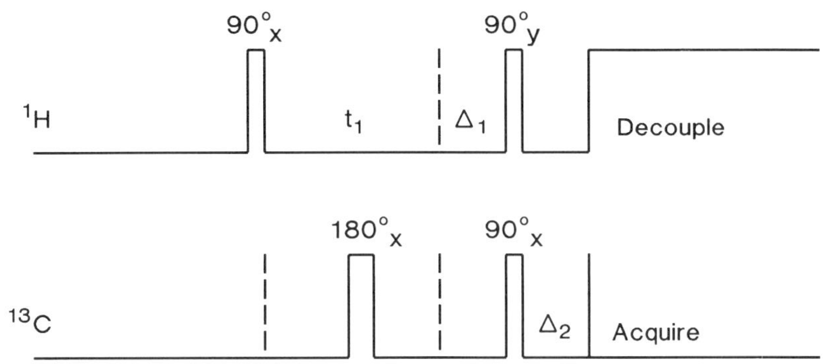

FIGURE 6. Pulse sequence for the heteronuclear correlation experiment.

D. Heteronuclear Shift Correlation[29-33]

Once the proton assignments have been established as far as possible with COSY and verified with decoupling, the next logical experiment to be done is the 2D heteronuclear shift correlation. The sequence, which has no commonly accepted name, is illustrated in Figure 6.

This 2D experiment requires more sample than the ones previously discussed because a series of ^{13}C spectra must be obtained. The experimental time thus depends on the minimum number of scans per ^{13}C spectrum which yields an adequate signal-to-noise ratio with appropriate chemical shift resolution in both dimensions. Figure 7 is the 2D heteronuclear correlation for *N*-methyl loline (**4**).

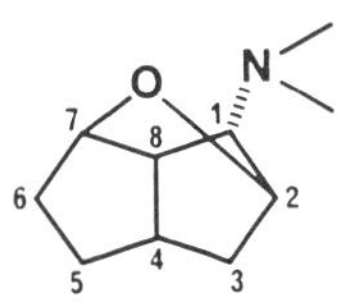

4

N-Methyl loline is derived from loline, an alkaloid found in tall fescue seed.[34] Responses of cattle grazing on tall fescue, such as decreased weight gain, increased respiration rate and temperature, and reduced intake of fescue may be related to the alkaloid content. The 2D heteronuclear shift correlated contour plot (Figure 7) typically has the ^{13}C shifts along the horizontal axis and the proton shifts appear along the vertical axis. As can be seen from Figure 7, knowledge of the proton assignments obtained by a COSY experiment combined with ^{13}C multiplicities obtained by the DEPT experiment enable ^{13}C assignments to be made quite readily without resorting to arguments based on chemical shift inferences. Cross-peaks in the spectrum indicate carbon-proton bonding. The H-1 and H-2 protons were assigned with the aid of a COSY spectrum. A DEPT spectrum identified all the methine carbons. The 2D heteronuclear shift correlation spectrum of 4 (Figure 7) establishes which carbon is C-1 and which is C-2, as well as all other carbon assignments in one elegant experiment.

As in many other 2D techniques variations of the heteronuclear correlation experiment occur. Nonprotonated carbons can be assigned by long-range proton-carbon couplings.[35-37] It has been known for some time that the sensitivity of the heteronuclear correlation experiment could be increased. Recent literature[38-40] has shown that use of inverse polarization transfer with proton detection can dramatically increase the sensitivity of the experiment. When the hardware to produce spectra with the new techniques becomes generally available, the 2D heteronuclear correlation will be a standard experiment to run on every sample requiring chemical shift assignments.

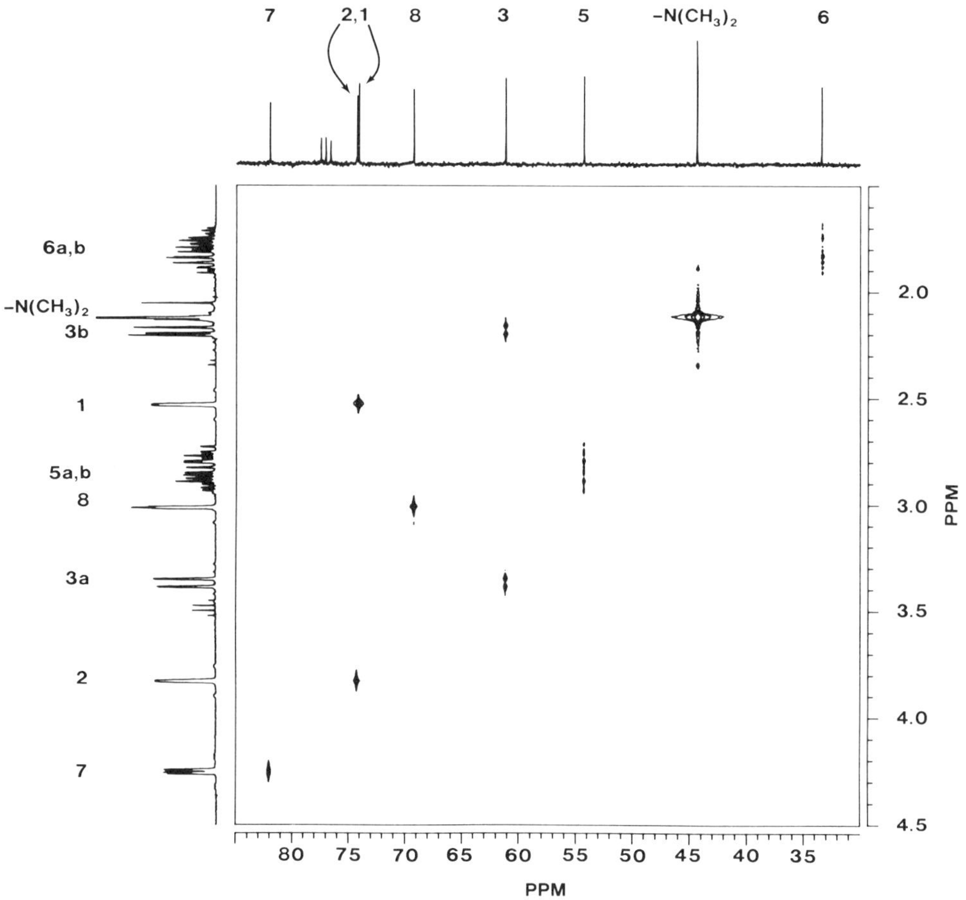

FIGURE 7. $^{13}C/^{1}H$ Correlation spectrum of *N*-methyl loline (Structure 4).

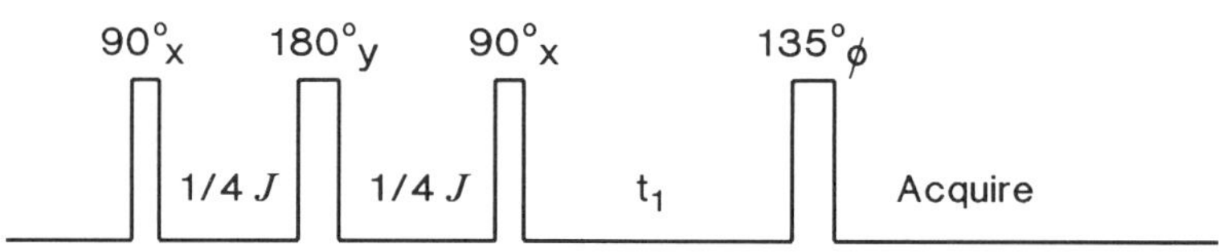

FIGURE 8. Pulse sequence for the INADEQUATE experiment.

E. INADEQUATE[41-43]

The INADEQUATE (incredible natural abundance double quantum transfer experiment) experiment as is true for many multipulse sequences has several variations. The 2D INADEQUATE sequence (Figure 8) is not a listed experiment in Table 1.

At present a 2D INADEQUATE experiment requires so much material that naturally occurring active compounds which are usually found in very low concentration are generally available in insufficient quantities to obtain a spectrum in a reasonable time. The reason that a great deal of material is necessary is because the experiment establishes carbon to carbon connectivity via observation of coupling between adjacent carbon-13 nuclei while suppressing the strong signals of the isolated carbon-13 nuclei. When only unenriched samples are available, approximately one in ten thousand molecules will contain adjacent ^{13}C nuclei. Thus the anomaly exists in that the potentially most useful experiment of all is hardly ever employable in situations

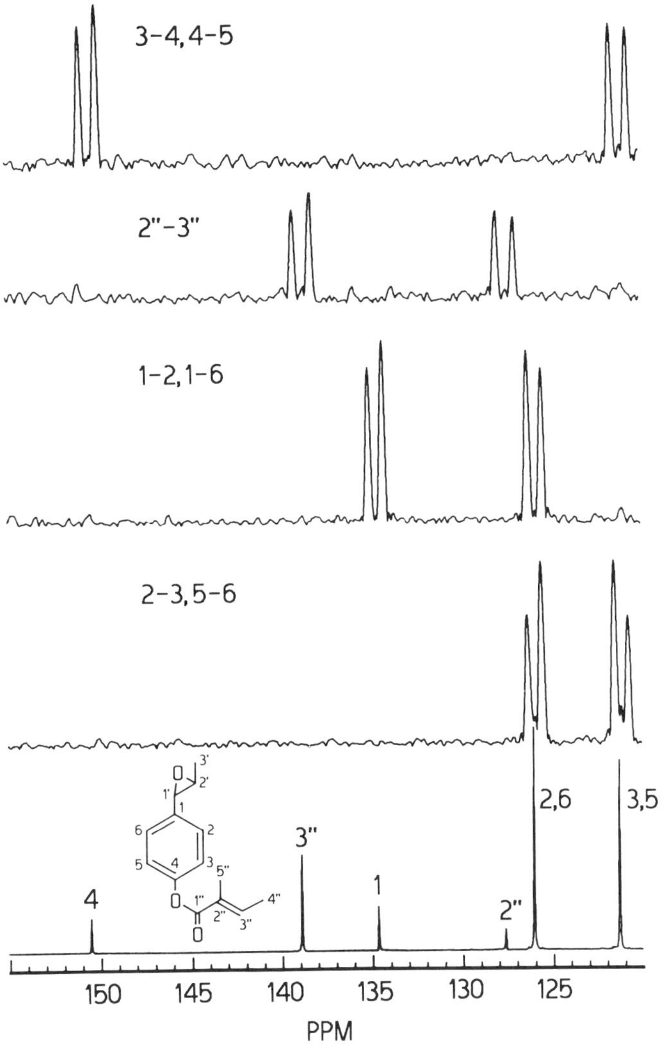

FIGURE 9. Cross-sections of the 2D INADEQUATE spectrum of (Structure 5).

where the structure is truly unknown. The way a 2D INADEQUATE experiment can be applied, to unambiguously determine the carbon skeleton when sufficient sample is available, is illustrated in Figure 9 for compound **5**.

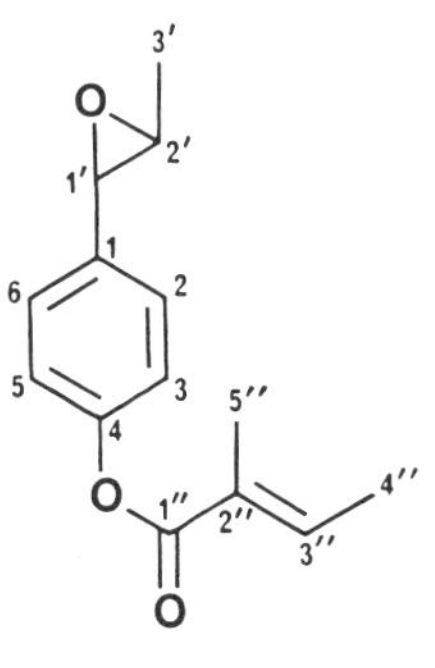

5

Compound **5** belongs to a group of phenylpropenoids isolated from various *Pimpinella* plant species. Compounds which occur with epoxy substituents have been shown to possess antigermination activity.[44] Because other compounds isolated from the same plant species had the two aromatic substituents in the *ortho* positions, rigorous proof was needed to establish that the substituents shown in structure **5** were indeed *para*. Figure 9 contains the cross-sections obtained from the contour plot of the 2D INADEQUATE spectrum of compound **5**. The oxygenated aromatic carbon is assigned to the resonance at 150.5 ppm and is the starting point for establishing the carbon skeleton. It can be seen in the top cross-section of Figure 9 that the lowest field aromatic carbon is connected to the carbons with shifts centered at 121.3 ppm. Because every aromatic carbon must by necessity be connected to two other carbons, the carbons centered at 121.3 ppm must be on either side of the oxygenated aromatic carbon and thus the substitution pattern must be *para* and not *ortho*.

IV. STRUCTURE DETERMINATION

To illustrate how the various modern NMR techniques interact we will consider the structure determination of the naturally occurring compounds which have been given the trivial names, drummondone A (**6**) and drummondone B (**7**).[45] Structures **6** and **7** were determined, almost exclusively, by NMR.

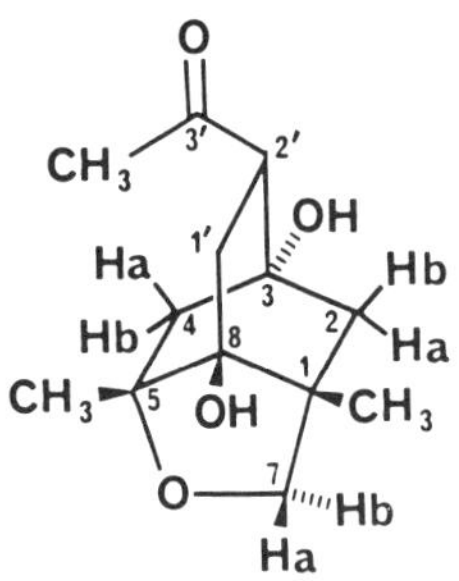

6

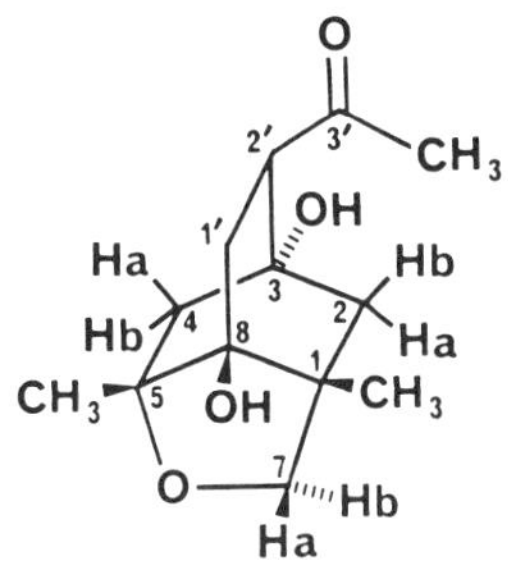

7

The drummondones were isolated from extracts of seeds of *Sesbania drummondii*, a perennial shrub or tree inhabiting the coastal plain from Florida to Texas. Drummondol (**8**) a previously isolated compound served as a model compound for the study of the drummondones. Seeds of *S. drummondii*, also known as coffee bean, rattlebush, and rattlebox, are toxic to livestock. Our interest in the seed extracts stemmed from the fact that pronounced antitumor activity was also found. The compound giving rise to the antitumor activity was identified as sesbanimide (**9**).[46] Abcisic acid (**10**), is an important plant growth regulator.[47] Vomifoliol (**11**), drummondol (**8**), drummondone A (**6**), and drummondone B (**7**) are catabolites of abcisic acid which have been isolated in our studies.

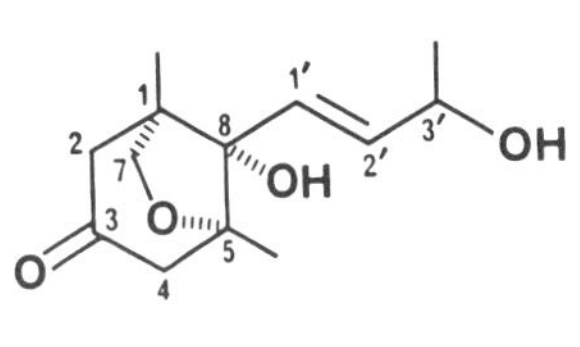

8

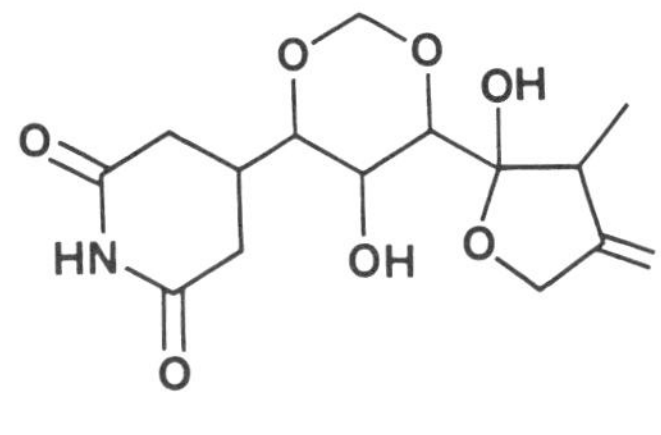

9

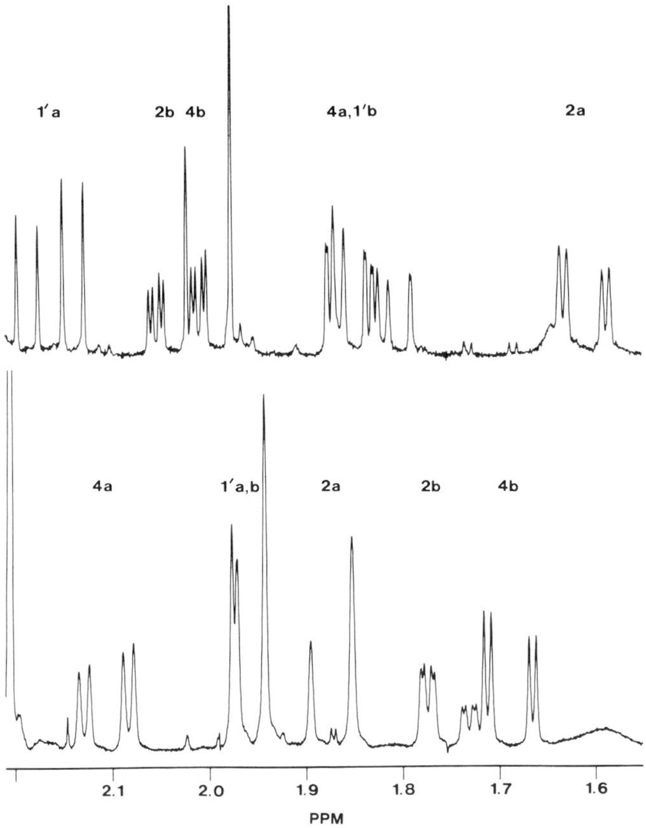

FIGURE 10. Comparison of the high-field region for proton spectra of (Structures 6 and 7).

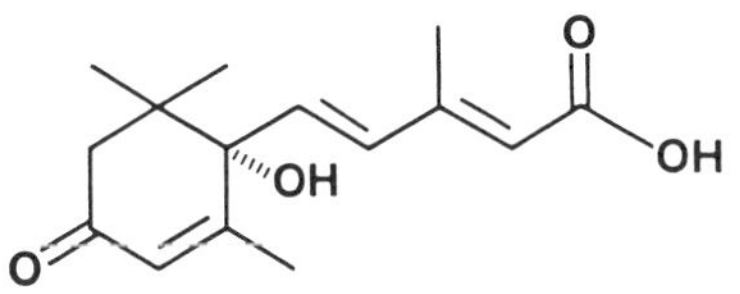

10

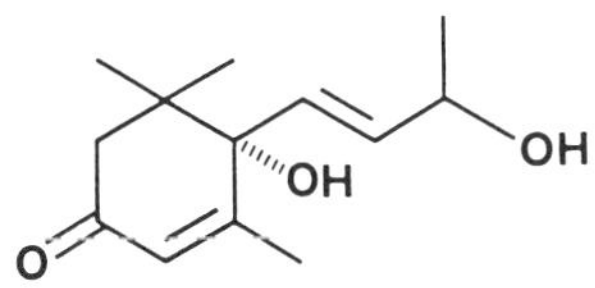

11

Since the molecular ions of structures **6**, **7**, and **8** were identical (m/z 240) and each exhibited 13 signals in their ^{13}C spectrum it was assumed that all three compounds were isomeric. Figure 10 shows a portion of the ^{1}H spectra of structures **6** and **7**.

Table 3 presents the ^{1}H NMR data for the two drummondones (**6**) and (**7**) compared with drummondol (**8**). The most dramatic difference in the NMR data occurs in the change of the methyl resonance which is found as a doublet at 1.33 ppm in structure **8** but appears as a singlet in structures **6** and **7** at 2.20 and 2.23 ppm, respectively. As a result of the change in multiplicity accompanied by the shift downfield it was initially assumed that the 3′ hydroxyl of structure **8** had become a carbonyl in structures **6** and **7**. Another assumption based on the data in Table 3 was that the double bond in structure **8** had been reduced.

Table 3 reveals the ^{1}H chemical shifts for structures **6** and **7** closely parallel each other. There are significant differences in some long-range coupling constants discernible in Table 3.

Table 3
NMR PROTON CHEMICAL SHIFTS AND COUPLINGS (*J*, Hz) FOR DRUMMONDONE A (6), DRUMMONDONE B (7), AND DRUMMONDOL (8)

Proton/compound	6	7	8
	Chemical Shifts		
2a	1.87 dd	1.61 ddd	2.60 dd
2b	1.75 dddd	2.03 ddd	2.35 d
4a	2.10 dd	1.86 ddd	2.61 s
4b	1.69 dd	2.00 d	2.61 s
7a	3.67 ddd	3.70 ddd	3.88 dd
7b	3.51 dd	3.56 dd	3.72 d
1′a	1.96 dd	2.16 dd	5.86 dd
1′b	1.95 ddd	1.84 ddd	—
2′	2.84 dddd	2.76 dddd	6.20 dd
3′	—	—	4.45 m
1-CH_3	1.06 s	1.05 s	1.00 s
5-CH_3	1.21 s	1.26 s	1.20 s
3′-CH_3	2.20 s	2.23 s	1.33 d
	Proton Couplings		
2a,2b	13.0	13.0	18
2a,7a	0.5	0.5	2.2
2a,2′	—	2.2	—
2b,4a	3.2	3.2	—
2b,7a	1.1	1.1	—
2b,2′	0.5	—	—
4a,4b	13.9	13.9	—
4a,2′	—	0.5	—
4b,2′	2.2	—	—
7a,7b	7.6	8.1	8.1
7b,1′b	0.5	0.5	—
1′a,1′b	14.0	14.0	—
1′a,2′	14.3	6.5	—
1′b,2′	4.3	12.0	—
1′,2′	—	—	15.1
1′,3′	—	—	1.1
2′,3′	—	—	6.2
3′,3′-CH_3	—	—	6.1

Note: Spectra were recorded at 300 MHz with $CDCL_3$ solutions.

In the ^{13}C NMR spectra of structures **6** and **7**, all the chemical shifts are very similar and the multiplicities are the same (Table 4).

As shown in Table 4, resonances of C-1, C-2, C-5, C-7, C-8, and methyl groups attached at C-1, C-5, and C-3′ in structure **8** were comparable to those found in structures **6** and **7**, implying the presence of a tetrahydrofuran moiety. Resonances observed for protons attached at C-7, and for methyl groups at C-1 and C-5, are very similar and further support the presence of the tetrahydrofuran ring. Overall, the ^{1}H NMR resonances for structures **6** and **7** are not as closely parallel to those of structure **8** as the corresponding ^{13}C resonances.

There are significant differences in the NMR spectra of structures **6** and **7** as contrasted with those of structure **8**. Compounds **6** to **8** all have one ketone function established by ^{13}C NMR resonances at 206.9 ppm for structure **8** and at 210.6 ppm for structures **6** and **7**. A carbon-carbon

Table 4
^{13}C NMR CHEMICAL SHIFTS OF DRUMMONDONE A (6), DRUMMONDONE B (7), AND DRUMMONDOL (8)

Carbon/compound	6	7	8
1	44.2 s	44.2 s	47.7 s
2	53.1 t	53.5 t	52.6 t
3	68.7 s	68.6 s	206.9 s
4	46.9 t	46.5 t	53.1 t
5	77.5 s	77.5 s	81.9 s
7	79.0 t	79.5 t	77.3 t
8	83.7 s	83.6 s	85.7 s
1′	26.3 t	25.8 t	124.1 d
2′	55.0 d	55.1 d	140.0 d
3′	210.6 s	210.6 s	68.0 d
1-CH_3	17.6 q	16.9 q	15.4 q
5-CH_3	19.7 q	20.7 q	18.9 q
3′-CH_3	31.3 q	31.1 q	24.2 q

Note: Spectra were obtained at 75.47 MHz with $CDCl_3$ solutions.

double bond in structure **8** is established by ^{13}C resonances at 124.1 and 144.0 ppm and by ^{1}H resonances at 5.86 and 6.20 ppm. None of the olefinic peaks are observed in the spectra of structures **6** and **7**. As mentioned above, the ^{1}H spectrum of structure **8** contains a methyl doublet at 1.33 ppm that is replaced by a methyl singlet in structures **6** and **7** at 2.20 and 2.23 ppm, respectively. The NMR data are consistent with a carbonyl group at C-3′ in structures **6** and **7** in place of the hydroxyl at that position in structure **8**. Since the ^{13}C spectra of structures **6** and **7** each exhibit only one carbonyl resonance, the carbonyl group at C-3 in structure **8** must be reduced to a hydroxyl in structures **6** and **7**. The accompanying structures show structures **6** and **7** with cagelike ring systems that are pseudosymmetrical about a plane passing through C-3, C-8, C-1′, and C-2′ and differ only in the configuration at C-2′. All three isomers (**6** to **8**) are probably formed via a common precursor.

In order to distinguish between isomers shown in structures **6** and **7** and to verify the structures a number of NMR experiments were undertaken. A few assumptions could be made in advance to guide the experiments. The protons attached at C-7, adjacent to an oxygen, were expected to have the lowest field ^{1}H chemical shifts. Protons at C-2 were more likely than those at C-4 to be coupled to the ones at C-7. The chemical shifts of protons 2a and 2b were pinpointed by a COSY experiment with drummondone A (**6**, Figure 11). It is evident that the proton farthest downfield, designated 7a, is coupled to protons 2a and 2b as well as to 7b. In addition to the expected vicinal couplings with the protons at 1′, proton 2′ is coupled to 4b, the methylene proton at highest field. With 4b located, its connection with 4a was readily apparent, and all the chemical shifts for structure **6** were established; similarly those of structure **7** were assigned by a separate COSY experiment (Figure 12).

In both structures **6** and **7**, protons 2b and 4a have the same observed coupling constant, J = 3.2 Hz and, therefore, they must occur in a planar, zigzag (W) orientation.[48]

A difference NOE[28] experiment (Figure 13) provided corroboration of the stereochemistry. When the C-1 methyl resonance of structure **6** was irradiated, signals for protons 7a and 2b were enhanced; this observation indicated that these three centers are in close proximity with a syn relationship. Similarly, irradiation at the C-5 methyl signal enhanced the 4a proton signal but not those of 7a or 7b.

Now that the ^{1}H chemical shifts are established they can be used in a 2D heteronuclear ^{1}H/^{13}C

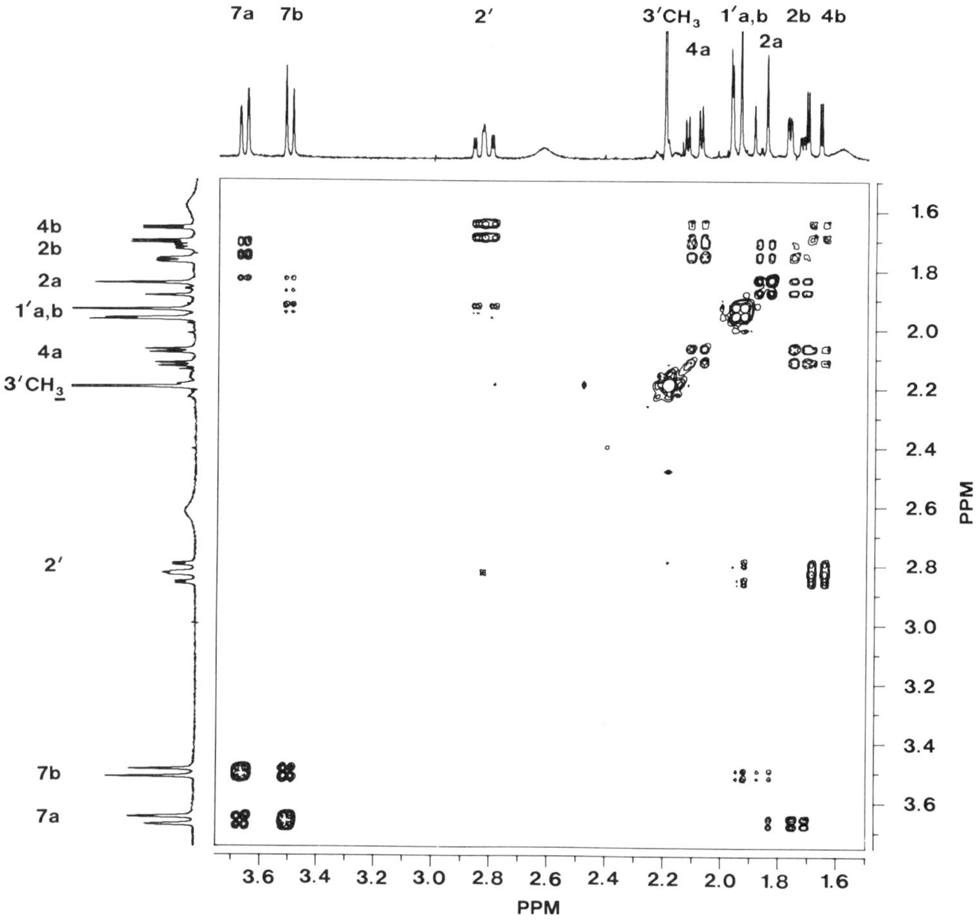

FIGURE 11. COSY spectrum of Structure 6.

shift correlation experiment to assign and confirm ^{13}C shifts in an unambiguous manner (Figure 14).

Using structure **6**, the resonances for protons 2a and 2b (1.87 and 1.75 ppm, respectively) correlate with the ^{13}C resonances at 53.1 ppm; protons 4a and 4b (2.10 and 1.69 ppm, respectively) correlate with the ^{13}C resonance peak at 46.9 ppm. Corresponding assignments for structure **7** were made by analogy.

With all the resonances assigned for structure **6**, the differences between structures **6** and **7** were readily apparent. In structure **6** there is an observed coupling of 2.2 Hz between protons 4b and 2′ but none between the protons 2a and 2′. These results require a planar, zigzag relationship between protons 4b and 2′ with the acetyl group syn to C4. In contrast, structure **7** has corresponding couplings appropriate for a zigzag relationship between protons 2a and 2′, thus placing the acetyl group syn to C-2.

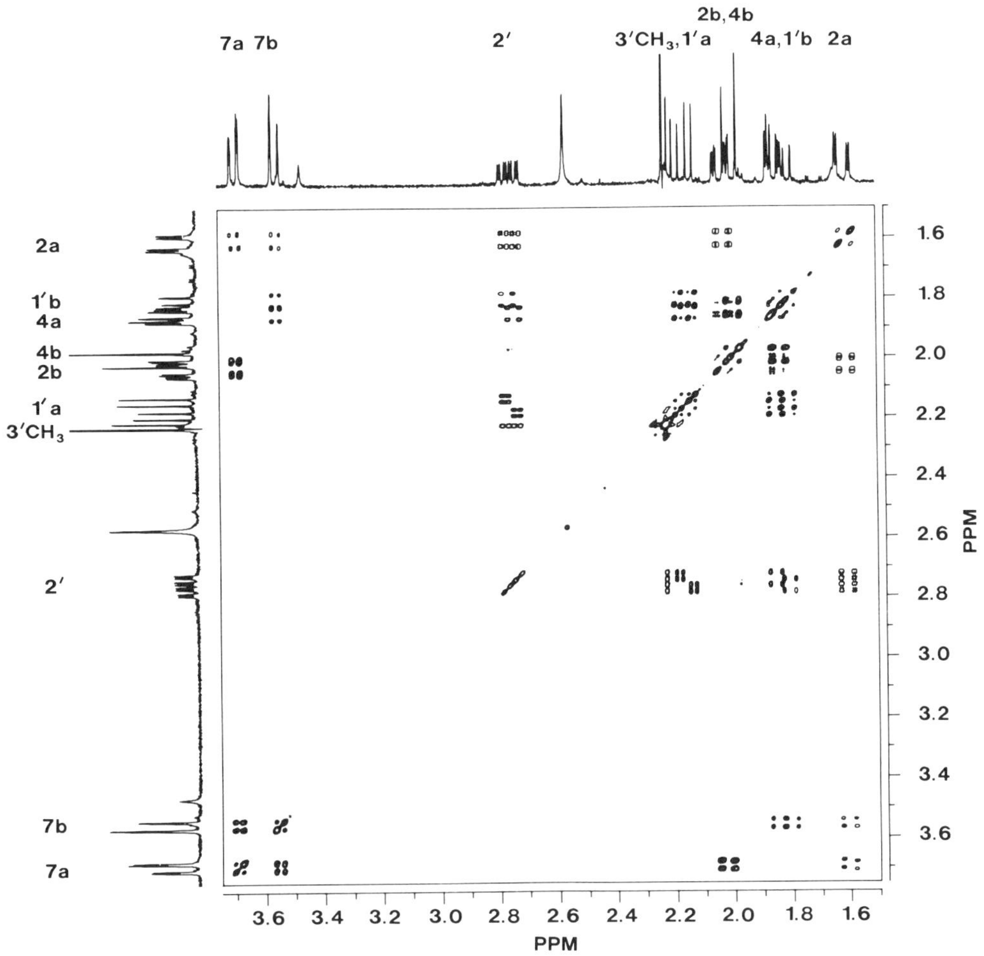

FIGURE 12. COSY spectrum of Structure 7.

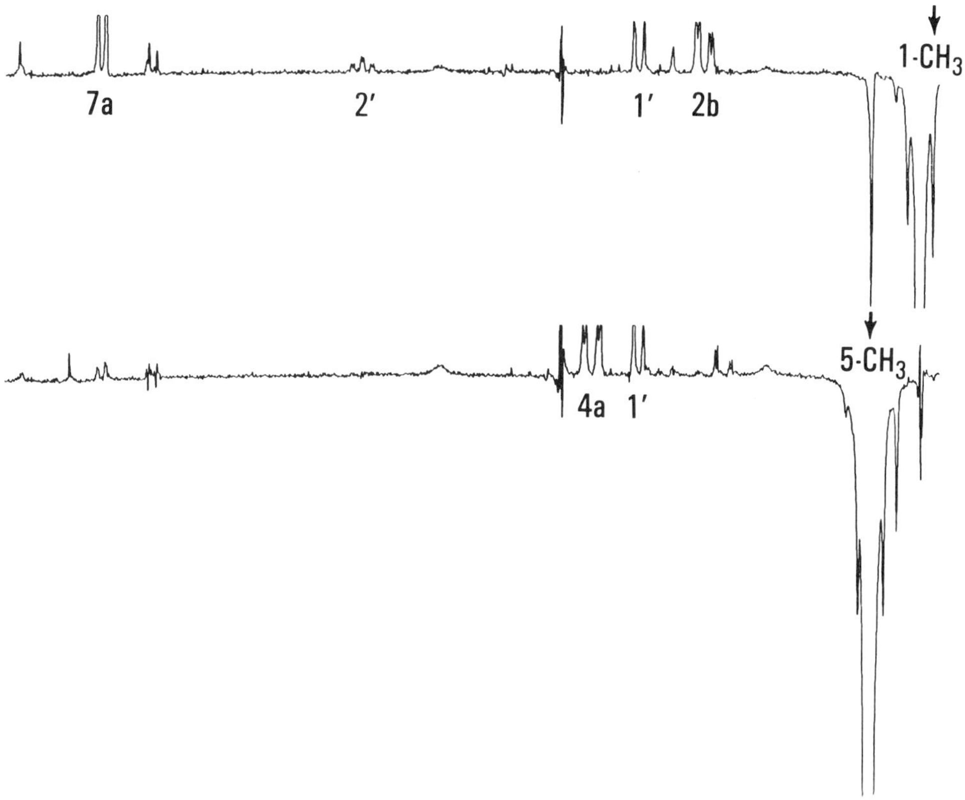

FIGURE 13. NOE difference spectra obtained with Structure 6.

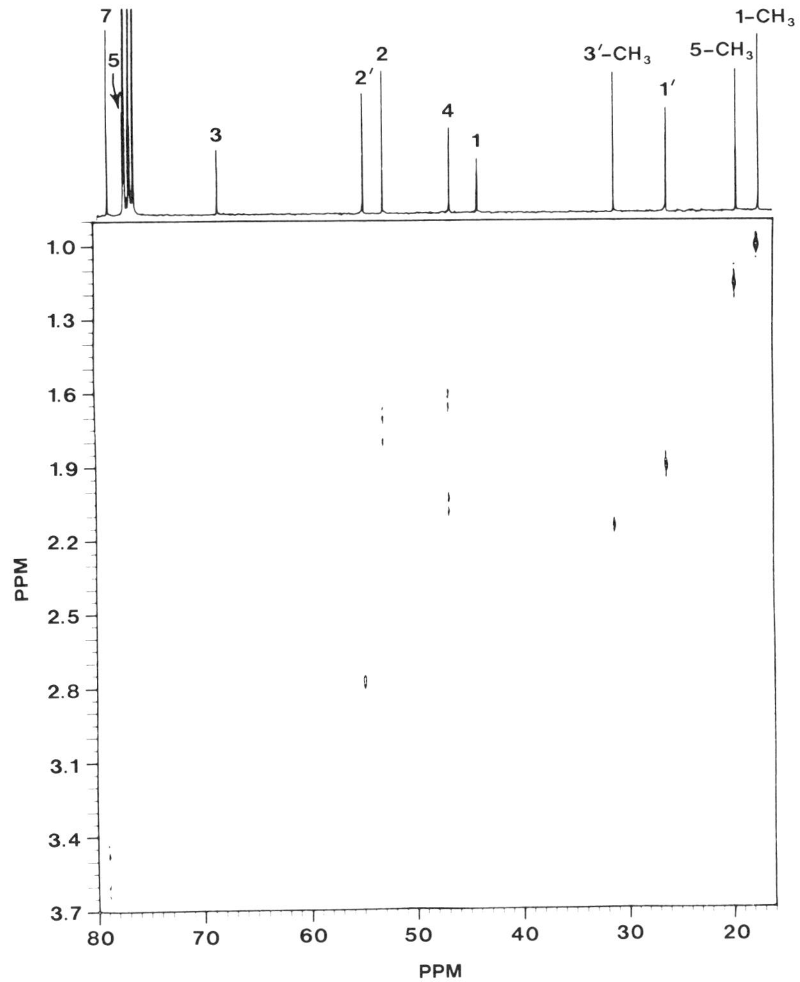

FIGURE 14. 2D $^{13}C/^{1}H$ heteronuclear shift correlation spectrum of drummondone A (Structure 6).

REFERENCES

1. **Wenkert, E. A., Clouse, A. O., Cochran, D. W., and Doddrell, D.,** Carbon-13 nuclear magnetic resonance spectroscopy of naturally occurring substances. I. A direct analysis of nonprotonated carbon centers, *J. Am. Chem. Soc.*, 91, 6879, 1969.
2. **Muller, L., Kumar, A., and Ernst, R. R.,** Two-dimensional carbon-13 NMR spectroscopy, *J. Chem. Phys.*, 63, 5490, 1975.
3. **Aue, W. P., Bartholdi, E., and Ernst, R. R.,** Two-dimensional spectroscopy. Application to nuclear magnetic resonance, *J. Chem. Phys.*, 64, 2229, 1976.
4. **Aue, W. P., Karhan, J., and Ernst, R. R.,** Homonuclear broad band decoupling and two-dimensional *J*-resolved NMR spectroscopy, *J. Chem. Phys.*, 64, 4226, 1976.
5. **Bodenhausen, G., Freeman, R., and Turner, D. L.,** Two-dimensional J spectroscopy: proton-coupled carbon-13 NMR, *J. Chem. Phys.*, 65, 839, 1976.
6. **Bodenhausen, G., Freeman, R., Niedermeyer, R., and Turner, D. L.,** High-resolution nmr in inhomogeneous magnetic fields, *J. Magn. Reson.*, 24, 291, 1976.

7. **Bax, A.,** *Two-Dimensional NMR in Liquids,* D. Reidel, Boston, 1982.
8. **Benn, R. and Gunther, H.,** Modern pulse methods in high-resolution NMR spectroscopy, *Angew. Chem. Int. Ed. Engl.,* 22, 350, 1983.
9. **Turner, C. J.,** Multipulse NMR in liquids, *Progress in Nuclear Magnetic Resonance,* Vol. 16, Emsley, J. W., Feeney, J., and Sutcliffe, L. H., Eds., Pergamon Press, Oxford, 1985, 311.
10. **Hall, L. D. and Sanders, J. K. M.,** Complete analysis of ^{1}H NMR spectra of complex natural products using a combination of one- and two-dimensional techniques. 1-Dehydrotestosterone, *J. Am. Chem. Soc.,* 102, 5703, 1980.
11. **Pegg, D. T., Bendall, R., and Doddrell, D. M.,** Heisenberg Vector model for precession via heteronuclear scalar coupling, *J. Magn. Reson.,* 44, 238, 1981.
12. **Doddrell, D. M., Pegg, D. T., and Bendall, M. R.,** Distortionless enhancement of NMR signals by polarization transfer, *J. Magn. Reson.,* 48, 323, 1982.
13. **Bendall, M. R., Doddrell, D. M., Pegg, D. T., and Hull, W. E.,** DEPT, *Bruker Brochure,* Karlsruhe, West Germany, 1982.
14. **Bendall, M. R. and Pegg, D. T.,** Complete accurate editing of decoupled ^{13}C spectra using DEPT and a quaternary-only sequence, *J. Magn. Reson.,* 53, 272, 1983.
15. **Bamburg, J. R., Riggs, N. V., and Strong, F. M.,** The structures of toxins from two strains of *Fusarium tricinctum, Tetrahedron,* 24, 3329, 1968.
16. **Bendall, M. R. and Pegg, D. T.,** ^{1}H-^{13}C Two-dimensional chemical shift correlation spectroscopy using DEPT, *J. Magn. Reson.,* 53, 144, 1983.
17. **Nakashima, T. T., John, B. K., and McClung, R. E. D.,** Selective 2D DEPT heteronuclear shift correlation spectroscopy, *J. Magn. Reson.,* 59, 124, 1984.
18. **Coxon, B.,** Two-dimensional DEPT *J* (CH)-resolved ^{13}C NMR spectrum editing, *J. Magn. Reson.,* 66, 230, 1986.
19. **Morris, G. A. and Freeman, R.,** Enhancement of nuclear magnetic resonance signals by polarization transfer, *J. Am. Chem. Soc.,* 101, 760, 1979.
20. **Stack, R. J., Stein, T. M., and Plattner, R. D.,** 4-*O*-[1-Carboxyethyl]-D-galactose: a component of the extracellular polysaccharide produced by *Butyrivibrio fibrisolvens* strain 49, in press.
21. **Bax, A. and Freeman, R.,** Investigation of complex networks of spin-spin coupling by two-dimensional NMR, *J. Magn. Reson.,* 44, 542, 1981.
22. **Plattner, R. D., Al-Hetti, M. B., Weisleder, D., and Sinclair, J. B.,** A new tricothecene from *Trichothecium roseum, J. Chem. Res.,* in press.
23. **Piantini, U., Sorensen, O. W., and Ernst, R. R.,** Multiple quantum filters for elucidating NMR coupling networks, *J. Am. Chem. Soc.,* 104, 6800, 1982.
24. **Shaka, A. J. and Freeman, R.,** Simplification of NMR spectra by filtration through multiple-quantum coherence, *J. Magn. Reson.,* 51, 169, 1983.
25. **Rance, M., Sorensen, O. W., Bodenhausen, G., Wagner, G., Ernst, R. R., and Wutrich, K.,** Improved spectral resolution in COSY ^{1}H NMR spectra of proteins via double quantum filtering, *Biochem. Biophys. Res. Commun.,* 117, 479, 1983.
26. **States, D. J., Haberkorn, R. A., and Ruben, D. J.,** A two-dimensional nuclear Overhauser experiment with pure absorption phase in four quadrants, *J. Magn. Reson.,* 48, 286, 1982.
27. **Marion, D. and Wutrich, K.,** Application of phase sensitive two-dimensional correlated spectroscopy (COSY) for measurements of ^{1}H-^{1}H spin-spin coupling constants in proteins, *Biochem. Biophys. Res. Commun.,* 113, 967, 1983.
28. **Sanders, J. K. M. and Mersh, J. D.,** Nuclear magnetic double resonance; the use of difference spectroscopy, in *Progress in Nuclear Magnetic Resonance Spectroscopy,* Vol. 15, Emsley, J. W., Feeney, J., and Sutcliffe, L. H., Eds., Pergamon Press, Oxford, 1983, 353.
29. **Maudsley, A. A., Muller, L., and Ernst, R. R.,** Cross-correlation of spin-decoupled NMR spectra by heteronuclear two-dimensional spectroscopy, *J. Magn. Reson.,* 28, 463, 1977.
30. **Bodenhausen, G. and Freeman, R.,** Correlation of proton and carbon-13 NMR spectra by heteronuclear two-dimensional spectroscopy, *J. Magn. Reson.,* 28, 471, 1977.
31. **Freeman, R. and Morris, G.,** Experimental chemical shift correlation maps in nuclear magnetic resonance spectroscopy, *J. Chem. Soc. Chem. Commun.,* 684, 1978.
32. **Bax, A. and Morris, G. A.,** An improved method for heteronuclear chemical shift correlation by two-dimensional NMR, *J. Magn. Reson.,* 42, 501, 1981.
33. **Hull, W. E.,** Two-Dimensional NMR, *Bruker Brochure,* Karlsruhe, West Germany, 1982.
34. **Tookey, H. L. and Yates, S. G.,** The alkaloids of tall fescue: loline (festucine) and perloline, *Ann. Quim.,* 68, 921, 1972.
35. **Reynolds, W. F., Enriquez, R. G., Escobar, L. I., and Lozoya, X.,** Total assignment of ^{1}H and ^{13}C spectra of kauradien-9(11),16-oic acid with the aid of heteronuclear correlated 2D spectra optimized for geminal and vicinal ^{13}C-^{1}H coupling constants: or what to do when "INADEQUATE" is impossible, *Can. J. Chem.,* 62, 2421, 1984.

36. **Kessler, H., Griesinger, C., Zarbock, J., and Loosli, H. R.,** Assignment of carbonyl carbons and sequence analysis in peptides by heteronuclear shift correlation via small coupling constants with broadband decoupling in t_1 (COLOC), *J. Magn. Reson.*, 57, 331, 1984.
37. **Bax, A. and Summers, M. F.,** ^{1}H and ^{13}C Assignments from sensitivity-enhanced detection of heteronuclear multiple-bond connectivity by 2D multiple quantum NMR, *J. Am. Chem. Soc.*, 108, 2093, 1986.
38. **Muller, L.,** Sensitivity enhanced detection of weak nuclei using heteronuclear multiple quantum coherence, *J. Am. Chem. Soc.*, 101, 4481, 1979.
39. **Bendall, M. R., Pegg, D. T., and Doddrell, D. M.,** Pulse sequences utilizing the correlated motion of coupled heteronuclei in the transverse plane of the doubly rotating frame, *J. Magn. Reson.*, 52, 81, 1983.
40. **Bax, A., Griffey, R. H., and Hawkins, B. L.,** Correlation of proton and nitrogen-15 chemical shifts by multiple quantum NMR, *J. Magn. Reson.*, 55, 301, 1983.
41. **Bax, A., Freeman, R., and Kempsell, S. P.,** Natural abundance ^{13}C-^{13}C coupling observed via double-quantum coherence, *J. Am. Chem. Soc.*, 102, 4849, 1980.
42. **Bax, A., Freeman, R., and Frenkiel, T. A.,** An NMR technique for tracing out the carbon skeleton of an organic molecule, *J. Am. Chem. Soc.*, 103, 2102, 1981.
43. **Bax, A., Freeman, R., Frenkiel, T. A., and Levitt, M. H.,** Assignment of carbon-13 NMR spectra via double-quantum coherence, *J. Magn. Reson.*, 43, 478, 1981.
44. **Kleiman, R., Plattner, R. D., and Weisleder, D.,** Anti-germination activity of phenylpropenoids from the genus *Pimpinella*, *J. Natl. Prod.*, 51, 249, 1988.
45. **Powell, R. G., Weisleder, D., Smith, C. R., and Boettner, F. E.,** Drummondones A and B: unique abscisic acid catabolites incorporating a bicyclo[2.2.2]octane ring system, *J. Org. Chem.*, 51, 1074, 1986.
46. **Powell, R. G., Smith, C. R., and Weisleder, D.,** Sesbanimide, a potent antitumor substance from *Sesbania drummondii* seed, *J. Am. Chem. Soc.*, 105, 3739, 1983.
47. **Walton, D. C.,** Structure-activity relationships of abscisic acid analogs and metabolites, in *Abscisic Acid*, Addicott, F. T., Ed., Praeger, New York, 1983, chap. 4.
48. **Jackman, L. M. and Sternhell, S.,** *Applications of Nuclear Magnetic Resonance Spectroscopy in Organic Chemistry*, 2nd. ed., Pergamon Press, Oxford, 1969, 334.

Chapter 6

HIGH-FIELD NMR STUDIES OF PHOTOSYNTHESIS

Ion C. Baianu

TABLE OF CONTENTS

I. INTRODUCTION

Photosynthesis in green plants is the process by which chlorophyll-containing organelles inside plant cells convert solar energy into photochemical energy, which is then used to form carbohydrates. The chlorophyll-containing organelles specialized in photosynthesis are called chloroplasts. A chloroplast is a 5- to10-mm lenticular-shaped particle with an envelope surrounding its stroma (protein- and water-containing matrix) and a number of sacklike membraneous disks called thylakoids, which appear stacked like a pile of coins in electron micrographs of chloroplast thin sections; the thylakoids membrane lamelae are clearly seen in such electron micrographs to form stacks (Figure 1). The overall reaction of photosynthesis could be oversimplified to:

$$2H_2O + CO_2 \xrightarrow[\text{thylakoids}]{\text{light}} O_2 + 4H^+ + 4e^- + \text{carbohydrate} + 112\ \text{kcal/mol}\cdot CO_2 \qquad (1)$$

Chlorophylls in thylakoid membranes absorb photons and trap the light energy so that it can be used in the subsequent photochemical reactions. Such "antenna" chlorophylls are organized into a photosynthetic unit that contains a redox reaction center; close to the reaction center there are present both an electron donor and an electron acceptor. The excitation energy migrates from chlorophylls to the reaction center where it sensitizes the redox reactions. The photosynthetic unit can be visualized as a group of 200 to 300 chlorophyll (Chl) molecules. Each photon absorbed by any Chl molecule of the unit is transferred through the complex until it reaches the specialized Chl a molecules called energy traps or reaction centers. When the excitation energy reaches the trap, the primary photoact facilitates the transfer of an electron (or proton) from the donor to the acceptor molecule. The trap molecules first undergo light-induced oxidation or reduction and subsequent reverse reactions in the dark, leading to changes in their light absorption.

It is now very well established that there are in fact two pigment systems and two light reactions needed for photosynthetic oxygen evolution to occur *in vivo*. This was originally proposed by Hill and Bendall;[1] later Duysens et al.[2] coined the terms pigment system I (PSI) and pigment system II (PSII) to denote the photosystems responsible for the photooxidation of cytochrome f (cyt f) and water, respectively; the light reaction I oxidizes cyt f and reduces $NADP^+$ (nicotinamide adenine dinucleotide phosphate), whereas the light reaction II oxidizes water to molecular oxygen (O_2). The two photosystems are linked, and operate in series, as schematically represented in Figure 2; they form together the photosynthetic electron transport chain running from water to $NADP^+$. PSII oxidizes water and produces a weak reductant (plastohydroquinone). PSI transfers electrons from plastohydroquinone to $NADP^+$, thereby producing the strong reductant NADPH. Coupled with the transport of electrons from water to $NADP^+$, there is a phosphorylating mechanism which converts ADP to ATP. NADPH and ATP diffuse to the stroma phase where they are used to drive the synthetic reactions by which CO_2 is fixed and carbohydrates, and other organic compounds, are produced. The primary acceptor, Q_A^-, of PSII is presumably also a plastoquinone, bound to the reaction center and reduced to the plastosemiquinone in the primary reaction.[3] The oxidation of Q_A^- by secondary acceptors can be selectively inhibited by a herbicide, 3-(3′,4′-dichlorophenyl)-1′,1-dimethylurea (DCMU), or diuron. This specific inhibitor is, therefore, useful as a means to separate PSI electron transport from PSII. The two photosystems are physically separated in the thylakoid membrane, with PSII located towards the lumen (interior) of the thylakoid and PSI located near the outer surface of the thylakoid membrane; therefore, both PSI and PSII "particles", or membrane fractions, could be separated by a combination of mechanical and detergent treatments of thylakoids. This electron transport model is called the "Z-scheme of photosynthesis". The traps in PSI and PSII

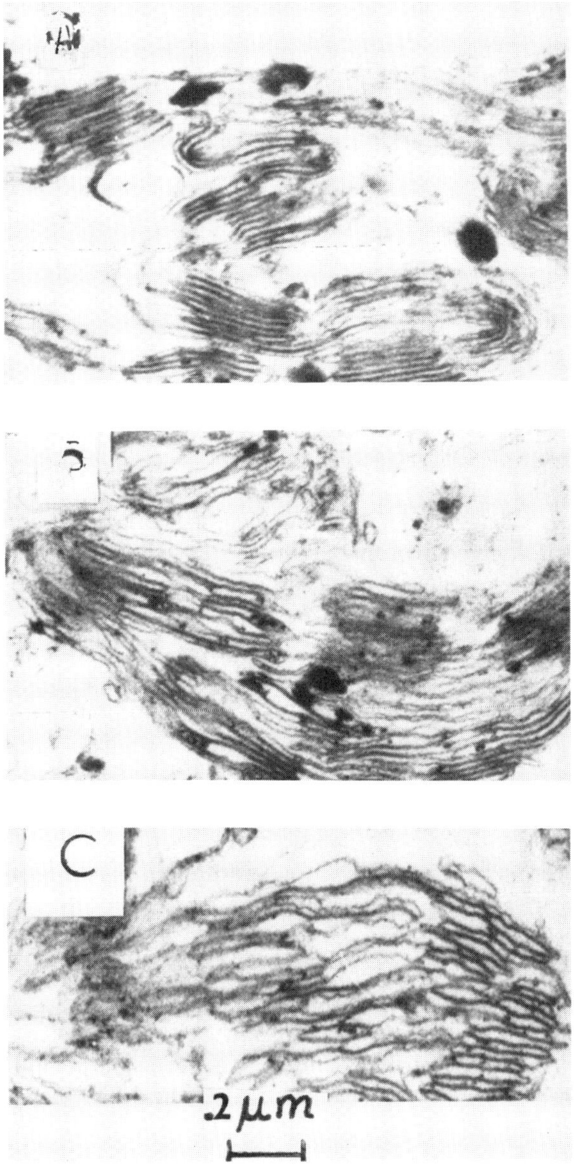

FIGURE 1. A, ^{35}Cl NMR spectra of thylakoid suspensions. B, Electron micrographs of halophytic plant thylakoids for separations at three different pH values: (A) pH 5.8, (B) pH 7.0, and (C) pH 9.0.[20]

contain specialized chlorophyll a molecules, Chl a_I and Chl a_{II}, respectively, that have different absorption maxima: 690 to 700 mm for PSI and 660 to 680 mm for PSII; therefore, the reaction center of PSI is called "P700" and the reaction center of PSII is called "P680" (Figure 2). The accepted view is that P700 molecules are dimers of Chl a_I; whereas, P680 contains Chl a_{II} monomers that might be associated with protein(s). The primary reactions of the two light reactions can be written as:

$$\text{P680} \cdot A_{II} \xrightarrow{h\upsilon_{II}} \text{P680}^* \cdot A_{II} \longrightarrow \text{P680}^+ A_{II}^- \tag{2}$$

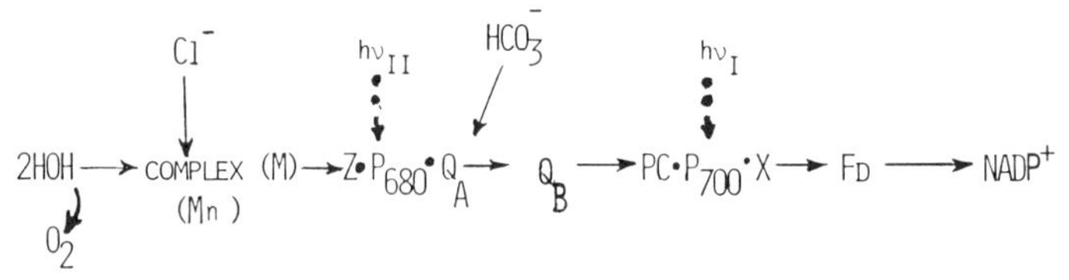

FIGURE 2. Electron flow in photosynthesis.[1] Here, P^{680} is the reaction center chlorophyll a (Chl) of PSII, I is the intermediate electron acceptor that includes pheophytin, Q_A and Q_B are the quinone electron acceptors, (Q_B or "R" is the 2-electron acceptor), PQ is plastoquinone molecule of the plastoquinone pool, Z is the electron donor to P680+, and M represents the charge accumulating species. The possible major sites of action of Cl^- and HCO_3^- are suggested to be at the M and Q_A levels, respectively.

$$P700 \cdot A_I \xrightarrow{h\upsilon_I} P700^* \cdot A_I \longrightarrow P700^+ A_I^- \tag{3}$$

where A_I and A_{II} are the primary electron acceptors and P_{700} and P_{680} are the primary electron donors of PSI and PSII, respectively. This review is concerned only with the mechanism(s) of photosynthetic water oxidation and is, therefore, focused on the processes taking place at the beginning of the electron transport chain in PSII. The evolution of molecular oxygen, O_2, from water involves the overall reaction:

$$2H_2O + \text{energy} \rightarrow O_2 + 4H^+ + 4e^-,\ E_O' = 0.8V \tag{4}$$

The energy required by this reaction is provided by the excitation energy $h\upsilon_{II}$, from light absorption in PSII, and the process may be catalyzed by a special enzyme located on the inside of the thylakoid membrane; such a catalyst is called the "oxygen-evolving" (OE) enzyme and appears to be part of a complex with other proteins (the oxygen-evolving complex, OEC). Since the OE enzyme has not yet been isolated in the active form, its characteristics are uncertain; educated guesses based on SDS-PAGE electrophoresis place its molecular weight between 23 and 47 kDa. However, it has been reasonably well established that photosynthetic oxygen evolution requires both Mn[4] and chloride[5] to function, that is, to catalyze water-splitting and molecular oxygen evolution. The Mn requirement is strict and estimated at ~4 to 6 Mn/400 Chl, in chloroplasts; whereas, the chloride (Cl^-) requirement is much more variable (Figures 3A and B), and oxygen evolution (albeit low) is observed even at low levels of Cl^-. If the thylakoids are depleted of Cl^-, and then exposed to light, however, an irreversible photoinhibition takes place, that is, thylakoids become unable to evolve oxygen even though sufficient Cl^- is present. Heat or Tris treatments also impair the oxygen evolving ability. Such experiments clearly indicate that the remainder of the electron transport chain is, however, unimpaired by the mild heat treatments[6] or by Tris,[7] and that the heat mainly affects oxygen evolution by impairing the function of the OEC.

The formation of an O_2 molecule as a result of photosynthetic water oxidation requires the removal of four electrons (Reaction 4), and in order to be able to oxidize water the average potential of the four oxidizing equivalents must be higher than 0.8 V. Since a single photoact transfers only one electron, the overall reaction ((Reaction 4) must involve at least four elementary photoacts. The four individual steps could very well be at different potential levels, involving different redox reactions. Because such photochemical reactions occur in rapid succession, their kinetics and relation to oxygen evolution can only be studied with short (2-μs, or less) flashes of light. Flash-induced oxygen evolution by thylakoids and green algae could be measured with very high sensitivity and reproducibility with a polarographic instrument designed and built by Joliot.[8] The results obtained from such measurements are summarized in an excellent review by Joliot and Kok [9] and form the basis for the currently accepted model of photosynthetic water oxidation kinetics.

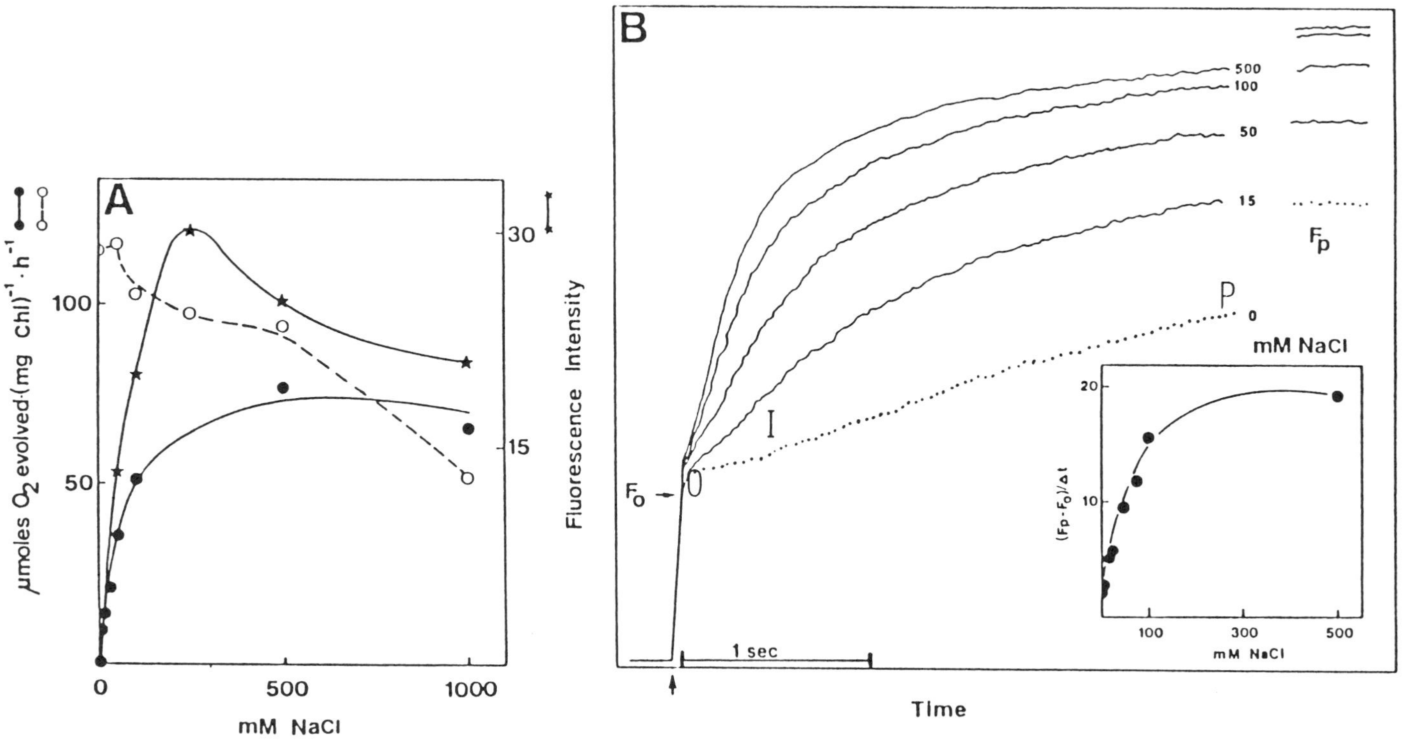

FIGURE 3. (A) Dependence of the Hill reaction rates of thylakoids from halophytic (bottom scale) and glycophytic (top scale) plants, as a function of the total [Cl^-]. (●), *Avi. germinans;* (☆), *Avi. marina;* and (○), *S. oleracea.* (B) The inset gives the rate of the variable fluorescence rise, O → P, in thylakoids (pH 7.6) from *Avi. germinans.* The figures shows tracings of the actual Chl^- a fluorescence induction curves (as a function of the total [Cl^-] up to 500 m*M*) from which the values in the insert were calculated; F_p, peak fluorescence intensity (or P level) and F_0, fluorescence intensity of the "O" level. (Modified from Critchley, C., Baianu, I. C., Govindjee, and Gutowsky, H. S., *Biochim. Biophys. Acta,* 436, 682, 1982.)

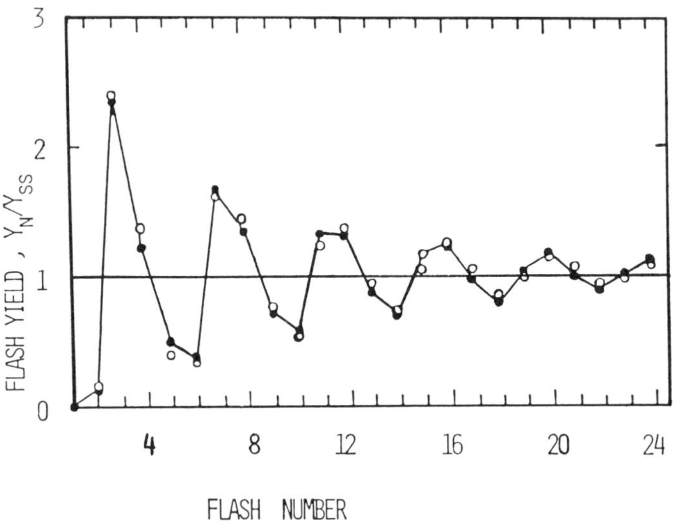

FIGURE 4. O_2 yields as a function of flash number in a suspension of thylakoids from spinach and peas. (o) O_2 yields observed after 40 min dark; (•) Calculated for $[S_0]_0 = 0.25$, $[S_1]_0 = 0.75$, $\alpha = 0.1$, and $\beta = 0.05$, with a being the "miss" parameter and β the "double hit" parameter.[66]

II. KOK'S O_2 CLOCK: KINETICS AND MECHANISM OF PHOTOSYNTHETIC WATER OXIDATION

In principle, a strong, 1-µs light flash excites all trapping centers at once, so that the subsequent events occur synchronously in all centers, and in a stepwise manner, provided that the system is uniform. Figure 4 shows an example of the yields of O_2 produced with a series of flashes given to a suspension of isolated spinach thylakoids which had been in darkness for 40 min prior to flashing. A characteristic flash pattern is observed: the flash yields, Y_n, normalized to the steady-state yields, Y_{ss}, after at least ~20 flashes (steady state), are close to zero for the first two flashes (Y_1 and Y_2) and are maxima for the 3rd (Y_3), 7th (Y_7), 11th (Y_{11}) flash, and so on. The flash yields show a dampened oscillation which is significantly faster in green algae than in thylakoids (Figure 5). Kok et al.[10] have shown that this pattern occurs because the trapping centers of PSII operate independently of each other in the oxidation of water. Each center and an associated catalyst (S, or "M") cycles — like a clock — through five oxidation states according to the following (Kok's O_2 clock) model:

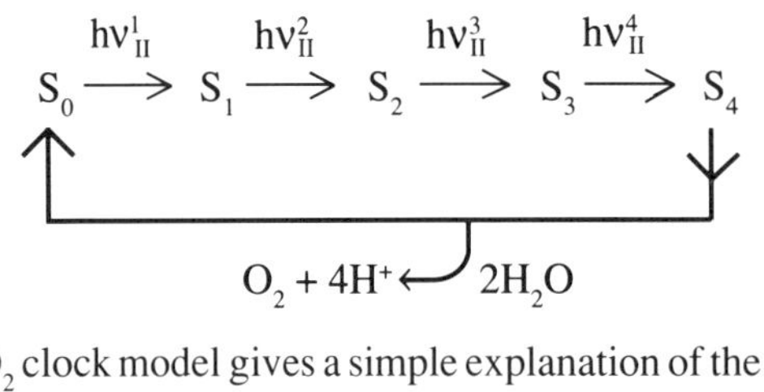

(5)

The O_2 clock model gives a simple explanation of the observed periodicity of four in the flash pattern of O_2 yields. To explain the high oxygen yield on the third flash it was assumed that in the dark-adapted thylakoids the state S_1 predominates, and that both S_0 and S_1 are stable in the dark; in dark adapted chloroplasts, the oxygen evolving system (OES) would be distributed in two states, S_0 and S_1, in the ratio ~1:3. The states S_2 and S_3 deactivate in darkness to S_1 and S_0; S_4 is unstable and does not require light to proceed to S_0 by splitting up water and releasing molecular oxygen and four protons. In the early model of Kok (Scheme 5), it was assumed for

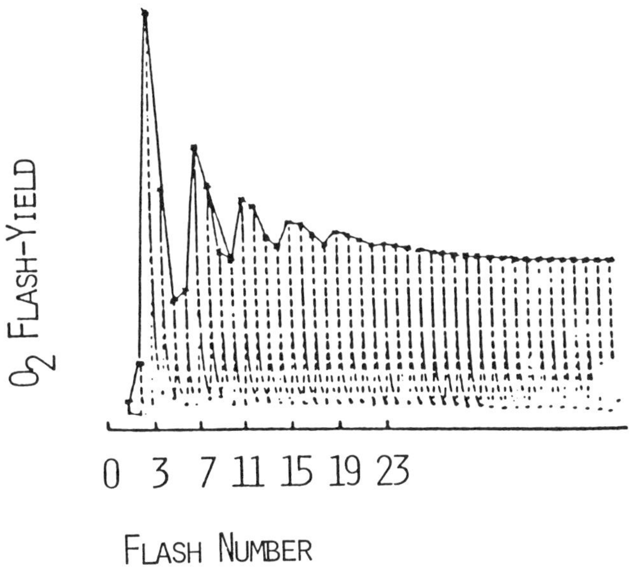

FIGURE 5. O_2 flash yields in green algae, observed at pH = 7.4 in 99%/D_2O/bicarbonate buffer with 0.1 *M* KCl added.[24]

simplicity that all four protons are released in the $S_4 \rightarrow S_0$ transition, that is, one has the proton-release pattern (0, 0, 0, 4). Recent data[11,12] indicate that the more probable scheme of proton-release is

$$S_0 \underset{1H^+}{\longrightarrow} S_1 \underset{(0)}{\longrightarrow} S_2 \underset{1H^+}{\longrightarrow} S_3 \underset{2H^+}{\longrightarrow} S_4 \longrightarrow S_0 \qquad (6)$$

or (1, 0, 1, 2). This suggests that water may undergo chemical changes prior to the last step. The details of the proton-release patterns still remain to be understood. The OEC would contain a charge accumulator, M, that accumulates up to four positive charges as the O_2 clock advances from S_0/S_1 to S_4. In Figure 2, Z is the electron donor to the reaction center Chl a P680 and Q_A is the electron acceptor; in addition, there is a 2-electron acceptor, called R, or Q_B.

With these notations, the PSII configurations corresponding to the five S states can be represented as:

1. $M \cdot Z \cdot P680 \cdot Q$ for S_0
2. $M^+ \cdot Z \cdot P680_+ \cdot Q^-$ for S_1
3. $M^{2+} \cdot Z \cdot P680 \cdot Q$ for S_2
4. $M^{3+} \cdot Z \cdot P680 \cdot Q$ for S_3
5. $M^{4+} \cdot Z \cdot P680 \cdot Q$ for S_4

The damping in the flash pattern with increasing flash number was explained by Kok to be due to the mixing of S states with time, and also due to a certain probability that the clock fails to advance in some PSII units, an event which is called a “miss”. Whenever on the second flash O_2 had evolved, the effect was explained by the presence of flashes that are too long, causing “double hits”[10] (i.e., one flash advances the O_2 clock two states, instead of only one state).

Kok's model assumes that the charges produced in one center are independent of all the other centers. Joliot and Kok[9] found that the flash pattern of relative O_2 yields remained unchanged when most of the centers were closed by the addition of the herbicide diuron (DCMU). If the centers were not independent of each other one would expect that the closed centers would transfer their energy to the open ones causing O_2 evolution even in the first or second flashes, which is not what they observed.

Experiments in which the removal of certain proteins from PSII results in blocking of oxygen evolution[12] appear to indicate that one or several proteins that contain bound Mn and Cl^- may be at the water oxidation site of OEC. However, it was not yet possible to prove that the cycling of S states occurs in such proteins, or to identify the role that Mn plays in oxygen evolution. Models were proposed, however, with some indirect evidence from low-field proton NMR relaxation measurements for the cycling of S states being associated with changes in the valence state of OEC-bound Mn; such measurements were made on suspensions of green plant thylakoids[13,14] or PSII membrane fractions,[15] following light flashes, and were based on the idea of using the paramagnetic relaxation enhancement of water protons by Mn^{+2}, and/or Mn^{+4} (hypothetical!), to probe the OEC(s) states.

The relaxation rate changes measured after a series of light flashes were, however, of the order of 5 to 15% for thylakoids[13] and 0.1 to 0.3% for PSII membrane fractions.[16] Furthermore, the flash patterns reported were different for thylakoids and the PSII membrane fractions, the reproducibility of the measurements was low, and flash patterns were variable, especially on the first two flashes.[12] There was no discernible "pattern" in the relaxation rate changes if the thylakoids were washed with a buffer solution containing 1 to 2 m*M* EDTA.[17,18]

^{1}H NMR measurements at high-field (220 MHz ^{1}H resonance) on thin films of hydrated spinach and halophytic thylakoids showed transverse relaxation changes between 20 to 25%, following 1 to 2 s after flashes, when the Mn content of the samples was in the range from 4.6 to 6 Mn per 400 Chl (as determined by neutron activation analysis), and the samples contained 50 m*M* $CaCl_2$ added to a 50 m*M* bicarbonate solution at pH 7.2.[19] For mangrove thylakoids, the flash-induced NMR transverse relaxation changes were as large as 30 to 35%, but the Mn content was as high as 7 Mn per 400 Chl.[20] Such changes were abolished by the addition of 25 μ*M* DCMU. Subsequent ^{1}H NMR observations were made at 360 MHz on spinach thylakoid suspensions with 100 m*M* phosphate buffer, 50 to 100 μs after series of 1 to 11 flashes; these showed, however, no significant transverse relaxation rate changes within ±1% of the dark-adapted sample.[21] The thylakoid samples in such experiments had, however, only between 3.8 and 4.6 Mn per 400 Chl, as determined by neutron activation analyses. At present, it is uncertain if the "functional" Mn in OEC is directly connected to the cycling of the S states in Kok's O_2 clock since there seem to be only about 4 "functional" Mn per 400 Chl in spinach thylakoids.[22,23] Further experiments are needed in this area to check out (albeit indirectly) the Mn valence-change models. There is, however, indirect evidence from high-field ^{1}H NMR and also ESR measurements[24] on *Chlamydomonas* algae treated with 10 m*M* hydroxylamine (NH_2OH), (with and without 50 μm DCMU), that Mn^{+2} is involved in the recovery of O_2-evolving ability of green algae in the dark; the transverse relaxation rate of water (HDO) in *Chlamydomonas* suspensions treated with 10 m*M* NH_2OH changed reversibly by 50%, and reproducibly, by alternating between light and dark at 10-min intervals, and repeating the measurements six times.[25] When 50 μM DCMU was added, however, there were no significant changes in the proton transverse relaxation rate, within ±2%. Because of the complex nature of the system studied it would be premature to imply that such experiments link Mn^{+2} to the cycling of the S states in Kok's O_2 clock model. The O_2 yield flash patterns of the untreated green algae and mangrove thylakoids had the characteristic periodicity of four (Figures 5 and 6A); the flash pattern of mangrove thylakoids treated with ≤1 m*M* NH_2OH in the presence of 0.25 *M* NaCl showed the characteristic two-flash delay (Figure 6B), previously observed for spinach and lettuce thylakoids.[13] The O_2 yield flash patterns of *Chlamydomonas* in a D_2O/bicarbonate buffer were the same as in the H_2O buffer, within the experimental variability of ±2%.[25] These results are in excellent agreement

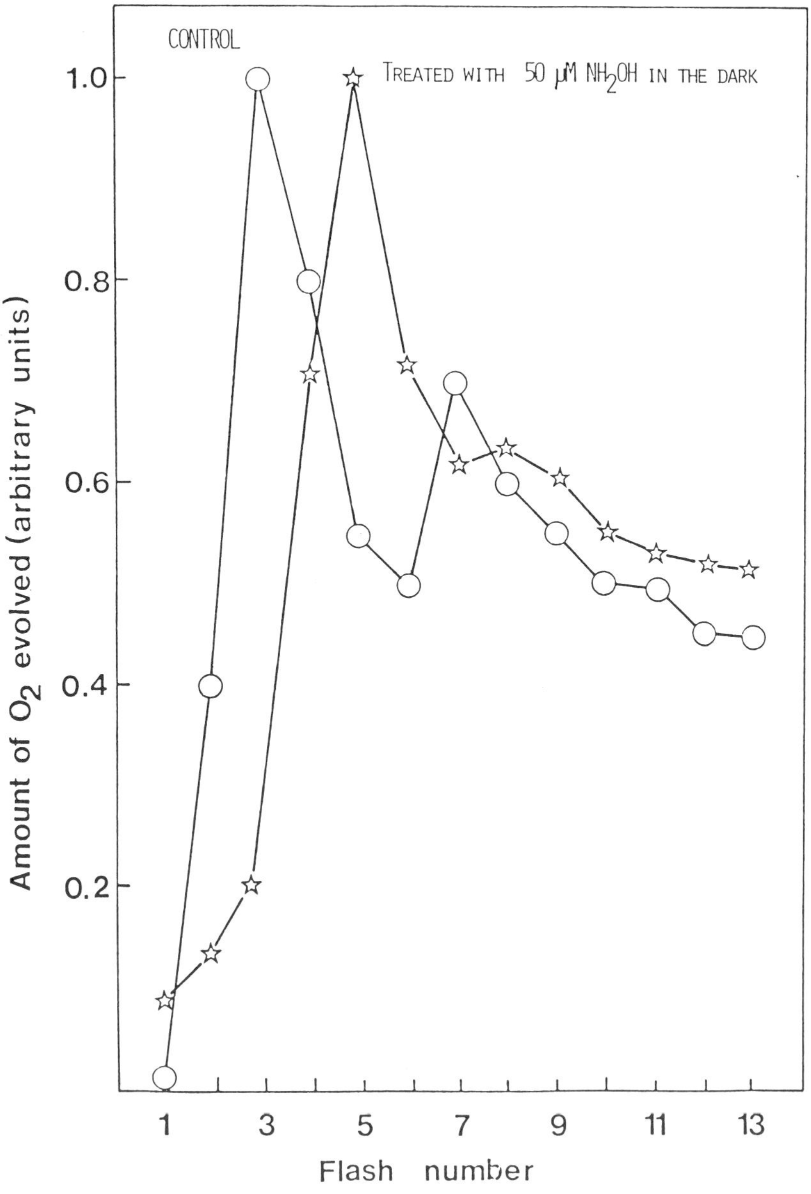

FIGURE 6. O_2 yields to light flashes (~3 µs) for thylakoids from halophytic plants *(A. germinans)*. A, ○, with 250 m*M* NaCl added and ☆ after the addition of 500 µ*M* NH_2OH in the dark.

with previous measurements by a different group,[26,27] which strongly suggested that the splitting of the OH bond of water is not rate limiting in photosynthetic oxygen evolution, in any of the S states, and that the reactions of the higher S states with water would be rapid.[27,28] A comparison of 30.7-MHz ^2H NMR spectra of D_2O in suspensions of green algae recorded at 20 and –40°C (Figures 7A and B, respectively) indicates that several cellular sites exchange hydrogen for deuterium from D_2O, as seen from the different quadrupole splittings in Figure 7B, at the lower temperature. At room temperature, the rate of this chemical exchange is fast on the ^2H NMR timescale. Measurements on green algae and thylakoids in suspension may provide, therefore, an improved means for investigating the possible role(s) of Mn in photosynthetic water oxidation; the ^2H NMR measurements would avoid problems of interpretation caused by cross-

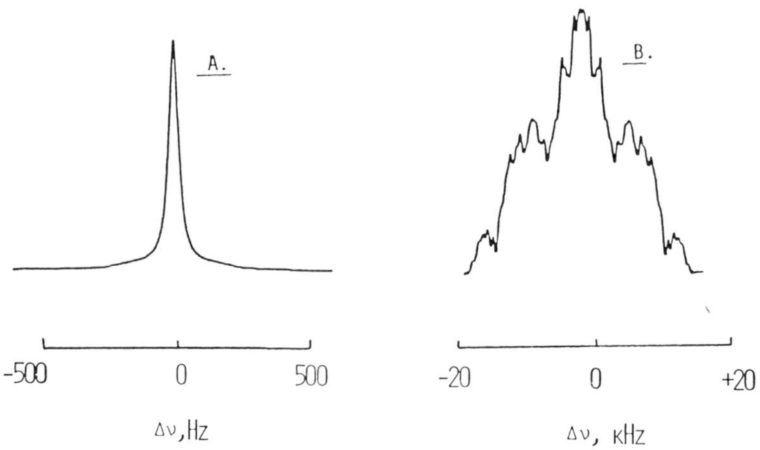

FIGURE 7. (A) 30 MHz ^{2}H NMR spectrum of a pellet of green algae *(Chlamydomonas s.)* in D_2O at 22°C and (B) 30 MHz ^{2}H NMR spectrum of a pellet of green algae at –30°C.

relaxation effects.[29] Furthermore, the resolved quadrupole splittings observed at the lower temperatures may provide additional information.

Summarizing the contents of this section, it can be stated that although the kinetics of the photosynthetic oxygen evolution process is well established, the mechanism governing this process, the nature, and roles of the participating components are yet to be found; the possible connection between OEC-bound Mn and the cycling of S states is only indirect, and tenuous at best, although much effort has been channeled into the investigation of this hypothesis. ^{1}H, ^{2}H NMR, and/or ESR measurements may still provide sufficient evidence to test critically such a role of Mn in photosynthetic water oxidation.

The other essential cofactor of OEC is chloride and its action does take place within PSII, near the water-splitting site.[30,34] Furthermore, the absence of Cl^- prevents the advancement of the O_2 clock beyond S_2 or S_3, and there is polarographic (flash-yield) evidence that Cl^- interacts strongly with the higher states, S_3 and S_4.[35] The experiments and available evidence for the possible role of Cl^- in photosynthetic water oxidation will be reviewed together with models that are now emerging from ^{35}Cl NMR studies of Cl^- binding to thylakoid membranes,[36-41] and more specifically to OEC.[41,42] Our data indicate the possibility of utilizing high-field NMR observations to study the mechanism of photosynthetic water oxidation in green algae, in addition to thylakoids and PSII particles.

III. CHLORIDE ROLE IN PHOTOSYSTEM II

Warburg and Luttgens[5] found that Cl^- plays a role in photosynthesis; its role in the photosynthetic electron transport chain which runs from H_2O to $NADP^+$ is, however, restricted to the O_2-evolving (or electron donor) side of PSII.[30-34] This role is to be contrasted with that of HCO_3^- which was shown to function on the electron acceptor side of PSII (see Figure 2).

Although the requirement of Cl^- on the O_2 evolving side (Figure 2) is established, its mechanism of action is unknown. We will, however, review the experiments in which thylakoids isolated from halophytic plants were used to show that Cl^- functions indeed on the electron donor side in PSII. We will also summarize the conclusions obtained through the application of ^{35}Cl NMR to study thylakoids isolated from halophytes and spinach.[40-42]

A. Photochemical Reactions: The Hill Reaction and Chl a Fluorescence Induction in Thylakoids of Halophytic Plants

Figure 3 shows the dependence of electron transport activities on the Cl^- concentration;

electron transport activities were measured as Hill reaction rates (Figures 3A and B) and from the Chl^- a fluorescence transients (Figure 3B) in thylakoids from two salt-tolerant higher plants, *Avi. germinans* and *Aster tripolium*. Duysens and Sweers[43] showed that the light-induced reduction of the primary electron acceptor, Q_A, (the quencher of fluorescence) of PSII causes an increase in the yield of Chl a_{II} fluorescence. This simple concept is the basis for the steady-state and kinetic fluorimetric assay of primary and secondary electron transfer processes in PSII.[44] When light is turned on, Chl a fluorescence undergoes transient changes with time. Fast changes in fluorescence yield that occur within a few seconds are related to the initial photochemical events of PSII. Thus, the rise of fluorescence from a low to a high level (the "variable" fluorescence, as opposed to the constant fluorescence, F_o, observed immediately when light is turned on), is caused by the accumulation of Q^-_A since the photons arriving at the reaction center can no longer drive the photochemical reactions and are then emitted through fluorescence. Ferricyanide, which can reoxidize Q^-_A also lowers the fluorescence level, whereas DCMU, which blocks the reoxidation of Q^-_A by the pool of secondary acceptors, causes the fluorescence to rise quickly to a high level, where it remains constant. In untreated thylakoids, the fast fluorescence component rises from an initial level O, (with fluorescence yield F_o) to a plateau, P (with fluorescence yield F_p). O represents the fluorescence level at which all the intersystem intermediates are oxidized. At the P level, a steady state is reached in which electrons are transferred to the intermediate pool by PSII and fed into PSI at equal transfer rates on both sides.[45] The fluorescence rise OI (Figure 3B) is almost synchronous with a rise in the rate of oxygen evolution as the water-splitting catalyst (OE enzyme) accumulates in its S_3 state. The amplitude P-O measures the redox potential difference spanned by the redox couple Q_A/Q_AH (reduced Q_A). Joliot and Joliot[46] were able to derive the following relationships for the variable fluorescence, F, (Figure 3B) and for the rate of oxygen evolution, V:

$$F = \frac{kI[Q_AH]}{1 - p[Q_AH]} \qquad \text{and } V = \frac{kI[Q]}{1 - p[Q_AH]} \tag{7}$$

where k is a rate constant, I is the excitation light intensity, and p is the probability of excitation exchange between a closed reaction center and a neighboring center, either open or closed. However, it was later found[45] that the variable Chl a fluorescence, which originates from the pigment population of PSII, is sensitive both to the rate of electron transport through the reaction centers, and to the changes in the thylakoid membrane ultrastructure that accompany the phosphorylating electron transport, as well as the size of the "antenna". In the case of halophytic thylakoids we find that $F \sim V$ or $V = BF$, with B being a constant. The apparent rate constant, K_{app}, for the equilibration of Q_A and $Q_B(R)$, turns out to be from Equation 7:

$$K_{app} = B \frac{[Q_B]}{[Q_B]}$$

noting that $V/F = [Q_A]/[Q_A^-]$.

In the case of thylakoids isolated in the presence of 50 m*M* NaCl from leaves of both species, optimal rates of photosynthetic O_2 evolution, measured with ferricyanide as the Hill acceptor, were obtained with 500 and 250 m*M* NaCl, respectively, (Figure 3A). In contrast, spinach thylakoids not only lacked this requirement of high $[Cl^-]$ but were adversely affected by salt concentrations in excess of about 150 m*M* NaCl (Figure 3A). Chl a fluorescence induction in thylakoids from *A. germinans* showed a salt requirement (Figure 3B, inset) very similar to that of Hill reaction; in this case, the variable part of the chlorophyll fluorescence induction curve (F_p-F_o) increases with added NaCl, the concentration dependence (Figure 3B and inset) paralleling that of the O_2 evolution (Figure 3A).

B. Tolerance of O_2 Evolution to High Salt Levels in Halophytes

Thylakoids from *Avi. germinans* not only required high Cl^- concentrations for O_2 evolution but also were unusually tolerant towards high levels of salt in the assay medium. Their O_2 evolution and Chl^- a fluorescence induction were not inhibited by NaCl concentrations as high as 3 *M* (data not shown). Despite these exceptional properties, thylakoids from *Avi. germinans* had the normal periodicity of four (Figure 6A) in flash-induced O_2 yield (O_2 yield per flash as a function of flash number) in the presence of sufficient NaCl (i.e., 250 m*M*), and showed (Figure 6B) typical changes in flash patterns[47] when treated with low concentrations of NH_2OH. Because an electron acceptor was added to the samples, the flash patterns showed a faster damping of the oscillations in O_2 yield than normally observed.[48]

C. Chloride-35 NMR of Chloroplast Suspensions

1. NMR Relaxation Mechanisms

In thylakoid suspensions, the ^{35}Cl and ^{81}Br NMR line shapes are broadened by the quadrupole relaxation associated with binding of the anions to these membranes.[40-42] By adding EDTA to the suspensions, by decoupling the protons, and by using ^2H_2O instead of H_2O as the medium, we have shown that relaxation by paramagnetic species[40] and by proton dipolar interactions, respectively, contributes less than 2 Hz to the ^{35}Cl NMR line width. Therefore, a ^{35}Cl nucleus (spin I = $^3/_2$) bound to a thylakoid membrane is predicted to have a Lorentzian line shape with a width determined by the quadrupolar interactions:[49-51]

$$\Delta\nu_b = (2/5)\pi^2\cdot(e^2qQ/h)^2\cdot\tau_c, \tag{8}$$

in which e is the electronic charge, eq is the electric field gradient at the nucleus, which has an electric quadrupole moment eQ, and τ_c is the correlation time characterizing the fluctuations of the electric field gradient; h is Planck's constant.

Fast chemical exchange of Cl^- between "bound" and "free" (but hydrated) states averages the different relaxation rates and line widths, and gives in the extreme narrowing limit[49-51] a single Lorentzian line of width

$$\Delta\nu_{obs} = P_b\cdot\Delta\nu_b + (1 - P_b)\cdot\Delta\nu_{free}, \tag{9}$$

in which P_b is the fraction of Cl^- that is bound. $\Delta\nu_b$ can be quite large so that the binding of even a small fraction of the Cl^- can affect the averaged line width by substantial amounts. Such averaging accounts for the single ^{35}Cl NMR lines found in most of our observations at various pH values (Figure 8).

However, composite line shapes similar to those in Figure 9, (curves b and c), were found in several instances. They tend to occur at higher Chl concentrations and acidic pHs, e.g., in *Avi. germinans* at [Chl] ≥ 6 mg/ml and pH 5.8. A study was made of the temperature dependence of the composite line shape for the latter; above 7°C the line shape became single Lorentzian and the temperature dependence of its width paralleled that for the single peak in Figure 8 for *Avi. germinans* in the less concentrated suspension, (2.9 mg of Chl per milliliter, Figure 9). This suggests that the broad component arises from the separation, under certain conditions, of the suspension into two "phases" between which the Cl^- exchange is slow. At the higher concentrations employed, the thylakoids are closely packed, or stacked, so such behavior is not unreasonable. A likely explanation is that the broad component arises from Cl^- "trapped" in and more tightly bound to the grana stacks of appressed membranes. Electron microscopy has shown that under the conditions in which the broader ^{35}Cl component is observed, there is also a larger proportion of appressed membranes[52-54] (glycophytes; Figure 1, halophytes); one would expect Cl^- diffusion through the grana stacks to be slower than through unappressed thylakoids. The ^{35}Cl NMR line width decreases with increasing [Cl^-], but "flattens out" (Figure 10) to a virtually

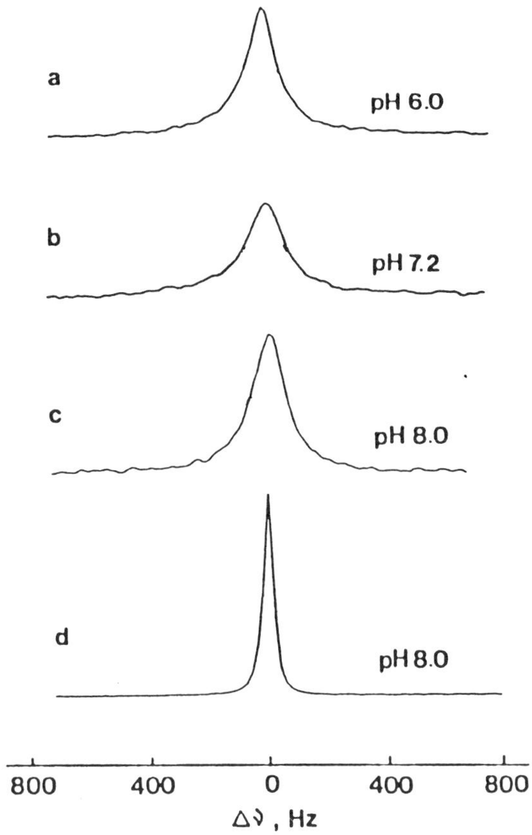

FIGURE 8. pH dependence of ^{35}Cl NMR spectra for *Avi. germinans* chloroplast suspensions with 600 m*M* NaCl added (a to c) and (d) ^{35}Cl NMR spectra of a 600 m*M* NaCl solution in D_2O. (Modified from Critchley, C., Baianu, I. C., Govindjee, and Gutowsky, H. S., *Biochim. Biophys. Acta,* 682, 436, 1982.)

constant value at high [Cl^-]. The inflection point is 0.55 m*M* Cl^- for 2.9 mg of Chl per milliliter and 0.95 *M* for 5 mg of Chl per milliliter, reflecting the dependence of $\Delta\nu_l$ upon [Chl].

2. Temperature Dependence of the ^{35}Cl NMR Line Width

The effect of temperature (0 to 17°C) upon Dn_l is shown in Figure 11 for the thylakoids of *Avi. germinans* at pH values of 7.2 and 5.8 with 2.9 mg of Chl per milliliter and 1 *M* NaCl. The broadening is less at the lower pH and increases for both pH values at lower temperatures. The increase is linear in 1/T between 7 and 17°C but deviates towards larger $\Delta\nu_l$ at the lower temperatures. In these experiments the line shape was a single Lorentzian at pH 7.2 for all temperatures. However, at pH 5.8 an incompletely resolved and broader peak was observed below 7°C when [Chl] was >5 mg/l. The center of the broader component appeared to have the same chemical shift as the narrow component and the NaCl solution.

Our analysis of the concentration dependence of the ^{35}Cl line width is based on Equation 9, which assumes that chemical exchange is fast enough for the extreme narrowing limit to apply. Usually, $\Delta\nu_b >> \Delta\nu_{free}$,[49] also $P_b << 1$, so to a good approximation,

$$\Delta\nu_l = \Delta\nu_{obs} - \Delta\nu_{free} = P_b \cdot \Delta\nu_b \tag{10}$$

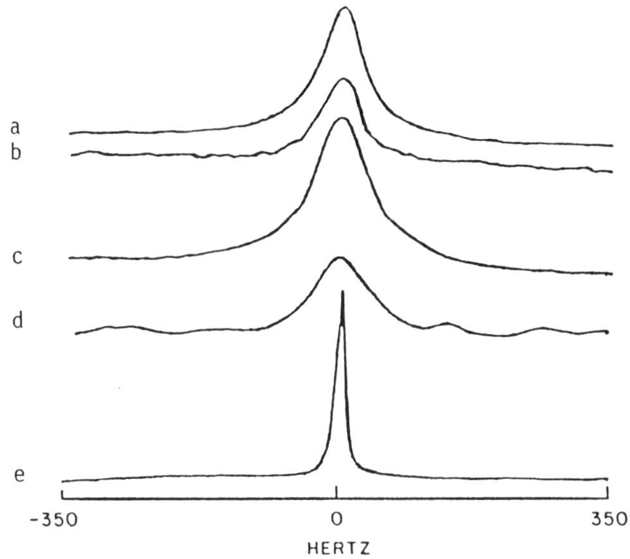

FIGURE 9. ^{35}Cl NMR spectra of thylakoid suspensions (a to d) and 500 m*M* NaCl (e) (Resonance at 24.5 MHz; spectra recorded at 5°C). a, *A. germinans* thylakoids at pH 7.4. with 250 m*M* NaC1. 1 m*M* $MgC1_2$ (2.9 mg Chl per milliliter; b, *Aster tripolium* thylakoids at pH 7.2 with 250 m*M* NaCl added (3 mg Chl per milliliter), c Spinach thylakoids at pH 7.2 with 10 m*M* NaCl added (4 mg Ch1 per milliliter); and d, Spinach thylakoids prepared with 0.2 m*M* NaCl at pH 7.2 (4 mg Chl per milliliter). a to d, 4000 scans and e, 200 scans. (Modified from Baianu, I. C., Critchley, C., Govindjee, and Gutowsky, H. S., *Proc. Natl. Acad. Sci. U.S.A.*, 81, 3713, 1984.)

When the chemical exchange is slow, the value for $\Delta\nu_b$ in Equation 10 is replaced by the more general term:[15]

$$1/(T_{2b} + \tau_{exc}) = 1/[(1/\Delta\nu_b) + \tau_{exc}] \tag{11}$$

in which T_{2b} is the transverse relaxation time of chlorine nuclear spins in the bound state; τ_{exc} decreases at the higher temperatures, so if it were significant, the broadening $\Delta\nu_l$ produced by the thylakoids would increase at higher temperatures. Alternatively, if τ_{exc} is negligible, the temperature dependence of $\Delta\nu_l$ is determined by that of τ_{exc} in Equation 11, which decreases at higher temperatures. In the actual case, $\Delta\nu_l$ decreases at higher temperatures (Figure 11), suggesting that the ^{35}Cl line width is dominated by the quadrupole relaxation. This conclusion is consistent with the earlier, detailed studies of Cl^- binding to hemoglobin,[51] where a similar concentration dependence to the lower curve in Figure 10 was observed, and the ^{35}Cl NMR line widths were in the region of 40 to 50 Hz.

3. Suspension Concentration Dependence of ^{35}Cl NMR Line Width

If the fraction P_b of the Cl^- bound to thylakoids is small, it may be expressed in terms of the binding constant K_b, the total concentration of binding sites [E], and the total Cl^- concentration $[Cl^-]$. Thereby, Equation 10 is converted to the following expression[51] for the concentration dependence of the resultant broadening:

$$\Delta\nu_l = P_b K_b [E]/(1 + K_b [Cl^-]) \tag{12}$$

For a fixed $[Cl^-]$, it predicts that $\Delta\nu_l$ will be a linear function of [E], and thereby of [Ch1].

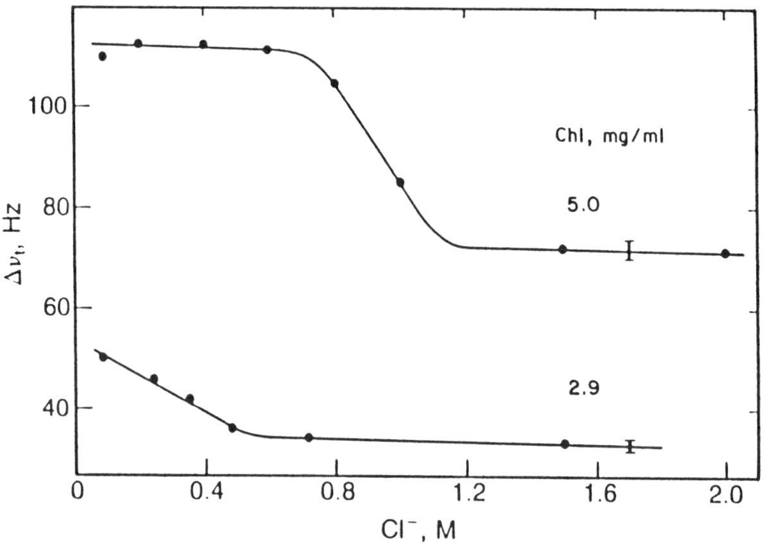

FIGURE 10. [35]Cl NMR line width dependence on [Cl⁻] for *Avi. germinans* chloroplast suspensions at pH 7.0 and 18°C. (From Baianu, I. C., Critchley, C., Govindjee, and Gutowsky, H. S., *Proc. Natl. Acad. Sci. U.S.A.*, 81, 3713, 1984. With permission.)

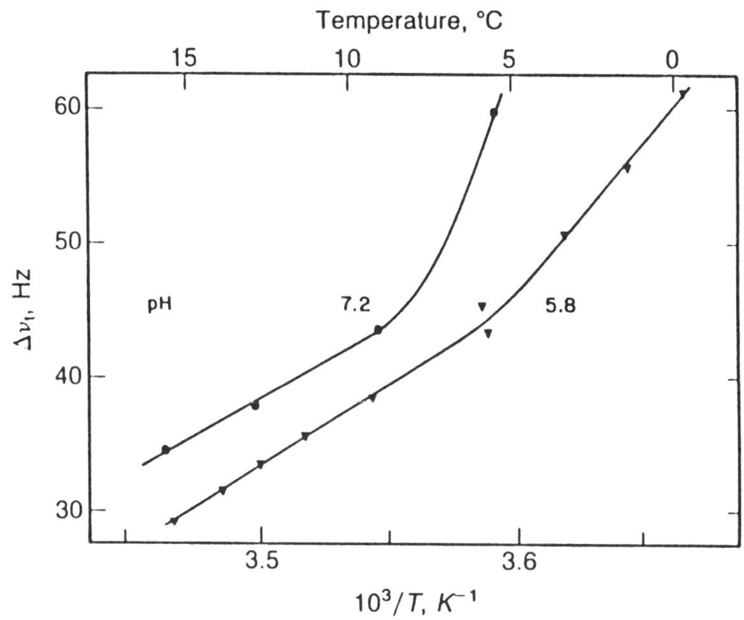

FIGURE 11. Temperature dependence of ^{35}Cl line broadening, $\Delta\nu_l$, by thylakoids of *Avi. germinans* at pH values of 7.2 and 5.8 with 1 *M* NaCl, 2.9 mg of Chl per milliliter, and a buffer of 50 m*M* Hepes per 1 m*M* $MgCl_2$. (From Baianu, I. C., Critchley, C., Govindjee, and Gutowsky, H. S., *Proc. Natl. Acad. Sci. U.S.A.*, 81, 3713, 1984. With permission.)

The dependence of $\Delta\nu_l$ upon [Chl] at a fixed Cl^- (0.6 *M*) is shown in Figure 12 for 5°C; it is linear except possibly for concentrations above 6 mg/ml.

For fixed [Chl], Equation 12 predicts that $\Delta\nu_l$ is largest at low [Cl^-] and drops to zero

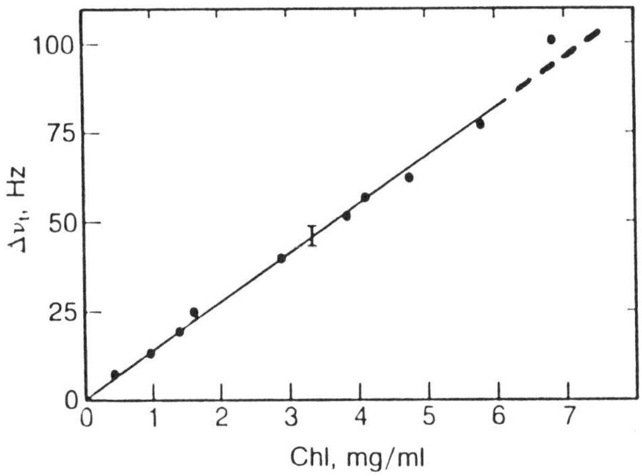

FIGURE 12. Dependence upon [Chl] of ^{35}Cl NMR line broadening, $\Delta\nu_l$, by thylakoids of *Avi. germinans* at pH 7.4, 5°C, 0.6 *M* NaCl, and buffer as for Figure 9, curve a. (From Baianu, I. C., Critchley, C., Govindjee, and Gutowsky, H. S., *Proc. Natl. Acad. Sci. U.S.A.*, 81, 3713, 1984. With permission.)

hyperbolically as the [Cl$^-$] becomes large and the fraction bound becomes small. The data for *Avi. germinans* in Figure 10 at Chl concentrations of 2.9 and 5.0 mg/ml are in only partial agreement with this prediction. At high [Cl$^-$], both curves approach relatively large $\Delta\nu_l$s, of 30 and 70 Hz, respectively, rather than zero. These results are similar to those reported for chloride binding to hemoglobin, which were interpreted in terms of two binding sites, one characterized by a high, and the other by a low binding constant.[51]

In our case, the near constancy of $\Delta\nu_l$ at high [Cl$^-$] leads us to suggest that it is a limiting value caused by nonspecific interactions of Cl$^-$ with membranes. The relaxation mechanism associated with such interactions could be the diffusional encounter by Cl$^-$ ions of the hydrophobic hydration spheres of nonpolar groups in the membranes such as that proposed to account for the strong relaxation of ^{81}Br in aqueous solution with organic solutes.[55] Support for this interpretation is provided by the effect of heating the thylakoids for 3.5 min at 45°C.[41] The treatment eliminates oxygen evolution and reduces the ^{35}Cl NMR line broadening to the 30 Hz found at higher [Cl$^-$] with unheated thylakoids, and is therefore attributable to the deactivation of the specific binding sites by heating. Note that such a heat treatment is known to leave intact the electron transport chain after M.[6,7]

K_b may be estimated graphically from the dependence of $\Delta\nu_l$ upon [Cl$^-$]. We subtracted the limiting value, $\Delta\nu_{lim}$, of $\Delta\nu_l$ found at high Cl$^-$ from the $\Delta\nu_l$ observed at lower Cl$^-$, defining $\Delta\nu_l' = \Delta\nu_l - \Delta\nu_{lim}$. The $\Delta\nu_l'$, should be the net line broadening caused by the binding of the Cl$^-$ to specific sites of the thylakoids. Introducing $\Delta\nu_l'$ into Equation 9 and rearranging it gives

$$\Delta\nu_b[E]\ (1/\Delta\nu_l') = [Cl^-] = (1/K_b) \tag{13}$$

Accordingly, the reciprocal of the net line broadening should be a linear function of the [Cl$^-$], with an intercept $(1/K_b)$ and the slope $1/(\Delta\nu_b[E])$.

Plots of this type are given in Figure 13 for suspensions of *Avi. germinans* at 5 and 25°C. They were obtained by adjusting $\Delta\nu_{lim}$ to give straight lines, the values used being 54 and 27 Hz, respectively. Intercepts of 10 and 7 M^{-1} were obtained for 5 and 26°C, respectively. The difference in K_b at these two temperatures corresponds to a binding energy ΔE_b of about 3 kcal/mol.

At this point, one can attempt to estimate the concentration of binding sites ([E]). [E] appears as a product with $\Delta\nu_b$, the linewidth for bound Cl$^-$, in the expression $1/\Delta\nu_b[E]$ for the slopes of

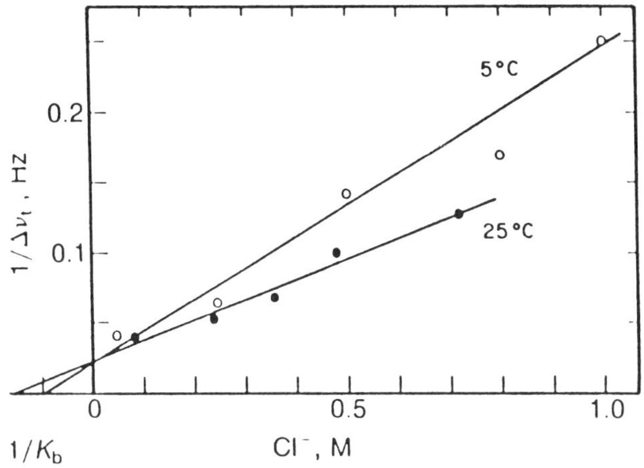

FIGURE 13. The inverse ($1/\Delta\nu_t$) of the broadening of the ^{35}Cl NMR line width, excluding that from the nonspecific interactions (see text) plotted vs. [Cl⁻] for suspensions of thylakoids from *Avi. germinans*, 2.9 mg of Chl per milliliter buffer. (From Baiano, I. C., Critchley, C., Govindjee, and Gutowsky, H. S., *Proc. Natl. Acad. Sci. U.S.A.*, 81, 3713, 1984. With permission.)

the lines in Figure 13. Such results for K_b are comparable with the value of 10 M^{-1} reported for the binding of Cl⁻ to hemoglobin.[51] In the latter, $[E]_t$ was known and thereby the concentration-dependence studies gave a value for the $\Delta\nu_b$ of Cl⁻ bound to hemoglobin of 3×10^4 Hz (for weak ionic binding), a result that provides us with an order of magnitude estimate of $\Delta\nu_b$ for OEC in thylakoids that appears to have a similar Cl⁻ binding curve to that of hemoglobin, (Figure 10, lower curve), (see also Section 6). The slope of the line in Figure 12 for 25°C is 0.15 $Hz^{-1}.M^{-1}$. Combining these two values leads to an $[E]_t$ of 0.2 m*M*; the Chl concentration of 3 mg/ml is 3.3 m*M*, which gives one binding site per 16 Chl (25 Cl⁻ per 400 Chl). This binding ratio is about four times that found necessary to activate the O_2-evolving system in spinach thylakoids, using the radioactive isotope ^{36}Cl.[56] (Such ^{36}Cl measurements are not yet available for halophyte thylakoids).

These estimates of 0.2 m*M* for [E] and 3×10^4 Hz for $\Delta\nu_b$ confirm the assumptions made earlier in the analysis that [Cl⁻] >> $[E]_t$.

4. Hill Reaction Rates of Spinach Thylakoids, Compared with Those of Halophytic Plants: Oxygen Evolution Kinetics and ^{35}Cl NMR

The dependence of the Hill reaction rate upon [Cl⁻] has been reported for thylakoids from several plant species. The results reported were for the halophytes *Avi. germinans* (Figures 1 and 2 in Reference 41) and *Avicennia marina* (Figure 2 of Reference 42), or for spinach. The Hill activities are given in terms of the maximal reaction rate V_{max} for each set of data. The [Cl⁻] scale for spinach is 100 times that for the halophytes and the data points obtained beyond 1.7 and 170 m*M* Cl⁻, respectively, are not shown. The concentrations given include any residual Cl⁻ in the thylakoids used (4), which are less than a few millimoles.

Figure 14 shows that the Hill activity of glycophytic and halophytic thylakoids increases with the [Cl⁻]. As pointed out by Kelley and Izawa,[57] Cl⁻ activates in some manner the O_2 evolving enzyme (E) of PSII. The simplest, schematic model for the activation is as follows:

$$E + Cl^- \underset{K_{-1}}{\overset{K_1}{\rightleftharpoons}} [ECl^-] \xrightarrow{h\upsilon_{II},\ H_2O} ;\ H_2O\cdot E + Cl^- + O_2 + 4H^+ + 4e^-$$

The kinetics of O_2 evolution by such a process are the same as the initial reaction rate of a

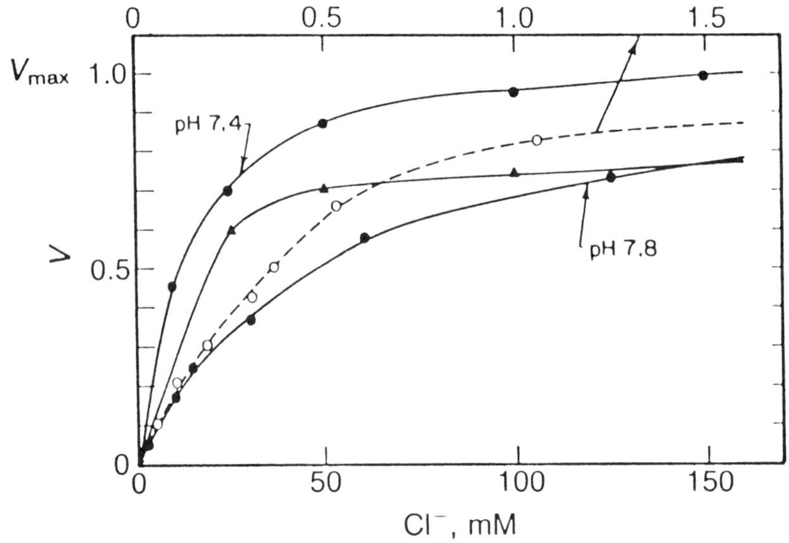

FIGURE 14. Dependence of the maximal reaction rates, V_{max}, of oxygen evolution (measured by the Hill reaction) on the Cl^- concentration in thylakoid suspensions. (○) Spinach thylakoids; *Avi. germinans* thylakoids isolated and tested at pH 7.4 (●) or at pH 7.8 (●); and (▲) *Aster tripolium* thylakoids tested at pH 7.4.[42] Halophytic plants (top scale) and a glycophytic plant (bottom scale).

unireactant enzyme for which Cl^- would be the (hypothetical) substrate and O_2 is the product.[58] If $[Cl^-] >> [E]$, the dependence of the Hill activity upon $[Cl^-]$ should be given by the Henry-Michaelis-Menten equation:

$$v/V_{max} = [Cl^-]/K_m + [Cl^-]) \tag{14}$$

For the reaction[11] the Michaelis-Menten constant is

$$K_m = (k^{-1} + K_p)/k_1 \tag{15}$$

The curves in Figure 14 are of the general form predicted by the Henry-Michaelis-Menten equation. The best fit to the Henry-Michaelis-Menten equation is given by recent results for spinach. They are presented in Figure 15 as a double-reciprocal plot of 1/Hill activity vs. $1/[Cl^-]$, the latter including "endogenous" Cl^-. The 1/v data have been renormalized to a V_{max} that is 35% larger than the rate reported for 25 m*M* Cl^-.[59] This gives a self-consistent linear plot in which the same value of 0.6 m*M* is obtained for K_m from the $1/v = 0$ intercept and from the $[Cl^-]$ at $V_{max}/2$, as predicted by Equation 12. The k_m from the $[Cl^-]$ for Izawa's $V_{max}/2$ is only 0.35 m*M*. An earlier analysis by Kelley and Izawa[57] of spinach thylakoid data gave a large K_m, 0.9 m*M*, at pH 7.2. However, the medium contained 2 mg of bovine serum albumin (BSA) per milliliter which might also bind Cl^-.

On the other hand, a linear, double-reciprocal plot (not shown), with a K_m of 25 m*M*, was obtained from the data in Figure 14 for *Avi. germinans* at pH 7.8. However, the double-reciprocal plots of the data for the other cases we tested exhibit curvature. Such data indicate large differences between thylakoids isolated from halophytic and glycophytic plants, as well as some variability in results caused by the addition of an extraneous protein (BSA, see Reference 57).

5. pH Dependence of ^{35}*Cl NMR Line Width and Oxygen Evolution*

The effect of Cl^- on O_2 evolution, as estimated by the ferricyanide, Hill reaction rates in

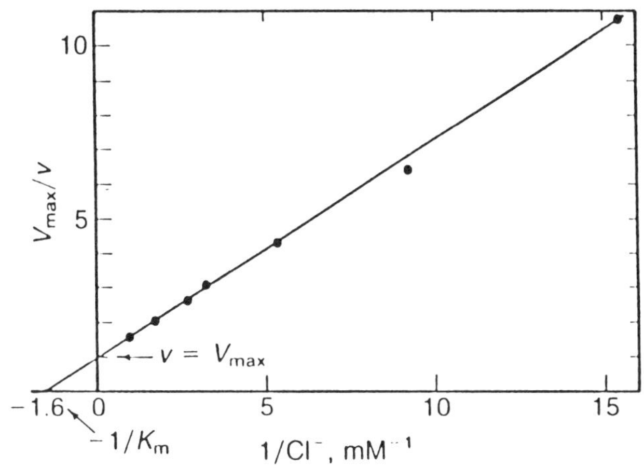

FIGURE 15. Double-reciprocal plot of the $1/O_2$-evolving rate (v) normalized to V_{max} (see text) vs. $1/[Cl^-]$ using data of Izawa et al.[32] for *S. oleracea* at 21°C with 10 μg of Chl per milliliter.[42]

thylakoids from *Avi. germinans* and *A. tripolium*, is pH dependent in a manner similar to that reported for *Avi. marina* thylakoids prepared at pH 7.6.[60]

The dependence of O_2 evolution (O) and ^{35}Cl NMR line width (❑) upon pH in the range 5.9 to 8.2 is shown in Figure 16A (from Reference 41) for thylakoids from *Avi. germinans*, prepared and kept at pH 7.6 prior to the assay. These samples were assayed at different pH values in the absence (❑) and in the presence of high NaCl concentrations (O, 0.5 *M* in Figure 16C, and 1 *M* in Figure 16B). In the absence of added salt, the O_2-evolution rates for pH values from 7.4 to 8.4 were unstable, and thus only the initial rates are plotted in Figures 16B and C. In the presence of high salt concentration all measured rates were stable and higher by approximately 30 μmol O_2 per milligram Chl per hour.

The rate of Chl^- a fluorescence rise, $(F_p\text{-}F_0)/\Delta t$, (equivalent to the initial slope) in *Avi. germinans* isolated at pH 7.0, with added salt, increases with increasing pH up to 7.6 (Figure 16C), while that without added Cl^- decreases. The pH dependence of O_2 evolution in *A. tripolium* thylakoids, prepared and kept at pH 7.0, and assayed with or without salt, is presented in Figure 16C (from Reference 41).

The photochemical efficiency,[61] $(1- F_o/F_p)$, of PSII showed the same trends with regard to Cl^- concentration and pH (data not shown) as those given in Figure 17B (from Reference 41).

The specific contribution to the ^{35}Cl NMR line broadening caused by the fraction of Cl^- involved in O_2 evolution was estimated by subtracting the ^{35}Cl NMR line widths for thylakoid suspensions that were heat inhibited.

Heat treatment of the thylakoids at 48°C for 3.5 min abolished completely the O_2 evolution (data not shown) and also the pH dependence of the ^{35}Cl NMR line width (Figure 18B), in this case, the line width for the heat-treated thylakoids (47 Hz), while 5 to 10 Hz less than that of the untreated thylakoids is still approximately 30 Hz greater than that (17 Hz) for aqueous Cl^-. This shows that the heat treatment does affect the Cl^- binding sites, the Cl^- binding to the treated thylakoids is extensive even when M (or OEC), is specifically inhibited.

Removal of the very loosely bound Mn^{+2} by EDTA washing [41] did not significantly alter the pH dependence (Figure 18A), indicating that any contributions by this paramagnetic species to the ^{35}Cl line broadening are small.

These results suggest (1) ^{35}Cl NMR may be used to study the role of Cl^- in O_2 evolution since it monitors Cl^- binding including binding to OEC; and (2) the amount of Cl^- involved in the function of O_2 evolving complex may be estimated by subtracting the amount of Cl^- bound after specific heat treatment from that of the untreated sample.

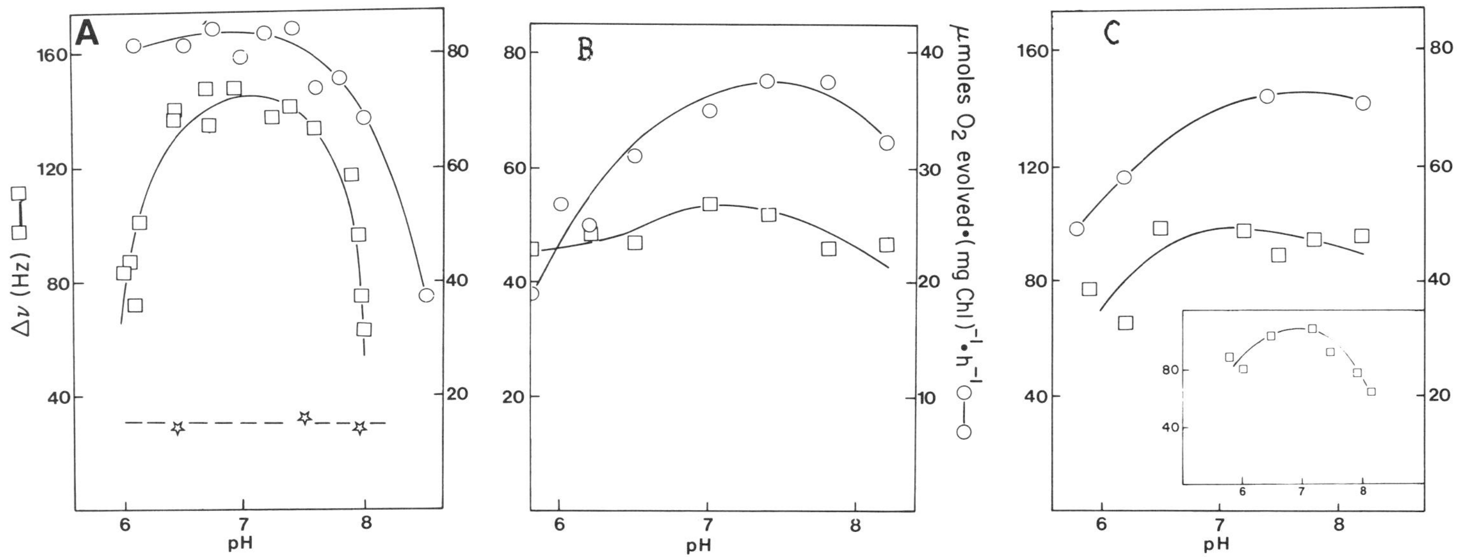

FIGURE 16. The pH dependence of O_2 evolution with ferricyanide as electron acceptor (open circles) and ^{35}Cl NMR line widths (open squares) in thylakoids. (A) from *Avi. germinans* isolated at pH 7.6 with 6 mg Chl per milliliter and 1 *M* NaCl. (B) from *Avi. germinans* isolated at pH 7.0 with 3 mg Chl per milliliter and 0.5 *M* NaCl. The points (☆) at the bottom of A are for 1*M* NaCl solution, without thylakoids. The inset in C is the pH dependence of the ^{35}Cl line width observed in the same samples 1 h after the first measurements. (C) Assays are as specified in the caption to Figure 3 of Reference 41 from *Aster tripolium,* isolated at pH 7.0, (3 mg Chl per milliliter), with 0.5 *M* NaC1 added. (From Critchley, C., Baianu, I. C., Govindjee, and Gutowsky, H. S., *Biochim. Biophys. Acta,* 682, 436, 1982. With permission.)

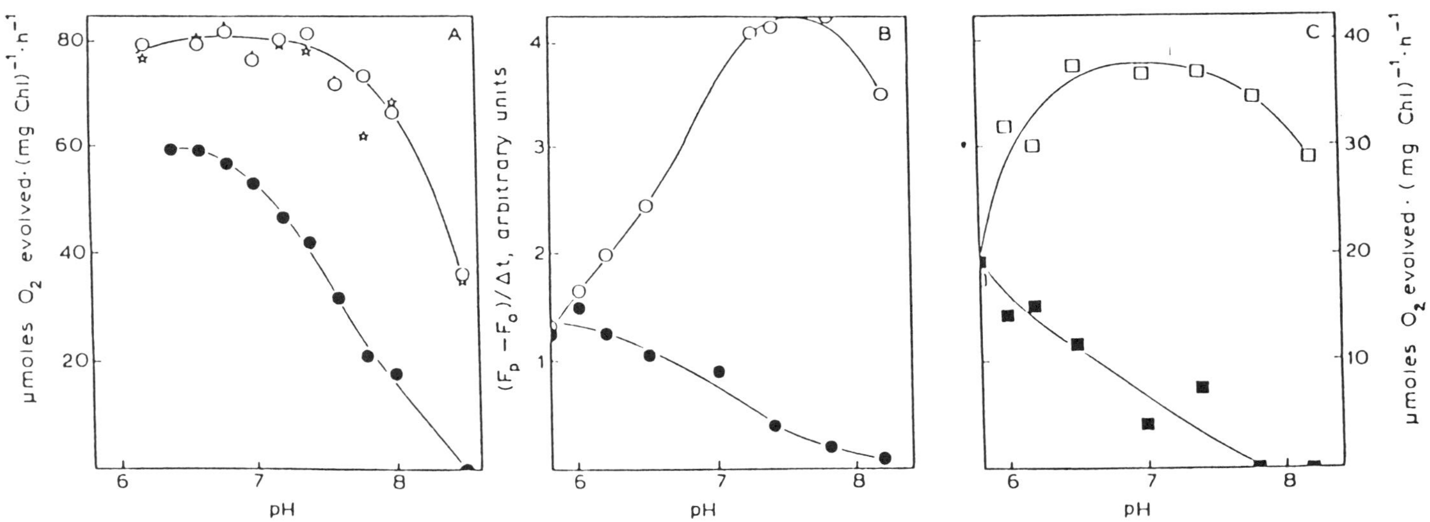

FIGURE 17. The pH dependence of electron transport activity in the absence (closed symbols), or presence (open symbols), of added 0.5 *M* (☆) or 1.0 *M* (○ and) NaCl: (A) O_2 evolution in thylakoids from *Avi. germinans* isolated at pH 7.6; (B) O_2 evolution in thylakoids from *Aster tripolium* isolated at pH 7.0; and (C) the initial rate of the O → P fluorescence rise in thylakoids from *Avi. germinans* isolated at pH 7.0. Buffers (50 m*M*): Mes (pH 5.8 to 6.7), Hepes (pH 6.8 to 7.8), and Tricine (pH 7.6 to 8.4) at the pH values indicated. Δt: time difference between the two points used for fluorescence measurements in C. (Data from Critchley, C., Baianu, I. C., Govindjee, and Gutowsky, H. S., *Biochim. Biophys. Acta*, 682, 436, 1982.)

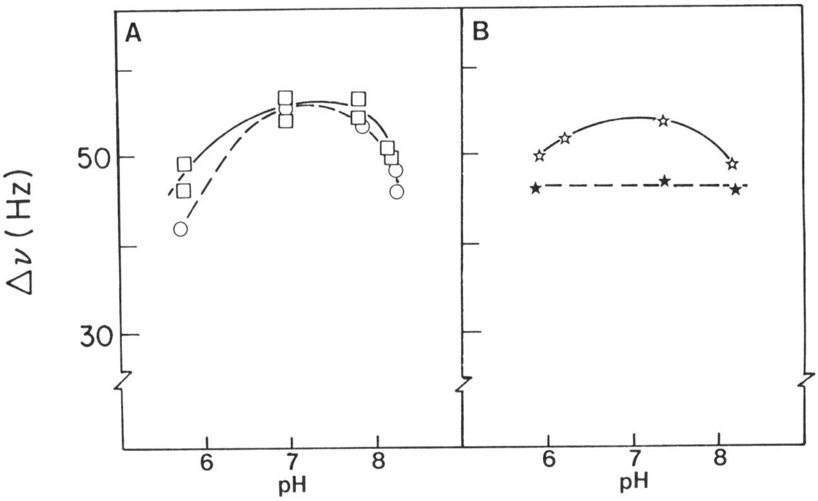

FIGURE 18. pH dependence of ^{35}Cl NMR linewidth in thylakoids from *Avi. germinans*: (A) isolated in the presence (○) or absence (□) of 2 m*M* EDTA in the grinding and washing media (Chl concentration, 3 mg/ml; NaCl, 0.5 m*M*); (B) before (☆) or after (★) heating to 48°C for 3.5 min, both in the presence of 2 m*M* EDTA (Chl concentration, 2.9 mg/ml, 2.9 mg/ml; NaCl, 1 *M*). (From Critchley, C., Baianu, I. C., Govindjee, and Gutowsky, H. S., *Biochim. Biophys. Acta*, 682, 436, 1982. With permission.)

6. Chloride Binding Parameters from ^{35}Cl NMR Measurements

The following information regarding Cl^- binding to thylakoids from halophytes has now been obtained.

a. Fast Chemical Exchange

Two observations suggest that there is a fast chemical exchange of Cl^- between the membrane sites and the bulk phase: (1) there is a single Lorentzian ^{35}Cl NMR line for most thylakoids; (only when thylakoids are stacked and Cl^- may be trapped, there is an additional broad line) and (2) as the temperature is increased, the ^{35}Cl NMR linewidth narrows suggesting that T_{2bound}, (the transverse relaxation time of the bound Cl^-) dominates the measurements. From these observations, the value of $1/T_{2bound}$ is estimated[62] to be >1000 s^{-1}.

b. Cl^- Binding Constant (K_b)

This was obtained[62] as follows: the inverse of the line width (y) for bound Cl^-, $y = (\Delta\nu_{observed} - \Delta\nu_{free})^{-1}$, was plotted as a function of $[Cl^-]$. The K_b was calculated by dividing the slope of the curve by the intercept of the plot since, for the fast exchange limit,

$$y = \frac{(\Delta\nu_{bound} - \Delta\nu_{free})^{-1}}{[sites]} [Cl^-] + \frac{(\Delta\nu_{bound} - \Delta\nu_{free})^{-1}}{K_b\,[sites]}$$

Such a calculation yielded a value of 1.13 M^{-1} for K_b in halophytes which is very similar to the value for Cl^- binding to site B of hemoglobin.[51]

c. Cl^- Binding Energy

Using the relation $K_b = A \exp(-E_{binding}/kT)$ (where k is the Boltzmann constant, A is another constant, and T = absolute temperature) and, by measuring K_b at two temperatures, one can calculate E from the following relationship:[63]

$$\Delta E = k \frac{\ln (k_{b2}/K_{b1})}{T(1)^{-1} - T(2)^{-1}}$$

where T(1) and T(2) are the two temperatures and K_{b1} and K_{b2} are binding constants at T(1) and T(2), respectively.

Using this method, a value of 3 kcal/mol was obtained for ΔE.[62] This value shows that the character of Cl^- binding is weakly ionic.

d. Estimate of the Number of Cl^- Sites per PSII

It is not straightforward to determine a firm number from the existing data. However, we were able to make an estimate with some assumptions (see below). This estimate for Cl^- sites in halophytes that may be related to the O_2 evolution activity, is 20 to 40 Cl^- per 400 Chl^- a; its order of magnitude compares favorably with a value of 10 Cl^- per 600 Chl, obtained in spinach by measuring ^{36}Cl binding.[56] The details of the method and the assumptions involved in obtaining the estimate for Cl^-/PSII follow. The inverse of $(\Delta\nu_{observed} - \Delta\nu_{free})$ was plotted as a function of the inverse of thylakoid concentration, in terms of $[Chl]^{-1}$ for a fixed $[Cl^-]$ of 0.6 *M*. The number of Cl^- binding sites is unknown, but it must be related to [Chl] by a fixed ratio: [sites] = [Chl]/f. The measured slope of the curve y as a function of $[Chl]^{-1}$ can be shown[64] to equal f $(\Delta\nu_{bound})^{-1} \cdot ([Cl^-] + 1/K^-)$. Since $\Delta\nu_{bound}$ is expected to be about 10^4 Hz (based on a comparison with the stronger binding site of Cl^- of hemoglobin (Reference 51, see Section 3), and since the other parameters are known ($[Cl^-] = 0.6\ M$; $K_b = 1.13\ M^{-1}$), one can estimate f to be about 4.[64] If we assume that PSII contains 400 Chl molecules, then there are 100 Cl^- sites per PSII. However, under complete inhibition of O_2 evolution activity, we had only 20 to 40% reduction of the line broadening; thus, the number of Cl^- binding sites associated with O_2 evolution is 20 to 40 Cl^- per 400 Chl in halophytes.[64] This number remains to be determined for other plants from ^{35}Cl NMR data.

The Cl^- associated with the O_2 evolution may be distributed between two distinct pools. In spinach, there are two reported values of the Henry-Michaelis-Menten constants, K_m, 0.35 and 0.9 m*M* at pH 7.2 (see Section 4), and the corresponding ones for halophytes[64] are 5 to 10 and 20 to 25 m*M* (at pH 7.8), or 12 and 1 m*M* (at pH 7.0); the last value is not too different from the 0.9 m*M* K_m value reported for spinach thylakoids at pH 7.2 by Kelley and Izawa.[57]

D. Site of Chloride Effect

Loss of Chl a $_{II}$ fluorescence induction is often due to inhibition of electron donation from water to the PSII reduction center. However, such a loss is also observed when the electron flow out of Q^-_A is enhanced by the addition of artificial electron acceptors. A choice between these two options can be based on the use of artificial electron donors to PSII. When an electron donor, e.g., catechol, diphenylcarbazide, or hydroxylamine was added to thylakoid samples in the absence of added Cl^- (Figure 4 in Reference 41), the variable fluorescence was reestablished to the same extent that it is by adding Cl^-, (Figure 19). This indicates that PSII electron transport is impaired in the vicinity of water-oxidation site because of the absence of sufficient Cl^-. (Note that Cl^- is not an electron donor to PSII.)

Earlier work located the site of Cl^- action on the electron donor side of PSII and supported the concept that the fast reversible binding of Cl^- might be necessary for the advancement of the S states.[35,65-67] Muallem et al.,[35] in reporting flash experiments on thylakoids, pointed to the Cl^- dependence of the $S_1 \rightarrow S_2$ step. Thus far, sensitivity limitations have restricted the ^{35}Cl NMR measurements to dark experiments. Recent improvements at high-field (11.5 T) NMR instruments should enable the observation of the ^{35}Cl NMR line widths at lower Cl^- and Chl concentrations, as well as of the possible effects of series of light flashes on the ^{35}Cl NMR spectra of thylakoid suspensions.

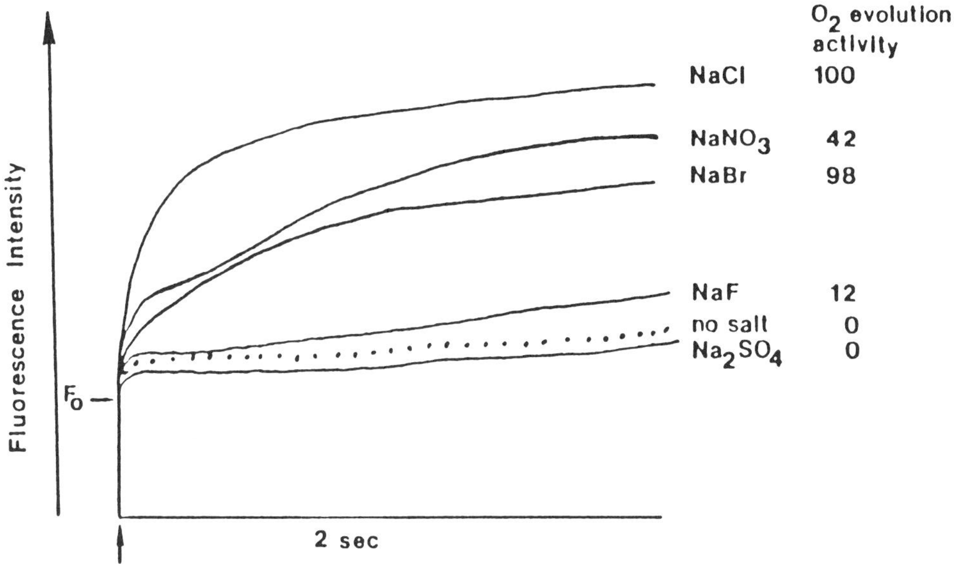

FIGURE 19. Chl a fluorescence transients at pH 7.8 in suspensions of *Avi. germinans* thylakoids (20 mg Chl per milliliter) in the absence of salt (dotted line), or upon the addition of various sodium salts (containuous lines). Monovalent salts, 500 m*M* and divalent salts, 250 m*M*. The corresponding rates of O_2 evolution (mmol/ml Chl per hour) are shown at the right for each anion. (From Critchley, C., Baianu, I. C., Govindjee, and Gutowsky, H. S., *Biochim. Biophys. Acta*, 682, 436, 1982. With permission.)

IV. ANION SPECIFICITY IN PHOTOSYNTHETIC OXYGEN EVOLUTION

The current hypothesis[41,64-65] is that the Cl^- related to O_2 evolution is bound to the OE enzyme; its location is thought to be on the inner side of the thylakoid membrane and the functional Cl^- binding site(s) may be expected to be in an enzyme "pocket" near or at the active site. The binding energy of Cl^- ions is ~3 kcal/mol, and "binding" is reversible (the exchange rate is >1000 s^{-1}). Cl^- ions must be weakly "bound" to some positively charged groups. F^- ions inhibit O_2 evolution. This inhibition may occur because F^- binds more tightly than Cl^- since its ionic field is about 1.8 times higher than that of Cl^-; however, this would not permit weak, reversible binding, which may be necessary for the cycling of the S states.[67] The inhibitory F^- effect in thylakoids from halophytic plants is illustrated in Figure 20. One problem with the above interpretation is the time course of the inhibition of O_2 evolution caused by F^- (Figure 20A) which is consistent only with a slow exchange of Cl^- by F^-, and would apparently contradict the fast exchange mechanism derived from the ^{35}Cl NMR data. An added difficulty is the relatively large $[F^-]$ concentration needed to completely inhibit oxygen evolution (Figure 20B). An alternative interpretation could involve irreversible binding of F^- to Mn^{+2} (or Mn^{+4}) buried in the thylakoid membrane, or to some other positively charged groups of OEC that are not exposed to water. Such a process could cause a slow denaturation of the OE enzyme, dissociation of the OEC, or both, and would be consistent with the relatively large concentrations of F^- needed to inhibit completely O_2 evolution.

The central peak observed in the ^{35}Cl NMR spectrum for *A. tripolium* and spinach (Figure 9, curves B and C), with 250 and 10 m*M* NaCl, 3.1 mg of Chl per milliliter, and pH 7.2, was about 50 and 70 Hz, respectively. Inclusion of 250 m*M* NaF in these suspensions and incubation for 10 min in the dark prior to measurements narrowed the lines to about 35 Hz for both suspensions (by 30 and 50%).

The mechanism of Cl^- action should be related to its specificity for PSII. For *Avi. marina* thylakoids, the specificity was very pronounced, such that Br^- and I^-, for example, could not

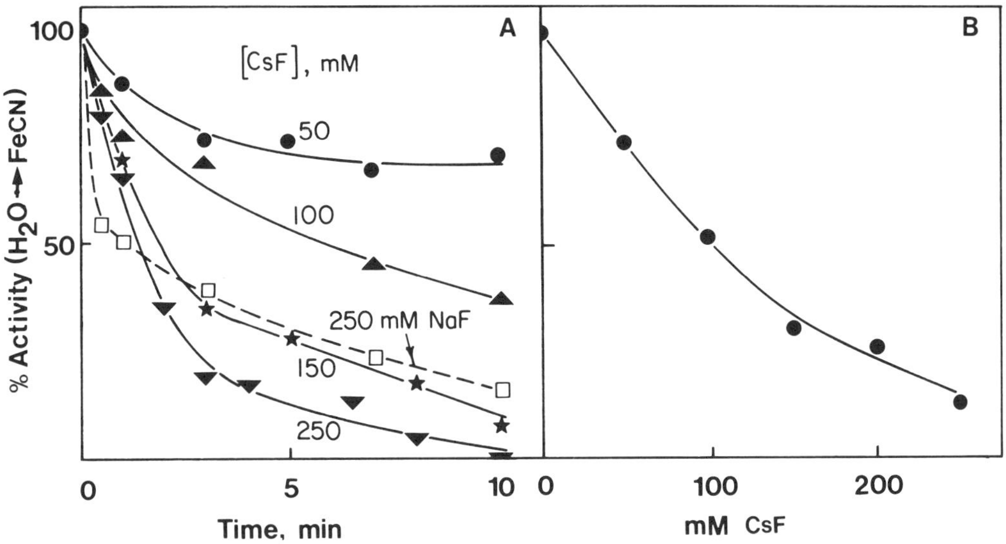

FIGURE 20. Effect of F^- on O_2 evolution activity in thylakoids from *Avi. germinans* at pH 7.5 in 250 m*M* NaCl: (A) as a function of time after adding various concentrations of F^-, and (B) as a function of [CsF] after a 3-min incubation time in the dark. (From Critchley, C., Baianu, I. C., Govindjee, and Gutowsky, H. S., *Biochim. Biophys. Acta*, 682, 436, 1982. With permission.)

substitute for Cl^-.[60] In thylakoids from *Avi germinans*, Br^- and NO_3^- were found to stimulate Chl a fluorescence transients and O_2 evolution to a substantial degree, as shown in Figure 19. On the other hand, both F^- and SO_4^{2-} have no significant stimulating effect on either the O_2 evolution or the Chl a fluorescence, and F^- is actually an inhibitor of both. For Br^- the fluorescence yield is low compared with the O_2 evolution suggesting that Br^- may somehow quench Chl a_{II} fluorescence, perhaps by substituting for Q^-_A. The effect of F^- occurs both on fluorescence and oxygen evolution (Figures 19 and 20).

One might expect the specificity of the anion to depend upon factors such as its size, shape, hydration, and charge. In Figure 21 we have plotted the O_2 evolution, in relative units, reported for spinach thylakoids in Reference 41 vs. the anion volume; the latter was calculated for spherically symmetric ions from the Pauling radius. A similar plot is given in Figure 21B for the mangrove (halophyte) data. Figure 21A shows that the Cl^- volume of 25 $Å^3$ is optimum for O_2 evolution in spinach, as well as in mangrove, thylakoids. The photochemical activity falls off progressively for smaller as well as for larger anions, with a somewhat more rapid fall-off for mangroves than for spinach chloroplasts with ions larger than Br^-.

The decreasing effectiveness of the large ions is consistent with steric effects, either at the binding site itself or in the access of the ion to the site through a channel, or pores, of restricted size.

The specificity of the requirement for Cl^- is quite sharp. Steric factors can account for the ineffectiveness of the large ions, but the inhibitory effects of the smaller ions require a different explanation. Although not tested for its inhibitory action in halophyte thylakoids, with spinach thylakoid membranes at 100 m*M* Cl^- and neutral pH, SO_4^{-2} is neither stimulatory nor inhibitory. This behavior of the "large" SO_4^{-2} ion is quite distinct from the inhibition by the "small" F^- of O_2 evolution in halophytic thylakoids (Figure 20). The simplest interpretation is that the size of SO_4^{-2} prevents it from affecting the membrane sites involved in the electron transfer. The situation with OH^- is more complex. Even though OH^- has approximately the same size as F^-, it inhibits O_2 evolution at much lower concentrations (3 m*M* vs. 100 m*M*) suggesting that OH^- and F^- must have different actions.

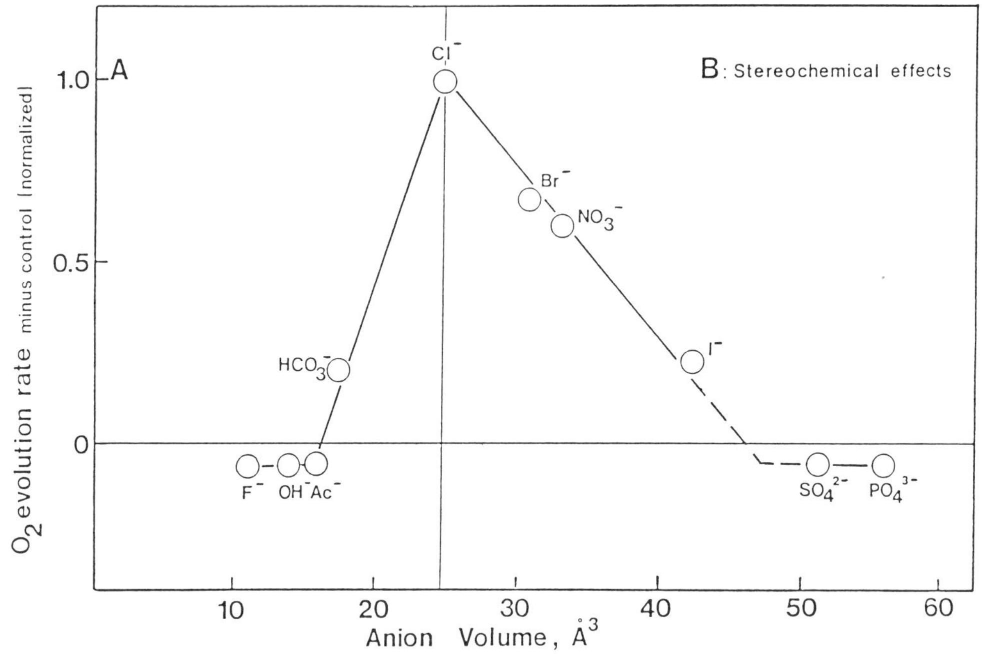

FIGURE 21. Hill reaction rates (O_2 evolution with ferricyanide as electron acceptor) vs. the anion volume calculated for a spherical ion of Pauling radius: (A) of spinach thylakoids at pH 7.2, O_2 data from Reference 5 and (B) of mangrove thylakoids at pH 7.8; data from Figure 17 plus I-oxide from Reference 32. The points in B are for the ions directly above in A. (Modified from Critchley, C., Baianu, I. C., Govindjee, and Gutowsky, H. S., *Biochim. Biophys. Acta*, 682, 436, 1982.)

Inhibition of the Hill reaction at alkaline pH values has been reported previously [68,69] and any efficient method for Cl^- depletion of thylakoids involves the use of an alkaline solution. The pH dependencies of the O_2 evolution and Cl^- binding shown in Figure 16A to C for mangrove thylakoids are a measure of the effects of OH^-, at least for the higher pH values. The data in Figure 15A show that both O_2 evolution and Cl^- binding are optimal at neutral pH values (approximately 7.2). Increasing the pH to above approximately 8 causes a sharp reduction in both O_2 evolution and Cl^- binding. This is probably caused by titration of Cl^- binding sites by OH^-, followed by release of Cl^- from such sites. (Mild heating can apparently cause similar Cl^- release.)[38] Since there is only a small number of such groups, low concentrations of OH^- are sufficient. The time course of the OH^- effect is also of interest as the effect occurs within 60 s.[69]

These differences suggest that changes in pH not only displace the Cl^- but also modify the charge distribution in the thylakoids and/or the binding sites. Moreover, in the case of spinach, evidence is found for conformational changes as well.[41]

OH^- may substantially modify the charge distribution in the thylakoids and/or the binding sites. In contrast, SO_4^{-2} and PO_4^{-3}, which are "large" anions, do not inhibit O_2 evolution in the presence of Cl^-; unlike F^- and OH^-, PO_4^{-3} and SO_4^{-2} do not replace Cl^- as they are too large (their anion volume is in the range of 55 Å^3 and they cannot enter the "Cl^- pocket"; the anion volume of Cl^- is ~25 Å^3). The order of effectiveness of those anions that can replace Cl^- is Br^-, NO_3^-, and I^-, with Br^- being the most effective.[41] Since Cl^- is the most effective anion in stimulating O_2 evolution, we have suggested that its molecular volume and ionic field are optimum for the function it performs. Its role remains to be found, but its rapid binding and release may facilitate the operation of the O_2 clock, and this is why a reversible binding and an appropriately low binding energy can be crucial to the function of Cl^- *in vivo*.

ACKNOWLEDGMENTS

The kind permission by several publishers to reproduce printed figures from our original papers in this review is gratefully acknowledged.

REFERENCES

1. **Hill, R. and Bendall, F.,** *Nature,* 186, 136, 1960.
2. **Duysens, L. N. M., Amesz, J., and Kamp, B. M.,** *Nature,* 190, 510, 1961.
3. **Velthuys, B. R.,** Charge Accumulation and Recombination in System II of Photosynthesis, Ph.D. thesis, Leiden University, The Netherlands, 1976.
4. **Cheniae, G. M. and Martin, I. F.,** Sites of function of manganese within photosystem II. Roles in O_2 evolution and system II, *Biochim. Biophys. Acta,* 197, 219, 1970.
5. **Warburg, P. and Luttgens, W.,** *Naturwissenchaften,* 32, 301, 1974.
6. **Homann, P. H.,** Effects of manganese on the fluorescence of chloroplasts, *Biochem. Biophys. Res. Commun.,* 33, 229, 1968.
7. **Blankenship, R. E., Babcock, G. E., and Sauer, K.,** Kinetic Study of Oxygen Evolution Parameters in Tris-washed, Reactivated Chloroplasts, *Biochim. Biophys. Acta,* 165, 387, 1975.
8. **Joliot, P.,** Polarographic Measurements of Oxygen Yields with Flashes in Green Algae, Ph.D. thesis, University of Paris, 1960.
9. **Joliot, P. and Kok, B.,** Oxygen evolution in photosynthesis, in *Bioenergetics of Photosynthesis,* Govindjee, Ed., Academic Press, New York, 1975, 388.
10. **Kok, B. et al.,** *Photochem. Photobiol.,* 11, 457, 1970.
11. **Fowler, C. F. and Kok, B.,** Proton evolution associated with the photoxidation of water in photosynthesis, *Biochim. Biophys. Acta,* 357, 299, 1974.
12. **Saphon, S. and Crofts, A. R.,** *Z. Naturforsch.,* 320, 617, 1977.
13. **Wydrzynski, T., Zumbulyadis, N., Schmidt, P. G., Gutowsky, M. S., and Govindjee,** *Proc. Natl. Acad. Sci. U.S.A.,* 73, 1196, 1976.
14. **Wydrzynski, T. et al.,** Nuclear magnetic relaxation by the manganese in aqueous suspensions, *Biochemistry,* 17, 2155, 1978.
15. **Sharp, R. R.,** Proton NMR relaxation changes in PS II particles in response to light flashes, Annu. Midwest Photosynthesis Conf., Michigan University, Indiana, 1987.
16. **Sharp, R. R.,** Paramagnetic intermediates in the catalytic cycle of photosynthetic oxygen evolution, *Biophys. J.,* 53(2), 614a, 1988.
17. **Sharp, R. R. and Yocum, C. F.,** Field dispersion profiles of the proton spin-lattice relaxation rate in chloroplast suspensions: effect of manganese extraction by EDTA, Tris, and hydroxylamine, *Biochim. Biophys. Acta,* 592, 185, 1980.
18. **Robinson, H. H., Sharp, R. R., and Yocum, C. F.,** On the origin of light-induced changes in the proton magnetic relaxation rate of chloroplast suspensions, *Arch. Biochem. Biophys.,* 207, 1, 1981.
19. **Baianu, I. C. et al.,** High-field proton NMR of algae and isolated chloroplasts. Annu. Meet. Am. Soc. Photobiology, WM2, Bethesda, 1981, 11.
20. **Critchley, C. and Baianu, I. C.,** unpublished results. University of Illinois, Urbana, 1982.
21. **Baianu, I. C. et al.,** First high-field ^1H NMR observations of light-flash effects in photosynthetic membranes, Res. Conf. in Photosynthesis, New Hampshire, 1982.
22. **Cheni, G. M. and Martin, I. F.,** *Plant Physiol.,* 47, 568, 1971.
23. **Yocum, C. F. et al.,** Stoichiometry, inhibitor sensitivity and organization of manganese associated with photosynthetic oxygen evolution, *Proc. Natl. Acad. Sci. U.S.A.,* 78, 7507, 1981.
24. **Baianu, I. C.,** Deuterium NMR and oxygen flash-yields in green algae, Annu. Midwest Photosynthesis Conference, Michigan University, Indiana, 1987.
25. **Baianu, I. C. and Coleman, W. J.,** Unpublished results, University of Illinois, Urbana, 1982.
26. **Sinclair, J. and Arnason, T.,** Studies on the rate-limiting reaction of photosynthetic oxygen evolution in the unicellular algae *Chlorella, Biochim. Biophys. Acta,* 368, 393, 1974.
27. **Arnason, T. and Sinclair, J.,** Studies of the rate-limiting reaction of photosynthetic oxygen evolution in spinach chloroplasts, *Biochim. Biophys. Acta,* 430, 517, 1976.

28. **Crofts, A. R. and Wood, P. M.,** Electron transport and proton pumps, in *Current Topics in Bioenergetics*, Part A, Vol. 7, Sanadi, D. R. and Vernon, L. P., Eds., Academic Press, New York, 1978, 175.
29. **Edzes, H. T. and Samulski, E. T.,** The measurement of cross-relaxation effects in the proton NMR spin-lattice relaxation of water in biological systems: hydrated collagen and muscle, *J. Magn. Reson.*, 34, 1978.
30. **Hind, G., Nakatani, H. Y., and Izawa, S.,** The role of Cl^- in photosynthesis. The Cl^- requirement of electron transport, *Biochim. Biophys. Acta*, 172, 277, 1969.
31. **Bove, J. M. et al.,** *Z. Naturforsch.*, 186, 683, 1963.
32. **Izawa, S., Heath, R. L., and Hind, G.,** The role of the chloride ion in photosynthesis. III, The effect of artificial electron donors upon electron transport, *Biochim. Biophys. Acta*, 180, 388, 1969.
33. **Cheniae, G. M.,** *Annu. Rev. Plant Physiol.*, 21, 467, 1970.
34. **Avron, M.,** The electron transport chain in chloroplasts, in *Bioenergetics of Photosynthesis*, Govindjee, Ed., Academic Press, New York, 1975, 381.
35. **Muallem, A. et al.,** Charge storage in Cl^- depleted chloroplasts. in *Photosynthesis II: Photosynthetic Electron Transport and Photophosphorylation*, Akoyunoglou, G., Ed., Balaban International Science Services, Philadelphia, 1981, 435.
36. **Critchley, C., Baianu, I. C., Govindjee, and Gutowsky, H. S.,** Chloride effects in photosystem II reactions from salt-tolerant higher plants, *Biophys. J.*, 37, 351a, 1982.
37. **Baianu, I. C. et al.,** Chloride-35 NMR of thylakoid suspensions, Annu. Midwest Photosynthesis Meeting, Argonne National Laboratory, Chicago, 1982.
38. **Coleman, W. J., Baianu, I. C., Gutowsky, H. S., and Govindjee,** The effect of chloride and other anions on the thermal inactivation of oxygen evolution in spinach thylakoids, in *Advances in Photosynthesis Research*, (Adv. Agric. Biotechnology Ser.), Sybesma, C., Ed., Martinus Nijhoff, The Hague, The Netherlands, 1984, I.3.283.
39. **Baianu, I. C. et al.,** High-field ^{35}Cl NMR and photosynthetic oxygen evolution, Annu. Midwest Photosynthesis Conference, Argonne National Laboratory, Chicago, 1983.
40. **Critchley, C., Baianu, I. C., Govindjee, and Gutowsky, H. S.,** Salt effects on oxygen evolution in isolated thylakoids. Aust. Soc. Plant Physiology Meeting, Perth, Australia, 1982.
41. **Critchley, C., Baianu, I. C., Govindjee, and Gutowsky, H. S.,** The role of chloride in O_2 evolution by thylakoids from salt-tolerant higher plants, *Biochim. Biophys. Acta*, 682, 436, 1982.
42. **Baianu, I. C., Critchley, C., Govindjee, and Gutowsky, H. S.,** NMR study of chloride ion interactions with thylakoid membranes, *Proc. Natl. Acad. Sci. U.S.A.*, 81, 3713,1984.
43. **Duysens, L. N. M. and Sweers, H. E.,** in *Studies of Microalgae and Photosynthetic Bacteria*, Ashida, T., Ed., University of Tokyo Press, Tokyo, 1963, 353.
44. **Kok, B. et al.,** Observations on the reducing side of the O_2-evolving photoact, *Brookhaven Symp. Biol.*, 19, 446, 1967.
45. **Papageorgio, G.,** Chlorophyll fluorescence: an intrinsic probe of photosynthesis, in *Bioenergetics of Photosynthesis*, Govindjee, Ed., Academic Press, New York, 1975, 319.
46. **Joliot, P. and Joliot, A.,** *C.R. Acad. Sci. Paris*, 258, 4622, 1964.
47. **Bouges, B.,** Action de faibles concentrations d'hydroxylamine sur l'emission d'oxygene des algues *Chlorella* et des cloroplastes d'epinards, *Biochim. Biophys. Acta*, 234, 103, 1971.
48. **Radmer, R. and Cheniae, G.,** in *Primary Processes of Photosynthesis*, Barber, J., Ed., Elsevier, Amsterdam, 1977, 303.
49. **Dwek, R. A.,** *Nuclear Magnetic Resonance (NMR) in Biochemistry*, Clarendon Press, Oxford, 1973, chap. 13.
50. **Stengle, Th. R. and Baldeschwieler, J. D.,** Halide ions as chemical probes for NMR studies of proteins, *Proc. Natl. Acad. Sci. U.S.A.*, 64, 1020, 1967.
51. **Chiancone, E. et al.,** NMR study of chloride binding to haemoglobin, *J. Mol. Biol.*, 70, 675, 1972.
52. **Leto, K. and Arntzen, C.,** Cation-mediated regulation of excitation energy distribution in chloroplasts lacking organized photosystem II complexes, *Biochim. Biophys. Acta*, 638, 107, 1981.
53. **Jennings, R. C. et al.,** Proton-induced grana formation in chloroplasts; distribution of chlorophyll-protein complexes and photosystem II photochemistry, *Biochim. Biophys. Acta*, 638, 100, 1981.
54. **Critchley, C.,** unpublished results.
55. **Hertz, H. G. and Holz, M.,** Hydrophobic hydration and quadrupole magnetic relaxation of ionic nuclei, *J. Phys. Chem.*, 78, 1002, 1974.
56. **Theg, S. M. and Homann, P. H.,** Light-, pH- and uncoupler-dependent association of chloride with chloroplast thylakoids, *Biochim. Biophys. Acta*, 679, 221, 1982.
57. **Kelley, P. M. and Izawa, S.,** The role of chloride ion in photosystem II; effects of chloride ion on photosystem II electron transport and on hydroxylamine inhibition, *Biochim. Biophys. Acta.*, 502, 198, 1978.
58. **Segel, I. H.,** *Enzyme Kinetics*, John Wiley & Sons, New York, 1975.
59. **Izawa, S.,** *Methods Enzymol.*, 69,413, 1980.
60. **Critchley, C.,** Chloride effect on oxygen evolution in halophytic plant thylakoids, *Nature*, 298, 483, 1982.

61. **Malkin, S. et al.,** Fluorescence determination of the photochemical efficiency of Photosystem II, *Plant Physiol*, 570, 67, 1981.
62. **Baianu, I. C. et al.,** Oxygen evolution and the interactions of anions with thylakoid membranes, *Annu. Meet. Am. Soc. Photobiol.*, Vancouver, Canada, 1982.
63. **Schultz, G. E. and Schirmer, R. H.,** *Principles of Protein Structure*, Springer-Verlag, New York, 1979, 30.
64. **Govindjee, Baianu, I. C., Critchley, C., and Gutowsky, H. S.,** Roles of chloride and bicarbonate in photosynthetic oxygen evolution, in *The Oxygen Evolving System of Photosynthesis*, Inoue, Y., Crofts, A. R., Govindjee, Murata, N., Renge, G., and Staoh, K., Eds., Academic Press, Tokyo, 1983, 303.
65. **Schröder, H. et al.,** in Photosynthesis Congress, Vol. 2, 2nd ed., Forti, G., Ed., 1972, 919.
66. **Wydrzynski, T. J.,** The role of manganese in photosynthetic oxygen evolution, Ph.D. thesis, University of Illinois, Urbana, 1977, 186.
67. **Wydrzynski, T. J. and Sauer, K.,** Periodic changes in the oxidation state of manganese in photosynthetic oxygen evolution upon illumination with flashes, *Biochim. Biophys. Acta*, 589, 59, 1980. (Diagram on page 69.)
68. **Maison-Peteri, B. et al.,** Modifications of the functioning of photosystem II, induced in the dark by alkaline pH in the presence of cations, *Biochim. Biophys. Acta*, 637, 202, 1981.
69. **Briantais, J. M. et al.,** Identification of S_2 as the sensitive state to alkaline photoinactivation of Photosystem II in chloroplasts, *Biochim. Biophys. Acta*, 461, 61, 1977.

Chapter 7

HIGH-RESOLUTION NMR STUDIES OF FOOD PROTEINS

Ion C. Baianu

TABLE OF CONTENTS

I. INTRODUCTION

The practical importance of cereals and especially wheat as a source of protein on a global scale is well recognized. In contrast with the rapid technological development in the uses of wheat derivatives in foodstuffs, the understanding of the basic physicochemical properties of wheat and wheat proteins has advanced rather slowly. Recently, however, new methods and techniques suitable for the investigation of chemical properties of wheat proteins have been tried and developed.[1] While major physical techniques such as X-ray diffraction and nuclear magnetic resonance (NMR) have been successfully applied to the study of a large number of polypeptides and proteins,[2-5] it was only recently that the first proton and carbon-13 (^{13}C) NMR spectra of wheat proteins were reported and assigned.[6,7] A significant sensitivity and dispersion increase was obtained for wheat proteins in the solid state by recording spectra at high magnetic fields.[1] This provides a direct approach to the problem of wheat variety identification, gluten microrheology, and nondestructive analysis of wheat proteins. The present limitations of the solid-state NMR technique in its application to the analysis of wheat are due to the complex composition of wheat and its components, and also to the incomplete averaging of the chemical shift anisotropy (which occurs at spinning rates of the samples below about 6 kHz). These limitations can, to a large extent, be removed by specific site labeling and by improved designs of NMR probes and/or pulse sequences. Nevertheless, the resolution of ^{13}C NMR spectra of wheat proteins in their solid state will still remain below the resolution obtainable in solution, at natural abundance and high fields (see Figure 25).

The detailed characterization of cereal proteins is a problem of considerable practical importance and, although there have been major advances by biochemical techniques in this direction, there are still basic limitations caused by the destructive character of the currently employed chemical techniques.[1] ^{13}C NMR spectroscopy has already been shown to be capable of nondestructive monitoring of individual atomic sites in protein solutions[2] and in solid proteins.[1] This is, therefore, a powerful method of protein characterization which can also monitor protein conformational changes, down to the level of individual atomic sites.[5]

Therefore, ^{13}C and ^{15}N NMR spectra of wheat protein solutions may be used to complement and advance the interpretation of the corresponding solid-state NMR spectra, and can be particularly helpful in the assignment of the chemical shifts to specific residues in the solid. Since in solution the chemical shift anisotropy is averaged out by fast molecular motions, increasing the magnetic field brings none of the problems encountered with solids; considerable sensitivity and dispersion increases are, therefore, readily obtainable for protein solutions at high magnetic fields, accompanied by a significant reduction in the time required for accumulating spectra and by reduced superposition of peaks. For small proteins in solution, ^{13}C-Fourier transform NMR studies have succeeded in identifying numerous individual ^{13}C sites and the technique is, therefore, uniquely suited for monitoring protein conformations.[8-10]

This approach was applied to wheat protein analysis and its possible implications for practical applications were considered. In conjunction with the NMR experiments, a simple preparation method was used which allowed the separation and fractionation of large quantities of wheat proteins soluble in water with 10 m*M* acetic acid after extraction.

II. EXPERIMENTAL PROCEDURES

A. Protein Preparation Techniques

1. Flinor Wheat Gliadins and Glutenins

Triticum aestivum cv. Flinor from the 1978 harvest was obtained from RHM agriculture as seed-grade undressed seed. The grain was milled in a Bühler MLU 302 laboratory mill to give a flour of 72% extraction. All wheat and flour samples were stored at –15°C until required.

The protein extract was prepared as follows: flour (1 kg) was suspended with gentle stirring in a 0.5 *M* sodium chloride solution (4 l) at 4°C. After 18 h the flour was separated by centrifugation (5000 g for 30 min). The cohesive mass was washed twice with water (4 l) and the majority of the starch removed by sieving (0.75 mm). The resulting glutenous mass was suspended with gentle stirring in 50% (v/v) propan-2-ol at 4°C. After 18 h the insoluble material was removed by centrifugation (5000 g for 30 min) and the clear extract was exhaustively dialyzed against distilled water at 4°C until all the propanol had been removed.[6]

After removal of the insoluble material by centrifugation (30,000 g for 2 h) the supernatant was lyophilized. The protein extract was characterized by amino acid analysis using a Rank Hilger Chromaspek amino acid analyzer after vacuum hydrolysis at 110°C for 21 h using constant-boiling hydrochloric acid.

Sodium dodecyl sulfate-polyacrylamide gel electrophoresis (SDS-PAGE) was carried out according to the method of Laemmli[11] using 10% slab gels with the following differences: (1) the reservoir buffer composition was 0.384 *M* glycine adjusted to pH 8.3 with Tris and 3.47×10^{-3} *M* SDS added and (2) dithiothreitol (DTT) (5×10^{-2}*M*) was added to the sample buffer in place of mercaptoethanol. Characteristic electrophoretograms of the samples which were obtained by this method are shown in Figure 1.

2. Flanders Wheat Gliadins

Wheat gliadins were separated from single variety Flanders flour by a simple, four-step preparation procedure at 4°C: (1) extraction of albumins and globulins with 1 *M* NaCl solution, (2) extraction of gliadins and certain glutenins with 50% propanol, (3) repeated dialysis of extracted gliadins and glutenins against distilled water, and (4) centrifugation at 50,000 g for 2 h to eliminate precipitate (mostly glutenins). The clear supernatant obtained at stage (4) was lyophilised and stored at –20°C. SDS-PAGE of the reduced supernatant indicates that the sample consists of a representative population of gliadins from Flanders flour, without any trace of glutenin subunits or other protein subunits, with molecular weights in excess of 75,000 Da (see Figure 2). This was confirmed by additional observations using conventional electrophoresis on starch gels; the latter technique also indicated a decrease in the relative amounts of ω-gliadins in the separated gliadins by comparison with the control.

3. Gluten Preparation

The wheat used in these experiments was high-yielding, hard-milling, European winter variety *(Triticum aestivum cv. Maris Huntsman)*. Flour (unbleached, untreated) and gluten were prepared from 1978 crop wheat as described by Booth and Ewart.[12]

4. Fractionation of Gluten

Freshly washed wet gluten (125 g) was cut into small pieces (approximately 0.5 cm in diameter) using scissors, and the gluten was suspended in distilled water (0.75 l) at 4°C. Solid urea (60 g) was added, the volume was made up to 1 l, and the suspension was stirred

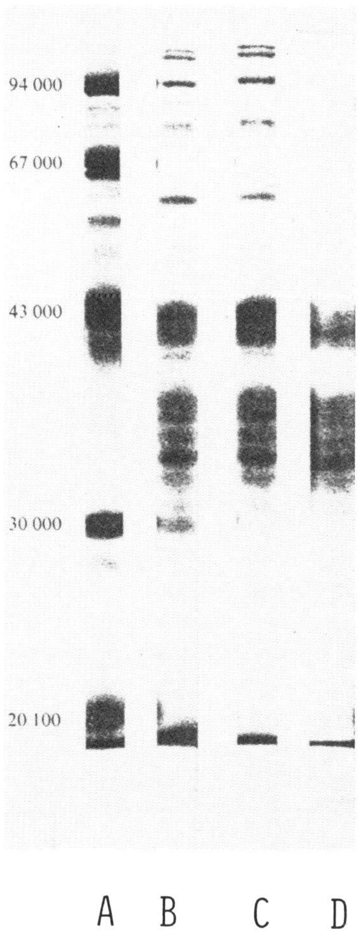

FIGURE 1. SDS-PAGE of Flinor proteins. (A) Pharmacia low-molecular-weight electrophoresis calibration proteins, (B) SDS-Tris extract of Flinor flour, (C) SDS-Tris extract of Flinor gluten, and (D) SDS-Tris extract of Flinor protein extract. Electrophoresis conditions were as follows: 10% PAGE cross-linked with 0.26% bis-acrylamide; the electrophoresis buffer was 0.382 *M* glycine adjusted to pH 8.3 with Tris containing 3.47×10^{-3} *M* SDS, and the electrophoresis was developed for 18 h at 15 mA; the gel was stained with 0.5% Coomassie Blue in 50% methanol containing 7.5% acetic acid and destained with 7.5% acetic acid in 5% methanol.

magnetically for 18 h at 4°C. After centrifugation at 10,000 × g for 1 h (4°C), the supernatant was dialyzed against ten volumes of distilled water at 4°C for a total of nine changes. The pellet was resuspended in 1 l of distilled water (4°C) and, after stirring for 2.5 h, the suspension was recentrifuged, and the pellet was again washed with water. The water washings were discarded. Both urea-soluble proteins in the supernatant and urea-insoluble proteins in the water-washed pellet were freeze-dried, and the dry products were ground in a Retsch Turbo Mill (0.5-mm sieve). Untreated gluten was freeze-dried immediately after preparation and ground in a Turbo Mill.

5. Maize Zeins

The zeins used in these experiments were provided by Colorcon, Inc. of West Point, PA and Sigma Co. These zein protein preparations were used because of the large quantities available

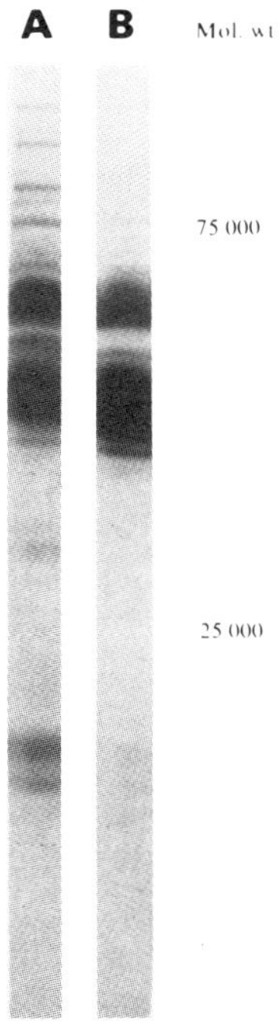

FIGURE 2. SDS-PAGE electrophoresis of protein extracts from Flanders single variety flour. Tracks: (A) 50% propanol extract and (B) standards.

for study, and because details of its preparation are readily available.[13] The protein content of the Colorcon zein was determined to be 90.4% by Kjeldahl analysis (N × 6.0).

The zein was extracted with propan-2-ol and water-saturated butan-1-ol to produce two subfractions. The propan-2-ol-soluble protein was prepared as follows: zein (12 g) was dissolved in aqueous 60% (v/v) propan-2-ol: (500 ml) at 4°C. After mixing for 1 h on a magnetic stirrer, the protein solution was dialyzed extensively against distilled water to remove the propan-2-ol. The protein was then lyophilized and stored at –17°C.

Treatment of the zein with butan-1-ol was done as follows: zein protein (10 g) was added to a two-phase butanol-water (250 ml:250 ml) system at 4°C. After vigorous mixing for 1 h, the suspension was allowed to stand overnight. The upper layer (the butan-1-ol phase) was poured off. This upper layer contained most of the pigment present in the zein and a small proportion (1% of initial weight) of the zein protein and was discarded. The bottom layer (primarily composed of water and protein) was then dialyzed against distilled water to remove any residual butan-1-ol. After removal of the excess butan-1-ol, the sample was lyophilized and stored at –17°C. (In this review the terms "maize" and "corn" will be used interchangeably, although "corn" is the wider concept.)

6. Soy Protein Isolates

The method of Tranh and Shibasaki[14] for the simultaneous isolation of the two major soybean storage proteins, glycinin (11S) and β-conglycinin (7S), is suitable for obtaining the relatively large amount of protein required for ^{13}C NMR spectroscopy. The purity of a 7S fraction prepared in this manner has been found to be about 86% by ultracentrifugation[14] or 70% by densitometry of SDS-PAGE gels although single radial immunodiffusion yielded a β-conglycinin[15] content as low as 52%, which is probably an underestimate.

SDS-PAGE under reducing conditions yielded a pattern that consisted of a very intense band at 64 kDa and two other major ones at 60 and 49 kDa which were identified as the α', α, and β subunits of β-conglycinin, respectively.[14] The measured molecular weights generally agree with those reported in the literature,[14] although they may represent an overestimate since β-conglycinin subunits are glycoproteins. The reason for the variability of the reported MW of β–conglycinin subunits, as estimated from SDS-PAGE, is not clear at present.

B. NMR Measurements

A Nicolet Technology NT-200 wide-bore NMR spectrometer (operating at 4.7 T) was used for ^{1}H, ^{13}C, and ^{15}N studies of wheat gliadins in solution. Proton NMR spectra were recorded with 5-mm diameter saddle probes; ^{13}C and ^{15}N NMR spectra were recorded with double-resonance, 20-mm diameter probes. Fourier transforms were carried out on-line with an NIC 1180 computer using an FT size of 16,384 points and an ADC of 16 bits. For the fully proton-decoupled ^{13}C NMR spectra, the ^{1}H irradiation ($H_1/2\pi$, ^{1}H-decoupling field of 5 kHz; 200.0670 MHz proton resonance frequency) was centered 6 ppm downfield from the ^{1}H resonance of tetramethylsilane (Me_4Si) and had a modulation bandwidth of about 10 ppm. Chemical shifts, line widths, integrated intensities, and difference spectra were recorded digitally. Samples contained 20% D_2O for ^{2}H-resonance lock and Me_4Si as the chemical shift standard.

The observation frequency of the ^{13}C resonance was 50.3122 MHz with a spectral width of ±5747.1 Hz. Free induction decays were recorded following 30-μs ^{13}C pulses (60° nutation angle) with a 0.713-s accumulation time and a pulse repetition rate of 1 s. Other experimental details are specified in the legends to Figures 1 to 8.

Samples for ^{13}C NMR measurements were prepared by dissolving lyophilized gliadins in a 5 m*M* CD_3COOD solution in D_2O (also containing 5% *n*-propanol), to a final protein concentration of 80 mg/ml^{-1}.

A laboratory-assembled NMR instrument, NSF-250 (operating at 5.85 T), was used for the ^{13}C NMR work on maize zeins. ^{13}C NMR spectra were recorded with laboratory-built, horizontal-spinning ^{13}C-{^{1}H} probes. Sample holders were 20 mm in diameter and contained up to 7 ml of solution. The sensitivity of such a probe is approximately twice that of most commercial probes, which employ vertical sample spinning and smaller sample quantities; it has, however, the disadvantage of difficult shimming. Fourier transforms were performed on-line with an NIC-1080, and more recently with an NIC-1280 computer using an FT size of 16,000 points for the 8000 data points acquired, and an ADC of 12 bits. For fully proton-decoupled ^{13}C NMR spectra, broad-band ^{1}H irradiation (250.160 MHz proton resonance frequency) was centered 4 to 7 ppm downfield from the ^{1}H resonance of (Me_4Si) and covered 10 ppm. The observation frequency of the ^{13}C channel was 62.908 MHz, with a spectral width of ±9 kHz. Free induction decays were recorded following 35 μs (45°) ^{13}C pulses with a pulse repetition rate of 1 to 4 s. For ^{13}C NMR measurements, the protein solutions contained 50% deuteriated pyridine-d_5:50% D_2O and Me_4Si as the chemical shift standard. The ^{13}C NMR spectra presented (Figures 3 and 4) were done in triplicate and found to be reproducible.

Because of the limited solubility of zein in aqueous media (due to its hydrophobic nature), a number of solvents with various physical properties was tested to establish the most efficient solvent for use in NMR work. Although initial testing involved the use of alcohol-containing solvents, the solubility of zein in 50% (v/v) pyridine-water proved to be over three times that of any of the other solvent mixtures with water that have been tested or used previously; a concentration of 40 mg/ml protein in solution could be readily obtained. This zein concentration in solution facilitated the NMR work. Zein solubility was also tested in pyridine (100%) and a concentration of 90 mg/ml could be obtained without precipitation. However, because of the toxicity of pyridine, the concentration of 50% (v/v) was used in NMR work. Pyridine has been used previously as a primary solvent in other systems, although it does not appear to have been used to solubilize zein.

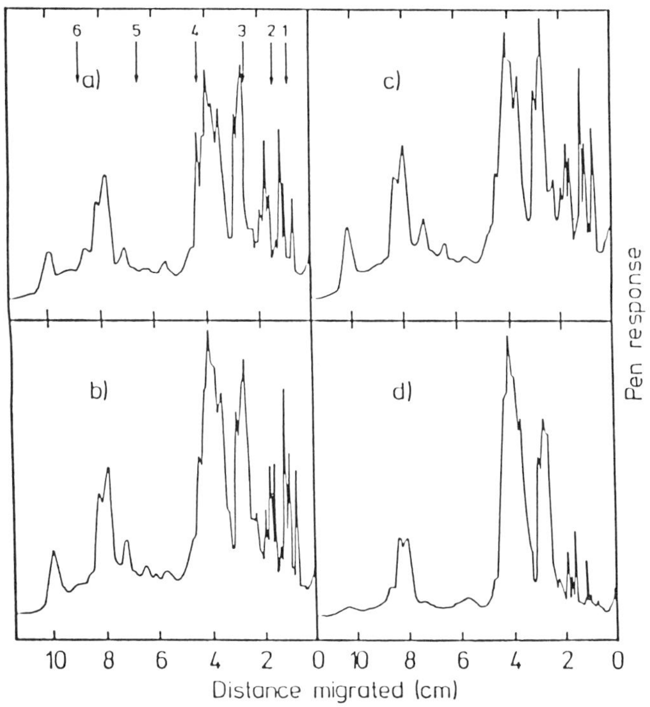

FIGURE 3. Densitometer traces of sodium dodecyl sulfate polyacrylamide gel electrophoretograms of reduced proteins from flour, gluten, and gluten subfractions. Flour (50 mg) and gluten fractions (5 mg) were solubilized in 1 cm^3 of sample buffer, 0.5 *M* dithioerythritol was included in the sample buffer. The samples were prepared for electrophoresis by heating at 100°C for 5 min. Migration distances are from the top of the separating gel. (a) flour, (b) gluten, (c) urea-insoluble fraction, (d) urea-soluble fraction. Arrows indicate the positions of molecular weight markers: 1, phosphorylase b, 94,000; 2, bovine serum albumin, 67,000; 3, ovalbumin, 43,000; 4, carbonic anhydrase, 30,000; 5, soybean trypsin inhibitor, 20,000; and 6, α-lactalbumin, 14,400.

III. PROTEIN ANALYSES

A. Wheat Proteins

1. SDS-PAGE

Electrophoresis on polyacrylamide gel slabs was performed in the presence of SDS according to the method of Laemmli,[11] with the exception that the reservoir buffer composition[13] was 0.384 *M* glycine/0.025 *M* Tris 3.47×10^{-3} *M* SDS, pH 8.3. The separating gel ($150 \times 150 \times 1.5$ mm) contained 154 g/l of total acrylamide monomer and 4 g/l of cross-linker, and the stacking gel ($150 \times 10 \times 1.5$ mm) contained 77 g/l of monomer and 2 g/l of cross-linker. After electrophoresis, done at 7.5 m*M* for approximately 18 h, the gels were stained with Coomassie brilliant blue and subsequently scanned densitometrically using a Joyce-Loebl Chromoscan 201 system at 520 nm (Figures 3 and 4).

2. Gel Filtration Chromatography

Gel filtration chromatography was performed on columns of Sepharose CL-4B in the presence of SDS.[16] Samples were prepared for chromatography by extraction into 0.069 *M* SDS/0.1 *M* Tris/HCl, pH 8.0, in "Oakridge"-type centrifuge tubes for 1 h under gentle rotation (28

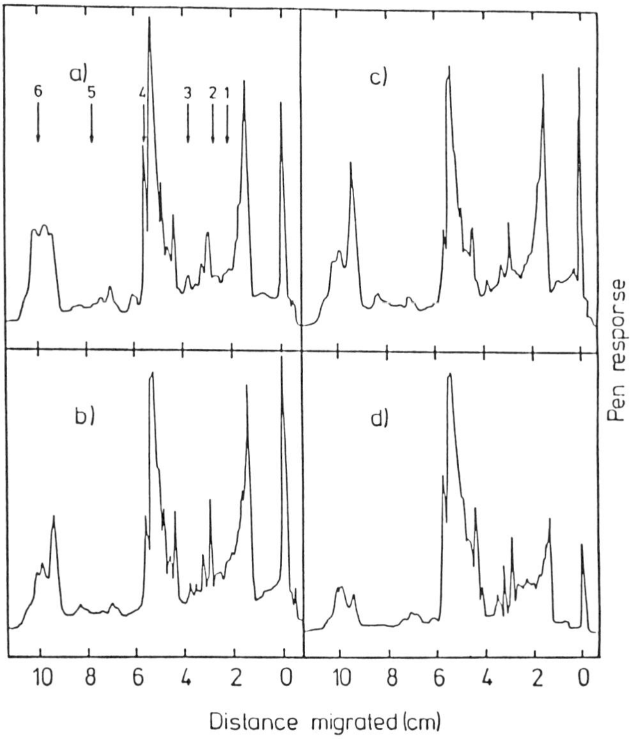

FIGURE 4. Densitometer traces of sodium dodecyl sulfate polyacrylamide gel electrophoregrams of unreduced proteins from flour, gluten, and gluten subfractions. Flour (50 mg) and gluten fractions (5 mg) were solubilized in 1 cm^3 of sample buffer; dithioerythritol was excluded from the sample buffer. The samples were prepared for electrophoresis by heating at 100°C for 5 min. Migration distances are from the top of the stacking gel. (a) flour, (b) gluten, (c) urea-insoluble fraction, and (d) urea-soluble fraction. Molecular weight markers as in Figure 3.

rotations per minute). After extraction hexane was added and the mixture was centrifuged for 1 h at 50,000 × g. The clear SDS extracts were carefully removed from the tubes and dialyzed against $3.47 \times 10^{-3} M$ SDS/0.1 *M* Tris/HCl, pH 8.0, for 4 h before application to the Sepharose column. The column (16 × 850 mm) was eluted with $3.47 \times 10^{-3} M$ SDS/0.1 *M* Tris/HCl, pH 8.0, at a flow rate of 12 cm^3/h^{-1}. Both extraction and chromatography were done at room temperature (20°C).

3. Extraction of Gluten Proteins Using 1 M Urea

Fractionation of flour and gluten proteins using aqueous urea solutions was done previously.[16-18] In our work, a urea concentration of 1 *M* was used because preliminary measurements of Rayleigh light scattering by isolated gliadins in solution had indicated that scattering was minimal at this concentration.[7] At urea concentrations of 1.5 *M* and 2 *M*, Rayleigh scattering increased markedly, suggesting that gliadins unfold at such urea concentrations. Cluskey and Wu[19] also reported α-helix loss by gluten proteins in solutions containing 3 *M* urea. We found that a single extraction using 1 *M* urea extracted 20% of the protein from *Maris Huntsman* gluten, similar to the 25% value obtained by Lee[20] for the amount of protein extracted from the flour using one extraction with 2 *M* urea. Total protein recovery in the urea-soluble and urea-insoluble fractions was 95%. The proportion of the gluten protein in each fraction and the recovery of protein were measured by the Kjeldahl nitrogen determinations.

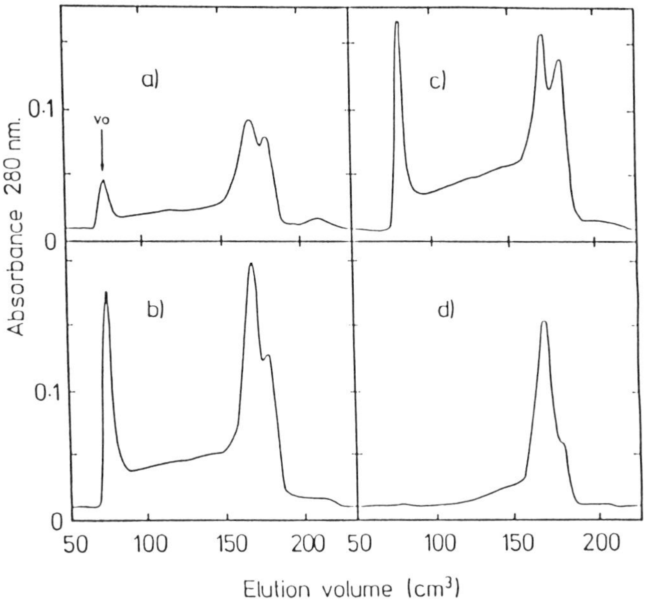

FIGURE 5. Gel filtration chromatography of proteins in *Maris Huntsman* flour, gluten, and gluten subfractions. Flour (0.75 g) and gluten fractions (0.15 g) were extracted in 15 cm^3 of extraction buffer, and 3 cm^3 of the extract was applied to the column. The column effluent was monitored continuously at 280 nm. (a) flour, (b) gluten, (c) urea-insoluble fraction, and (d) urea-soluble fraction.

4. Molecular Size Distribution of Proteins in Gluten and Gluten Subfractions

Proteins from the *Maris Huntsman* vs. wheat flour, gluten, and gluten subfractions were solubilized in SDS buffer and chromatographed on Sepharose CL-4B to determine the molecular size distribution of the proteins (Figure 5). Previous work showed that approximately 95% of flour and gluten proteins are extracted by SDS. By contrast, under our extraction conditions the more commonly used acetic acid/urea acetyltrimethylammonium bromide solvent (AUC) extracts only 85% of flour proteins.

The glutenins extracted from flour or gluten have extremely polydisperse molecular size and are eluted over a large area of the chromatogram from the void volume (75 cm^3) to an elution volume of approximately 155 cm^3. Assessing absolute molecular weights for these components is difficult because of the lack of adequate standards with which to calibrate Sepharose gels, but they apparently range from about 10^3 to values in excess of 5×10^6. The peak between 155 and 175 cm^3 comprises the gliadin fraction, with molecular weights of 3 to 7×10^4, whereas the peak between 175 and 185 cm^3 comprises albumins and globulins of low molecular weight (approximately 1.5×10^4). These different fractions are distinct from each other not only in terms of their molecular sizes but also in terms of their polypeptide subunit distributions as determined by SDS-PAGE.

Examination of *Maris Huntsman* gluten (Figure 5b) indicated that glutenins accounted for approximately 40% of the protein, and gliadins accounted for approximately 37%, as determined by measuring the areas beneath the respective regions of the chromatogram and expressing these as a percentage of the total area (based on absorbance at 280 nm of the column effluent). For the urea-insoluble fraction (Figure 5c), glutenin comprised about 52% of the total protein, and gliadin comprised about 28%; for the urea-soluble fraction (Figure 5d), gliadins were the predominant species, accounting for about 70% of the protein, with glutenins comprising only 17%. The relatively small proportion of glutenins present was difficult to remove.

The accuracy of quantification by this method depends on the assumption that the contents of aromatic amino acids in the different fractions are equivalent. Thus, the area values given above may not accurately reflect the actual amounts of protein in the different regions of the chromatograms. Nevertheless, such measurements do indicate the comparative compositions of the different samples. They also indicate that the urea-soluble fraction of gluten is considerably enriched in gliadins, and that the urea-insoluble fraction is enriched to a lesser extent in relatively high molecular weight glutenin species.

5. Polypeptide Subunit Distributions in Gluten and Gluten Subfractions

Polypeptide subunit distributions in the various samples were compared by SDS-PAGE of the reduced proteins. Previous work has shown that the subunit compositions of the glutenin and gliadin fractions prepared by gel filtration chromatography are quite distinct.[16,17] Subunits of molecular weights between 8×10^4 and 1.4×10^5 (migration distances 1.35 to 0.70 cm, Figures 4 and 5) were found only in the glutenin fractions, and these are referred to here as high molecular weight glutenin subunits (HGS). Those subunits having molecular weights between 3.9 and 4.6 $\times 10^4$ (migration distances: 2.95 to 2.30 cm, Figures 6 and 7) were distributed similarly, and these are referred to as low molecular weight glutenin subunits (LGS). Components with molecular weights between 2.9 and 3.6×10^4 (migration distances 4.50 to 3.35 cm) were largely α-, β-, and γ-gliadin (GL) subunits and were considerably enriched in GL subunits compared with the gluten (Figures 6, 7, and Table 1). The glutenins in the urea-soluble fraction were atypical of most of the glutenins in the parent gluten in that the ratio of HGS to LGS was much lower for this fraction than for the gluten.

When the relative levels of total glutenin subunits (HGS + LGS) and total gliadin subunits (GL + WGL) were calculated by adding the relevant areas beneath densitometer peaks, the ratio of glutenins to gliadins was similar for flour and gluten (0.72 and 0.75, respectively); it was somewhat higher for the urea-insoluble fraction (0.85) but much lower for the urea-soluble fraction (0.45). The accuracy of this quantification of subunit distributions from SDS-PAGE depends on the basic amino acids, particularly lysine, contained in each polypeptide type. Because the content of such residues in different types of subunits varies,[18] the values quoted cannot be regarded as absolute, but they do give useful information when comparing different fractions. For the same reason, the values derived by SDS-PAGE and by gel filtration chromatography for the relative proportions of glutenins and gliadins cannot be expected to yield the same values; quantification from gel filtration chromatographs depends on the aromatic amino acid contents of the different polypeptide types.

The various samples were also examined by SDS-PAGE in the unreduced state (i.e., DTT was omitted from the sample preparation buffer), and the results are presented in Figures 6 and 8. HGS and LGS components were no longer observed, but a considerable proportion of the protein was present as a trailing smear from the top of the gel. This observation indicates that the glutenin subunits are linked by disulfide bonds originally and that the unreduced disulfide-linked glutenins are highly polydisperse with respect to molecular size. The patterns for flour, gluten, and the urea-insoluble fraction were quite similar, with protein staining being observed at the tops of both the separating and stacking gels (Figure 8A, B, and C). However, the urea-soluble fraction was different (Figure 8D); the densitometer peak at the stacking gel top was caused by the gel/destaining solution interface and not by protein staining. This again indicates that the glutenins in this fraction have a lower average molecular weight than those in the other fractions.

B. Corn Zeins

1. Solubility

Basic solubility data for commercial zeins is at present unavailable in the literature in spite of the importance of solubility for analytical studies of zein. Because the solubility of zein is rather limited in aqueous media (due to its hydrophobic nature), a number of solvents with

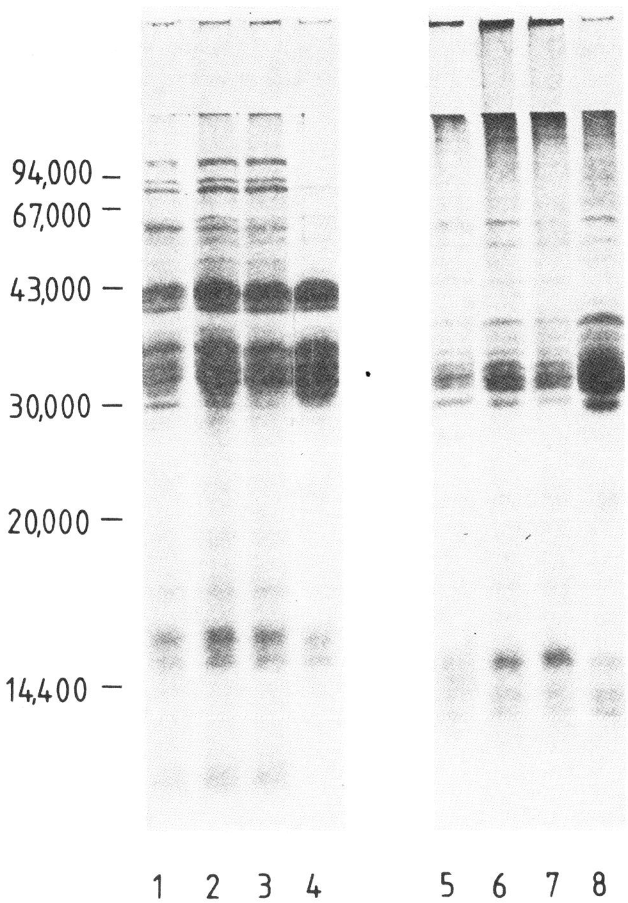

FIGURE 6. SDS PAGE of reduced and unreduced proteins from flour, gluten, and gluten subfractions. Four (50 mg) and gluten fractions (5 mg) were solubilized in 1 cm^3 of sample buffer. When the samples were examined in the reduced state, 0.05 *M* dithioerythritol was included in the sample buffer. The samples were prepared for electrophoresis by heating at 100°C for 5 min. 1, flour, reduced; 2, gluten, reduced; 3, urea-insoluble fraction, reduced; 4, urea-soluble fraction, reduced; 5, flour, unreduced; 6, gluten, unreduced; 7, urea-insoluble fraction, unreduced; and 8, urea-soluble fraction, unreduced.

various dielectric constants were tested to establish the most efficient solvent for use in NMR work. Initial testing involved the use of alcohol-containing solvents.

Alcohols of various chain lengths as well as other organic solvents were used in mixtures with water, and protein solubility differences were recorded.[21] Protein solubility was determined by examining the maximum protein content obtainable before the measured "absorbance" at 420 nm reached 2.0 (Figure 9).

The first solvent used for zein was an ethanol/water mixture. The solubility in this mixture proved to be greatest at the ethanol concentration of 70% v/v (12.0 *M*). The limit of solubility, however, was somewhat low, 15 mg protein per milliliter (2.0% w/v), in comparison with other organic solvents. Second, we examined zein solubility in propan-2-ol/water mixtures. The optimum propan-2-ol concentration was determined to be 60% v/v (7.8 *M*) in water. The 60% v/v propan-2-ol mixture dissolved up to 15 mg/ml of zein (3.0% w/v). The next alcohol in the series was butan-1-ol. Butanol has the important property of being immiscible with water and,

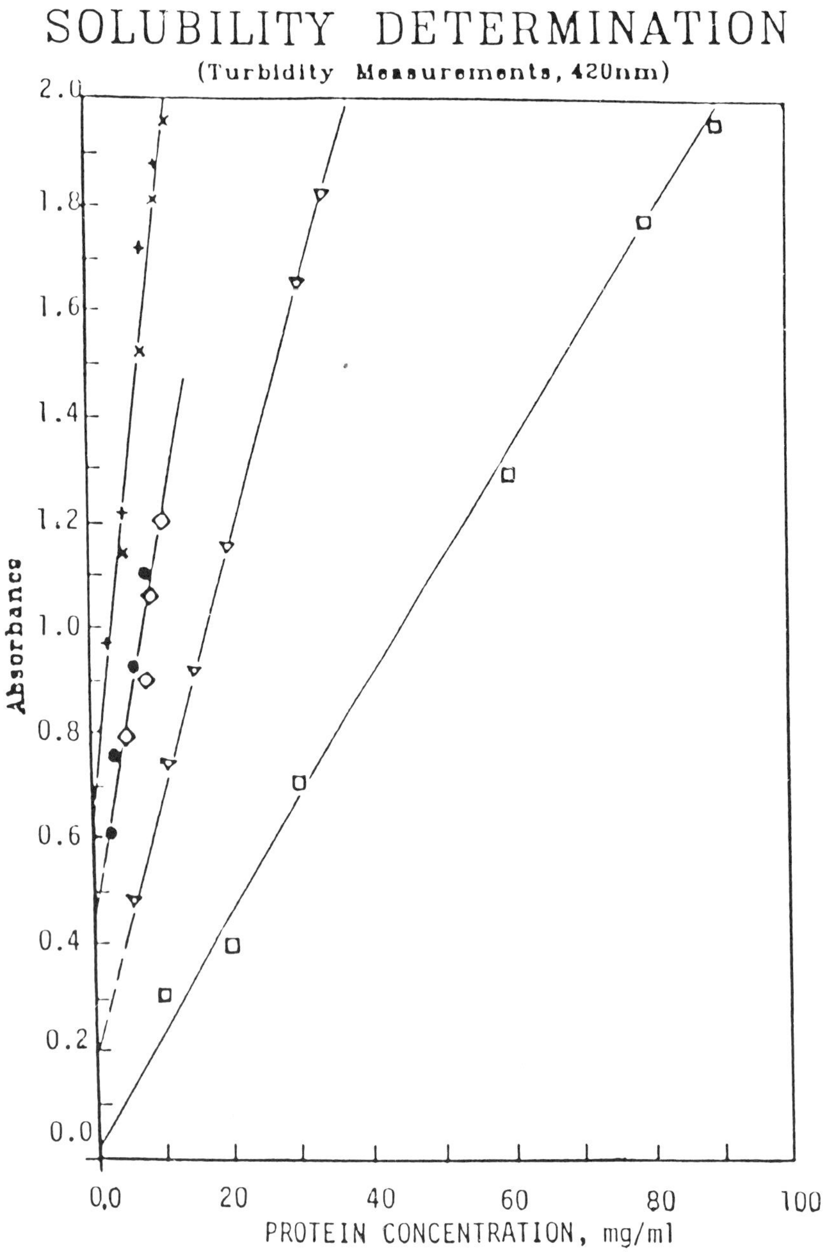

FIGURE 7. Plot of turbidity data for zein in organic solvents obtained at 420 nm. ❑ = DMSO and pyridine (100%); ∇ = 50% pyridine; ◊ = propan-2-01 (60%); ● = ethanol; ✖ = acetone (70%); and ✚ = pH 12.00.

therefore, can be used to phase-separate the most hydrophobic zein components. This procedure also largely removed the yellow-colored pigments into the butanol phase.

Zein solubility in acetone/water mixtures was also tested. At a concentration of 70% v/v, water/acetone proves to be only as good in its ability to solubilize zein proteins as propan-2-ol.

For comparison with the solubility of other related cereal proteins such as wheat gliadins, we investigated the solubility of zein in acetic acid. Addition of acetic acid to a water suspension of zein solubilizes very little protein; at a final concentration of 0.96 *M* acetic acid (pH = 2.5), only 0.2 mg/ml (0.02% w/v) of protein was dissolved. As expected, when the pH was adjusted to pH 11.4, zein solubility was good, dissolving up to 10 mg/ml.

Table 1
DISTRIBUTION OF POLYPEPTIDE SUBUNITS AS DETERMINED BY SODIUM DODECYL SULFATE POLYACRYLAMIDE GEL ELECTROPHORESIS OF REDUCED SAMPLES

	Glutenin subunits		Gliadin subunits				
	High molecular weight (HGS)	Low molecular weight (LGS)	α-Gliadin (WGL)	Other (GL)	HGS/ LGS	HGS/ GL	LGS/ GL
Flour	16	26	12	46	0.62	0.35	0.45
Gluten	16	27	10	47	0.59	0.34	0.47
Urea-insoluble fraction	19	27	11	43	0.70	0.44	0.50
Urea-soluble fraction	4	27	8	61	0.15	0.07	0.39

Note: Values are expressed as a percentage of the total area beneath the densitometer peaks for all glutenin and gliadin subunits.

From Schofield, J. D. and Baianu, I. C., *Cereal Chem.*, 59(4), 240, 1982.

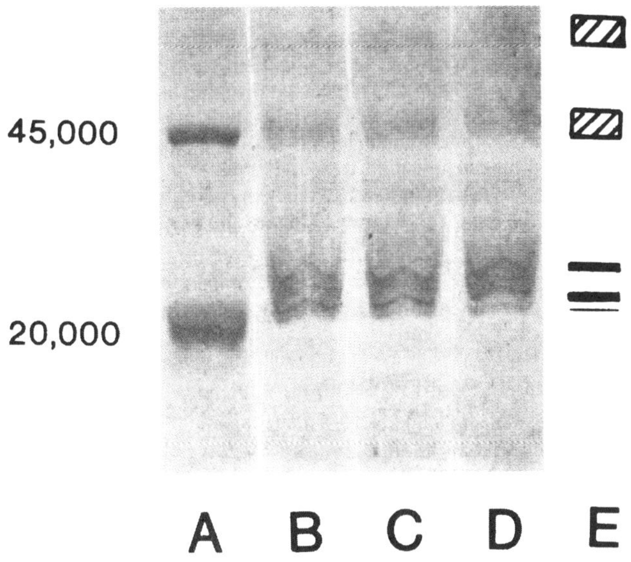

FIGURE 8. SDS-PAGE electrophoresis of unreduced zeins: (A) Sigma Chemical Co., low molecular weight electrophoresis standards (proteins), (B) untreated zeins, unreduced, (C) zeins following extraction with propan-2-ol, unreduced, (D) zein proteins following pretreatment with butan-1-ol, and (E) schematic representation of the electrophoresis band present in slots B, C, and D.

Solubility of zein was also tested using various concentrations of dimethyl sulfoxide (DMSO). Zein solubility was greatest in 100% DMSO. In 100% DMSO, zein concentrations of 100 mg/ml could be easily obtained. Zein solubility was also tested in pyridine (100%) and a zein concentration of 90 mg/ml could be obtained without precipitation. Because of pyridine toxicity the concentration of 50% (v/v) was used in the NMR work.

In the presence of more than 5% water in DMSO, however, zein aggregated and precipitated.

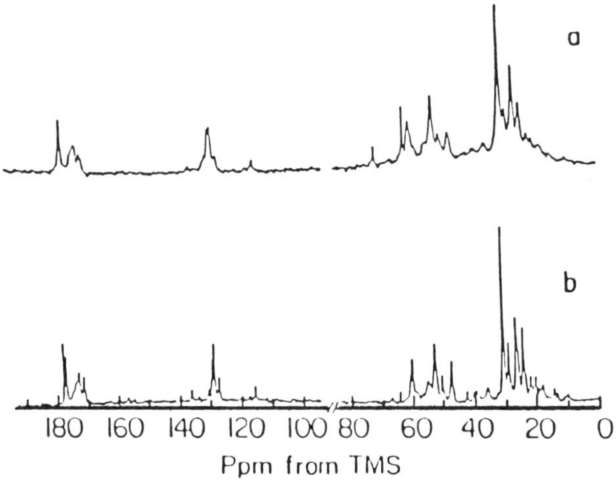

FIGURE 9. The 50.3-MHz ^{13}C NMR spectra of wheat gliadins in 10 m*M* CD_3COOD solutions in D_2O (99%) at pD 3.4 and 22°C; NMR accumulation times: 20 min. Proton broadband decoupling was centered at 4.7 ppm and covered the entire spectrum. (a) at 32% (w/v) wheat gliadin concentration and (b) at 15.8% wheat gliadin concentration; both spectra were recorded nonspinning.

Pyridine has been previously used as a primary solvent in other systems, but it has not been used to solubilize zein, perhaps because of toxicity. The solubility of zein in pyridine/water proved to be over three times that of any of the other solvent mixtures with water, tested or used previously; a concentration of 40 mg/ml (10% w/v) protein in solution could be readily obtained; this zein concentration in solution facilitated the NMR work.

Butane-1-ol as a solvent, however, exhibits entirely different characteristics. Because butane-1-ol is nonpolar (ε = 17.8), its interactions with zein components are limited to the hydrophobic groups near the surface of zein proteins. These characteristics make butane-1-ol a rather poor solvent for most zein proteins.

2. SDS-PAGE

SDS-PAGE was performed according to a modification of the method of Laemmli.[11] Gels consisting of 10% acrylamide were used to provide greater separation of the zein protein subunits. Sample buffers were prepared as follows: (1) 0.0625 *M* Tris-HCl, 2% SDS, 10% glycerol, 6 *M* urea, (2) 50% pyridine, 10% glycerol, (3) 60% 2-propanol, 10% glycerol, and (4) 70% ethanol, 20% glycerol. The reservoir buffer consisted of a Tris/glycine (pH 8.3). Electrophoresis was performed on zein with and without the presence of a reducing agent in the sample buffer. The nonreducing sample buffer consisted of Tris-HCl (0.0625 *M*), SDS (2%), glycerol (10%), bromophenol blue (0.1%), and urea (6 *M*). Samples were prepared with a protein concentration of 5 mg/ml. Protein samples were incubated for 4 h at 37°C prior to electrophoresis. Electrophoresis was carried out over a period of 4 h at 15 mA. The protein bands were fixed in 10% TCA for 2 h, and then the gels were stained for 2 h (50 ml), methanol (250 ml), and distilled water (250 ml). The gels were then destained until most of the background stain was removed using a destaining solution consisting of 7.5% acetic acid.

SDS-PAGE of zein separated four distinguishable bands in the reduced state and only three subfractions in the absence of a reducing agent (such as β-mercaptoethanol or DTT), in the solvent buffers. In the reduced state, protein bands occur at 16,000, 22,000, 24,000, and 48,000 Da. In the unreduced state the 22,000-, 24,000-, and 48,000-Da bands were the only bands present if SDS is used, regardless of the organic solvent in which zein was extracted. This indicates that these solvents had no effect on the overall composition of the zein protein. In the pyridine buffer system, however, three additional bands could be separated; in this system up to six bands could be distinguished with two major bands at 12,000 and 32,000 Da, under nonreducing conditions[13] (Figure 8). These observations are consistent with the possibility of zein aggregation in the other solvents.

C. Soy Globulins

1. Gel Electrophoresis

SDS-PAGE was carried out in a continuous buffer system using a T = 10%, C = 2.6% horizontal slab gel (250 × 115 × 2 mm). An LKB 2117 Multiphor system (Bromma, Sweden) operating with an LKB 219 regulated voltage/current/power supply was used for electrophoresis.[22] The cooling water was thermostated at 15 ± 2°C. A mixture of protein standards (MW-SDS-70L, Lot 55F6048 from Sigma, St. Louis, MO) was run in parallel with soy protein samples.

After preelectrophoresis (100 mA for 20 min), 100 µg of soy protein and 300 µg of SDS in 10-µl sample buffer (0.045 *M* Tris·HCl pH = 8.9, 0.1% SDS, 0.05% bromophenol blue), placed in boiling water for 2 min and allowed to cool, were loaded into the gel sample slots. Reduced samples contained 100 µg of DDT. Electrophoresis was performed in a 0.375 *M* Tris·HCl pH 8.9 buffer system at a constant current of 20 mA for 30 min and then at 100 mA for 4 h. Gels were fixed in a 12% trichloroacetic acid-3.5% sulfosalicylic acid solution in 30% aqueous CH_3COOH for 1 h, stained in a 0.25% solution of Coomassie Brilliant blue R250 in H_2O-CH_3OH-CH_3COOH (6:6:1) for 4 h, diffusion-destained with several changes of 30% C_2H_5OH-10% CH_3COOH aqueous solution over 24 h, and finally soaked in a preserving 30% C_2H_5OH-10% CH_3COOH-10% glycerol aqueous solution for 1 h.

SDS-PAGE under reducing conditions yielded a pattern that consisted of three major bands at 20, 39, and 71.5 kDa and four less intense bands at 34.8, 54.3, 64.5, and 81.5 kDa. The electrophoretograms were typical of soy proteins, similar to those found in the literature[23,24] and there was no evidence of sample proteolysis.

SDS-PAGE under nonreducing conditions[21] showed that the majority of protein components had a molecular weight in excess of 54 kDa, with major bands at 54.3, 59, 71.5, 81.5, and 91.5 kDa. Some of the acidic (34.8 and 39 kDa) and some of the basic (20 kDa) glycinin subunits existed in the reduced state, despite the absence of a reducing agent; this is presumably a result of the protein isolate manufacturing process.

2. Gel Filtration Chromatography

A laboratory-assembled gel filtration system was employed that consisted of a valveless piston pump (model RP-SY-1CSC from FMI Fluid Metering, Inc., Orchard Bay, NY), a Teflon 4-way rotary valve (type 50 from Rheodyne, Cotati, CA) for sample injection through a 0.5-ml sample loop, and a 2.5 × 46 cm column of Fractogel TSK-HW-55S (Merck; 25- to 40-µm particle size and 1000 to 700,000 molecular weight fractionation range). The column was packed[23,29] at 2.5 ml/min flow rate and at a final pressure of 60 psi. It was then equilibrated with 10 m*M*, 3-(cyclohexylamino)-propane-sulfonic acid (CAPS) and 0.02% NaN_3, pH 10.5; 0.5 ml of a 0.5% dispersion of soy protein isolate in this buffer was loaded onto the column through a 0.45-µm Nylon-66 membrane filter and was eluted with the same buffer at a flow rate of 0.45 ml/min at 10 psi, at room temperature. The pH and the ionic strength of the buffer were chosen to be similar to those used for the ^{13}C NMR spectroscopy of the soy isolate. The eluent absorbance was continuously monitored at 280 nm (0.01 AUFS) with an UV-Vis absorption detector (Model Spectroflow 757 from Kratos Analytical Instruments, Ramsey, NJ) connected to a recorder (Model 8376-20 from Cole-Palmer Instrument Co., Chicago, IL). A mixture of protein standards (MW-GF-1000, Lot 125F6005 from Sigma, St. Louis, MO) was run under the same conditions.

The gel filtration chromatography of the soy protein isolate indicated the presence of insolubilized protein that could not go through the 0.45-µm filter and that of smaller aggregates with molecular weights in excess of 700,000 Da. The chromatogram also contained two broad overlapping peaks that correspond to apparent molecular weights of 100,000 and 84,000 Da.

Our results indicate that the soy protein isolate studied consists of a complex mixture of globulins with a molecular weight distribution from 20,000 to over 700,000 Da and major components at 59,000, 71,500, 84,000, 180,000, and higher than 700,000 Da, with the last three molecular species dominating.

IV. NMR STUDIES OF FOOD PROTEINS IN AQUEOUS SOLUTIONS: SPECTRAL ASSIGNMENTS

A. Assignments of ^{13}C NMR Spectra of Wheat Gliadins in Solutions at pD 3.4

Although ^{13}C NMR can currently identify numerous single nuclei resonances even in systems as complex as proteins, the spectral assignment problem is only partially solved even for small proteins such as myoglobin — due to considerable overlap of resonances in the carbonyl and aliphatic regions of the spectra. The assignment problem would seem even more complex for heterogeneous mixtures of wheat protein fractions with molecular weights between 28,000 and 60,000 in dilute acetic solutions. For wheat protein fractions in such solutions there are three major simplifications to this problem. (1) The conformations of wheat proteins are highly disordered [as observed by X-ray diffraction (Baianu, I. C., unpublished results)] and, therefore, the chemical shift values of ^{13}C nuclei in random copolypeptides and amino acid solutions can be used for assignment.[8] (2) The amino acid compositions of various gliadin fractions are quite similar, and gliadin terminal sequences have been determined which show considerable homology between these gliadin fractions (some ω-gliadins are a notable exception to this rule since they do not contain any cysteine[12]). (3) A major proportion of the amino acids in wheat proteins consists of glutamic acid/glutamine + proline (~70%) and serine (~10%), while the cysteine content is typically between 1 and 2%.

It is possible therefore, to begin the assignment of ^{13}C resonances of wheat protein fractions in the 15.8% (w/v) gliadin solutions (Figures 9 and 10) by comparing their chemical shifts with those of glutamic acid/glutamine, proline, and serine groups in other diamagnetic proteins and in small polypeptides. The assignments for nonprotonated carbon resonances are in an even better position since these are known to yield sharp ^{13}C resonances with stable chemical shifts — comparable with those of the corresponding amino acid solutions.[4] This is not the case for the ^{1}H resonances (Figure 11) which are overlapping and cannot be readily assigned in this one-dimensional (1D) NMR spectrum.

The ^{13}C chemical shifts associated with glutamines and prolines occur in the carbonyl and aliphatic regions of the NMR spectra; within these regions, the ^{13}C resonances of glutamines and prolines have rather distinct chemical shift values from the other amino acids. While, at present, it is not possible to identify single ^{13}C resonances in these regions of the spectra, it is possible to make collective assignments of the observed peaks to glutamines, prolines, or aromatic amino acids. Such assignments are listed in Table 2 and are typical for protein solutions, as discussed in References 2 and 4. One of the important points to emerge from these collective assignments is that the wheat protein spectra are dominated by the peaks from glutamines, serines, and prolines. As expected, the aromatic amino acids give well-separated peaks of lower intensities than those of glutamines and serines/prolines. It is possible, therefore, to follow by ^{13}C NMR conformational and mobility changes in protein regions containing polar amino acids and hydrogen bonds, as well as in those protein regions containing nonpolar amino acids.

B. Irreversible Spectral Changes Following Heating Treatments of Concentrated Gliadin Solutions

^{13}C and ^{15}N NMR spectra of concentrated [32% (w/v)] gliadin solutions were recorded[7] at temperatures up to 95°C. A series of ^{13}C NMR spectra is shown in Figure 12, in the temperature

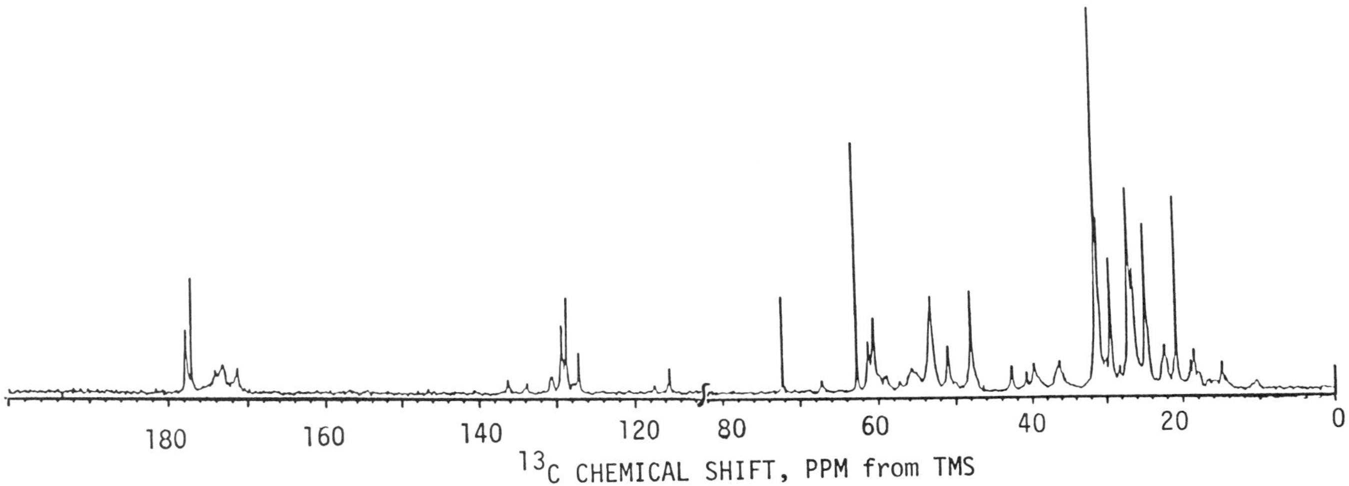

FIGURE 10. The 125-MHz ^{13}C NMR spectrum of a solution of 15.8 Flinor *var.* wheat gliadins. Experimental conditions as for Figure 9, except for the accumulation time which was 13.7 min at 11. 75 T.

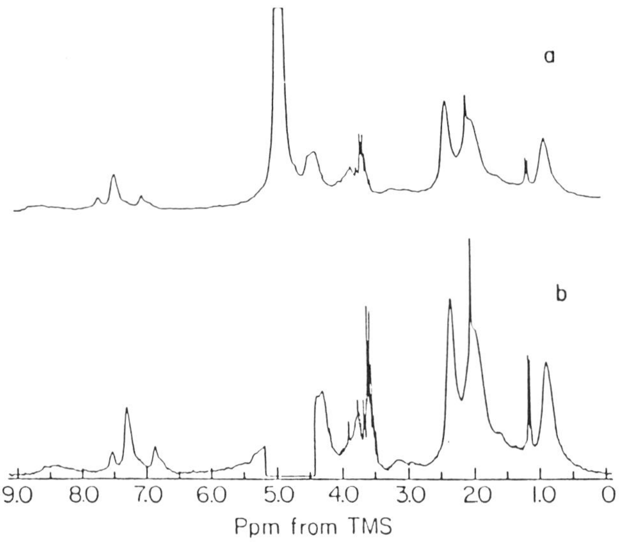

FIGURE 11. The 200-MHz proton NMR spectra of wheat protein solutions at 22°C: (a) at 32% (w/v) concentration of protein in 10 m*M* acetic acid solutions, and (b) at 15.8% (w/v) concentration of protein in 10 m*M* acetic acid solution.

Table 2
^{13}C CHEMICAL SHIFTS OF WHEAT PROTEINS IN SOLUTION

Chemical shift (ppm from Me_4Si)	Collective assignments	Chemical shift (ppm from Me_4Si)	Collective assignments
11.1	Ile δCH_3	57.7	Phe αCH, Trp αCH
12.0	Ile δCH_3	59.0	He α-CH, Val α-CH
15.1	Ala βCH_3	60.4	He α-CH, Val α-CH, Pro α–CH, Thr α-CH
15.8	Ile γCH_3		
17.1	Ile γCH_3, Val γCH_3	61.8	Pro αCH, Ser βCH_2
18.2	Ala βCH_3, Val $C\gamma_2/C\gamma_1$	62.9	Pro α-CH
18.9	Ala CH_3, Val $C\gamma_1/C\gamma_2$	67.5	Thr βCH_2
19.3	Ala CH_3, Val γCH_3, Thr γCH_3	72.6	—
20.4	Leu δCH_3, Thr γCH_3	80.0	—
21.6	Leu δCH_3	110.8	Trp Cγ
22.6	Leu δCH_3, Pro γCH_2 (*cis*)	112.8	Trp Cγ
24.3	Leu CH_3, Pro γCH_2 (*trans*)	115.8	Trp $C\varepsilon_3$, $C\eta_2$; Tyr C_3
26.5	*trans* Pro γCH_2, Gln βCH_2	118.0	Tyr C_3, His C-4 ($C\delta_2$)
27.0	Gln βCH_2, His βCH_2	127.5	Trp $C\delta_2$
29.8	Arg, His, Trp: $\beta CH_2\Delta$, Lys δCH_2, Pro, γCH_3, βCH_2	128.6	Trp $C\delta_2$/Tyr C_1/Phe C_4
		129.5	Trp $C\delta_2$/Tyr Co_1 (Cγ)/His Cγ
31.7	Gln γCH_2, Pro βCH_2, Val βCH_2	130.8	His Cγ, Tyr Cγ, Trp δ_2
36.6	Tyr βCH_2/Cys-6 βCH_2/Phe βCH_2	136.4	His Cγ, Trp $C\delta_2$, Phe Cγ
40.1	Lcu βCH_2/Cys-1,6 βCH_2/Lys εCH_2 Ala α-CH, Gly α-CH	140.0	Phe Cγ, (C_1), Trp $C\varepsilon_2$
		155.1	Tyr C_4 (Cζ)l
41.0	Arg δCH_2	157.2	Tyr C_4, Arg Cζ
42.7	Gly αCA/Cys-1 βCH_2	171.6	Gly C=O, Cys C=O, His C=O
43.3	Pro δCH_2	173.0	Pro δC=O, Gln CO=O, Ser C=O

From Baianu, I. C., Johnson, L. F., and Waddell, D. K., *J. Sci. Food Agric.*, 33, 373, 1982.

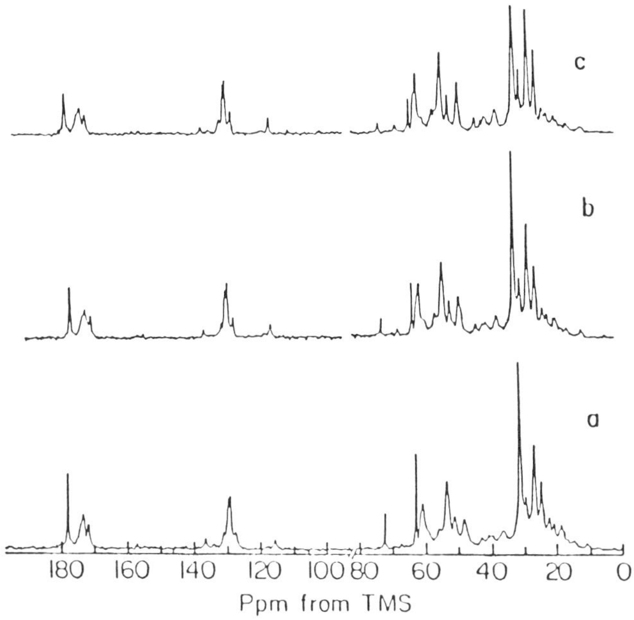

FIGURE 12. Changes with temperature in the ^{13}C NMR spectra of wheat proteins: (a) initially at 22°C for 32% (w/v) wheat protein concentration, (b) spectrum recorded at 45°C [32% (w/v)], and (c) spectrum recorded at 60°C [32% (w/v)], (From Baianu, I. C., Johnson, L. F., and Waddell, D. K., J. *Sci. Food Agric.*, 33, 373, 1982.)

range 22 to 60°C. The spectral changes due to raising the temperature were readily observed by calculating difference spectra (Figure 13), which showed that there are present not only intensity changes for particular chemical shift values but also chemical shift changes in the carbonyl region of the spectra. Furthermore, the group A of difference peaks in Figure 13c is due to an intensity decrease of the peaks at these chemical shift values upon increasing the temperature. On the other hand, the remaining peaks at higher fields (i.e., for chemical shifts <62 ppm)—which were assigned to the aliphatic groups

$$-\overset{|}{\underset{|}{C}}H,\ \overset{|}{\underset{|}{C}}H_2 \text{ and } -CH_3$$

exhibited a narrowing and increase of peak heights upon increasing the temperature. A summary of these spectral changes, together with the assignments of the corresponding chemical groups, is presented in Table 2.

The presence of the group A of peaks in Figure 13c is inconsistent with a deaggregation process (upon heating of gliadins) involving certain tyrosines, histidines, tryptophans, phenylalanines, and threonines because these become less mobile upon heating (as indicated by the broadening and loss in intensity of the corresponding peaks in group A). This is, however, consistent with a process in which some intermolecular hydrogen bonds are broken and/or ion pairs are disrupted at residues belonging to different gliadins; this is followed by the formation of some intramolecular hydrogen bonds upon cooling the sample after heating to 45°C. These intramolecular bonds could involve glutamine-NH_2 groups, carboxyl groups/glutamic acid residues, serine βOH groups, and NH groups. The motions of tyrosine, tryptophan, phenylalanine, histidine, and threonine residues can become more restricted as a result of intramolecular bonding in the regions surrounding these residues, directly involving these residues, or by further involvement of nonpolar interactions between neighbor aromatic amino acids. Other aromatic peaks in the chemical shift range from 129 to 158 ppm, however, exhibit slight

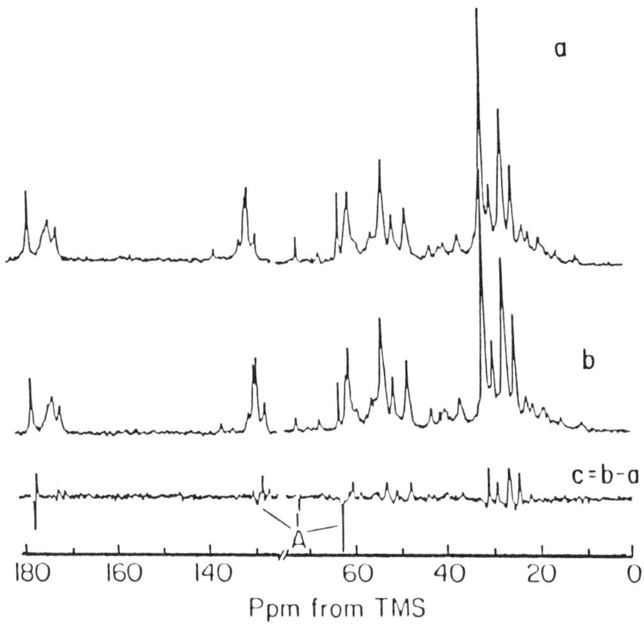

FIGURE 13. Determination of a difference spectrum of the ^{13}C resonances of wheat protein solutions between recordings at two temperatures: (a) spectrum recorded at 45°C, (b) spectrum recorded at 60°C, and (c) difference spectrum of (a) and (b) spectra (other experimental conditions are as described in the legend to Figure 10). (From Baianu, I. C., Johnson, L. F., and Waddell, D. K., J. *Sci. Food Agric.*, 33, 373, 1982.)

increases in height upon heating, and this reflects their greater mobility. A similar effect is seen for peaks in the chemical shift range from 11 to 61 ppm. It is interesting that these are assigned (compare Tables 2 and 3) mostly to glutamine and proline residues although they also include a small contribution from nonpolar residues such as Leu CH_3 and Ile αCH. These nonpolar amino acids could be part of terminal sequences of gliadins and could be freed upon disruption of polar interactions between different gliadins. That glutamines and prolines make up some 70% of the total amino acid content of gliadins makes it probable that a number of these residues are statistically distributed in the terminal part of the primary sequences.

The balance between hydrogen bonding and the disruption of hydrogen bonds/ion pairs between gliadin molecules must be, however, in favor of the latter since most ^{13}C peaks are narrow and increase in intensity upon raising the temperature. The associated decrease in the macroscopic viscosity of the gliadin samples is also consistent with this interpretation.

The ^{15}N NMR spectra, recorded in a heating cycle up to 60°C, followed by cooling to 22°C, show a considerable increase in the intensity of the peaks assigned to $-NH_2$ and $\rangle$NH. Both the ^{13}C and ^{15}N NMR spectra recorded in this heating cycle show that the spectral changes are irreversible (see Figures 13 and 14, respectively). These observations are also consistent with the mechanism proposed earlier. The observed correlation between the sample viscosity and the irreversible spectral changes points towards further possible connections between NMR measurements[7] and rheological testing.

C. Corn Zeins

1. Assignments of ^{13}C Resonances

The assignment of chemical shifts to ^{13}C resonances is often difficult in multicomponent systems such as proteins. In the case of zein dispersions, the ^{13}C NMR spectra are further complicated by the overlap of peaks. However, the peak assignments are facilitated somewhat

Table 3
SPECTRAL CHANGES AFTER A HEATING CYCLE UP TO 60°C AND COOLING TO 22°C AS OBSERVED BY ^{13}C AND ^{15}N NMR OF WHEAT PROTEIN SOLUTIONS

Chemical shift	Integrated intensity difference[a,b,c]		Assigned chemical group(s)
C-13 nucleus (ppm from Me_4Si)			
26.5	0.22 ± 0.02		*trans*-Pro γCH_2, Gln βCH_2
27.0	0.22 ± 0.02		Gln βCH_2
29.8	0.30 ± 0.02		Gln CH, Pro γCH_2
31.7	0.14 ± 0.02		Gln CH_2, Pro γCH_2
48.3	0.46 ± 0.02		Pro δCH_2
53.5	0.18 ± 0.02		Gln αCH, Leu αCH, Cys-αCH, Arg αCH
60.4	0.20 ± 0.02		Ile αCH
61.8	0.20 ± 0.02		Pro αCH
62.9	−0.82 ± 0.02		Pro αCH, Ser βCH_2
72.6	−0.50 ± 0.02		—
129.5	0.30 ± 0.02		Trp $C\delta_2$/TyrC_1($C\gamma$)/His $C\gamma$
173.0	0.25 ± 0.02		Pro C=O/Gln C=O/Ile C=O/Ser C=O
179.0 and 178.3 shifted to 178.1	—		Gln dC=O/Pro-$CONH_2$
N-15 nucleus (ppm from liquid NH_3)			
71.0	4.0 ± 0.1	$-NH_2$	Both peaks have increased, negative NOE[d] in the spectra recorded at 22°C after heating at 60°C
82.9	0.7 ± 0.2	NH	

[a] Accumulation times for ^{13}C NMR spectra were typically 8 to 20 min with shimming at all temperatures under computer control (Compushim accessory, Nicolet Technical Co.); sidebands were completely eliminated by recording spectra nonspin; the field stability and homogeneity of the superconducting magnet was sufficiently high so that no further improvement in resolution was possible by sample spinning of the concentrated protein solutions.
[b] Accumulation times of the ^{15}N NMR spectra were typically 10 h at 22°C and 2 h after heating to 60°C.
[c] Normalized to the initial peak intensity and shown with the corresponding SD.
[d] NOE = Nuclear Overhauser enhancement.

From Baiani, I. C., Johnson, L. F., and Waddell, D. K., *J. Sci. Food Agric.*, 33, 373, 1982.

by the fact that the ^{13}C NMR spectra of zeins are dominated by a small group of amino acids.[28] In addition, the more random coil or denatured the protein becomes, the more the protein residues behave like free amino acids, giving sharper ^{13}C resonances and better defined chemical shift values. The ^{13}C NMR spectra of zeins in various solvent systems are presented in Figures 15A to E.

Assignments of chemical shifts to the ^{13}C resonances of zein were done by comparing chemical shift values for specific L-amino acid residues with previously published data,[4,8-10] and also using the amino acid analysis of acid hydrolysates of zein by ion-exchange chromatography (Table 4). The ^{13}C assignments of zein and its subfractions are summarized in Tables 5 and 6, respectively. Chemical shift values for glutamic and aspartic acid residues indicated that these are present in the amide forms (see References 13 and 25). Amino acid sequencing of the amino terminal chain also indicates predominantly the presence of the amide forms.[25]

Analysis of ^{13}C NMR spectra for zein and its two subfractions indicated that 20 ^{13}C resonances

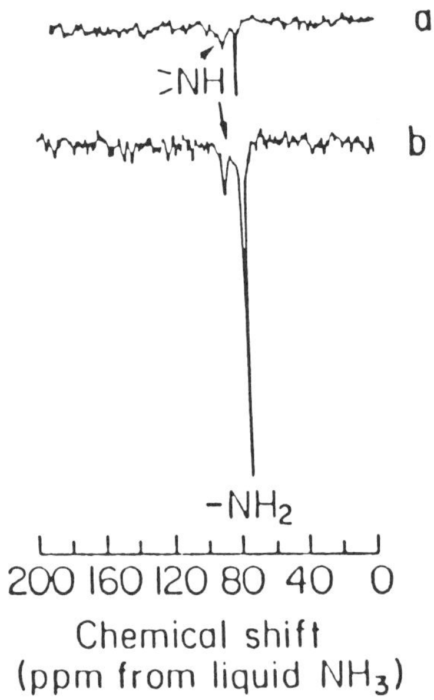

FIGURE 14. Irreversible changes induced by heating in wheat protein[27] conformations observed by ^{15}N NMR: (a) at 22°C before heating [for the 32% (w/v) wheat protein concentration] and (b) at 22°C after 10 min heating cycle up to 60°C [the same concentration as in (a)]. [It is interesting and significant that the peaks in spectrum (b) have large, negative nuclear Overhauser enhancements (NOE); the NOE values calculated from the spectra correspond to correlation times characteristic of the $-NH_2$ groups and neighboring residues shorter than about 1×10^{-9} s (1 ns)].[7]

could be resolved in the aliphatic region and, furthermore, 5 resonances were resolved in the aromatic region.[13,21] Examination of the spectra also showed that the carbonyl region was unresolved for all the protein samples; more recent ^{13}C NMR measurements at 11.75 T resolved, however, several carbonyls[26] (Figure 16). SDS-PAGE analysis of the zein preparation showed that zeins consisted primarily of subunits having molecular weights in the range 22 to 24 kDa in the unreduced state, in agreement with other SDS-PAGE studies of zein maize proteins.[27] This suggests that the sharpness of the resolved ^{13}C peaks is probably due to the relatively fast tumbling rate of zeins in solution, as well as to the relatively high mobility of amino acid side chains.[13]

The collective assignments presented in Tables 2 and 3, respectively, were made by using the information from the amino acid analysis[28] of zein together with previous ^{13}C NMR assignments for amino acids and proteins;[4,7-10] these indicate that the ^{13}C resonances in the aliphatic region of the spectra are dominated by leucine, glutamine, alanine, and proline. The resolved peaks present in the aromatic region are readily assigned to phenylalanine, histidine, and tyrosine, in view of the fact that tryptophan, which otherwise would complicate interpretation, is in too low a concentration to be detected. (Our amino acid data could not be used to determine tryptophan since this is chemically decomposed by the acid hydrolysis procedure used.) These peaks, although resolved, are somewhat obscured by peak overlap and also possibly by the reduced mobility of the corresponding residues. Recent work at 11.75 T increased considerably the resolution of the ^{13}C resonances of the aromatic amino acid residues[26] (Figure 17).

The carbonyl region of all the protein samples at 5.85 T had two broad peaks — one occurring at approximately 177 ppm and the other, less broad peak at approximately 180 ppm. The peak at approximately 180 ppm was assigned[13] to the C-terminal carbonyl of proline and the δC=O group of glutamines.[7] The carbonyl region also shows the effect that pyridine has upon chemical

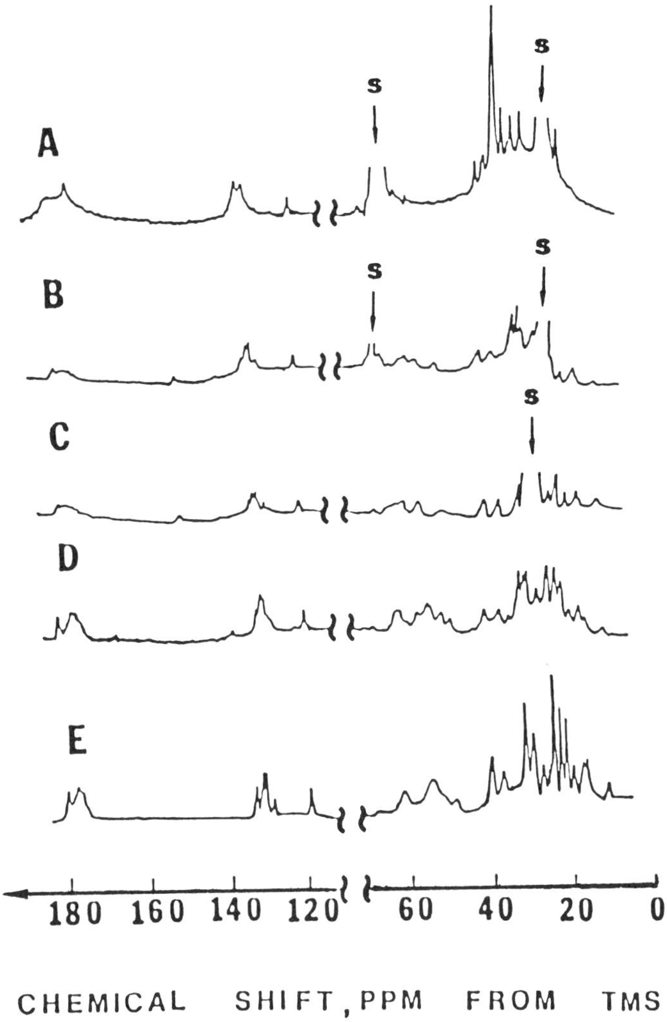

FIGURE 15. Proton broadband-decoupled, 62.908-MHz ^{13}C NMR spectra of zeins[21] in various organic solvents: A, 70% ethanol solution; B, 60% 2- propanol solution; C, 70% acetone solution; D, aqueous solution adjusted to pH 11.5; and E, 50% pyridine solution. S denotes peaks assigned to solvent carbons.

shift values; because pyridine is a strong proton acceptor, the effect of pyridine upon the protein spectra is similar to that exhibited in alkaline solutions.[21]

2. Spectral Differences between Zein Samples after Organic Solvent Treatments

Significant differences in the spectra of each of the three protein samples were observed. The spectra presented for native zeins and the propan-2-ol subfraction were broadly similar but the mobility of most residues, as indicated by the sharpness of the peaks, differed slightly (Figure 18). The ^{13}C NMR spectrum for the zeins following the butan-1-ol treatment (Figure 18b), however, showed several sharp peaks that were not observed either for native zein or for the propan-2-ol-extracted subfractions when examined in 50% (v/v) pyridine. However, the butan-1-ol treatment did *not* affect the amino acid composition significantly, and neither did propan-2-ol extraction. These observations could be explained either by differences in the subfraction compositions (which is not supported by the electrophoresis data) or by changes in the conformations of zeins caused by the treatments with alcohols.

Table 4
AMINO ACID COMPOSITION OF ZEIN[a]

Amino acid	%[b]
Asp/Asn	3.4
	3.3
Thr	2.0
	1.8
Ser	3.2
	2.6
Glu/Gln	22.8
	26.1
Pro	11.9
	10.6
Gly	2.6
	1.5
Ala	13.2
	9.1
Cys ($^1/_2$)	0.3
	0.3
Val	3.7
	3.5
Met	1.5
	1.7
Ile	4.0
	4.0
Leu	20.3
	20.8
Tyr	3.6
	5.0
Phe	5.6
	7.1
His	1.0
	1.2
Lys	< 0.1
	< 0.1
Arg	1.3
	1.8

[a] ±5% experimental error of values shown.
[b] Percentage of total amino acid composition, moles residue per 100 moles of recovered amino acid residues.

The ^{13}C NMR spectrum of zeins following the treatment with butan-1-ol showed marked differences, as discussed above. All differences appear in the aliphatic region of the spectrum and are labeled by 1, 2, and 3, respectively, in Figure 18. Assignments of these peaks were as follows:[21] peak 1, alanine βC; peak 2, leucine αC; and peak 3, leucine αC and glutamine αC.

Because of the complexity of the zein mixtures, certain ^{13}C NMR peaks, such as those in the carbonyl region or a number of peaks in the aliphatic region, were more difficult to assign to specific amino acid sites; there was a considerable overlap of peaks. Resolving and assigning peaks to amino acid sites might be improved further by two-dimensional Fourier transform (2D-FT) NMR techniques combined with deconvolution; such experiments for the characterization of zeins and other maize products are in progress. In addition to 2D-FT NMR of solutions, solid-state NMR might also be of some help for elucidating the structure of zeins, although severe

Table 5
CHEMICAL SHIFTS OF ZEINS IN SOLUTION

Chemical shift (ppm)	Collective assignments
13.2	Ile δC[a]
17.6	Ile εC
18.2	Ala βC, Val γ_1/γ_2[a]C
19.4	Ala βC, Val γ_1/γ_1C
21.5	Thr γC
22.1	Leu δC
23.8	Leu δC
25.3	Pro γC, Arg βC, His βC
27.8	Glu βC,
29.7	Glu βC, Val βC, Arg βC, Met γC
32.1	Pro βC, Met βC
34.4	Gln/Glu γC
36.8	Phe βC, Ile βC, Asn/Asp βC, Tyr βC[a]
39.5	Arg δC, Phe βC[a]
42.9	Leu βC
45.9	Gly αC

Chemical shift (ppm)	Collective assignments
50.6	Pro δC, Asn/Asp αC
54.1	Ala αC, Asn/Asp αC, Met αC, Arg αC
57.1	Gln/Glu αC, Leu αC, Met αC, Arg αC, Phe αC, Cys αC
57.9	Asn/Asp αC, Ser αC
64.7	Pro αC, Ile αC, Val αC, Ser αC
70.0	Thr βC[a]
119.1	His C-4, Tyr C-3
130.6	Phe o, m, C; His C-5, Tyr C_1, Cγ[a]
132.4	Phe o, m, C[a]
132.9	Phe γC, His C-2[a]
134.1	Phe γC, His C-2[a]
159.4	Arg εC, Asn/Asp ζC[a]
177.1	δC=O:Gln, Leu, Pro, Ala, Phe, Val, Ile, Tyr, Asn/Asp, His, Ser, Cys, Met[a]
180.5	Gln δC, Pro C=O[a]

[a] Chemical shift values and collective assignments are identical for these peaks in the zein subfractions, and are therefore omitted in Table 3.

Table 6
CHEMICAL SHIFTS OF ZEIN SUBFRACTIONS IN SOLUTION

Propan-2-ol-subfraction

Chemical shift (ppm)	Collective assignments
16.7	Ile εC, Met εC
21.1	Leu δC, Thr γC
24.3	Leu δC, Pro γC, Ile γC, Arg γC
26.6	Gln βC, His βC
28.8	Pro βC, Met γC, Val βC, Arg βC
53.1	Ala αC, Asn/Asp αC, Met αC, Arg αC
56.2	Gln αC, Leu αC, Met αC, Arg αC, Phe αC, Cys αC
59.2	Ile αC, Val αC
60.7	Ile αC, Ser βC, Val αC, Thr αC
63.8	Pro αC, Val αC, Ser βC

Protein following butan-1-ol extraction

Chemical shift (ppm)	Collective assignments
12.9	Ile δC, Ile εC, Met εC
27.1	Gln βC, His βC, Cys βC
29.8	Pro βC, Arg βC, Met γC, Val βC, Arg βC
32.9	Gln γC, Pro βC, Met βC
35.1	Leu βC, Ile βC, Asn/Asp βC, Tyr βC
40.9	Leu βC, Arg δC
50.3	Ala αC, Asn/Asp αC
63.4	Pro αC, Val αC

overlap of peaks occurs in the solid-state caused by a broad distribution of chemical shift anisotropies (Figure 19).

There are also a number of noticeable differences between the ^{13}C NMR spectra of 3% zeins in various other organic solvents. The region which exhibited the greatest spectral differences is the aliphatic region. In alcohol and acetone, the resolution of the aliphatic region is also restricted by the presence of the CH_3/CD_3 and CH_2/CD_2 groups from the solvents used. The solvent effect on the zein aliphatic residues is best exemplified by monitoring the line width of

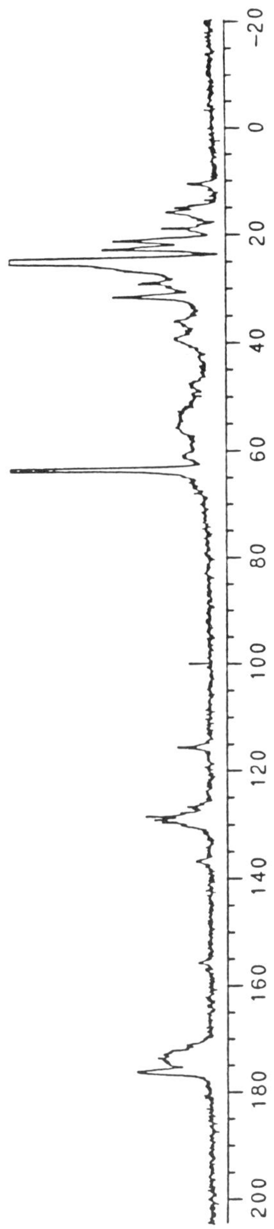

FIGURE 16. Proton broadband-decoupled, 125-MHz ^{13}C NMR spectrum of an 8% (w/w) solution of corn zeins in D_2O at pD 11.6.

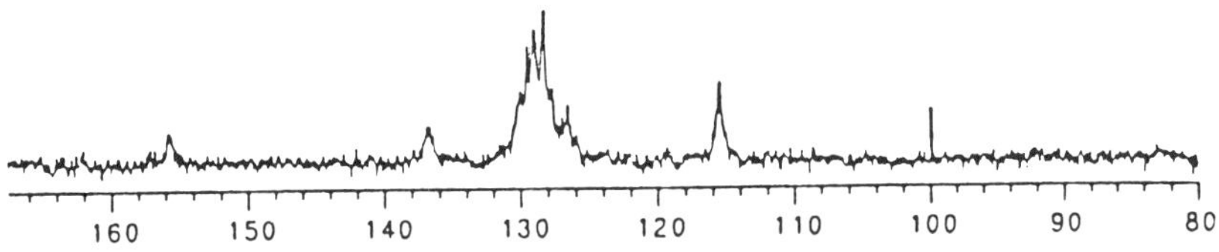

FIGURE 17. Expanded aromatic region of the 125-MHz ^{13}C NMR spectrum in Figure 16, showing the increased resolution caused by the increased dispersion at 11.75 T in comparison with the spectra recorded at 62.9-MHz (5.4 T) (Figures 15 and 18).

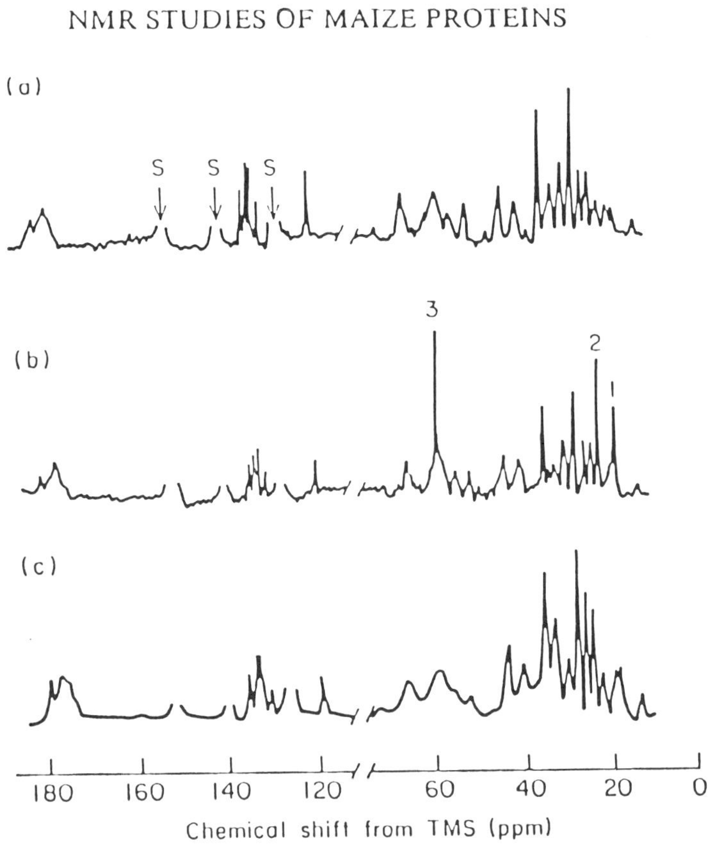

FIGURE 18. Proton broadband-decoupled, 62.908-MHz ^{13}C NMR spectra of zeins in deuteriated pyridine d_5/D_2O (1:1) solutions.[13] (a) Propanol-2-ol subfraction, (b) zeins after treatment with butan-1-ol, and (c) untreated zeins. S denotes solvent carbon peaks.

the leucine βC peak located at 43.2 ppm. The broadening of this peak (~100 Hz at half height) is caused by the low mobility of the corresponding groups in most of the solvents studied, including the pH 11.5 solution. The βC peak, however, shows much greater mobility in the pyridine solution, as indicated by the narrowing of the 43.2 ppm peak to less than 50 Hz at half height. The peak corresponding to the leucine βC in the ethanol solution is also relatively sharp (~30 Hz); this peak, however, is only part of a broader band which covers all of the region between 15 and 50 ppm. This was considered to indicate that at least some of the leucine residues in zein have a relatively high mobility in solution,[21] perhaps as a result of broken hydrogen bonds in the protein structure in the presence of ethanol. The aliphatic groups of zeins showed overall

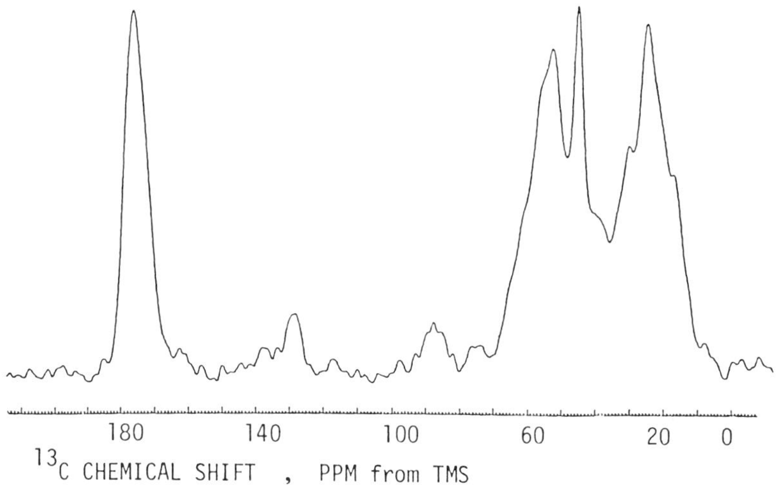

FIGURE 19. CP-MAS, 50.3-MHz ^{13}C NMR spectrum of a zein powder recorded with a 4.5-kHz sample spinning rate using a CXP200 NMR spectrometer; sample temperature was 22°C, the rf proton decoupler power was 1 kW into a 50-ohm load and the rf observe pulse power was carefully matched. Note, however, the substantially lower resolution in this spectrum compared with those in solution (Figures 15 to 18).

a much higher mobility in the pyridine solution than in any other solvent (line widths measured at half height are of the order of 10 to 50 Hz only in pyridine), and they were sharper, indicating that at least some of the aliphatic residues of zeins are possibly exposed to the solvent.

Another important region of the 62.90 MHz ^{13}C NMR spectra which shows significant differences is the aromatic one (between 110 and 140 ppm). The resolution in the aromatic region can be a good indicator of the protein interactions with the solvent. For all zein samples, with the exception of zeins in pyridine, the aromatic region lacks resolution at 5.85 T, but not at 11.75 T. When contrasted with the pyridine/water, the peaks in the aromatic region of zein in the other organic solvents are much broader (~200 to 300 Hz) at 5.85 T than for zeins dispersed in 50% pyridine/water (line widths of 20 Hz at 5.85 T).

The last region which needs to be discussed is the carbonyl region (170 to 181 ppm). This region exhibited some interesting chemical shift differences observed at 5.85 T between the samples in pyridine or the pH 11.5 solution and in the other organic solvents. The carbonyl region for zeins in the acetone, ethanol, and 2-propanol/water appeared to be similar, exhibiting a broad, unresolved band, except for the spectrum in ethanol which did show a sharper carbonyl peak. The carbonyl regions of zeins in the pH 11.5 and pyridine solutions showed a major difference in comparison with the other organic solvents in that they had a sharp (~30 Hz) resolved peak at about 181 ppm. Interestingly, this carbonyl peak displays a chemical shift change of 1 ppm from the values recorded for zeins in propanol and acetone; we previously assigned this peak to the δC=O of glutamic acid residues[30] and to the carbonyl of the C-terminal prolines. At pH 11.5 this peak occurs at 181.0 ppm and shifts to ~180.5 ppm in pyridine. This shift is probably caused by the pH effect upon carbonyl groups;[8] as the pH of the solution is increased, the shielding effect of the proton on the carboxylic acid is decreased causing the peak to shift downfield. The shift is also present in the zeins spectra in pyridine, most likely due to the strong proton acceptor characteristics of pyridine, creating a similar effect to increased pH. The difference in the chemical shift values for this peak is significant because it does indicate that glutamic acid is present to an appreciable level in zeins. Nevertheless, the amide form of the glutamic residue is the one predominant in zeins.[21,25]

3. Importance of the SDS-PAGE Data for the NMR Analysis

In solution, the line widths of the ^{13}C peaks can be significantly decreased either by the high

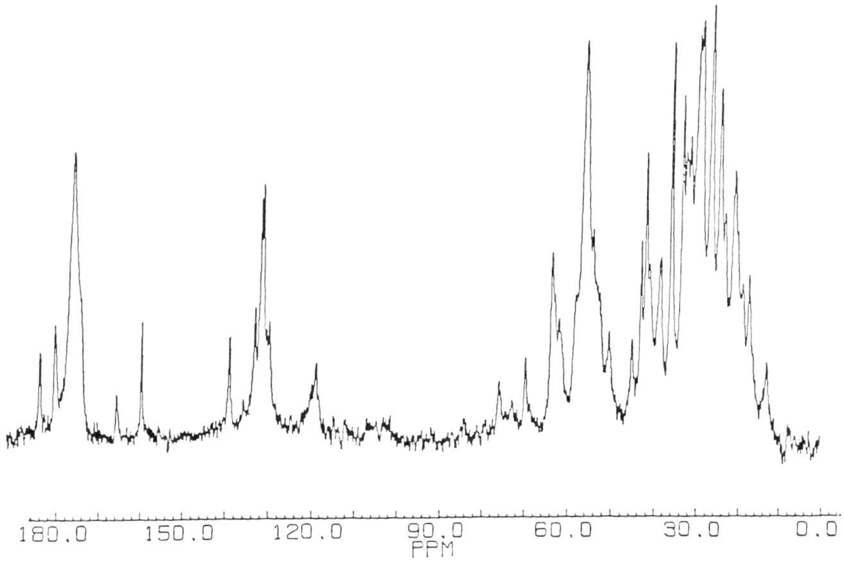

FIGURE 20. WALTZ-(proton)-decoupled, 50.3-MHz ^{13}C NMR spectrum of a soy protein isolate in solution at pH 10.5, obtained with a CXP200 NMR spectrometer. Rf decoupling power: 1 W; 16,000 data points and Fourier transform size; 16 μs, 90° pulses. (Data from Kakalis, L. and Baianu, I. C., *J. Agric. Food Chem.*, submitted.)

mobility of the amino acid side chains or by the fast tumbling of small proteins in solution. In the absence of aggregation, the tumbling rate of the protein is primarily determined by the molecular weight of that protein and the viscosity of the solution PAGE and SDS-PAGE electrophoresis were used to determine the molecular weights of corn zeins at a given concentration (5 mg/ml) after treatment with various organic solvents.

Such results showed that, in the absence of SDS by utilizing ethanol or 2-propanol for dispersing unreduced zeins, only a streak could be obtained on the gel (data not shown), therefore strongly indicating protein aggregation. The two major bands of 22,000 and 24,000 Da, as seen in the unreduced zein with SDS, appear to be of greatest intensity and are, therefore, the major subfractions of unreduced zeins. However, when electrophoresis was performed using a 50% pyridine buffer system, the unreduced zein patterns showed six bands with the three additional bands having molecular weights of approximately 12,000, 16,000, and 32,000 Da, indicating that causes other than disulfide bonding influence the aggregation of zeins in alcohols; these are likely to be the hydrophobic interactions.

D. Soy Proteins

^{13}C NMR spectra of soy protein isolates in solution that exhibit more than 40 peaks were observed[29] reproducibly (Figure 20). The aliphatic region (below 70 ppm) contains many overlapping peaks, whereas the aromatic region (110 to 165 ppm) is well resolved in comparison with the hydrolizate spectrum. The carbonyl region (above 170 ppm) contains one broad peak that represents the envelope of several carbonyl resonances, a peak at 178.3 ppm assigned to Gln δC=O and Asp δC=O, and a peak at 178.7 ppm assigned to the Glu δC=O groups. Tentative assignments of the peaks were made by comparing the observed chemical shifts with those of other proteins, peptides, and free amino acids,[4,7-10,13,21] as well as by taking into account the hydrolizate spectrum and the amino acid analysis of the sample.[28] The chemical shift values and proposed assignments[29] for soy proteins are given in Table 7.

Tabulated chemical shift values in the literature usually refer to neutral solutions; in assigning the peaks of ionizable groups we used the values for the ionic form(s) expected to be present at

Table 7
^{13}C CHEMICAL SHIFTS OF SOY PROTEIN ISOLATES IN SOLUTION

Chemical shift (ppm from TMS)	Proposed assignments
11.0	Ile δCH_3 (11.6)[a]
15.6	Ile γCH_3 (14.8)
17.1	Val γCH_3 (17.9), Ala βCH_3 (16.0)
19.2	Ala βCH_3, Val γCH_3, Thr γCH_3
21.5	Leu $\delta_1$$CH_3$
22.9	Leu $\delta_2$$CH_3$, Pro γ$CH_2$, Lys γ$CH_2$
25.0	Leu γCH (24.0), Arg γCH_2 (24.6), Pro γCH_2
26.2—28.9	Ile γCH_3, Cys βCH_2, Lys δCH_2, Arg γCH_2, Gln βCH_2, His βCH_2, Trp βCH_2
30.0	Pro βCH_2 (30.2), Arg βCH_2 (27.5), Met γCH_2
31.0	Lys βCH_2, Met βCH_2, Val βCH
31.8	Pro βCH_2, Lys δCH_2, Val βCH, Gln γCH_2
34.3	Glu γCH_2 (30.2)
36.8	Asn βCH_2, Ile βCH (36.2), Phe βCH_2 (36.6), Tyr βCH_2 (36.4)
39.3	Asp βCH_2, Cys_2 βCH_2
40.2	Lys εCH_2 (39.9)
41.3	Lys εCH_2, Leu βCH_2, Arg δCH_2 (41.2)
43.3	Gly αCH (41.2)
48.5	Pro δCH_2 (47.2)
51.2	Ala αCH (49.0), Asn αCH
52.4	Ala αCH, Asp αCH, Cys_2 αCH, His αCH (49.9)
54.5	Arg αCH (54.7), Glx αCH (53.2), Lys αCH (52.2), Leu αCH
56.2	Phe αCH (55.3), Ser αCH, Trp αCH, Tyr αCH, Cys αCH
60.0	Ile αCH (57.9), Val αCH, Ser αCH, Pro αCH, Thr αCH
62.5	Pro αCH, Ser βCH_2, Val αCH
67.7	Thr βCH_2
71.9	Carbohydrate impurity and/or a glycoprotein component
114.4	Trp γC
116.8	Trp γC, Tyr εCH (116.7)
117.7	Trp γC, His δCH (116.7)
127.8	Trp δ_1C, Trp ζC, Tyr γC (128.7)
128.4	Trp δ_2C, Tyr γC
129.4	Phe εCH (129.9), Tyr γC, Trp δ_2C
129.8	Phe δCH (130.1), Tyr γC, Trp δ_2C
131.0	His εCH (131.5), Tyr δCH
136.8	His γC (126.2), Phe γC, Trp ε_2C
157.5	Arg ζC (157.4)
163.4	Tyr ζC (155.7)
172.2—175.2	Ala δC=O (172.9), Gly δC=O (170.4), Arg δC=O (171.0), Leu δC=O (172.8), Glx δC=O (172.1), Asn δC=O, δAsn C=O, Thr δC=O, Ile δC=O, Ser δC=O, His δC=O (170.1), Cys δC=O, Met δC=O, Phe δC=O, Pro δC=O (172.0), Lys δC=O, Val δC=O (171.6), Tyr δC=O
178.3	Gln δC=O, Asp δC=O
181.7	Glu δC=O (176.8)

[a] Corresponding chemical shifts for the soy protein hydrolysate.

pH 11.5. The ^{13}C chemical shifts of nonionizable groups are relatively insensitive to changes in the ionic state of nearby residues unless they are actually H-bonded to them, or if pH-induced conformational changes took place. In view of the relatively high molecular weights of the soy isolate protein components (Sections III.A and B), the presence of sharp peaks in the ^{13}C NMR spectra of the soy protein isolate (Figures 20 and 21) requires further investigation. By measuring the line width of symmetric, well-resolved, and individually assigned resonances,

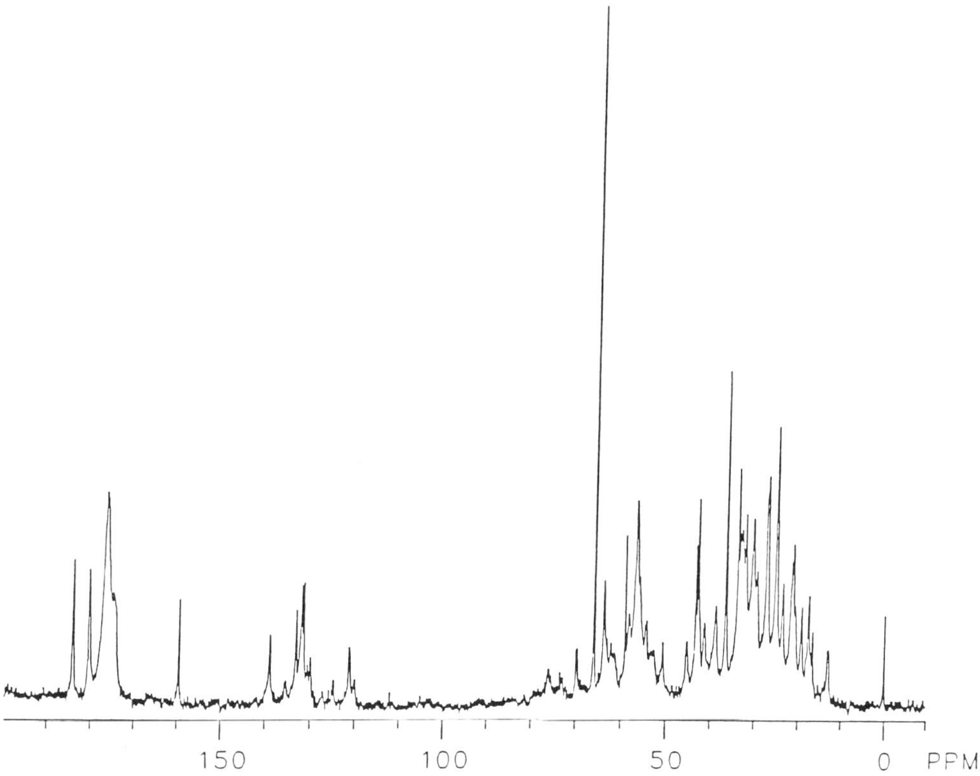

FIGURE 21. The 50.3-MHz, natural abundance ^{13}C NMR spectrum of a soy protein isolate (5% w/w) in solution (H_2O:D_2O; 2:1) at pH 11.5 and 18°C; other experimental conditions as in Figure 20.

after correcting for an estimated 2-Hz broadening due to instrumentation, one can calculate the effective correlation time τ_{eff} that characterizes the isotropic rotation of the corresponding chemical groups.

In the aliphatic region such groups are Glu βCH_2 at 34.3 ppm and Thr βCH_2 at 67.7 ppm. Assuming that the relaxation of protonated carbons is dominated by the ^{13}C-^1H dipolar interaction, τ_{eff} was determined to be 1.8 and 2.1 ns, respectively.[29] Interestingly, the first value is close to the reported one of 1.63 ns for bovine pancreatic ribonuclease.[30] These values are significantly lower than the τ_{eff} that characterizes the tumbling of even the smallest protein component (20 kDa) of the soy isolate and are most likely due to internal local motions.[31,32]

In the aromatic region, the Arg ζC resonance at 157.5 ppm has a line width of 6 Hz that corresponds to a τ_{eff} of 53 ns, assuming that at a magnetic field of 4.7 T the relaxation of nonpronated carbons is dominated by chemical shift anisotropy.[9] This value is reasonable for the tumbling of protein molecules and suggests that Arg side chains may have no internal motion faster than the overall molecular rotation.[31,32]

Based on fluorescence polarization data for globular proteins,[33] and assuming a spherical shape for soybean proteins, one can estimate a molecular weight of 173,770 that corresponds to τ_{eff} = 53 ns; this value is close to 180,000, the molecular weight that of one of the dominating molecular species.

In the case of the soy protein hydrolizate the number of resolved peaks exceeded 60 (Figure 22). This is due to the sharper peaks of the hydrolizate, a consequence of their faster tumbling in solution than that of the much larger protein molecules (even in the presence of considerable segmental chain motions in the latter). Assignments of the hydrolizate peaks[4,7-10,13,21] were made by comparison of the observed chemical shifts with those of small peptides and free amino acids,

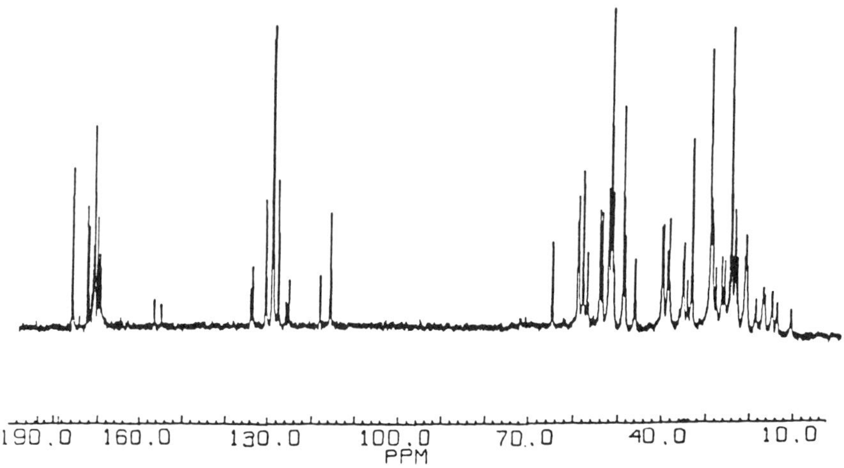

FIGURE 22. 50.3-MHz ^{13}C NMR spectrum of a 10% soy protein hydrolizate in D_2O at pD ≅1.0 and 18°C; 50° nutation angle pulses were 25-μs long; 2-s recycle time was used, together with 12-kHz sweep widths, 32 k Fourier transform size, and 2-Hz exponential line broadening.

at the same pH, taking into account the effect of amino acid incorporation into peptides on the chemical shift values and the fact that Trp and Lys are being degraded during acid hydrolysis. Although the isolate and the hydrolizate spectra were recorded at different pH values, their comparison can provide additional information about certain components of the broad protein peaks as shown in Figure 23.

The ^{13}C NMR spectra of soy proteins appear equally resolved to those of gliadins, at 4.7 T[7] and more resolved than those of α_{s1} casein,[34] corn zeins,[21] and myosin,[35] opening the possibility of probing soy protein conformations in solution by means of ^{13}C chemical shifts, line widths, and spin-lattice relaxation times of sharp resonances, particularly in the spectral region of aromatic carbons which is well resolved at 4.7 T.

V. SOLID-STATE NMR

A large number of practically important materials are solids. In agriculture, cereal grains and products derived from them are in this category. For technological applications it is becoming increasingly important to be able to analyze chemically such materials without their alteration or destruction. NMR techniques that were developed over the last 10 years are potentially capable of such chemical analysis, based on chemical shift differences, in a manner similar to the approach presented in the previous sections. The methodology employed is, however, substantially different (Figure 24) from the NMR of liquids; this methodology is explained in Section B. Some of the principles upon which such methodology is based will be briefly presented and illustrated by solid-state NMR spectra that we have obtained.

A. Magic-Angle Spinning and High-Power Proton Decoupling

There are four major obstacles to obtaining highly resolved solid-state NMR spectra in complex agricultural materials:

1. The NMR peaks are severely broadened (~20 to 40 kHz) by dipolar interactions among the nuclear spins.
2. The longitudinal relaxation times (T_1) of low-γ spin-$^1/_2$ nuclei are long and, therefore, the accumulation periods required for observing the low-γ, spin-$^1/_2$ nuclei would be long.
3. The presence of chemical shift anisotropies (CSA) which are dependent upon the orientation of a given bond axis (e.g., a carbon-carbon bond) with respect to the applied, static magnetic field.

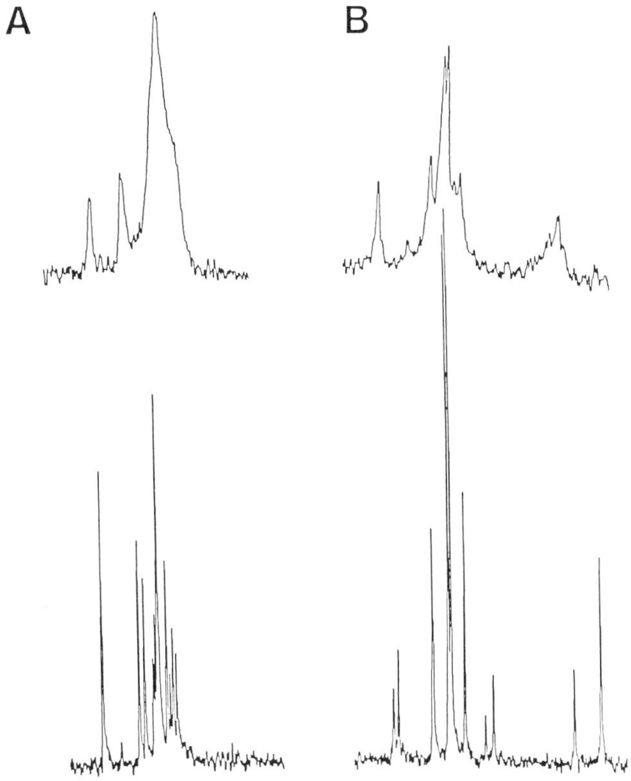

FIGURE 23. Comparison between (upper) and the soy protein hydrolizate (lower) traces, expanded from the corresponding ^{13}C NMR spectra in Figures 20 and 22.

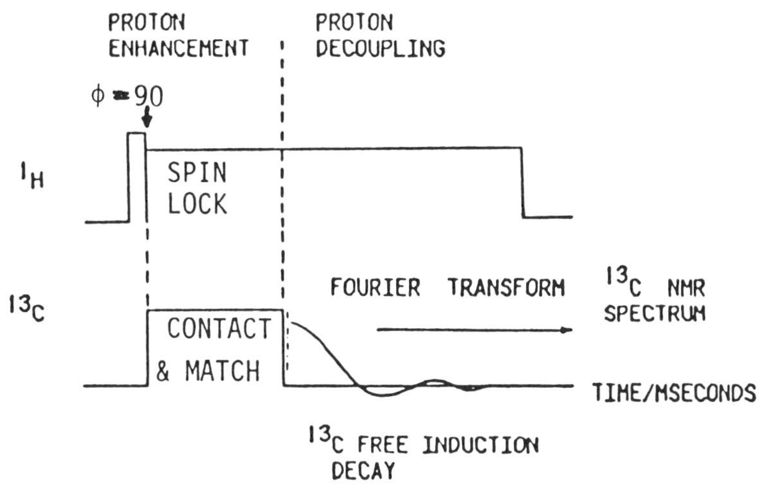

FIGURE 24. Basic scheme for cross-polarization (or polarization transfer) in solids, shown here for ^{1}H-{^{13}C}. The spin-lock pulse is 90° phase-shifted after the first 90° pulse (with zero phase), and is employed for polarization transfer to the ^{13}C spins; after mixing (or contact), the long pulse in the proton channel is employed for high-power decoupling of protons from ^{13}C. The maximum signal-to-noise enhancement (4:1) is obtained for rigid solids when the rf power level in the ^{13}C channel matches (Hartmann-Hahn) that in the proton channel, during spin-lock of the protons, or contact time.

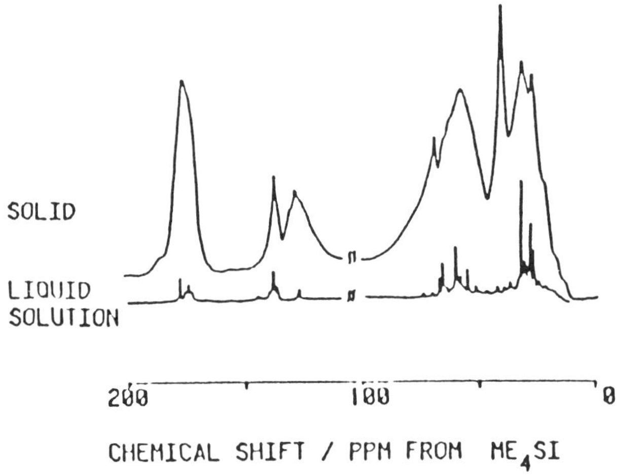

FIGURE 25. CP-MAS, 75-MHz ^{13}C NMR spectrum of lyophilized wheat gliadins (top spectrum) compared with the proton broadband-decoupled ^{13}C NMR spectrum (bottom) of a 15.8% (w/w) solution of wheat gliadins with 10 m*M* CD_3COOD in 99% D_2O, at pD = 3.4 and 22°C. Sample spinning rates were 3.5 kHz for the top spectrum and 0 for the solution spectrum. The H_1 decoupling field value for the spectrum was 30 G. (From Baianu, I. C. and Forster, H., *J. Appl. Biochem.*, 2, 347, 1980.)

4. The presence of a distribution of CSAs caused by the presence of a range of distinct conformations in the solid; such conformations would be rapidly averaged in solution by molecular motions.

Dipolar interactions between any two nuclear spins have a characteristic dependence[36] upon the orientation of the internuclear vector with respect to the external magnetic field H_0, which is of the form $(3\cos^2\Theta-1)$, where Θ is the angle between the internuclear vector and the external magnetic field direction. The expression $(3\cos^2\Theta-1)$ becomes zero for an angle $\Theta_M = 54° 44'$ called the "magic-angle". If a sample is spun rapidly around an axis of rotation which makes a constant angle $\Theta_M = 54° 44'$ with the applied static magnetic field, the dipolar broadenings are reduced to zero for spin-$^1/_2$ nuclei[37] and the CSA is reduced to a relatively small value. To obtain such narrowing it is necessary that the sample spinning rate is higher than the static line width (in kilohertz). For proton-proton dipolar interactions (~100 kHz) this is not readily achieved for rigid solids. Furthermore, "broadening" caused by the overlap of *different* CSAs will, of course, remain — as illustrated in Figure 25; this figure presents the magic-angle spinning ^{13}C NMR spectrum of a wheat gliadins powder sample. Note, however, that this spectrum was acquired in the presence of proton-enhancement and decoupling Section B.

Even with magic-angle spinning (MAS), the NMR sensitivity for dilute spins at natural abundance, such as ^{13}C, is insufficient for routinely obtaining an acceptable signal-to-noise ratio.

B. The Cross-Polarization Process

The presence of long T_1s adds to the difficulty of obtaining NMR spectra of the dilute spins because of the long accumulation periods required. The solution to this problem is based upon an idea of Hartmann and Hahn,[38] that has been renamed "cross-polarization" (CP).[39] Essentially, the abundant spins (protons) are first polarized by a 90° pulse and then their polarization is transferred to the excited dilute spins (such as ^{13}C) under well-defined rf-power matching conditions[38] for the two rf channels. Subsequent to the proton polarization by a 90° pulse, the proton spins are spin locked (SL) by a long, 90° phase-shifted pulse (Figure 24); this long pulse is continued after polarization transfer, or CP, with the purpose of high-power, proton decoupling from the dilute spins. The entire pulse sequence takes place under conditions of fast sample spinning at the magic-angle, Θ_M, and the procedure is, therefore, referred to as CP-MAS. The top spectrum of a wheat gliadin powder sample in Figure 25 was obtained by CP-MAS. If the spinning rate was insufficient to completely average out CSA, the spectrum would contain side bands separated by the spinning frequency, each side of the central peak.

C. Comparison of CP-MAS NMR Spectra with the Corresponding Solution Spectra

Whenever it is possible to solubilize a material without altering its chemical structure, a comparison should be made between the solution and the CP-MAS NMR spectra. Such a comparison would reveal differences between the solution and the solid-state if they are present, or else provide help with the assignment of the solid-state spectrum. It was recently claimed[40] that ^{13}C CP-MAS NMR spectra can be used to determine secondary structure of noncrystalline, solid proteins if there is short-range order within the protein domains, and the secondary structure is well defined (e.g., α-helical segments). In the case of cytochrome *c*, however, the ^{13}C CP-MAS NMR spectrum was broad and the secondary structure could not be determined from the NMR spectra of a powder.[40] This was alleged (but not proved) to be caused by a distribution of slightly different loops in the peptide backbone of cytochrome *c* that are resolved in the three-dimensional structure of cytochrome *c* crystals by X-ray diffraction. It was, therefore, proposed that the ^{13}C NMR chemical shifts in CP-MAS spectra "sum up the spectroscopic effect of the local conformations...as defined by the rotational angles Φ, ψ, ω, and χ, and an 'amorphous' type spectrum would be expected for the 'non-periodic' regions with secondary structure of a globular protein such as cytochrome *c*." In the case of elastin, however, both α-helical and β-sheet regions were derived from the NMR, whereas the X-ray powder pattern suggests an amorphous structure. The ^{13}C resonances that were employed for such a characterization of secondary structures of solid proteins were those of the carbonyls (C_α and C_β corresponding respectively to an α-helix and β-sheets), and — when sufficiently resolved — those of the methyl, methylene, and –CH groups. Resolution enhancement was essential in several cases. The intensity of the α-CHs (signal *"l"* at ~52 ppm and "m" at ~56 ppm) of several helix-promoting amino acids such as Ala, Ser, Glu, and Asp was low for proteins that had predominantly (80 to 90%) β-pleated sheet secondary structures. The chemical shifts of the two carbonyl and β-CH_3 (Ala) peaks for the α-helix and β-sheet structures differed by 4 to 7 ppm, corresponding to the large differences between the (Φ, ψ) Ramachandran angles for the two types of regular secondary structures. Additionally, the β- and δ–CH_2 peaks of Pro *(trans)* (at 28.7 and 24 ppm, respectively), as well as the β-CH_2 of Pro *(cis)* were considered to be informative. If one applied these concepts to the ^{13}C CP-MAS NMR spectrum of wheat gliadins (Figure 25A), which is just as broad and unresolved as that of cytochrome *c* (Figure 3 in Reference 40), one may conclude that there is a wide distribution of local, nonperiodic secondary structures of wheat gliadins, or else that their structure is close to a random coil. Such a conclusion is consistent with our previous ^{13}C NMR analysis of wheat gliadins in solution (Figure 25B), and with the expected secondary structure from the large proportion of Pro present (~30%).

The CP-MAS ^{13}C NMR spectrum of the wheat gliadin powder in Figure 25 (top curve) is much less resolved than that of wheat gliadins in solution (lower curve), presumably because of the broad distribution of CSA patterns in the solid-state. Somewhat similar spectra were observed for wheat gluten powders[1] (Figures 26 and 27). Although the spinning rate was increased from 3.0 kHz at 7 T (Figure 26) to 4.5 kHz at 4.7 T, there was no significant increase in resolution (Figure 27), supporting the interpretation that the resolution in the solid-state is limited by a distribution of CSA patterns characteristic of noncrystalline materials. Significant differences are, however, observed between the wheat gliadins-enriched (Figure 28a) and glutenin-enriched (Figure 28b) subfractions,[16] as well as the wheat gluten (Figure 28c). Differences were observed both in the aliphatic (Figures 29a and b) and aromatic regions (Figures 28a to c). It is interesting that the differences in the aliphatic region can also be observed under low-power decoupling conditions for hydrated (~50% w/w moisture) samples (Figures 30B and C). For comparison, the ^{13}C NMR spectrum of a gliadin acid-hydrolysate solution is also shown in Figure 30D. The hydrated wheat grains spectrum (Figure 30A) contains additional ^{13}C NMR peaks from carbohydrates and lipids. In the dry state, the low-power proton-decoupled ^{13}C NMR spectra are completely unresolved in the absence of MAS (data not shown).

The carbonyls in the ^{13}C CP-MAS NMR spectra of wheat gliadins and wheat glutens in (Figures 27 and 28) do not appear significantly different. However, as indicated above, there are significant differences in the aliphatic region, between 20 and 55 ppm, with the wheat glutenins spectrum including several sharp and resolved peaks; the βCH_2 peaks of Leu at ~39.3 ppm would be consistent with a significant proportion of α-helix being present, as would be the 35 ppm (βCH_2) peak of Asp. On the other hand, the sharp peaks from 29 to 33 ppm assigned to βCH_2 of Pro, βCH of Val, and βCH_2 of Glu are consistent with the presence of β-pleated sheets in the secondary structure of wheat glutenins. The sharpness of the peaks in the aliphatic region would imply that *locally*, such regions of wheat glutenins are *well ordered*. There is also an indication of a partial splitting of the carbonyl envelope of glutenins (Figure 28b, at arrows) which would be consistent with a mixture of α-helix and β-pleated sheets (in almost equal proportions), perhaps superimposed on a much broader spectrum from irregular regions in these proteins. Such structural differences between gliadins and glutenins would be related to the markedly different functionalities of these two major groups of wheat proteins. However, a systematic, CP-MAS study of a large group of proteins and homopolypeptides of known structures is needed to establish the analysis proposed in Reference 40, and to confirm the proposed assignments. The assignments discussed above are consistent with those for wheat gliadins in solution,[7] and with known chemical shifts of amino acids (Figure 30D) and random polypeptides.[8]

VI. TWO-DIMENSIONAL AND MULTIPLE-QUANTUM NMR

In recent years two-dimensional (2D) and multiple-quantum (MQ) NMR techniques are becoming increasingly popular because of the substantially improved resolution and versatility in handling complex assignment problems.

A group of techniques collectively named "correlation spectroscopy" are in widespread use for chemical and biochemical analyses; their development stems from Jeener's idea[41] for 2D NMR. The pulse sequences employed for correlation spectroscopy can be represented in the following sequence:

$$90° - t_1 - \beta°_{PH2} - FID_{PH3} \qquad (1)$$

$$90° - t_{1/2} - \beta°_{PH2} - t_{1/2} - FID_{PH3} \qquad (2)$$

$$90° - t_1 - \beta\text{-}_{PH1}\ t_d - \beta°_{PH2} - t_d - FID_{PH3'} \qquad (3)$$

where β is the pulse width and PHs are rf phases. The first sequence is essentially Jeener's and its use for 2D is indicated schematically in Figure 31. The 2D NMR experiment includes four distinct processes: (1) preparation, (2) evolution, (3) mixing, and (4) detection. At the preparation stage, the experimenter prepares the spin system under investigation in some well defined state which is the same for all set values of the t_1 time interval; such a selected initial

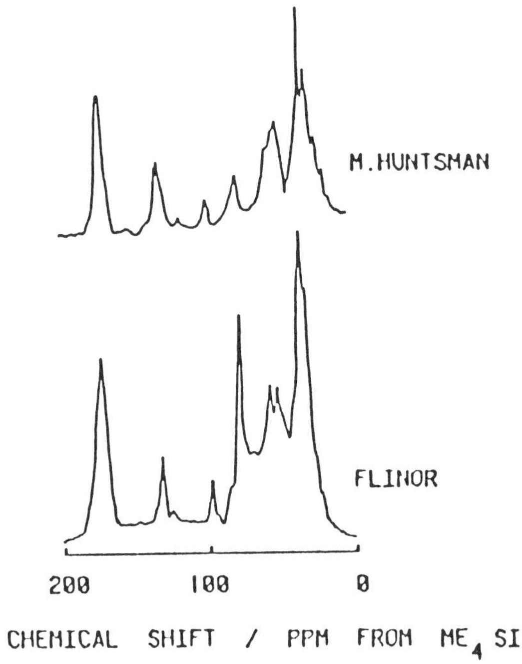

FIGURE 26. Comparison of CP-MAS, 75-MHz ^{13}C NMR spectra of wheat glutens extracted from two different English wheat varieties: var. *Flinor* and var. *M. huntsman*. Other conditions were as in Figures 24 and 25. (From Baianu, I. C. and Forster, H., *J. Appl. Biochem.*, 2, 347, 1980.)

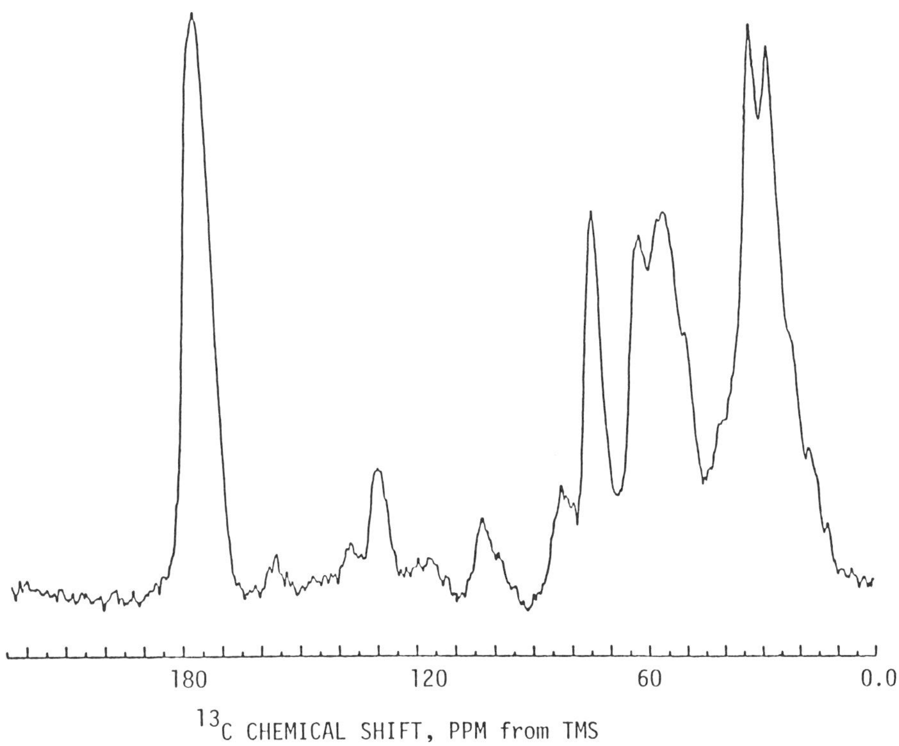

FIGURE 27. CP-MAS, 50.3-MHz ^{13}C NMR spectrum of *Flinor* var. wheat gluten, recorded at 4.7 T with a sample spinning rate of 4.5 kHz; the 90° pulse width at 200 MHz was ~2 μs.

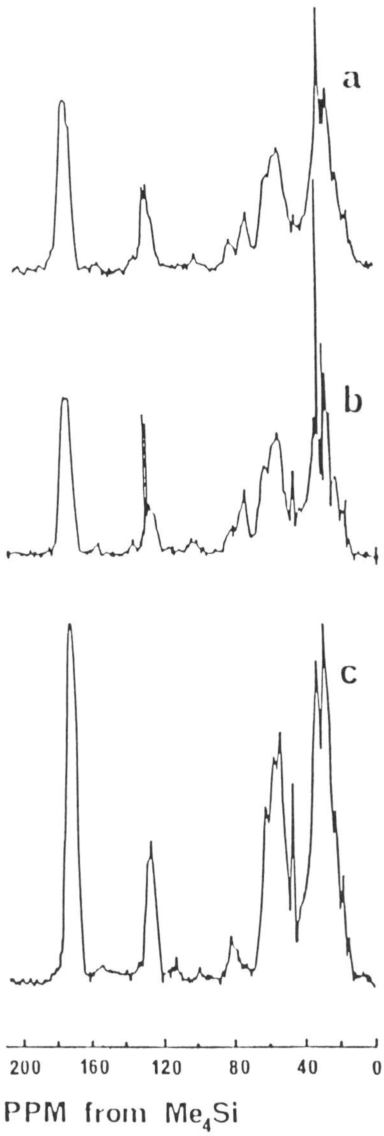

FIGURE 28. The 75-MHz, CP-MAS, ^{13}C NMR spectrum of *M. Huntsman* var. wheat proteins. Sample spinning rate was 3.5 kHz; sidebands removed by subfraction. (a) *M. Huntsman* var. wheat gluten powder, (b) urea-insoluble. *M. Huntsman* var. wheat protein fraction in powder form, and (c) urea-soluble, *M. Huntsman* var. wheat protein fraction (gliadins enriched), in powder form. (From Schofield, J. D. and Baianu, I. C., *Cereal Chem.*, 59(4), 240, 1982.)

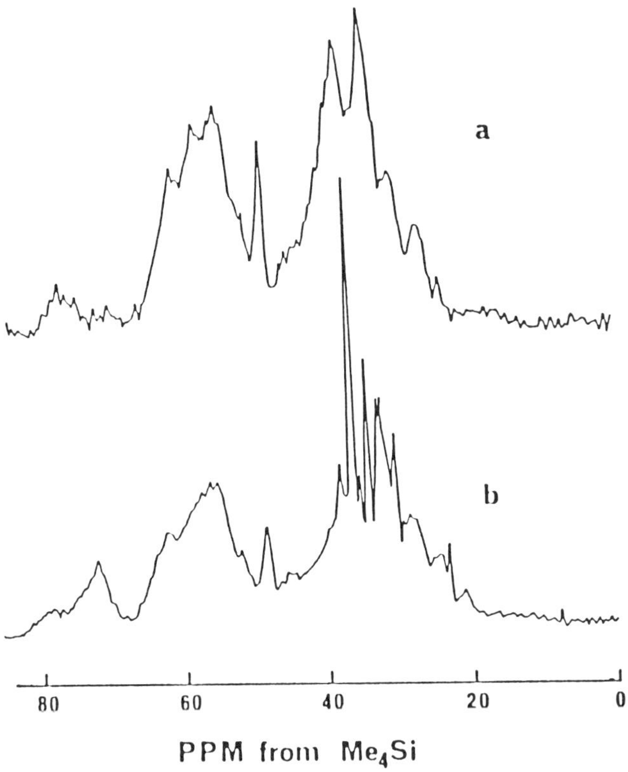

FIGURE 29. Expanded plot of the aliphatic region of Figures 28b and 28c: (a) spectrum of the urea-soluble (gliadins-enriched) protein fraction of *M. Huntsman* var. gluten and (b) spectrum of the glutenins-enriched fraction.

condition of the spin system could be, for example, thermal equilibrium or saturation of one/several spins. Usually, the preparation period ends with a pulse that generates transverse magnetization or polarization transfer.

During the evolution period, the precession of the nuclear spin magnetization is affected, or "modulated", by couplings or interactions, such as J-spin coupling or dipolar interactions, that are being explored in the 2D NMR experiment. The mixing pulse(s) causes mixing or separation of coherences by converting one spin transition into others. The detection period (t_2) involves the recording of the free-induction decay (FID), or NMR spin echo, of the observe nucleus, starting at "0" time after the last pulse in the sequence and ending at a time t_{2max} which is constant for each t_1 value. Following acquisition of the data, the FIDs are first Fourier transformed in the regular fashion, then the files are stored as a matrix, followed by conversion of matrix rows into columns (transposition), and the resulting "pseudo FIDs", or interferrograms, are again Fourier transformed (Figure 31). The result is a 2D plot, or square box, with an F_2-vertical axis which is usually the chemical shift of the observe nucleus and an F_1, horizontal axis whose meaning is determined by the interactions present during the evolution period; the latter are determined, or selected, by the pulse sequence employed during the evolution period. The pattern of interactions is, therefore, labeled or indexed by the chemical shift on the F_2 axis. An excellent, pictorial account of the mechanics of action of nuclear spins in a 2D NMR experiment is available in a recent 2D NMR review by Gray,[42] which helps one understand the basic actions of rf pulses on the nuclear spins, allows one to decompose the 2D NMR experiment into standard packages of pulses that have predictable results and, therefore, lets one tailor a specific pulse sequence to yield a desired result. The practical, "nuts and bolts" aspects of 2D NMR were also

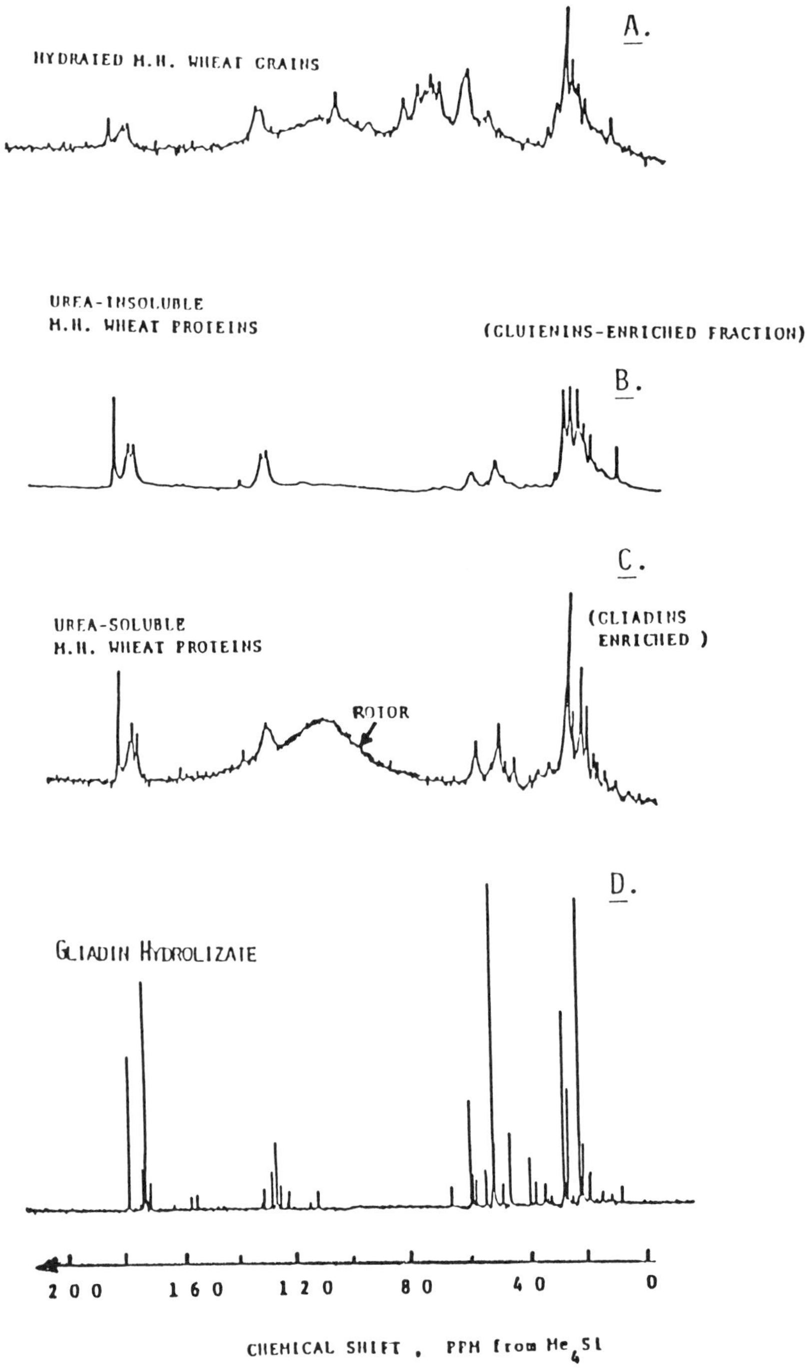

FIGURE 30. Comparison of 62.5-MHz ^{13}C NMR spectra of hydrated wheat grains (50% water, w/w) of *M. Huntsman* var. (A), and its storage protein components (B and C), and acid hydrolizate of wheat gliadins (D). Low-power, proton broadband-decoupling was employed in all cases, centered at the water resonance frequency. Spectra C and D are for solutions, whereas A and B are for hydrated solids containing about 50% water (w/w); recycling times were 2 s for the spectra A to C and 5 s for spectrum D; 16,000 data points were accumulated using 90° (60 μs) pulses for spectra A to C and 45° pulses for spectrum D. Spectra A and C were accumulated (50,000 scans) with a sideways (horizontal)-spinning rotor that contained two Teflon caps (central, broad ^{13}C absorption band at ~100 ppm).

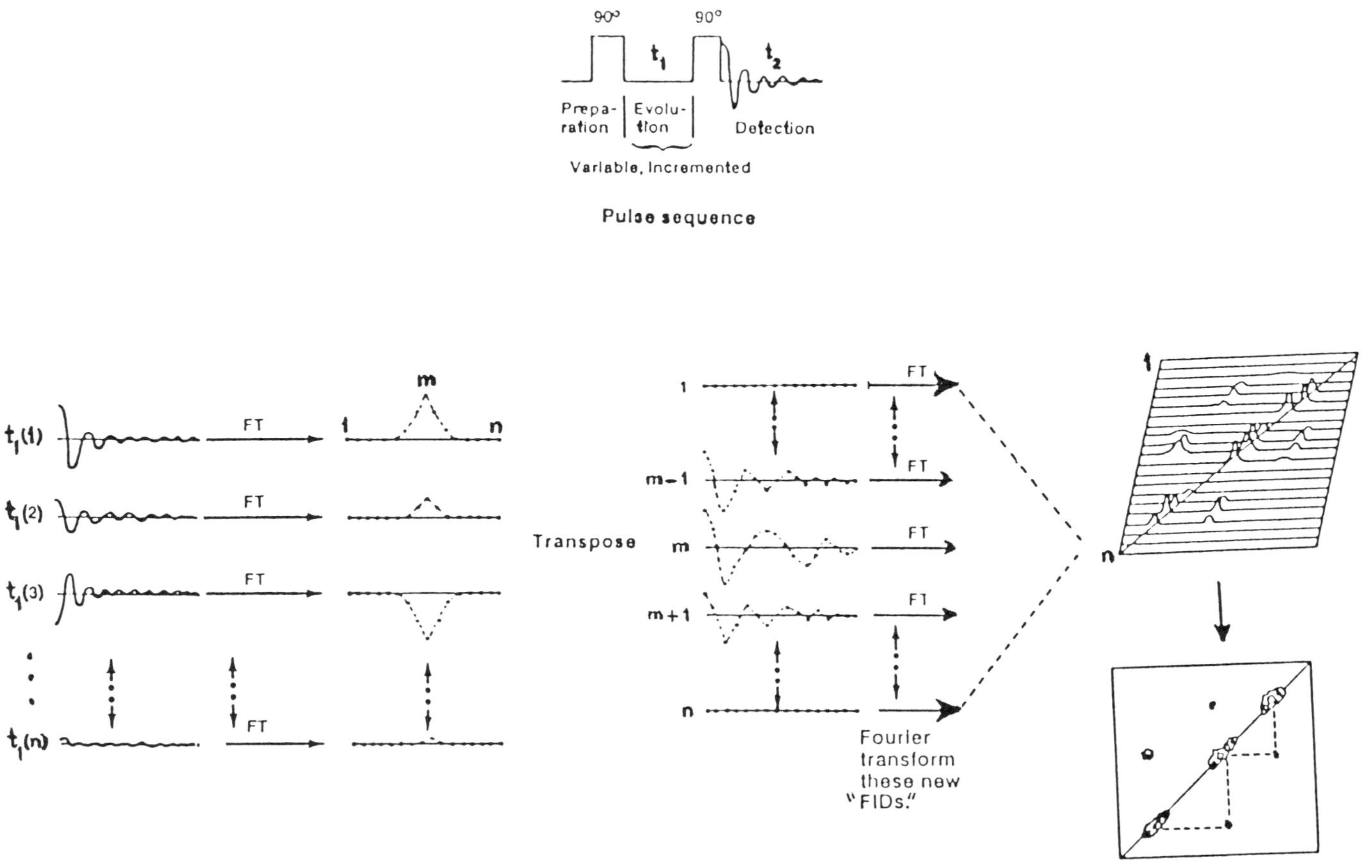

FIGURE 31. Schematic representation of the 2D NMR procedures. Using the pulse sequence shown at the top, *n* free induction decays (FIDs) are collected for a liquid sample and stored; these FIDs are then Fourier transformed and stored. A transposition of the latter spectra then yields n interferograms, or "pseudo-FIDs", is performed; these are then Fourier transformed to produce the stacked, or contour, plots of the 2D NMR spectrum.

the subject of a comprehensive review by Hull,[43] and the reader with little experience of 2D NMR is urged to look it over before embarking on a 2D enterprise.

The purpose of this section is to illustrate only a few of the 2D and MQ NMR applications to the characterization of proteins and to provide the reader with an understanding of the approach, as well as the methods employed to analyze 2D NMR plots and to extract molecular information. There are at least two (related) applications of 2D NMR techniques: (1) the improvement of spectral assignments for regions that are very crowded, or unresolved in 1D NMR spectra and (2) determination of structure and conformations of complex molecules.

The first of these applications, the improvement of spectral assignments (such as the assignment of chemical shifts to nuclei belonging to specific chemical groups), is readily related to the NMR characterization of food proteins discussed in the previous sections. Especially relevant for such applications is the heteronuclear chemical shift correlation (HETCOR) 2D NMR experiment in which one can resolve along the F_2 axis ^{13}C or ^{15}N chemical shifts, and ^1H chemical shift along the F_1, axis, for example. If one wishes to simplify the spectra and decouple the spin multiplets, then one can use the following (HETCOR) pulse sequence:

^1H 90° – τ – 90° – τ 180°– t – 90° – t′ – 90° – BROADBAND DECOUPLING

^{13}C 1/J 180°– ...– t_1 90° – t_2 – FID

The 180° pulses in this sequence refocus the chemical shifts, which appear in a chemical shift-correlated map (CSCM) that relates a ^{13}C chemical shift of a specific nucleus to the chemical shift of the protons directly bonded to that ^{13}C; the directly bonded nuclei have large J_{CH} couplings which provide the improved separation in the HETCOR experiment. HETCOR is especially useful for macromolecules with fast internal motions; we expect, therefore, that HETCOR would provide much improved separation for soy storage proteins and wheat gliadins in solution. Two examples of HETCOR experiments that were successful in resolving most of the residues in a protein molecule, the turkey ovomucoid, are shown in Figures 32 and 33. An improved version of HETCOR employs the indirect, or reverse, detection methods which utilizes the high ^1H sensitivity to detect indirectly the large ^{13}C, or ^{15}N, chemical shifts. This is especially useful in the cases when the amounts of purified proteins are minute (≤20 mg), and for ^{15}N NMR studies.

In the more common, complex cases encountered with large protein molecules (M_w > 20 kDa), with/without internal motions (such as soy storage proteins), two other 2D NMR correlation spectroscopy experiments, COSY and NOESY, are perhaps harder to perform than HETCOR and may be almost impossible to interpret. For protein molecules less than about 20,000 Da, however, both COSY and nuclear Overhauser enhancement and exchange spectroscopy (NOESY) provide valuable spin connectivity and structural information. The simplest COSY (or JEENER 2D) experiment (HOMCOR) utilizes Jeener's 2D pulse sequence (1) with β° = 90°. The COSY sequence allows one to find which protons are scalar coupled to each other by correlating the ^1H chemical shifts of the coupled spins. Coherence transfer is induced by the pulse sequence (1), and one can actually see the coherence transfer echo in a COSY experiment. The contour plot of a COSY spectrum has a characteristic diagonal from left to right which contains the autocorrelation ^1H couplings, reproducing, therefore, the 1D NMR spectrum. An example is given in Figure 34 for an 8-m*M* lysozyme in solution. The presence of off-diagonal peaks is proof of spin coupling between the protons at the chemical shift value in F_1. In the COSY experiment *all* possible off-diagonal interactions (intergroup couplings) will be generated; it leads therefore, to a *complete* description of the coupling partners. To find the connectivity, or coupling between two protons, one can draw a parallel line to the F_1 axis, starting at the selected off-diagonal peak and ending on the diagonal; then, one draws two lines parallel to the F_2 axis: one parallel line through the diagonal intersection point obtained at the previous step, and the

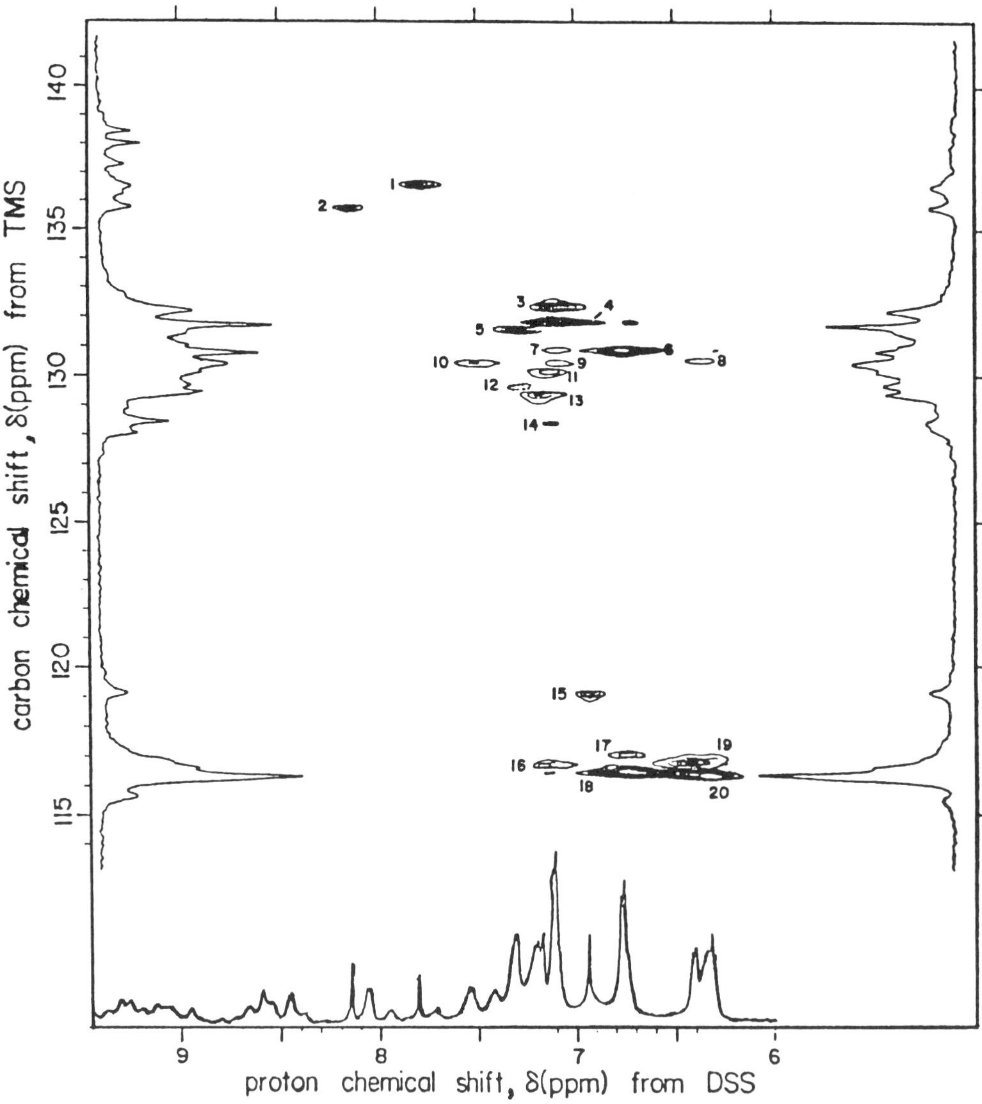

FIGURE 32. Chemical shift-correlated, ^1H-{^{13}C} NMR spectrum of turkey ovomucoid. (From Markley, J. L., *Protein Engineering*, Alan R. Liss, New York, 1987, 15. With permission.)

other through the off-diagonal peak of interest. The two parallel lines intersect the F_1 axis at the peaks corresponding to the proton, or proton groups, that are scalar coupled. The protons with the largest scalar couplings would be the ones most remote from the diagonal. In relatively large molecules, such as proteins, one useful feature of the COSY contour plot is that it breaks up in "subpatterns" characteristic of different amino acid residues, thereby simplifying the assignment. Note such a characteristic pattern for the Tyr residues of lysozyme in Figure 34. COSY does not, however, allow one to measure J-couplings except when the diagonal multiplets are resolved; such couplings can be, however, determined with a 2D J pulse sequence (2), with $\beta = 180°$. Another limitation of the simple COSY experiment is the crowding of peaks on the diagonal which may obscure the small couplings that are close to the diagonal and that are of interest. A solution to this problem is the COSY-45 experiment for which $\beta = 45°$ in the pulse sequence (Sequence 1). COSY-45 minimizes the diagonal contributions, and the cross-peaks of the AB type are more readily seen; this also allows coherence transfer among multiplet lines of

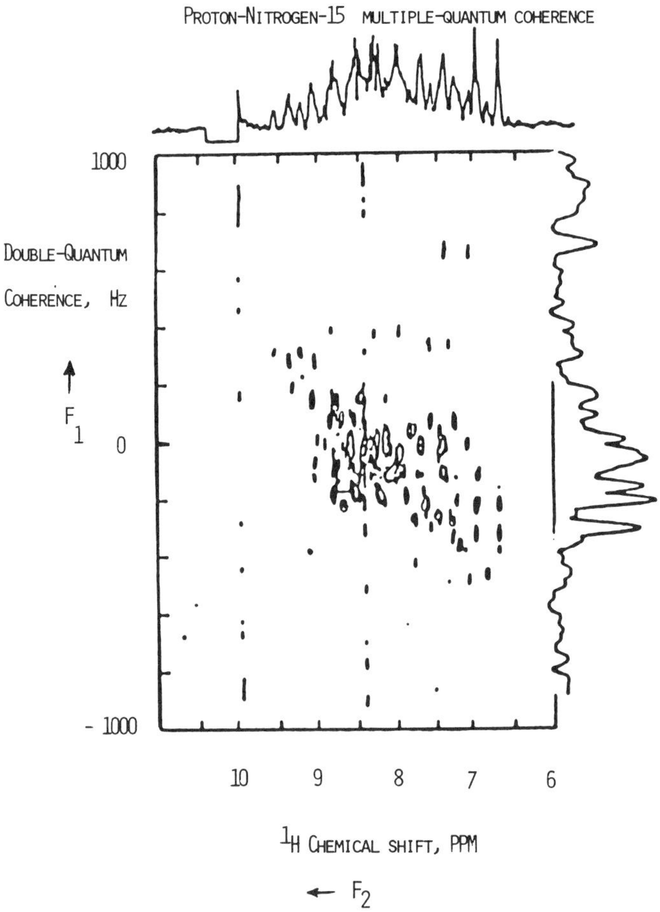

FIGURE 33. 2D, ^{1}H-{^{15}N}, DQF NMR spectrum of a solution of 15-m*M* turkey ovomucoid third domain (OMTKY3) at pH 3.3, 25°C. The F_1 dimension represents the DQ frequencies. A total of 48 t_1 increments of 500 ms each were collected in 512 data points. (Other details in Reference 44.)

a single spin (as in 2D J-resolved) to be resolved on the diagonal. For long-range ^{1}H-^{1}H scalar couplings, a delayed COSY experiment (COSYLR) has been successfully carried out with the pulse sequence (3). The resolution of a COSY plot is vastly improved in a phase-sensitive (PS) COSY experiment (with positive and negative peaks!), and the "smearing" on the diagonal caused by tailing off from intense peaks is removed using a double-quantum "filter" (DQF) COSY sequence:

$$90° - t_1 - 90°_{PH2} - t_{DQ} - 90°_{PH3} - FID_{PH4} \qquad (4)$$

The second pulse in this sequence generates a number of quantum coherences, (transitions for which $\Delta m = \pm n$, where n = integer); the coherences with $n \neq 1$ would remain undetected if the sequence stopped after the second pulse. The third pulse, with appropriate phasing, converts double-quantum (2Q) coherences into single-quantum (n = 1), observable magnetization. Such 2Q coherences arise from any coupled spin pair. By appropriate phase cycling,[42,43] the single-spin signals (such as the H_2O peak) are removed from the spectrum, or greatly reduced in amplitude, because the 2Q and 1Q coherences have different rf phase dependences. 3Q or higher MQ coherences also afford further spectral simplification of COSY 2D plots. The resolved

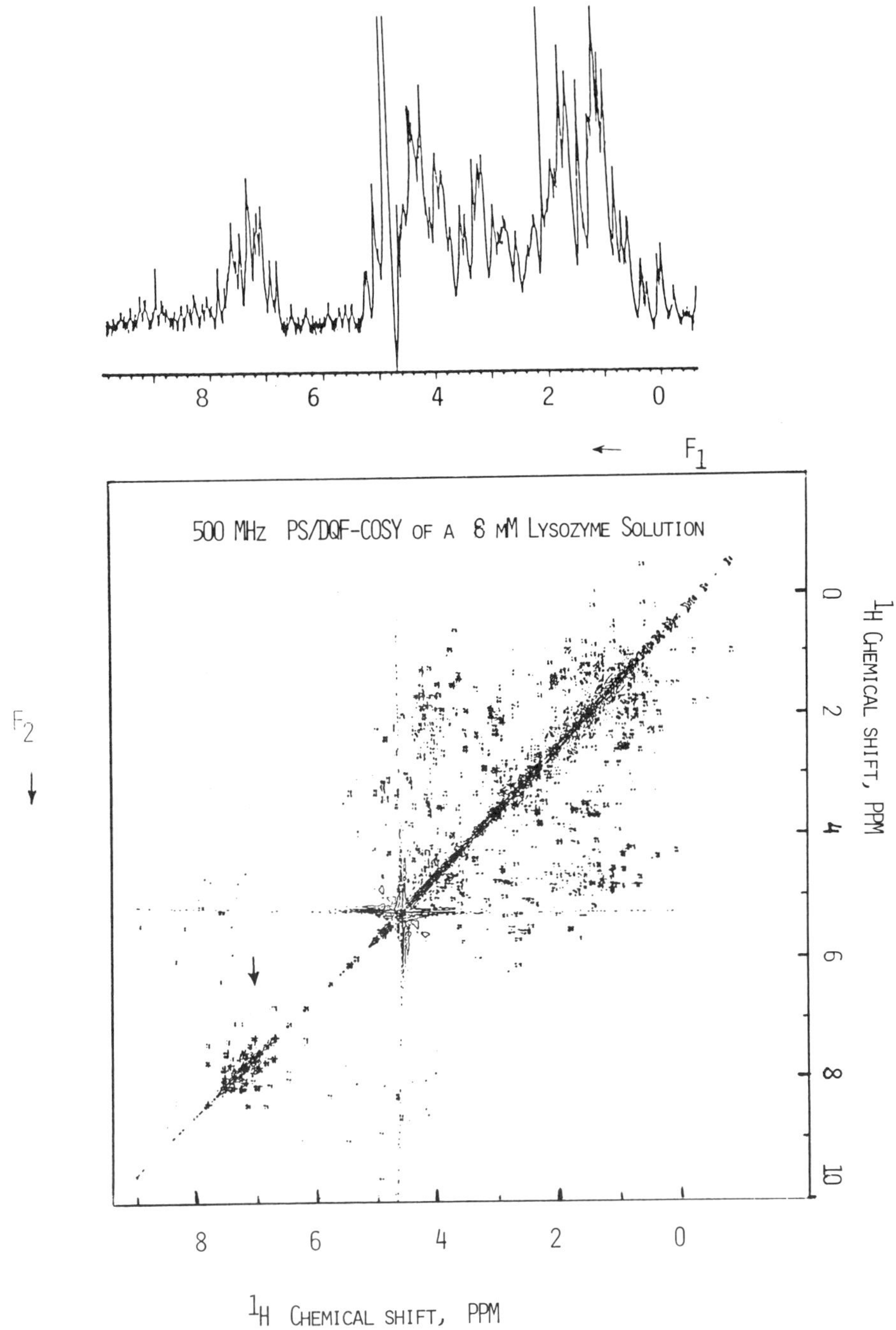

FIGURE 34. PS/DQF, COSY 2D (500-MHz) NMR spectrum of an 8-m*M* lysozyme solution in D_2O at pD = 4.5.

COSY 2D spectrum of a lysozyme solution in Figure 34 was actually obtained with DQF, in a (PS) experiment and provides a good illustration of the advantages of PS/DQF COSY for small proteins. To determine distances between coupled protons, after their coupling patterns or connectivities were unraveled by COSY, one can use a 2D technique called NOESY, that employs incoherent magnetization transfer and monitors the *dipolar couplings*, or chemical exchange, between protons. The NOESY pulse sequence (5) may appear similar to the DQF-

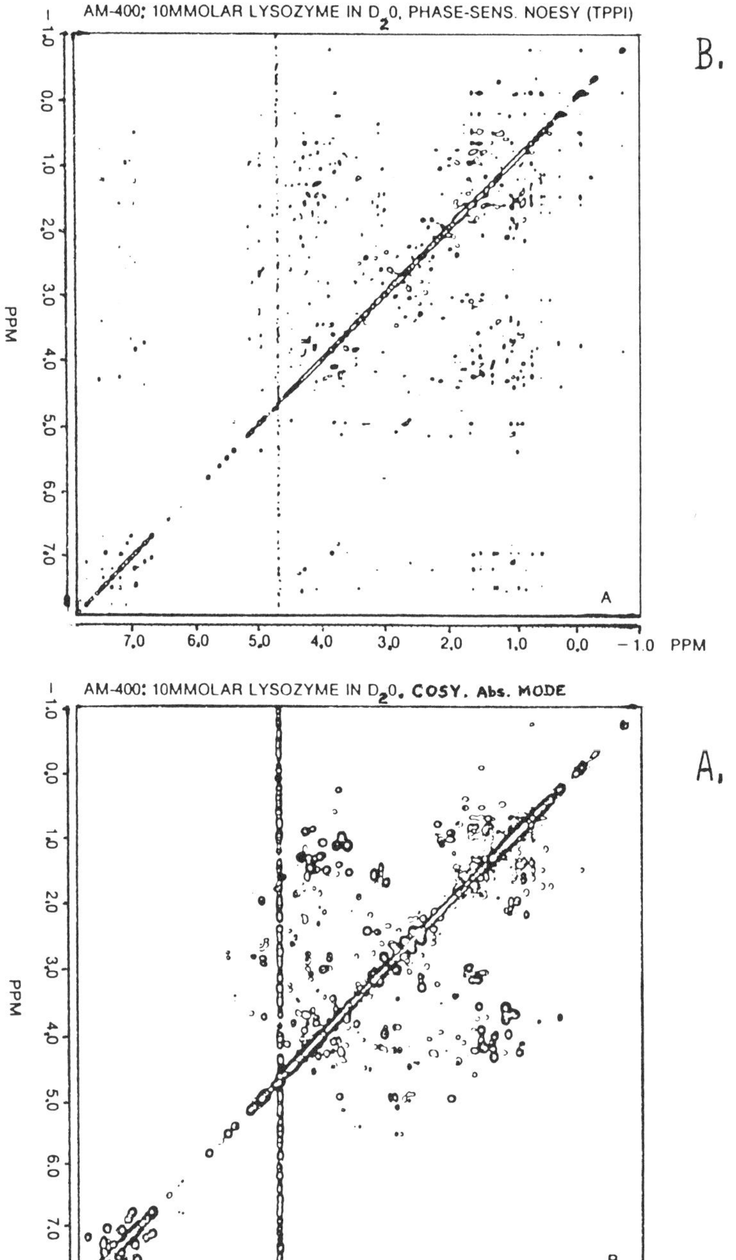

FIGURE 35. Comparison of 400-MHz ^{1}H NOESY-TPPI (A) and COSY (magnitude mode), (B) spectra. The sample is 10-mmol lysozyme in D_2O at 303 K. (A) Phase-sensitive NOESY via the TPPI technique: SW2 = 6024 Hz, TD2 = SI2 = 2K, TD1 = 512, SI1 = 1K, HZ/PT2 = 5.9, HZ/PT1 = 11. 8, D1 = 1. 8 s, AQ = 0.17 s, t_m = 0.1 s (constant), t^0_1 = 3 μs, NS = 16, DS = 4, total time about 6 h, WDW2 = SSB^2 (π/6), and WDW1 = SSB^2 (π/10). The residual water signal was suppressed by presaturation during all periods except detection. (B) A standard magnitude mode COSY-90 experiment has been performed with conditions identical to (A) except that LG windows were used with LB = –250, GB = 0.2 in t_2, and LB = –150, GB = 0.4 in t_1. The same absolute contour levels are plotted in both cases, and the improved resolution in the phase-sensitive TPPI mode is quite apparent. TPPI = time propositional phase increment. (From Hull, W. E., *Two-Dimensional NMR Spectroscopy. Applications for Chemists and Biochemists,* Croasmun, W. R. and Carlson, R. M. K., Eds., VCH, New York, 67, 1987. With permission.)

COSY, but the rf phases and the phase-cycles involved are different; the MQ coherences which lead to COSY-like signals are suppressed by appropriate phase cycling with the exception of the zero-quantum signals which *cannot be* eliminated.

$$90°_{PH1} - t_1 - 90°_{PH2} - t_m - 90°_{PH3} - FID_{PH4} \qquad (5)$$

In the 2D chemical exchange correlation experiment, the 2D matrix has the same form as that for the COSY experiment and in the PS mode the chemical exchange correlation signals have the same phase characteristics as the diagonal, autocorrelation (COSY) peaks. Thus, for positive autocorrelation peaks, all the components of a correlation multiplet will also be positive when chemical exchange dominates or when cross-relaxation approaches the spin diffusion limit ($\omega_o\tau_c >> 1$), such as the case of large molecules or slow tumbling, (negative NOE!). Negative correlation peaks will appear only for cross-relaxation in the extreme-narrowing limit (positive NOE). For small molecules, the chemical exchange and NOE effects have, therefore, opposite signs and are readily distinguished in the PS-NOESY plot. The NOESY 2D plot (Figure 35A) looks like the COSY one, with a left-to-right diagonal and off-diagonal peaks; the F_1 and F_2 axes also represent ^1H chemical shifts. When compared with the simple (magnitude mode) COSY spectrum of lysozyme (Figure 35B), the PS-NOSEY is much better resolved; however, the PS/DQF COSY (Figure 34) has comparable resolution to NOESY (Figure 35A). Since cross-peak amplitudes in the NOESY spectra depend on a complicated function of two or more exponentials, extracting distance information (r_{AB}^{-6}) is not straightforward but for very short t_m (Sequence 5); for very short t_m the cross-peak amplitudes are proportional to the individual cross-relaxation rates R_C that are then proportional to $\tau_c(r_{AB}^{-6})$ for isotropic tumbling molecules, or groups. If anisotropic tumbling or internal motions are present, the situation is substantially more complex, but not hopeless; with a knowledge of the molecular motions involved and individual proton relaxation times of a biopolymer, absolute distances $r_{AB} < 5Å$ can be determined with an accuracy of ±0.5 Å. As an example, in a protein molecule of $M_w < 20{,}000$ Da, such distances could be measured between the αCH and the neighbor > NH, or more significantly between αCH and the protons in the side chain residues next to the αCH. Progressing from one end of a polypeptide chain, the 3D arrangement of the polypeptide protons can be, thus, mapped out using an algorithm that is implemented on a computer. For proteins, the rotating-frame version of NOESY, called ROESY, is needed since groups that have zero NOE could still be observed in the rotating frame. There is a 2Q detection experiment, DQC INADEQUATE, which is simpler to interpret than both COSY or NOESY:

$$90° - \tau - 180° - \tau - 90° - t_1 - 90° - t_2 \qquad (6)$$

The delay is chosen to maximize conversion of single-quantum magnetization to 2Q magnetization. The double-quantum coherence (DQC) evolves during t_1, spreading in frequency at a rate which is determined by the *sum* of the chemical shifts of the two nuclei involved. DQC is converted to single-quantum magnetization by the last 90° pulse in the sequence (6) for the purpose of detection by the receiver coil. By adding multiple transients acquired with appropriate phase cycling one can detect single-quantum signals in t_2 that originate only from DQC in t_1. Other signals are cancelled by the phase-cycling scheme. The resulting 2D plot has carbon chemical shifts in F_2 and 2Q frequency in F_1. Each peak represents a directly bonded carbon. Two carbons that are directly bonded will share a common frequency in F_1, and therefore, one can immediately determine the direct connectivity of all carbons. The method is not restricted to carbons; it can also be applied to protons. For protons, production of DQC does *not* require a direct bond; all that is needed is a J-coupling that can be used to set up a favorable delay between the first two pulses. Therefore, for protons, this experiment would allow a COSY-like experiment but with the main diagonal absent, which would be convenient for resolving very small scalar couplings.

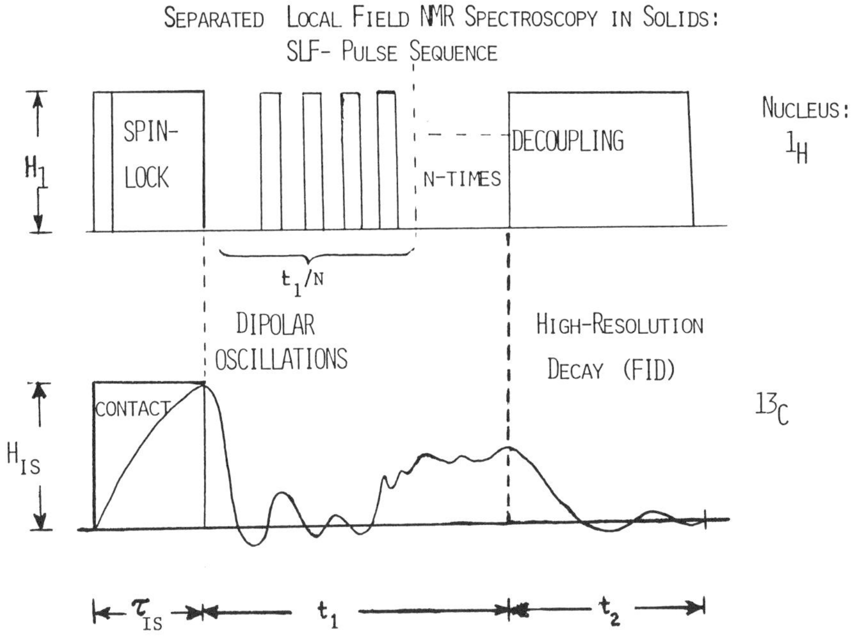

(I)

The 2D NMR techniques discussed above are only a few of the large number of pulse sequences available for 2D spectroscopy of solutions. These would not work in an organic solid because of the dominating, very large dipolar interactions present in such a solid. MQ NMR experiments have been, however, designed for solids.[50,51] The instrumental requirements are much more strict for 2D of solids in order to achieve dipolar decoupling; a residual broadening of ~100 Hz remains even with the best designed multipulse sequences. An illustration of a 2D pulse sequence designed for solids is presented in Diagram I. Note that the beginning two pulses in the proton (I) channel are the same as in the CP sequence of Figure 24, as is the ^{13}C "contact" (polarization transfer) pulse. The decoupling scheme in Diagram I for separated local-field (SLF) spectroscopy is, however, different from that in Figure 24; the homonuclear, $^{1}H...^{1}H$, dipolar couplings are selectively removed by a multipulse, WAHUHA sequence,[46] whereas the ^{13}C magnetization is modulated by $^{13}C...^{1}H$ dipolar couplings during t_1 (the evolution period). At the end of the t_1 intervals, high-power decoupling is employed to obtain a high-resolution ^{13}C FID, without any further dipolar modulation. The resulting 2D plot has ^{13}C chemical shift on the F_2 axis and dipolar frequency (coupling), in kilohertz on the F_1 axis. This 2D NMR experiment allows, therefore, the determination of the dipolar $^{13}C...^{1}H$ couplings for different chemical groups in solids, which can be converted into bond distances and group orientations (for single crystals).[46-54] Notably, bond distances can be obtained even from powders.[45] A relatively simple application of the SLF sequence for a single crystal of ammonium hydrogen malonate ($NH_4HC_3H_2O_4$) is illustrated in Figure 36A and B (from Reference 45). In this simple molecule, the dipolar splittings of the carbonyls, as well as of the CH_2 group, are clearly resolved and the observed dipolar spectrum (Figure 3\ B) along the F_1 axis agrees very well with the calculated spectrum (Figure 36A). For crystallizable proteins, the SLF technique may provide interesting results. Other means to order protein microcrystals may also be employed.[47-49]

Overall, however, the 2D NMR of protein solutions has provided much detailed information that is both interesting and useful for the characterization of proteins. The solid-state 2D NMR techniques lack the superb resolution obtainable by 2D NMR of protein solutions. Nevertheless, for very large molecules, the solids and solution 2D NMR may converge to provide information that is now unavailable from any other technique. MQ NMR[50,51,57] holds special promise for solid

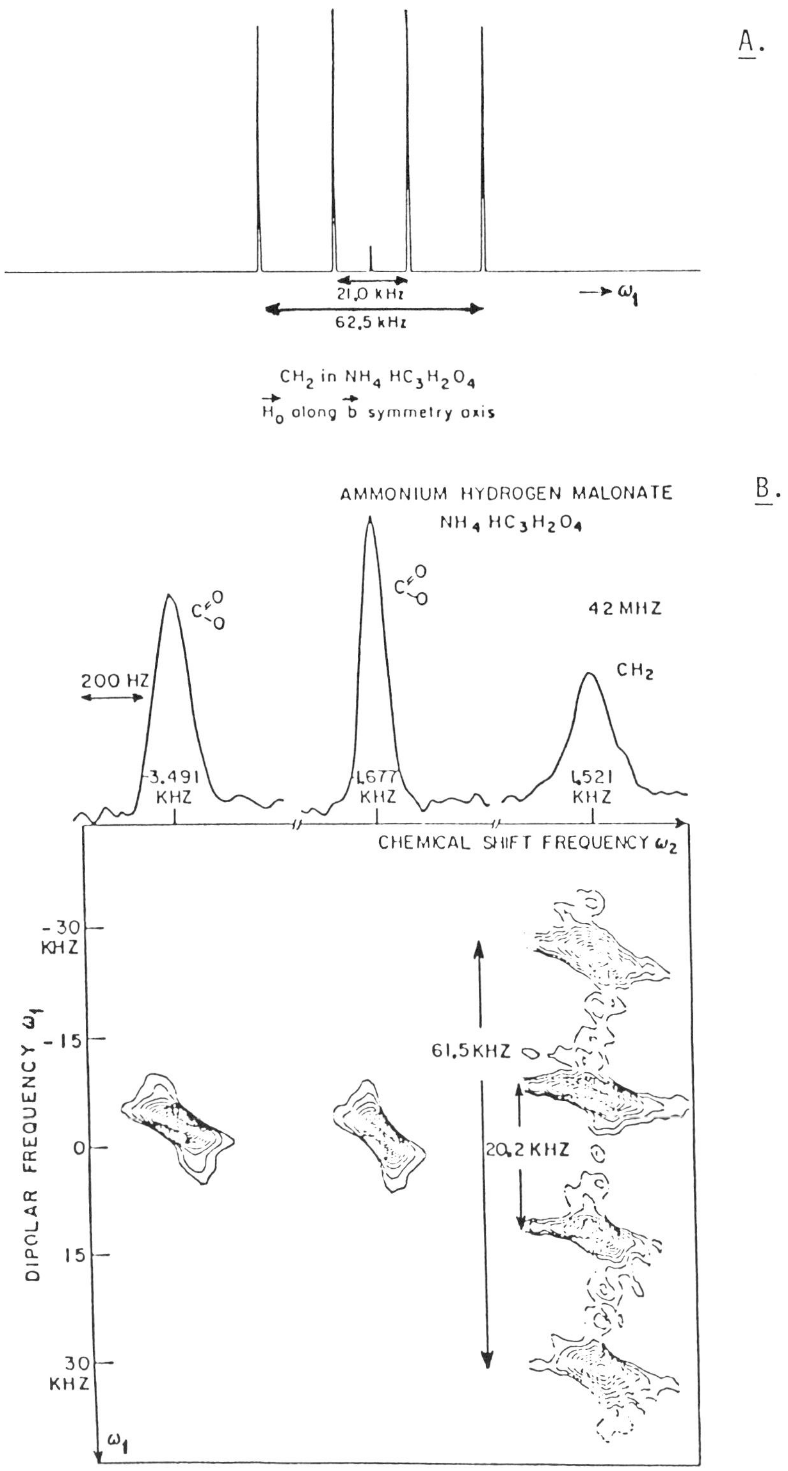

FIGURE 36. 2D, ^{13}C-^{1}H,SLF NMR spectrum (stacked plot) of a single crystal of ammonium hydrogen malonate.[46] (A) Calculated ^{13}C-^{1}H dipolar spectrum of CH_2 in an ammonium hydrogen malonate crystal for H_0 along the *b* symmetry axis of the crystal, as determined by X-ray diffraction, and by assuming r = 1.09Å. (B) Contour plot of the SLF, 2D NMR spectrum of an ammonium hydrogen malonate single crystal aligned with the b axis along the static magnetic field direction; the C-H bond angles of the CH_2 group relative to the b axis were 78 and 164°. (From Hester, P. K., Ackerman, J. L., Neff, B. L., and Waugh, J. S., *Phys. Rev. Lett.*, 36(18), 1081, 1976. With permission.)

biopolymers and very large molecules. The applications of these "new" techniques[45-50,54,55,58,59] to food proteins are yet to be developed.

VII. DISCUSSION AND CONCLUSIONS

The physical and chemical characterization of complex food proteins is a problem of practical importance. Such a characterization is greatly aided by NMR spectroscopy. Progress has recently been made with the characterization of wheat,[7,26] soy,[29,53] and corn[13,21,26] proteins by NMR, especially by ^{13}C NMR techniques. Wheat gliadins and soy globulins yielded high-resolution spectra in solution that could be interpreted and assigned. In all cases studied, however, the resolution was substantially lower in the solid-state[1] (using CP-MAS) than in solution (obtained under conditions of low-power, broadband proton decoupling). Because of the chemical complexity and diversity of food proteins, the NMR assignments of resolved peaks depend on the availability of additional information from "standard" analyses such as amino acid composition analysis, amino acid sequencing, and molecular weight distribution of protein components and subunits by gel filtration and electrophoresis (especially SDS-PAGE). Most promising are, however, the prospects of combining NMR analyses with high performance liquid chromatography preparation and analysis. ^{1}H NMR and direct, proton broadband-decoupled ^{15}N appear — at this stage — much less useful than ^{13}C for the study of complex food proteins.

However, the application of recently developed, 2D NMR and MQ techniques such as HETCOR, phase-sensitive COSY, and NOESY/ROESY, as well as indirect, proton detection of the dilute nuclei (^{15}N and ^{13}C) is likely to provide powerful means for the characterization of food proteins and other food systems. The application of novel, 2D NMR techniques to solid food systems may also alleviate some of the severe problems encountered with resolution in 1D NMR spectra of these solid systems.

Last but not least, specific labeling[10] with ^{13}C and deuterium[47-49] at selected sites of food proteins may provide improved means for extracting valuable information regarding the conformation, dynamics, and functionality of food proteins. Selective labeling could be combined with heteronuclear, chemical shift-correlated 2D NMR techniques to greatly simplify NMR spectra of food proteins and their peak assignments.

In a relatively short period since the beginnings of its application to food systems (in the late 1970s), NMR spectroscopy has contributed substantially to our understanding of the molecular properties of such complex systems. The new NMR techniques (2D, MQ excitation) are still developing[43,52,54-59] and hold even greater promise for the study of food systems.

ACKNOWLEDGMENTS

The kind permission of several publishers to reprint original figures is gratefully acknowledged. This work was supported in part by ICMB Grant # 1-6-3555 from the Illinois Corn Marketing Board. The recording of 125-MHz ^{13}C NMR spectra by Dr. LeRoy F. Johnson of General Electric Co., Fremont, CA, using a GE-500, is also gratefully acknowledged.

REFERENCES

1. **Baianu, I. C. and Förster, H.,** Cross-polarization, high-field carbon-13 NMR techniques for studying physicochemical properties of wheat grain, flour, starch, gluten and wheat protein powders, *J. Appl. Biochem.*, 2, 347, 1980.

2. **Dwek, R. A., Campbell, I. D., Richards, R. E., and Williams, R. J. P., Eds.,** *NMR in Biology*, Academic Press, London, 1977.
3. **Wilson, H. R.,** *Diffraction of X-Rays by Proteins, Nucleic Acids and Viruses*, Edward Arnold, London, 1966.
4. **Meienhoffer, J., Ed.,** *Chemistry and Biology of Peptides*, Ann Arbor Science, Ann Arbor, MI, 1972.
5. **Allerhand, A., Childers, R. F., and Oldfield, E.,** Carbon-13 Fourier transform NMR studies in 20 mm sample tubes: observation of individual carbon sites in proteins, *Ann. Natl. Acad. Sci. U.S.A.*, 222, 765, 1972.
6. **Baianu, I. C.,** Carbon-13 and proton nuclear magnetic resonance studies of wheat proteins, *J. Sci. Food Agric.*, 32, 309, 1981.
7. Baianu, **I. C., Johnson, L. F., and Waddell, D. K.,** High-resolution protein, carbon-13 and nitrogen-15 nuclear magnetic resonance studies of wheat proteins at high magnetic fields: spectral assignments, changes with concentration and heating treatments of Flinor gliadins in solution. Comparison with gluten spectra, *J. Sci. Food Agric.*, 33, 373, 1982.
8. **Wütrich, K.,** *NMR in Biological Research: Peptides and Proteins*, North-Holland, Amsterdam, 1976.
9. **Norton, R. S., Clouse, A. O., Addleman, R., and Allerhand, A.,** Studies of proteins in solution by natural-abundance carbon-13 nuclear magnetic resonance. Spectral resolution and relaxation behavior at high magnetic field strengths, *J. Am. Chem. Soc.*, 99, 79, 1977.
10. **Brown, D. T. et al.,** Studies of macromolecular structure by carbon-13 nuclear magnetic resonance. II. A specific labeling approach to the study of histidine residues in proteins, *J. Am. Chem. Soc.*, 95, 1316, 1973.
11. **Laemmli, U. K.,** Cleavage of structural proteins during the assembly of the head of bacteriophage T4, *Nature*, 227, 680, 1970.
12. **Booth, M. R. and Ewart, J. A. D.,** Studies of four components of wheat gliadins, *Biochim. Biophys. Acta*, 81, 226, 1969.
13. **Augustine, M. and Baianu, I. C.,** High-resolution carbon-13 nuclear magnetic resonance studies of maize proteins, *J. Cereal Sci.*, 4, 371, 1986.
14. **Tranh, H. and Shibasaki, K.,** Major proteins of soybean seeds. A straightforward fractionation and their characterization, *J. Agric. Food Chem.*, 24, 1117, 1976.
15. **Iwabuchi, S. and Yamauchi, F.,** Determination of glycinin and β-conglycininin soybean proteins by immunological methods, *J. Agric. Food Chem.*, 35, 200, 1987.
16. **Schofield, J. D. and Baianu, I. C.,** Solid-state, cross-polarization magic-angle spinning carbon-13 nuclear magnetic resonance and biochemical characterization of wheat proteins, *Cereal Chem.*, 59(4), 240, 1982.
17. **Payne, P. I. and Corfield, K. G.,** Subunit composition of wheat glutenin proteins isolated by gel filtration in a dissociating medium, *Planta*, 145, 83, 1979.
18. **Kasarda, D. D. et al.,** Wheat proteins, in *Advances in Cereal Science and Technology*, Vol. 1, Pomeranz, Y., Ed., American Association Cereal Chemists, St. Paul, 1977, 158.
19. **Cluskey, J. E. and Wu, V.,** Optical rotatory dispersion of wheat gluten, gliadin and glutenin in acetic acid and aluminum lactate systems, *Cereal Chem.*, 43, 158, 1966.
20. **Lee, J. W.,** Preparation of gliadin by urea extractions, *J. Sci. Food Agric.*, 19, 15.3, 1968.
21. **Augustine, M. and Baianu, I. C.,** Basic studies of corn proteins for improved solubility and future utilization: a physicochemical approach, *J. Food Sci.*, 52(3), 649, 1987.
22. **Fehrstrom, H. and Moberg, U.,** LKB Appl. Note 306, LKB Company, Bromma, Sweden, 1977.
23. **Koshyiama, I.,** Purification and physico-chemical properties of 11S globulin in soybean seeds, *Int. J. Pept. Protein Res.*, 4, 167, 1972.
24. **Fukushima, D.,** Internal structure of 7S and 11S globulin molecules in soybean proteins, *Cereal Chem.*, 45, 203, 1968.
25. **Bietz, J. A., Paulis, J. W., and Wall, J. S.,** Zein subunit homology revealed through amino-terminal sequence analysis, *Cereal Chem.*, 56(4), 327, 1979.
26. **Baianu, I. C.,** Proton broadband decoupled 125 MHz carbon-13 NMR spectra of corn zeins in alkaline (pD 11.2) solutions and wheat gliadins in acidic solutions (pD 3.4), unpublished results, 1987; in collaboration with L. F. Johnson.
27. **Paulis, J. W. et al.,** *J. Agric. Food Chem.*, 23, 197, 1975.
28. **Augustine, M. E. and Baianu, I. C.,** Analysis of amino acid composition of cereal and soy proteins by carbon-13 NMR and ion-exchange chromatography, *Proc. Fed. Am. Soc. Exp. Biol.*, 43(3), 672, 1984.
29. **Kakalis, L. and Baianu, I. C.,** High-resolution carbon-13 NMR studies of soy protein isolates, *J. Agric. Food Chem.*, submitted for publication.
30. **Glushko, V., Lawson, P. J., and Gurd, F. R. N.,** Conformational states of bovine pancreatic ribonuclease *A* observed by normal and partially relaxed carbon-13 NMR, *J. Biol. Chem.*, 247, 3176, 1972.
31. **Howarth, O. W. and Lilley, D. M. J.,** Carbon-13 NMR of peptides and proteins, *Prog. Nucl. Magn. Reson. Spectrosc.*, 12, 1, 1978.
32. **Komorosky, R. A. et al.,** ^{13}C NMR of biopolymers, in *Topics in Carbon-13 NMR Spectroscopy*, Levy, G. C. J., Ed., John Wiley & Sons, New York, 1976.
33. **Demchenko, A. P.,** in *Ultraviolet Spectroscopy of Proteins*, Springer-Verlag, Berlin, 1986, 204.

34. **Watanabe, M. and Arai, S.**, in *Modification of Proteins: Food, Nutritional and Pharmacological Aspects*, (Advances in Chemistry Series, Vol. 198), Feeney, R. E., and Whitaker, J. R., Eds., American Chemical Society, Washington, D.C., 1982, 202.
35. **Eads, T. M. and Mandelkern, L.**, Backbone and side-chain motion in myosin, subfragment 1, and rod, determined by natural abundance carbon-13, *J. Biol. Chem.*, 259, 10689, 1984.
36. **Abragam, A.**, *The Principles of Nuclear Magnetism*, Oxford University Press, New York, 1961.
37. **Andrews, E. R. and Eades, R. G.**, The narrowing of NMR spectra of solids by high speed specimen rotation, *Prog. Nucl. Magn. Reson. Spectrosc.*, 8, 2, 1962.
38. **Hartmann, S. R. and Hahn, E. L.**, Nuclear double resonance in the rotating frame, *Phys. Rev. Lett.*, 128, 2042, 1962.
39. **Pines, A., Gibby, M. G., and Waugh, J. S.**, Proton enhanced NMR of dilute spins in solids, *J. Chem. Phys.*, 59, 569, 1973.
40. **Kricheldorf, H. R. and Müller, D.**, Secondary structure of peptides. Characterization of proteins by means of ^{13}C NMR CP/MAS spectroscopy, *Colloid. Polym. Sci.*, 262, 856, 1984.
41. **Jeener, J.**, Ampére International. Summer School, Baske Poljé, Yugoslavia, 1971.
42. **Gray, G. A.**, Introduction to two-dimensional NMR methods, in *Two-Dimensional NMR Spectroscopy: Applications for Chemists and Biochemists*, Vol. 9, Croasmun, W. R. and Carlson, R. M. K., Eds., VCH, NewYork, 1987, 1.
43. **Hull, W. E.**, Experimental aspects of two-dimensional NMR, in *Two-Dimensional NMR Spectroscopy: Applications for Chemists and Biochemists*, Croasmun, W. R. and Carlson, R. M. K., Eds., VCH, New York, 1987, 67.
44. **Markley, J. L.**, in *Protein Engineering*, Alan R. Liss, New York, 1987, 15.
45. **Vaughan, R. W.**, High-resolution, solid state NMR, *Annu. Rev. Phys. Chem.*, 29, 397, 1978.
46. **Hester, P. K., Ackerman, J. L., Neff, B. L., and Waugh, J. S.**, Separated local-field spectra in NMR: determination of structure of solids, *Phys. Rev. Lett.*, 36(18), 1081, 1976.
47. **Oldfield, E. and Rothgeb, T. M.**, NMR of individual sites in protein crystals, magnetic-ordering effect, *J. Am. Chem. Soc.*, 102, 3635, 1980.
48. **Rothgeb, T. M.** and **Oldfield, E.**, NMR of heme protein crystals. General aspects, *J. Biol. Chem.*, 256, 1432, 1981.
49. **Baianu, I. C., Gutowsky, H. S., and Oldfield, E.**, Origin and behavior of deuteron NMR spin echoes in selectively labeled amino acids, myoglobin microcrystals, and purple membranes, *Biochemistry*, 23, 3105, 1984.
50. **Vega, S., Shattuck, P. W., and Pines, A.**, Fourier transform double quantum NMR in solids, *Physiol. Rev. Lett.*, 37(1), 43, 1976.
51. **Vega, S. and Pines, A.**, Operator formalism for double-quantum NMR, *J. Chem. Phys.*, 66(12), 5624, 1977.
52. **Boyd, J. and Redfield, C.**, Analysis of cross-peak fine structure in multiple-quantum filtered COSY spectra, *J. Magn. Reson.*, 68, 67, 1986.
53. **Kakalis, L.**, High Field NMR Studies of Soy Protein Molecular Dynamics, Hydration and Conformations, Ph.D. thesis, University of Illinois, Urbana, 1988.
54. **Waugh, J.**, Uncoupling of local field spectra in nuclear magnetic resonance: determination of atomic positions in solids, *Proc. Natl. Acad. Sci. U.S.A.*, 73(5), 1394, 1976.
55. **Rybaczewski, E. F., Neff, B. L., Waugh, J. S., and Sherfinski, J. S.**, High resolution ^{13}C NMR in solids: ^{13}C local fields of CH_1, CH_2 and CH_3, *J. Chem. Phys.*, 67, 1231, 1977.
56. **Bodenhausen, G., Stark, R. E., Ruben, P., and Griffin, R. G.**, Separation of dipolar and quadrupolar splittings in single crystal nitrogen-14 NMR, *Chem. Phys. Lett.*, 67(2), 397, 1979.
57. **Schnur, G., Kimmich, R., and Winter, F.**, Limits of {^{14}N} ^{1}H double-quantum polarization-transfer spectroscopy in solids, *J. Magn. Reson.*, 66, 295, 1986.
58. **Bodenhausen, G.**, Multiple-quantum NMR, *Prog. Nucl. Magn. Reson Spectrosc.*, 14, 137, 1981.
59. **Freeman, R. and Morris, G. A.**, Two-dimensional Fourier transformation in NMR, *Bull. Magn. Reson.*, 1(1), 5, 1979.

Chapter 8

RELAXATION AND CROSS RELAXATION: NMR DETERMINATIONS OF WATER INTERACTIONS WITH PROTEINS

Thomas F. Kumosinski and Helmut Pessen

TABLE OF CONTENTS

I. INTRODUCTION

The importance of water interactions with naturally occurring biological macromolecules to the scientific community is well documented.[1] It is thought that the quantitation of this interaction will go a long way toward enabling investigators to understand the energetics of protein secondary, tertiary, and quaternary structure as well as changes in the relevant structural parameters with varying environmental conditions. However, to date not even the amount of water bound to a globular protein (hydration) can be determined with confidence, much less can one directly observe the constituent groups on the protein responsible for the binding of water. Only the concept of hydrophobic and hydrophilic interaction, as quantitated by solubilities of the individual amino acids, allows some qualitative extrapolation to the macromolecular system. To complicate the problem further, the hydration of a protein has been found to be dependent on the methodology used in the study. Traditionally, the hydrodynamic methods, i.e., sedimentation, translational and rotational diffusion, and viscosity, have been used to measure the hydration of proteins, with some success; with these, however, unlike with small angle X-ray scattering (SAXS), an assumption of the overall shape of the unhydrated protein must be made.

Recently it has been shown that, in investigating hydration, the hydrodynamic approach cannot stand alone.[2,3] At best, it requires supplementation by a structural method, preferably SAXS, to give unequivocal information on hydration. It has also been demonstrated that sedimentation and SAXS information can be correlated to such an extent that sedimentation parameters are fairly predictable and, in effect, do not constitute an independent method. It would appear, then, that for the purpose of investigating hydration, SAXS could be considered a more useful approach than sedimentation, and that it might replace the latter, were it not for the circumstance that, in contemporary laboratories, SAXS instruments are even rarer than analytical ultracentrifuges.

As a practical matter, the SAXS method suffers from the lack of a substantial published data base. It would of course be helpful if investigators reporting SAXS results on a protein in each instance undertook the routine labor of calculating and reporting a full set of the parameters their data are capable of generating[4] instead of merely those engaging their immediate attention. Further, it is to be hoped that those interested in hydration and structural information obtain access in larger numbers to instrumentation enabling them to use this powerful and versatile method which, with the commercial availability in recent years of position-sensitive detectors, has been a very productive method as well.

In lieu of SAXS data, it will be shown in this chapter that NMR is capable of giving information on hydration. While the absolute values of NMR hydration are not completely independent of assumptions regarding mechanism, relative values are largely independent of such assumptions and can serve to assess changes in structure or biochemical behavior as a function of varying environmental conditions. The interpretation of such changes of course will depend on the system studied.

One way of using NMR relaxation to obtain this kind of information has been by means of frequency dependence.[5] There are, however, few instruments available that allow dispersive measurements. Attention must also be paid to the frequencies chosen: if measurements are made at two different frequencies to eliminate one of the modes of relaxation employed here (see Equations 16a and b), care must be taken that the frequencies are sufficiently different to exhibit a substantial enough dispersion for adequate relaxation statistics. In any case, the frequencies must be high enough (and so must the molecular weight, on which the correlation time depends) to escape the line-narrowing region, i.e., the product $\omega_o^2\tau_c^2$ (where ω_o is the Larmor frequency and τ_c the correlation time) must be clearly larger than 1. Otherwise, the system of requisite simultaneous equations will either have no solution, or the solution will be very imprecise.

In place of frequency dependence, concentration dependence (readily accessible with any NMR instrument) may be emphasized to advantage. Two modes of relaxation, spin-lattice and spin-spin, have been used with deuteron resonance to avoid the cross relaxation encountered with spin-lattice relaxation of protons. ^{17}O relaxation has also been used, for similar reasons.[6,7] In a different method, proton spin-lattice relaxation has been employed, but cross-relaxation effects were determined and allowed for by the joint examination of two genetic variants of the same protein, which differed only in the known extent of an association reaction.[8] Again, the combination of protein molecular weight and resonant frequency must satisfy the condition for avoiding the line-narrowing mentioned.

In addition to hydration, NMR relaxation can give other valuable information that can be correlated with structural and biochemical changes. Correlation times are related to molecular size and shape and are relevant to hydrodynamics. Virial coefficients, for which concentration dependence data are indispensable, can be evaluated directly from a concentration plot. It is true for either mode of relaxation that the observed relaxation rates are linear functions of hydration, concentration, and an exponential containing both the concentration and the virial coefficient B_0. It can be seen that nonlinear regression applied to this relationship will give B_0 directly. Virial coefficients, related to the average net charge carried by molecules, provide a particularly useful measure of molecular interaction.

II. THEORY

A. General Nonideal Effects

Proteins, by their very nature as polyelectrolytes with large charge-to-mass ratios, are susceptible to severe interaction effects in solution. Although the consequent deviations from ideality are among the most important effects in solution theory, they have not received sufficient consideration in NMR studies of proteins. A prime example is the nonideality caused by repulsion due to net positive or negative charges, which has been demonstrated to have serious effects on measurements other than NMR (light scattering, osmotic pressure, sedimentation equilibrium and velocity, translational diffusion, and even pH titration[1,9-13]). Thus Pedersen,[11] on the basis of the theory of Tiselius,[12] as early as 1940 showed that high net charges on a macromolecule result in the creation of a potential gradient in a sedimenting solution during both equilibrium and velocity runs; Booth[13] has quantitated this behavior by a general theory for charged spheres. Sedimentation coefficients of bovine serum albumin at acid pH are lower by at least half when compared with the same protein in the presence of 0.1 *M* salt;[14] the salt thus minimizes the high repulsive effect of the molecular charges. Light scattering and osmotic pressure measurements of aqueous proteins carrying high net charge indicate decreased apparent weight-average molecular weights with increasing concentration, in consequence of intermolecular repulsion,[9,15-17] according to the equation:

$$M_{w,app}^{-1} = M_{w,o}^{-1} (1 + c \text{ dln } \gamma/dc) \tag{1}$$

where γ is the activity coefficient of the protein at a concentration c, $M_{w,app}$ is the experimentally derived apparent weight-average molecular weight, and $M_{w,o}$ is the true molecular weight at infinite dilution of the protein. From the virial expansion for osmotic pressure follows

$$\text{dln } \gamma/dc = 2B_0 + 3B c + \ldots \tag{2}$$

where the B quantities are the second and higher virial coefficients; for a repulsive effect usually dln $\gamma/dc = 2 B_0 > 0$ (i.e., $\gamma > 1$), in accord with general electrolyte solution theory.

Beyond simple electrostatic repulsion, one needs to be concerned with a less obvious phenomenon, namely the charge fluctuations treated in the theory of Kirkwood and Shumaker.[18] Extensive experimental evidence for this theory has been furnished by many investigations using light scattering; for proteins in water under isoionic conditions, the apparent weight-average molecular weight increases with increasing concentration.[19-22] Ordinarily, this phenomenon might be interpreted simply as an aggregation of the molecule. However, Kirkwood and Shumaker have shown that, since proteins are polyampholytes rather than simple polyelectrolytes, attractive forces arise in isoionic protein solutions from statistical fluctuations, both in charge and in charge distribution; these in turn are associated with fluctuations in the number and configurations of protons bound to the protein molecule. Briefly stated, one or more virial coefficients in $c^{n/2}$ powers must be added to Equation 2, and these coefficients, due to progressive ionization of the macromolecule, have negative values. Moreover, the virial coefficients of the c^n terms should also usually be negative. Experimental results show the virtual elimination of such charge-fluctuation effects in the presence of a moderate amount of salt. Since virial effects are found for other solution parameters, such as sedimentation coefficients, intrinsic viscosities, and linear as well as rotational diffusion coefficients, charge fluctuations can be expected to affect these hydrodynamic parameters also.

Specifically, the translational diffusion coefficient D_t and the rotational diffusion coefficient D_r exhibit the influence of the activity coefficient, γ, which expresses the combined result of all such effects, according to

$$D_t = (8/6)D_r = [kT/(6\pi\eta r_s)](1 + c \text{ dln } \gamma/dc) \tag{3}$$

where k is Boltzmann's constant, 1.3806×10^{-16} erg °K^{-1}, T is the temperature in K, η is the viscosity of the solvent, and r_s is the Stokes radius of the macromolecule. Accordingly, electrostatic repulsion and charge fluctuation will affect these, as well as other hydrodynamic parameters.

Measurements of protein hydration by NMR relaxation techniques using the two-state model of Zimmerman and Brittin[23] have yielded surprisingly large values for the correlation time of bound water when cross-relaxation effects were avoided by experiments in D_2O.[6,24] When cross-relaxation effects were not taken into account (e.g., in proton NMR of water), correlation times obtained from dispersion data also were high,[5,6,24,25] even though Andree[26] had shown that apparent τ_c values from frequency dependence of spin-lattice relaxation should be smaller as a result of cross-relaxation. Also, recent ^{17}O NMR T_1 and T_2 data of Halle et al.,[7] give τ_c values larger than would be predicted by the Debye-Stokes-Einstein equation, a result attributed by the authors to long-range coulombic protein-protein interactions. However, these measurements as well as many others were made, again, on isoionic protein solutions with no added salt. However, since rotational diffusion is basic to these measurements (i.e., the bound water rotates with substantially the same τ_c as the protein) and is related to τ_c(i.e., $\tau_c = r_s^2/6D_r$), it can be seen from Equation 3 that

$$\tau_c = [4\pi\eta r_s^3/(3kT)](1 + c \text{ dln } \gamma/dc)^{-1} \tag{4}$$

and, therefore, not only electrostatic effects due to the high charge-to-mass ratio of the protein but also charge-fluctuation effects should be encountered, both as a result of the methodology of these measurements. The absence of salt and the consequently large negative virial coefficients thus account for the excessive apparent τ_c values observed. The addition of salt should then result in correlation times more in accord with those calculated for the protein from structural considerations directly. In fact, two recent studies,[8,27] using different means to deal with the complicating effects of cross-relaxation, but both using salt, obtained τ_c values in good agreement with the known structure.

B. Derivation of Requisite Expressions

In order to address this problem, one clearly should work with solutions containing salt. The theory now needs to be expanded to accommodate a three-component system. Expressions for NMR relaxation rates are rederived here in terms of the multicomponent theory of Casassa and Eisenberg,[28] since the salt components can compete with water for binding sites on the protein molecule (e.g., charged side chains).

The model used as a point of departure for this derivation is a fast-exchange, two-state, three-component system. The initial assumptions are (1) only one correlation time exists for the bound-water state; (2) there is competition between bound water and salt for the interaction sites on the surface of the protein; but (3) there is no competition between salt and protein for water (since experiments were performed here only on protein solutions with added salt, not on insoluble or powdered samples). Hence, a general equilibrium expression for the binding is

$$P + iW + jX \rightleftarrows PW_iX_j \tag{5}$$

where P, W, and X represent protein, water, and salt at free concentrations p, w, and x, bound concentrations p_b, w_b, and x_b, and total concentrations

$$P = p + p_b \tag{5a}$$

$$W = w + w_b \tag{5b}$$

and

$$X = x + x_b \tag{5c}$$

respectively. Associated with the formation of each species PW_iX_j is an apparent macroscopic association constant K_{ij}. Then

$$p_b = p \sum_{i=0}^{q} \sum_{j=0}^{r} K_{ij} w^i x^j \tag{6a}$$

where q and r are any positive integers such that q + r = n, the total number of binding sites per molecule. At the same time,

$$w_b = p\Sigma\Sigma iK_{ij}w^ix^j \tag{6b}$$

and

$$x_b = p\Sigma\Sigma jK_{ij}w^ix^j \tag{6c}$$

The average number of molecules of water or salt bound to a molecule of protein, i.e., the Scatchard hydration, $\bar{\nu}_W$, and salt binding $\bar{\nu}_X$, respectively, are then given by

$$\bar{\nu}_W = w_b/P \tag{6d}$$

and

$$\bar{\nu}_X = x_b/P \tag{6e}$$

which may be also written in terms of the Equations 6a to c. (We designate here as Scatchard hydration all protein-water interactions characterized by multiple equilibria. Another type of water binding, preferential hydration, will be encountered further on. For a particularly clear exposition of the relationships between these concepts, see Reference 29.)

The concentrations are commonly expressed in moles per liter but may, when more convenient, equally well be expressed in moles per 1000 g water, as will be the case in the following. While this will change the numerical values and dimensions of association constants, $\bar{\nu}_W$ and $\bar{\nu}_X$ will remain unaffected. Strictly speaking, all terms should be expressed as the corresponding activities rather than as molar or molal concentrations. (However, in view of the generally large molecular weights of proteins, even fairly concentrated solutions are usually no more than 10^{-13} *M* and, under the conditions of many experiments, virial coefficients are negligible, i.e., the protein has a very small net charge. Deviations from this assumptions will be discussed specifically later.)

The experimentally derived quantity to be considered is the slope of the plot of either the spin-lattice (R_1) or the spin-spin (R_2) relaxation rate of the water vs. the concentration of protein present. In analogy to other quantities encountered in solution theory, the slope $(dR/dP)_\mu$ may be termed the relaxation increment; the subscript μ indicates that solutions are at constant chemical potential and, in terms of experimental procedure, that exhaustive dialysis against buffer was performed prior to the proton NMR relaxation measurements.

Considering now the relaxation increment of an aqueous solution of protein in the presence of salt as a concentration-dependent function, and expressing the total relaxation rate of the water in the three-component system in terms of the solution components, one may write, on the basis of fast exchange of water between two fractions of rates R_b (for bound water) and R_f (for free water), and in view of Equation 6b,

$$R = \frac{w_b}{W} R_b + \frac{w}{W} R_f = \frac{(p\Sigma\Sigma iK_{ij}w^ix^j)R_b + wR_f}{p\Sigma\Sigma iK_{ij}w^ix^j + w} \tag{7}$$

From R as a function of P, W, and X, one has the total derivative

$$\left(\frac{dR}{dP}\right)_\mu = \left(\frac{\partial R}{\partial P}\right)_{W,X} + \left(\frac{\partial R}{\partial W}\right)_{P,X}\left(\frac{dW}{dP}\right)_\mu + \left(\frac{\partial R}{\partial X}\right)_{P,W}\left(\frac{dX}{dP}\right)_\mu \tag{8}$$

The coefficients of the three terms can be evaluated by partial differentiation of Equation 7 with the use of Equations 6a to c. More directly, from Equations 7 and 5b (where w, because of the exhaustive dialysis, is a constant),

$$R = \frac{(W - w)R_b + wR_f}{W} = R_b - (R_b - R_f)\frac{w}{W} \tag{9}$$

and from this,

$$\left(\frac{\partial R}{\partial P}\right)_{W,X} = 0, \qquad \left(\frac{\partial R}{\partial W}\right)_{P,X} = \frac{(R_b - R_f)w}{W^2}, \qquad \left(\frac{\partial R}{\partial X}\right)_{P,W} = 0 \tag{9a}$$

and, therefore,

$$\left(\frac{dR}{dP}\right)_{\mu} = \left(\frac{\partial R}{\partial W}\right)_{P,X}\left(\frac{dW}{dP}\right)_{\mu} = \frac{(R_b - R_f)w}{W^2}\left(\frac{dW}{dP}\right)_{\mu} \tag{10}$$

From Equations 5b and 6d, $W = w + w_b = w + \bar{\nu}_W P$ and, with $dw/dP = 0$,

$$\left(\frac{dW}{dP}\right)_{\mu} = \bar{\nu}_W \tag{11}$$

Since $w_b << W$ implies $w \cong W$ ($\cong$55.6), Equation 10 becomes

$$\left(\frac{dR}{dP}\right)_{\mu} = \frac{R_b - R_f}{W}\bar{\nu}_W \tag{12}$$

Because protein molecular weights are not always precisely known, it is convenient to replace the molality P by the concentration c (in grams of protein per grams of water) and the hydration $\bar{\nu}_W$ (in moles per mole) by $\bar{\nu}'_W$ (in the conventional units of grams bound water per gram protein). Then $c = PM_P/1000$ and $\bar{\nu}'_W = \bar{\nu}_W M_W/M_P$, where M_P and M_W are the molecular weights of protein and water, respectively. Also, $dP/dc = 1000/M_P$ and, $1000/WM_W$ being unity,

$$\left(\frac{dR}{dc}\right)_{\mu} = \left(\frac{dR}{dP}\right)_{\mu}\left(\frac{dP}{dc}\right) = (R_b - R_f)\bar{\nu}'_W \tag{13}$$

In the cited work,[24] the relationship of R to c (as shown in Figure 1 for both R_1 and R_2 results) was linear. One may write Equation 9 alternatively as

$$R = R_f + (R_b - R_f)\frac{w_b}{W} = R_f + (R_b - R_f)\frac{P\bar{\nu}_W}{W} = R_f + (R_b - R_f)c\bar{\nu}'_W \tag{13a}$$

and from Equations 13 or 13a, with a constant relaxation increment,

$$(dR/dc)_{\mu} = (R_b - R_f)\bar{\nu}'_W = k \tag{13b}$$

or

$$R = R_f + (R_b - R_f)c\bar{\nu}'_W = R_f + kc \tag{13c}$$

Absence of such linearity might be due to polydispersity and consequent changes in the (R_b – R_f) factor, which would have to be allowed for. The other factor in the relaxation increment, $\bar{\nu}'_W$, is generally taken to be independent of protein concentration (see Reference 30). If it is not actually constant, an additional term accounting for its concentration dependence would have to be included in Equations 13b and c, unless a change from concentration units to activities will remove the nonlinearity, as discussed in Section III.A.2.

If it is desired to express protein concentrations as c_2 in grams protein per milliliter solution, the solution density ρ and the concentration of the third component, c_3, need to be known, whereupon

$$c = \frac{c_2}{\rho - c_2 - c_3} \tag{14}$$

and

$$\left(\frac{dR}{dc_2}\right)_\mu = k \frac{\rho - c_3}{(\rho - c_2 - c_3)^2} \tag{14a}$$

The need to know solution densities may be eliminated for all except very high protein concentrations with a knowledge of the partial specific volume $\bar{v}$ and the solvent density ρ_o. From the definition of $\bar{v}$, it can then be shown that

$$c \simeq \frac{c_2}{\rho_0(1 - \bar{v}c_2) - c_3} \tag{14'}$$

and

$$\frac{dR}{dc_2} \simeq k \frac{\rho_0 - c_3}{[\rho_0(1 - \bar{v}c_2) - c_3)^2} \tag{14'a}$$

In cases of high protein concentration, the more general Equations 14 and 14a may have to be used. At low protein concentration, where $\rho_o > 1.0$ while $c_2, c_3 << \rho$, Equations 14a and 14′a reduce to expressions formally identical to Equation 13b with c_2 in place of c; a corresponding remark is true for Equations 14 and 14′ with respect to Equation 13c. From Equation 13b, $R_b = k/\nu\text{-}'_W \mp R_f$ for either R_1 or R_2, and, therefore,

$$R_{2b}/R_{1b} = (k_2 + R_{2f}\,\bar{v}'_W)/(k_1 + R_{1f}\,\bar{v}'_W) \tag{15}$$

However, generally $R_f\,\bar{v}'_W << R_b$; hence, from the definition of k, one obtains the useful approximation

$$R_{2b}/R_{1b} \simeq k_2/k_1 \tag{15a}$$

The parameters k_1, R_{1f}, k_2, and R_{2f} are experimentally accessible as the slope k and intercept R_f from R_1 or R_2 vs. c_2 plots, respectively, such as those of Figure 1.

C. Evaluation of Data

Use of these relationships in conjunction with the Kubo-Tomita-Solomon equations,[31,32] which relate R_{1b} and R_{2b} to the correlation time of the bound water, will give the requisite number of equations to permit simultaneous solution for $\bar{v}'_W$ and τ_c. These equations may be written as

$$R_{1b} = 2K\tau_c[(1 + \omega_0^2\tau_c^2)^{-1} + 4(1 + 4\omega_0^2\tau_c^2)^{-1}] \tag{16a}$$

and

$$R_{2b} = K\tau_c[3 + 5(1 + \omega_0^2\tau_c^2)^{-1} + 2(1 + 4\omega_0^2\tau_c^2)^{-1}] \tag{16b}$$

where $\omega_o = 2\pi\upsilon_o$ is the nuclear angular precessional frequency for the nuclide observed, in radians per second, and K is a measure of the strength of the nuclear interaction, i.e.,

$$K_{deuterons} = (3/80)(e^2qQ/\hbar)^2(\eta^2/3 + 1)^{-1}S_{deut}^2 \tag{16c}$$

and

$$K_{protons} = (3/20)\hbar^2\gamma^4r^{-6}S_{prot}^2 \tag{16d}$$

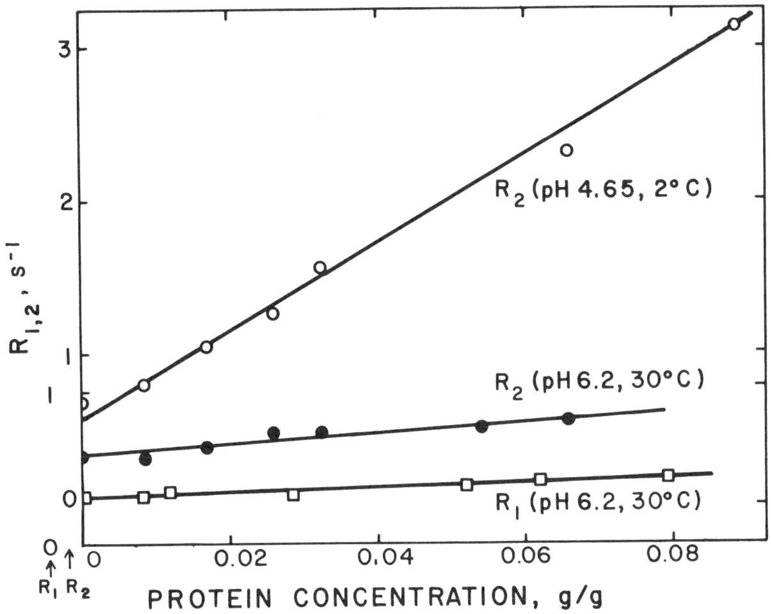

FIGURE 1. Dependence of proton relaxation rates on β-Lg A concentrations (g protein per gram water) in H_2O. Transverse relaxation rates R_2 at pH 4.65 and 2°C (-○-○-) and at pH 6.2 and 30°C (-●-●-). Longitudinal relaxation rates R_1 (-□-□-) at pH 6.2 and 30°C (shown for comparison only; not used in calculations). Points represent experimental values; lines represent least-squares fits. All points show linear relationship of relaxation rates to concentration at pH 4.65 and 6.2. (From Kumosinski, T. F. and Pessen, H., *Arch. Biochem. Biophys.*, 218, 286, 1982.)

Here e is the electronic charge, 1.6022×10^{-19} C, q is the electric field gradient, Q is the nuclear electric quadrupole moment, $\hbar$ is Planck's constant divided by 2π, 1.0546×10^{-27} erg·s; η is a dimensionless parameter measuring the deviation from axial symmetry;[33] γ is the gyromagnetic ratio for the proton, 2.6752×10^4 rad·Gauss^{-1}s^{-1}; r is the internuclear proton distance for water, 0.1526 nm; and S_{deut} and S_{prot} are respective order parameters.[7] It may be noted that these expressions assume no particular model for the relaxation mechanism accompanying NMR hydration. The thermodynamic theory can be used whether isotropic relaxation (S = 1) or anisotropy of the bound water (S < 1) is hypothesized, where in the latter case the "bound" should be understood in the sense of "hydrodynamically influenced layers" or "surface-induced probability distribution of water molecules".[7,34,35] The above treatment can easily be extended to a model postulating three or more states.

III. RELAXATION: ^{2}H NMR

The nuclide chosen for these particular studies[24] was deuterium. Its use eliminates the effect of cross relaxation of water protons with protein protons,which had been shown by Edzes and Samulski[36] to have a significant effect on spin-lattice relaxation rates of water protons with varying protein concentrations.

A. Influence of Protein Hydrophilic Self-Association: β–Lactoglobulin A

The previously derived relationships were tested on bovine β-lactoglobulin A (β-Lg A), the major whey protein in milk, under various environmental conditions. At pH 5, β-Lg A exists as a dimer of molecular weight 36,200. At pH 4.65, β-Lg A undergoes a dimer-to-octomer self-association involving hydrophilic groups. ^{2}H NMR relaxation measurements made at the above conditions served to test whether the notion of increased hydration with increasing hydrophilic self-association is valid.

1. Experimental Procedures

a. Preparation of Solutions

The following procedures are described to illustrate methods that have led to satisfactory results in studies of β-Lg in solution.[27] Protein solutions to be used for proton resonance measurements, prepared 1 d before use, were exhaustively dialyzed overnight against buffer at 0 to 5°C; dilutions for the concentration series to be studied were made with the appropriate dialyzate. Solutions to be used for deuteron resonance measurements were made up from a stock solution prepared by partial deuterium exchange. A suitable amount of crystalline protein in a stoppered vial was allowed to equilibrate repeatedly for 24-h periods at 4°C as a slurry with a small quantity of D_2O, followed by high-speed centrifugation and addition of fresh D_2O, for a total of five times. The solutions were buffered by direct addition of solid potassium phosphate, and the pH was adjusted by addition of 0.1 *N* NaOD in D_2O. Concentrations of β-Lg were determined spectrophotometrically from an absorption coefficient of 0.96 $ml{\cdot}mg^{-1}{\cdot}cm^{-1}$ at 278 nm.[37]

b. Relaxation Measurements

Resonance relaxation spectra were obtained by pulse Fourier transform spectroscopy with a JEOL FX60Q spectrometer operating at a nominal frequency of 60 MHz. The frequency of observation for protons was 59.79 MHz; for deuterons, 9.17 MHz. Raw data were in the form of relative intensities as calculated by the JEOL 980B computer.

Since the high concentration of water in a dilute solution produces an intense signal, a single accumulation at the particular sample temperature (2, 10, or 30 ± 1°C) was sufficient for each spectrum. Even then, care was necessary to avoid exceeding the dynamic range of the computer with consequent truncation. To this end, as well as to economize on the limited amount of the A variant of the protein available, small sample volumes were employed by use of a microcell assembly with an expendable 35-μl sample bulb, available from Wilmad Glass Company, Inc. The protein solution was introduced very slowly into the spherical bulb by means of a fine-gauge syringe needle inserted through its capillary neck, to avoid the inclusion of any air bubbles. Bubbles trapped below the neck could lead to vortex formation in the spinning sample bulb and vitiate the necessary assumption of spherical sample geometry.

The bulb, suspended by its neck from a chuck attached to a plastic cap, was positioned snugly inside a precision 5-mm OD sample tube which, initially, contained also the lock-signal solvent. The small amount of this solvent in the residual annular space outside the bulb was not always sufficient to assure maintenance of the lock; occasional failure of the lock during a lengthy series of automatic measurements resulted in loss of usable data. A second arrangement was then used in which the 5-mm tube, containing the sample bulb but no solvent, was positioned by means of fluorocarbon plastic spacers concentrically within a precision 10-mm OD sample tube accommodating a much larger quantity of lock-signal solvent. Incidental advantages of this arrangement were that the outside of the sample bulb was thus kept dry, and that the solvent could be sealed within the annular space between the two tubes and so kept from contamination for a greatly extended time. Except for these advantages, either arrangement resulted in the same measurements. The cell assembly, in either case, was positioned in the JEOL FX60Q 10-mm ^1H/^{13}C dual probe insert.

Longitudinal (spin-lattice) relaxation rates R_1 were measured by the inversion-recovery method,[38] where the repetition time T in the pulse sequence [...T...π...τ...π/2...] was chosen to be at least five times T_1 ($\equiv R_1^{-1}$) and the values of the variable delay time τ ranged from 10 ms to 3 s, for a total of between 5 and 20 τ-values, depending on the detail desired. From the Bloch equations[39] under the conditions of this method, the relation of the peak intensity A_τ to the pulse delay time τ becomes

$$A_\tau = A_\infty[1 - 2\exp(-R_1\tau)] \tag{17}$$

where A_∞ is the limiting peak intensity for $\tau \rightarrow \infty$. Independent measurement of A_∞, a source of irreducible error, can be dispensed with, and the problem of weighting the data points in the conventional linear plot (logarithm of a function of relative peak heights vs. τ) can be eliminated, by fitting directly to the data points (τ, A_τ) by least-squares an exponential of the form of Equation 17, from which the two parameters A_∞ and R_1 can be obtained.

The factor 2 preceding the exponential in Equation 17 is based on theory which predicts $A_\infty = -A_0$, where A_0 is the peak intensity for $\tau = 0$. It is recognized that in point of fact this equality rarely holds exactly because of slightly imperfect adjustment of the flip angle π. When Equation 17 is used in its logarithmic form, A_∞ must be determined by explicit mesurement, and A_0 then is usually seen, from the ordinate intercept of the straight-line plot of $\ln(A_\infty - A_\tau)$ vs. τ, to differ somewhat from its theoretical value. This, however, is essentially without relevance with respect to the only parameter of intrinsic interest, namely the R_1 obtained from the slope, except possibly insofar as agreement of the intercept with $\ln 2A_\infty$ would be some measure of the quality of the data as a whole.

When the equation is used, as it was here, in the exponential form with A_∞ not explicitly determined, the assumption $A_\infty = -A_0$ implies a two-parameter exponential fit, whereas $A_\infty \neq -A_0$ would imply a three-parameter exponential fit. The latter has been advocated or practiced in the past by various authors.[40-43] Such a practice, however, has ignored the definitive treatment of this matter by Leipert and Marquardt,[44] concurred in by Becker et al.,[45] which has shown conclusively that introduction of a third adjustable parameter, so far from improving the statistics, leads to a significant loss of precision in the estimate of R_1 because of an undesirable correlation between what should be statistically independent parameters, R_1 and A_∞. The effect, on the other hand, of even a considerably misadjusted flip angle on the value of R_1 obtained was shown to be insignificant for values of $R_1 > 0.1$ s, a condition generally satisfied.

The fitting of the two-parameter exponential was carried out by computer by means of an iterative program. For each sample, R_1 was determined at least four times and the results were averaged. This procedure was repeated at each concentration; a minimum of six concentrations were used under each set of conditions of temperature and pH at which the resonance relaxation of each nuclide was examined.

Transverse (spin-spin) relaxation rates R_2 were determined by spin-locking measurement[46] of $R_{1\rho}$, the longitudinal relaxation rate in the rotating frame. R_1 equals R_2 in dilute solutions of low viscosity whenever the magnitude of $R_{1\rho}$ is independent of $H_{1\rho}$, the spin-locking radio-frequency field in the rotating frame; this was the case, within the limits of experimental error, in this work. R_2 was evaluated as described above for R_1, except that the relation between peak intensity A_τ and decay time τ derived from the Bloch equations in this case[39] becomes

$$A_\tau = A_0 \exp(-R_2\tau) \tag{18}$$

where the initial intensity A_0 replaces A_∞ as the maximum peak intensity. Again, a least-squares two-parameter exponential fit to the data points was performed by an iterative computer program, from which A_0 and R_2 were obtained.

For each sample, R_2 was determined with the same number of replications as R_1. Measurements of one mode of relaxation were made on the identical samples and immediately following the completion of measurements of the other mode, or at latest the next day. In this manner, measurements for proton relaxation at 59.75 MHz were made at pH 6.2 at 2 and 30°C, at pH 4.65 at 2 and 30°C, and at pH 2.7 at 10°C. Measurements for deuteron relaxation at 9.17 MHz were made at pH 6.2 and 4.65, both at 2 and 30°C.

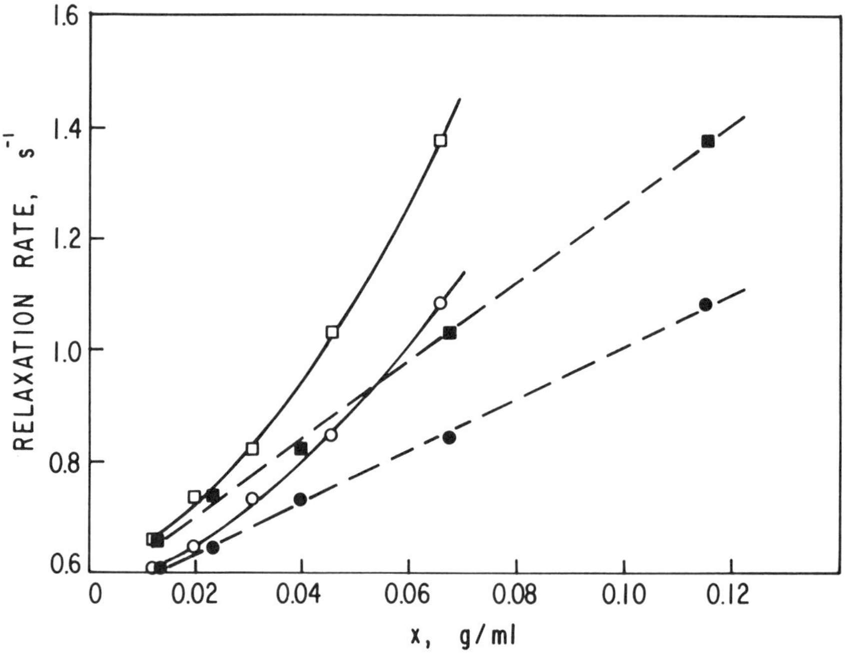

FIGURE 2. Dependence of water proton relaxation rates of b-Lg A on both concentration and activity, at pH 2.7 and 10°C. Transverse relaxation rates (squares) as function of X (-❑-❑-), concentration and (-■-■-), activity. Longitudinal relaxation rates (circles) as function of X (-❑-❑-), concentration and (-●-●-), activity. All points represent experimental values; lines represent least-squares polynominal fits of highest degree to make a statistically significant contribution to goodness of fit (F-test). Dependence on concentration is found to be of second degree as consequence of charge effects at pH 2.7. (Polynominal coefficients for R_1, $a_o = 0.562$, $a_1 = 2.926$, and $a_2 = 75.86$; for R_2, $a_0 = 0.606$, $a_1 = 3.468$, and $a_2 = 125.9$.) Activities take these effects into account, and dependence on activities is linear. (Coefficients of straight line: for R_1, $a_0 = 0.541$, $a_1 = 4.691$; for R_2, $a_0 = 0.560$, $a_1 = 7.061$). (From Kumosinski, T. F. and Pessen, H., *Arch. Biochem. Biophys.*, 218, 286, 1982.)

2. Analysis of Data

a. Activities and the Multicomponent Expression

At this point, an observation regarding predictions resulting from the derived multicomponent expressions above may be in order. Since these are based ultimately on equilibrium constants, the mass terms should be properly expressed as activities instead of concentrations. Consequently, relaxation rate vs. concentration curves should be expected to be nonlinear whenever there are appreciable protein activity coefficients. Figure 1 shows concentration dependences of R_1 and R_2 for β-Lg A at pH 6.2 and 30°C, and of R_2 at pH 4.65 and 2°C. Under these conditions, the charge-to-mass ratio and the second virial coefficient are relatively small ($B_0 = 0.9$ ml/g).[37] In fact, all these data exhibit linear relationships over a concentration range from 0 to 0.08 g protein per gram water, where γ does not differ greatly from unity. Figure 2, on the other hand, shows corresponding plots for pH 2.7 and 10°C, where the net charge is approximately 40 and the second virial coefficient is 8.5 ml/g.[47] The plots in terms of concentration here are clearly nonlinear. A change in the concentration scale to grams protein per gram water had no significant effect on the nonlinearity of the plots. (Low-temperature data were used at this pH for experimental reasons: the protein under these conditions undergoes a dissociation from dimer to monomer, but at low temperature, the amount of monomer at concentrations in excess of 0.01 g/ml is negligible.[47])

A polynominal curve-fitting program was used with all data on Figures 1 and 2, as well as all other concentration-dependent data, in order to determine linearity or nonlinearity, the degree of the polynomial being determined by goodness of fit as judged by the standard F-test. (The regression program used selects the lowest-degree expression for which the sum of squares due to the addition of one higher degree is statistically insignificant.) The two concentration plots of Figure 2 give polynomials of degree two. Protein concentrations, c, were then transformed into activities, a, by means of the relationship $a = \gamma c$, where the activity coefficient $\gamma = \exp(2B_0c)$ was obtained (see Equation 2) from the second virial coefficient $B_0 = 8.5$ ml/g, as cited above. Activity plots corresponding to the concentration plots are also shown in Figure 2; it is evident that these are linear by the same criteria.

Under the conditions of Figure 2, γ is sensibly larger than unity and, even in the presence of salt, failure to treat R as a function of activity rather than concentration will evidently lead to excessively high values of hydration (see Equation 13c). The opposite would be true under conditions when, in the absence of salt, charge fluctuations and consequent intermolecular attraction exist. With γ less than unity, a concentration plot in place of the correct activity plot for R must lead to inordinately low values of hydration. Alternatively, protein concentrations could continue to be used, provided the right-hand side of Equation 13 is multiplied by $(da/dc)_\mu$.

These results may be taken to demonstrate the validity of the multicomponent expression. This expression is used in the following in analyzing the data to describe the hydration of the genetic A variant of β-Lg, under nonaggregation conditions, as well as under the well-characterized dimer-to-octamer association.[48-52] (This phenomenon, because it involves a four-fold association of dimer under conditions where dissociation to monomer is negligible, is referred to in the following as "tetramerization".) It has been demonstrated that here the virial coefficients are small,[37,47] so that it is permissible to simplify the treatment by using protein concentrations.

b. Isotropic Binding, Two-State Model

i. Determination of Correlation Times

Proton spin-spin relaxation measurements of water in solutions of β-Lg A gave results indicating a single relaxation rate, whereas spin-lattice results could be fitted only by the sum of two exponential functions (Figure 3A). This behavior is consistent with the work of Edzes and Samulski[36] and others,[26,53] who found a cross relaxation mechanism between water protons and protein protons to make a significant contribution to the spin-lattice relaxation rate in water-collagen systems. Subsequently, Koenig et al.[54] showed that cross relaxation also exists in spin-lattice relaxation (T_1 processes) in globular protein solutions, i.e., the cross-relaxation rate disperses as a T_1 process and not as a T_2 process. For the solution illustrated in Figure 3A, with the notation of Edzes and Samulski, the parameters in the equation $m(t) = c^+ \exp(-R^+_1 t) + c^- \exp(-R^-_1 t)$ are $R^+_1 = 15.0$, $R^-_1 = 0.45$, $c^+ = 0.01$, and $c^- = 1.01$, where the reduced magnetization m(t) $\equiv (A_\infty - A_\tau)/2A_\infty$. The statistics here were poor because of instrumental limitations.[55] (Cross relaxation, calculated for β-Lg in a novel way,[8] was found to contribute to the observed apparent spin-lattice relaxation to the extent of about 90%.) As an alternative approach, protein solutions were made up in D_2O, and the dependence of both R_1 and R_2 on protein concentration was measured by deuteron NMR at 9.17 MHz which, in effect, eliminated cross relaxation from the T_1 measurements (Figure 3B). The concentration dependence of R_2 of protons in solutions of β-Lg A in H_2O under the same environmental conditions was also measured (see Figure 1).

Concentration plots of R_1 and R_2 for D_2O (Figure 4) showed no evidence of nonlinearity, at either pH and either temperature, over the concentration range studied. This agrees with the low virial coefficient of β-Lg A under these conditions.[37] The relaxation increments k_1 and k_2, together with the corresponding intercepts R_{1f} and R_{2f} (Figure 4), were used in Equations 13 and 16 to determine the bound-water correlation times τ_c shown in Table 1. As can be seen, τ_c increased as the temperature decreased; this is in quantitative agreement with the requirement of Stokes' equation that τ_c increase with both increasing viscosity and decreasing temperature. Furthermore, τ_c also increased when the pH was lowered from 6.2 to 4.65, as would be predicted from the work of Timasheff and Townend,[56] which showed that β-Lg A associates at the lower, but not at the higher pH, and that the association increases with decreasing temperature.

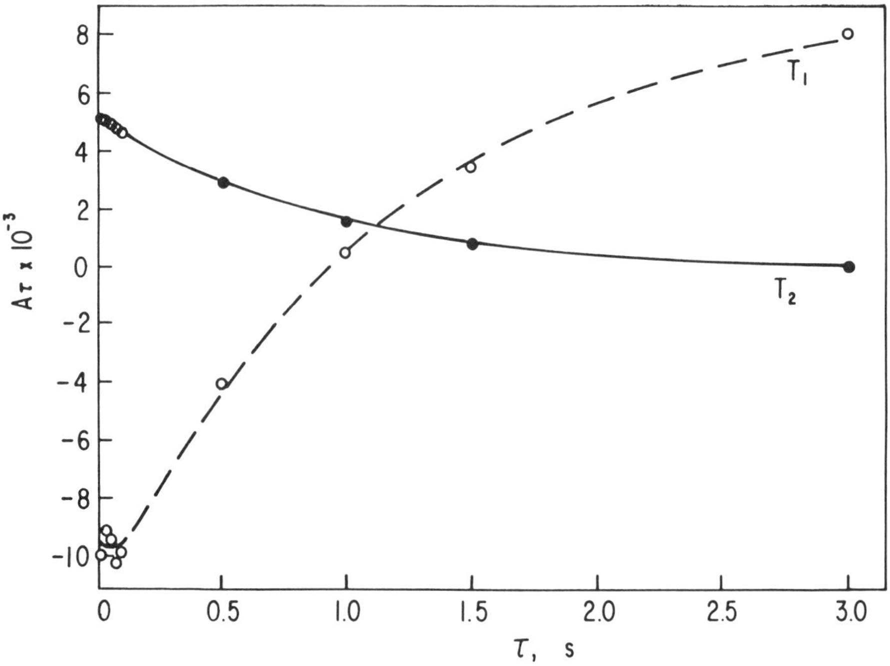

A

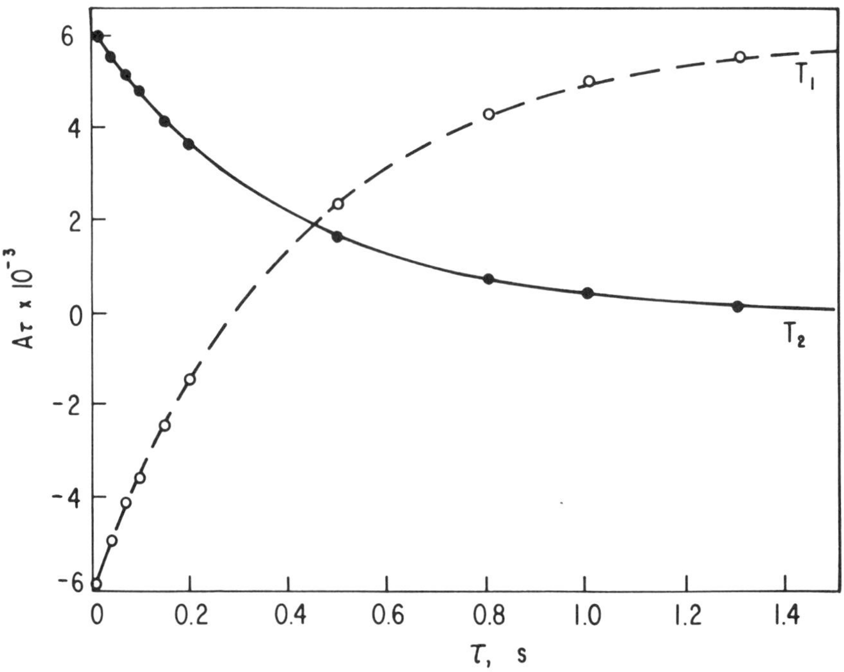

B

FIGURE 3. (A) Proton resonance peak intensities *A* as functions of time t, for solutions of β-Lg A in H_2O (6.09×10^{-2} g protein per gram H_2O) at pH 6.2, 30°C. Intensities for transverse relaxation (T_2, -●-●-) as function of decay time, from spin-locking measurements of $T_{1\rho}$; intensities for longitudinal relaxation (T_1, -○-○-) as function of delay time, from inversion-recovery measurements. Points represent experimental values. Line for T_2 represents two-parameter exponential fit. Line for T_1 represents four-parameter double-exponential fit; need for this fit shown by initial course of this line, indicative of cross relaxation. (B) Deuteron resonance peak intensities *A* as function of time t, for solutions of β-Lg A in D_2O (2.86×10^{-2} g protein per gram D_2O) at pH 6.2, 30°C. In this case, the indication of cross relaxation in T_1 is absent and a two-parameter exponential shows excellent fit to the experimental points even at shortest times. (From Kumosinski, T. F. and Pessen, H., *Arch. Biochem. Biophys.*, 218, 286, 1982.)

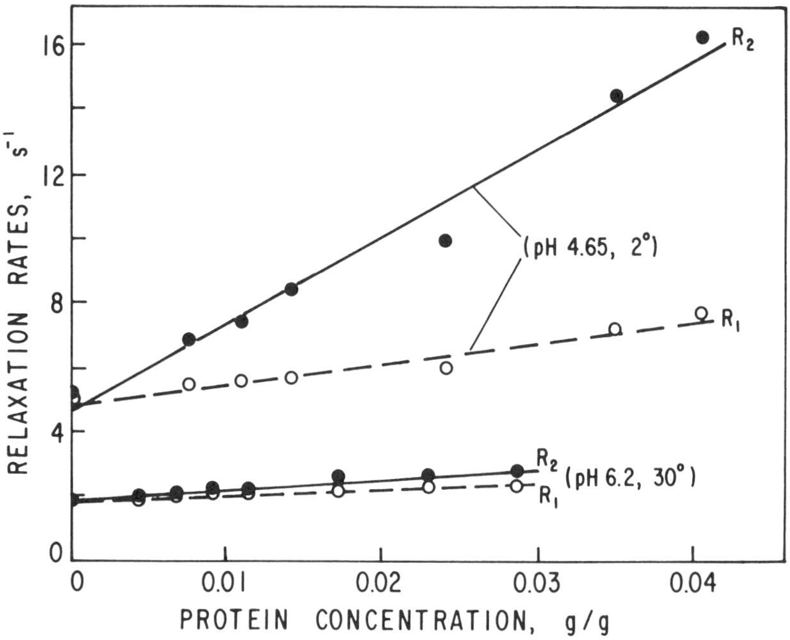

FIGURE 4. Dependence of deuteron relaxation rates on β-Lg A concentrations (g protein per gram water) in D_2O. Transverse relaxation rates R_2 (-●-●-) at pH 4.65, 2°C and at pH 6.2, 30°C. Longitudinal relaxation rates R_1 (-○-○-) under the same two sets of conditions. Points represent experimental values; lines represent least-square fits. Points at all concentrations show linear relationship of relaxation rates to concentration, at both sets of pH and temperature conditions, and for both modes of relaxation. (From Kumosinski, T. F. and Pessen, H., *Arch. Biochem. Biophys.*, 218, 286, 1982.)

Table 1
DEUTERON NMR RELAXATION AND HYDRODYNAMIC RESULTS FOR β-LACTOGLOBULIN IN SOLUTION[a]

pH	T (°C)	k_1[a]	k_2	R_{1f} (s^-)	R_{2f} (s^{-1})	τ_c (ns)	r_{NMR}[b] (nm)	r_{sed} (nm)	$(\tau_c)_{calc}$ (ns)
6.2	30	20.7	31.9	1.93	1.93	10.0	2.32	2.70	10.2
		±0.9	±1.9	±0.01	±0.04	±2.7	0.19		
	2	17.9	58.9	5.46	5.25	25.6			22.5
		±2.1	±5.2	±0.24	±0.28	±4.1			
4.65	30	25.4	73.2	2.01	1.79	22.5	3.04	4.35	65.9
		±1.9	±5.0	±0.04	±0.12	±3.3	±0.15		
	2	63.9	274.2	4.90	4.67	32.2			145.0
		±7.1	±19.0	±0.16	±0.44	±4.6			

[a] Adapted from Reference 27.
[b] Error terms in this and subsequent tables represent the standard error of the parameter.
[c] Spherical model assumption.

ii. Determination of Hydration Parameters

Since the extent of a possible intermolecular contribution, R'_{2b}, to the spin-spin relaxation of bound-water protons and the number of protein protons so contributing are unknown, determination of Scatchard hydrations was attempted by three different methods and the results were compared.

Table 2
HYDRATION AND THERMODYNAMICS FOR β-LACTOGLOBULIN FROM DEUTERON NMR USING $(e^2qQ/h) = 215.6$ kHz[a,b]

pH	T (°C)	υ'_w (g H_2O per g protein)	ΔG (kcal)	−ΔH (kcal)	−ΔS (e.u.)
6.2	30	0.0063 ±0.0008	0.90 ±0.08		
				0.8 ±3.0	6 ±10
	2	0.0072 ±0.0020	0.74 ±0.15		
4.65	30	0.0095 ±0.0002	0.65 ±0.01		
				6.9 ±1.6	24.8 ±5.8
	2	0.0301 ±0.0003	−0.044 ±0.006		

[a] Adapted from Reference 27.
[b] Method I, as described in Section III.A.2.b.

Method I — Only the deuterium NMR relaxation increments were used, with a value of $e^2qQ/\hbar/ = 215.6$ kHz[57] and with the asymmetry parameter η assumed to be zero. This type of calculation gives low values of $\bar{v}'_W$, as shown in Table 2; however, low values could be expected because the relaxation increment probably samples only a percentage of the total hydration of a protein, since at 9.17 MHz any bound water with τ_c values less than 6 ns would have a T_1/T_2 ratio of unity. At pH 4.65, where tetramerization occurs, the hydration markedly increases with decreasing temperature, whereas at pH 6.2, where none occurs, the hydration is lower and independent of temperature. This is consistent with the findings of Timasheff and co-workers[48,58] from small-angle X-ray scattering that the geometry of the octamer must include a large central cavity in which trapped water could reside.

Method II — A combination of the τ_c values found by deuteron NMR at 9.17 MHz (Table 1) and the k_2 values found from proton NMR at 59.75 MHz was used (see Figure 1). The reason for this procedure is that the quadrupole coupling constant for the bound water should actually decrease as hydrogen bonding increases.[59] Here R_{2b} can be calculated from Equation 16b at 59.75 MHz, and the $\bar{v}'_W$ values can be easily obtained from the simple relationship $\bar{v}'_W = k_2/(R_{2b} - R_{2f})$, derived from Equation 13b. Such hydration values (Table 3) are slightly higher than those from deuteron relaxation measurements only (Table 2), but show the same temperature and pH dependence. (Combination experiments of this kind would be best performed at the same Larmor frequency; however, the availability of only a single spectrometer with no variable frequency capability would preclude this possibility. These experiments could shed light on such problems as the constancy of the hydrogen-bond distance under various conditions and the existence of a distribution of correlation times in the total hydration shell.)

Method III — This is a combination procedure also, with an extra intermolecular interaction term R'_{2b} added to the proton spin-spin relaxation rate R_{2b}. Based on the small-angle X-ray scattering results for β-Lg A of Witz et al.,[48] in conjunction with the known molecular weight and amino acid composition, a simple consideration of the molecular geometry shows that, on the average, each proton in the protein will have six neighboring protons at a distance of 0.261 nm. From this average intermolecular distance, together with the τ_c values of Table 1 and the relationship $\bar{R}_{2b} = R_{2b} + 12\,R'_{2b}$ (where $\bar{R}_{2b}$ is the total spin-spin relaxation rate), the hydration can be calculated as $\bar{v}'_W = k_2/(\bar{R}_{2b} - R_{2b})$. These values (Table 3) are in close agreement with those in Table 1. Altogether, no great difference exists in the $\bar{v}'_W$ values from all three methods.

Table 3
HYDRATION AND THERMODYNAMICS FOR β-LACTOGLOBULIN DERIVED FROM τ_c VALUES OF TABLE 1 BY METHODS II AND III[a]

				R_{2b} (s^{-1})		υ'_W (g H_2O per gram protein)		ΔG (kcal)		−ΔH (kcal)		−ΔS (e.u.)	
pH	T (°C)	k_2	R_{2f} (s^{-1})	II[b]	III[c]	II[b]	III[c]	II[b]	III[c]	II[b]	III[c]	II[b]	III[c]
6.2	30	3.5	0.34	228	337	0.0152	0.0103	0.35	0.59				
		±0.5	±0.02			±0.0021	±0.0007	±0.07	±0.04				
										0	0	1.2	1.9
	2	8.1	0.70	530	784	0.0152	0.0103	0.32	0.54			±0.2	±0.1
		0.4	±0.02			±0.0008	±0.0014	±0.03	±0.07				
4.65	30	9.2	0.27	468	694	0.0197	0.0133	0.20	0.43				
		±0.8	±0.04			±0.0016	±0.0011	±0.05	±0.04				
										4.62	4.62	15.9	16.7
	2	28.2	0.56	660	978	0.0428	0.0289	−0.25	−0.032	±0.73	0.70	±2.6	±2.4
		±1.1	±0.05			±0.0017	±0.0011	±0.02	±0.021				

[a] Adapted from Reference 27.
[b] Method II uses τ_c from Table 1 and proton k_2.
[c] Method III uses, in addition to the procedure of Method II, the assumption $\overline{R}_{2b} = R_{2b} + 12 R'_{2b}$, with the intermolecular proton distance for β-Lg A calculated from the partial specific volume as 0.261 nm, as described in Section III.A.2.

iii. Comparison of Results with Other Structural Information

Dynamics of β-Lg dimer — With the τ_c values calculated from k_2/k_1 from deuteron NMR spin-spin and spin-lattice relaxation increments, dR_2/dc and dR_1/dc, a Stokes radius r_{NMR} for the bound water can be calculated from the Stokes-Einstein relation[60] on the basis of a spherical model (Table 1). At pH 6.2, where β-Lg A exists as the unassociated dimer, the value of r_{NMR} is slightly lower than the one for the protein itself derived from hydrodynamic data, r_{sed}.[61] This discrepancy could be due to the spherical approximation inherent in the use of the Stokes-Einstein equation, inasmuch as the β-Lg dimer actually has an axial ratio of approximately 2:1.[37] Moreover, the Stokes radius of the protein obtained from sedimentation includes the water of hydration and should, therefore, be larger than the Stokes radius of the bound water calculated from the ^2H NMR relaxation data. What should be compared with the τ_c of the bound water is the τ_c of the protein without any contribution from hydration. For the latter, values of 10.2 ns at 30°C and 22.5 ns at 2°C (Table 1) can be calculated for the protein with the use of its partial specific volume, 0.751 ml/g, and an asymmetry factor of 1.168 [62] to account for the dimer axial ratio of 2:1. These values are in excellent agreement with the experimental τ_c of the bound water at pH 6.2 at 30°C and 2°C (see Table 1).

Hydration and dynamics of β-Lg octamer — At pH 4.65, where the protein exists to a large extent as the octamer even at 30°C at concentrations above 0.01 g/ml,[49] the Stokes radius of the bound water is about 30% less than the Stokes radius of the octamer itself. However, this value is still much closer to the theoretical value than those obtained by other investigators for other proteins.[5,6,24-26] Furthermore, the 422-symmetry model for the octamer according to Timasheff and Townend[58] possesses a large central cavity which could accommodate trapped water; if the NMR experiment observed this trapped water, the τ_c value found would be less than that of the protein. Also, if the assumption is made that the NMR hydration of the octamer itself at 2°C equals $(\bar{v}'_W)_{pH 4.65} - (\bar{v}'_W)_{6.2}$, values from 0.019 to 0.028 g H_2O per gram protein can be calculated by the three methods described above. The total volume of the cavity, approximated by an internal sphere tangent to the subunits on the basis of known structural parameters,[58] amounts to about 6.5 nm^3. Taking the specific volume of water as unity and thus its molecular volume as 0.03 nm^3 per molecule, this would correspond to about 220 moles H_2O per mole of octamer, or 0.027 g H_2O per gram protein, which is within range of the NMR-derived hydration values for the octamer.

Since the derived NMR correlation times are number-average values, the hypothesis that the increase in hydration accompanying octamer formation is largely due to trapped water may be tested by calculating a number-average correlation time from the relationship $(\bar{v}'_W)_{4.65}\,\tau_c = (\bar{v}'_W)_{6.2}\,(\tau_c)_0 + [(\bar{v}'_W)_{4.65} - (\bar{v}'_W)6.2](\tau_c)_{cc}$, where $(\tau_c)_0$ is the correlation time of the octamer at 2°C (i.e., 145 ns, see Table 1), $(\tau_c)_{cc}$ is the correlation time of the central cavity of volume 6.5 nm^3 (i.e., 1.4 ns), and $(\bar{v}'_W)_{6.2}$ and $(\bar{v}'_W)_{4.65}$ are the NMR hydration values at pH 6.2 and 4.65, respectively. Calculation of τ_c from ^2H NMR hydration values by method I at 2°C gives 36 ns, in fair agreement with the ^2H NMR experimental value of 32.2 ± 4.6 ns at pH 4.65 and 2°C. While the results of this calculation furnish an indication of the reasonableness of the approach, they do not, however, show any exact mechanism of increased hydration accompanying octamer formation.

In contrast to method I, method II assumes no constant quadrupole coupling constant and therefore may serve, incidentially, to calculate values of $e^2qQ/\hbar$ (assuming $\eta = 0$) from the $\bar{v}'_W$ values and the relaxation increments obtained by deuteron NMR, together with the experimentally derived τ_c. The quadrupole coupling constants calculated for the respective methods range from 120 to 160 kHz. Hunt and MacKay[59] have correlated O...^{2}H...O and N...^{2}H...O hydrogen-bond distances with values of quadrupole coupling constants. From their relationships and the above $e^2qQ/\hbar$ values, one obtains distances for O...^{2}H...O from 0.15 to 0.17 nm, and for N...^{2}H...O from 0.16 to 0.20 nm. These compare with linear hydrogen bond lengths of 0.181 to

0.187 nm recently reported by Ceccarelli et al.[63] in an extensive review of neutron diffraction data.

c. Contrast of NMR Hydration with Preferential Hydration

To contrast these NMR hydration results with results from another physical method which measures water-protein interactions, preferential hydrations were obtained by Wyman's theory of linked functions,[30,64] from which it follows that for a tetramerization reaction (dimer → octamer, in the case of β-LgA)

$$\begin{aligned} \mathrm{dlnk_T/dln}\ a_{X,T} &= (\bar{\nu}_{X,T})_{pref} - 4(\bar{\nu}_{X,M})_{pref} \\ &= -(W_T/X_T)[(\bar{\nu}_{W,T})_{pref} - 4(\bar{\nu}_{W,M})_{pref}] \end{aligned} \tag{19}$$

where the preferential interactions are defined by

$$(\bar{\nu}_X)_{pref} = \bar{\nu}_X - \left(\frac{X}{W}\right)\bar{\nu}_W \quad \text{and} \quad (\bar{\nu}_X)_{pref} = \bar{\nu}_W - \left(\frac{W}{X}\right)\bar{\nu}_X \tag{19a}$$

Here K_T is the association constant, a_X is the activity of salt, $(\bar{\nu}_{X,T})_{pref}$ and $(\bar{\nu}_{X,M})_{pref}$ are the preferential salt binding of octamer (subscript T) and dimer (subscript M), and $(\bar{\nu}_{W,T})_{pref}$ and $(\bar{\nu}_{W,M})_{pref}$ are the preferential hydrations, respectively. Preferential interaction parameters can thus be readily obtained from the slope of a plot of association constants at various salt concentrations vs. the activity of the salt at the corresponding concentrations.

Association constants can be calculated by the use of Gilbert's theory for rapidly reequilibrating association in a sedimenting boundary.[65,66] Gilbert has shown that for a reversible association there exists a minimum concentration, c_{min}, above which bimodality of a schlieren ultracentrifuge pattern appears. Furthermore, the area of the slow peak remains constant as the loading concentration is increased well above c_{min}, which is related to the equilibrium constant of the association. Thus, for a tetramerization (in this case, dimer → octamer),[48-50]

$$K_T = M_M^3\ \delta[1 + \delta/4(1 - \delta)]^3/[16(1 - \delta)c^3] \tag{20}$$

where $\delta \equiv (s - s_1)/(s_4 - s_1)$, ($s$, s_1, and s_4 being the sedimentation coefficients for the leading edge of the boundary, for the dimer, and for the octamer, respectively), K_T is the association constant, c is the total loading concentration, and M_m is the dimer molecular weight. Since $\delta_{min} = (n - 2)/3(m - 1)$ and $\delta = 2/9$ for n = 4 (where n is the degree of association), $K_T = 1.087 \times 10^{12}/c^3_{min}$. The product of the percentage of the slow peak and the loading concentration equals c_{min}, and K_T can thus be evaluated.

In this way, sedimentation velocity data for β-Lg A at pH 4.65 give a linked-function plot (Figure 5B), which clearly shows a negative slope. From the least-squares value of this slope, a preferential solvation of –3.78 moles salt per mole octamer is calculated. With the assumption that the tetramerization does not release any salt, a value of 0.258 g H_2O per gram protein for the preferential hydration is obtained, presumed to be equal to $(\bar{\nu}'_W)_{pH4.65} - (\bar{\nu}\beta'_W)_{pH6.2}$ from NMR at 2°C. The linked-function preferential hydration thus appears to differ significantly from the NMR hydration difference.

However, as previously noted, NMR probably samples only a certain percentage of the total hydration of a protein. For example, if a major portion of the bound water has a τ_c value of less than 1 ns, its contributions to the spin-lattice and spin-spin relaxations of the bound state would be equal since at 9.17 MHz the T_1/T_2 ratio for this fast-tumbling water would be unity. Therefore, the correlation time and hydration values calculated from the NMR results by use of a two-state

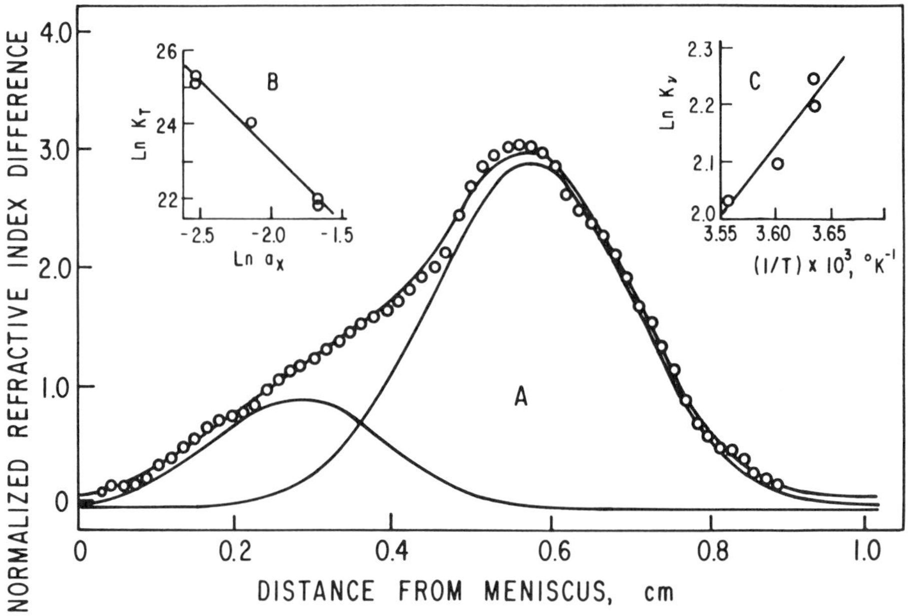

FIGURE 5. (A) Sedimentation of β-Lg A under association conditions. Gilbert pattern (open circles) and decomposition into two Gaussian peaks (solid lines). (B) Linked-function plot. Plot of logarithm of tetramerization equilibrium constant vs. logarithm of activity of salt. The slope is a measure of preferential hydration. (C) van't Hoff plot. Plot of logarithm of equilibrium constant of water binding vs. reciprocal of absolute temperature. The slope gives a value –5.5 ± 1.2 kcal for the enthalpy of hydration. (From Kumosinski, T. F. and Pessen, H., *Arch. Biochem. Biophys.*, 218, 286, 1982.)

model approximation would yield number-average values weighted toward the slow-tumbling component. It would probably be more realistic to compare the enthalpies of hydration derived by NMR with those from linked functions by the van't Hoff relationship (Figure 5C). This is possible by assuming a thermodynamic model involving the transfer of free water to bound water as protein is added to the solution, i.e., $\Delta G = RT\ln(\bar{v}_W/55.6)$. Values of ΔG, ΔH, and ΔS of hydration for each of the three NMR methods are shown in Tables 2 and 3.

From the slope of the van't Hoff plot of Figure 5C, a ΔH of –5.5 ± 1.2 kcal/mol dimer is obtained. ΔS values calculated at each temperature average –14.1 ± 0.03 e.u. The small SD for ΔS indicates constancy of the entropy with respect to temperature and lends support to the model. Furthermore, ΔH of hydration values from NMR range from approximately –6 to –9 kcal/mol dimer, seemingly in agreement, within experimental error, with the value of –5.5 derived from linked functions. Nevertheless, it needs to be remembered that the linked-function method measures the difference between the total hydrations of octamer and dimer, whereas the NMR method yields a quantity proportional to the total hydration of the octamer. Direct comparison of the temperature dependence of these two methods, by assuming that the linked-function hydration is proportional to the difference in the NMR hydration at pH 4.65 and 6.2, is not feasible since the proportionality constant itself should change as a function of temperature. However, the total hydration of the dimer appears to be temperature independent, as indicated by essentially constant NMR hydration values at pH 6.2 for 30 and 2°C. These considerations lead to an alternative evaluation of the NMR data in terms of a three-state model.

i. Isotropic Binding, Three-State Model

For the reasons indicated, an attempt was made to evaluate the ^{2}H NMR relaxation increment on the basis of a three-state model (i.e., free water, fast- and slow-tumbling bound water), as used

Table 4
TOTAL HYDRATION FOR β-LACTOGLOBULIN DERIVED FROM A THREE-STATE MODEL (^{2}H NMR)[a,b]

pH	T (°C)	$(\bar{v}'_w)_f$	$(\bar{v}'_w)_s$	$(\tau_c)_s$ (ns)	$-(\Delta H)_s$ (kcal)
6.2	30	0.1	0.0054	11.0	
			±0.0008	±2.7	
					0
	2	0.1	0.0056	33.9	
			±0.0020	±4.1	
4.65	30	0.201	0.0085	28.8	
			±0.0002	±3.3	
					6.0
					±1.6
	2	0.358	0.0235	42.3	
			±0.0003	±4.6	

[a] Assumptions: $(\tau_c)_f = 48$ ps; $(R_{1b})_f = (R_{2b})_f = 31.94\ s^{-1}$ at 30°C for pH 6.2 and 4.65; $(R_{1b})_f = (R_{2b})_f = 70.27\ s^{-1}$ at 2°C for pH 6.2 and 4.65. Subscripts f and s refer to fast-tumbling and slow-tumbling fractions, respectively.

[b] Adapted from Reference 27.

by Cooke and Kuntz in treating lysozyme data,[67] with the following assumptions. First, the fast-tumbling hydration component of dimer was assumed to be 0.1 g water/per gram protein on the following grounds. The lysozyme crystal has been determined to contain 80 molecules of water per protein molecule, or 0.1 g H_2O per gram protein,[68] and 100 to 300 water molecules per crystalline protein molecule distributed over the molecular surface have generally been reported.[67] For β-Lg, 200 moles water per mole protein is equivalent to 0.1 g water per gram protein. Also, Teller et al.[69] have shown that experimentally derived frictional coefficients from sedimentation results are in agreement with those calculated from known X-ray crystallographic structures provided water molecules are added to each charged side chain on the protein surface. The β-Lg dimer has 90 charged amino acids;[70] a value of 0.1 g water per gram protein corresponds, therefore, to about two water molecules bound to each charged side-chain amino acid.

Second, since from Teller's work, this water would be bound to charged side chains, and since Brown and Pfeffer[71] have shown that deuterium-modified lysine groups tumble at 48 ps, a τ_c value of 48 ps can be assumed. Any water bound to lysine should then have a τ_c value no lower than that of the side chain itself. It was assumed also that, because of the fast segmental motion of the side chain, any increase in association of the protein would not affect this τ_c unless the side chain was directly involved in the interaction site. However, the viscosity and temperature effect on the Stokes-Einstein relationship was taken into account when the temperature changes from 30 to 2°C. Finally, the fast-tumbling hydration values at pH 4.65 were increased, by 0.101 (30°C) and 0.258 (2°C) g H_2O per gram protein, respectively, in line with the linked-function results, which indicate such preferential hydrations at these temperatures.

Subtraction of the fast-tumbling contribution from the ^{2}H NMR spin-lattice and spin-spin relaxation increments yields new values from which the correlation time of the slow component, $(\tau_c)_s$, and its corresponding hydration, $(\bar{v}'_W)_s$, can be calculated. Table 4 shows that the $(\tau_c)_s$ are slightly larger than the τ_c of Table 1, and the $(\bar{v}'_W)_s$ are slightly larger than the $\bar{v}'_W$ of Table 2 for the two-state model. However, the corresponding values are probably within experimental error, as are the derived enthalpy of hydration of the slow component (Table 4) and the enthalpy of hydration derived from the two-state model (Table 2). Calculation of a number-average correlation time of the slow component of the octamer, by assuming that the increase in hydration upon octamer formation is due to water trapped in the cavity of the octamer, yields 37 ns. This is in reasonable agreement with the experimentally derived value of $(\tau_c)_s$ at pH 4.65 and 2°C of 42.3 ± 4.6 ns (Table 4).

ii. Anisotropic Binding Mechanism

The preceding calculations assume an isotropic relaxation mechanism, as detailed above. In the presence of salt, all the relaxation at pH 6.2 can be accounted for by a slow-tumbling and a fast-tumbling component, amounting to 13 and 204 mol H_2O per mole dimer, respectively, and increasing at pH 4.65 and 2°C to 61 and 730 mol H_2O per mole dimer; these may be considered reasonable values for the hydration of a protein.[1,72]

This does not, however, eliminate the possibility of an anisotropic relaxation mechanism for hydrodynamically bound water. The present results may be interpreted equally well on the basis of the three-component derivation in conjunction with either a two- or three-state model and an appropriate order parameter $S < 1$ (see Equations 16c and d). Here a three-state model is defined, as for the isotropic mechanism (Table 4), as comprising free-motion water, a slow-motion component (i.e., $\tau_c > 5$ ns), and a fast-motion component (i.e., $\tau_c \cong 48$ ps, as assumed in Table 4). For the latter, under extreme-narrowing conditions the factor S^2 attached to Equations 16c and d is changed to $(1 - S^2)$.[7] The slow motion, in either the two- or three-state anisotropic mechanism, may be due to such processes as protein reorientation, internal motion of the protein, or translational diffusion of water along the protein surface.[7,34]

Reported values of S = 0.06 from ^{17}O relaxation[7] have been derived by applying to a protein the line-splitting data obtained for a liquid crystal, on the assumption that three to six water molecules are bound to carboxyl groups and one to three to hydroxyl groups. Theoretical results of Walmsley and Shporer[34] give relationships for S (termed the scaling factor by these authors) based on ^1H, ^2H, and ^{17}O relaxation. From these it follows that a value of S = 0.06 for ^{17}O would imply S = 0.12 for ^2H. At pH 6.2 (30 and 2°C) and pH 4.65 (30 and 2°C) one obtains (from Equations 13a to c and the ^2H NMR data of Table 1 for the two-state, and those of Table 4 for the three-state model) hydrations, all in units of g H_2O per gram protein, of 0.483, 0.500, 0.660, and 2.090 for the two-state, and 0.298, 0.295, 0.509, and 1.250 for the three-state model, respectively. A more reasonable estimate is obtained from the theoretical relationships of Walmsley and Shporer, together with the experimental results of Koenig et al.,[5,6,24] which give S = 0.23 and corresponding hydrations, in the same units,of 0.119, 0.136, 0.180, 0.569, and 0.102, 0.105, 0.163, 0.435. The ΔH of hydration at pH 4.65 is found to be –6.8 for the two-state and –5.9 kcal for the three-state model. These results agree with the isotropic mechanism (Tables 2 and 4), since S enters simply as a factor in the Kubo-Tomita-Solomon equations.

Alternatively, equating the preferential hydration from linked functions with the difference between the 2°C ^2H NMR hydrations at pH 4.65 and pH 6.2, one obtains S = 0.30 for the two-state and S = 0.26 for the three-state model, both not far from the 0.23 predicted from the theory of Walmsley and Shporer. Furthermore, with the above values of S, the pH 4.65 enthalpies of hydration are –6.8 and –5.9 kcal for the two- and three-state models, respectively. Thus, the increase in hydration as well as the corresponding enthalpy change attendant on octamer formation are the same for either assumption of relaxation mechanism.

B. Influence of Protein Hydrophobic and Electrostatic Self Association: Bovine Casein

From the previous section, it appears that hydrophilic self-association of proteins can increase hydration values through changes in quaternary structure. An extension of this work would be to study the influence of protein-protein interactions involving hydrophobic and electrostatic groups on the hydration of proteins as evaluated by NMR relaxation measurements. The model system chosen was casein, a family of phosphoproteins which are the major components of milk.[73,74] Casein monomers undergo hydrophobic self-associations at pH 7 which increase with increasing temperature.[75,76] The associated state is commonly referred to as the submicellar form.[75,77] Upon addition of calcium, casein polymers further associate via calcium phosphate salt bridges into a colloidal state referred to as the micellar form.[73,77] Although the chemical and physical properties of the individual caseins are well documented, few studies have addressed the association properties of the whole complex as found in milk. For these reasons, NMR relaxation measurements of water were made with varying concentrations and temperatures of casein under both submicellar and micellar conditions.[78]

1. Experimental Procedures

a. Preparation of Solutions

Casein micelles were isolated from 2 l of fresh warm milk to which 1 g of phenylmethyl sulfonyl fluoride had been added to retard proteolysis. The milk was centrifuged at 4000 × g for 10 min to remove the cream fraction; 400 ml of this skim milk was centrifuged for 1 h at 88,000 × g (37°C). The pellets were washed twice in D_2O containing 25 m*M* piperazine-*N,N'*-bis(2-ethanesulfonic acid) (PIPES) (pH 6.75), 20 m*M* $CaCl_2$, and 80 m*M* KCl. The final protein concentration was fixed at about 100 mg/ml (total volume of 5 ml). Subsequent dilutions were made with the same buffer. To produce submicelles, sodium caseinate prepared from the same skim milk was dialyzed and lyophilized at pH 7.2; the lyophilized protein was dissolved in D_2O, in the same PIPES-KCl buffer without $CaCl_2$, but with added dithiothreitol to promote self-association of κ-casein.[76] These procedures were designed to minimize the concentration of H_2O in the D_2O solutions and thus to eliminate any significant contribution to the relaxation rates from deuterium exchange. Casein concentrations were determined spectrophotometrically on samples diluted 1/50 to 1/100 in 0.1 *N* NaOH; an absorptivity of 0.850 $ml \cdot mg^{-1} \cdot cm^{-1}$ at 280 nm was used for whole casein.[76]

b. Relaxation Measurements

Measurements were made essentially as described under β-Lg A above, except that only deuteron resonance relaxation was investigated at 9.17 MHz, on samples at pH 7.0, at 2, 15, and 30 ± 1°C. These rates were measured in D_2O to avoid cross-relaxation effects between water and protein protons, such as have been observed by Edzes and Samulski[36] and by Koenig et al.[54]

2. Analysis of Data

a. General Considerations

Bovine casein is composed of four major proteins, α_{s1}-, α_{s2}-, β-, and κ-casein, in the approximate ratios of 4:1:4:1.[79] α_{s1}-casein contains 8 phosphoserines; β-casein contains 5 phosphoserines; κ-casein contains on the average 1 phosphoserine, while α_{s2} is variable, containing 8 to 11.[74] All caseins are generally considered to have little or no ordered secondary structure and to contain a large number of hydrophobic residues. With the above assumptions, the average molecular weight of monomeric casein is estimated to be 23,300; the average partial specific volume, $\bar{v}$, is 0.736 ml/g; and the weight-average number of phosphate groups is 6.6 per 23,300-Da monomer unit.

At pH 6.75 with no calcium present, studies of the individual caseins have shown that they undergo mainly hydrophobically driven self-associations which increase with increasing temperature and ionic strength.[77] Studies on whole casein are limited but show similar results.[76,77] Figure 6A shows one proposed structure of this limiting polymer, commonly referred to as the submicellar form of casein. Here, the hydrophobic core is considered to be composed mostly of the hydrophobic portions of α_s- and β-caseins, while κ-casein resides mostly at the surface because of its ability to keep α_s- and β-caseins from precipitating at 37°C in the presence of calcium. All charged groups, including the serine phosphates, are located on the surface of the submicellar structure. In this model, the κ-casein content of the submicelles is variable.

Upon addition of calcium, these submicellar spherical particles are thought to self-associate by way of calcium phosphate salt bridges to form a large colloidal spherical particle of approximate radius of 65 nm, called the micellar form of casein (see Figure 6B). The integrity

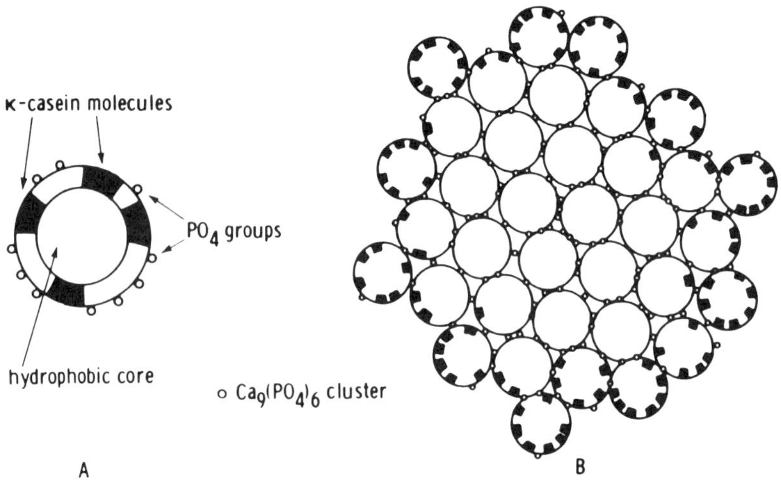

FIGURE 6. Quaternary structural forms of casein. A, submicellar form and B, micellar form upon addition of calcium. (From Schmidt, D. G., *Developments in Dairy Chemistry*, Vol. 1, Fox, P. E., Ed., Applied Science, Essex, England, 1982. With permission.)

of the submicelle is thought to be preserved upon the addition of calcium if enough calcium is used and no phosphate buffer is present to compete with the serine phosphate interaction sites. Moreover, it has been conjectured by several investigators that trapped water exists within the micellar structure.[1] The characteristic of κ-casein to be present predominantly on the surface of the micelle has been shown by electron microscopy coupled with gold-labeled κ-casein,[77] or with ferritin conjugate and anti-κ-casein.[80]

For the above reasons, ^{2}H NMR relaxation measurements, both spin-lattice, R_1, and spin-spin, R_2, of D_2O with varying concentrations of casein were made on casein, with and without calcium, at 30, 15, and 2°C. Figure 7 shows R_1 and R_2 measurements at 15°C under submicellar and micellar conditions. All data were fitted by Equation 13a, modified by replacing the concentration, c, by the protein activity, a_P, with $a_P = c \exp(2B_0 c + \ldots)$, where B_0 is the second virial coefficient of the protein. Least-square fits were obtained by means of a Gauss-Newton nonlinear regression program. The experimental data and the data calculated from the model employed are in excellent agreement, as shown by the solid line in Figure 7. Under these and all other conditions investigated, the nonlinear portion of the curves yielded a virial coefficient of 0.0032 ± 0.0003 ml/mg, indicating the consistency of the experimental results. The linear portions of the curves were evaluated with a propagated SE of about 4%; they contain the product of the relaxation of the bound water, hydration, and finally the asymmetry parameter, S. These parameters are separated and each is discussed in the following section.

b. Hydration and Dynamics: Isotropic Model

From the linear portion of spin-lattice and spin-spin relaxation results, Equation 13a, and the Kubo-Tomita-Solomon equations, Equations 16a and b, the following parameters were calculated at the various environmental conditions of the caseins: correlation times, τ_c, hydration values, ν_w, for an isotropic tumbling model (S = 1), and the relaxation rates of the bound water, R_{1b} and R_{2b}. The results are shown in Table 5. Here, $\bar{\nu}_w$ values increased from 0.00652 to 0.01201 g water per gram protein and τ_c values decreased from 38.9 to 29.8 ns as the temperature decreased from 30 to 2°C for casein in the submicellar form; propagated standard errors were about 8% for τ_c and 6% for $\bar{\nu}_w$. The same temperature dependence of τ_c and $\bar{\nu}_w$ was exhibited under micellar conditions, although at all temperatures their absolute values were larger for the micellar than for the submicellar form.

At this point it may be noted that although the caseins are self-associating, one needs to consider here only the aggregated form. The concentrations used were high enough so that the

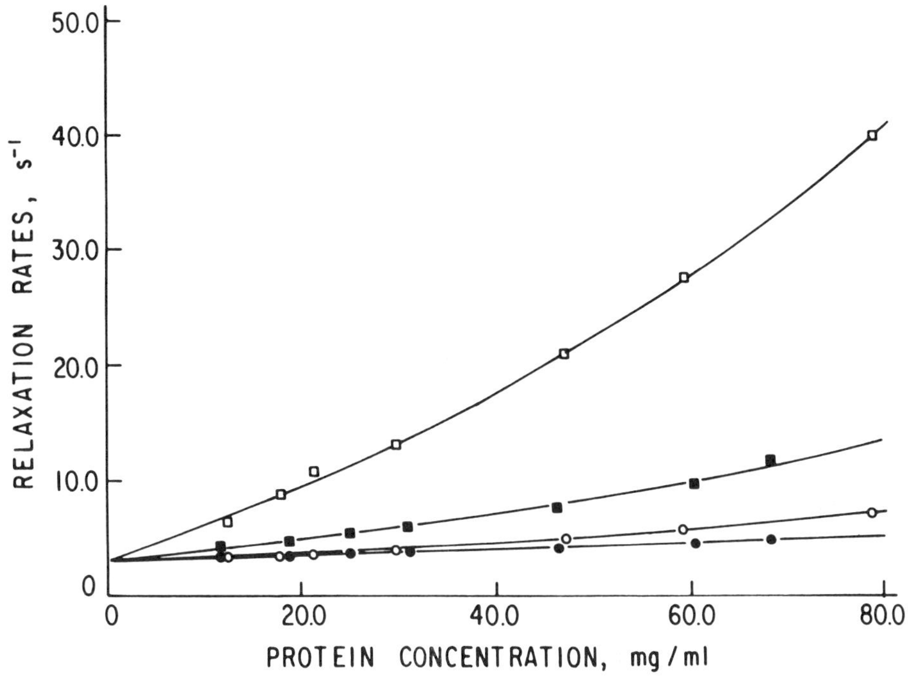

FIGURE 7. Dependence of deuteron relaxation rates of water on casein concentrations in D_2O at pH 6.75 in 0.2 *M* PIPES-KCl buffer at 15°C. -●-●-, R_1 measurements, submicellar form; -❑-❑-, R_2 measurements, submicellar form; and -○-○-, R_1 measurements, micellar form; -❑-❑-, R_2 measurements, micellar form. (From Reference 78.)

Table 5
HYDRATION AND DYNAMICS OF BOUND WATER FOR CASEINS

	Temperature (°C)	τ_c (ns)	υ_w (g H_2O per gram protein	R_{1b} (s^{-1})	R_{2b} (s^{-1})
Submicelle	30	38.9	0.00652	1904	10,510
	15	34.7	0.00824	2080	9,840
	2	29.8	0.01201	2323	9,070
Micelle	30	63.6	0.0165	1249	14,790
	15	51.1	0.0225	1515	12,570
	2	45.1	0.0282	1689	11,530

association equilibrium favors polymer formation.[75,77] For both micelles and submicelles, no significant differences in hydration would result from protein concentration-dependent dissociation effects at 30°C. (At lower temperatures, this must be qualified as discussed further on.) Also, these $\bar{v}_w$ values will in most probability show only a fraction of the total hydration, since at 9.17 MHz any bound water with τ_c <6 ns would have an R_2/R_1 ratio of unity and would not be observable by this methodology.

Since it had been shown previously that the τ_c values derived from NMR relaxation results are those for the unhydrated rather than the hydrated form of the protein,[27,81] the Stokes radius, r, calculated from τ_c values using the Stokes-Einstein relationship would indeed be a representation of the quaternary structure for the unhydrated protein. Such r values were calculated for the caseins from all τ_c results and are listed in Table 6.

A Stokes radius of 3.64 nm (Table 6) found at 30°C is at the lower limit of radii reported for

Table 6
MOLECULAR PARAMETERS OF CASEINS DERIVED FROM DATA OF TABLE 5

	Temperature (°C)	r (nm)	M_P	$(\upsilon_w)_r$	S	$(\upsilon_w)_{S = 0.237}$
Submicelle	30	3.64	165,000			0.116
	15	3.05	97,200			0.147
	2	2.55	56,800			0.214
Micelle	30	4.29	270,700	0.469	0.188	0.294
	15	3.48	144,500	0.357	0.251	0.400
	2	2.93	86,200	0.380	0.272	0.502

submicelles, whose sizes range from 4 to 9 nm depending on the method of measurement.[76,77,82,83] (It should be recalled that direct comparison between Stokes radii derived by this NMR method and those calculated from hydrodynamic or small-angle scattering data would be inappropriate since these latter include water of hydration, whereas the NMR values pertain to the anhydrous protein.[27,81]) Results for the submicelles show a decrease in hydration value (Table 5) and an increase in Stokes radius (Table 6) with increasing temperature. This suggests that hydrophobic interactions are involved in the formation of the submicelle, since, as the temperature is raised, water is excluded from the hydrophobic interface during an association process.

Although the absolute value of the Stokes radius calculated for the micelle was on the same order of magnitude as that of the submicelle, it was not as large as might expected, evidently because of experimental limitations. These limitation are due to the large size of the casein micelle (r = 65 nm), which would result in a τ_c value of nearly 200 μs. Such a slow motion would yield a spin-spin relaxation rate too large to be seen by these NMR experiments at 9.17 MHz. In one sense, what the data may show is the average hydration of the caseins within the micelle, since the fastest motions dominate relaxation data. The micelle exhibits the same temperature dependence as the submicelle, showing hydrophobic interactions, in agreement also with earlier investigators who theorize that micelles are formed by aggregation, via Ca^{2+} salt bridges, of submicelles.[73,77] The slight increase in r from submicelle to micelle is probably due to a gradual increase in internal hydration (trapped water) as the submicelle is incorporated into the micelle. In the course of micelle formation, the electrostatic forces involving Ca^{2+} and phosphate or carboxyl groups on the exterior of the submicelle are in competition with, and finally outweigh, the hydrophobic effects within the submicelle.

c. Derived Molecular Parameters of the Protein

Since it has been shown that the Stokes radius of the bound water derived from NMR relaxation results can be related to the anhydrous volume,[27,81] a molecular weight of the caseins can be calculated from

$$M_P = 4/3\ \pi\ r^3\ N_A/\bar{v}_P \tag{21}$$

where r is the Stokes radius (Table 6), N_A is Avogadro's number, and $\bar{v}_p$ is the average partial specific volume of the caseins, taken here to be 0.736. The results are shown in Table 6. Here the increase in M_p for both the submicelle and the micelle, as the temperature is increased, is a qualitative indication of hydrophobic self-association not only for the submicelle, but also within the micelle structure itself.

To quantitate this temperature-dependent variation of M_p, apparent equilibrium constants K_A were calculated from $K_A = M_p/23{,}300$, where 23,300 is the average monomer molecular weight of casein. (This relationship is reasonable since the measurements were performed at high

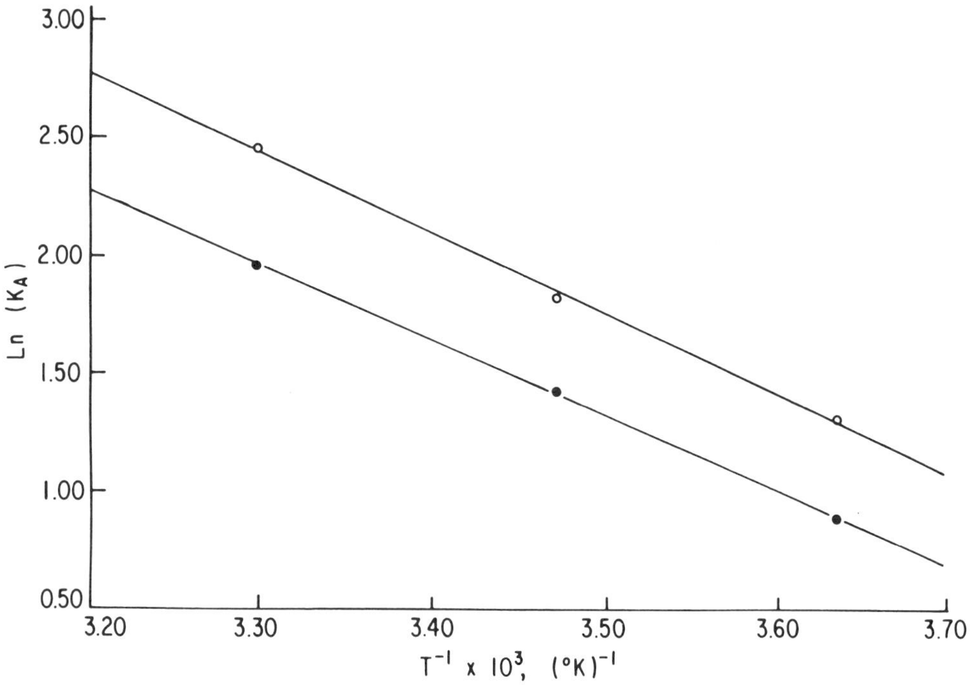

FIGURE 8. van't Hoff plots for temperature dependence of the self-association of casein. -●-●-, submicellar form and -○-○-, micellar form. (From Reference 78.)

concentrations of casein, where the equilibrium is driven nearly completely toward the aggregated form.) The ln K_A and the inverse temperature (1/T) were then used with the van't Hoff expression (see Figure 8) to calculate the apparent enthalpy of self-association, ΔH, for submicellar and micellar casein. As can be seen in Figure 8, the van't Hoff plots for the two forms of casein are essentially parallel. In fact, ΔH values for the submicelle formation were 6.34 ± 0.11 kcal, and for the self-association to the micelle only slightly higher at 6.81 ± 0.28 kcal. These values are in good agreement with ΔH of 4.67 found for the association of purified α_{s1}- and κ-caseins.[84] This quantitation of the temperature variation of the self-association strongly suggests that the integrity of the submicelle is at least partly preserved when it is incorporated into the micellar form by Ca-phosphate salt bridges. Moreover, extrapolation of the van't Hoff plot to 37°C yields an apparent molecular weight of 210,000 which is in agreement with results from other investigations.[77,82,85]

Finally, an average charge $\mathcal{Z}$ per average monomer molecular weight M_p = 23,300 can be evaluated from the virial coefficient B_0 of 0.0032 ml/mg using the following expression:[86]

$$2\,B_0 = \frac{\mathcal{Z}^2}{4m_s\,M_p} + \bar{v}_p/1000 \qquad (22)$$

where m_s is the molarity of salt used. The value of ±8.1, calculated from the NMR relaxation measurements, is not in good agreement with the average value of −16.1, which was derived from the amino acid sequence,[74] average pK values from the literature, and the assumption that at pH 6.75, where these experiments were performed, the serine phosphates have a charge of −2. However, Arakawa and Timasheff[87] have shown that another term must be added to Equation 9, to take into account the preferential interactions of salt and water at the protein interface, namely the quantity $-(\partial g_s/\partial g_p)^2/m_s$, where $(\partial g_s/\partial g_p)$ is the preferential binding term. If this is added to Equation 9, with $\mathcal{Z}$ chosen as –16.1, and B_0 = 0.0032 ml/mg, a value of 0.046 g salt per gram protein is found for the preferential binding term. For micellar casein this value is reasonable, if one translates Ca^{2+} binding of 0.043 g salt per gram protein into 8.5 mol Ca^{2+} per mole of 23,300-Da protein. This is consistent with the notion of Ca^{2+} binding to phosphate groups since, as stated above, the weight-average value for all the caseins is 6.6 phosphates per monomer.

d. Hydration: Anisotropic Tumbling Model

Up to this point, all hydration values were calculated using an isotropic motion mechanism of the bound water (S = 1). However, the motion of the bound water may in fact be anisotropic (S < 1) if the correlation times are long with respect to the Larmor frequency. Such may be the case for the casein micelles where water may be trapped at the surfaces of submicelles as they self-associate via calcium-phosphate salt bridges into micelles. In the following, an attempt, although speculative in nature, is made to calculate the asymmetry factor S for casein.

In Table 6, r values at all temperatures are somewhat larger for the micellar form of casein than for the submicellar form. This consistently larger Stokes radius (also represented by the molecular weight, M_p, in Table 6) could be due to increased hydration either by weakening of hydrophobic interactions within the submicellar form, caused by calcium phosphate salt bridge formation, or to trapped water at the surfaces of the submicelles as they are incorporated into the micelle. This increased hydration value, $(\bar{v}_w)_r$, may easily be calculated from the Stokes radii of the micelle, r_m, and submicelle, r_{sm}, by

$$(\bar{\nu}_w)_r = \bar{v}_p \left(\frac{4/3\ \pi\ r_m^3}{4/3\ \pi\ r_{sm}^3} - 1 \right) \tag{23}$$

where $\bar{v}_p$ = 0.736 for an average partial specific volume of the caseins. The results, shown in Table 6, range between 0.357 and 0.469 g water per gram protein at these temperatures. Asymmetry values, S, can now be calculated from $S = (\bar{v}_w/(\bar{v}_w)_r)^{1/2}$ and the $\bar{v}_w$ and $(\bar{v}_w)_r$ values from Tables 5 and 6, respectively. These S values are listed in Table 6 and average 0.237 ± 0.033, in good agreement with the value of 0.23 predicted by Walmsley and Shporer.[34] New hydrations $(\bar{v}_w)_{s=0.237}$ can now be calculated for an anisotropic motion mechanism using the $\bar{v}_w$ values of Table 5 and the average S of 0.237. These are listed in the last column of Table 6. The absolute values of these new hydrations, ranging from 0.116 to 0.502 g water per gram protein for submicellar and micellar casein, are closer to the expected hydration values derived by other methods.[27,81]

It should be stressed that, although the above calculation is not proof of the existence of water with anisotropic motion bound to casein, it does furnish valuable information. What is important here is the variation of hydration with quaternary structural changes of the casein rather than its absolute value. It may be that the absolute value of the hydration derived from NMR relaxation results will be obtained only some time in the future when the controversy regarding the isotropic vs. anisotropic nature of water binding to proteins in solution is resolved. At any rate, water bound at the surface of the micelles or influenced by the slow motion of the large particle would not be sensed at the frequencies used in these studies.

Here it has been found that the hydration value, no matter what absolute value is used, decreases and that r increases with increasing temperature for submicellar casein, in agreement with the general notion of hydrophobic protein self-association. Moreover, this same temperature-dependent variation of $\bar{v}_w$ and r exists for micellar casein, indicating preservation of the submicelle structure within the micelle. The consistent increase in r and $\bar{v}_w$ for the micellar form over the submicellar form clearly demonstrates that trapped water exists in the micellar form of casein, as had been predicted by Kuntz and Kauzmann.[1]

IV. CROSS RELAXATION

A. Influence of Environmental Changes on β-Lactoglobulins

Much controversy exists in the literature concerning the use of proton NMR relaxation rates for measuring the hydration of proteins. In fact, there has been little unanimity on this subject because of the complex nature of hydration phenomena and their relationships to NMR relaxation, and because of complicating factors such as cross relaxation between protein and water protons. Examination of water-protein interactions by measurement of proton spin relaxation of water in a well-defined system of a protein capable of undergoing structural changes should, however, afford opportunity to observe correlations between these changes and the measured relaxations.

A suitable protein for this purpose is β-lactoglobulin, which occurs in two genetic variants, A and B, possessing physical properties nearly identical except for the extents of specific structural changes.[50,51,58,88-91] Pessen et al.[8] have reported measurements of the longitudinal (R_1) and transverse (R_2) proton relaxation rates of water in buffered solutions as a function of protein concentration, with pH and temperature varied to allow examination of several of the protein structural states for concomitant differential behavior between the two variants. These data were evaluated first in terms of very simple model assumptions for the water-macromolecule interaction, with the aim of determining (1) whether significant changes in this interaction can be found for dilute solutions of this model protein as a result of environmental changes and (2) whether such changes can be shown to reflect the respective molecular states of the protein, as reported in the literature from studies by other methods under the same conditions. Next, the consequences of considering cross relaxation were examined. As a special alternative to such other methods as those using magnetic field dependence of NMR relaxation[24,92] or measurements on solvent ^{17}O and ^{2}H nuclei,[24,27,54] the combined use of proton NMR data from the two genetic variants was explored, making use of their differing association behavior. The results of cross relaxation evaluated in this way can be compared with the previous results from ^{2}H NMR relaxation as well as with other protein structural information obtained from hydrodynamics, X-ray diffraction and small-angle X-ray scattering.

1. Experimental Procedures

a. Preparation of Solutions

β-Lactoglobulins A and B (β-Lg A, β-Lg B) were the recrystallized lyophilized products, prepared from the milk of homozygous A/A and B/B cows by the method of Aschaffenburg and Drewry.[93] To exclude the possibility that observations might be affected by paramagnetic enhancement due to the presence of heavy-metal ions, some experiments were carried out both with and without prior treatment of the preparations with EDTA. In no case was a significant difference found. Distilled water from an all-glass still was used without further treatment. The buffers employed were pH 4.65, 0.1 *M* acetate (β-Lg B) and 0.3 *M* acetate (β-Lg A); pH 6.2, 0.1 *M* phosphate (β-Lg A) and 0.1 *M* acetate (β-Lg B); pH 8.0, 0.1 *M* glycylglycine plus 0.1 *M* KCl, adjusted to pH by addition of KOH. All buffer salts were the potassium salts of the respective acids and were Baker Analyzed reagents. Glycylglycine was purchased from Calbiochem, and benzene-d_6 and acetone-d_6 (99.5 atom %) were purchased from Wilmad Glass Co., Inc.

Protein solutions, prepared 1 d before use, were exhaustively dialyzed overnight against buffer at 0 to 5°C, except for the study at pH 8.0. In that case, all manipulations, including filling of the sample cells, were carried out at room temperature, and the low-temperature measurements were made last. All dilutions were made with the appropriate dialyzate. Protein concentrations were determined using an absorption coefficient of 0.96 ml·mg^{-1}·cm^{-1} at 278 nm.[37]

b. Relaxation Measurements

Measurements were made essentially as described under β-Lg A above, except that spectra were obtained from two different spectrometers (R_1 measurements, at 90 MHz, from a Bruker WH-90; R_2 measurements, at 59.79 MHz, from a JEOL FX60Q), at 2° and 30 ± 1°C. To avoid exceeding the dynamic range of the computer with consequent signal truncation, it was necessary to provide the Bruker WH-90 with a 20-db attenuator in the probe preamplifier, in addition to reducing the sample volume for both instruments by the use of a microcell assembly with an expendable 35-μl sample bulb (Wilmad Glass Co., Inc.). The assembly included this microbulb, filled with the sample and inserted into either a 10- or 5-mm OD sample tube containing deuterated solvent (benzene-d_6 or acetone-d_6) to provide an external heteronuclear lock signal.

2. Analysis of Data

a. Relaxation Increments and Free-Water Relaxation Rates

In the following, it will be convenient to supplement the notation of Equations 13 a to c by the introduction of a few modified quantities. Letting $h \equiv \bar{v}_w$, (in grams bound water per gram dry protein) and $\Delta R \equiv R_b - R_f$ (i.e., the difference in relaxation rates of protein-influenced and free states of water, or the total excess relaxation rate), it is noted from Equations 13a and 13b that the relaxation increment, $(dR_{obs}/dc)\mu = k$, becomes identical to the specific excess relaxation rate, $(R_{obs} - R_f)/c$, if the concentration dependence of R_{obs} is linear. Furthermore, the quantity h ΔR, termed the hydration product, equals these other two quantities only in the absence of cross relaxation; i.e., k is a hydration product uncorrected for cross relaxation effects. The appropriate correction is made in conjunction with the topic of cross relaxation below. (The R and k terms throughout may be subscripted with 1 or 2, depending on whether they pertain to longitudinal or transverse relaxation.)

Typical concentration dependences for R_1 and R_2 data are shown in Figures 9A and 9B, respectively, and are seen to be linear, in agreement with the original theory. Values of k_1 and R_{1f}, and of k_2 and R_{2f}, calculated by linear regression of the respective dependence of relaxation rates on protein concentration (Equation 13c), are listed in Table 7. As expected,[60] the relaxation rates of bulk water protons, R_{1f} and R_{2f}, increased at the lower temperature in each of these systems; the magnitude of the change did not differ significantly among the solvent buffers used. The relaxation increments k_1 and k_2 also changed in the same direction.

Values of k_1, k_2, R_{1f}, and R_{2f} so found were used, as described, in the calculation of apparent degrees of hydration h, uncorrected for cross relaxation (also listed in Table 7), as well as apparent rotational correlation times and longitudinal and transverse relaxation times for water in the bound state, R_{1b} and R_{2b} (not listed). In each case, R_{2b} increased while R_{1b} decreased at the lower temperature. This temperature effect is consistent with the absence of an exchange contribution, as is implicit in Equations 16a and b.[94]

b. Structural States and Hydration

The temperatures at which measurements were made had been chosen because of known properties of the protein. At both pH 4.65 and 8.0, it is known to undergo significant structural changes, mainly in the cold; at pH 6.2, no change in structure accompanies the same temperature change. On inspection of the temperature and pH dependence of the various parameters in Table 7, it is noticeable that not only h but also the hydration product k_2 appears to be suitable as at least a preliminary measure of hydration: allowing for the temperature effect (exhibited by itself at pH 6.2), k_2 and h on going from 30 to 2°C at pH 4.65 display increases by factors of 2.7 and 3.1 for the A variant and of 2.3 and 2.5 for the B variant, while at pH 8.0, the A variant shows increases by factors of 1.7 for k_2 and 2.0 for h. These indications of pronounced increases in apparent hydration in the cold at both pH 4.65 and 8.0 may be interpreted in terms of the known molecular behavior of β-Lg.

This protein has been reported to occur naturally in five genetic variants.[74] The A and B forms employed here undergo a variety of changes in conformation and state of association, summarized in Figure 10. The investigation described here focused on two of these: a slow, irreversible

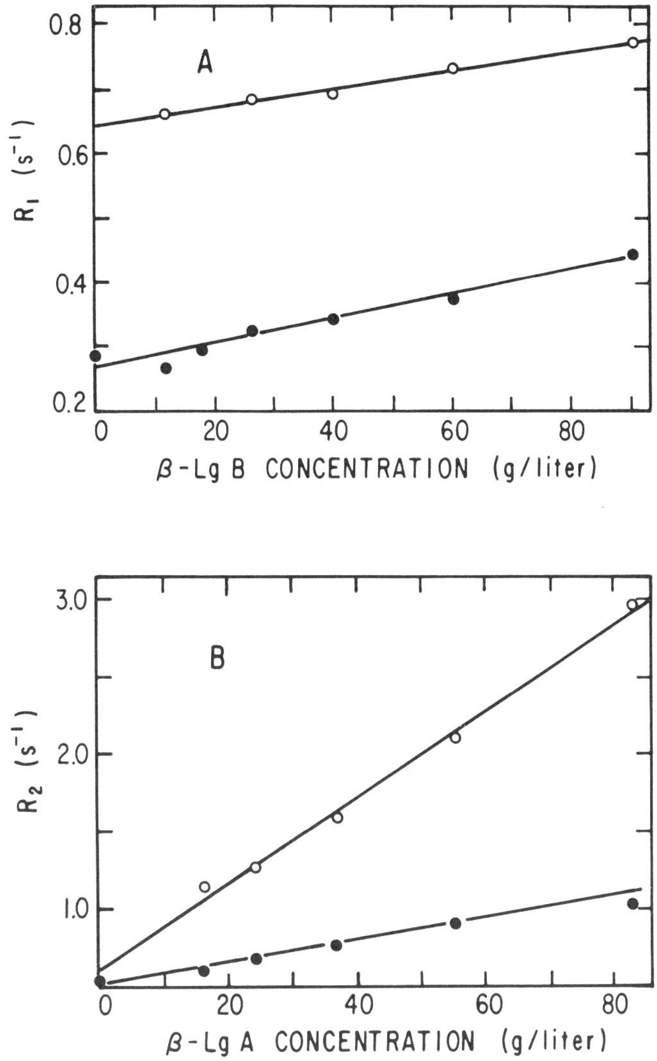

FIGURE 9. Proton relaxation rates of water in β-Lg solutions as a function of protein concentration. The slope (k) is the relaxation increment, and the intercept (R_f) is the relaxation rate of free water, as described in Section III. A.2. (A) Proton longitudinal relaxation rates (R_1) of water as function of β-Lg B concentration at pH 6.2. Temperatures: (-●-●-), 30°C ($k_1 = 1.94 \pm 0.18$, $R_{1f} = 0.271 \pm 0.009$) and (-○-○-), 2°C ($k_1 = 1.49 \pm 0.11$, $R_{1f} = 0.643 \pm 0.006$). (B) Proton transverse relaxation rates (R_2) of water as function of β-Lg A concentration at pH 8.0. Temperatures: (-●-●-), 30°C ($k_2 = 7.0 \pm 0.2$, $R_{2f} = 0.53 \pm 0.01$) and (-○-○-), 2°C ($k_2 = 27.9 \pm 1.3$, $R_{2f} = 0.61 \pm 0.06$). (From Reference 8.)

denaturation[91] and a rapid dimer ⇔octamer equilibrium,[50,51,58,89,90] both occurring primarily in the cold. The 2-subunit, 36,700-Da dimer is the kinetic unit persisting over a wide range of moderate conditions of pH from 3 to about 7;[89] its structure has been reported in detail.[95,96]

Above pH 6.5, the dimer is known to begin to dissociate[97-99] and then to denature irreversibly[98,100,101] while at first remaining in solution. This time-dependent denaturation, referred to as "cold-denaturation" since it is accelerated in the cold as compared to room temperature, occurs about three times faster with β-Lg A than with β-Lg B.[91] This correlates well with the observed indication of increased hydration of β-Lg A on exposure to the cold, since both the concurrent

Table 7
RELAXATION INCREMENTS (LONGITUDINAL, K_1, AND TRANSVERSE, K_2) AND SOLVENT RELAXATION RATES (R_{1f} AND R_{2f}) FOR SOLUTIONS OF TWO VARIANTS OF β-LACTOGLOBULIN[a]

			Relaxation increment ($s^{-1} \cdot g_{water} \cdot g^{-1}_{prot}$)		Solvent relaxation rate (s^{-1})		Uncorrected hydration ($g_{water} \cdot g^{-1}_{prot}$)
pH	Variant	Temp. (°C)	k_1	k_2	R_{1f}	R_{2f}	h
4.65	A	30	1.52 ± 0.25	8.9 ± 1.5	0.307 ± 0.011	0.23 ± 0.07	0.070 ± 0.012
		2	3.56 ± 0.23	55.3 ± 4.4	0.603 ± 0.010	0.70 ± 0.24	0.298 ± 0.021
4.65	B	30	1.26 ± 0.26	5.2 ± 0.4	0.314 ± 0.004	0.34 ± 0.02	0.047 ± 0.008
		2	2.18 ± 0.24	27.7 ± 2.6	0.644 ± 0.011	0.60 ± 0.16	0.163 ± 0.017
6.2	A	30	1.92 ± 0.26	4.5 ± 0.2	0.301 ± 0.007	0.31 ± 0.01	0.056 ± 0.005
		2	—	10.4 ± 1.2	—	0.89 ± 0.06	0.077 ± 0.007
6.2	B	30	1.94 ± 0.18	—	0.271 ± 0.009	—	
		2	1.49 ± 0.11	—	0.643 ± 0.006	—	
8.0	A	30	1.77 ± 0.13	7.0 ± 0.2	0.264 ± 0.005	0.53 ± 0.01	0.064 ± 0.004
		2	2.73 ± 0.51	27.9 ± 1.3	0.514 ± 0.015	0.61 ± 0.06	0.178 ± 0.023

[a] Longitudinal measurements (at 90.0 MHz) and transverse measurements (at 59.79 MHz), made at two temperatures and three values of pH, as indicated. Values of the parameters are the results of linear least-square fits to points at five to seven concentrations, each representing the mean of ±quadruplicate relaxation rate determinations based on measurements at five time values each. Error terms indicate standard errors of the parameter.

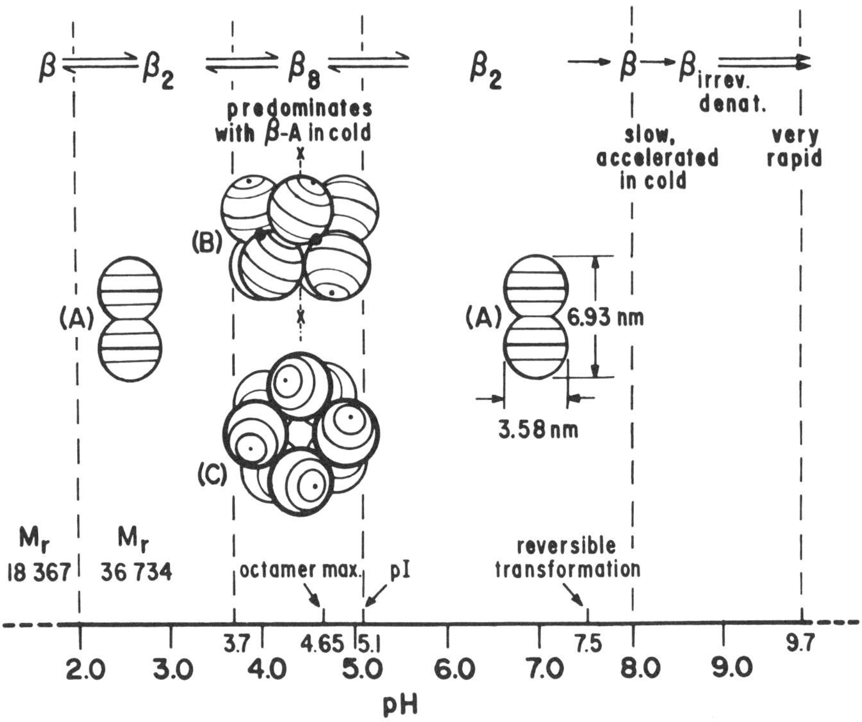

FIGURE 10. Schematic representation of changes in the structure of β-lactoglobulins as a function of pH. β:β-lactoglobulin. Insets (molecular models): (A), dimer; (B), octamer, with square decahedral faces on top and bottom; 0, octamer bonds; X-X, tetrad axis; circular lines indicate monomer equators and parallels perpendicular to the dimer axes; and (C), octamer, with square faces in front and back, tetrad axis perpendicular to plane of paper. (Adapted from Reference 8. Based on data of References 50, 51, 58, 88 to 91, 95 to 101, and 108.)

dissociation and subsequent denaturation presumably involve increased availability of water-binding sites on the denatured protein molecule.

Between pH 3.7 and 5.1, self-association of β-Lg dimer to octamer occurs as temperature decreases. This is a rapid equilibrium process that is also well characterized;[50] it occurs to a greater extent with the A variant than with the others. Since self-association involves protein-protein interactions which should remove some of the potential water-binding sites on the dimer surface, a decrease in hydration might be expected. The observed increased hydration requires a more detailed examination of the molecular geometry. The octamer, as determined by Timasheff and Townend,[58] is a closed ring, consisting of four dimers associated symmetrically about a tetrad axis; its general shape is that of a decahedron with a substantial central cavity (Figure 10, insets B and C). Because of a sizable hole (approximately 1.4 nm diameter) in each of the two square faces of the decahedron (Figure 10, inset C), the water in the interior of the octamer is readily accessible for fast exchange with the bulk water on the outside, as was assumed by the theory above.

An estimate of the amount of water contained in the cavity may be obtained from the geometric parameters,[58] which indicate that an inscribed sphere, tangent to the interior van der Waals surfaces of the octamer, would have a volume of about 6.45 nm^3. (This estimate neglects the spaces in corners and crevices of the cavity; these will be largely offset by the dimer-dimer contact areas between the eight monomer units, which tetramerization makes unavailable to water binding at the 12 new protein-protein contact sites.) From the molecular weight and volume of water and the molecular weight of the protein,[96] one finds that this cavity corresponds to 0.0264 g H_2O per gram protein. It can, therefore, accommodate a considerable amount of water detectable by NMR, giving rise to the increased hydration observed under the conditions where octamer formation occurs. It may now also be seen that the relative enhancements of the hydration parameters noted above (i.e., factors of 3.1 for h and 2.7 for k_2 for the A variant vs. 2.5 and 2.3, respectively, for the B variant) are consistent with the known higher degree of octamer formation by β-Lg A compared with B.[50]

There is, thus, qualitative agreement between the hydrations shown in Table 7 and the known structure of the corresponding states of the protein, but it remains to be examined how this agreement may be affected by a consideration of cross relaxation.

c. Dynamics and Hydration from Transverse Relaxation

Values of apparent correlation times calculated from Equations 5b, 6, and 7 range from 1.4 to 7.1 ns, considerably lower than corresponding estimates obtained for β-Lg from the Debye-Einstein relation[60] for dielectric relaxation (15 to 53 ns[102]) or fluorescence depolarization (20 to 78 ns[62]), or from NMR by means of a method which obviates the effects of cross relaxation (10 to 32 ns[27]). The lowered apparent correlation times may be considered a clear indication of cross relaxation,[53] not clearly evidenced here in the form of nonexponential relaxation only because of the vast excess of bulk water protons in dilute solutions. Instead, it shows its presence by its effect on the apparent τ_c and by increasing the apparent water proton relaxation rate, R_1,[53,103] and, thus, the values of k_1. These, as well as the validity of the uncorrected Equation 13c, require further examination, especially as there is definite evidence for the general occurrence of cross relaxation in protein systems.[36,53,54,103]

Since, however, cross relaxation affects primarily longitudinal relaxation,[36,54] one may start by dealing separately with the transverse relaxation, for which Equation 13c will continue to serve in conjunction now with the added information supplied by the availability of data for both genetic variants at pH 4.65. For this purpose, the extent of tetramerization for each variant at this pH at 2°C can be calculated from light-scattering data of Kumosinski and Timasheff[50] as 91.2% (β-Lg A) and 31.0% (β-Lg B).

This information allows the values of k_2 from Table 7 to be expressed in terms of the relative contributions of dimer and octamer and, together with the knowledge of the amount of water in the cavity and Equation 13c, provides two simultaneous equations for the two variants. In this way, a number of parameters of interest not affected by cross relaxation can be obtained (Appendix, a). These include the hydrations of dimer, $h_D = 0.0276$ g/g, and octamer, $h_O = 0.0540$ g/g, various correlation times, and the Stokes radius of the octamer, $R_{s,O} = 3.82$ nm. Repetition of this calculation for 30°C with the separately determined values of k_2 of Table 7 at that temperature, combined with the information[50] that the extents of tetramerization here are 26.9% (β-Lg A) and <0.02% (β-Lg B), gives $h_D = 0.0223$ g/g, $h_O = 0.0487$ g/g, and $R_{s,O} = 3.75$ nm. The values of $R_{s,O}$ from measurements at the two temperatures thus are in close agreement, as would be expected since the Stokes radius should be nearly temperature independent. They are in general agreement also with literature values of 4.33 nm, from sedimentation;[61] 4.44 nm, from small-angle X-ray scattering;[48,61] and 3.04 nm, from ^{2}H NMR.[27]

d. Cross Relaxation and Longitudinal Relaxation

Longitudinal relaxation data can be utilized in similar manner when Equation 13c is modified by decomposition into a corrected term, $k'_1 \equiv h'\Delta R$ (where h' is a hydration corrected for cross relaxation), and a term for a cross-relaxation increment, k_x, to account for the contribution of cross relaxation to k_1. The justification for such a procedure derives from a consideration of a cross-relaxation model which treats the water and protein protons as separate thermodynamic systems in magnetic interaction[54,74] (Appendix, b). This results in values at 2°C (see Appendix, c) for the k_x terms for octamer and dimer, respectively, of $k_{x,O} = 3.71$ and $k_{x,D} = 1.39$. Since spin diffusion, which is the basis for cross relaxation, should be dependent on distance from particle center of mass to surface (see Reference 54), the pertinent distance for the fairly isotropic octamer should be approximately its Stokes radius, 4.33 nm,[61] whereas for the dimer, consisting of two tangent, nearly spherical monomer subunits, it should be approximately the monomer radius, 1.79 nm.[58] The ratio of these distances is 2.42, while the ratio $k_{x,O}/k_{x,D} = 3.71/1.39$ is 2.67, and thus in good agreement with the value estimated from the known geometry. (The 30°C values, based on k_1 values and extents of tetramerization much smaller, and, therefore, containing larger relative errors, yield a ratio $k_{x,O}/k_{x,D} = 1.86$, which, however, is still of the correct order of magnitude.)

To examine the phenomena at pH 8.0, a first assumption was made that k_x at 30°C may be approximated by its value for the dimer at pH 4.65, at the opposite side of the isoelectric point of 5.1, since additional cross relaxation due to higher pH should be largely offset by decreased spin diffusion with the here more disordered protein. For pH 4.65, the modified Equation 5b, $k_1 = h_D\Delta R_1 + k_x$, with $h_D = 0.0223$ obtained from k_2, and with the k_1 of Table 7, gives $k_x = 1.088$. This value, substituted in the same equation applied to the pH 8.0 conditions, gives $h_D\Delta R_1 = 0.69$ and, from Equation 6, $\tau_c = 5.36$ ns, or a Stokes radius of 1.88 nm. Compared to the approximately 1.79 nm Stokes radius of the monomer,[58] this implies that at this alkaline pH, even at room temperature, a substantial fraction of the protein exists as monomer. This is in agreement with the findings of Townend et al.[97] and Georges et al.,[98] according to which the dimer at room temperature above about pH 7.0 begins to dissociate before alkaline denaturation takes place. Since this might weaken the above assumption regarding k_x, one may look, alternatively, to the pH 6.2, 2°C data for confirmation. Used as before, these yield a k_x of 1.43 which, applied to the pH 8.0, 2°C data, gives $\tau_c = 8.67$ ns, or a Stokes radius of 1.68 nm, and again the distinct indication is of a prevalence of monomer, in accord with the literature.

Cross-relaxation increments and hydrations calculated in this manner for the various forms of the protein under the different conditions, as well as the respective correlation times and Stokes radii, are listed in Table 8 (where h' designates values obtained by taking cross relaxation explicitly into account; see Appendix, b). Correlation times here range from 10.2 to 51.2 ns, more in keeping with the literature values cited above.[27,62,102] In view of the various approximations made and the high contribution of k_x to k_1, the values of h' listed must, however, be considered to be somewhat less accurate than their standard errors would indicate.

Table 8
PARAMETERS CALCULATED FOR β-LACTOGLOBULIN SOLUTIONS WITH CROSS RELAXATION TAKEN INTO ACCOUNT[a]

pH	Variant or species	Temp (°C)	k_x ($s^{-1} \cdot g_{water} \cdot g^{-1}_{prot}$)	h' ($g_{water} \cdot g^{-1}_{prot}$)	τ_c (ns)	R_{Stokes} (nm) This work	R_{Stokes} (nm) Literature
4.65	A	30	1.34 ± 0.25	0.029 ± 0.001	12.5 ± 1.5		
		2	3.50 ± 0.23	0.052 ± 0.009	51.2 ± 4.6		
	B	30	1.09 ± 0.26	0.022 ± 0.002	10.2 ± 0.1		
		2	2.11 ± 0.24	0.036 ± 0.008	33.0 ± 1.6		
	Pure dimer	30	(1.09 ± 0.26)[b]	(0.022 ± 0.002)[b]	10.2 ± 0.1		
		2	1.39 ± 0.32	0.028 ± 0.010	23.6 ± 0.1		
	Pure octamer	30		0.049 ± 0.002	18.6 ± 5.6 (42.2 ± 12.4)[b]	3.75	4.33,[f] 4.44,[g] 3.04[h]
		2	3.71 ± 0.25 2.03 ± 1.29[c]	0.054 ± 0.010	53.8 ± 5.0 (102.6 ± 38.4)[b]	3.82	
6.2	A or B	30	1.79 ± 0.18	0.019 ± 0.001	10.2 ± 0.1		
		2	3.52 ± 0.19	0.021 ± 0.002	23.6 ± 0.1		
8.0	A	30	1.09 ± 0.26	0.047 ± 0.009	5.4 ± 0.3	1.88[d]	1.79[i]
		2	2.08 ± 0.51	0.103 ± 0.007	12.4 ± 0.6	1.68[e]	

[a] Based on data of Table 7 and Equations 16 and 29. For statistics, see legend to Table 7.
[b] Pure octamer without cavity contribution.
[c] Parameters of B variant.
[d] From cross relaxation at pH 4.65.
[e] From cross relaxation at pH 6.2.
[f] From Reference 61.
[g] From Reference 61, as calculated from Reference 48.
[h] From Reference 27.
[i] From Reference 58.

The contributions of cross relaxation indicated by k_x (Table 8) comprise the major parts of the values of k_1 (Table 7): for either variant, under all three pH conditions, k_x at 30°C amounts to about 90% of k_1. This could be expected, inasmuch as solutions of proteins ranging in molecular weight from 30,000 to 100,000, under various conditions of concentration and temperature, have been found to exhibit zero-field cross relaxation rates already roughly equal to water proton relaxation rates.[54] At higher frequencies, the ratio of cross relaxation to total proton relaxation must increase further, since cross relaxation, particularly for proteins of molecular weight above 20,000 and at resonant frequencies near 100 MHz, increases substantially with frequency,[53,104] while the bound-water proton relaxation rate decreases (see Equation 16a). This expectation should hold for the present data relating to a 37,000-Da protein examined at 90 MHz. It is notable, however, that k_x at 2°C, which for pH 4.65 and 6.2 amounts to more than 97% of k_1, for pH 8.0 shows a marked drop to 76%, in agreement with the predicted loss of spin diffusion in the more disordered state under these denaturation conditions.

e. Significance of Parameters

In regard to the significance of the hydration parameter h, it is well recognized[1,29,94] that any technique used to measure hydration of proteins implies an operational definition of the water observed as "bound" that pertains to that particular technique, as well as to the model used to analyze the data thus obtained. The results described here pertain to rotationally as well as irrotationally bound water,[15] together with the effects of inter- as well as intramolecular contributions to relaxation. Cross relaxation effects have already been mentioned. Anisotropic surface effects also may well play a role, as dealt with by Walmsley and Shporer[34] and Halle et al.[7]

Using the approach of these authors, one can relate the hydration values shown in Table 8 to hydrodynamic hydration values in the literature by taking into account the effects of the anisotropic surface environment of the protein molecule. This involves the use of a scaling factor[34] or order parameter S,[7] which is nearly independent of the specific protein, and which enters Equations 16a and b so as to change the R_b value by a factor of S^2, and consequently h (related to R_b by Equation 13c) by very nearly $1/S^2$. From the data of Koenig et al.,[6] and the arguments of Walmsley and Shporer[34] as embodied in Figure 3 of these authors, one obtains an estimate of S as approximately 0.21. Applying this, as indicated, to the values of h' of Table 8 gives conventional hydrations from 0.43 to 0.50 for the pH 4.65 and 6.2 dimer at room temperature. This is squarely within the 0.30 to 0.54 range, clustering around 0.46, reported for β-Lg in the literature.[1]

Without the use of S, the h' values of Table 8, ranging from 0.019 for dimer at 30°C, pH 6.2, to 0.052 for the A variant at 2°, pH 4.65, may be directly compared with results from a method based on deuteron NMR to avoid the effects of cross relaxation, which gave corresponding values from 0.015 to 0.043 (Table 3 of Reference 27).

While the hydration values of Table 8 are smaller than the uncorrected ones of Table 7 (roughly half, or less), in either set the relationships between them under the various conditions are not greatly different. It is seen, therefore, that no matter in which way one prefers to interpret the data of Table 7, one can obtain definite correlations with known structural information, and that, contrary to doubts expressed in the literature,[24,54] relaxation measurements at the frequencies employed here contain considerable structural information relating to the solute protein. Furthermore, because one should be concerned mainly with changes rather than absolute values of h, other effects not specifically evaluated may largely cancel. Although absolute values obtained for a hydration parameter may be very dependent on model assumption, observed changes in such a parameter are less dependent on assumptions and can be equally useful in correlations with structural changes.

Returning to the questions posed at the outset, one sees that it is possible by a simple procedure to obtain useful hydration parameters which can account, at least in a qualitative fashion, for the effects of (1) cold-denaturaion of β-lactoglobulin A at alkaline pH and (2) of octamer formation of β-lactoglobulin at pH 4.65; (3) give a quantitative account of the effect on relaxation of the marked difference between variants A and B in octamer formation; (4) provide quantitative cross relaxation information; and (5) confirm such conjectures as that of Beall et al.[105] regarding correlation of water proton relaxation rate changes with changes in molecular states of a protein. It thus indicates that observed proton relaxation rate changes can be used by way of the parameter h or h' to evaluate concurrent structural changes in a given system.

3. Appendix

a. Parameters from Transverse Relaxation Increments and Solvent Relaxation Rates

Summing the contributions to k_2 of dimer and octamer for each variant according to Equation 13c, one has, from Table 7, the two simultaneous equations

$$k_{2,A} = 55.3 = 0.912\, h_O \Delta R_{2,O} + (1 - 0.912) h_D \Delta R_{2,D} \tag{24a}$$

and

$$k_{2,B} = 27.7 = 0.310\, h_O \Delta R_{2,O} + (1 - 0.310) h_D \Delta R_{2,D} \tag{24b}$$

where h_O, the hydration of the octamer, is that of the dimer, h_D, augmented by the contribution of the cavity i.e., $h_O = h_D + 0.0264$; and where $\Delta R_{2,O}$ and $\Delta R_{2,D}$ are the total excess transverse relaxation rates for octamer and dimer, respectively. $R_{2,D}$ can be obtained by way of an estimate of the correlation time of the dimer, τ_D, from the known dry volume of the latter, suitably corrected for its deviation from spherical shape.

The Einstein diffusion equation, expressing the rotary diffusion coefficient D_{rot} in terms of spherical volume V, provides a correlation time equal to $V\eta/kT$,[60] where k is Boltzmann's constant, 1.381×10^{-16} erg/°K, and η is the viscosity of solvent, approximated sufficiently by that of water, 0.01673 and 0.00801 P at temperatures T of 275.2 and 303.2°K, respectively. V equals vM/N_A, where v, the partial specific volume of β-Lg, is 0.751 ml/g,[106] M is $2 \times 18{,}370$, and N_A is Avogadro's number, 6.022×10^{23} mol^{-1}. The correlation times for spheres then would be 20.18 ns at 2°C and 8.77 ns at 30°C. Application of the shape factor 1.168 for the elongated β-Lg dimer[62] gives values for τ_D of 23.57 ns at 2°C and 10.24 ns at 30°C; from Equation 24, together with the values of R_{2f} from Table 7, $\Delta R_{2,D}$ becomes 488.6 s^{-1} at 2°C and 231.8 s^{-1} at 30°C.

Equations 8a and b can now be solved for the remaining unknowns, which are found to be $h_D = 0.0276$ and $\Delta R_{2,O} = 1096.8$. It follows that $h_O = 0.0276 + 0.264 = 0.0540$. From $\Delta R_{2,O}$ and R_{2f} at 2°C from Table 7, one obtains $R_{2b,O}$ and, by means of Equation 16b, the correlation time $\tau_{O,h}$ of the hydrated octamer, 53.82 ns. The latter may be regarded as the sum of weighted correlation times of dry octamer, τ_O, and the cavity water, $\tau_{cav} = V_{cav}\eta/kT$. With the values of V_{cav} and η above, τ_{cav} equals 2.839 at 2°C and 1.234 ns at 30°C. Thus, at 2°C, 53.82 = (0.0276/0.0540)τ_O + (0.0264/0.0540)(2.839), and τ_O = 102.6 ns. However, from $\tau_O = V\eta/kT = 4\pi R^3_{s,O}/3kT$, where $R_{s,O}$ is the Stokes radius of the octamer, follows $R_{s,O}$ = 3.82 nm.

b. Cross-Relaxation Increment

The cross-relaxation model results in a set of simultaneous differential equations for the coupled magnetization decays of water and protein protons in terms of: corresponding reduced magnetizations, $M_W(t)$ and $M_P(t)$, defined as $M(t) \equiv (A_\infty - A_t)/2A_\infty$ (see Equation 17); corresponding longitudinal relaxation rates in the absence of cross relaxation, R_W and R_P; the rate of magnetization transfer from water to protein protons, R_T; and the ratio of water to protein protons, n_W/n_P. Standard methods lead to a double-exponential solution of the form[23,36,54]

$$M_{W,P}(t) = C^{+}_{W,P}\exp(-R^{+}t) + C^{-}_{W,P}\exp(-R^{-}t) \tag{25}$$

where R^+ and R^- refer to two apparent relaxation rates, fast and slow respectively, which are the same for both kinds of proton and are given by

$$2R^{\pm} = R_P + R_W + R_T n_W/n_P + R_T$$
$$\pm [(R_P - R_W + R_T n_W/n_P - R_T)^2 + 4R_T^2 n_W/n_P]^{1/2} \tag{26}$$

Judging from the data of Koenig et al.,[54] where neglecting the $4R^2_T n_W/n_P$ term in Equation 26 in comparison with the terms preceding it resulted in less than 5% error, one may approximate the component rates of the double-exponential relaxation by

$$R^{+} \simeq R_P + R_T n_W/n_P \quad \text{and} \quad R^{-} \simeq R_W = R_T \tag{26a}$$

In dilute solutions, $n_W >> n_P$, and R^+ becomes so large that the first right-hand term of Equation 25 is small compared to the second term. (This, together with a pulse-width dependence, was the reason why the double exponential was not detectable in graphs of the original data.) The observed relaxation rate then becomes $R \cong R_W + R_T$ and, since the observed proton longitudinal relaxation can be expressed as the sum of contributions from the two states as well as from cross relaxation,

$$R_{obs} \simeq p_b R_b + p_f R_f + R_T \tag{27}$$

Here p_b and p_f are the fractions of protons in the respective state, so that $p_b = hc$, and $p_f = 1 - p_b = 1 - hc$.

Since for dilute solutions R_T should be proportional to the protein concentration c, or $R_T\ k_x c$, Equations 13b and c can be rewritten as

$$R_{obs} = R_f + [(h'\ \Delta R) + k_x]c = R_f + (k_1' + k_x)c \tag{28}$$

where

$$k_1' = h'\ \Delta R, \qquad k_1 = k_1' + k_x \tag{28a}$$

and the observed relaxation increment now equals the sum of the hydration product and the cross relaxation increment. The term k_x, it should be noted, is an approximate representation of R_T, reflecting in addition the combined effects of the simplifying assumptions made above.

c. Parameters from Longitudinal Relaxation Increments and Solvent Relaxation Rates

In analogy to Equations 24a and b, one has

$$k_{1,A} = 3.56 = (0.912)(0.0540)\Delta R_{1,O} + (1 - 0.912)(0.0276)\Delta R_{1,D} + k_{x,A} \tag{29a}$$

and

$$k_{1,B} = 2.18 = (0.310)(0.0540)\Delta R_{1,O} + (1 - 0.310)(0.0276)\Delta R_{1,D} + k_{x,B} \tag{29b}$$

Here, $\Delta R_{1,O}$ and $\Delta R_{1,D}$ are found from Equation 13a with $\tau_{O,h} = 53.82$ and $\tau_D = 23.57$ ns, together with the appropriate R_{1f} from Table 7, as 0.969 and 2.936/s^{-1}, respectively. With these values, $k_{x,A} = 3.50$ and $k_{x,B} = 2.11$. Considering each of these, in turn, to be made up of contributions from octamer, $k_{x,O'}$, and dimer, $k_{x,D'}$, one has the two equations

$$k_{x,a} = (0.912)k_{x,O} + (1.088)k_{x,D} \tag{30a}$$

and

$$k_{x,b} = (0.310)k_{x,O} + (0.690)k_{x,D} \tag{30b}$$

which yield $k_{x,O} = 3.71$ and $k_{x,D} = 1.39$.

B. Overview

To relate resonance relaxation behavior to protein structural states, pulse Fourier transform NMR was employed to obtain water proton longitudinal and transverse relaxation rates (R_1 and $R_{1\rho} = R_2$) of bovine β-Lgs A and B in buffered solutions. Measurements at concentrations from 5 to 100 mg/ml were made at pH 4.65, 6.2, and 8.0, at 30° and 2°C, to monitor specific structural changes. The parameters characterizing the concentration dependence of the observed R_1 and R_2 were used to derive a number of quantities relating to protein-influenced water, including a hydration parameter h. Changes in h under the different sets of conditions were correlated with (1) the irreversible denaturation of this protein at pH 8.0, 2°C and (2) the dimer ⇔octamer association at pH 4.65, 2°C. Corresponding correlation times, however, were low, indicating cross relaxation which had not manifested itself as nonexponential relaxation because of the large amount of water present. Differences in the extent of the dimer octamer association between genetic variants A and B allowed an evaluation of dynamics and extent of hydration from R_2 alone, assuming the absence of intermolecular interactions. Derived parameters were in agreement with hydrodynamic and X-ray values in the literature. Cross relaxation was likewise evaluated and was found to contribute to R_1 to a large extent. The results show that changes in proton relaxation rates in solutions of a globular protein occurring as genetic variants with different physical properties (such as β-Lg) can be utilized to detect variations in hydration corresponding to changes in molecular association and conformation, as well as to obtain cross relaxation and structural data.

Therefore, the conclusions of Edzes and Samulski that the protons of the bound water cross-relax with the protein protons (i.e., methyl proton group magnetic sinks), and that this effect is exhibited in the spin-lattice rather than the spin-spin relaxation data, appears to be correct. However, although these findings are in relatively good agreement with those from the previously presented ^2H NMR relaxation, it must be emphasized that the use of proton NMR complicates an already cumbersome problem. The appearance of a new cross relaxation term in the spin-relaxation data necessitates either additional NMR experiments or a creative use of different genetic variants of the same protein, as in the cited study.[8] Clearly, the use of ^2H NMR or ^{17}O NMR relaxation experiments for proteins dissolved in D_2O, whereby cross relaxation is virtually eliminated in the T_1 process because of the large difference in the magnetic moments of these nuclei with protons, is much easier and more prone to success.

V. CONCLUDING REMARKS

The foregoing considerations add up to considerable agreement between certain theoretically and experimentally derived quantities. However, none of the above arguments should be interpreted as proof of any particular NMR mechanism or model, nor of the identity of the particular groups on the protein surface that interact with water. Even without such conclusions, and in place of the quest for absolute values of hydration, it can be useful to scrutinize relative changes in hydration when these can be taken as functions of changes in secondary, tertiary, or quaternary structure of a protein.

However, it must be stressed that the use of frequency-dependent NMR relaxation instruments in lieu of protein concentration-dependent relaxation measurements at a minimum of two frequencies can yield erroneous results. High virial effects can indeed be present in protein solutions, especially if salt is not added to the system in order to minimize the protein-protein electrostatic potential. However, by use of nonlinear regression analysis, the virial coefficient of the protein, which is related to its net charge, can also be extracted from these protein concentration-dependent relaxation data.

Furthermore, although the NMR relaxation mechanism controversy still exists between the isotopic binding model and the intermediate asymmetry model, it can be concluded from all the above results that the "hydrodynamically influenced water" model of Koenig et al. can finally be placed to rest. Only a model whereby water directly binds to proteins can account for the water relaxation rate experiencing the charge of the macromolecule in protein solutions and for the protein structural results presented in this chapter. In addition, the model of water moving from site to site on the asymmetric surface of a macromolecule with a fast-motion component equal to the diffusion coefficient of unbound water, is also unlikely in protein solution. With the abundance of free water (i.e., 55.6 mol/l of water) relative to the protein concentration (approximately 10^{-3} mol/l), such a phenomenon would be thermodynamically unsound. On the other hand, in experiments involving protein powders and other two-phase biological systems, where only a limited amount of water is added to a solid or amorphous material and where surface adsorption of water to an insoluble second phase can exist, the above mechanism of asymmetric motion along the macromolecular surface most likely does predominate.

Therefore, in a true protein solution, only the isotropic two- or three-state binding mechanism or the water binding with a fast motional component to a macromolecule with a slow anisotropic motion are the most reasonable models of choice at this time.

One point with the above results should be noted here. The use of the ^{2}H NMR experiments which, of course, yield a sharp and well-defined water peak, can yield an extra relaxation resulting from fast-exchangeable protein protons, such as those arising from arginine, lysine, and even aspartic and glutamic acid side chains. However, in most cases, a calculation from the amino acid sequence of the protein in question can show the percentage contribution of this term to the results, and in most cases it will be small. Nevertheless, in the above studies, this effect would still not account for the change in hydration with temperature as observed for casein self-association, the β-lactoglobulin tetramerization, and the gelation of β-lactoglobulin at high pH. However, investigators may also perform the ^{2}H NMR relaxation experiments using ^{17}O NMR relaxation of the samples of protein dissolved in D_2O, if they are still plagued by the magnitude of this contribution to the hydration term.

Finally, the above absolute hydration results using a two- or three-state isotropic binding model are much lower than hydration values determined by small-angle X-ray scattering and sedimentation velocity experiments. It appears that only an asymmetric binding model could bring the order of magnitude in these hydration results in line with SAXS and hydrodynamic values. However, SAXS and hydrodynamic results may also contain another contribution not considered by some investigators. In fact, there may also be a dynamic contribution to the hydration, since it is measured by the difference between two volumes. The concept of protein "breathing" has been emphasized before,[71] and the entire topic of protein dynamics has been reviewed recently.[3,107] The effects of dynamic changes such as fluctuations (e.g., ring flipping and domain hinge bending) on packing volumes and accessible surface areas remain unclear. Progress on these questions may come from dynamic modeling by computer simulation.

ACKNOWLEDGMENT

The authors acknowledge a debt of gratitude to Norma C. Pessen for undertaking the demanding task of collating and checking the references, and for proofreading and correcting the successive versions of the entire manuscript.

REFERENCES

1. **Kuntz, I. D. and Kauzmann, W.,** Hydration of proteins and polypeptides, *Adv. Protein Chem.*, 28, 239, 1974.
2. **Kumosinski, T. F. and Pessen, H.,** Estimation of sedimentation coefficients of globular proteins: an application of small-angle X-ray scattering, *Arch. Biochem. Biophys.*, 219, 89, 1982.
3. **McCammon, J. A. and Karplus, M.,** The dynamic picture of protein structure, *Acc. Chem. Res.*, 16, 187, 1983.
4. **Kumosinski, T. F. and Pessen, H.,** Structure and mechanism of action of riboflavin-binding protein: small-angle X-ray scattering, sedimentation, and circular dichroism studies on the holo- and apoproteins, *Arch. Biochem. Biophys.*, 214, 714, 1982.
5. **Koenig, S. H. and Schillinger, W. E.,** Nuclear magnetic relaxation dispersion in protein solutions, *J. Biol. Chem.*, 244, 3283, 1968.
6. **Koenig, S. H., Hallenga, K.,** and **Shporer, M.,** Protein-water interaction studied by solvent ^{1}H, ^{2}H, and ^{17}O magnetic relaxation, *Proc. Natl. Acad. Sci. U.S.A.*, 72, 2667, 1975.
7. **Halle, B., Andersson, T., Forsén, S., and Lindman, B.,** Protein hydration from water oxygen-17 magnetic relaxation, *J. Am. Chem. Soc.*, 103, 500, 1981.
8. **Pessen, H., Purcell, J. M., and Farrell, H. M., Jr.,** Proton relaxation rates of water in dilute solutions of β-lactoglobulin. Determination of cross relaxation and correlation with structural changes by the use of two genetic variants of a self-associating globular protein, *Biochim. Biophys. Acta*, 828, 1, 1964.
9. **Tanford, C.,** *Physical Chemistry of Macromolecules*, John Wiley & Sons, New York, 1961, 227, 293, 352, 563.
10. **Schachman, H. K.,** *Ultracentrifugation in Biochemistry*, Academic Press, New York, 1959, 226.
11. **Pedersen, K. O.,** Detailed theory, in *The Ultracentrifuge*, Svedberg, T. and Pedersen, K. O., Eds., Oxford University Press, London, 1940, 16.
12. **Tiselius, A.,** Über den Einfluss der Ladung auf die Sedimentations geschwindigkeit von Kolloiden, besonders in der Ultrazentrifuge, *Kolloid-Z.*, 59, 306, 1932.
13. **Booth, F.,** Sedimentation potential and velocity of solid spherical particles, *J. Chem. Phys.*, 22, 1956, 1954.
14. **Pedersen, K. O.,** On charge and specific ion effects on sedimentation in the ultracentrifuge, *J. Phys. Chem.*, 62, 1282, 1958.
15. **Timasheff, S. N. and Kronman, M. J.,** The extrapolation of light-scattering data to zero concentration, *Arch. Biochem. Biophys.*, 83, 60, 1959.
16. **Kronman, M. J. and Timasheff, S. N.,** Light scattering investigation of ordering effects in silicotungstic acid solutions, *J. Phys. Chem.*, 63, 629, 1959.
17. **Timasheff, S. N.,** The application of light scattering and small-angle X-ray scattering to interacting biological systems, in *Electromagnetic Scattering-I.C.E.S.*, Kerker, M., Ed., Pergamon Press/Macmillan, New York, 1963, 337.
18. **Kirkwood, J. G. and Shumaker, J. B.,** Forces between protein molecules in solution arising from fluctuations in proton charge and configuration, *Proc. Natl. Acad. Sci. U.S.A.*, 38, 863, 1952.
19. **Kirkwood, J. G. and Timasheff, S. N.,** The effect of ionization on the light scattering of isoionic proteins, *Arch. Biochem. Biophys.*, 65, 50, 1956.
20. **Timasheff, S. N., Dintzis, H. M., Kirkwood, J. G., and Coleman, B. D.,** Light scattering investigation of charge fluctuations in isoionic serum albumin solutions, *J. Am. Chem. Soc.*, 79, 782, 1957.
21. **Timasheff, S. N. and Tinoco, I.,** Light scattering of isoionic conalbumin, *Arch. Biochem. Biophys.*, 66, 427, 1957.
22. **Timasheff, S. N. and Coleman, B. D.,** On light-scattering studies of isoionic proteins, *Arch. Biochem. Biophys.*, 63, 87, 1960.
23. **Zimmerman, J. R. and Brittin, W. E.,** Nuclear magnetic resonance studies in multiple phase systems: lifetime of a water molecule in an adsorbing phase on silica gel, *J. Phys. Chem.*, 1328, 1957.
24. **Hallenga, K. and Koenig, S. H.,** Protein rotational relaxation as studied by solvent ^{1}H and ^{2}H magnetic relaxation, *Biochemistry*, 15, 4255, 1976.
25. **Lindstrom, T. R. and Koenig, S. H.,** Magnetic-field-dependent water proton spin-lattice relaxation rates of hemoglobin solutions and whole blood, *J. Magn. Reson.*, 15, 344, 1974.
26. **Andree, P. J.,** The effect of cross relaxation on the longitudinal relaxation times of small ligands binding to macromolecules, *J. Magn. Reson.*, 29, 419, 1978.
27. **Kumosinski, T. F. and Pessen, H.,** A deuteron and proton magnetic resonance relaxation study of β-lactoglobulin A association: some approaches to the Scatchard hydration of globular proteins, *Arch. Biochem. Biophys.*, 218, 286, 1982.
28. **Casassa, E. F. and Eisenberg, H.,** Thermodynamic analysis of multicomponent solutions, *Adv. Protein Chem.*, 19, 287, 1964.
29. **Na, G. C. and Timasheff, S. N.,** Interaction of calf brain tubulin with glycerol, *J. Mol. Biol.*, 151, 165, 1981.
30. **Tanford, C.,** Extension of the theory of linked functions to incorporate the effects of protein hydration, *J. Mol. Biol.*, 39, 539, 1969.

31. **Kubo, R. and Tomita, K.,** A general theory of magnetic resonance absorption, *J. Phys. Soc. Jpn.*, 9, 888, 1954.
32. **Solomon, I.,** Relaxation processes in a system of two spins, *Physiol. Rev.*, 99, 559, 1955.
33. **Abragam, A.,** *The Principles of Nuclear Magnetism,* Oxford University Press (Clarendon), London, 1961, chap. 8.
34. **Walmsley, R. H. and Shporer, M.,** Surface-induced NMR line splittings and augmented relaxation rates in water, *J. Chem. Phys.*, 68, 2584, 1978.
35. **Berendsen, H. J. C. and Edzes, H. T.,** The observation and general interpretation of sodium magnetic resonance in biological material, *Annu. N.Y. Acad. Sci.*, 204, 459, 1973.
36. **Edzes, H. T. and Samulski, E. T.,** The measurement of cross relaxation effects in the proton NMR spin-lattice relaxation of water in biological systems: hydrated collagen and muscle, *J. Magn. Reson.*, 31, 207, 1978.
37. **Townend, R., Winterbottom, R. J., and Timasheff, S. N.,** Molecular interactions in β-lactoglobulin. II. Ultracentrifugal and electrophoretic studies of the association of β-lactoglobulin below its isoelectric point, *J. Am. Chem. Soc.*, 82, 3161, 1960.
38. **Vold, R. L., Waugh, J. S., Klein, M. P., and Phelps, D. E.,** Measurement of spin relaxation in complex systems, *J. Chem. Phys.*, 48, 3831, 1968.
39. **Bloch, F.,** Nuclear induction, *Physiol. Rev.*, 70, 460, 1946.
40. **Gerhards, R. and Dietrich, W.,** The analysis of spin-lattice relaxation time experiments, *J. Magn. Reson.*, 23, 21, 1976.
41. **Sass, M. and Ziessow, D.,** Error analysis for optimized inversion recovery spin-lattice relaxation measurements, *J. Magn. Reson.*, 25, 263, 1977. .
42. **Kowalewski, J., Levy, G. C., Johnson, L. F., and Palmer, L.,** A three-parameter non-linear procedure for fitting inversion-recovery measurements of spin-lattice relaxation times, *J. Magn. Reson.*, 26, 533, 1977.
43. **Brunetti, A.,** Computation of T_1 from pulsed NMR data using a generalized and rapid least-squares formula, *J. Magn. Reson.*, 28, 289, 1977.
44. **Leipert, T. K. and Marquardt, D. W.,** Statistical analysis of NMR spin-lattice relaxation times, *J. Magn. Reson.*, 24, 181, 1976.
45. **Becker, E. D., Ferretti, G. A., Gupta, R. K., and Weiss, G. H.,** The choice of optimal parameters for measurement of spin-lattice relaxation times. II. Comparison of saturation recovery, inversion recovery, and fast inversion recovery experiments, *J. Magn. Reson.*, 37, 381, 1980.
46. **Farrar, T. C. and Becker, E. D.,** *Pulse and Fourier Transform NMR: Introduction to Theory and Methods,* Academic Press, New York, 1971, 92.
47. **Timasheff, S. N. and Townend, R.,** Molecular interactions in β-lactoglobulin. VI. The dissociation of the genetic species of β-lactoglobulin at acid pH's, *J. Am. Chem. Soc.*, 83, 470, 1961.
48. **Witz, J., Timasheff, S. N., and Luzzati, V.,** Molecular interactions in β-lactoglobulin. VIII. Small-angle X-ray scattering investigation of the geometry of β-lactoglobulin A tetramerization, *J. Am. Chem. Soc.*, 86, 168, 1964.
49. **Townend, R. and Timasheff, S. N.,** Molecular interactions in β-lactoglobulin. III. Light scattering investigation of the stoichiometry of the association between pH 3.7 and 5.2, *J. Am. Chem. Soc.*, 82, 3168, 1960.
50. **Kumosinski, T. F. and Timasheff, S. N.,** Molecular interactions in β-lactoglobulin. X. The stoichiometry of the β-lactoglobulin mixed tetramerization, *J. Am. Chem. Soc.*, 88, 5635, 1966.
51. **Timasheff, S. N. and Townend, R.,** Molecular interactions in β-lactoglobulin. V. The association of the genetic species of β-lactoglobulin below the isoelectric point, *J. Am. Chem. Soc.*, 83, 464, 1961.
52. **Timasheff, S. N. and Townend, R.,** β-lactoglobulin as a model of subunit enzymes, *Protides Biol. Fluids,* 16, 33, 1969.
53. **Kalk, A. and Berendsen, H. J. C.,** Proton magnetic relaxation and spin diffusion in proteins, *J. Magn. Reson.*, 24, 343, 1976.
54. **Koenig, S. H., Bryant, R. G., Hallenga, K., and Jacob, G. S.,** Magnetic cross relaxation among protons in protein solutions, *Biochemistry,* 17, 4348, 1978.
55. **Kumosinski, T. F., Pessen, H., and Purcell, J. M.,** Use of proton NMR relaxation rates in the determination of the thermodynamics of primary hydration in an associating protein system (β-lactoglobulin), *Fed. Proc. Fed. Am. Soc. Exp. Biol.*, 39, 1673, 1980.
56. **Timasheff, S. N. and Townend, R.,** Molecular interactions in β-lactoglobulin. I. The electrophoretic heterogeneity of β-lactoglobulin close to its isoelectric point, *J. Am. Chem. Soc.*, 82, 3157, 1960.
57. **Waldstein, P., Rabideau, S. W., and Jackson, J. A.,** Nuclear magnetic resonance of single crystals of D_2O ice, *J. Chem. Phys.*, 41, 3407, 1964.
58. **Timasheff,. S. N. and Townend, R.,** Structure of the β-lactoglobulin tetramer, *Nature,* 203, 517, 1964.
59. **Hunt, M. J. and Mackay, A. L.,** Deuterium and nitrogen pure quadrupole resonance in deuterated amino acids, *J. Magn. Reson.*, 15, 402, 1974.
60. **Poole, C. P. and Farach, H. A.,** *Relaxation in Magnetic Resonance,* Academic Press, New York, 1971, 65.
61. **Pessen, H., Kumosinski, T. F., and Timasheff, S. N.,** Small-angle X-ray scattering, *Methods Enzymol.*, 27, 151, 1973.

62. **Wahl, P. and Timasheff, S. N.,** Polarized fluorescence decay curves for β-lactoglobulin A in various states of association, *Biochemistry*, 8, 2945, 1969.
63. **Ceccarelli, C., Jeffrey, G. A., and Taylor, R.,** A survey of O-H...O hydrogen bond geometries determined by neutron diffraction, *J. Mol. Struct.*, 70, 255, 1981.
64. **Wyman, J., Jr.,** Linked functions and reciprocal effects in hemoglobin: a second look, *Adv. Protein Chem.*, 19, 223, 1964.
65. **Gilbert, G. A.,** General discussion on the physical chemistry of enzymes. Characterization and physical properties, *Discuss. Faraday Soc.*, 20, 68, 1955.
66. **Gilbert, G. A.,** Sedimentation and electrophoresis of interacting substances. I. Idealized boundary shape for a single substance aggregating reversibly, *Proc. R. Soc. London, Ser. A*, 250, 377, 1959.
67. **Cooke, R. and Kuntz, I. D.,** The properties of water in biological systems, *Annu. Rev. Biophys. Bioeng.*, 3, 95, 1974.
68. **Hagler, A. T. and Moult, J.,** Computer simulation of the solvent structure around biological macromolecules, *Nature*, 272, 222, 1978.
69. **Teller, D., Swanson, E., and De Haën, C.,** The translational friction coefficient of proteins, *Methods Enzymol.*, 61, 103, 1979.
70. **Whitney, R. McL., Brunner, J. R., Ebner, K. E., Farrell, H. M., Jr., Josephson, R. V., Morr, C. V., and Swaisgood, H. E.,** Nomenclature of the proteins of cow's milk: fourth revision, *J. Dairy Sci.*, 59, 785, 1976.
71. **Brown, E. M. and Pfeffer, P. E.,** Relative exposure of lysine groups in β-lactoglobulin, *Fed. Proc. Fed. Am. Soc. Exp. Biol.*, 40, 1615, 1981.
72. **Schoenborn, B. P. and Hanson, J. C.,** The determination of structural water by neutron protein crystallography: an analysis of the carbon monoxide myoglobin water structure, in *Water in Polymers*, Rowland, S. P., Ed., American Chemical Society, Washington, D.C., 1980, 217.
73. **Farrell, H. M., Jr. and Thompson, M. P.,** The caseins as calcium binding proteins, in *Calcium Binding Proteins*, Thompson, M. P., Ed., CRC Press, Boca Raton, FL, 1986.
74. **Eigel, W. N., Butler, J. E., Ernstrom, C. A., Farrell, H. M., Jr., Harwalkar, V. R., Jenness, R., and Whitney, R. McL.,** Nomenclature of proteins of cow's milk: fifth revision, *J. Dairy Sci.*, 67, 1599, 1984.
75. **Slattery, C. W. and Evard, R.,** A model for the formation and structure of casein micelles from subunits of variable composition, *Biochim. Biophys. Acta*, 313, 529, 1973.
76. **Pepper, L. and Farrell, H. M., Jr.,** Interactions leading to formation of casein submicelles, *J. Dairy Sci.*, 65, 2259, 1982.
77. **Schmidt, D. G.,** Association of caseins and casein micelle structure in *Developments in Dairy Chemistry*, Vol. 1, Fox, P. E., Ed., Applied Science, Essex, England, 1982.
78. **Kumosinski, T. F., Pessen, H., Prestrelski, S. J., and Farrell, H. M., Jr.,** Water interactions with varying molecular states of bovine casein: ^{2}H NMR relaxation studies, *Arch. Biochem. Biophys.*, 257, 259, 1987.
79. **Davies, D. T. and Law, A. J. R.,** Content and composition of protein in creamery milks in South-West Scotland, *J. Dairy Res.*, 47, 83, 1980.
80. **Carroll, R. J. and Farrell, H. M., Jr.,** Kappa-casein: an immunological approach to its location in the casein micelle using electron microscopy, *J. Dairy Sci.*, 66, 679, 1983.
81. **Pessen, H. and Kumosinski, T. F.,** Measurements of protein hydration by various techniques, *Methods Enzymol.*, 117, 219, 1985.
82. **Capalja, G. G.,** An electron microscope study of the ultrastructure of bovine and human casein micelles in fresh and acidified milk, *J. Dairy Res.*, 35, 1, 1968.
83. **Schultz, B. C. and Bloomfield, V. A.,** Structure of bovine casein oligomers, *Arch. Biochem. Biophys.*, 173, 18, 1976.
84. **Clarke, R. and Nakai, S.,** Investigation of κ-α_{s1}-casein interaction by fluorescence polarization, *Biochemistry*, 10, 3353, 1971.
85. **Pepper, L.,** Casein interactions as studied by gel chromatography and ultracentrifugation, *Biochim. Biophys. Acta*, 278, 147, 1972.
86. **Tanford, C.,** *Physical Chemistry of Macromolecules*, John Wiley & Sons, New York, 1961, 236.
87. **Arakawa, T. and Timasheff, S. N.,** Preferential interactions of proteins with salts in concentrated solutions, *Biochemistry*, 21, 6545, 1982.
88. **Timasheff, S. N. and Townend, R.,** Structural and genetic implications of the physical and chemical differences between β-lactoglobulins A and B, *J. Dairy Sci.*, 45, 259, 1962.
89. **Timasheff, S. N.,** The nature of interactions in proteins derived from milk, in *Symposium on Foods: Proteins and Their Reactions*, Schultz, H. W. and Angelmeier, A. F., Eds., AVI Publishing, Westport, CT, 1964, 174.
90. **Townend, R.,** β-lactoglobulins A and B: the environment of the Asp/Gly difference residue, *Arch. Biochem. Biophys.*, 109, 1, 1965.
91. **Waissbluth, M. D. and Grieger, R. A.,** Alkaline denaturation of β-lactoglobulins. Activation parameters and effect on dye binding site, *Biochemistry*, 13, 1285, 1974.

92. **Grösch, L. and Noack, F.,** NMR relaxation investigation of water mobility in aqueous bovine serum albumin solutions, *Biochim. Biophys. Acta,* 453, 218, 1976.
93. **Aschaffenburg, R. and Drewry, J.,** Occurrence of different β-lactoglobulins in cow's milk, *Nature,* 176, 218, 1955.
94. **Cooke, R. and Kuntz, I. D.,** The properties of water in biological systems, *Annu. Rev. Biophys. Bioeng.,* 3, 95, 1974.
95. **Green, D. W., Aschaffenburg, R., Camerman, A., Coppola, J. C., Dunnill, P., Simmons, R. M., Komorowski, E. S., Sawyer, L., Turner, E. M. C., and Woods, K. F.,** Structure of bovine β-lactoglobulin at 6Å resolution, *Mol. Biol.,* 131, 375, 1979
96. **Dayhoff, M. O., Ed.,** *Atlas of Protein Sequence and Structure,* Vol. 5, Suppl. 1, National Biomedical Research Foundation, Washington, D.C., 1973, S-83.
97. **Townend, R., Weinberger, L., and Timasheff, S. N.,** Molecular interactions in β-lactoglobulin. IV. The dissociation of β-lactoglobulin below pH 3.5, *J. Am. Chem. Soc.,* 82, 3175, 1960.
98. **Georges, C., Guinand, S., and Tonnelat, J.,** Etude thermodynamique de la dissociation réversible de la β-lactoglobuline B pour des pH supérieurs à 5,5, *Biochim. Biophys. Acta,* 59, 737, 1962.
99. **McKenzie, H. A. and Sawyer, W. H.,** Effect of pH on β-lactoglobulins, *Nature,* 214, 1101, 1967.
100. **Roels, H., Préaux, G., and Lontie, R.,** Polarimetric and chromatographic investigation of the irreversible transformation of β-lactoglobulin A and B upon alkaline denaturation, *Biochimie,* 53, 1085, 1971.
101. **Brown, E. M. and Farrell, H. M., Jr.,** Interaction of β-lactoglobulin and cytochrome c: complex formation and iron reduction, *Arch. Biochem. Biophys.,* 185, 156, 1978.
102. **Oncley, G. L.,** The electric moments and the relaxation times of proteins as measured from their influence upon the dielectric constants of solutions, in *Proteins, Amino Acids, and Peptides,* Cohn, E. J. and Edsall, J. T., Eds., Reinhold, New York, 1943, 557, 563.
103. **Shirley, W. M. and Bryant, R. G.,** Proton-nuclear spin relaxation and molecular dynamics in the lysozyme-water system, *J. Am. Chem. Soc.,* 104, 2910, 1982.
104. **Daskiewicz, O. K., Hennel, J. W., Lubas, B., and Szczepkowski, T. W.,** Proton magnetic relaxation and protein hydration, *Nature,* 200, 1006, 1963.
105. **Beall, P. T., Hazlewood, C. F., and Rao, P. N.,** Nuclear magnetic resonance patterns of intracellular water as function of HeLa cell cycle, *Science,* 192, 904, 1976.
106. **Lee, J. C. and Timasheff, S. N.,** Partial specific volumes and interactions with solvent components of proteins in guanidine hydrochloride, *Biochemistry,* 13, 257, 1974.
107. **Petsko, A. and Ringe, D.,** Fluctuations in protein structure from X-ray diffraction, *Annu. Rev. Biophys. Bioeng.,* 13, 331, 1984.
108. **Tanford, C., Bunville, L. G., and Nozaki, Y.,** The reversible transformaion of β-lactoglobulin at pH 7.5, *J. Am. Chem. Soc.,* 81, 4032, 1959.

Solid-State NMR of Intact Agricultural Polymers and Matrices

Chapter 9

INTRODUCTION TO SOLID-STATE NMR SPECTROSCOPY AND ITS APPLICATION TO THE STUDY OF SOILS AND RELATED SYSTEMS

Walter V. Gerasimowicz and Philip E. Pfeffer

TABLE OF CONTENTS

I. GENERAL PRINCIPLES OF SOLID-STATE NMR

A. Introduction

Over the past 25 years, high-resolution liquid-state NMR has become the most widely utilized technique employed in the elucidation of molecular structure, identification of chemical species, and in the study of nuclear spin dynamics. NMR spectroscopy (as seen in the other chapters of this book) is an invaluable tool for the investigation of phenomena, such as chemical shift and proton coupling in the liquid or solution state. Due to the fact that the experimentalist can obtain high-resolution spectra, one can observe weak interactions, such as those associated with spin-spin coupling and with the chemical shift. Such interactions are a reflection of the local nuclear spin environment, and may be used as a diagnostic probe of that phenomenological level. However, NMR spectroscopists have continued to cope with several serious limitations. Lack of NMR sensitivity due to sample size limitation is being overcome by increases in magnetic field strength, electronic development, and computer (data processing) technology. Many scientists viewed the requirement (until recently) of liquid samples as the greatest limitation with respect to NMR analysis. Although solid-state physicists have conducted wide-line studies since the early days of NMR, these results have been of little practical interest to chemists, biochemists, agricultural, and soil scientists. Developments over the last few years have resulted in a dramatic and ever-expanding growth in this area of research.

This chapter is written with several purposes in mind. We intend to provide the reader with a concise overview of the practical limitations as well as the major problems associated with solid-state NMR research. Our emphasis will center upon ^{13}C cross-polarization magic angle spinning (^{13}C-CPMAS) with other examples as appropriate. However, many of the solutions applied to the procurement of solid-state carbon spectra can be analogously utilized in other heteronuclear pulse experiments of this type. Most of the subject matter is too complex to be fully discussed either theoretically or mathematically within the intended scope of this volume. Literature references will be cited so that the reader can later pursue various areas of personal interest in depth. Selected topics and papers (but not exhaustive) are presented to give the agricultural scientist examples from the literature which may stimulate an awareness for the vast potential, which is as yet untapped, in this "fertile field" of spectroscopy.

Obviously, solid-state NMR is having an enormous impact on agricultural science from the standpoint of the study of agricultural materials, their structure, polymeric properties, structure-function relationships, and chemical and biological properties which only represent a partial list of the areas which could be mentioned. Furthermore, several profound reasons could be given for the study of agricultural substances in the solid state. The solid may be insoluble or intractable. Additionally, the scientific question of interest may concern the properties of the material in the solid state itself. One can never be absolutely certain that data obtained while molecules are in solution is necessarily a reflection of the solid state. Molecular structure may change markedly when the species is in different phases. High-resolution NMR in the solid state can be used (in certain situations) to confirm the validity of X-ray structural determinations, conformations, and molecular configurations. The technique is non destructive. The sample utilized in the solid-state NMR study is retained intact and unmodified, and has not been subjected to solvent interactions. Valuable and rare samples are thus preserved and do not require extensive recovery from or removal of solvent.

By convention ^{1}H (abundant spin) will be designated as I and ^{13}C (rare spin) will be designated as S. At times the I·S notation will be employed to remind the reader of the generality of this discussion *i. e.,* the fact that we are discussing spin systems in which two different spin species are present or heteronuclear interactions are being treated.

B. Basis of Nuclear Spin Interactions in Solids

A summary of the basic nuclear spin interactions which occurs in solids is depicted in Figure 1. Formal and detailed descriptions of the physics and quantum mechanics may be found in the

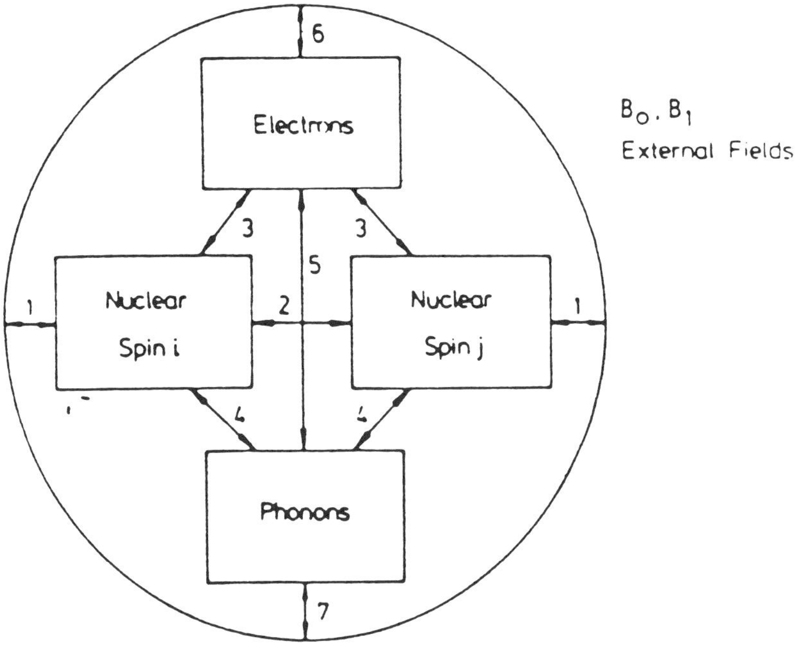

FIGURE 1. The sevenfold way a nuclear spin system can interact with its surrounding. (1) Zeeman interaction of spins, (2) direct spin interactions, (3) nuclear spin-electron interaction and indirect spin interaction, (4) direct spin-lattice interaction, (3 to 5) indirect spin lattice interaction via electrons, (3 to 6) shielding and polarization of nuclear spins by electrons, and (4 to 7) coupling of nuclear spins to sound fields, etc. (From Mehring, M., *High Resolution NMR in Solids*, 2nd ed., Springer-Verlag, Berlin, 1983. With permission.)

excellent texts of Mehring [1], Slichter,[2] Haeberlein,[3] Gerstein and Dybowski,[4] and of course in Abragam.[5] The nuclear spin interactions can be expressed first in terms of the effects between internal and external magnetic fields (B_0 and B_1).

$$\mathcal{H} = \mathcal{H}_{ext} + \mathcal{H}_{int} \tag{1}$$

where

$$\mathcal{H}_{ext} = \mathcal{H}_0 + \mathcal{H}_1 \tag{2}$$

$\mathcal{H}_0$ and $\mathcal{H}_1$ are the Hamiltonian operators which represent the Zeeman interactions corresponding to the internally induced fields due to B_0 and B_1. The internal Hamiltonian for the interaction of two different spins (I and S) with corresponding magnetogyric ratios of γ_I and γ_S is shown in equation (3):

$$\mathcal{H}_{int} = \mathcal{H}_{II} + \mathcal{H}_{SS} + \mathcal{H}_{IS} + \mathcal{H}_Q + \mathcal{H}_S + \mathcal{H}_L \tag{3}$$

$\mathcal{H}_{II}$ and $\mathcal{H}_{SS}$ represent direct and indirect interactions among the various spins (path 2 and 3) as shown in Figure 1. Paths 2 and 3 are also involved in the I·S interactions with the corresponding $\mathcal{H}_{IS}$ operator. $\mathcal{H}_Q$ is the quadrapolar operator. The shielding Hamiltonians are depicted by $\mathcal{H}_S$, i.e., Knight and chemical shifts (paths 3 and 6 in the Figure). $\mathcal{H}_L$ represents the spin-lattice interactions as seen in paths 3, 4, and 5 of Figure 1. A summary of the other terms in Equation 3 and a tabulation of the explicit forms of the Hamiltonian are shown in Table 1.[1]

Table 1
INTERACTION HAMILTONIANS

Interaction	Form of the Hamiltonian
Chemical shift	$\mathcal{H}_s = \gamma I \cdot \tilde{\sigma} \cdot B_0$
Dipole-dipole	$\mathcal{H}_D = \sum_{i<j} \hbar\gamma_i\gamma_j r_{ij}^{-3}\left(I_i \cdot I_j - \frac{3(I_i \cdot r_{ij})(I_j r_{ij})}{r_{ij}^2}\right)$
	$\mathcal{H}_D = \sum_{i=j} I_i \cdot \tilde{D} \cdot I_j$
	$D_{\alpha\beta} = \hbar\gamma_i\gamma_i r_{ij}^{-3}(\sigma_{\alpha\beta} - 3e_\alpha e_\beta)$
	$\alpha,\beta = x,y,z$; e_α: α-component of the unit vector along r_{ij}
J-coupling	$\mathcal{H}_J = \sum_{i \neq j} I_i \cdot \tilde{J} \cdot I_j$
Spin-rotation	$\mathcal{H}_{CR} = \sum_i I_i \cdot \tilde{C}_i \cdot J$
Quadrupole	$\mathcal{H}_Q = \frac{eQ}{2I(2I - 1)\hbar} I \cdot \tilde{V} \cdot I$
	$\tilde{V} = \{V_{\alpha\beta}\}$; $\alpha,\beta = x,y,z$

C. Solid-State NMR Methodology in Solids

Attempts to obtain solid-state ^{13}C NMR spectra by using ordinary pulse Fourier transform techniques and instrumentation as employed for routine liquid samples will normally not even result in a spectrum which can be detected. The major (although not the exclusive) reason for the lack of resolution in the spectra of typical solids is the dipole-dipole interaction exerted by protons. Furthermore, rare spins are generally dipolar-coupled to abundant spins in the solid state. Due to their low natural abundance (1.1% in the case of ^{13}C), the signal which can be detected from a single experiment is generally small, and signal averaging must be implemented. The repetition rate, however, is dependent upon the time (the ^{13}C spin-lattice relaxation time) necessary for the ^{13}C nuclei to recover (for the most part) to their equilibrium state and orientation after being subjected to an rf pulse at or near the resonance frequency of the nuclei. Chemical shift anisotropy (CSA) also results in line broadening. The chemical shift observed for a nucleus is a function not only of its identity, but also depends on its orientation within the molecule relative to the magnetic field. As such, in powders and solid samples where isotropic motion and rapid and random tumbling is not present (as in solutions where sharp lines are seen due to rapid motional averaging), a broad band of resonance frequencies is observed for individual carbon types which can be as broad as the typical frequency range over which ^{13}C is normally observed.

1. Cross-Polarization of Dilute Spins

Due to their low natural abundance, rare spins do not exhibit significant homonuclear dipole-dipole interactions. This advantage is reduced by the inherent low signal-to-noise which is due to the low population of nuclei which can be observed. One can attempt to enrich the sample isotopically. Such procedures may be costly and may affect the characteristics of the sample

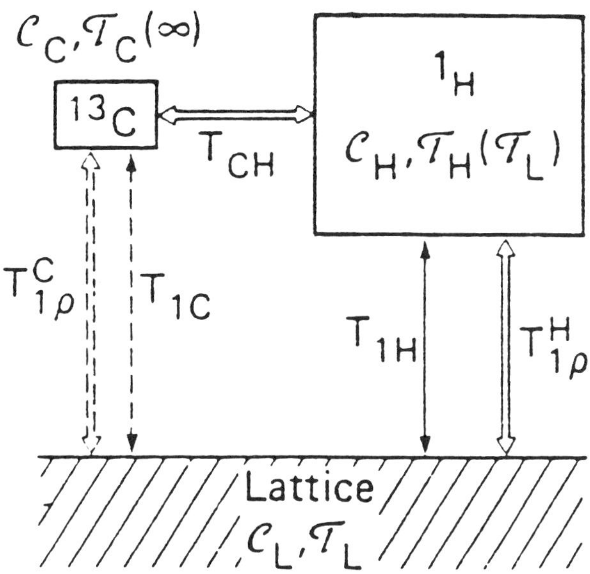

FIGURE 2. Large ^1H, small ^{13}C, and infinite lattice reservoir, characterized by a heat capacity, C (vide infra), and temperature T. Temperatures in parentheses are those at the start (t = 0) of the CP sequence. (From Yannoni, C. S., *Acc. Chem. Res.*, 15, 201, 1982. With permission.)

under study. Other methods have been devised which permit observation of dilute nuclei in a solid by taking advantage of the strong dipole-dipole interactions among abundant and rare spins. This process is commonly referred to as cross-polarization (CP) or polarization transfer. The history of polarization transfer can be traced back to the 1950s and the 1960s.[6-10] Several research groups demonstrated that they could detect communication between spins in a spin-locked system which was externally perturbed. Hartmann and Hahn[6] were able to detect rare spins by means of a double resonance experiment.

The basic manner in which CP is carried out in order to observe ^{13}C nuclei at natural abundance in the presence of an abundant proton spin population depends upon spin temperature. When a solid organic sample experiences a magnetic field (H_0), the individual protons comprising the sample will align parallel and antiparallel to H_0. They will populate Zeeman energy levels with energies of $\pm(\gamma_H/2\pi)\ H_0/2$. These energy levels correspond to spin alignments which are either parallel or antiparallel to H_0. The protons will undergo mutual spin flips by exchanging energy with the lattice until a state of internal equilibrium is attained (~100 μs). Upon reaching this equilibrium state, they can be treated as a thermodynamic reservoir which requires several T_{1H}s (proton spin-lattice relaxation times) to equilibrate with the lattice at some temperature T_L.[11] This process is depicted in Figure 2. At this point, the proton magnetization may be defined as follows:

$$M_{0H} = C_H H_0/T_L \tag{4}$$

$$E = C_H H_0^2/T \tag{5}$$

$$\mathcal{L} = (C_H H_0^2/T^2) + \text{constant term} \tag{6}$$

E is the energy, L is the entropy, and C_H is the Curie constant defined in Equation 7 as:

$$C_H = N_H \gamma_H^2 \hbar^2/4k \tag{7}$$

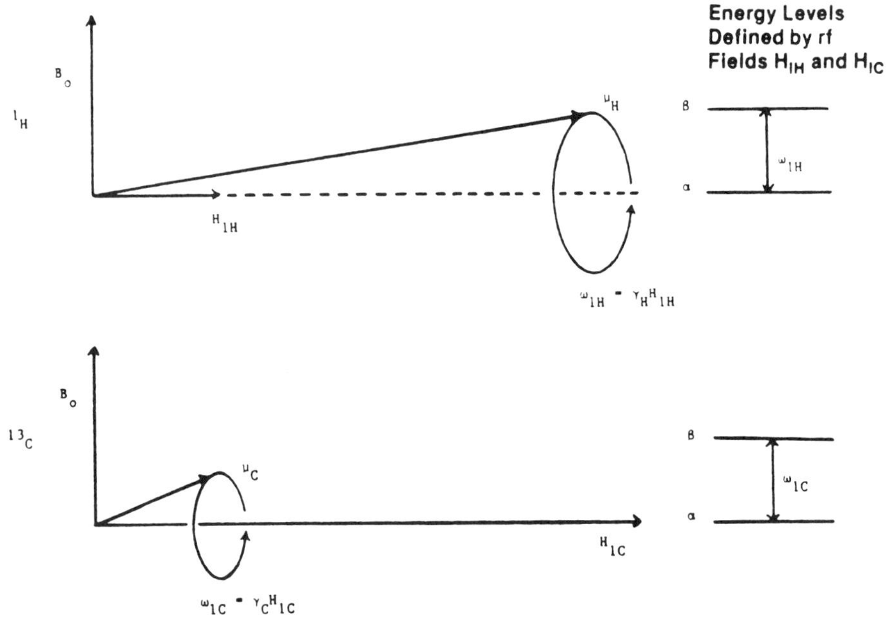

FIGURE 3. Matched rotating frame energy levels for ^{1}H and ^{13}C spin species during the Hartmann-Hahn condition. In the diagram, H_{1H} and H_{1C} are applied rf fields, γ_H and γ_C are the respective gyromagnetic ratios, and ω_{1H} and ω_{1C} are the energy levels (in units of rad s^{-1}) induced by the rf fields. Since $\omega_{1H} = \omega_{1C}$ under the match condition, efficient CP can occur. (From O'Donnell, Bartuska, V. J., Palmer, A. R., and Sindorf, D. W., *Am. Lab.*, 104, 1986. With permission.)

N_H is the number of proton spins and k is the usual Boltzmann constant.[12] Not only can the I spin system be considered a reservoir, but also the S spin system can form a reservoir both of which are in contact with one another and with the lattice as well. The rate at which polarization is transferred from the I to the S reservoir is described kinetically by the cross-relaxation time T_{IS} or in this case T_{CH}.

Three steps are involved in the CP process. The first step involves cooling or reduction of the S spin temperature. The magnetization of the carbon nuclei is rather small initially due to the fact that T_C, the carbon spin temperature is large:

$$M_{OC} = C_C H_O / T_C \tag{8}$$

and

$$C_C = N_C \gamma_C^2 \hbar^2 / 4k \tag{9}$$

where C_C is the Curie constant associated with the ^{13}C nuclei and the other terms are defined analogously to Equation 7 above. In order to actually transfer polarization, thermal contact is established between the I and S spin reservoirs by means of applying a spin lock in the rotating frame. The S spins are first irradiated on resonance and the rf field is adjusted until the conditions in Equations 10 and 11 are fulfilled:

$$\gamma_H H_{1H} = \gamma_C H_{1C} \tag{10}$$

$$\omega_H^{eff} = \omega_C^{eff} \tag{11}$$

These equations define the Hartmann-Hahn match and are further illustrated by Figure 3.[6,13]

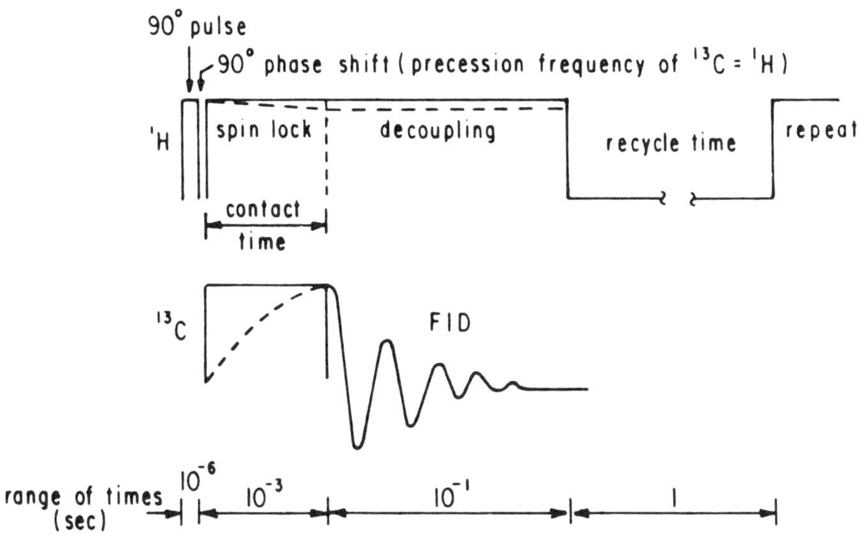

FIGURE 4. ^{13}C solid-state CP sequence. (From Pfeffer, P. E., *J. Carbohydr. Chem.*, 3, 613, 1984. With permission.)

Therefore, an efficient mechanism has been established for magnetization transfer between the ^{1}H and ^{13}C reservoirs. Accordingly, as seen in Equation 8, the ^{13}C spin temperature is reduced, and the ^{13}C nuclear magnetization (M_{0C}) increases. The process is depicted schematically in Figure 4 during the contact time (CT) period. Typical time regimes are also shown. Furthermore, the repetition rate of the experiment is now governed by the shorter ^{1}H relaxation time (ca. 0.3 to 1.0 s), rather than by the much longer ^{13}C relaxation time. Therefore, signal averaging and repetition of the basic experiment are much more efficient. Due to the CP process, ^{13}C magnetization is enhanced by a factor of ca. 4 over that observed by natural relaxation mechanisms. In summary, Figure 4 depicts:

1. Establishment of I spin polarization in H_0
2. Spin-locking of I spins along $H_1{}^I$
3. Establishment of I and S spin contact via the Hartmann-Hahn matching condition
4. Turning off $H_1{}^S$
5. Rccording S-spin FID
6. Repeating sequence n times

2. High-Power Proton Decoupling

Proton dipolar broadening of ^{13}C resonances in solids can be removed by means of high-power decoupling techniques. The protons are irradiated at their Larmor frequency. Effectively this irradiation stirs up the proton spins and eliminates the influence of the proton dipolar interaction on the ^{13}C nuclei. The experiment is exactly analogous to scalar decoupling of protons in the liquid state, but requires much higher power. Dipolar couplings in a solid may be several thousand Hz as opposed to the much smaller J couplings found in the liquid state. Therefore, in order to effectively decouple protons in solids, ^{1}H irradiation fields and power requirements are orders of magnitude larger (12 to 18 G) than those employed in liquids. Decoupling is normally carried out throughout the spin-locking and acquisition of the FID (Figure 4). Note the spectrum in Figure 5A where powdered crystalline sucrose is examined under solution state conditions as compared with Figure 5B where high-power proton decoupling is employed.[14] Although little detail is evident, some signal is beginning to emerge in Figure 5b.

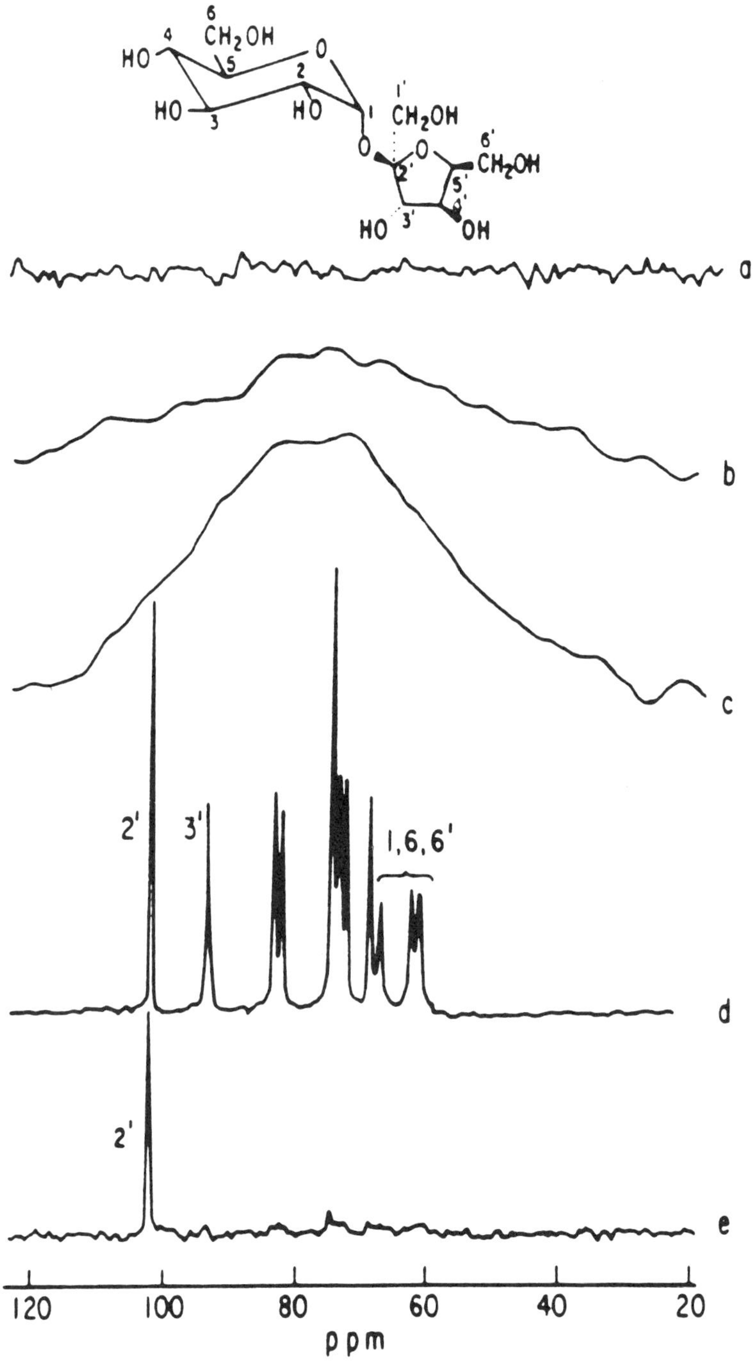

FIGURE 5. The 15-MHz ^{13}C solid-state spectra of powdered crystalline sucrose taken under the following experimental conditions: (a) solution spectrum conditions, 50° pulse angle, 8-s pulse delay, and 5-kHz proton decoupling; (b) same conditions as a except with high-power decoupling of 45 kHz; (c) same as b except with proton-carbon CP, CT of 2 ms; (d) same as c except sample was spun at 2.0 kHz at the magic angle of 54.7°; and (e) same as d except a 40-μs delay without decoupling was inserted into the pulse sequence following the CT. (From Pfeffer, P. E., *J. Carbohydr. Chem.*, 3, 613, 1984. With permission.)

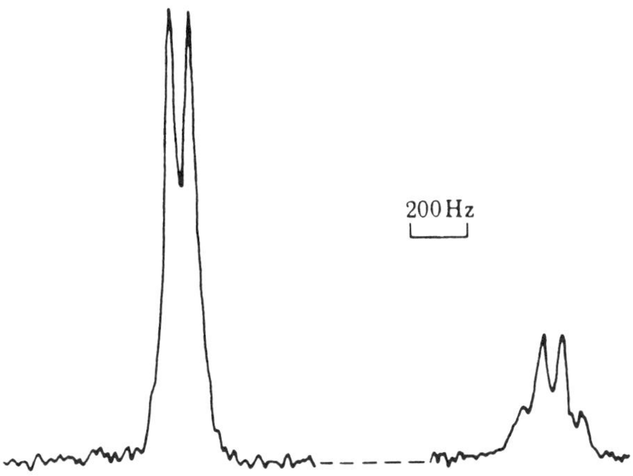

FIGURE 6. ^{31}P-NMR spectrum in polycrystalline P_4S_3 at 420 K (melting point 446 K). The spectrum is strongly narrowed by molecular motion in the solid and displays an AB_3 type fine structure.(From Andrew, E. R., *Philos. Trans. R. Soc. London, Ser. A*, 299, 505, 1981. With permission.)

3. Magic Angle Spinning

The two techniques mentioned above (CP and high-power dipolar decoupling) would only provide limited insight(s) into the spin dynamics of solids if magic angle spinning (MAS) did not further increase the resolution seen in solid state NMR spectra. Most agricultural materials are intractable solid powders. Note in Figures 5b and 5c that high-power decoupling alone or CP does not appreciably enhance resolution where structural detail similar to liquid state spectra can be obtained. Complex powder spectra are much broader than those seen for liquids due to the overlap of many resonances. An excellent example of this phenomenon is provided in the case of the water molecule. The typical proton NMR line width of water at room temperature is on the order of 0.1 Hz or less. The line width of ice at low temperatures is approximately six orders of magnitude larger or about 10^5 Hz. This behavior has been well studied and is due to static anisotropic nuclear interactions in the solid state. In a liquid, rapid and random tumbling of the molecules averages out this anisotropy, and it is removed from the spectrum resulting in the nice, sharp resonances one is accustomed to seeing. In some limited cases, sufficient motion is present in solids to result in NMR line narrowing. Figure 6 depicts such an example. An AB_3 in the ^{31}P spectrum could be resolved in the solid. The doublet was attributed to the three basal phosphorus nuclei in the molecule and the quartet to the apical nucleus.[16,17]

In most cases, however, sufficient motion is not present in solids to average the dipolar interactions. One alternative to this problem, MAS, was first introduced by Lowe and Andrew et al.[18-21] Effectively, one can emulate nature by mechanically imposing motion on the nuclei. Andrew's early studies also indicate that MAS affects other spin interactions, such as the chemical shift.[21] The combination of high power decoupling, CP, and MAS for rare spins was first shown by Schaefer et al.,[22,23] and for abundant spins by Gerstein.[24] Because of these developments, modern commercial spectrometers used for solid-state work are normally equipped with MAS capabilities.

The truncated dipolar Hamiltonian, $\mathcal{H}_d$, for all nuclear spin pairs in a solid is defined as

$$\mathcal{H}_d = \sum_{i<j} 1/2\gamma_i\gamma_j\hbar^2 r_{ij}^{-3}(I_i \cdot I_j - 3I_{iz}I_{jz})(3\cos^2\text{Ø}_{ij} - 1) \tag{12}$$

In this equation, r_{ij} is the internuclear displacement, Ø_{ij} is the angle between the internuclear vector r_{ij} and the Zeeman field H_0, and the remaining terms are defined as previously specified. The geometry of the MAS experiment is shown in Figure 7.[15] The broadening due to chemical

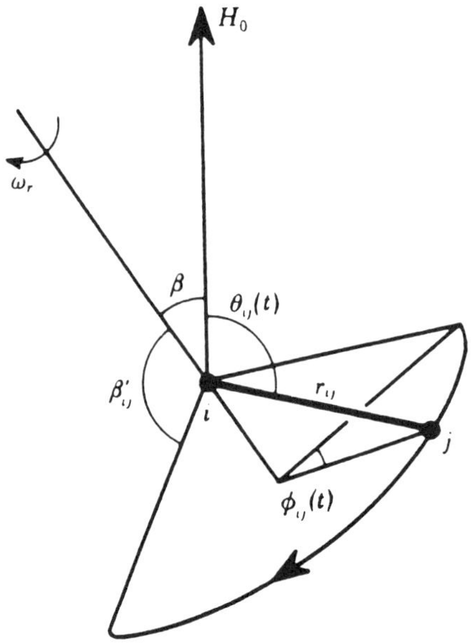

FIGURE 7. Diagram illustrating the motion of a typical internuclear vector r_{ij} when a solid is rotated with angular velocity ωr about an axis inclined at angle B to H_0. (From Andrew, E.R., *Philos. Trans. R. Soc. London, Ser. A*, 299, 505, 1981.With permission.)

shift anisotropy is eliminated (for the most part) by spinning at the "magic angle" of 54.7° with respect to H_O (Figure 7). This is the angle which causes the function ($3Cos^2Ø_{ij}$ –1) to vanish (Equation 12) and H_d = 0.

A summary of solution vs. high-resolution solid-state NMR spectroscopy is shown in Table 2.[25] Proper execution of the CPMAS experiment results in the spectrum shown in Figure 5d. The reader is referred to the literature references for a detailed discussion of the mathematical and theoretical treatments associated with these spectroscopic techniques.

II. DIRECT CPMAS MEASUREMENTS OF WHOLE SOILS

The first solid-state ^{13}C NMR spectra of whole soil were reported in 1981 by Barron and Wilson.[26] Ideally, the tool or technique employed in the analysis of soil organic matter should be nondestructive. Until the advent of CPMAS, structural group analysis, characterization of soil humic and fulvic substances, and other soil components could not be performed adequately without chemical alteration during pyrolysis and extraction procedures. Therefore, a degree of doubt existed as to how representative the measurements were in terms of the actual soil, *in situ* structure and composition. Particularly, a controversy exists even now regarding the aromaticity of soil organic matter.[26-28] A number of reasons for such controversy will be discussed below. Additionally, precautions will be stated which should be adhered to in order to prevent similar mistakes based upon "ideal?" methods, such as CPMAS.

Preston and Ripmeester published the spectra of several whole soils and some of their fractions as shown in Figure 8.[29] The ^{13}C-CPMAS spectra of the various organic soils show some common features. Resonance peaks were identified at 30 ppm and assigned to unresolved methyl, methylene, and methine carbon nuclei. Methoxyl carbon is seen at approximately 58 ppm. Carbon attributed to carbohydrate is indicated by resonances around 74 ppm (carbohydrate ring structures) and at 104 ppm (the anomeric region). Further assignments include the region from 50 to 70 ppm where N-substituted carbon, possibly from amino acid residues, gives a small

Table 2
SOLUTION-STATE VS. HIGH RESOLUTION SOLID-STATE NMR SPECTROSCOPY

Total interaction	=	Zeeman	+	Dipolar	+	Scalar	+	Chemical shift
Solution state		50 MHz		0		200 Hz		Isotropic, single frequency
Solid state		50 MHz		50 kHz		200 Hz		200 ppm-wide chemical shift anisotropy
Solid state NMR technique used to overcome broadening				High-power (dipolar) proton decoupling		High-power decoupling also removes J-coupling		Magic-angle spinning gives the isotropic line

Note: This is a specific example showing ^{13}C nuclear spin interactions in a 4.7-T field.

From Jelinski, L. W., *Chemtech*, March, 1986, 188. With permission.

contribution.[30] Somewhat lower levels of intensity are evident in the aromatic region ranging from about 110 to 160 ppm. Resonances can be assigned at approximately 130 (C- and H-substituted C) and at 150 ppm (O- and N- substituted C). The peak at 174 is indicative of carbonyl type carbon and may be assigned to carboxyl, amide, and ester group linkages. The results of various groups indicate that studies of soil organic matter are being performed and may be potentially correlated to variations in soil formation, function, and vegetation coverages. Further information is available in the references already cited as well as the others which are listed.[31,32]

A. Experimental Conditions — Spin Dynamic Considerations

1. NMR Spectrometer Settings

To obtain reasonable information on such complex materials, certain experimental and instrumental parameters must be considered. Current high-field, high-resolution NMR instruments are stable in terms of the B_0 field and also with regard to magic angle settability and spinning speeds. Problems are encountered however, in initially getting samples to spin at 4 to 6 kHz, but these are generally not insurmountable. Samples suitable for analysis are generally packed into rotors which are then spun by light gases in stators of various design. An excellent discussion of rotor and stator design has been given by Fyfe. He presents the advantages and disadvantages of such equipment as well as design and material modifications which have transpired since the original designs were reported.[33] Modern MAS devices are based upon the designs of Andrew and Lowe.[20,34] Their designs are depicted schematically in Figure 9.[33] Variable temperature MAS is also possible, and devices have been introduced for this purpose as well.[35] Upon completion of the experiment, the sample is returned to the scientist intact, unmodified, and preserved for further experimental study by other techniques.

2. Degree of Cross-Polarization Enhancement

During the CP process, a number of relaxation processes (mechanisms) must be considered in order to establish that spectral intensities are not distorted. Of the techniques described for obtaining CPMAS spectra, matching of the abundant ^{1}H spin and the dilute ^{13}C spin rotating frame Larmor frequencies to induce CP energy tranfer is the most critical for controlling the relative magnetizaton or intensity response exhibited by the ^{13}C resonances. Because the abundant protons in organic solids are close together, rapid spin diffusion occurs within this population. This process allows all protons within an individual component to have the same

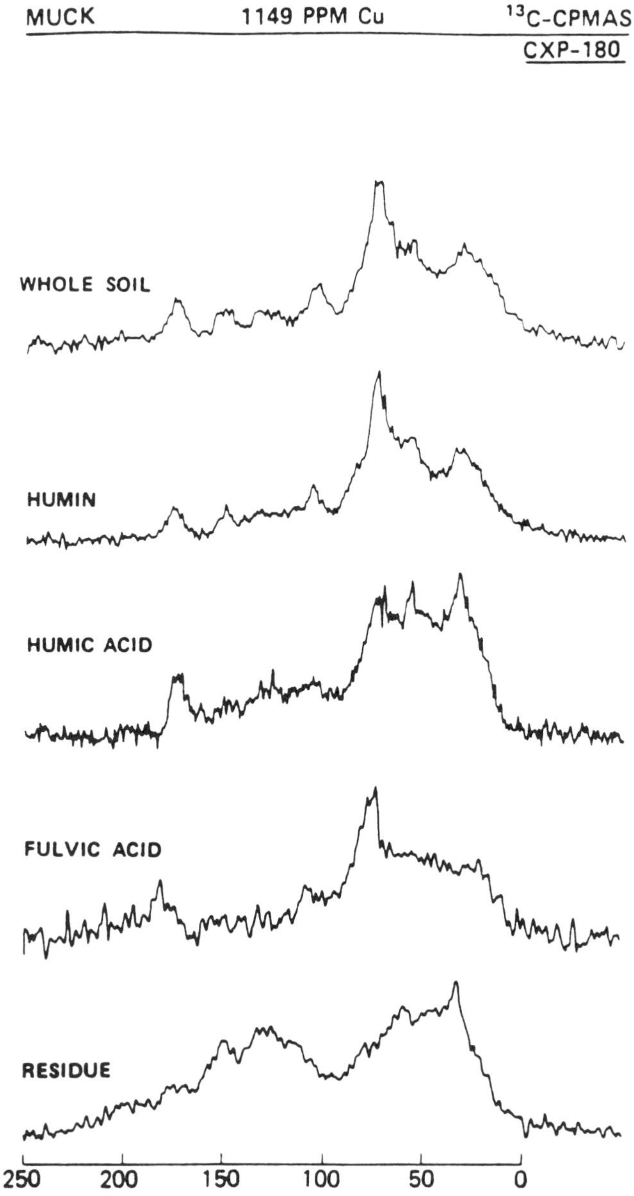

FIGURE 8. Solid-state CPMAS ^{13}spectra for organic soil A, humin, HA, fulvic acid, and residue of acid hydrolysis. (From Preston, C. M. and Ripmeerster, J. A., *Can. J. Spectrosc.*, 27, 700, 1982.)

relaxation behavior. If spin diffusion is not uniform within a solid due to heterogeneity, the proton relaxation processes can differ for each component. Amorphous heterogeneous samples usually have T_{1H} values between 0.1 and 0.2 s as a result of local motions and amplitude variations. The first relaxation process which must be considered is called the contact time (CT). The CT is a period (millisecond range) when the proton and carbon spin reservoirs are in thermal contact with each other. The CP rate depends on the internuclear distance (r_{CH}^{-6}) between carbon and hydrogen. Clearly, nonprotonated carbon nuclei will cross polarize at a slower rate than will protonated species.[37,38] The time constant associated with the carbon-proton CP process is referred to as the T_{CH}. In Figure 4, the buildup of ^{13}C magnetization (dashed line) at the expense

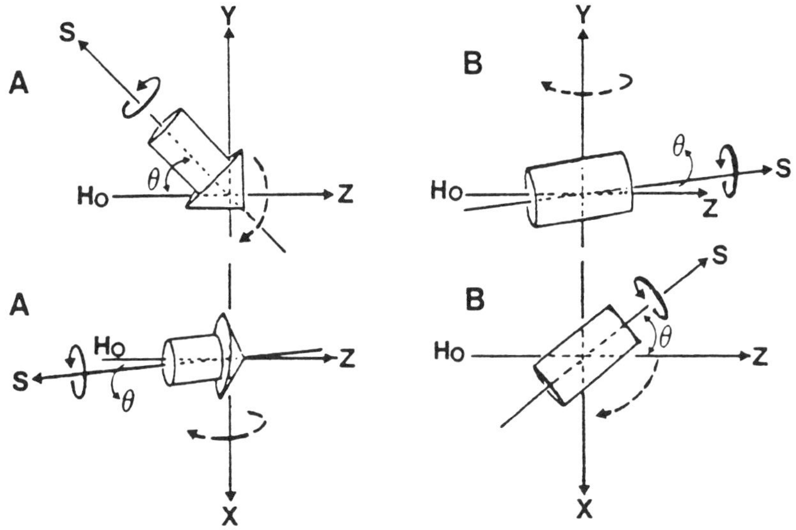

FIGURE 9. Schematic representation of the geometric arrangements for MAS using (A) the Andrew-Beams design and (B) the Lowe design. (From Fyfe, C. A., *Solid State NMR for Chemistry*, CFC Press, Guelph, Ontario, 1983, 272. With permission.)

of the ^{1}H spin population is seen. However, a competitive relaxation is also occurring, $T_{1\rho H}$, known as the proton spin-lattice relaxation time in the rotating frame. This time constant reflects the decay or loss of proton polarization as a function of time. Typically, T_{CH} values are in the range of 50 μs to 1 ms. Because of this dependence on T_{CH} not all carbons are equivalent. Some carbons are isolated from neighboring protons (e.g., carbonyl carbons, tetra-substituted double bond carbons, and quaternary carbons) while others have directly bonded protons (e.g., methylene, methine, and methyl group carbons). For quantitative work, the observed carbon resonance intensities must be proportional to the number of carbons of any given type. This means that the carbons that cross-polarize slowly must be given adequate time to equilibrate with the spin-locked protons which are concurrently decaying during the time impacted upon by the $T_{1\rho H}$ process. To satisfy this requirement, it is essential that the value of $T_{1\rho H}$ be greater (order of magnitude) than the corresponding T_{CH} for each carbon population in the system. Figure 10 illustrates this point. In the early part of the contact period when the proton population has had minimal time to decay (proton population with a long $T_{1\rho H}$), optimal energy exchange is achieved and a large ^{13}C signal is generated. However, for those carbons having T_{CH} values closer to the proton $T_{1\rho}$, a significant amount of proton magnetization has been drained off before sufficient energy is interchanged. The result is a much diminished ^{13}C signal, even when the experiment is optimized at a somewhat longer CT. Finally, when the protons have decayed to the point where there is little or no magnetization left ($T_{CH} >> T_{1\rho}$), virtually no ^{13}C signal is observed. In general, this situation is not observed unless the carbons under consideration are significantly isolated from any protons, as has been found in some selected coal samples.[37]

Assuming the criteria of proton $T_{1\rho} >> T_{CH}$ is met, one must also consider the relationship between the experimental parameter, CT, and these decay times. In general, the CT should be much larger than the largest T_{CH} value in the sample (approximately a factor of five to ten), and much smaller than the proton $T_{1\rho}$ (approximately one fifth the magnitude) to generate optimum responses.

In solid-state CP experiments, power levels are also very critical. Modulation of proton interactions (decoupling) in a solid requires approximately ten times more power than employed in solution state spectroscopy to achieve acceptable responses. Deviations in the power levels of the ^{1}H or ^{13}C transmitters can be very deleterious. T_{CH} values can be increased if the CP

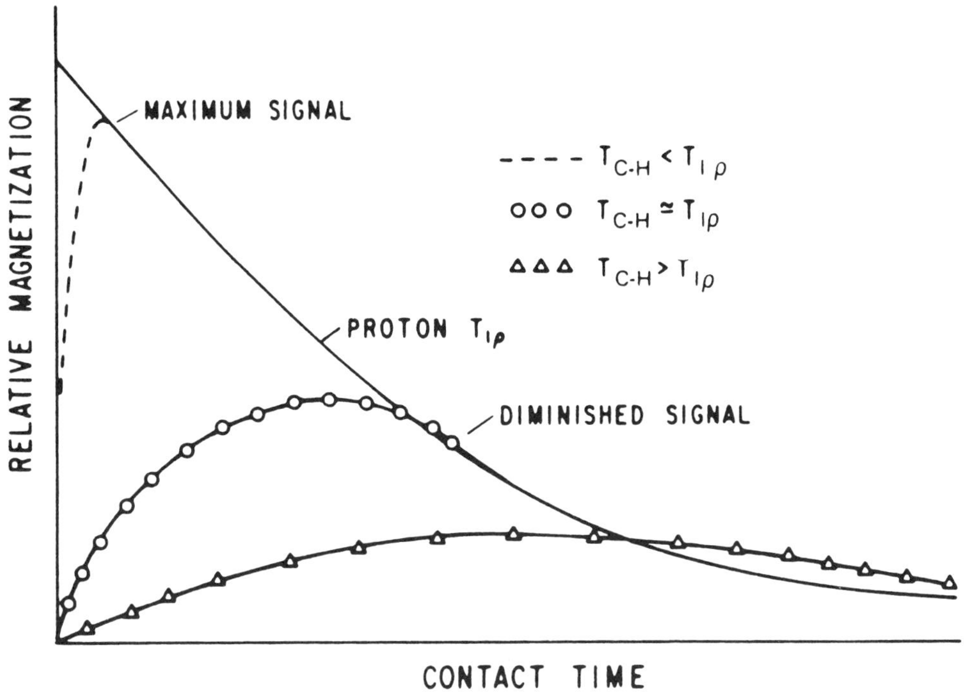

FIGURE 10. Competing relaxation processes during CP. (From Pfeffer, P. E., *J. Carbohydr. Chem.*, 3, 613, 1984. With permission.)

Hartmann-Hahn matching condition is not met. In the case of weak C-H and H-H dipolar coupling (e.g., in systems with rapid motional changes occurring in the solid), signal intensities can be changed markedly if the rf output power of the ^{1}H or ^{13}C transmitter changes by even ± 0.1 dB.[36] Such deviations would affect the resonance signal area. Quantitation would not be feasible through integration of such peaks.

3. Model Response System for Soils, Mucks, Sludges, and Humic and Fulvic Systems

The importance of these response factors for predominant carbon types present in soil and soil related systems has been illustrated.[39] To establish the relative response of various components in a heterogeneous matrix such as soil, we initially examined a model multicomponent mixture. We do not wish to imply that the components which make up the model mixture are the only components present in soils and/or sludges. These model constituents were selected to represent the predominant carbon types evident in the spectra of the materials selected for this study. Prior to making these measurements, the relaxation values (proton T_1 and $T_{1\rho}$) of the individual compounds used to prepare the model system were evaluated (Figure 11 shows the ^{13}C-CPMAS spectrum of each component used). The range of relaxation parameters was also measured in the presence of paramagnetic Fe^{+3} (in concentrations approximating sludge) to note its affect on each type of component. Organic free radicals and metal ions are present in soils and soil components. If these paramagnetic centers are removed, the structure of the sample will be modified. In solution, metal ions can exchange rapidly. In the solid however, different peaks may be broadened to different extents and even lost entirely from the spectrum. Note however, that in solution, cases of preferential metal ion complexation can cause different degrees of line broadening as well. Table 3 lists the proton T_1 and $T_{1\rho}$ values for the individual components in both the presence and absence of Fe^{+3}. All of these samples were prepared in a somewhat amorphous state by lyophilization from water. The final state of the lipid fatty acid component

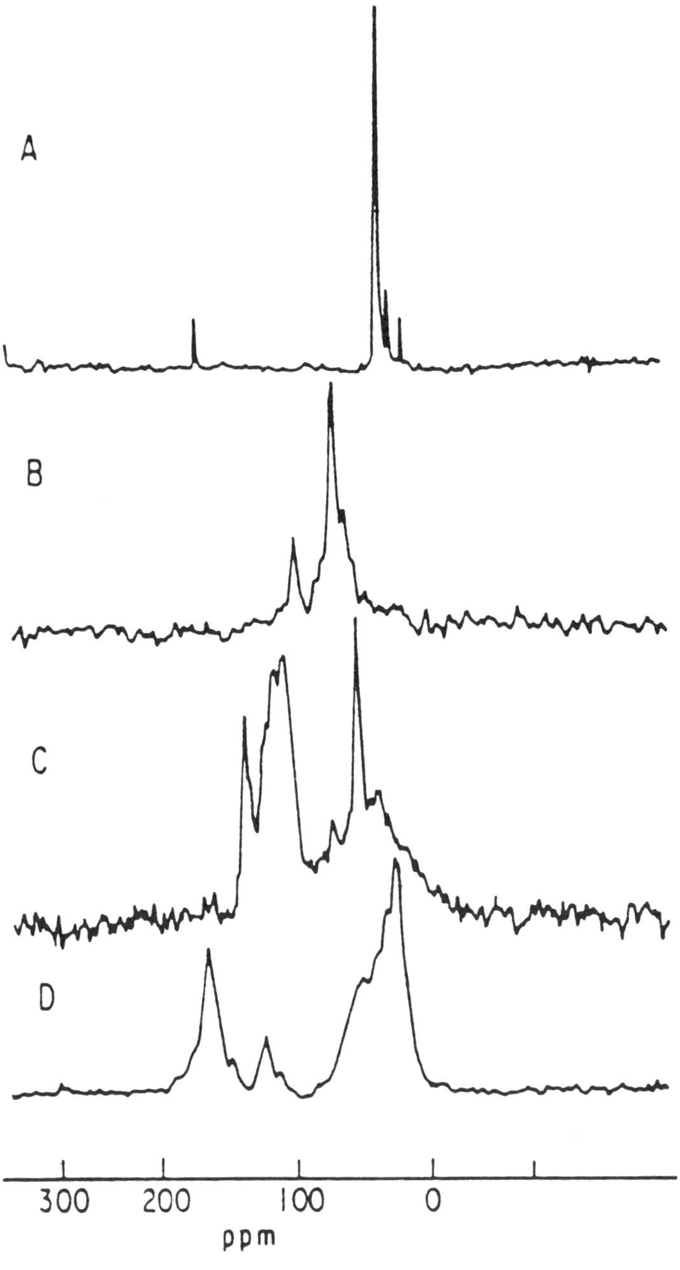

FIGURE 11. CPMAS ^{13}C NMR spectra of reference compounds: (A) stearic acid, (B) wood pulp, (C) lignin, and (D) keratin. Each spectrum was obtained from 3000 acquisitions with a pulse delay of 1.5 s and a CT of 0.5 ms and 10- to 15-Hz line broadening. (From Pfeffer, P. E., Gerasimowicz, W. V., and Piotrowski, E. G., *Anal. Chem.*, 56, 734, 1984. With permission.)

in the model mixture appears to be somewhat more ordered than that of the other model compounds as evidenced by its extremely long T_1 and $T_{1\rho}$ values. The T_1 and $T_{1\rho}$ values associated with this hydrophobic component exhibit little change when the paramagnetic-amended system is compared to the analogous non-iron-containing mixture. Similarly, the hydrophobic nature of lignin does not foster intimate spatial communication with the Fe^{+3} since its T_1 has only diminished by half while its $T_{1\rho}$ has essentially remained the same. The more

Table 3
RELAXATION VALUES[a] OF MODEL COMPONENTS

Component	Fe^{3+}[b] T_{1H}	Fe^{3+}[b] $T_{1\rho H}$	Without Fe^{3+} T_{1H}	Without Fe^{3+} $T_{1\rho H}$
Wood pulp	3.0	2.2	78.8	7.0
Lipids	2.5×10^3	18.6	5.0×10^3	21.0
Protein	9.1	3.8	83.0	4.7
Lignin	73.0	7.3	164.0	7.0
Microcrystalline cellulose	2.0	7.3	505.0	

[a] Milliseconds ± 10%.
[b] 1.9%.

From Gerasimowicz, W. V., Hicks, K. B., and Pfeffer, P. E., Macromolecules, 17, 2597, 1984. With permission.

hydrophilic carbohydrate and protein constituents of the model do, however, show signs of significant interaction with the paramagnetic centers. Iron-induced shortening of T_1 is evident in both the protein and carbohydrate. Researchers have exploited this concept through the addition of small quantities of paramagnetic species to pure peptides thereby reducing their proton T_1s and recycling times while simultaneously enhancing the efficiency of the CPMAS experiment.[40,41] Such preferential shortening of T_1 can have a pronounced effect on ^{13}C sensitivity when one uses a recycling time of 4 to 5 T_1s. In our sample, the large spread of proton T_1 values necessitates the use of pulse delay values of 20 and 10 s, respectively, in the absence and presence of 1.9% Fe^{+3} to ensure reestablishment of nearly total proton equilibrium magnetization.

The rate of proton relaxation, $T_{1\rho}$, presents a different problem when trying to establish equal carbon population responses. A CT must be selected which will be fast enough to facilitate carbon-proton energy transfer for the components with rapidly relaxing protons (short $T_{1\rho}$s) as well as those with long $T_{1\rho}$ values.[42,43] This question has been addressed in a paper concerning the analytical reliability of the solid-state ^{13}C spectra of coals.[42] Yoshida et al. examined the effect of pulse repetition rates and CTs on the ^{13}C solid-state spectra of a coal sample and a hexane-soluble fraction.[44] Optimal CTs and pulse delays were required to obtain reliable estimates of carbon aromaticity in their sample.[45] Unlike the heterogeneous models and sludges examined in this work, the coal samples contained compounds of limited diversity, e.g., aliphatic and aromatic hydrocarbons, with a relatively narrow range of $T_{1\rho}$ values. As seen in Table 3, the dynamic range of $T_{1\rho}$ values in the present study is exceedingly large even for the metal-free components (~7.0 ms to 21.0 ms). In order to maximize the absolute ^{13}C responses of the model mixture, a study of the contact time vs. intensity of our model mixture in the presence and absence of Fe^{+3} as found in sludge mixtures was performed. In these systems, with and without Fe^{+3}, the rapid growth of the carbohydrate resonances at ~70 ppm and 102 ppm as well as the aliphatic resonances at ~30 ppm in the range of CTs 0.03 ms to 0.5 ms was observed. From 0.5 ms to 7.0 ms there is a rapid loss of signal from the carbohydrate peaks while the aliphatic peak, having long $T_{1\rho}$ values (21.0 ms no Fe^{+3}, 18.6 ms Fe^{+3}), clearly persists. Cross-relaxation effects on the protein and lignin resonances are more difficult to assess because of their overlap with each other and with the aliphatic region. For both the metal- and nonmetal-containing samples, the maximum response for both components (lipid and carbohydrate) is found at a contact time of about 0.5 ms. Figure 12 illustrates the generation of carbon magnetization for the Fe^{+3}-containing model system as a function of variable CT. At the shorter contact times (<0.5 ms) necessary to overcome the rapid proton relaxation process, the relationship of $T_{CH} << CT$

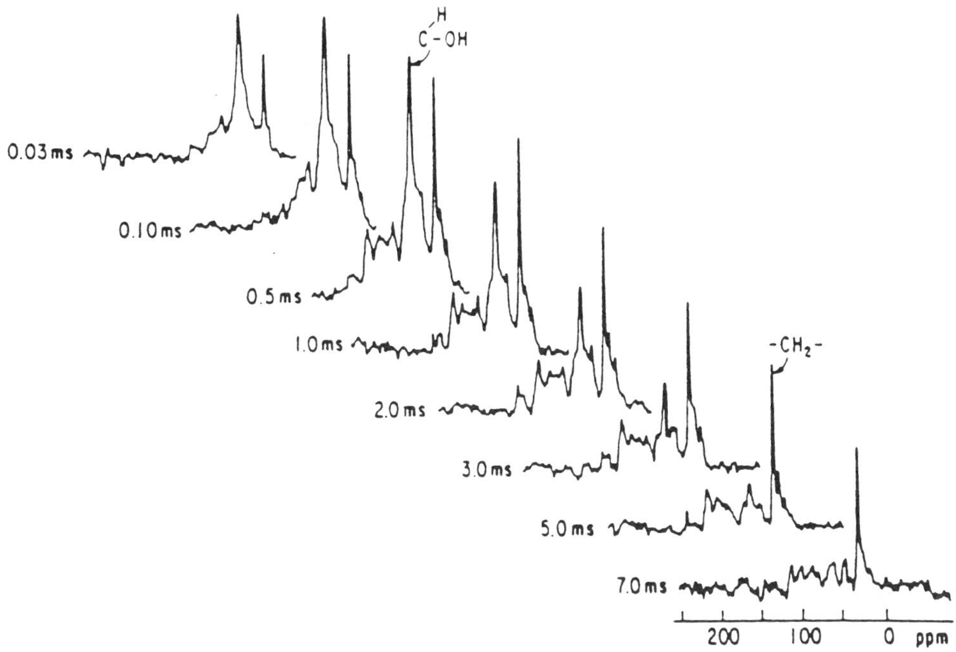

FIGURE 12. CT study of model sludge mixture containing 1.9% Fe^{3+}. Each spectrum was obtained after 5000 scans, a pulse delay of 1.5 s, CTs of 0.03 to 7.0 ms, and 15-Hz line broadening. (From Pfeffer, P. E., Gerasimowicz, W. V., and Piotrowski, E. G., *Anal. Chem.*, 56, 734, 1984. With permission.)

becomes invalid and loss of absolute ^{13}C intensity is apparent. This phenomenon is clearly illustrated when the spectra of the model mixture is examined under optimum conditions with and without metal as seen in Figure 13. In spectrum 13A we see that the ratio of the carbohydrate resonances at ~102 ppm and ~70 ppm (corrected for lignin content) to the aliphatic region at ~30 ppm is (2.2/1.0) ± 10% (theory 2.1/1). Spectrum 13B, in the presence of 1.9% Fe^{+3} with the aliphatic region normalized to the aliphatic region of 13A, shows a ratio of carbohydrate/lipid of only (1.1/1.0) ± 10%. From these data it is obvious that even under these most optimum conditions of short CTs of 0.5 ms and long recycling times of 20 s, 48% of the carbohydrate signal has been lost. At these high Fe^{+3} concentrations (1.9%), preferential loss of proton polarization associated with the carbohydrate species is observed. The Fe^{+3} concentration dependence of the phenomenon can be seen in Table 3.

The lost ^{13}C intensity is due to the loss of magnetization for those carbons that have severely diminished $T_{1\rho}$ values relative to their corresponding cross-relaxation times (T_{CH}s). While ^1H spin polarization is being transferred to all of the components in the model, the short proton $T_{1\rho}$ associated with the carbohydrate component upon paramagnetic doping causes a loss in its signal intensity much more rapidly than any decay seen in the other constituents with their longer $T_{1\rho}$s. The regime of $T_{CH} \ll CT \ll T_{1\rho H}$ is difficult to meet in our model system for the carbohydrate component in the presence of large quantities of paramagnetic iron. This effect is reflected in the wide dynamic range of proton $T_{1\rho}$ values seen in Table 3. If one cannot account for all carbon in a sample in an absolute sense, the fraction of carbon resonances must at least be representative of all carbon in the sample. The preferential localization of Cu in the carbohydrate or hydrophobic portions of soil organic matter has been reported. However, the effects are less dramatic than those shown here. These authors further concluded that carbohydrates such as uronic acids may account for the metal-binding effects. The importance of such soil components and their ability to bind metals is well established.[47,48]

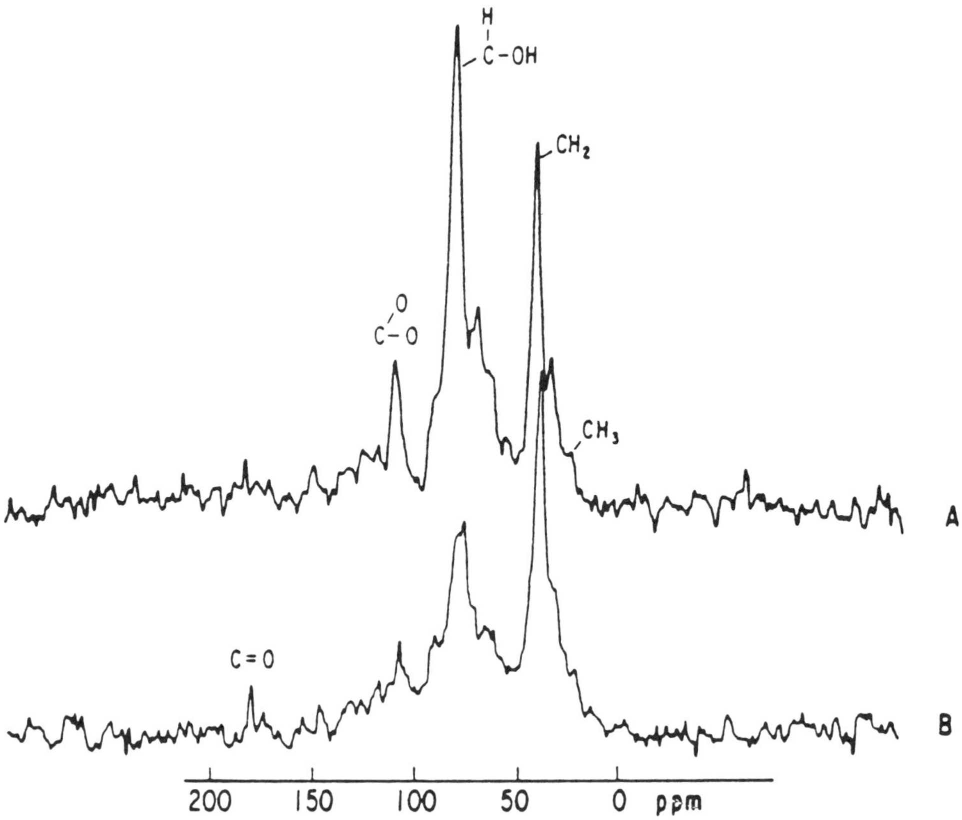

FIGURE 13. Comparison of ^{13}C-CPMAS NMR spectra of model sludge mixtures (A) with no Fe^{3+} and (B) containing 1.9% Fe^{3+}. Each spectrum was obtained under optimized conditions, 14,000 scans, 16-s pulse delays, a 0.5-ms contact time, and 20-Hz line broadening. (From Pfeffer, P. E., Gerasimowicz, W. V., and Piotrowski, E. G., *Anal. Chem.*, 56, 734, 1984. With permission.)

4. Humic Substances

Several recent reviews are available on solid-state NMR and humic matter.[49-53] Our purpose is not to provide another review but rather to cite some further examples and techniques which will be useful to the soil scientist in applications to agricultural systems, and to stimulate further developments in these areas. Humic materials are known to be involved in a number of interrelated, chemical processes in soil, and to influence plant growth directly as well as indirectly. Indirectly, humic acids (HAs) may improve soil fertility by increasing the availability of plant nutrients or by sequestering toxic heavy metals in the soil environment. Kononova suggested early on (1966) that plant roots can assimilate these naturally occurring heteropolymers thereby directly affecting plant growth and metabolism.[54]

Humic substances derived from terrestrial plants differ in chemical structure from those originating from aquatic plants.[55-58] One extraction scheme is depicted in Figure 14.[59,60] Figure 15 shows some representative spectra of HAs.[50] Signals are evident from 110 to 160 ppm (aromatic/olefinic and carboxyl groups are seen from 160 to 190 ppm. Paraffinic and carbohydrate signals constitute the largest area in Figure 15. Carbohydrates originate from the fact that at this soil horizon nondegraded plant litter is present. Therefore, during the treatment with base, hemicelluloses and other carbohydrates are being extracted. Hatcher comments further on the notion that large amounts of paraffinic carbon in HAs may affect the determination of aromaticity for these substances.[50] The relative intensities of the paraffinic resonance ranges from 7.1 to 36% in this case. Aromaticities determined from chemical degradation methods can range from 61 to 100% and are generally higher than those ascertained from NMR methods.[61]

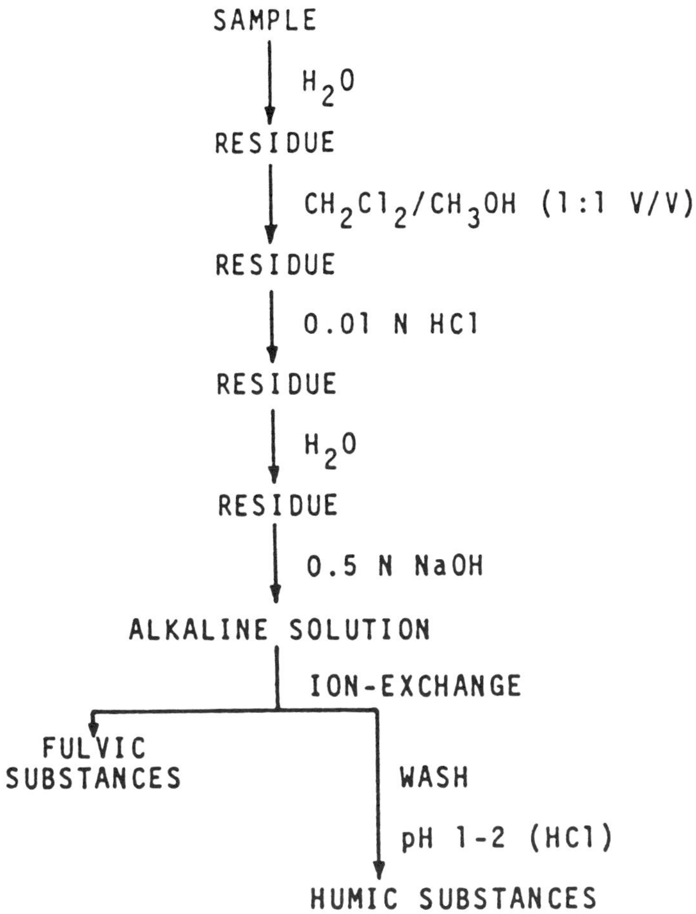

FIGURE 14. Humic substance extraction scheme (all extractions were performed in an N_2 atmosphere). (From Gerasimowicz, W. V. and Byler, D. M., *Soil Sci.*, 139, 270, 1985. With permission.)

Controversy, however, does exist as to the accuracy and detection of all components based upon the considerations mentioned above, and it is still not clear as to which of the methods (traditional or solid-state NMR) is the more accurate. The problem of detection is discussed at length below.

In order to establish stable cropping systems, soil fertility and structure must be maintained. Soil organic matter is vital to this stability in that the organic matter serves as a source of nutrient elements which can be released through mineralization. Cultivation of a soil previously supporting pasture or native vegetation generally leads to reduced levels of soil organic matter.[62,63] Decreases in organic carbon from five Vertisol soils ranged from 31 to 59% after 25 years of continuous cultivation for cereal crops in southern Queensland.[62,63]

Skjemstad et al. studied a soil type which had been under cultivation for as long as 45 years.[64] These soils were Langlands-Logie clays used primarily for wheat production in Queensland. Samples were collected from fields (and fractionated as described in the reference) which varied in their time of continuous cultivation from 20, 35, and 45 years.[64] A sample was also examined from an adjacent area which had never been cultivated and which was under virgin brigalow *(Acacia harpophylla)* scrub. Over the same time course as the cultivation process, organic matter levels dropped by 50, 61, and 66% (as determined by chemical analysis).[64] However, ^{13}C-CPMAS spectroscopy performed directly on these samples, and on the uncultivated virgin soil does not reveal significant differences in the *in situ* organic.

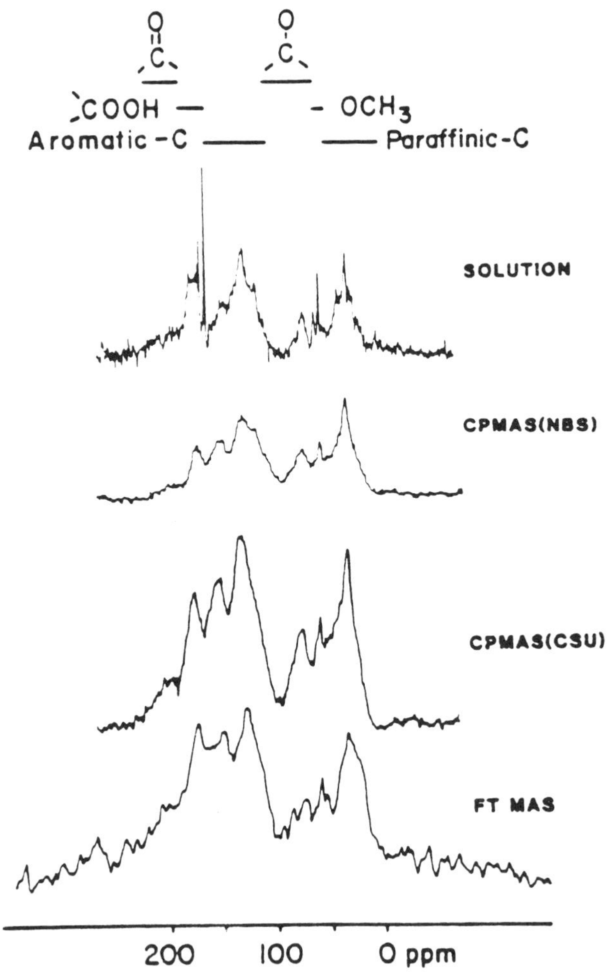

FIGURE 15. ^{13}C-NMR spectra of humic acids from the Rice Lake peat sample obtained in solution (0.5 *N* NaOD) and in solid state by CPMAS and by conventional Fourier transform (FT) with MAS. (From Hatcher, P. G., Berger, I. A., Dennis, L. W., and Maciel, G. E., *Aquatic and Terrestrial Humic Materials*, Christman, R. F. and Gjessing, E. T., Eds., Ann Arbor Science, Ann Arbor, MI, 1983, 37. With permission.)

Such discrepancies may be attributed to lack of CP as well as paramagnetics which may, for example, affect the ratio of aromatic/carbonyl functional groups. Furthermore, the HA fractions were shown to decrease as a function of cultivation time, while the relative concentration of clay-associated matter increased. The authors state that UV-visible and IR analysis agreed with the NMR. The investigators concluded that the most resistant, and therefore stable, organic components are those physically associated with the inorganic components of the soil. As such, these materials were inaccessible to enzymatic or chemical substances, rendering them inert to degradation. In a ^{13}C-NMR solution study, Cartoux and Schnitzer reported the characteristics of the organic matter in particle size fractions separated from an Aquoll.[65] The samples were from the surface horizon of the cultivated Aquoll. HAs and residual humic substances (obtained from the residue of the HAs by treatings with HCl-HF and reextracting) were analyzed. The chemical analyses showed that the five silt fractions had the highest carbon and nitrogen concentrations. The ^{13}C-NMR spectra showed that the HAs derived from the clays were richer in aliphatics than HAs derived from larger classes of particles. The residual HAs were more homogeneous in

Table 4
DETAILS OF SAMPLES USED

Sample	Parent material
Ornithogenic soil	Guano deposits collected from several locations in Antarctica (i.e., Cape Adare, Cape Hallet, Cape Bird, Cape Crozier, and Inexpressible Island)
Antarctic moss	Under dead mosses consisting of comminuted (whole) and visible organic matter of *Bryum antarcticum* and *Bryum argenicum* mosses
Antarctic peat	Polytrichum moss from Signy Island, British Antarctic Peninsula

From Wilson, M. A., Goh, K. M., Collin, P. J., and Greenfield, L. G., *Org. Geochem.* With permission.

composition than the HAs. Thus, their structures are more stable than the HAs. So, these data also appear to be consistent with the idea that the organic matter associated with inorganic soil components is more resistant to environmental decomposition than the unassociated fraction.

In all of the studies cited, the reader should bear in mind that the results apply only to the systems under investigation. One should not necessarily attempt to apply interpretations or trends observed in one system to another system.

Clearly, NMR analysis has demonstrated the variablilty in the structure of organic matter in soils and extracts from the soils. The fraction of aromatic carbon to total carbon has been a particularly sticky problem, where values ranging from 10 to 74% have been reported.[58,61,66] Similarly, large variations in carbohydrate content have been reported.[31,58,61] Specific reasons for such differences have not been clearly elucidated. Furthermore, little information is available as to the reason(s) for relatively high carbohydrate content in soils. Thus, Wilson examined humic substances from the Antarctic continent (terrestrial HAs) where lignin is a negligible portion of the vegetation.[67] The soil humification rate should be very slow compared to warmer climates, thus preserving organic matter in its native state.

The samples obtained are described in Table 4.[67] Solid-state ^{13}C-CPMAS spectra are shown in Figure 16.[67] The Antarctic peat sample appears to be composed predominately of carbohydrate. Resonances are seen at 99.8 ppm from C-1 dioxygenated carbons and at 78.8 and 63 ppm from hydroxylated carbons. The HA fraction also shows carbohydrate at 71 ppm. The overall spectrum of the HA is more complex than that of the peat sample. Resonances at 56 ppm may be assigned to methoxyl carbons or even amino carbons. With dipolar dephasing (see discussion below), the 56 ppm resonance persists, therefore it arises from methoxy as opposed to amino-like moieties.

The spectrum of the fulvic acid fraction from the moss sample (not shown) suggests that the fulvic acid consists of two carbohydrates.[67] The authors assign these carbohydrates to a mixture of α and β glucose units. The fulvic acid fraction appears to be derived from hydrolysis of cellulosic components.

Furthermore, the spectra shown indicate that while aromatic components are evident in soils without appreciable lignin, lignin is not necessarily the precursor for the aromatics. Thus, other factors, such as microbial activity, may contribute and alter aromaticity. Temperature changes have an important environmental impact on humus formation, and when organic turnover rates are slowed down, carbohydrate content is high when compared to soil systems from warmer environments.

Soil humic materials are thought to be a source for some of the dissolved organic carbon (DOC) found in natural waters. Many components and compound types have been identified, such as amino acids, carbohydrates, and fatty acids.[68] The difficulties associated with understanding the chemistry of terrestrial humates are compounded in the case of freshwater and marine environments due to the problem of isolating sufficient quantities of organic matter from such sources. Yet, marine environments are the second largest reserve of organic carbon on this

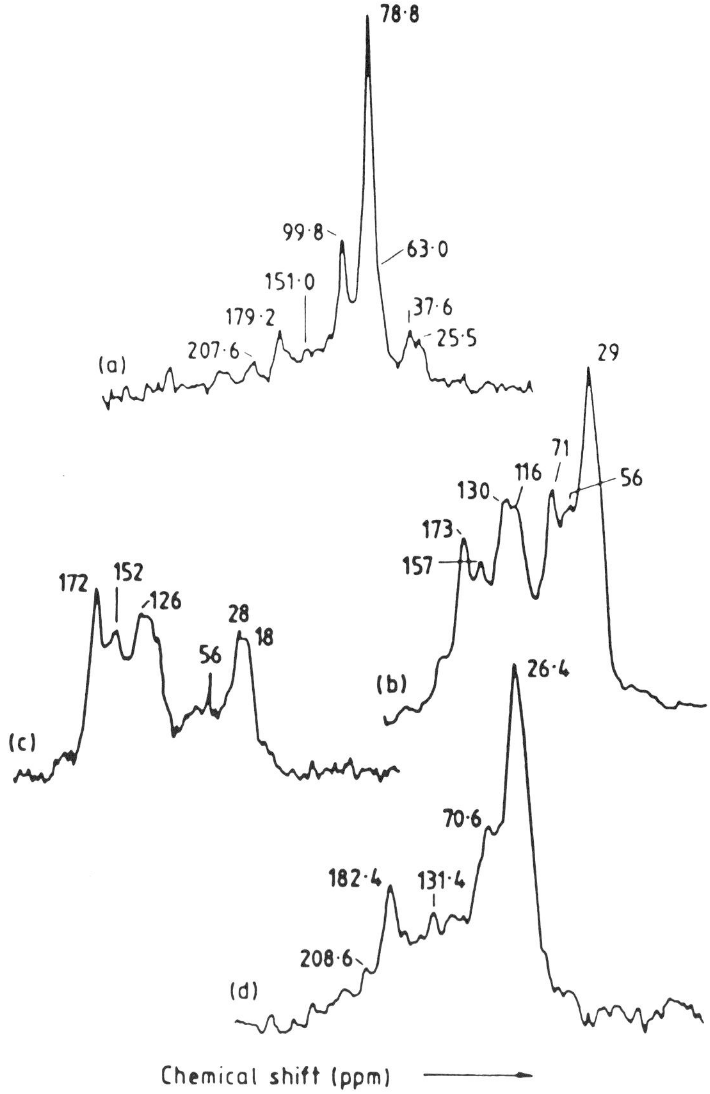

FIGURE 16. ^{13}C CPMAS spectra of (a) Antarctic peat, (b) HA from Antarctic peat, (c) 40-μsec dipolar dephased spectrum of HA from Antarctic peat, and (d) HA from soil under moss.(From Wilson, M. A., Goh, K. M., Collin, P. J., and Greenfield, L. G., *Org. Geochem.*, 9, 225, 1986. With permission.)

planet, (sediments contain the largest reserves) with an estimated 800×10^9 tons, 89% of which is dissolved.[69] Some of the factors which influence humate chemistry and structure are shown in Table 5.[70] Geography, environment, vegetation, and microbial activity all act synergistically to affect soils and humic substances. Transport of humic substances which are water soluble can also affect the composition or mixture found at any particular site or location.

A study has been reported which attempted to link the relationship of dissolved marine humics with phytoplankton.[71] The reader is referred to the reference for a detailed discussion of the isolation procedures. ^{13}C-CPMAS spectra are shown in Figure 17. The usual ^{13}C resonance assignments can be made

1. 0 to 60 ppm, aliphatic carbon

Table 5
FACTORS INFLUENCING THE STRUCTURE OF HUMIC SUBSTANCES

Primary factor	Secondary factor	Influence
Geographical location/climate	Type of plant life	Produces variations in proporations of groups of chemical compounds, e.g., lignins
	Microbial activity	Transformation (biodegradation/ polymerization) or organic materials
	Transport of humic substances	Structure of aquatic humic substances may reflect terrestrial input

From Gillman, A. H. and Wilson, M. A., *Org. Geochem.*, 8, 15, 1985. With permission.

2. 60 to 95 ppm, carbon singly-substituted with electron withdrawing elements such as oxygen or nitrogen in the case of humic substances
3. 95 to 110 ppm, dioxygenated carbon
4. 110 to 160 ppm, aromatic and olefinic carbon
5. 160 to 190 ppm, ester, amide, carbonyl, and salts thereof

Further analytical information is shown in Table 6 for the HA as well as for intra- and extracellular material and exudates of *P. tricornutum*. The marine humic material contains a large proportion of highly branched alkyl chains, and a smaller amount of aromatic. The authors also point out the presence of appreciable carbohydrate (uronic acid(s)).[71] GC-MS analysis of the *P. tricornutum* exudates allows identification of furans (from carbohydrates), pyrroles, nitrites, phenols, and methylphenols.[71] Therefore, one can see the similarities between the exudates and the dissolved marine humic substances.

Dissolved organic matter (DOM) in pore waters from sediments of different depositional environments has also been examined by ^{13}C-CPMAS.[72] The DOM appears to exist mostly as two major types: one type is largely composed of carbohydrates and paraffinic functionality, whereas the second type is dominated by paraffinic and aromatic groups.[72] The authors conclude that the dominance of one DOM structural type over the other is dependent upon the oxidizing and/or reducing capacity of the sediment from which the materials are derived. Under one set of conditions (anaerobic) microbes can degrade carbohydrates and accumulate in the pore waters. Aerobic conditions show reduced carbohydrate and enhanced aromaticity, probably from lignin components.[72]

In the case of one particular aquatic humic substance, aromatic ketone groups were found to comprise the major portion of the carbonyl functionality.[73] Spectra of underivatized HAs and fulvic acids were compared and contrasted with chemically modified derivatives. The authors postulate that the mechanism of aromatic-ketone group formation could be attributed to photolytic rearrangement and oxidation of hydrocarbon and phenolic ester precursors.

The literature contains additional discrepencies in such measurements. For example, a recent comparison of solution and solid state spectra of HAs shows a nonconformity when functional group composition is determined for samples from Ap horizons of Typic Borolls near Indian Head, Saskatchewan.[74] These discrepancies were particularly emphasized in terms of the CO_2H and phenolic regions.

Chemical degredation of four different humus types from Norway and analysis by ^{13}C-CPMAS NMR confirmed that 40 to 50% of the soil carbon is unaccounted for from among the degredation products.[75] More recently, spin counting experiments have been utilized in an attempt to account for the carbon in the HA and fulvic acid of geochemical samples, and the HA and fulvic acid fractions of soil organic matter.[76,77] The spin counting experiments involved obtaining spectra on mixtures of sample and glycine at varying contact times.[76,77] These spin

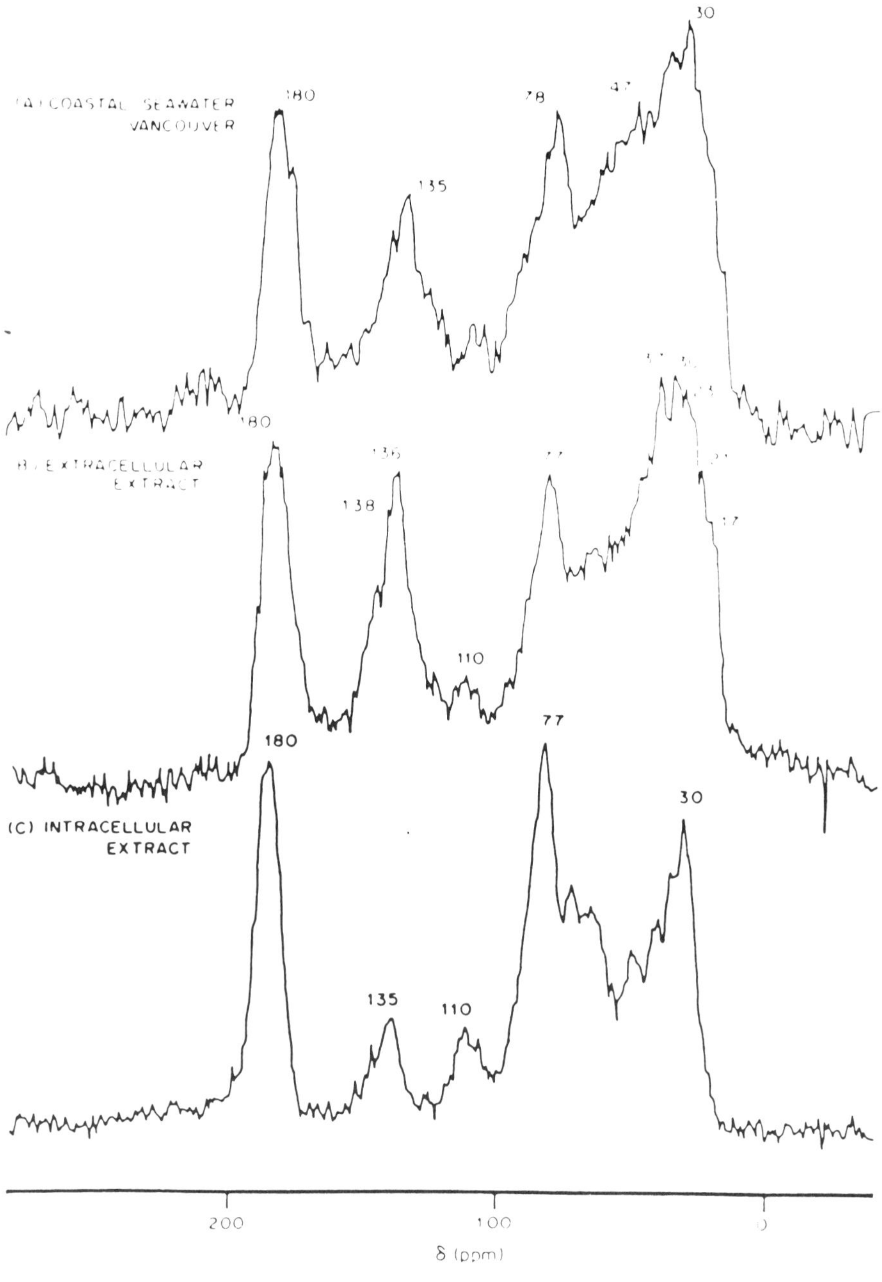

FIGURE 17. ^{13}C-CPMAS spectra of humic materials. Resonances identified in Table 6. (From Gillman, A. H. and Wilson, M. A., *Org. Geochem.*, 8, 15, 1985. With permission.)

counting experiments demonstrated that less than 50% of the carbon from the various fractions examined was detectable in the NMR experiment. The differences were apparent even after correction for differential relaxation. Once again, the amount of aromatic and noncarbonyl aliphatic carbon was not in agreement with that determined by calculation methods.[76,77]

The question of aromaticity in soil humates has also been raised recently in a different light.[78] ^{13}C-CPMAS studies were carried out on soil samples from the A horizons of a Typic Rendoll, Typic Xerorthent and two Typic Xerochrept(s) on different parent rocks (granite and shale, respectively).[78] The researchers found that the humic fractions were more aromatic than the

Table 6
RELATIVE AMOUNTS OF CARBON AND HYDROGEN TYPES IN HUMIC MATERIALS

Dissolved Seawater Humic Substances

	% Hydrogen types					% Carbon types					
		Aryl		*O*-Alkyl							
Sample	Method	H_{AR} (6.1—9.0)	H_{R-0} (3.0—4.7)	H[a] (1.8—3.0)	H[b] (0.5—1.8)	Method	Carboxyl[c] (160—190)	Aryl[d] (110—160)	Acetal (95—110)	*O*-Alkyl (60—95)	Alkyl (0—60)
Coastal sea water	—[e]	7	19	32	42	—[f]	14	18	2	18	49
P. tricornutum Extracellular material	—[e]	5	22	27	46	—[f]	15	19	3	15	49
P. tricornutum Intracellular material	—[e]	6	30	26	37	—[f]	21	6	7	42	25

[a] Largely hydrogen attached to carbon α to aromatic rings or electron-withdrawing groups, such as carboxyl.
[b] Mainly alkyl protons β or further from aromatic rings or electron-withdrawing groups.
[c] Includes salt, ester, and amide groups.
[d] Includes olefinic carbon.
[e] By solution F.T. NMR.
[f] By CP/MAS-NMR.

From Gillman, A. H. and Wilson, M. A., *Org. Geochem.*, 8, 15, 1985. With permission.

humin fractions. The humic substances appear to have lower polysaccharide contents when the spectra are compared to those of the humins. The results were explained on the basis of the procedure employed in isolating the fulvic acids. The fulvic acids were isolated by using Polycar AT (a polyvinyl pyrrolidone polymer). Polycar AT preferentially retains polyphenols while partially excluding the polysaccharide fraction present in the extract at that point.[78] Fulvic acids isolated by means of adsorption onto charcoal exhibit higher polysaccharide concentrations than analogous Polycar AT preparations. Thus, the manner of isolation can impact on the final result observed spectroscopically.

Melanoidins have been synthesized from amino acids and sugars in various ratios and spectroscopically (^{13}C-CPMAS) compared with humic substances from the Hula Basin in Israel.[79] The spectra showed significant similarities between the melanoidins and humics. The authors suggest that the Maillard reaction may be more important in the formation of humic materials than lignin-derived sources as traditionally assumed.[79]

One other important aspect of this report is the fact that aromaticities calculated from NMR data of the melanoidins are quite close to the aromaticities of both marine and terrestrial humic substances.[79] Thus, the issue of aromaticity persists.

Earl et al. have addressed relaxation processes which occur in fulvic acids and the effect of lowering the temperature in order to assure a quantitative response in CPMAS NMR spectra of this type.[80] Generally, researchers in this area have selected contact times of 1.0 to 2.0 ms with the belief that the correct relative intensities will be obtained even though the total response may not represent fully cross-polarized carbon. Such assumptions do not necessarily hold for heterogeneous mixtures such as fulvic acid. Other authors point out that CPMAS spectra of certain fulvic acids do not agree with their analogous high resolution solution spectra taken under quantitative conditions.[81,82] The most likely cause of such discrepencies was attributed to the CP and relaxation dynamics of the solid.

Figure 18 shows a series of ^{13}C spectra of Armadale fulvic acid taken under a number of conditions.[80] Spectrum A is the high-resolution solution data where the intensities are thought to be representative of the actual carbon content of the sample. Spectrum B illustrates the response upon Fourier transformation of simple Bloch decay data using MAS and high-power decoupling (no CP) at room temperature. Spinning sidebands are a cause of intensity distortion in this spectrum, making it relatively useless for quantitative evaluation. Figure 18C is a CPMAS spectrum (room temperature) of the sample. Intensity distortions are readily obvious due to varying relaxation domains as well as carbon-protein distortions in this material. Figure 18D depicts the spectrum under the same conditions as for C other than the temperature (lowered to –57 °C). Under such conditions several effects are assumed:

1. All extreme motions are frozen out
2. $T_{CH} << T_{1\rho H}$
3. S/N is increased by a factor of three when compared to the room temperature data

The authors also point out that various chemical exchange reactions may be ongoing in these complex systems. Proton exchange could result in line broadening and unexpected chemical shifts. Line shapes are also a direct function of temperature.

These results strongly suggest that a combination of VT experiments, Bloch decay, and CT measurements are necessary to address the problems of molecular mobility and physical heterogeneity. A combination of such techniques is needed if one desires to advance beyond simple organic molecules in order to study complex natural products.

Clay minerals which are classified as smectites have the propensity to adsorb and take up small organic and polymeric species into their interlayers.[83] This intercalation or interlayer complex formation is generally seen to cause an increase in the basal spacing to distances of 10 Å or more. Through a combination of X-ray diffraction and ^{13}C-CPMAS along with chemical

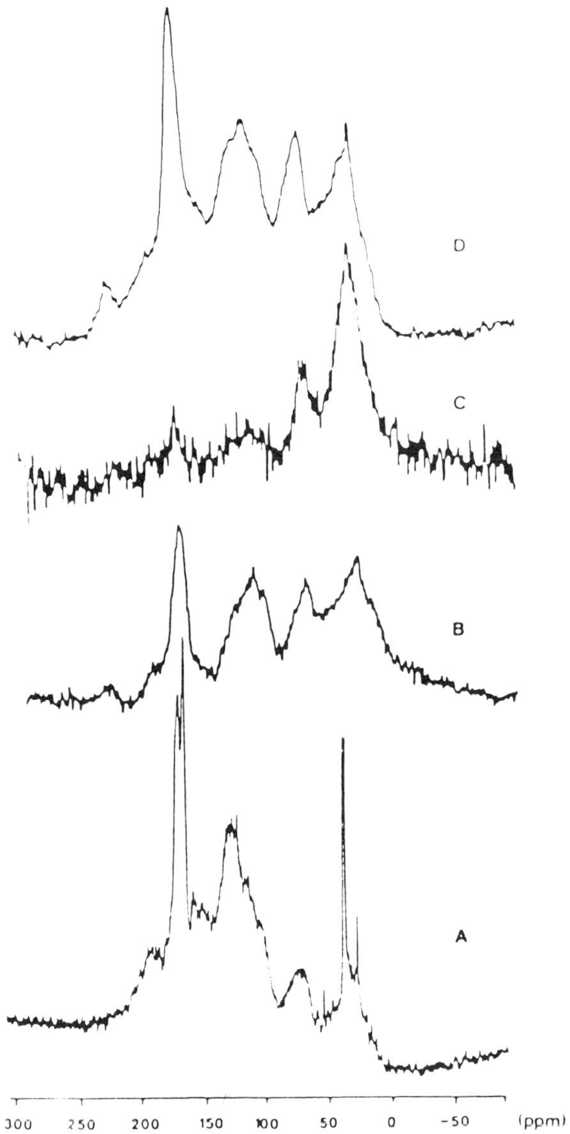

FIGURE 18. Four ^{13}C spectra of the Armadale fulvic acid taken under different conditions. Spectrum A is the 75-MHz solution spectrum taken with 7500 scans using π/2 pulse, a total interpulse delay of 10.2 s and inverse gated decoupling on a solution of 300 mg of sample dissolved in 2 ml of $_{13}C$ depleted DMSO-d_6 which is the sharp, truncated peak at 39.5 ppm. Spectrum B is the 50-MHz Fourier transform of a Bloch decay taken with 8192 scans at 20°C using a π/2 pulse and a decoupling field strength of 60 kHz with MAS. Spectrum C is the 50-MHz CP spectrum at 20°C with 4096 scans and 0.5 ms CP CT. Spectrum D is the CP spectrum taken at –57°C with 4568 scans and a contact time of 1.5 ms. (From Earl, W. L., Wershaw, R. L., and Thorn, R. A., *J. Magn. Reson.*, 74, 264, 1987. With permission.)

and heat pretreatments, reasonable evidence has been presented for the occurrence of interlayer clay-organic complexes in several New Zealand soils.[84] The clay-mineral can be described as a mica-smectite. The organic species in question is characteristic of a humic substance with a polymethylene chainlike structure.[84] The organic complex material was resistant to pyrophosphate extraction, decomposition, and peroxidation as seen in the NMR spectra. Furthermore, in the complex form, thermal stability up to 400°C was observed. Temperatures had to be raised above these levels in order to observe extensive decomposition.[84] Certain soil qualities were necessary in order to achieve such stability. These factors include a minerology of the clay smectite type, low microbial perturbation, and a highly acid soil environment.[84]

5. Soil/Plant Components: Lignin-Carbohydrates — Interrupted Decoupling

Many terrestrial humic and soil components are aromatic. Much of this aromaticity is derived from vascular plant debris consisting of lignin and lignin analogue structures which impart the aromaticity to these systems. ^{13}C-^{1}H interrupted decoupling and enzymatic degradation can be employed to infer details of structure in such constituents. A wood pulp sample *(Piceae glauca)* whose analytical characteristics have been previously described will be utilized to illustrate some of the salient features of interrupted decoupling.[85] The chemical composition of *Piceae glauca* pulp (PGP) consists predominantly of two major components; lignin and cellulose. Several ^{13}C-CPMAS NMR studies of lignin and cellulose in the solid state have been published.[86-92] These studies have focused on the assignment of resolvable resonances for these wood constituents. The assignments of the various peaks and signals from these reports on lignin, cellulose, and wood will be utilized in the interpretation of the ^{13}C-CPMAS spectra presented here. Although some of the referenced papers contain ^{13}C NMR solid-state spectra of lignin and cellulose from instruments operating at higher ^{13}C magnetic field strengths than that in our laboratories, the resolution obtained (upon careful examination of the published results) is not significantly enhanced over that observed here.

Figure 19A depicts the ^{13}C-CPMAS NMR spectrum of the PGP. Several resonance lines are clearly resolved in this spectrum. However, these bands are relatively broad and reflect the complex chemical composition of the sample, as well as its amorphous, noncrystalline structure. The rather broad resonance line widths also reflect the thermomechanical treatment, subsequent grinding, and preparation of the samples. Amorphous samples, such as those examined here, generally produce fairly wide NMR lines in solid-state ^{13}C-CPMAS spectra. The resonance at 105 ppm may be slightly narrower than similar resonances in ball-milled wood spectra.[93] This slight narrowing may reflect a small degree of crystallinity or nonamorphous regions in our samples. However, the extensive ball-milling as well as the spectral line widths observed essentially define the morphology of the PGP utilized in this work as amorphous. The spectrum shows that local molecular environments are nonequivalent within this polymeric network. The resolution achieved, however, is still adequate for identifying the major components of the PGP. The ^{13}C NMR spectrum is essentially dominated by resonances that are attributed to components arising from cellulose and hemicellulose. Carbon nuclei from these carbohydrate moieties give rise to signals between 60 and 110 ppm. The peak at approximately 105 ppm corresponds to the anomeric or C1 carbon from β-D-glucopyranose repeating units of cellulose. This resonance is particularly well resolved. The broad signals from the cellulose are superimposed on those from the hexose and pentose units of the hemicellulose compounds.

The spectrum also contains features which are less intense. The shoulder at 56.5 ppm may be assigned to methoxyl groups from lignin and also from hemicellulose. The aromatic components of lignin are seen between 130 and 155 ppm. These less intense features and the ^{13}C-CPMAS spectra cited and discussed so far only provide limited, qualitative information about the structure of these compounds. Other chemical and NMR methodologies are necessary to examine the chemical and structural relationship (connectivity) between lignin and carbohydrate in PGP and to enhance the underlying and unresolved components in the spectra obtained.

In order to enhance the spectral resolution of the noncarbohydrate components of the PGP, the ^{13}C-^{1}H dipolar dephasing interaction was utilized.[94] Many examples now exist in the literature where dipolar dephasing provides spectra consisting only of resonances from

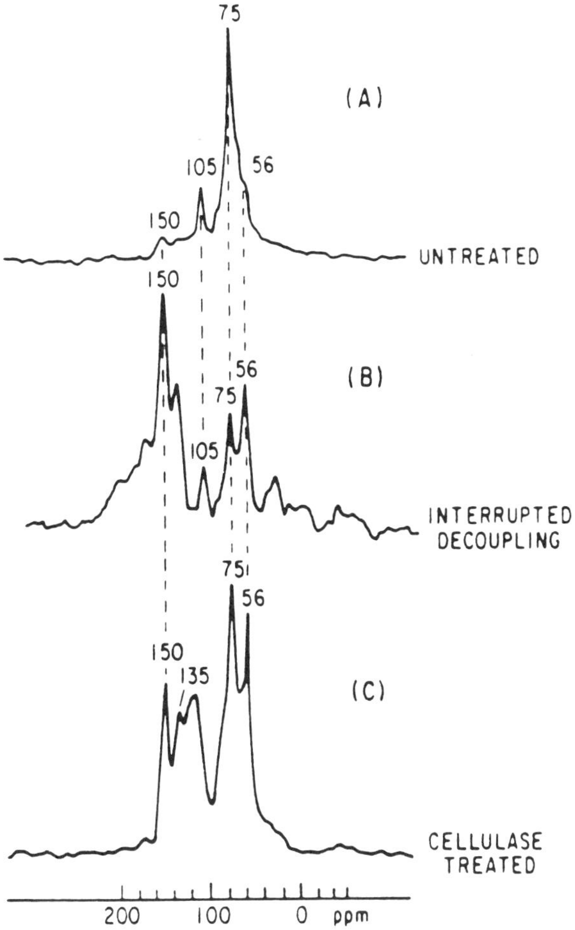

FIGURE 19. *Piceae glauca* pulp ^{13}C-CPMAS spectra. (From Gerasimowicz, W. V., Hicks, K. B., and Pfeffer, P. E., *Macromolecules*, 17, 2597, 1984. With permission.)

nonprotonated carbon nuclei and methyl groups.[94-98] Certain functional groups, such as methyl groups, undergo rapid, internal rotations, even in the solid state. Such motion attenuates the ^{13}C-^{1}H dipolar relaxation mechanism rather markedly, with the result that resonances from such carbon nuclei are attenuated but not entirely eliminated from the spectrum. Even though the internal motion of the methyl group reduces the dipolar effect because of angular averaging, such groups usually have distinct chemical shifts and are easily assigned in solid-state CPMAS ^{13}C NMR spectra. Although the interrupted decoupling technique is a somewhat phenomenological method and not totally quantitative, the method can effectively increase the resolution one may achieve in ^{13}C-CPMAS studies. In essence, dipolar dephasing is a very promising spectral editing method for the solid-state.

Figure 19B depicts the spectrum of the PGP after a decoupling delay time of 40 μs. Almost all of the ^{13}C resonance responses from the carbohydrate portion of the sample have been attenuated or eliminated entirely from the spectrum. Some loss in the signal-to-noise ratio (approximately 25 to 30%) is apparent in such an experiment as would be expected due to some dipolar dephasing of the remaining nonprotonated and methyl resonance signal intensities. ^{13}C resonances are still evident at 105 ppm and some response is still seen at 75 ppm. When the magnitude of the signal at 75 ppm is compared to that at 150 ppm in Figure 19A, the domination

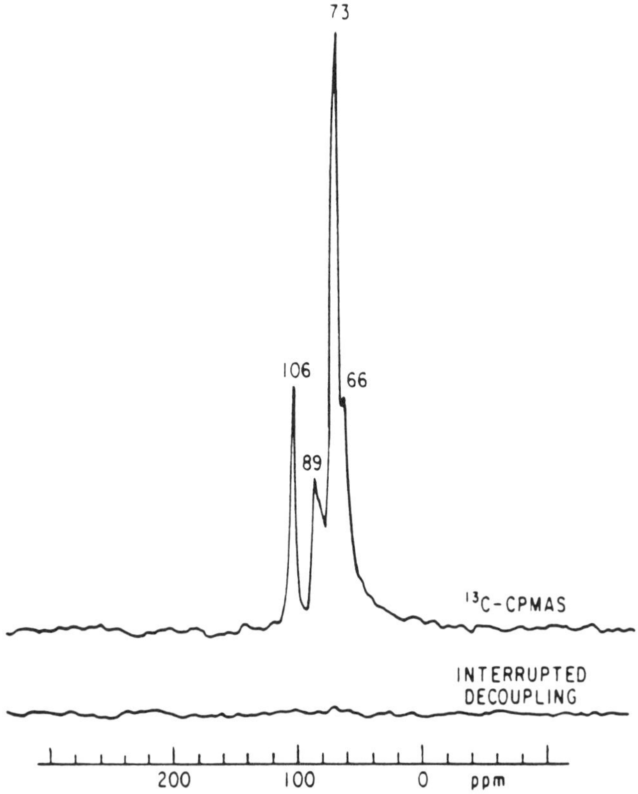

FIGURE 20. Microcrystalline cellulose ^{13C}CPMAS spectra. (From Gerasimowicz, W. V., Hicks, K. B., and Pfeffer, P. E., *Macromolecules*, 17, 2597, 1984. With permission.)

of the former is noticed. The 75 ppm resonance is only seen as a residual signal when compared with the 150 ppm resonance in Figure 19B. Therefore, this residual signal is probably from structures that are very low in concentration and does not adversely affect the values of the relaxation times (determined by standard T_{1H} methodology) for the 75 ppm resonances in Figure 19A. In order to determine whether those residual signals at 75 ppm were from ordered cellulose molecules, a microcrystalline sample (Avicel) was examined under standard ^{13}C-CPMAS conditions and under identical conditions but with interrupted ^{13}C-^{1}H dipolar dephasing (Figure 20). When the microcrystalline cellulose is examined with dipolar dephasing, no signals are evident either for the cellulose ring or for the anomeric carbon nuclei. Thus, the peaks in Figure 19B at approximately 75 and 105 ppm are not from ordered, crystalline cellulose constituents in the PGP matrix. In several reports of the ^{13}C NMR solution spectra of lignins, investigators have proposed that ^{13}C signals between about 102 and 108 ppm have as much likelihood of being due to ^{13}C resonances from sinapyl structures as from anomeric carbon nuclei of glycosides.[99,100] Thus, the ^{13}C resonance at 105 ppm in Figure 19B may be associated with lignin polymer, especially due to the corresponding strong intensity at 150 to 155 ppm. Another possible assignment for this resonance could be C-2 nonprotonated carbon nuclei of associated keto sugar polymers. However, the noncarbohydrate and other previously unresolved components are no longer obscured as seen when Figure 19B is compared with Figure 19A. The features that were previously hidden, such as the lignin methoxy at 56 ppm and the aromatic carbon moieties between 130 and 155 ppm, now dominate the spectrum in Figure 19B.

In order to measure the T_{1H} values for these components, we have combined the T_{1H} inversion-recovery sequence with the ^{13}C-^{1}H dipolar dephasing mechanism. Under very careful Hartmann-

Table 7
PICEAE GLAUCA PULP COMPONENT T_{1H} RELAXATION VALUES[a] (INTERRUPTED DECOUPLING)

	^{13}C chemical shift[b]			
$[Fe^{3+}]$[c]	**150**	**135**	**74**	**55**
153	36.9	36.5	51.4	42.1
184	23.0	25.7	44.2	30.1
419	23.3	24.7	31.3	27.8
2860	14.1	12.3	17.2	14.3
25400	4.5	5.9	9.2	7.7

[a] ms ± 10%.
[b] ppm.
[c] Concentration in ppm.

From Gerasimowicz, W. V., Hicks, K. B., and Pfeffer, P. E., *Macromolecules*, 17, 2597, 1984. With permission.

Table 8
RELAXATION VALUES[a] OF *PICEAE GLAUCA* PULP[b]

$[Fe^{3+}]$[c]	T_{1H}	$T_{1\rho H}$	T_{CH}[d]
153[e]	78.8 (82.0)[f]	9.9	0.095
184	79.0 (82.6)	9.2	
419	53.2 (55.9)	7.8	
2860	13.1 (13.8)	6.8	
25400	2.9 (2.9)	2.7	

[a] ms ± 10%.
[b] Carbohydrate region of ^{13}C spectra at about 75 ppm.
[c] Concentration in ppm.
[d] ±25%.
[e] Iron content of zero-doped *Piceae glauca* pulp.
[f] Values in parentheses represent T_{1H} values for anomeric carbohydrate resonances at about 105 ppm.

From Gerasimowicz, W. V., Hicks, K. B., and Pfeffer, P. E., *Macromolecules*, 17, 2597, 1984. With permission.

Hahn matching conditions for spectral acquisition, the T_{1H} relaxation values were determined for these now-visible components. The results are listed in Table 7. Unlike the homogeneous relaxation behavior observed for the wood pulp in Table 8, the relaxation of the various resonances is heterogeneous. Proton spin-lattice relaxation effects are not uniform throughout the sample.

Fe^{+3} induces shortening of proton spin-lattice relaxation times across the board in homogeneous samples. The residual resonances seen at a ^{13}C chemical shift of 74 ppm at an Fe^{+3} concentration of 153 ppm do not have the same T_{1H} as observed for the overall wood pulp carbohydrate seen in Table 8 under the same conditions. This result suggests that these structures may be isolated and different in nature than the majority of the cellulose or hemicellulose previously observed under the normal noninterrupted decoupling conditions. Furthermore,

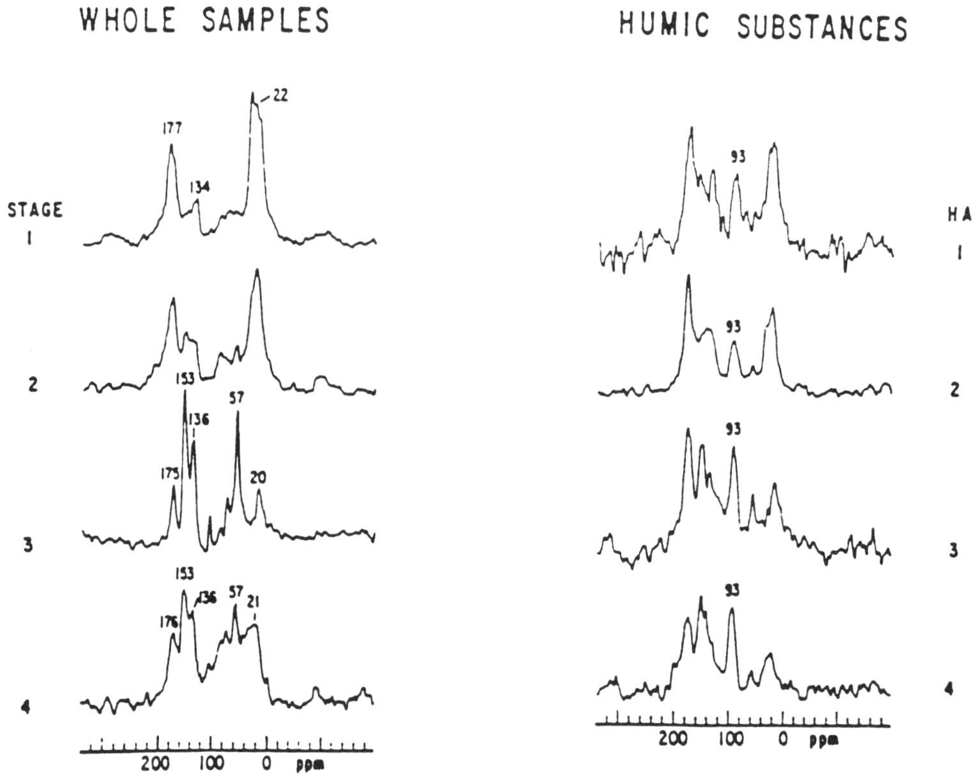

FIGURE 21. ^{13}C-CPMAS spectra of whole sludges and humic substances with ^{1}H-^{13}C interrupted decoupling. (From Gerasimowicz, W. V. and Byler, D. M., *Soil Sci.*, 139, 270, 1985. With permission.)

these structures may be somewhat more mobile in the solid than more rigid cellulose-like domains resulting in localized attenuation of the ^{13}C-^{1}H dipolar interaction. Such differences in native wood carbohydrate, lignin, and aromatic relaxation times would not be inconsistent with the fact that molecular weight distributions of polysaccharides in typical wood and plant polymers point to their polydisperse nature, as well as their very large size.[101] Zumbulyadis has shown that multicomponent proton relaxation behavior is useful in examing crystalline and amorphous domains in poly(ethylene)terephthalate, and has suggested that proton spin-lattice relaxation times may be used to estimate domain sizes in heterogeneous systems, particularly polymers.[102] Our work extends the utility of this concept to naturally occurring solid hetero-polymers.

One final aspect of this system revolves around the data shown in Figure 19C. This figure represents the standard ^{13}C-CPMAS NMR spectrum of cellulase-treated PGP. A remarkable qualitative similarity is noticeable in the resolvable ^{13}C chemical shifts when Figure 19B is compared with 19C. Due to their nonprotonated nature (OCH_3 excepted), the interrupted decoupling technique seems to differentiate the noncarbohydrate components without any chemical or enzymatic degradation of the cellulose. The utility of this method is thus further exemplified.

Another example of this technique is shown in Figure 21. To facilitate comparison, this figure depicts the ^{13}C-CPMAS spectra of both whole sludge samples and the HAs extracted from these materials.[39,59] One important consideration concerning the spectra shown in Figure 21A is that these data were obtained with no chemical modification whatsoever of the original, whole sample. The interrupted-decoupling technique permits some resolution enhancement of the substituted aromatic and remotely protonated carbon nuclei of the HA components. (See

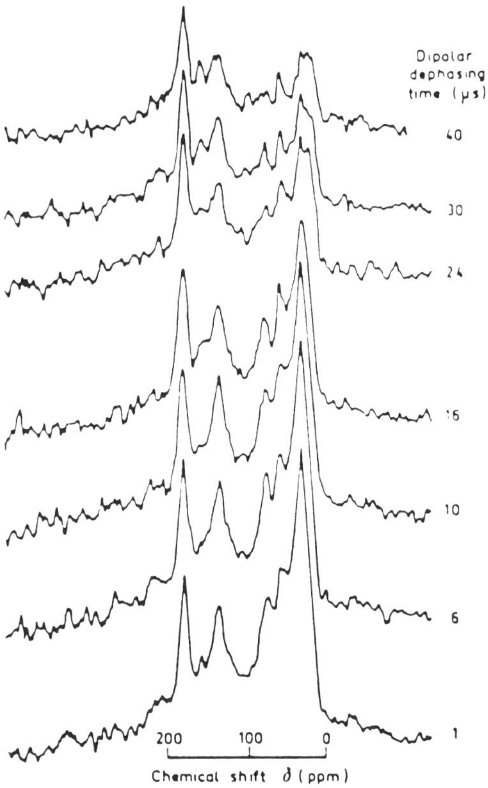

FIGURE 22. Effects of varying the dipolar dephasing time T_{DD} on spectral features in the NMR spectrum of sedimentary HA. (From Gillam, A. H., Wilson, M. A., and Collin, P. J., *Org. Geochem.*, 11, 91, 1987. With permission.)

specifically the spectral frequency from about 135 to 155 ppm.) One can now more readily discern peaks centered around 153 ppm, which are O- or N-substituted aromatic C nuclei, as well as resonances at approximately 136 ppm, which may be assigned to aromatic or olefinic nuclei that are not electronegatively substituted (Figure 21). The aliphatic, or paraffinic region centered around 22 ppm is also more clearly resolved in all cases shown in Figure 21 than in any of the standard ^{13}C-CPMAS spectra in Figure 15, for example. The envelope of resonance peaks centered around 22 ppm may also be indicative of carbon-to-carbon branch points, rather than straight-chain methylene groups.

^{13}C-^{1}H dipolar dephasing NMR has been employed in a structural analysis of sea loch sedimentary humic material.[103] A sea loch is an enclosed, highly productive area along a sea coast. Quite often these areas are fjordic in nature. Increasingly, commercial operations, such as fish farms and paper pulping mills are being established in such areas. Therefore, the natural environment may be disrupted over time. Knowledge of the structure and composition of sedimentary humic substances from sea lochs which have not been polluted or disturbed would be beneficial in establishing the conditions of the habitat at natural equilibrium. Furthermore, one could then assess possible future interactions of the sediment with cellulosic materials.

Such a study has been reported on sediment samples collected from Loch Thurnaig, approximately 1000 m in diameter and semienclosed.[103] The effect of varying the dipolar dephasing time on the NMR spectrum of the humic acid is shown in Figure 22. The resonance at 155 ppm is more prominent, showing nonprotonation. This signal has been attributed to phenols or an aryl ether functionality. A means of distinguishing between methoxy and amino acid carbon is to study rates of signal loss as a function of interrupted decoupling time. The rate of decay for the resonance at 55 ppm is shorter than that for methoxyl groups alone.[103] Although one cannot rule out the presence of methoxy carbon in this system, certainly unresolved amino acid appears to contribute to that area of the spectrum. However, one must be aware that errors arise in any attempt to quantify carbon types by measurements of this type.[95-98] Aromaticity was found to be higher in the sedimentary HA fraction when compared to the sedimentary fulvic acid fraction. Furthermore, the DOM appeared to resemble marine humic matter more closely than terrestrial derivatives.[103]

Similarly, oil shale deposits can be examined in order to evaluate the potential of such resources. These same spectroscopic techniques have been employed in examining various kerogen concentrates.[104] The NMR was able to distinguish between highly aliphatic Type I algal kerogens and kerogens derived from carbonaceous shales which have significant aromatic carbon fractions and resemble Type III kerogens.[104]

Spectra of the carbonaceous kerogens appear similar to those of lignite which are suggestive of algal and terrestrial sources. The interrupted decoupling spectra reveal oxygen-substituted carbons in the carbonaceous shales. This structural characteristic may be related to its ability to undergo CO_2 evolution at low temperature during heating, and to subsequent conversion levels upon pyrolysis to oil and gas products.[104]

Decomposition of leaves in soil has also been studied by ^{13}C-NMR with and without interrupted decoupling, as well as chemical structure of fossil leaves.[105] The results show that fossil leaves and rhizomes retain carbohydrates, and on the basis of carbon content, the concentrations of carbohydrate in fossil rhizomes and their living counterparts are similar. Fossilized leaves, however, contain much less carbohydrate than do their living counterparts.[105] The organic matter associated with the fossilized leaves was quite different in structure than the fossilized leaves themselves. This associated organic matter seemed to resemble humic substances.[105] Thus, the leaves and their associated organics may be subject to rather different transformation processes in their natural habitat.

B. Some Other Nuclei in CPMAS Studies of Soils

1. ^{31}P NMR Studies in Soil Systems

Several examples of ^{31}P-CPMAS spectra of whole soils have now appeared in the literature. In 1981, Williams reported a ^{31}P spectrum of whole soil.[106] Recently Frye has published ^{31}P-CPMAS spectra of HAs and fulvic acids.[107] These spectra are shown in Figures 23 and 24.[107] The authors concluded that the single pulse spectra (those acquired with MAS and high-power decoupling but without CP) are as easily acquired as the CP data. However, the overall spectral resolution is still very poor. Thus, difficulties remain in the interpretation of such spectra due to line broadening. The broadening effects may be caused by heterogeneity inherent in the samples themselves.

Turner et al. have obtained high-resolution MAS ^{31}P-NMR spectra of a variety of orthophosphates.[108] Table 9 shows a listing of the compounds, chemical shift ranges, average P-O bond length, and other information is tabulated. These workers were able to develop an empirical correlation between the isotropic chemical shift and $Z\sqrt{r}$, where Z is the cationic charge and r is the cationic radius.[108]

Many of the compounds displayed rather small chemical shift anisotropies, and many of the spectra were dominated by the main isotropic peak (see Figure 25).[108] The CSA was demonstrated to vary in an approximately linear fashion with P-O bond length, and with the deviation of the O-P-O bond angle from tetrahedral.[108]

Such model systems may prove useful in testing phosphate bonding in soils and other heterogeneous or amorphous systems. Clearly however, better resolution in the ^{31}P spectra of soils must be achieved first in order to apply the techniques described. Thus, ^{31}P-CPMAS, although possible, is still not nearly as useful as ^{13}C-CPMAS in the analysis of soils. However,

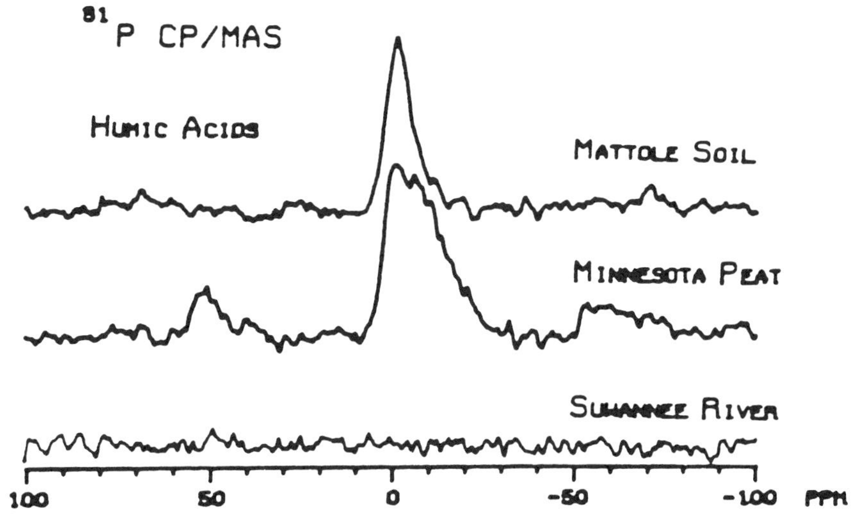

FIGURE 23. ^{31}P-CPMAS spectra of HAs. Approximately 10,000 scans were taken on a modified Nicolet NT-150 spectrometer, using a 2-ms CT, 2-s repetition time, and a spinning rate of 3.5 kHz. (From Frye, J. S., Bonnimann, C. E., and Maciel, G. E., *NMR of Humic Substances and Coal*, Wershaw, W. L. and Makita, M. A., Eds., Lewis Publishers, Chelsea, MI, 1987, 33. With permission.)

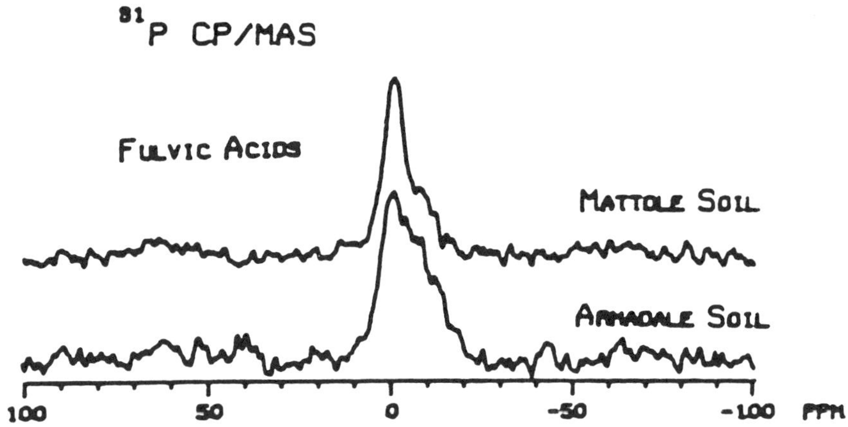

FIGURE 24. ^{31}P-CPMAS spectra of fulvic acids. Experimental conditions as in Figure 23. (From Frye, J. S., Bonnimann, C. E., and Maciel, G. E., *NMR of Humic Substances and Coal*, Wershaw, W. L. and Makita, M. A., Eds., Lewis Publishers, Chelsea, MI, 1987, 33. With permission.)

direct observation of the ^{31}P nuclei and small differences between samples is feasible and adds another tool to the arsenal of the soil scientist.

2. ^{15}N NMR Studies in Soil Systems

^{15}N-CPMAS NMR studies of enriched humic soil components and ^{15}N-labeled melanoidins have appeared.[109,110] Figure 26 depicts ^{15}N-CPMAS spectra of these substances. The three ^{15}N-CPMAS NMR spectra of the melanoidins and the HA show a predominant peak at 97 ppm. This resonance is characteristic of nitrogen from secondary amide structures. The band at 155 ppm is due to pyrrolelike nitrogen, and the 128 band can be assigned either to a pyrrole or secondary amide structure and to indole nitrogen. ^{15}N-NMR (thus far) has only shown utility where isotopic enrichment has been performed. ^{15}N at natural abundance is not yet very promising due to the

Table 9
^{31}P-NMR AND STRUCTURE PARAMETERS

^{31}P shifts in orthophosphates

Compound	δ_1 (ppm)	δ_{11} (ppm)	δ_{22} (ppm)	δ_{33} (ppm)	$\lvert\Delta\delta\rvert$ (ppm)	r^a_{P-O} (Å)	$\frac{\Sigma\lvert 109.5 - \theta_{O-P-O}\rvert}{n}$	n_r	Ref.[b]
$NH_4H_2PO_4$	0.9	–15.4	–5.6	23.6	34.1	1.537	1.17	1.23	17
$(NH_4)_2HPO_4$	1.5	–33.6	–9.9	48.0	69.8	1.539	2.38	1.23	18
Li_3PO_4	10.8	NM[c]	NM	NM	NM	1.546	0.68	1.17	19
$Na_3PO_4 \cdot 12H_2O$	7.8	NM	NM	NM	NM	1.526	0.15	1.19	20
$NaH_2PO_4 \cdot H_2O$	2.3	61.1	23.1	–77.3	119.4	1.552	3.87	1.23	21
Na_2HPO_4	6.6	–48.9	–5.7	74.5	101.8	1.550	3.17	1.20	22
$NaNH_4HPO_4 \cdot 4H_2O$	5.1	–36.9	–4.2	56.5	77.0	1.541	2.73	1.21	23
KH_2PO_4	4.3	–15.6	3.9	24.8	30.4	1.538	0.925	1.21	24, 25
$K_2HPO_4 \cdot 3H_2O$	2.1	–45.0	–6.5	57.8	83.6	NA[d]	NA	1.23	—
$Be_3(PO_4)_2$	–9.2	NM	NM	NM	NM	NA	NA		—
$Mg_3(PO_4)_2 \cdot 8H_2O$	4.6	NM	NM	NM	NM	1.538	0.53	1.21	26
$Ca_3(PO_4)_2$	3.0	NM	NM	NM	NM	1.531	2.51	1.22	27
$CaHPO_4$	–0.6	NM	NM	NM	NM	1.541	21.0	1.24	28
$BaHPO_4$									
Site 1	–0.5	–40.9	–7.8	47.3	71.7	NA	NA	1.24	—
Site 2	–2.8	–39.1	–14.9	45.7	72.7	NA	NA	1.26	—
BPO_4	–29.5	NM	NM	NM	NM	NA	NA	1.43	—
$AlPO_4$	–24.5	NM	NM	NM	NM	1.516	0.70	1.40	29
$GaPO_4$	–9.8	NM	NM	NM	NM	1.56	NA	1.30	30
YPO_4	–0.9	NM	NM	NM	NM	1.50	NA	1.25	31
$LaPO_4 \cdot nH_2O$	–3.3	NM	NM	NM	NM	NA	NA	1.26	—
$P_2O_5 \cdot 24WO_3 \cdot nH_2O$	–15.1	NM	NM	NM	NM	NA	NA	1.34	—

[a] Average P–O bond length.
[b] Structure reference.
[c] NM, not measured.
[d] NA, not available.

From Turner, G. L., Smith, K. A., Kirkpatrick, R. J., and Oldfield, E., *J. Magn. Reson.*, 70, 408, 1986. With permission.

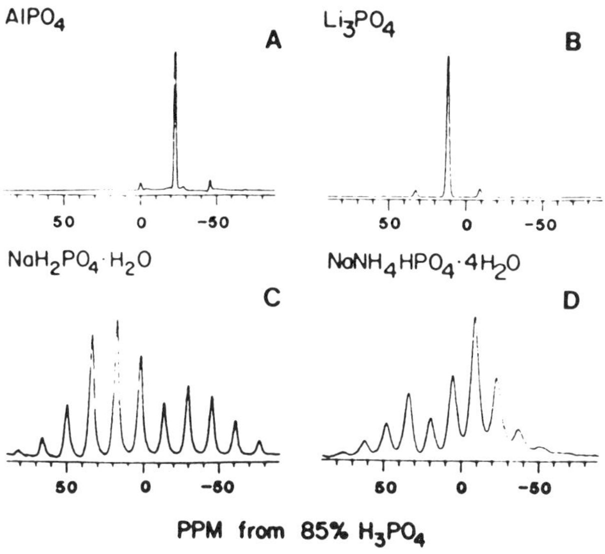

FIGURE 25. Representative ^{31}P-NMR spectra of several orthophosphates. (A) $AlPO^4$, 4.6 kHz MASS, 20 scans, 4-μs pulse (20° flip angle). (B) Li_3PO_4, 4.2 kHz MASS, 444 scans, 9-μs pulse (45° flip angle). (C) $NaH_2PO_4{\cdot}H_2O$, 3.2 kHz MASS, 60 scans, 9-μs pulse (45° flip angle). (D) $NaNH_4HPO_4{\cdot}4H_2O$, 2.8 kHz MASS, 28 scans, 9-μs pulse (45° flip angle). All spectra were recorded using a sweep width of 36.4 kHz, a 10-s recycle time, 10-Hz Gaussian line broadening, and ^{1}H dipolar decoupling was routinely applied for all systems. (From Turner, G. L., Smith, K..A., Kirkpatrick, R. J., and Oldfield, E., *J. Magn. Reson.*, 70, 408, 1986. With permission.)

low natural abundance of this species (0.37%). Very little signal is available at such levels. Furthermore, the relative sensitivity of ^{15}N (due to its small magnetogyric ratio) as compared with proton is only about 0.1%. Therefore, the Hartmann-Hahn matching or CP condition is difficult to achieve under the high-power conditions of the CPMAS experiment.

3. High-Resolution Solid-State Proton NMR of Soil Materials

Just as solid-state NMR studies of ^{13}C and other nuclei have been stimulated by the availability of CPMAS spectrometers and technology, sophisticated pulse sequences have also been developed to observe ^{1}Hs in solids.[111,112] The technique employed is known as combined rotation and multiple pulse spectroscopy (CRAMPS). A full exposition of the theory behind this method is beyond the scope of this paper. Suffice it to say that the large homonuclear couplings among protons in solids can be reduced to manageable levels by means of the CRAMPS experiment in certain cases. Figures 27 and 28 illustrate the only spectra in the literature of such a type on soils with which we are familiar.[107] Quantitation (integration of signal areas) is still not very accurate. However, some boundary conditions can now be placed on the proton distribution and environment in soils by solid state means. The CRAMPS method is subject to certain advantages and limitations. Advantages include the fact that protons afford high sensitivity due to their natural abundance and large magnetogyric ratios. Fewer spinning side bands are apparent thus causing less overlap with isotropic resonance positions. Disadvantages include more severe hardware requirements, smaller sample size, and smaller rotors for these experiments. The spectra presented indicate the potential for qualitative analysis at the very minimum.

4. Oxygen-17 and Other Nuclei

Oxygen-17 is now being actively pursued as a probe, and is being used to study the structure

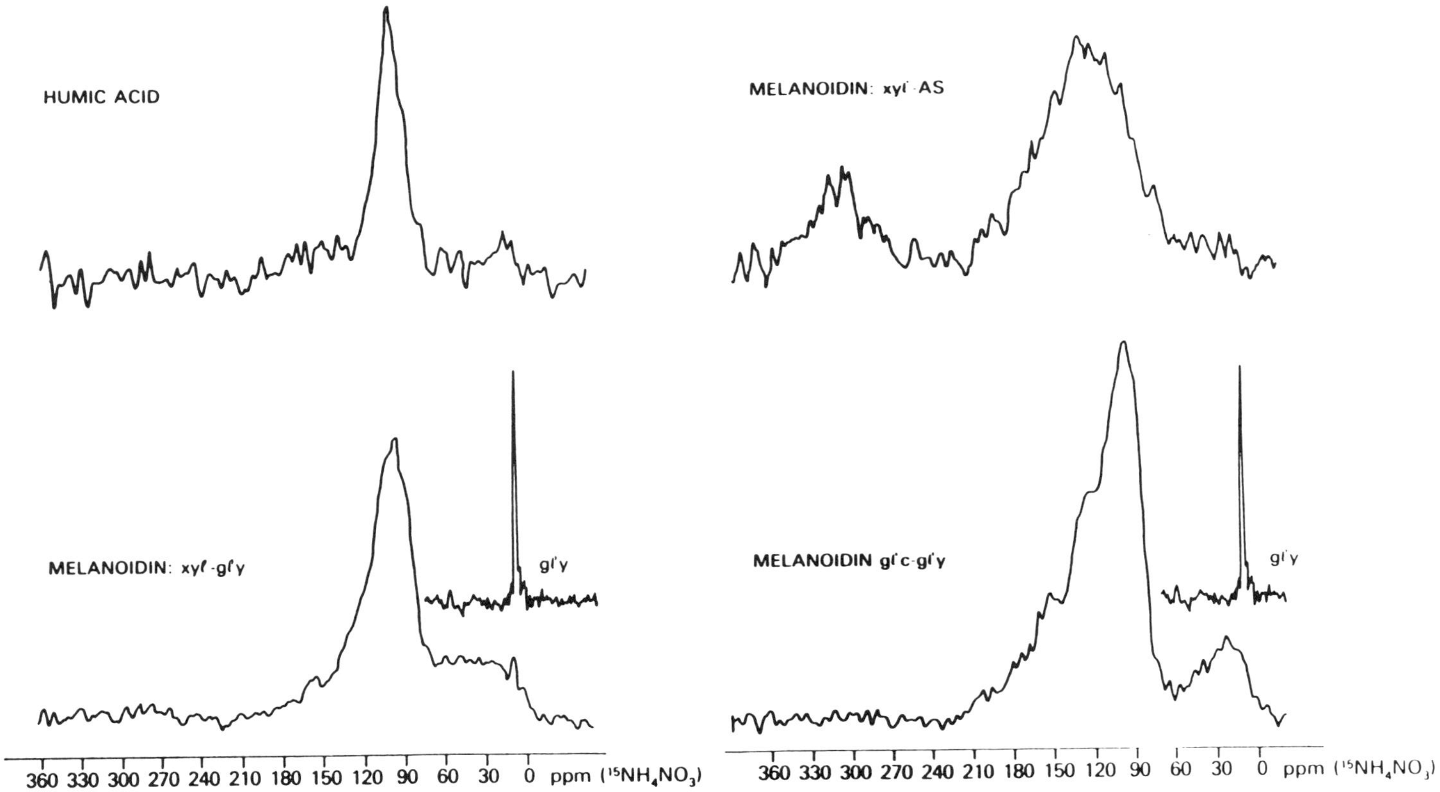

FIGURE 26. ^{15}N-CPMAS spectra of ^{15}N-labeled HA and melanoidins 1 (Xyl-Gly), 2 (Gly-Gly), 3 (Xyl-AS). Spectrum of glycine-^{15}N is shown as an insert. (From Benzing-Purdie, L., Ripmeester, J. A., and Preston, C., *J. Agric. Food Chem.*, 31, 913, 1983. With permission.)

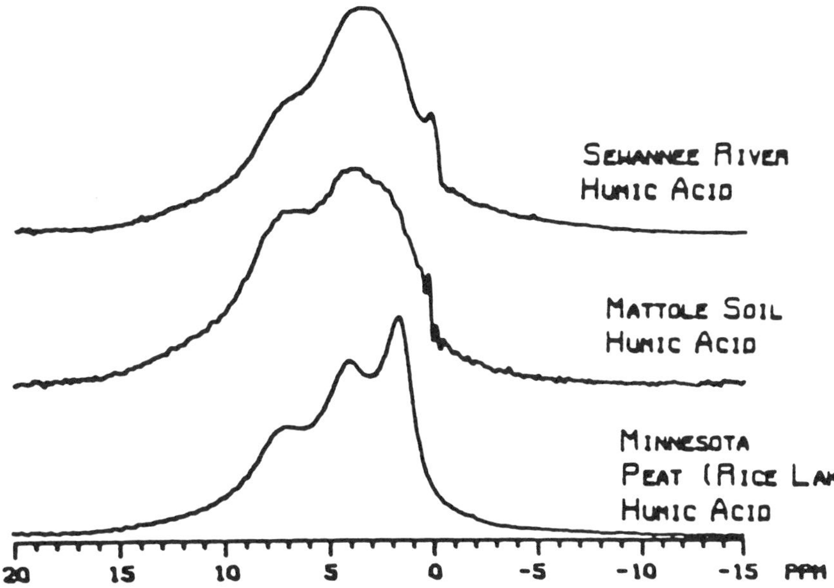

FIGURE 27. ¹H CRAMPS spectra of HAs. 500 scans were taken. (From Frye, J. S., Bonnimann, C. E., and Maciel, G. E., *NMR of Humic Substances and Coal,* Wershaw, W. L. and Makita, M. A., Eds., Lewis Publishers, Chelsea, MI, 1987, 33. With permission.)

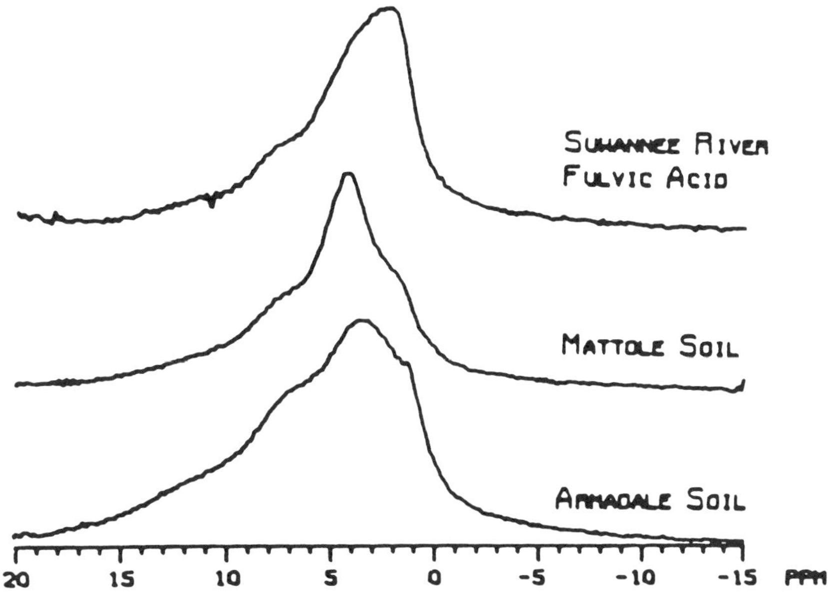

FIGURE 28. ¹H CRAMPS spectra of fulvic acids. Experimental conditions as for Figure 27. (From Frye, J. S., Bonnimann, C. E., and Maciel, G. E., *NMR of Humic Substances and Coal,* Wershaw, W. L. and Makita, M. A., Eds., Lewis Publishers, Chelsea, MI, 1987, 33. With permission.)

and reactivity of inorganic solids, glasses, catalysts, ceramics, and minerals. This NMR-active nuclide has an I = 5/2, a natural abundance of 0.037%, and a magnetogyric ratio of –0.5793 KHz/G. Walter et al. have recently reported results on a series of CP experiments on ^{17}O-labeled inorganic solids.[113] ^{17}O spectra were obtained on $Mg(OH)_2$, $Ca(OH)_2$, boehmite (AlO(OH)), talc ($Mg_3Si_4O_{10}(OH)_2$), $(C_6H_5)_3SiOH$, and amorphous SiO_2.[113]

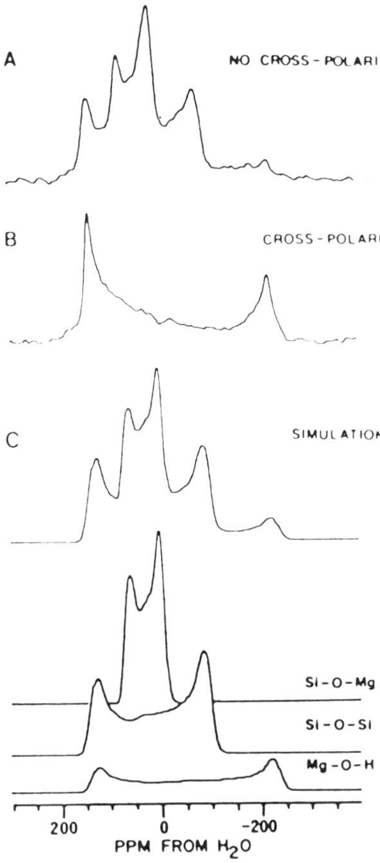

FIGURE 29. Static ^{17}O-NMR spectra of polycrystalline talc ($Mg_3Si_4O_{10}(OH)_2$) obtained at 67.8 MHz. (A) Proton-decoupled spectrum without CP: 500 scans, 10-s recycle time. (B) Cross-polarized spectrum: 500 scans, 33-μs CT, 10-s recycle time. (C) Simulated spectrum using three second-order quadrapolar powder patterns with the following parameters: (1) $e^2qQ/h = 3.2$ MHz, $\nu = 0$, $\delta i = 40$ ppm, relative intensity, 2.0; (2) $e^2qQ/h = 5.8$ MHz, $\nu = 0$, $\delta i = 50$ ppm, relative intensity 3.0; (3) $e^2qQ/h = 7.3$ MHz, $\nu = 0$, $\delta i = 0$ ppm, relative intensity 1.0. (From Walter, T. W., Turner, G. L., and Oldfield, E., *J. Magn. Reson.*, 76, 106, 1988. With permission.)

Spectra of the mineral talc are shown in Figure 29. The ratio of chemically distinct oxygen sites is 2:3:1 and corresponds to Si-O-Mg, Si-O-Si, and Mg-O-H. The nuclear quadrapole coupling constant (e^2qQ/h) and the assymetry parameter of the electric field gradient tensor, and the isotropic chemical shift are sensitive to the environment. Although Figure 29A shows overlapping powder patterns, the authors were able to simulate three axially symmetric second-order quadrapolar powder patterns (2:3:1 ratio) as seen in Figure 29C.[113] Furthermore, Figure 29B shows how CP can be used to observe hydroxyl groups even when the spectra suffer from extensive overlap. In this same paper, the authors point out methods of spectral "editing" by means of differences in cross-relaxation ratios, and selective observation of resonances from surface hydroxy groups that are rather difficult (or impossible) to detect by conventional ^{17}O NMR.

The subject of solid-state NMR of clays and mineral systems is another treatise unto itself. For further information on ^{27}Al and ^{29}Si solid-state NMR in clays and minerals see Reference 33.

The nuclei and techniques are simply pointed out as potentially applicable to whole soil systems themselves.

III. CONCLUSIONS

The current explosion in new NMR developments, hardware, software, and pulse sequences is producing a rapid acceleration in the ability of the capable spectroscopist to perform complex solid-state experiments on soils. Structure and a variety of nuclei can easily be examined as well as the dynamics within soil matrices. However, the NMR scientist still needs very well-characterized samples as a basis upon which to interpret his spectra. Over the next 10 years, a combination of NMR and advanced wet chemical techniques should be explored in order to fully exploit and discover the vast amount of information yet untapped in these complex, heterogeneous, environmental matrices.

REFERENCES

1. **Mehring, M.,** *High Resolution NMR in Solids,* 2nd. ed., Springer-Verlag, Berlin, 1983.
2. **Slichter, C. P.,** *Principles of Magnetic Resonance,* Springer-Verlag, Berlin, 1978.
3. **Haeberlen, U.,** High Resolution NMR in Solids. Selective Averaging, in *Advances in Magnetic Resonance,* Supplement 1, Waugh, J., Ed., Academic Press, New York, 1976.
4. **Gerstein, B. C. and Dybowski, C. R.,** *Transient Techniques in NMR of Solids,* Academic Press, Orlando, FL, 1985.
5. **Abragam, A.,** *The Principles of Nuclear Magnetism,* Clarendon Press, Oxford, 1961.
6. **Hartmann, S. R. and Hahn, E. L.,** Nuclear double resonance in the rotating frame, *Phys. Rev.,* 128, 2042, 1962.
7. **Lurie, F. M. and Slichter, C. P.,** Spin temperature in nuclear double resonance, *Phys. Rev.,* 133, A1108, 1964.
8. **Slusher, R. E. and Hahn, E. L.,** Sensitive detection of nuclear quadrapole interactions in solids, *Phys. Rev.,* 166, 332, 1968.
9. **Redfield, A. G.,** Pure nuclear electric quadrapole resonance in impure copper. *Phys. Rev.,* 130, 589, 1963.
10. **Kaplan, D. E. and Hahn, E. L.,** Experience de double irradiation en resonace magnetique par la methode d'impulsions, *J. Phys. Radium,* 19, 821, 1958.
11. **Yannoni, C. S.,** High resolution NMR in solids: the CPMAS experiment, *Acc. Chem. Res.,* 15, 201, 1982.
12. **Pines, A., Gibby, M. G., and Waugh, J. S.,** Proton-enhanced NMR of dilute spins in solids, *J. Chem. Phys.,* 59, 569, 1973.
13. **O'Donnell, D. J., Bartuska, V. J., Palmer, A. R., and Sindorf, D. W.,** Modern NMR techniques for solid samples, *Am. Lab.,* 104, 1986.
14. **Pfeffer, P. E.,** High resolution solid state ^{13}C NMR and its applications in carbohydrate chemistry, *J. Carbohydr. Chem.,* 3, 613, 1984.
15. **Andrew, E. R.,** Magic angle spinning in solid state NMR spectroscopy, *Philos. Trans. R. Soc. London, Ser. A,* 299, 505, 1981.
16. **Andrew, E. R., Hinshaw, W. S., and Jasinski, A.,** High-resolution NMR spectra in the solid state, *Chem. Phys. Lett.,* 24, 399, 1974.
17. **Andrew, E. R., Hinshaw, W. S., Hutchins, M. G., and Jasinski, A.,** A nuclear magnetic resonance investigation of solid tetraphosphorus trisulphide, *Proc. R. Soc. London, Ser. A,* 364, 553, 1978.
18. **Andrew, E. R., Bradbury, A., and Eades, R. G.,** Nuclear magnetic resonance spectra from a crystal rotated at high speed, *Nature,* 182, 1659, 1958.
19. **Andrew, E. R., Bradbury, A., and Eades, R. G.,** Removal of dipolar broadening of nuclear magnetic resonance spectra of solids by specimen rotation, *Nature,* 183, 1802, 1959.
20. **Lowe, I. J.,** Free induction decays of rotating solids, *Phys. Rev. Lett.,* 2, 285, 1959.
21. **Andrew, E. R.,** *Nuclear Magnetic Resonance in Rapidly Rotating Solids,* Coogan, C. K., Ham, N. S., Stuart, S. N., Pilbrow, J. R., and Wilson, G. H. H., Eds., Plenum Press, New York, 1970, 163.
22. **Schaefer, J. and Stejskal, E. O.,** Carbon-13 nuclear magnetic resonance of polymers spinning at the magic angle, *J. Am. Chem. Soc.,* 98, 1031, 1976.

23. **Schaefer, J., Stejskal, E. O., and Buchdahl, R.,** Magic-angle carbon-13 NMR analysis of motion in solid glassy polymers, *Macromolecules,* 10, 384, 1977.
24. **Gerstein, B. C., Pembleton, R. G., Wilson, R. C., and Ry, L. M.,** High resolution NMR in randomly oriented solids with homonuclear dipolar broadening: combined multiple pulse NMR and magic angle spinning, *J. Chem. Phys.,* 66, 362, 1977.
25. **Jelinski, L. W.,** NMR of plastics, *Chemtech,* March 1986, 188.
26. **Barron, P. F., Wilson, M. A., Stephens, J., Cornell, B. A., and Tate, K. R.,** Cross-polarization ^{13}C NMR spectroscopy of whole soils, *Nature,* 286, 585, 1981.
27. **Wilson, M. A., and Goh, G. M.,** Proton-decoupled pulse Fourier-transform ^{13}C nuclear magnetic resonance of soil organic matter, *J. Soil Sci.,* 28, 645, 1977.
28. **Barron, P. F. and Wilson, M. A.,** Humic soil and coal structure with magic-angle spinning carbon-13 CP-NMR, *Nature,* 289, 275, 1981.
29. **Preston, C. M. and Ripmeester, J. A.,** Application of solution and solid-state 13C NMR to four organic soils, their humic acids, fulvic acids, humins, and hydrolysis residues, *Can. J. Spectrosc.,* 27, 100, 1982.
30. **Preston, C. M., Mathur, S. P., and Rauthan, B. S.,** The distribution of copper, amino compounds, and humus fractions in organic soils of differing copper contents, *Soil Sci.,* 131, 344, 1981.
31. **Wilson, M. A., Barron, P. F., and Goh, K. M.,** Differences in structure of organic matter in two soils as demonstrated by ^{13}C cross-polarization nuclear magnetic resonance spectroscopy with magic angle spinning, *Geoderma,* 26, 323, 1981.
32. **Wilson, M. A., Pugmire, R. J., Zilm, K. W., Goh, K. M., Heng, S., and Grant, D. M.,** Cross-polarization ^{13}C-NMR spectroscopy with magic angle spinning characterizes organic matter in whole soils, *Nature,* 294, 648, 1981.
33. **Fyfe, C. A.,** *Solid-State NMR for Chemists,* CFC Press, Guelph, Ontario, 1983, 272.
34. **Schaefer, J., Stejskal, E. O., and Buchdahl, R.,** Magic-angle carbon-13 NMR analysis of motion in solid polycarbonate, *J. Macromol. Sci., Phys.,* B13, 665, 1977, and references cited therein.
35. **Fyfe, C. A., Mossburger, H., and Yannoni, C. S.,** A simple magic-angle spinning apparatus for routine use in NMR spectroscopy of solids, *J. Magn Reson.,* 36, 61, 1979.
36. **Stejskal, E. O., Schaefer, J., and Waugh, J. S.,** Magic-angle spinning and polarization transfer in proton enhanced NMR, *J. Magn. Reson.,* 28, 105, 1977.
37. **VanderHart, D. L. and Retcovsky, H. L.,** Estimation of coal aromaticities by proton-decoupled carbon-13 magnetic resonance spectra of whole coals, *Fuel,* 55, 202, 1976.
38. **Bartuska, V. J., Maciel, G. E., Schaefer, J., and Stejskal, E. O.,** Prospects for carbon-13 nuclear magnetic resonance analysis of solid fossil fuel materials, *Fuel,* 56, 345, 1977.
39. **Pfeffer, P. E., Gerasimowicz, W. V., and Piotrowski, E. G.,** Effect of paramagnetic iron on quantitation in carbon-13 cross-polarization magic angle spinning nuclear magnetic resonance spectrometry of heterogeneous environmental matrices, *Anal. Chem.,* 56, 734, 1984.
40. **Ganapathy, A. N. and McDowell, C. A.,** Paramagnetic doping as an aid in obtaining high-resolution carbon-13 NMR spectra of biomolecules in the solid state, *J. Am. Chem. Soc.,* 103, 6011, 1981.
41. **Brown, C. E.,** Effects of chemical modification and cobalt (II) chloride addition on the carbon-13 NMR spectra of carnosine as a solid powder, *J. Am. Chem. Soc.,* 104, 5608, 1982.
42. **Sullivan, M. J. and Maciel, G. E.,** Structural resolution in the carbon-13 nuclear magnetic resonance spectrometric analysis of coal by cross-polarization and magic-angle spinning, *Anal. Chem.,* 54, 1610, 1982.
43. **Earl, W. L. and VanderHart, D. L.,** Observations in solid polyethylene by carbon-13 nuclear magnetic resonance with magic angle spinning, *Macromolecules,* 12, 762, 1979.
44. **Yoshida, T., Maekawa, Y., and Fujita, T.,** Cross-polarization and magic angle spinning carbon-13 nuclear magnetic resonance spectrometry for quantitative analysis of coal and coal-derived solids, *Anal. Chem.,* 55, 388, 1983.
45. **Gerstein, B. C., Ryan, M. L., and Murphy, P. D.,** A tentative identification of average aromatic ring size in an Iowa vitrain and a Virginia vitrain, *Prepr. Pap., Am. Chem. Soc., Div. Fuel Chem.,* American Chemical Society, Washington, D.C., 24, 90, 1979.
46. **Preston, C. M., Dudley, R. L., Fyfe, C. A., and Mathur, S. P.,** Effects of variations in contact times and copper contents in a ^{13}C CPMAS study of samples of four organic soils, *Geoderma,* 33, 245, 1984.
47. **Lowe, L. E.,** Carbohydrate, in *Soil Organic Matter,* Schnitzer, M. and Khan, S. U., Eds., Elsevier, New York, 1978, 65.
48. **Rendlemann, J., Jr.,** Complexes of alkali metals and alkaline-earth metals with carbohydrates, *Adv. Carbohydrate Chem.,* 21, 209, 1966.
49. **Wilson, M. A.,** Applications of NMR spectroscopy to the study of the structure of soil organic matter, *J. Soil Sci.,* 32, 167, 1981.
50. **Hatcher, P. G., Berger, I. A., Dennis, L. W., and Maciel, G. E.,** Solid-state ^{13}C NMR of sedimentary humic substances: new revelations on their chemical composition, in *Aquatic and Terrestrial Humic Materials,* Christman, R. F. and Gjessing, E. T., Eds., Ann Arbor Science, Ann Arbor, MI, 1983, 37.

51. **Lindberg, J. J. and Hortling, B.,** Cross-polarization magic-angle spinning NMR studies of carbohydrates and aromatic polymers, *Adv. Polymer Sci.,* 1, 66, 1985.
52. **Hatcher, P. G., Breger, I. A., Maciel, G. E., and Szeverenyi, N. M.,** Geochemistry of humin, in *Humic Substances in Soil, Sediment, and Water,* Aiken, G. R., McKnight, D. M., Wershaw, R. L., and MacCarthy, P., Eds., John Wiley & Sons, New York, 1985, 275
53. **Wershaw, R. L.,** Applications of NMR spectroscopy for determining functionality in humic substances, in *Humic Substances in Soil, Sediment, and Water,* Aiken, G. R., McKnight, D. M., Wershaw, R. L., and MacCarthy, P., Eds., John Wiley & Sons, New York, 1985, 561.
54. **Kononova, M. M.,** *Soil Organic Matter,* Pergamon Press, Oxford, 1966.
55. **Forsman, J. P.,** Geochemistry of kerogen, in *Organic Geochemistry,* Berger, I. A., Ed., Pergamon Press, Oxford, 1963, 148.
56. **Steurmer, D. H., Kaplan, I. R., and Peters, K. E.,** Source indicators of humic substances and proto-kerogen: stable isotopic ratios, elemental compositions and electron spin resonance spectra, *Geochim. Cosmochim. Acta,* 42, 989, 1978.
57. **Hatcher, P. G., Rowan, R., and Mattingly, M. A.,** ^{1}H and ^{13}C NMR of marine humic substances, *Org. Geochem.,* 2, 77, 1980.
58. **Hatcher, P. G., Maciel, G. E., and Dennis, L. W.,** Aliphatic structure of humic acids: a clue to their origin, *Org. Geochem.,* 3, 43, 1981.
59. **Gerasimowicz, W. V. and Byler, D. M.,** Carbon-13 CPMAS NMR and FTIR spectroscopic studies of humic acids, *Soil Sci.,* 139, 270, 1985.
60. **Gerasimowicz, W. V., Byler, D. M., and Piotrowski, E. G.,** Infrared spectroscopic characterization of sludges and sludge extracts: comparison of several mild extraction procedures, *Soil Sci.,* 136, 237, 1983.
61. **Hatcher, P. G., Schnitzer, M., Dennis, L. W., and Maciel, G. E.,** Aromaticity of humic substances in soils, *J. Soil Sci. Soc. Am.,* 45, 1089, 1094, 1981.
62. **Dalal, R. C. and Mayer, R. J.,** Long term trends in Fertility of soils under continuous cultivation and cereal cropping in southern Queensland. I. Overall changes in soil properties and trends in winter cereal yields, *Aust. J . Soil Res.,* in press.
63. **Dalal, R. C. and Mayer, R. J.,** Long term trends in fertility of soils under continuous cultivation and cereal cropping in Southern Queensland. II. Total organic carbon and its rate of loss from the soil profile, *Aust. J. Soil Res.,* in press.
64. **Skjemstad, J. O., Dalal, R. C., and Barron, P. F.,** Spectroscopic investigations of cultivation effects on organic matter of vertisols, *Soil Sci. Soc. Am. J.,* 50, 354, 1986.
65. **Catroux, G. and Schnitzer, M.,** Chemical, spectroscopic, and biological characteristics of the organic matter in particle size fractions separated from an Aquoll, *Soil Sci. Soc. Am. J.,* 51, 1200, 1987.
66. **Skjemstad, J., Frost, R. L., and Barron, P. F.,** Structural units in humic acids from South-Eastern Queensland soils as determined by ^{13}C-NMR spectroscopy, *Aust. J. Soil. Res.,* 21, 539, 1983.
67. **Wilson, M. A., Goh, K. M., Collin, P. J., and Greenfield, L. G.,** Origins of humus variation, *Org. Geochem.,* 9, 225, 1986.
68. **Christman, R. F. and Gjessing, E.,** Priorities in humic research, in *Aquatic and Terrestrial Humic Materials,* Christman, R. F. and Gjessing, E., Eds., Ann Arbor Science, Ann Arbor, MI, 1983, 517.
69. **Horne, R. A.,** *Marine Chemistry,* Wiley Interscience, New York, 1969, 229.
70. **Gillman, A. H. and Wilson, M. A.,** Application of ^{13}C NMR spectroscopy to the structural clucidation of dissolved marine humic substances and their phytoplanktonic precursors in aquatic and terrestrial humic materials, Christman, R. F. and Gjessing, E., Eds., Ann Arbor Science, Ann Arbor, MI, 1983, 25.
71. **Gillman, A. H. and Wilson, M. A.,** Pyrolysis-GC-MS and NMR studies of dissolved seawater humic substances and isolates of a marine diatom, *Org. Geochem.,* 8, 15, 1985.
72. **Orem, W. H. and Hatcher, P. G.,** Solid-State ^{13}C-NMR studies of dissolved organic matter in pore waters from different depositional environments, *Org. Geochem.,* 11, 73, 1987.
73. **Lenheer, J. A., Wilson, M. A., and Malcolm, R. L.,** Presence and potential significance of aromatic-ketone groups in aquatic humic substances, *Org. Geochem.,* 11, 273, 1987.
74. **Schnitzer, M. and Preston, C. M.,** Analysis of humic acids by solution and solid-state carbon-13 nuclear magnetic resonance, *Soil Sci. Soc. Am. J.,* 50, 326, 1987.
75. **Ogner, G.,** A comparison of four different raw humus types in Norway using chemical degradations and CPMAS ^{13}C-NMR spectroscopy, *Geoderma,* 16, 343, 1985.
76. **Wilson, M. A., Vassallo, A. M., Perdue, E. M., and Reuter, J. H,** Compositional and solid-state nuclear Magnetic resonance study of humic and fulvic acid fractions of soil organic matter, *Anal. Chem.,* 59, 551, 1987.
77. **Vassallo, A. M., Wilson, M. A., Collin, P. J., Oades, J. M., and Waters, A. G,** Structural analysis of geochemical samples by solid-state nuclear magnetic resonance spectrometry. Role of paramagnetic material, *Anal. Chem.,* 59, 558, 1987.
78. **Saiz-Jimenez, J., Hawkins, B. L., and Maciel, G. E.,** Cross-polarization, magic-angle spinning ^{13}C nuclear magnetic resonance spectroscopy of soil humic fractions, *Org. Geochem.,* 9, 277, 1986.

79. **Kan, R. I., Rubinsztain, Y., Ioselis, P., Aizenshtat, Z., Pugmire, R., Anderson, L. L., and Woolfenden, W. R.,** Carbon-13 Cross-polarized magic-angle sample spinning nuclear magnetic resonance of melanoidins, *Org. Geochem.*, 9, 199, 1986.
80. **Earl, W. L., Wershaw, R. L., and Thorn, K. A.,** The use of variable temperature and magic-angle sample spinning in studies of fulvic acids, *J. Magn. Reson.*, 74, 264, 1987.
81. **Preston, C. M. and Blackwell, B. A.,** Carbon-13 nuclear magnetic resonance for a humic and a fulvic acid: signal-to-noise optimization, quantitation, and spin-echo techniques, *Soil Sci.*, 139, 88, 1985.
82. **Lobartini, J. C. and Tan, K. H.,** Differences in humic acid characteristics as determined by carbon-13 nuclear magnetic resonance, scanning electron microscopy, and infrared analysis, *Soil Sci. Soc. Am. J.*, 52, 125, 1988.
83. **Theng, B. K. G.,** *The Chemistry of Clay-Organic Reactions,* Higler, London, 1974.
84. **Theng, B. K. G., Churchman, G. J., and Newman, R. H.,** The occurrence of interlayer clay-organic complexes in two New Zealand soils, *Soil Sci.*, 142, 262, 1986.
85. **Gerasimowicz, W. V., Hicks, K. B., and Pfeffer, P. E.,** Evidence for the existence of associated lignin-carbohydrate polymers as revealed by carbon-13 CPMAS solid-state NMR spectroscopy, *Macromolecules,* 17, 2597, 1984.
86. **Atalla, R. H., Gast, J. C., Sindorf, D. W., Bartuska, V. J., and Maciel, G. E.,** Carbon-13 NMR spectra of cellulose polymorphs, *J. Am. Chem. Soc.*, 102, 3249, 1980.
87. **Earl, W. L. and VanderHart, D. L.,** High resolution, magic angle spinning carbon-13 NMR of solid cellulose I, *J. Am. Chem. Soc.*, 102, 3251, 1980.
88. **Robert, D. and Gagnaire, D.,** Quantitative analysis of lignins by carbon-13 NMR, *Int. Symp. Wood Pulping Chemistry,* 1, 86, 1981.
89. **Earl, W. L. and VanderHart, D. L.,** Observations by high-resolution carbon-13 nuclear magnetic resonance of cellulose I related to morphology and crystal structure, *Macromolecules,* 14, 570, 1981.
90. **Maciel, G. E., Kolodziesjki, W., Bertran, M., and Dale, B.,** Carbon-13 NMR and order in cellulose, *Macromolecules,* 15, 686, 1982.
91. **Horii, F., Hirai, A., and Kitamaru, R.,** Solid-state high-resolution carbon-13 NMR studies of regenerated cellulose samples with different crystallinities, *Polym. Bull.*, 8, 163, 1982.
92. **Nimz, H. H. and Nemr, M.,** Carbon-13 NMR spectra of lignins. 9. Spin-lattice relaxation times (T1) and determination of interunit linkages in three hardwood lignins, *(Alnus glutinosa, Corylus avellanus* and *Acer pseudoplatanus), J. Wood Chem. Technol.*, 2, 371, 1982.
93. **Kolodziesjki, W., Frye, J. S., and Maciel, G. E.,** Carbon-13 nuclear magnetic resonance spectrometry with cross-polarization and magic-angle spinning for analysis of lodgepole pine wood, *Anal. Chem.*, 54, 1419, 1982.
94. **Opella, S. J. and Frey, M. H.,** Selection of non-protonated carbon resonances in solid-state nuclear magnetic resonance, *J. Am. Chem. Soc.*, 101, 5854, 1979.
95. **Opella, S. J., Frey, M. H., and Cross, T. A.,** Detection of individual carbon resonances in solid proteins, *J. Am. Chem. Soc.*, 101, 5856, 1979.
96. **Alemany, L. B., Grant, D. M., Pugmire, R. J., Alger, T. D., and Zilm, K. W.,** Cross-polarization and magic angle sample spinning NMR spectra of model organic compounds. I. Highly protonated molecules, *J. Am. Chem. Soc.*, 105, 2133, 1983.
97. **Alemany, L. B., Grant, D. M., Pugmire, R. J., Alger, T. D., and Zilm, K. W.,** Cross-polarization and magic angle sample spinning NMR spectra of model organic compounds. II. Molecules of low or remote protonation, *J. Am. Chem. Soc.*, 105, 2142, 1983.
98. **Alemany, L. B., Grant, D. M., Alger, T. D., and Pugmire, R. J.,** Cross-polarization and magic angle sample spinning NMR spectra of model organic compounds. III. Effect of the ^{13}C-^{1}H dipolar interaction on cross-polarization and carbon-proton dephasing, *J. Am. Chem. Soc.*, 105, 6697, 1983.
99. **Ludemann, H.-D. and Nimz, H.,** Carbon-13 nuclear magnetic resonance spectra of lignins, *Biochem. Biophys. Res. Commun.*, 52, 1162, 1973.
100. **Himmelsbach, D. S. and Barton, F. E.,** Carbon-13 nuclear magnetic resonance of grass lignins, *J. Agric. Food Chem.*, 28, 1203, 1980.
101. **Sjostrom, E.,** *Wood Chemistry Fundamentals and Applications,* Academic Press, New York, 1981, 56.
102. **Zumbulyadis, N.,** Selective carbon excitation and the detection of spatial heterogeneity in cross-polarization magic-angle spinning NMR, *J. Magn. Reson.*, 53, 486, 1983.
103. **Gillam, A. H., Wilson, M. A., and Collin, P. J.,** Stuctural analysis of sea loch sedimentary humic substances by ^{13}C-^{1}H dipolar dephasing NMR, *Org. Geochem.*, 11, 91, 1987.
104. **Miknis, F. P., Lindner, A. W., Gannon, A. J., Davis, M. F., and Maciel, G. E.,** Solid-state ^{13}C-NMR studies of selected oil shales from Queensland, Australia, *Org. Geochem.*, 7, 239, 1984.
105. **Wilson, M. A., Verheyen, T. V., Vassallo, A. M., Hill, R. S., and Perry, G. J.,** Selective loss of carbohydrates from plant remains during coalification, *Org. Geochem.*, 11, 265, 1987.
106. **Williams, R. J. P., Giles, R. G. F., and Posner, A. A. M.,** Solid-state phosphorus NMR spectroscopy of minerals and soils, *J. Chem. Soc. Chem. Commun.*, 20, 1051, 1981.

107. **Frye, J. S., Bronnimann, C. E., and Maciel, G. E.,** Solid-state NMR of humic materials, in *NMR of Humic Substances and Coal,* Wershaw, W. L. and Mikita, M. A., Eds., Lewis Publishers, Chelsea, MI, 1987, 33.
108. **Turner, G. L., Smith, K. A., Kirkpatrick, R. J., and Oldfield, E.,** Structure and cation effects on phosphorus-31 NMR chemical shifts and chemical-shift anisotropies of orthophosphates, *J. Magn. Reson.,* 70, 408, 1986.
109. **Benzing-Purdie, L., Ripmeester, J. A., and Preston, C.,** Elucidation of the nitrogen forms in melanoidins and humic acid by nitrogen-15 cross-polarization magic angle spinning nuclear magnetic resonance spectroscopy, *J. Agric. Food Chem.,* 31, 913, 1983.
110. **Benzing-Purdie, L., Cheshire, M. V., Williams, B. L., Sparling, G. P., Ratcliffe, C. I., and Ripmeester, J. A.,** Fate of [^{15}N] Glycine in peat as determined by ^{13}C and ^{15}N CP-MAS NMR spectroscopy, *Agric. Food Chem.,* 34, 170, 1986.
111. **Burum, D. P. and Rhim, W. K.,** Analysis of multiple pulse NMR in solids. III, *J. Chem. Phys.,* 71, 944, 1979.
112. **Pembleton, R. G., Ryan, L. M., and Gerstein, B. C.,** NMR probe for combined homonuclear multiple pulse decoupling and magic angle spinning, *Rev. Sci. Instrum.,* 48, 1286, 1977.
113. **Walter, T. W., Turner, G. L., and Oldfield, E.,** Oxygen-17 cross-polarization NMR spectroscopy of inorganic solids, *J. Magn. Reson.,* 76, 106, 1988.

Chapter 10

THE STRUCTURE OF CELLULOSE AS STUDIED BY CP/MAS ^{13}C NMR SPECTROSCOPY

Fumitaka Horii

TABLE OF CONTENTS

I. INTRODUCTION

Cellulose, which is a polymer of D-glycosyl residues joined by β-1,4-glycosidic linkages (Figure 1), was one of the first polymers whose molecular structures were determined at the earliest stage of the modern history of macromolecules. Since then, subsequent developments in syntheses, characterization, and properties of synthetic polymers have greatly been promoted by cellulose research, in both experimental and theoretical fields. In contrast, however, the detailed morphological structure of cellulose itself is still unknown despite much research effort.[1] Although X-ray diffraction is also a powerful technique for cellulosic materials, the question of parallel vs. antiparallel chain packing is still controversial. The conformations about the β-1,4-glycosidic linkages as well as about the exocyclic C5-C6 bonds, which are defined in terms of torsion angles ϕ, ψ, and χ, respectively, as shown in Figure 1, have not been defined very precisely even in the crystalline regions. In such a situation, one would expect that the structure of the noncrystalline region is poorly understood until the present time, because there was almost no means suitable for characterizing this region.

Recent CP/MAS ^{13}C-NMR spectroscopy, which is a combined technique of high-power ^{1}H dipolar decoupling, magic-angle sample spinning (MAS), and ^{1}H-^{13}C cross-polarization (CP), is a promising method for studying these unsolved problems of cellulose because this method is very sensitive to local molecular conformations and dynamics. In this chapter, recent progress in structural studies of cellulose by this method is described, particularly considering the contents of the crystalline and noncrystalline regions separately. Some reports of cellulose derivatives and woods are also briefly reviewed. An earlier review of the application of CP/MAS ^{13}C-NMR techniques to cellulose chemistry has been published elsewhere.[2]

II. CP/MAS ^{13}C-NMR SPECTRA OF CELLULOSE

A. Assignments of Resonance Lines

Figure 2 shows a 50-MHz CP/MAS ^{13}C-NMR spectrum of ramie cellulose, which was obtained in the dried state, and a stick-type ^{13}C-NMR spectrum of low molecular weight cellulose (DP ~10) in deuterated dimethyl sulfoxide solution.[3] The assignments of the resonance lines (except for the C2, C3, and C5 lines) in the CP/MAS spectrum can be readily made as shown in Figure 2, based on analogies with the solution state spectrum.[4,5]

In the CP/MAS ^{13}C-NMR spectrum, the following three remarkable features are recognized: (1) as is seen in the C4 and C6 resonances, these lines split into two components, sharp downfield and broader upfield components. (2) Although these two components do not explicitly appear in the C1 line, this line contains another sort of smaller splitting. Such fine structure will be more clearly visible for the C1 line as well as C4 and C6 lines when the spectrum is measured in the presence of water, as described later. (3) The C1, C4, and C6 lines as well as their multiplets are shifted downfield by 2.3 to 9.6 ppm in the solid state, compared to the solution state. These features are commonly observed for other native and regenerated celluloses, but the details change from sample to sample.[3-20] Therefore, the crystal structure and morphology of the samples must be reflected in these features.

Earl and VanderHart[6] first tried to explain the cause of the larger splittings appearing in the C4 and C6 sites. They compared the mass fraction of the downfield component in the C4 line with the degree of crystallinity determined by the X-ray diffraction method for cotton, ramie, and hydrolized cotton. However, these samples exhibited little differences in the C4 relative intensity and in the appearance of the broad component, whereas the previously reported crystallinities were varied from 80 to 96% for these samples. Therefore, they proposed that the broad upfield components of C4 and C6 carbon resonances were not attributed to the noncrystalline component, but could be ascribed to anhydroglucoses on the surface of elementary fibrils.

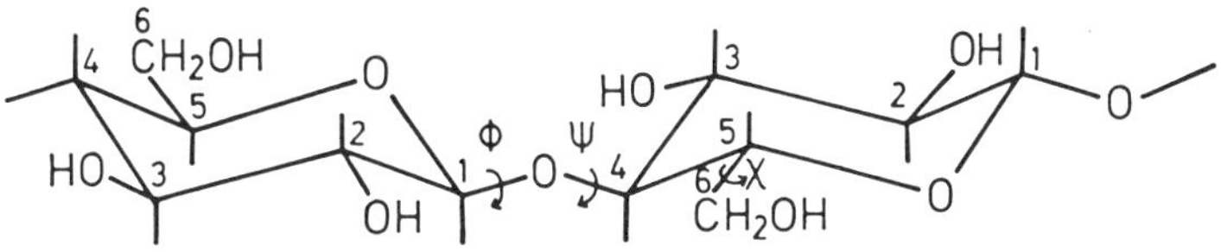

FIGURE 1. Schematic structure of a cellulose molecule.

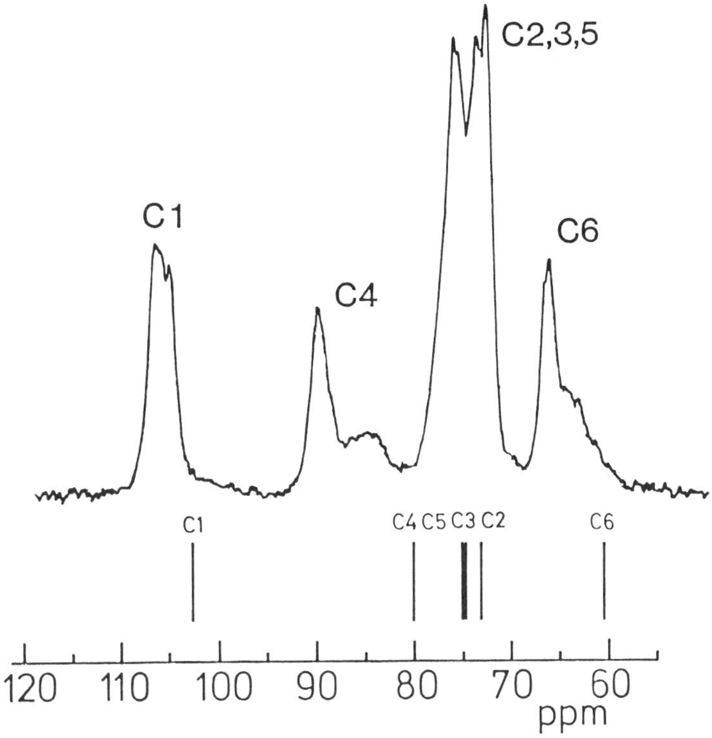

FIGURE 2. The 50-MHz CP/MAS ^{13}C-NMR spectrum of ramie cellulose and stick-type scalar-decoupled ^{13}C-NMR spectrum of low molecular weight cellulose in DMSO-d_6 solution. (From Horii, F., Hirai, A., and Kitamaru, R., *Polymers for Fibers and Elastomers*, ACS Symp. Ser., No. 260, Arthur, J. C., Jr., Ed., American Chemical Society, Washington, D.C., 1984, 27. With permission.)

In contrast, we estimated the crystallinities to be 0.5 to 0.6 for cotton, ramie, and hydrolized cotton by an X-ray analysis based on a modified Hermans' method.[3,7] These values correspond well to the fractions of the downfield components of C4 and C6 sites. Furthermore, it has been found, as shown in Figure 3, that a linear relationship exists between the mass fraction (f_{nmr}) of the sharp C4 component and the crystallinity (f_x) determined by the X-ray analysis for the different samples including both native and regenerated celluloses. It is, therefore, concluded that the sharp downfield and broad upfield components of the C4 and C6 lines are the crystalline and noncrystalline components, respectively. Recent studies[11] also suggest that chains in three-dimensionally disordered regions were included in the broad component, although the contribution of the surface chains on crystallites still seems to be overestimated.

It should be noted here that, in the case of cellulose samples, the relative carbon intensities of the respective resonances are proportional to the number of carbons at least in the dried state, even if the cross-polarization technique is employed.[9] Such quantitative measurements are probably based on the rapid ^1H spin diffusion throughout the crystalline and noncrystalline regions.

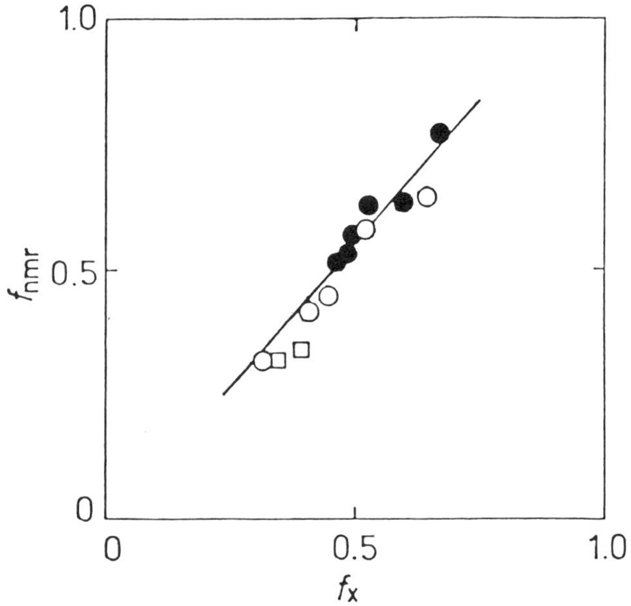

FIGURE 3. Integrated fraction f_{nmr} of the downfield component of the C4 resonance line vs. degree of crystallinity f_x determined by X-ray analyses. ●: native cellulose, ❍: regenerated cellulose, and ❑: mercerized cotton and ramie. (From Horii, F., Hirai, A., and Kitamaru, R., *Polymers for Fibers and Elastomers,* ACS Symp. Ser., No. 260, Arthur, J. C., Jr., Ed., American Chemical Society, Washington, D.C., 1984, 27. With permission.)

The second feature mentioned above is undoubtedly associated with the crystalline component, which will be described later in detail. The third one may be mainly due to the difference in molecular chain conformations in the solid and solution states. This interpretation will be briefly discussed in the following section where several factors affecting ^{13}C isotropic chemical shifts in solids are described.

B. Effects of Water

Since cellulose is a hydrophilic polymer, it is very important to measure CP/MAS ^{13}C spectra in the presence of water. However, the large centrifugal force produced by rapid sample spinning readily removes water from hydrated samples packed in a conventional MAS rotor and, thus, only a few reports[14,21-23] have been published on polymers containing some water. Recently a new type of MAS rotor with an O-ring seal has been developed by modifying a commercial bullet-type rotor.[14,23] This rotor can be rotated at a uniform rate of 3 to 4 kHz even for long-term measurements without any significant loss of water.

Figure 4 shows 50-MHz CP/MAS ^{13}C-NMR spectra of cotton cellulose with different water contents (g H_2O per gram cellulose × 100), which were recorded using the MAS rotor with an O-ring seal.[14] The fine splitting of the C1 resonance line is more clearly observed with increasing water content. The multiplicities in the C4 and C6 sites are also visible for the hydrated samples, although it is not so evident in this figure. On the other hand, the noncrystalline line of the C4 carbons becomes much narrower with increasing water content and as a result this line is well separated from the crystalline component at the downfield side. The separation of both components in the C6 resonance is also clearly observed for the hydrated samples owing to the upfield shift of the noncrystalline components. Similar effects of water are recognized for other cellulose samples. It is, therefore, essential to add enough water to cellulose samples to obtain highly resolved spectra.

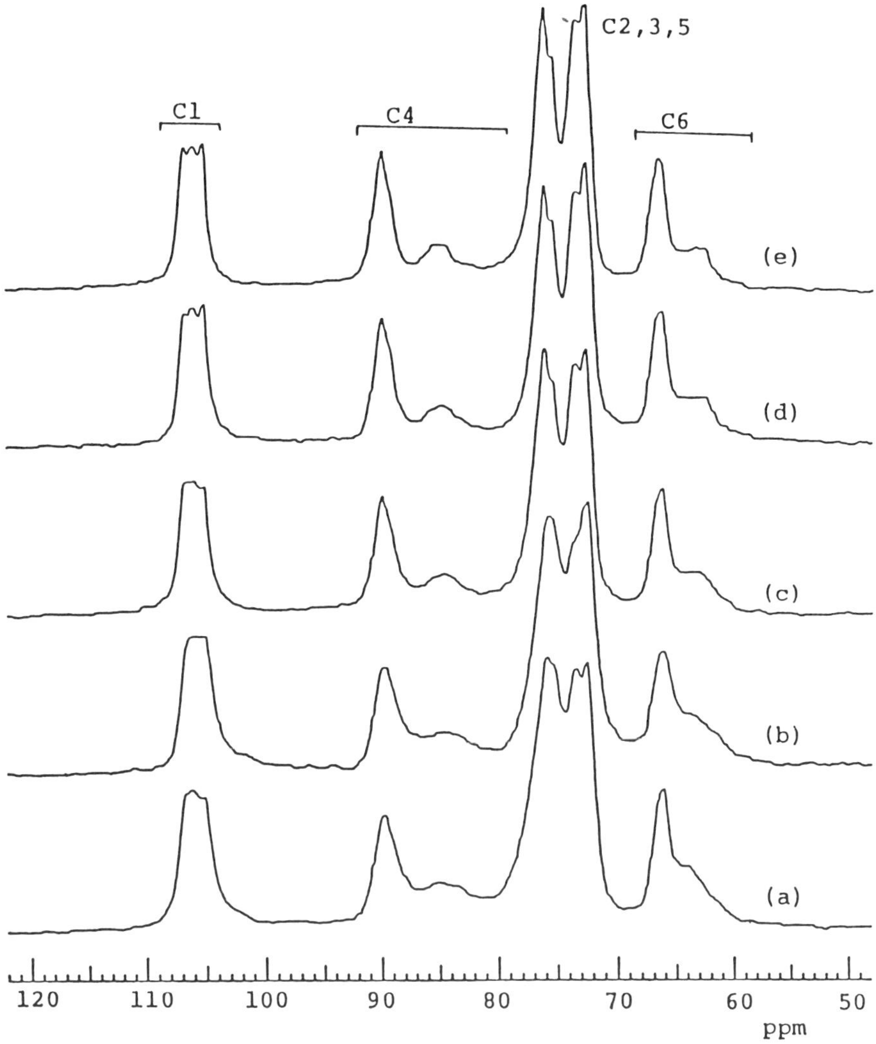

FIGURE 4. The 50-MHz CP/MAS ^{13}C-NMR spectra of cotton cellulose with different water contents: (a) 0%, (b) 4.2%, (c) 8.4%, (d) 19.1%, and (e) 161%. (From Horii, F., Harai, A., Kitamaru, R., and Sakurada, I., *Cellul. Chem. Technol.*, 19, 513, 1985. With permission.)

III. FACTORS AFFECTING ^{13}C CHEMICAL SHIFTS IN SOLIDS

An isotropic chemical shift obtained by CP/MAS ^{13}C-NMR spectroscopy reflects the averaged strength of the magnetic shielding field which is produced by the interaction between the static magnetic field and the electron clouds surrounding a given carbon nucleus. This shielding field is thought to change mainly depending on variations in bond lengths, bond angles, torsion angles, hydrogen bonding, and intermolecular contributions. Of these effects, the contributions of bond lengths and bond angles are normally difficult to estimate, but they are presumed to be negligibly small compared with other contributions, as long as there are no extreme deviations from their standard values.

Contributions of torsion angles, hydrogen bonding, and intermolecular interaction (which are frequently referred to as molecular packing) have been experimentally elucidated. For example, cycloalkane rigid crystals exhibit five ^{13}C resonance lines assigned to the folding parts and an enhanced peak attributed to the CH_2 sequences in the crystalline core.[24] These six resonances can be well correlated to six different conformations which are defined in terms of

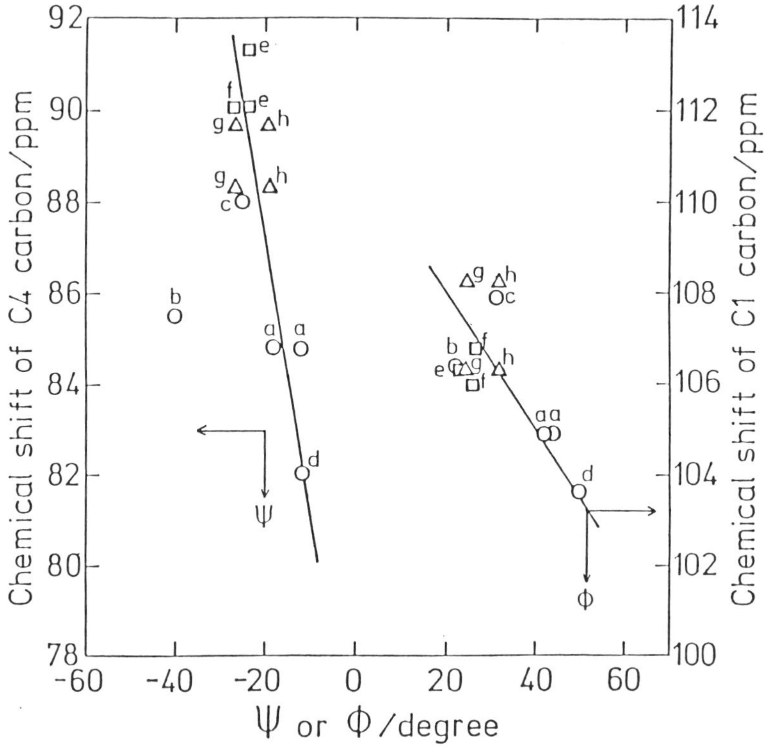

FIGURE 5. [13]C chemical shifts of C1 and C4 sites vs. torsion angles φ and ψ. a: β-cellobiose, b: β-methylcellobioside·CH_3OH, c: α-lactose·H_2O, d: β-lactose, e and f: cellulose I and g; and h: cellulose II. (From Horii, F., Hirai, A., and Kitamaru, R., *Bull. Magn. Reson.*, 5, 233, 1983. With permission.)

the successive two torsion angles on both sides of the carbon in question, *gt-gg, gt-tg, gg-tt, tg-gt, gt-tt,* and *tt-tt,* where *t* and *g* indicate torsion angles of about 180 and 60°, respectively. On the other hand, it has been found that a relatively small downfield shift of 1.3 ppm appears for the central CH_2 resonance of triclinic *n*-$C_{20}H_{42}$ crystals compared to the corresponding line for *n*-paraffin crystals with monoclinic and orthorhombic crystal forms.[25] Since the molecular conformations are almost the same *tt-tt* in those crystals, such a downfield shift seems to be due to the difference in molecular packing. More significant downfield shifts are observed for carbons chemically bonded to hydroxyl groups, when the hydrogen bonding is formed between those hydroxyl groups and other appropriate groups.[26,27] However, this sort of shift would be drastically decreased with increasing O•••O distance in the hydrogen bonding, and almost no effect appears for cases where O•••O distances are more than 2.7 Å.

In an attempt to estimate the effects of molecular packing and hydrogen bonding for cellulose, ^{13}C solid-state chemical shifts were compared with the corresponding solution-state values for monosaccharides and disaccharides related to cellulose.[3] In this case, the comparison was made only for ring carbons that are not allowed to adopt different conformations. As a result, it was suggested that the superimposed effects of packing and hydrogen bonding would occasionally produce a downfield or upfield shift less than at most 2.5 ppm for cellulose samples.

It has also been examined, using different disaccharides containing β-1,4-glycosidic linkages, as to whether or not the ^{13}C chemical shifts of C1 and C4 carbons are correlated to the torsion angles φ and ψ about the β-1,4-glycosidic linkages.[3,28] In Figure 5 the C1 and C4 chemical shifts are plotted against the torsion angles φ and ψ, respectively. These angles were determined by X-ray crystal analyses. Since there are not many appropriate disaccharides, the data of the cellulose crystals are also plotted there. Although the data points are somewhat scattered, a linear

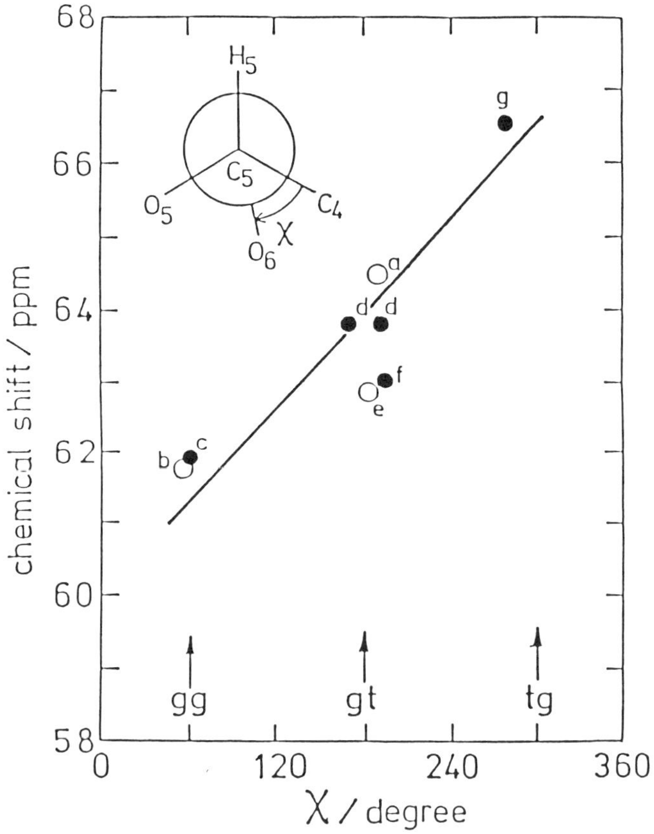

FIGURE 6. ^{13}C chemical shifts of C6 carbons vs. torsion angles χ, a: α-D-glucose, b: α-D-glucose·H_2O, c: β-D-glucose, d: β-D-cellobiose, e: α-D-lactose·H_2O, f: β-D-lactose, g: cellulose I, ○: α-D-glucose and α-D-glycosyl residues, and ●: β-glucose and β-D-glycosyl residues. (From Horii, F., Hirai, A., and Kitamaru, R., *Polymers for Fibers and Elastomers*, ACS Symp. Ser., No. 260, Arthur, J. C., Jr., Ed., American Chemical Society, Washington, D.C., 1984, 27. With permission.)

relationship seems to exist between the C1 and C4 chemical shifts and the torsion angles ϕ and ψ, respectively.

A similar linear relationship was also obtained between the C6 chemical shifts and the torsion angles χ about the exocyclic C5-C6 bonds for α- and β-glucoses and the glycosyl residues of different disaccharides, as shown in Figure 6.[3,28,29] Since the CH_2OH groups of low-molecular-weight glucosides preferably adopt two conformations of *gauche-gauche* and *gauche-trans*, the result of cellulose I crystals is also plotted as only one case of *trans-gauche* conformation. Here, for example, *gauche-trans* indicates that the C6-O6 bond is *gauche* to the C5-O5 bond and *trans* to the C4-C5 bond. According to these results, three preferred conformations, *gauche-gauche, gauche-trans*, and *trans-gauche*, are well correlated to the chemical shifts of 61 to 62 ppm, 62.7 to 64.5 ppm, and about 66 ppm, respectively.

The origins of the scattering of the data seen in Figures 5 and 6 are not clear at present. One possible origin may be the effects of molecular packing and hydrogen bonding mentioned above. Another cause would be due to neglecting the contributions from other torsion angles related to these carbons. Similarly to the case of the cycloalkanes described above, the C1 and C4 chemical shifts may be associated with variations in both ϕ and ψ although the other two torsion angles in the pyranose rings are constant. It seems, however, more appropriate for the detailed elucidation of this problem to first deal with carbohydrates containing α-1,4-glycosidic linkages, because much more precise experimental results can be obtained for the α-1,4 linked cycloamylose crystals.[30-32]

IV. CRYSTALLINE AND NONCRYSTALLINE CONTRIBUTIONS

A. Spectral Resolution of Components

As is clearly seen in Figures 2 and 4, the noncrystalline components of the C4 and C6 carbons are considerably well separated from the crystalline lines on the downfield sides. Thus, computer analyses would be employed for resolving those resonances into the two components if Lorentzian curves could be assumed for the respective components. However, this technique is not applicable in the case of the C1 line, because the number of overlapped lines cannot be precisely estimated from the spectrum. In such a case, the differences in nuclear magnetic relaxation rate, which are based on the different molecular mobilities, can be utilized to afford better spectral resolution between the crystalline and noncrystalline regions.

One of the most reliable ways to implement this method is to use the difference in ^{13}C spin-lattice relaxation time T_{1C}. In fact, the crystalline components of the cellulose samples have much longer T_{1C} values than those of the noncrystalline components,[3,9,12,14,15,20] e.g., 140 to 233 s for the former and 7 to 20 s for the latter in the case of hydrated cotton cellulose under the static magnetic field of 4.7 T.[14,15,20] Furthermore, the T_{1C} values are greatly increased for bacterial and valonia celluloses, which are more than 300 and 800 s, respectively, probably reflecting the more ordered and rigid crystal structure.[33] The pulse sequence used for the selective observation of the crystalline component is the one for T_{1C} measurements, which was proposed by Torchia,[34] where the magnetization of each ^{13}C species decays from the value produced by the CP process to zero due to the longitudinal relaxation. In the system containing at least two components with different T_{1C} values, therefore, the spectrum of the longer T_{1C} component can be observed selectively by recording the residual magnetization after the shorter T_{1C} component decays to zero.

Figure 7b shows the spectrum of the longer T_{1C} component of cotton cellulose soaked in H_2O, which was obtained by the pulse sequence mentioned above.[14] The delay time employed to suppress the noncrystalline contribution was 100 s. Although the conventional CP/MAS ^{13}C spectrum (Figure 7a) exhibits rather broad noncrystalline resonances at about 85 and 63 ppm, such lines completely disappear in Figure 7b. This indicates that Figure 7b represents the resonance profile of the crystalline component, although the relative intensities of the observed lines are not always proportional to the number of the corresponding carbons because of the different T_{1C} values for the respective lines. On the other hand, the noncrystalline spectrum can be obtained by subtracting Figure 7b from Figure 7a, as shown in Figure 7c. Similar recording of the respective spectra are also possible for other native celluloses, such as ramie, bacterial, and valonia celluloses as well as regenerated celluloses.[35] On the basis of those respective spectra the structure for the crystalline and the noncrystalline regions are discussed below.

The suppression of the noncrystalline contribution also seems to be possible by using the difference in ^{13}C and ^{1}H spin-lattice relaxation times $T_{1\rho C}$ and $T_{1\rho H}$ in the rotating frame, respectively.[9,36] However, it would be technically impossible in these cases to completely remove the noncrystalline contribution because of the difficulty in the use of longer spin-locking times.

B. Crystal Structure

On the basis of the crystalline spectra, which were obtained by the method described above, new information about the crystal structure of cellulose can be obtained. Figure 8 shows the crystalline spectra of the representative native cellulose,[2] and in Table 1, ^{13}C chemical shift data are summarized for the respective carbons of the crystalline components of native and regenerated celluloses.[35] As is clearly seen in the figure, cotton and ramie celluloses produce

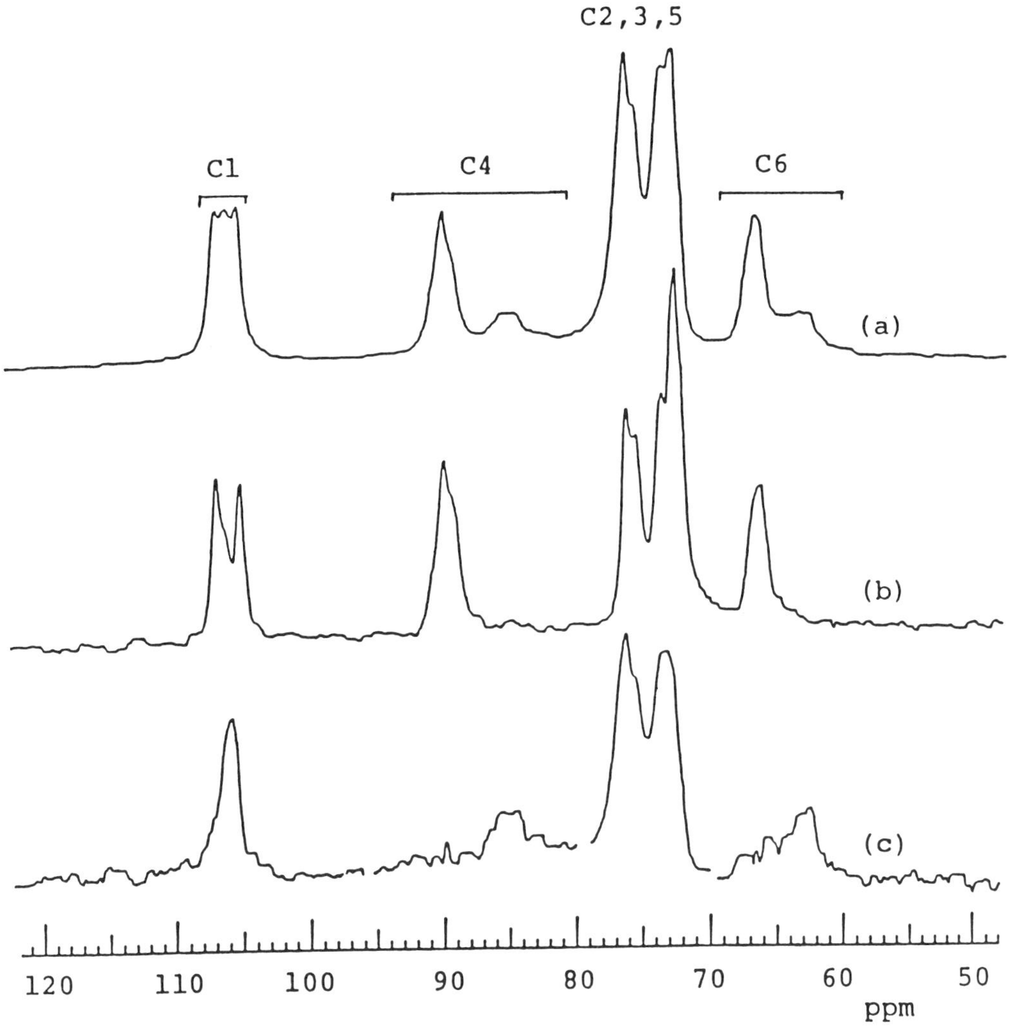

FIGURE 7. The 50-MHz CP/MAS ^{13}C-NMR spectra of cotton cellulose in the hydrated state: (a) whole spectrum, (b) crystalline component, and (c) noncrystalline component. (From Horii, F., Hirai, A., and Kitamaru, R., *The Structures of Cellulose. Characterization of the Solid-States*, ACS Symp. Ser., Atalla, R. H., Ed., American Chemical Society, Washington, D.C., 1987, 119. With permission.)

almost the same crystalline spectrum, whereas bacterial and valonia celluloses give another different type of crystalline spectrum. The most prominent features of these two types of crystalline spectra are recognizable in the C1 resonances; a predominant doublet with a weak singlet appears for cotton and ramie celluloses, while a triplet composed of an enhanced singlet and a minor doublet is observed for bacterial and valonia celluloses. It is, therefore, concluded that the crystal structures of the native celluloses are classified into two groups, bacterial-valonia type and cotton-ramie type. Although these spectra are referred to as cellulose I_a and I_b, respectively, both spectra seem to be composites of different crystal forms in accord with the previous proposals. Atalla and VanderHart[10,11] suggested that all native celluloses are a mixture of two crystalline modifications, cellulose I_α and I_β, which exhibit different fine multiplicities. Cael et al.,[16] also proposed a similar idea that the ^{13}C NMR spectra of native cellulose are linear combinations of two spectra. However, they assumed that these two spectra were associated with the resonances from two- and eight-chain unit cell regions of cellulose samples. Although

Table 1
^{13}C CHEMICAL SHIFTS OF THE CRYSTALLINE COMPONENTS OF NATIVE AND REGENERATED CELLULOSE IN THE HYDRATE STATE

Sample	C1			C4			C6			C2, 3, 5					
Native cellulose															
Valonia	106.9	106.2	105.2	91.0	90.1	89.2	66.7	66.4	66.2	75.8	75.3	73.7	72.9	72.4	72.0
Bacterial	106.9	106.3	105.2	91.0	90.1	89.5		66.4		75.8	75.3	73.8	72.9	72.5	71.9
Ramie	106.9	106.2	105.1		90.0	89.2	66.7		66.2		76.1	75.3	73.6	72.5	
Cotton	107.0	106.3	105.2		89.9	89.4		66.3	66.2		76.2	75.4	73.5	72.5	
Degenerated cellulose															
Cupra rayon	108.4	106.1			89.9	88.6		64.2	63.6		77.9	76.0	74.1	73.2	

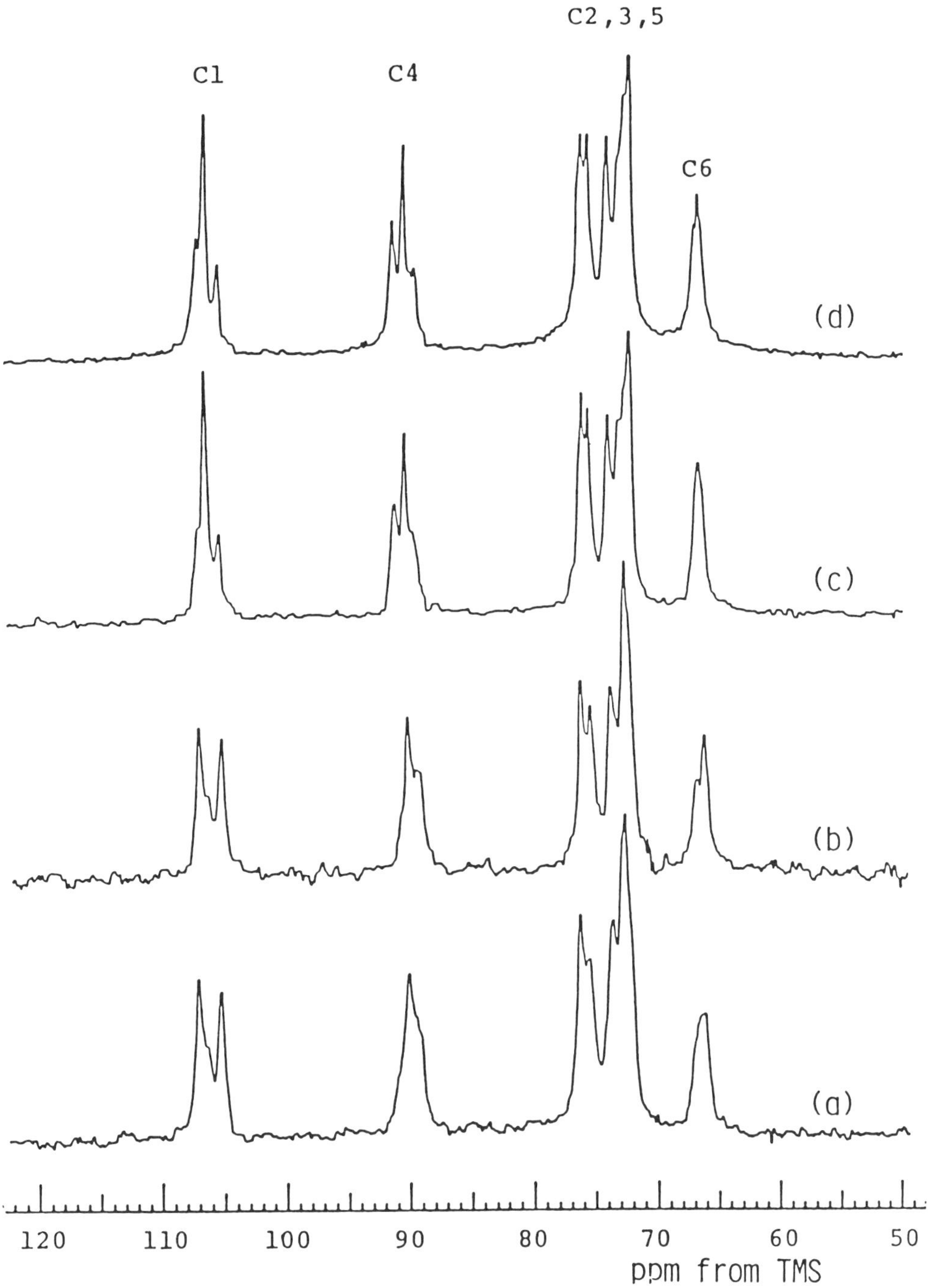

FIGURE 8. The 50-MHz CP/MAS ^{13}C-NMR spectra of the crystalline components included in native celluloses: (a) cotton, (b) ramie, (c) bacterial, and (d) valonia celluloses. (From Horii, F., Hirai, A., and Kitamaru, R., *Macromolecules*, 20, 2117, 1987. With permission.)

their models are very interesting, some significant modifications may be necessary. In fact, the relative intensities of the C1 and C4 triplets were not in accord with those predicted by a computer line shape analysis based on their models.[35]

Although it is very difficult at present to find the origin of the fine multiplicities, further information about the crystal structure of cellulose can be obtained from the experiments of crystal transformations. For example, it is well known that solid-state triacetylation followed by saponification produces crystal structure cellulose I under a low swelling condition. By use of these chemical reactions, it has been found that the bacterial-valonia type and cotton-ramie type crystal forms are transformed into cellulose I_a' which is newly defined but very similar to cotton-ramie type crystal form (cellulose I_a).[37] Similar crystal transformation has also been observed for bacterial cellulose by regeneration from cellulose III_I,[38] which is prepared by treating the sample with an aqueous solution of ethylene diamine and washing with anhydrous methanol. More direct transformations from I_b and I_a to I_a' have been realized for bacterial, valonia, and cotton celluloses by the treatment of these materials with saturated steam at high temperatures.[39] These findings may indicate that cellulose I_a' is a more stable crystal form than celluloses I_a and I_b.

Cellulose II, which is another well known crystal form, is obtained by regenerating cellulose samples through either the cuprammonium or the viscose process. The alkali treatment known as mercerization also converts cellulose I to cellulose II as described later in somewhat greater detail. Based upon the results of X-ray diffraction studies, the packing of the molecular chains is completely different in celluloses I and II; all chains are packed parallel to each other in cellulose I, while corner and center chains in the unit cell are antiparallel in cellulose II. The possibility of such different packing models as well as the possibility of a drastic change from parallel to antiparallel packing during mercerization are still under consideration.

Figure 9 shows the CP/MAS spectrum of the crystalline component cuprammonium rayon in the hydrated (Figure 9b) and dried (Figure 9c) states.[14] Here, the conventional CP/MAS spectrum (Figure 9a) of the hydrated sample is also depicted to show that the isolation of the resonance representing the crystalline component is essential for the detailed characterization of this material. The water enhances the resolution of each line, but the fine structure remains unchanged; the C1 and C4 lines split into a doublet and the relative intensities of the constituent lines are almost the same. Recent work[40] also shows that the cluster of C2, C3, and C5 carbon resonances can be resolved into three doublets with equivalent intensities. These results suggest that there are two inequivalent glycosyl residues in the unit cell. Atalla et al.[4,11] suggested the nonequivalence of successive glycosidic linkages along the chain on the basis of their spectroscopic evidence. In contrast, Dudley et al.[41] have found that cellotriose, which has three glycosidic linkages in one molecule, still exhibits a 1:1 doublet for the C1 sites. Therefore, they proposed that there are two independent chains in the unit cell. This means that the corner chains have different chain conformations than those of the center chains. Although such a deviation from the equivalent chains may be relatively small, the acquisition of CP/MAS spectra for related materials, including other polymorphs of cellulose, will help us to determine the exact crystal structures of cellulose. Some results have already been reported for samples which contain the crystal forms cellulose III and IV,[8,38,41] but the detailed analysis of the CP/MAS spectra of these materials should be performed by spectral editing described above to isolate the spectra of the crystalline component.

C. Noncrystalline Component

The CP/MAS ^{13}C-NMR method is also powerful for characterizing the structure of the noncrystalline regions. Above the glass transition temperature (T_g) the interfacial region, which is defined as a transition zone between the crystalline and the amorphous region, can be isolated spectroscopically from the amorphous regions by using the difference in ^{13}C spin-spin relaxation times (T_{2C}) of the two regions.[42] In cellulosic materials, such an analysis is not easy because the T_g is very high. However, by swelling with appropriate solvents, we can examine the characteristic structure of the noncrystalline regions of cellulose samples. Figure 10 shows the spectra of noncrystalline cuprammonium rayon in the dried and hydrated states.[14] These spectra were obtained by the subtraction method mentioned above. Each resonance line does not narrow

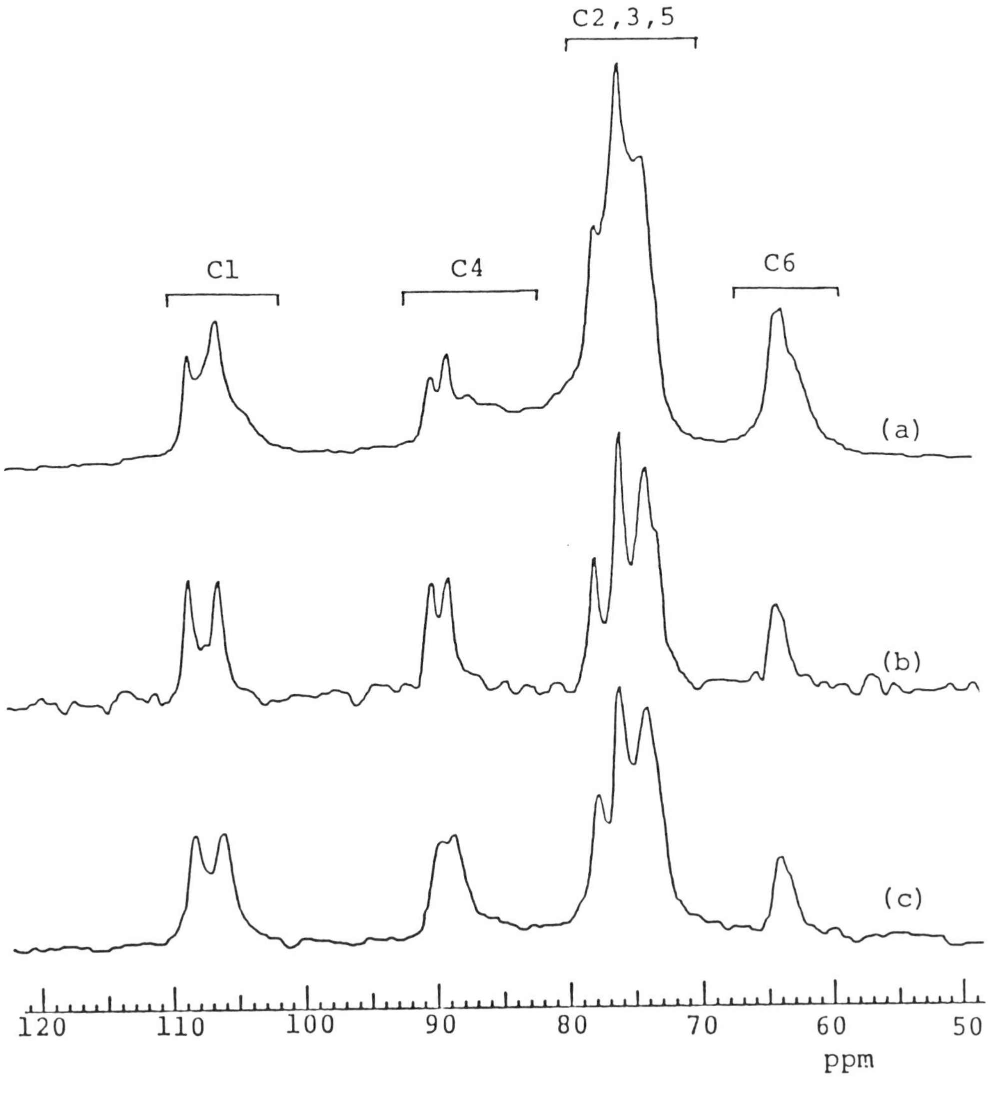

FIGURE 9. The 50-MHz CP/MAS ^{13}C-NMR spectra of (a) where spectrum in thc hydrated state, (b), (c) crystalline spectra in the hydrated and dried states, respectively of cupra rayon. (From Horii, F., Hirai, A., Kitamaru, R., and Sakurada, I., *Cellul. Chem. Technol.*, 19, 513, 1985. With permission.)

significantly upon the addition of water and the T_{1C} values were similar in both states. This indicates that water does not enhance the segmental mobilities of cellulose even in the noncrystalline region. This finding is very interesting, because the molecular chain conformation may be widely distributed and randomized, which is suggested from the relatively broad line widths in Figure 10.

Figure 11 shows the noncrystalline spectra of cotton cellulose in the corresponding two states.[14] In this case, the line widths of the C1 and C4 sites become markedly narrow for the hydrated cotton, while the chemical shifts remain unchanged. Such a marked narrowing is not ascribed to the enhanced motion of cellulose as in the case of rayon, but to the narrowing of the distribtuion of conformations giving spectral responses. Compared to the line widths of rayon, cotton has much narrower lines in the dried state, suggesting also the existence of a narrow distribtuion of conformations. Such a narrow distribution must be further reduced by water in the hydrated state. Therefore, it is assumed that the chains may undergo some distortion upon

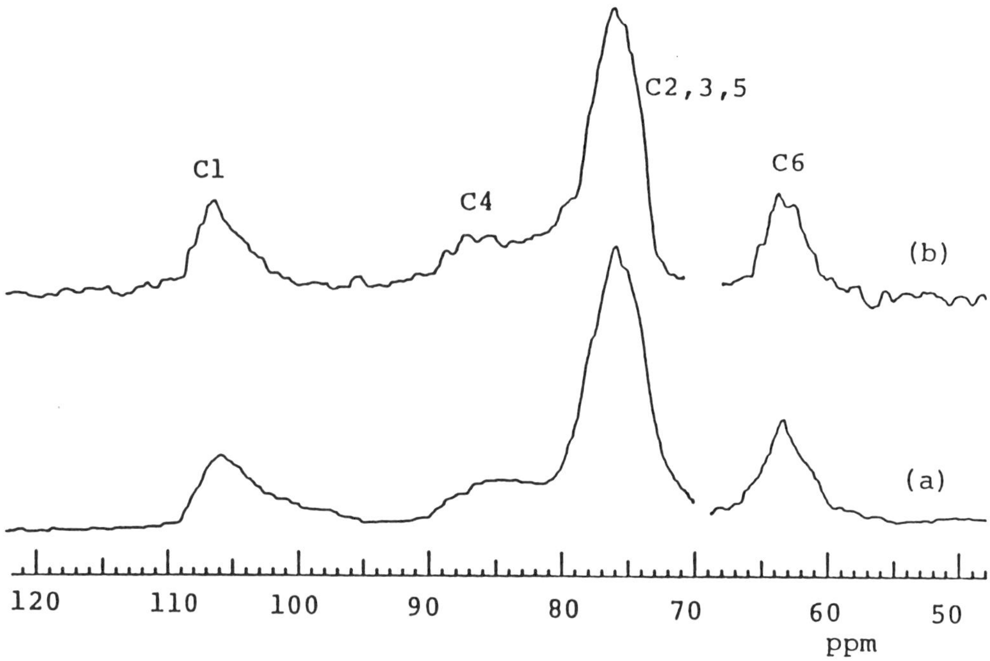

FIGURE 10. The 50-MHz CP/MAS ^{13}C-NMR spectra of the noncrystalline components of cupra rayon with the water contents of 0% (a) and 158% (b). (From Horii, F., Hirai, A., Kitamaru, R., and Sakurada, I., *Cellul. Chem. Technol.*, 19, 513, 1985. With permission.)

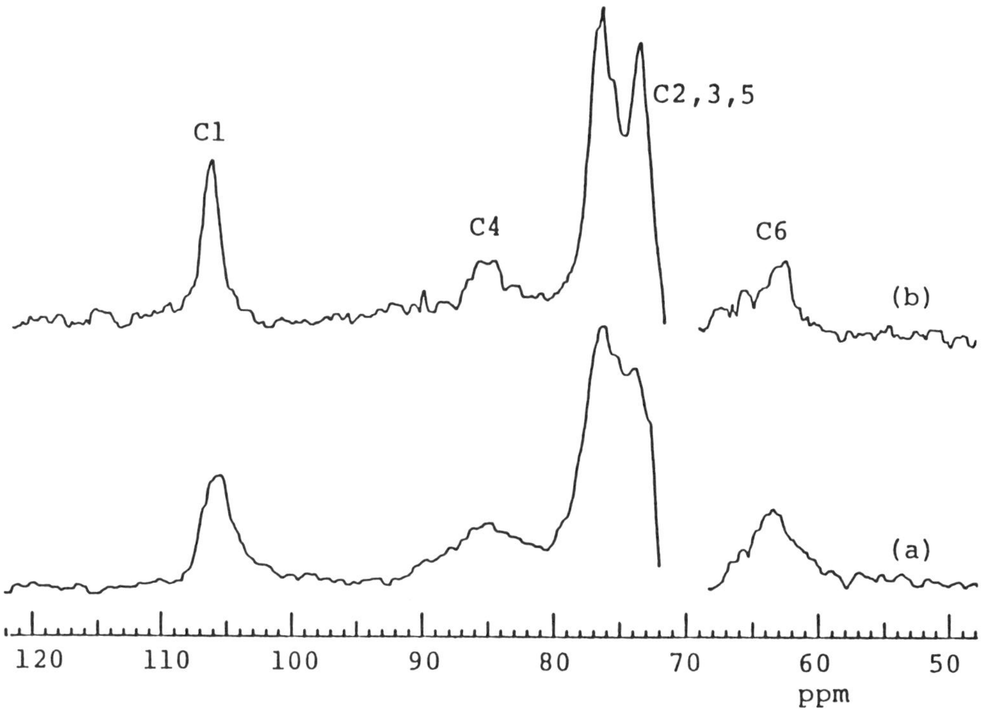

FIGURE 11. The 50-MHz CP/MAS ^{13}C-NMR spectra of the noncrystalline components of cotton cellulose with the water contents of 0% (a) and 161% (b). (From Horii, F., Hirai, A., Kitamaru, R., and Sakurada, I., *Cellul. Chem. Technol.*, 19, 513, 1985. With permission.)

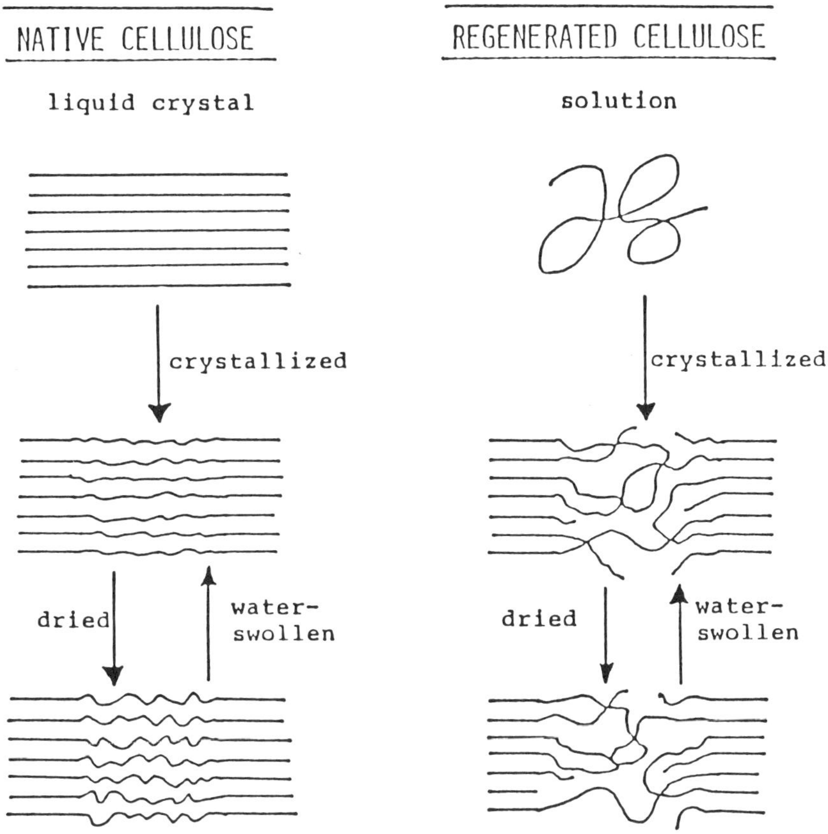

FIGURE 12. Schematic structural models of native and regenerated celluloses. (From Horii, F., Hirai, A., and Kitamaru, R., *The Structures of Cellulose. Characterization of the Solid-States*, ACS Symp. Ser., No. 340, Attala, R. H., Ed., American Chemical Society, Washington, D.C., 1987, 119. With permission.)

drying due to the restricted conformations, but water will relax this sort of distortion to some extent by activating local motions.

On the basis of these results, schematic structural models shown in Figure 12 have been proposed for native and regenerated celluloses.[15,20] In the incubation of *Acetobacter xylinum*, it has been found that the nascent fibrils are composed of bundles of chains in the tactoidal and noncrystalline states.[43] Furthermore, in that case, the cellulose chains are not randomly oriented in the cross-section of the fibril, but are in the form of monomolecular layers corresponding to the (110) plane of cellulose I.[44] Therefore, it is assumed that native cellulose including cotton and ramie may be crystallized from the ordered state like a "liquid crystal". After the crystallization, the noncrystallizable chains must still be relatively ordered as shown in Figure 12. These sorts of noncrystalline chains will undergo some distortion upon drying, but absorbed water must relax the distortion of the chains again. In the case of rayon, however, the fibers are spun from a concentrated solution where the molecules form almost random coils and are severely entangled with each other. Thus, the chains are not well extended by conventional drawing. In addition, some drawn chains may be already relaxed, by soaking with hot water during the purification process. Therefore, the noncrystalline chains are considerably disordered in rayon and, as a result, hardly distorted upon drying as shown in Figure 12. These differences in structure of the noncrystalline regions of native and regenerated celluloses must be reflected in the line widths of spectra shown in Figures 10 and 11.

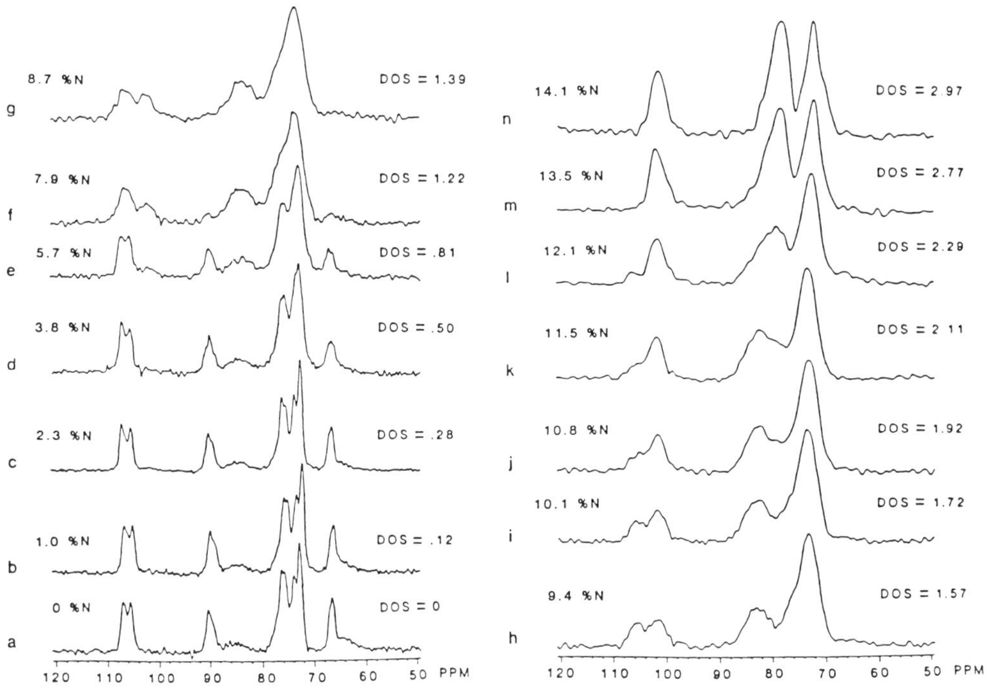

FIGURE 13. The 37.7 MHz CP/MAS NMR spectra of cotton cellulose and cellulose nitrates with different degrees of substitution. (From Patterson, P. M., Patterson, D. J., Blackwell, J., Koenig, J. L., and Jamieson, A. M.,. *J. Polym. Sci. Polym. Phys. Ed.*, 23, 483, 1985. With permission.)

V. CELLULOSE DERIVATIVES AND ADDITION PRODUCTS

Cellulose derivatives, such as cellulose nitrates and cellulose acetates are also useful products because they are soluble in conventional organic solvents and readily converted to fibers, plastics, films, etc. These derivatives are commercially prepared by heterogeneous chemical reactions and the properties of the products are influenced by the degrees of substitution (DOS) as well as by the sequence distributions. Therefore, recent CP/MAS ^{13}C-NMR work[36,45,46] of cellulose derivatives has focused mainly on the examination of the processes of the substitution reactions in the heterogeneous state.

Figure 13 shows CP/MAS ^{13}C-NMR spectra of cotton linters and cellulose nitrates with different degrees of substitution, where the heterogeneous reactions were performed at 0 to 4°C for 3 d using mixtures of anhydrous nitric acid and dichloromethane in different proportions.[45] The first nitration appears to occur in the C6 sites of the noncrystalline regions, because the upfield shoulder of the C6 line diminishes with increasing DOS. After the completion of the reaction at DOS = 0.50, further substitution of the C6 sites proceeds by inducing decrystallization, as indicated by the decreases in C4 and C6 resonances at 90 and 66 ppm, respectively, which are both assigned to the crystalline components as described in Section II. For the C1 sites, a new resonance appears at 101 ppm for DOS = 0.28 and, in contrast, the original C1 doublet of cotton cellulose disappears at DOS = 2.65. This change is the second process which is ascribed to the nitration of the C2 sites. The C3 sites are finally subjected to the substitution reaction, which is observed as an enhancement of the intensity at 82 ppm at DOS = 1.92. Thus we can successfully follow the nitration of cellulose by CP/MAS ^{13}C-NMR spectroscopy and can differentiate between substitution sites in both the crystalline and noncrystalline regions.

The heterogeneous acetylation process of cotton linters was also examined by almost the

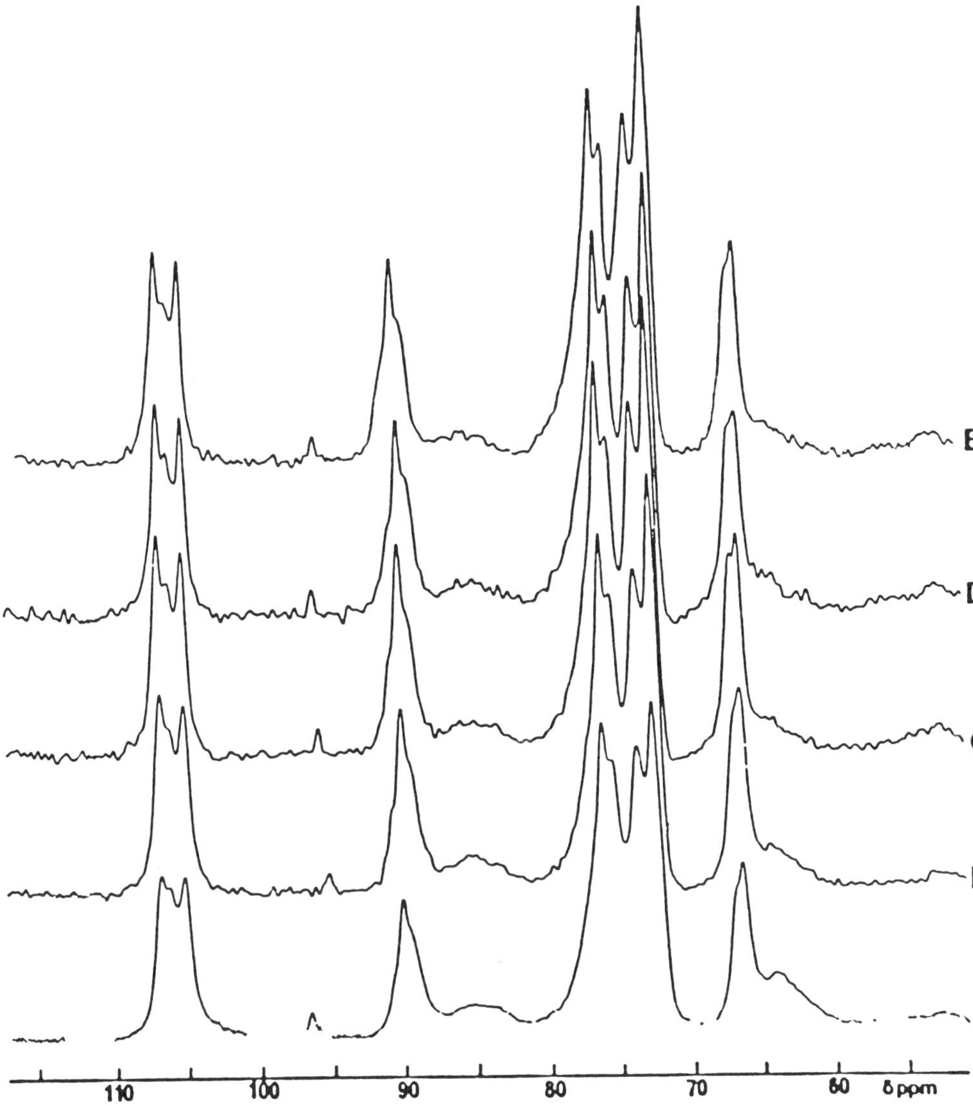

FIGURE 14. The 50-MHz CP/MAS ^{13}C-NMR spectra of cotton cellulose labeled by trimethylsilyl groups at different DOS: (A) 0, (B) 0.10, (C) 0.18, (D) 0.32, and (E) 0.45. (From Gagnaire, D., Saint-Germain, J., and Vincendon, M., *Polym. Bull.*, 13, 365, 1985. With permission.)

same method.[36] Although the results were not definitive as the case of nitrates due to the overlapping of the resonances, significant changes occur in the C4 regions of the spectrum. The initial acetylation starts in the noncrystalline regions, and hence retention of a peak at 89 ppm indicates the existence of the crystalline component of cellulose. However, this peak disappears for DOS = 2.4 or greater, suggesting the occurrence of acetylation in the crystalline regions with concomitant structural changes at the molecular level in these regions.

Morphological effects on heterogeneous reactions were also found for chemical substitution of accessible hydroxyl groups to a lesser extent by deuterated methoxyl (-OCD_3) and trimethylsilyl (-OTMS) groups of sonicated cotton cellulose.[46] Figure 14 shows 50-MHz CP/MAS ^{13}C-NMR spectra of cotton cellulose labeled with trimethylsilyl groups at different DOS. The main change in the spectrum is characterized by the decrease in the relative intensities of the noncrystalline resonances of C4 and C6 sites as well as of the central line in the C1 triplet. These

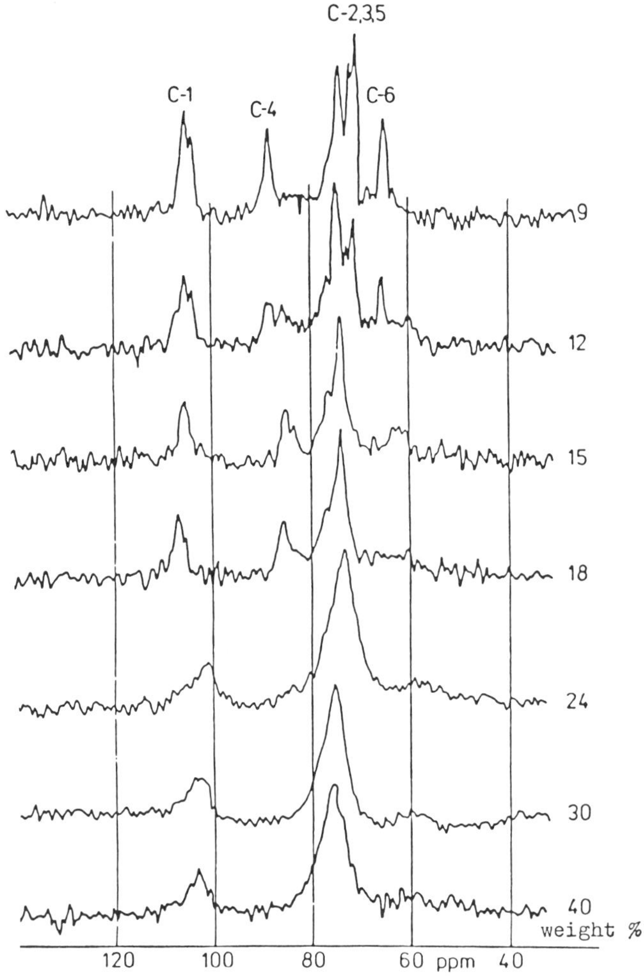

FIGURE 15. The 15 MHz CP/MAS ^{13}C-NMR spectra of cellulose treated with different aqueous solutions of NaOH. (From Kunze, J., Ebert, A., Schröter, B., Frigge, K., and Philipp, B., *Polym. Bull.*, 5, 399, 1981. With permission.)

decreases in the original resonances of the cotton should correspond to the appearance of new lines at 103 and 61.5 ppm attributed to C1 and C4 carbons of trimethylsilylated cotton. However, this is not obvious in these spectra. As a result, the spectrum for DOS = 0.45 is very similar to the crystalline spectrum shown in Figure 7b.

It is well known that cellulose forms different swelling complexes or addition products with bases, inorganic acids, inorganic salts, amines, etc. These treatments produce inherently characteristic X-ray diffraction patterns. Among these complexes, soda celluloses have been most extensively studied because they are commercially important intermediates for converting native celluloses to mercerized celluloses and to cellulose ethers, such as carboxymethylcellulose and methylcellulose. However, no one completely knows the mechanism of formation and morphological structure of soda celluloses as well as the dependence of the reactivities on the morphology. CP/MAS ^{13}C-NMR spectroscopy is also useful for solving these problems, although it seems technically difficult to use the high power ^{1}H decoupling field in such a highly dielectric system.

Figure 15 shows 15-MHz CP/MAS ^{13}C-NMR spectra of cotton linters treated with aqueous

NaOH solution of different concentrations.[47] The spectrum remains unchanged up to a steeping lye concentration of 9 wt%, in good accord with X-ray diffraction results which indicate no lattice conversion in the same range of concentration. With increasing lye concentration, new peaks corresponding to the C1, C4, and C6 carbons appear at about 108, 87, and 63 ppm, which are assigned to the resonances of soda cellulose I, and the pure resonance profile of that crystal form can be obtained at 18 wt% where mercerization and chemical modifications are normally performed. Furthermore, another transition of the crystal form appears to occur in the concentration range between 18 and 24 wt%.

Another recent study[48] also indicated that, compared to the corresponding chemical shifts of original wood pulp, the C1 resonance of soda cellulose I shifts downfield while the values of C4 and C6 carbons shift upfield. On the basis of the examination of possible shifts induced by selective coordination of sodium ions to the hydroxyl oxygens, it was suggested that the sodium ions selectively coordinate to the hydroxyl oxygens at the C2 sites. In contrast, an X-ray diffraction study[49] suggests that all sodium ions form secondary ionic bonds with O2 as well as O6 oxygens. Moreover, it has been found that the T_{1C} values of C1, C4, and C6 carbons are 2.4, 2.1, and 0.4 s under the static magnetic field of 4.7 T.[50] This indicates that the molecular mobilities of cellulose chains are extremely enhanced in soda cellulose I. Therefore, more detailed interpretation of the chemical shifts, particularly considering the effects of the coordination of sodium ions and molecular motions, is still necessary for this system.

VI. WOODS AND THEIR CONSTITUENTS

Typical conifer and deciduous woods are composed of cellulose, hemicellulose, lignin, and a small amount of extractives. The chemical structure of these materials has been studied after separation into the respective components by destructive methods. Therefore, these structures are probably changed during preparation. This is particularly true for lignin, which is easily modified. Real morphological structures of woods, involving lignin-carbohydrate complexes, would also be impossible to understand from studies of the individual components. Recent studies[51-62] of woods and the related materials have proved CP/MAS ^{13}C-NMR spectroscopy to be extremely effective as a nondestructive means for analyzing the chemical and morphological structures of intact woods as well as pulping processes.

Figure 16 shows a 50-MHz CP/MAS ^{13}C-NMR spectrum of ash wood.[55] The spectrum clearly reflects the contributions from cellulose, hemicellulose, and lignin. (For comparison, the resonance profiles of hemicellulose [extracted fraction B] and enzymatic lignin [dioxane fraction] are shown in Figure 17, which are obtained by successive separation treatments from lodgepole pine wood.)[54] The resonances between 65 and 110 ppm can be assigned to cellulose and hemicellulose components. In particular, the line of the anomeric carbons of both components is well recognized at about 105 ppm. Hemicellulose components also exhibit several other less intense resonances; acetyl methyl and acetyl carbonyl lines for the *O*-acetyl (4-*O*-methyl-glucurono) xylan at 26 ppm and 175 ppm, respectively, and a minor peak at 180 ppm which may result from the carboxyl groups of 4-*O*-methylglucuronic acid residues attached to the xylan chain. The lignin components give rise to several signals: the resonance at 60 ppm is assigned to the methoxyl substituents, while a broad cluster of resonances at 120 to 160 ppm may result from the aromatic carbons of lignin. Although other considerably intense resonances assignable to the aliphatic carbons of lignin are in the region of 50 to 95 ppm, they are completely overlapped by the contributions from cellulose and hemicellulose.

The same technique as mentioned in Section IV was applied for white birch and Japanese cypress woods to obtain their crystalline spectra.[62] However, the resolution was not high and there was no evidence that the contribution of the hemicellulose was completely removed. It is, therefore, difficult at present to conclude which type of crystal form, cellulose I_a or I_b, is formed in woods. This difficulty may be partly related to the low level of crystallinity, which is

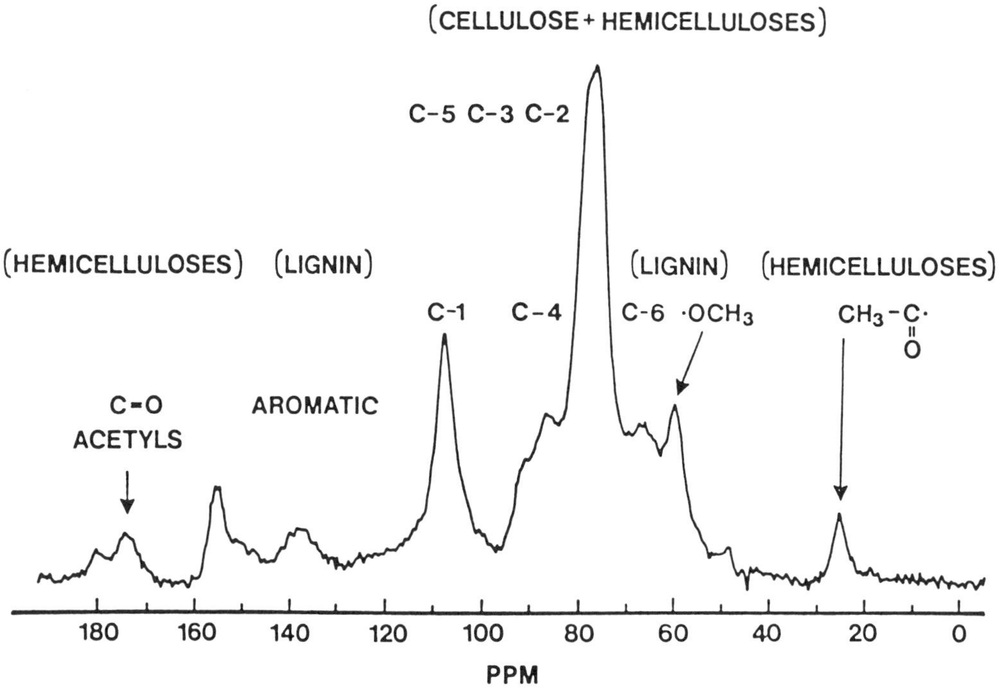

FIGURE 16. The 50 MHz CP/MAS ^{13}C-NMR spectrum of solid ash wood. (From Taylor, M. G., Deslandes, Y., Bluhm, T., Marchessault, R. H., Vincendon, M., and Saint-Germain, J., *Tappi*, 66, 92, 1983. With permission.)

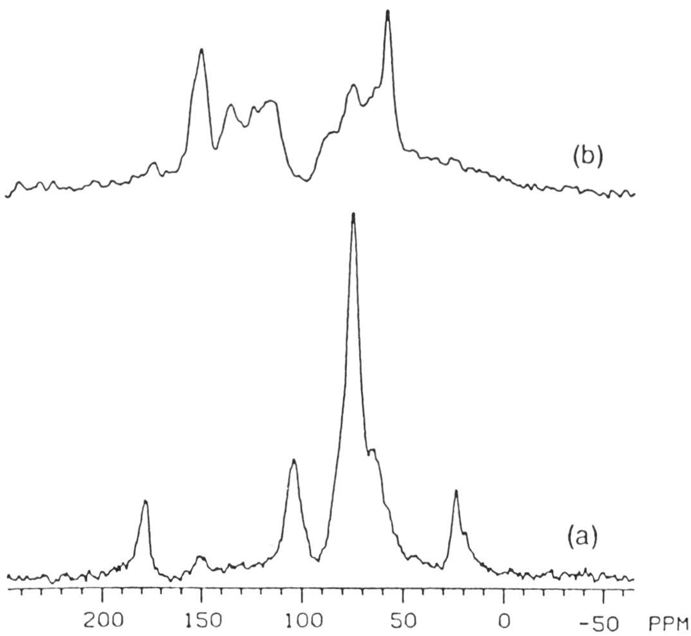

FIGURE 17. The 37.7-MHz CP/MAS ^{13}C-NMR spectra of hemicellulose (a) and enzymatic lignin (dioxane fraction) (b). (From Kolodzieski, W., Fyfe, J. S., and Maciel, G. E., *Anal. Chem.*, 54, 1419, 1982. With permission.)

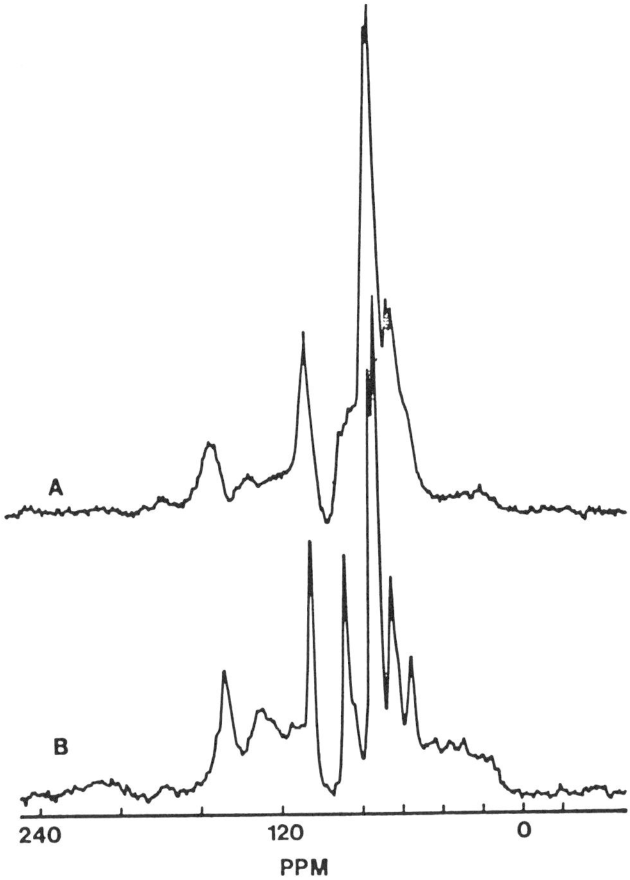

FIGURE 18. 50-MHz CP/MAS ^{13}C-NMR spectra of spruce wood (A) and the product (B) treated by the steam explosion process. (From Taylor, M. G., Deslandes, Y., Bluhm, T., Marchessault, R. H., Vincendon, M., and Saint-Germain, J., *Tappi*, 66, 92, 1983. With permission.)

suggested by the less intense peak at 90 ppm in Figure 16, compared to the more pronounced one for other native celluloses (see Figure 2). Howevere, such low crystallinity appears to provide indirect evidence for a supermolecular structures including lignin-carbohydrate complexes.

Figure 18 presents the effects of the steam explosion process for spruce wood.[55] A pronounced narrowing of the resonances occurs for the exploded wood, and the drastic increase in the C4 line at about 90 ppm is obtained possibly resulting from the increased crystallinity of cellulose. Separate X-ray analyses confirmed the increase in crystallinity and lateral size as well as the increased orientation in the crystallites.[55] Similar effects of the steam explosion process on wood composition and structure were observed for *Pinus radiata* and *Eucalyptus regnans*.[59] It may be, however, still difficult to examine the fine multiplicities of the C1 and C4 sites of cellulose by conventional CP/MAS spectra. In contrast, the selective observation method of the crystalline spectra was also very effective for the exploded white birch and Japanese cypress woods: interestingly, their crystalline spectra clearly indicated that both exploded woods contain cellulose I_a', which is very similar to the cotton-ramie type crystal form.[62]

The existence of the lignin-carbohydrate complex was examined by the subtraction methods for spectra of extractives-free lodgepole pine wood and the chlorine gas holocellulose, which are free from lignin components by a NaOC1 treatment.[54] Thus the subtracted spectrum indicates that some cellulose and hemicellulose are removed with the lignin during holocellulose preparation. This result may provide indirect evidence for the lignin-carbohydrate complex. The existence of the complex was also confirmed by the examination of the proton spin diffusion processes for the wood pulp constituent in conjunction with induced paramagnetic relaxation, ^{13}C-^{1}H interrupted-proton decoupling, and enzymatic degradation.[56]

VII. CONCLUDING REMARKS

As described above, CP/MAS ^{13}C-NMR spectroscopy is a powerful method for investigating the structure of the crystalline and noncrystalline regions of cellulose. The overlapping resonances in the conventional CP/MAS spectra were successfully resolved into the contributions of these two regions by using the differences in spin-lattice relaxation times T_{1C}. The relaxation edited spectra thus obtained for native cellulose were classified by the differences in the fine multiplicities of C1 and C4 resonances into two groups, cotton-ramie type and bacterial-valonia type which are referred to as cellulose I_a and I_b, respectively. However, the causes of the existence of the two types as well as the fine multiplicities are not clear at present. It is hoped that these problems will be solved through a better understanding of the crystal structure of cellulose based on further CP/MAS ^{13}C-NMR studies in conjunction with X-ray diffraction and electron microscopy. Some reliable aids will also be provided by similar investigations of other polymorphs of cellulose including cellulose II, III, and IV as well as by CP/MAS ^{13}C-NMR studies of amylose and cycloamylose which are composed of D-glycosyl residues joined by α-1,4-glycosidic linkages.

The noncrystalline structure of native and regenerated celluloses was also discussed on the basis of the difference spectra which were obtained by subtracting the crystalline spectra from the conventional CP/MAS spectra. The molecular chains are considerably confined in conformation for native cellulose, suggesting that cellulose molecules would be crystallized from some ordered state like a liquid crystal in nature. On the other hand, the molecular chain conformation of regenerated celluloses is widely distributed and almost randomized, probably reflecting that these moelcules are crystallized from a concentrated solution where they adopt random coils and are heavily entangled with each other. It is, however, greatly necessary for further progress in structural studies of the noncrystalline regions to well understand the origins of ^{13}C isotropic chemical shifts and their distribution.

CP/MAS ^{13}C-NMR has also been proved to be a promising new technique in following the heterogeneous reactions to produce cellulose derivatives and addition products as well as in characterizing the structure of intact woods and their constituents. At this stage, however, more quantitative analyses seem considerably difficult to perform for the individual components; for example, original cellulose and reaction products in cellulose derivatives or cellulose, hemicellulose, and lignin in woods, because of the serious overlapping of their resonances. Further developments in pulse techniques would be important for enhanced resolution in combination with studies of isolated components.

REFERENCES

1. **Marchessault, R. H. and Sundararajan, P. R.,** Cellulose, in *The Polysaccharides,* Vol. 2, Academic Press, New York, 1983, chap. 2.
2. **Fyfe, C. A., Dudley, R. L., Stephenson, P. J., Deslandes, Y., Hawer, G. K., and Marchessault, R. H.,** Application of high-resolution solid-state NMR with cross-polarization/magic-angle spinning (CP/MAS) techniques to cellulose chemistry, *J. Macromol. Sci. Rev. Macromol. Chem. Phys.,* C23(2), 187, 1983.

3. **Horii, F., Hirai, A., and Kitamaru, R.,** Cross-polarization/magic-angle spinning ^{13}C-NMR study. Molecular chain conformations of native and regenerated cellulose, in *Polymers for Fibers and Elastomers,* ACS Symp. Series, No. 260, Arthur, J. C., Jr., Ed., American Chemical Society, Washington, D.C., 1984, 27.
4. **Atalla, R. H., Gast, J. C., Sindorf, D. W., Bartuska, V. J., and Maciel, G. E.,** ^{13}C NMR spectra of cellulose polymorphs, *J. Am. Chem. Soc.,* 102, 3249, 1980.
5. **Earl, W. L. and VanderHart, D. L.,** High resolution, magic angle sample spinning ^{13}C NMR of solid cellulose I, *J. Am. Chem. Soc.,* 102, 3251, 1980.
6. **Earl, W. L. and VanderHart, D. L.,** Observation by high-resolution carbon-13 nuclear magnetic resonance of cellulose I related to morphology and crystal structure, *Macromolecules,* 14, 570, 1981.
7. **Horii, F., Hirai, A., and Kitamaru, R.,** Solid-state high-resolution ^{13}C-NMR studies of regenerated cellulose samples with different crystallinities, *Polym. Bull.,* 8, 163, 1982.
8. **Kunze, J., Scheler, G., Schroter, B., and Philipp, B.,** ^{13}C high-resolution solid-state NMR studies on cellulose samples of different physical structure, *Polym. Bull.,* 10, 56, 1983.
9. **Horii, F., Hirai, A., and Kitamaru, R.,** CP/MAS carbon-13 NMR study of spin relaxation phenomena of cellulose containing crystalline and noncrystalline componenets, *J. Carbohydr. Chem.,* 3, 641, 1984.
10. **Atalla, R. H. and VanderHart, D. L.,** Native cellulose: a composite of two distinct crystalline forms, *Science,* 223, 283, 1984.
11. **VanderHart, D. L. and Atalla, R. H.,** Studies of microstructure in native celluloses using solid-state ^{13}C NMR, *Macromolecules,* 17, 1465, 1984.
12. **Teeäär, R. and Lippmaa, E.,** Solid-state ^{13}C-NMR of cellulose. A relaxation study, *Polym. Bull.,* 12, 315, 1984.
13. **Kamide, K., Okajima, K., Matsui, T., and Kowsaka, K.,** Study on the solubility of cellulose in aqueous alkali solution by deuteration IR and ^{13}C NMR, *Polym. J.,* 16, 857, 1984.
14. **Horii, F., Hirai, A., Kitamaru, R., and Sakurada, I.,** Cross-polarization/magic-angle spinning ^{13}C NMR studies of cotton and cupra rayon with different water contents, *Cellul. Chem. Technol.,* 19, 513, 1985.
15. **Hirai, A., Horii, F., and Kitamaru, R.,** Solid-state high-resolution carbon-13 NMR studies of celluloses, *Bull. Inst. Chem. Res. Kyoto Univ.,* 63, 340, 1985.
16. **Cael, J. J., Kwoh, D. L. W., Bhattacharjee, S. S., and Patt, S. L.,** Cellulose crystallites: a perspective from solid-state ^{13}C NMR, *Macromolecules,* 18, 819, 1985.
17. **Atalla, R. H., Whitmore, R. E., and VanderHart, D. L.,** A highly crystalline cellulose from *Rhizocloninum hieroglyphicum, Biopolymers,* 24, 421, 1985.
18. **Kamide, K., Okajima, K., Kowsaka, K., and Matsui, T.,** CP/MAS ^{13}C NMR spectra of cellulose solids: an explanation by the intramolecular hydrogen bond concept, *Polym. J.,* 17, 701, 1985.
19. **Nagavi, B. G., Mithal, B. H., Chawla, J. S., and Dhar, K. L.,** CP/MAS solid-sample ^{13}C NMR of microcrystalline cellulose and amylose, *Tappi,* 69, 132, 1986.
20. **Horii, F., Hirai, A., and Kitamaru, R.,** CP/MAS carbon-13 NMR approach to the structural analysis of cellulose, in *The Structures of Cellulose. Characterization of the Solid States,* ACS Symp. Ser., No. 340, Atalla, R. H., Ed., American Chemical Society, Washington, D.C., 1987, 119.
21. **VanderHart, D. L.,** Deductions about the morphology of wet and wet beaten celluloses from solid-state ^{13}C-NMR, National Bureau of Standards Report (NBSIR), 1982, 82.
22. **Fyfe, C. A., Stephenson, P. J., Taylor, M. G., Bluhm, T. L., Deslandes, Y., and Marchessault, R. H.,** Hydration effects in the ^{13}C CP/MAS NMR spectra of solid (1→3)-β-D-glucans, *Macromolecules,* 17, 501, 1984.
23. **Horii, F., Hirai, A., and Kitamaru, R.,** CP/MAS ^{13}C NMR spectroscopy of hydrated amyloses using a magic-angle spinning rotor with an O-ring seal, *Macromolecules,* 19, 930, 1986.
24. **Moller, M., Gronski, W., Cantow, H.-J., and Hocker, H.,** Discrimination of rotational isomeric states in cycloalkanes by solid-state CP/MAS ^{13}C NMR spectroscopy, *J. Am. Chem. Soc.,* 106, 5093, 1984.
25. **VanderHart, D. L.,** Influence of molecular packing on solid-state ^{13}C chemical shifts: the n-alkanes, *J. Magn. Reson.,* 44, 117, 1981.
26. **Imashiro, F., Maeda, S., Takegoshi, K., Terao, T., and Saika, A.,** Hydrogen-bonding and conformational effects on ^{13}C NMR chemical shifts of hydroxybenzaldehydes in the solid-state, *Chem. Phys. Lett.,* 99, 189, 1983.
27. **Terao, T., Maeda, S., and Saika, A.,** High-resolution solid-state ^{13}C NMR of poly(vinyl alcohol): enhancement of tacticity splitting by intramolecular hydrogen bonds, *Macromolecules,* 16, 1535, 1983.
28. **Horii, F., Hirai, A., and Kitamaru, R.,** Relationships between carbon-13 chemical shifts and conformations of oligosaccharides and cellulose in the solid state, *Bull. Magn. Reson.,* 5, 233, 1983.
29. **Horii, F., Hirai, H., and Kitamaru, R.,** Solid-state ^{13}C-NMR study of conformations of oligosaccharides and celluloses. Conformation of CH_2OH groups about the exocyclic C-C bond, *Polym. Bull.,* 10, 357, 1983.
30. **Saito, H., Izumi, G., Mamizuka, T., Suzuki, S., and Tabeta, R.,** A ^{13}C cross polarization-magic angle spinning (CP/MAS) N.M.R. study of crystalline cyclohexa-amylose inclusion complexes. Conformation-dependent ^{13}C chemical shifts are related to the dihedral angles of glycosidic linkages, *J. Chem. Soc., Chem. Commun.,* 1387, 1982.

31. **Inoue, Y., Okuda, T., and Chujo., R.,** A high-resolution C.P.-M.A.S. ^{13}C-NMR study of solid-state cyclomaltohexaose inclusion-complexes: chemical shifts and structure of the host cyclomaltohexaose, *Carbohydr. Res.*, 141, 179, 1985.
32. **Veregin, R. P., Fyfe, C. A., Marchessault, R. H., and Taylor, M. G.,** Correlation of ^{13}C chemical shifts with torsional angles from high-resolution ^{13}C C.P.-M.A.S.N.M.R. studies of the crystalline cyclomalto-oligosaccharide complexes, and their relation to the structures of the starch polymorphs, *Carbohydr. Res.*, 160, 41, 1987.
33. **Hirai, A., Horii, F., and Kitamaru, R.,** CP/MAS ^{13}C NMR study of solid structure of the dry and hydrated cellulose samples, *Polym. Prepr., Jpn.*, 34, 2473, 1985.
34. **Torchia, D. A.,** The measurement of proton-enhanced carbon-13 T_1 values by a method which suppresses artifacts, *J. Magn. Reson.*, 30, 613, 1978.
35. **Horii, F., Hirai, A., and Kitamaru, R.,** CP/MAS ^{13}C NMR spectra of the crystalline components of native celluloses, *Macromolecules*, 20, 2117, 1987.
36. **Doyle, S., Pethrick, R. A., Harris, R. K., Lane, J. M., Packer, K. J., and Heatley, F.,** ^{13}C nuclear magnetic resonance studies of cellulose acetate in the solution and solid-states, *Polymer*, 27, 19, 1986.
37. **Hirai, A., Horii, F., and Kitamaru, R.,** Transformation of native cellulose crystals from cellulose I_b to I_a through solid-state chemical reactions, *Macromolecules*, 20, 1440, 1987.
38. **Chanzy, H., Henrissat, B., Vincendon, M., Tanner, S. F., and Belton, P. S.,** Solid state ^{13}C-N.M.R. and electron microscopy study on the reversible transformation cellulose I→cellulose III_I in *Valonia*, *Carbohydr. Res.*, in press.
39. **Horii, F., Yamamoto, H., Kitamaru, R., Tanahashi, M., and Higuchi, T.,** Transformation of native cellulose crystals induced by saturated steam at high temperatures, *Macromolecules*, 20, 2946, 1987.
40. **Chen, Y. Y., Horii, F., and Kitamaru, R.,** Observation of ^{13}C spin exchanges in solids by two-dimensional NMR spectroscopy, Preprints of the 25th NMR Symposium, Japan, 1986, 42.
41. **Dudley, R. L., Fyfe, C. A., Stephenson, P. J., Deslandes, Y., Harmer, G. K., and Marchessault, R. H.,** High-resolution ^{13}C CP/MAS NMR spectra of solid cellulose oligomers and the structure of cellulose II, *J. Am. Chem. Soc.*, 105, 2469, 1983.
42. **Kitamaru, R., Horii, F., and Murayama, K.,** Phase structure of lamellar crystalline polyethylene by solid-state high-resolution ^{13}C NMR: detection of the crystalline-amorphous interphase, *Macromolecules*, 19, 636, 1986.
43. **Haigler, C. H., Brown, R. M., and Benziman, M.,** Calcofluor white ST alters the *in vivo* assembly of cellulose microfibrils, *Science*, 210, 903, 1980.
44. **Kai, A.,** The structure of the nascent fibrils produced by *Acetobacter xylinum:* the X-ray diffraction diagram of cellulose produced in the presence of a fluorescent brightener, *Macromol. Chem. Rapid Commun.*, 5, 310, 1984.
45. **Patterson, P. M., Patterson, D. J., Blackwell, J., Koenig, J. L., and Jamieson, A. M.,** High-resolution solid-state ^{13}C NMR spectoscopy of cellulose nitrates, *J. Polym. Sci. Polym. Phys. Ed.*, 23, 483, 1985.
46. **Gagnaire, D., Saint-Germain, J., and Vincendon, M.,** High-resolution solid state ^{13}C NMR study of heterogeneous reactions on cotton cellulose, *Polym. Bull.*, 13, 365, 1985.
47. **Kunze, J., Ebert, A., Schröter, B., Frigge, K., and Philipp, B.,** ^{13}C- and ^{23}Na-NMR investigations on alkali cellulose, *Polym. Bull.*, 5, 399, 1981.
48. **Kamide, K., Kowsaka, K., and Okajima, K.,** Determination of intramolecular hydrogen bonds and selective coordination of sodium cation in alkalicellulose by CP/MASS ^{13}C NMR, *Polym. J.*, 17, 707, 1985.
49. **Nishimura, H., Okano, T., and Sarko, A.,** Mercerization of cellulose. V. The crystal structure of Na-cellulose I, *J. Appl. Polym. Sci.*, in press.
50. **Yokata, H., Sei, T., Kitamaru, R., and Horii, F.,** CP/MAS ^{13}C NMR of alkali cellulose, *Preprints of the Fall Meeting of the Society of Fiber Science and Technology, Japan*, 1986, 104; to be published.
51. **Bartuska, V. J., Maciel, G. E., Bolker, H. I., and Fleming, B. I.,** Structural studies of lignin isolation procedures by ^{13}C NMR, *Holzforschung*, 31, 214, 1980.
52. **Maciel, G. E., O'Donnell, D. J., Ackerman, J. J. H., Hawkins, B. H., and Bartuska, V. J.,** *Macromol. Chem.*, 182, 2297, 1981.
53. **Schaefer, J., Sefcik, M. D., Stejskal, E. O., McKay, R. A., and Hall, P. L.,** Characterization of the catabolic transformation of lignin in culture using magic-angle carbon-13 nuclear magnetic resonance, *Macromolecules*, 14, 557, 1981.
54. **Kolodzieski, W., Fyfe, J. S., and Maciel, G. E.,** Carbon-13 nuclear magnetic resonance spectrometry with cross-polarization and magic-angle spinning for analysis of lodgepole pine wood, *Anal. Chem.*, 54, 1419, 1982.
55. **Taylor, M. G., Deslandes, Y., Bluhm, T., Marchessault, R. H., Vincendon, M., and Saint-Germain, J.,** Solid-state ^{13}C NMR characterization of wood, *Tappi*, 66, 92, 1983.
56. **Gerasimowicz, W. V., Hicks, K. B., and Pfeffer, P. E.,** Evidence for the existence of associated lignin-carbohydrate polymers as revealed by carbon-13 CP/MAS solid-state NMR spectroscopy, *Macromolecules*, 17, 2957, 1984.

57. **Haw, J. F. and Maciel, G. E.,** Carbon-13 nuclear magnetic resonance study of the rapid steam hydrolysis of red oak, *Holzforschung*, 38, 327, 1984.
58. **Haw, J. F. and Schultz, T. P.,** Carbon-13 CP/MAS NMR and FT-IR study of low-temperature lignin pyrolysis, *Holzforschung*, 39, 289, 1985.
59. **Hemmingson, J. A. and Newman, R. H.,** A CP/MAS ^{13}C NMR study of the effect of steam explosion processes on wood composition and structure, *J. Wood Chem. Technol.*, 5, 139, 1985.
60. **Leary, G. J., Morgan, K. R., and Newman, R. H.,** A CP/MAS NMR comparison of wood fractions from spruce, *Holzforschung*, 40, 221, 1986.
61. **Willis, J. M., Jagannathan, N. R., and Herring, F. G.,** The potential of ^{13}C CP/MAS NMR in the study of kraft pulping kinetics., *J. Wood Chem. Technol.*, 6, 249, 1986.
62. **Horii, F., Hirai, A., Kitamaru, R., Tanahashi, M., and Higuchi, T.,** Polymorphs of native cellulose: an approach by high-resolution solid-state ^{13}C NMR, Preprints of the Fall Meeting of the Society of Fiber Science and Technology, Japan, 1986, 103.

Chapter 11

THE STRUCTURE AND DEGRADATION OF INTACT PLANT CELL WALL MATRICES BY ^{13}C CPMAS NMR AND RELATED TECHNIQUES

Peter L. Irwin

TABLE OF CONTENTS

I. INTRODUCTION

At the most general level of our understanding, the primary cell wall and middle lamellar complex of higher plant tissues[1] is considered to be a three phase system which is made up of a "crystalline" component (cellulose), a "matrix" component (the acid sugar-containing polymers or pectic polysaccharides and hemicellulose, etc.) as well as a "packing" component (bound H_2O). The architecture of the cell wall system, beyond simple primary structure, is not well understood,[2] albeit great advances have been made with regard to the three dimensional arrangement of cellulose microfibrils in certain plant tissues.

Over the last 5 years, work in the author's laboratory has centered on acid sugar-containing polysaccharides (e.g., polygalacturonans) as they exist, intact, in the cell wall complex.[3-7] Specifically, we have been addressing the process of cell wall separation,[8] which is believed to be due to the dissolution of the middle lamella (the electron dense region between adjacent cell walls) in certain agriculturally important commodities (e.g., *Malus pumila* fruit), because the physicochemical mechanism of and the ability to modulate this process are obscure. In many climacteric fruit crops (those which have a marked increase in ethylene synthesis and respiratory activity during fruit ripening), the degradation of the middle lamella or cell wall separation process is caused, at least in part, by the action of specific random-cleaving polygalacturonan hydrolases (endoPG). The apple system is unique because cell separation occurs without a detectable increase in endoPG activity[9,10] even though there is a marked increase in cell wall polyuronide solubility.[11]

In this chapter, ^{13}C cross-polarization and magic angle spinning (CPMAS) NMR and other magnetic resonance experiments are presented which address questions regarding the higher order structure, molecular motion, and degradation of polygalacturonan blocks in intact cell wall polymers as a function of the cell separation process and specific enzymatic degradation. These molecular spectroscopic techniques are the major methods available for studying the dynamic and structural properties of individual components in heterogeneous solid materials.[4,6,12] Such experiments are of interest because bulk mechanical properties of polymeric materials, such as a cell wall complex, can be related to polymer chain dynamics and molecular motions.[12-14]

II. SPECTROSCOPIC PROBLEMS ASSOCIATED WITH PLANT CELL WALLS

Before proceeding with a discussion of higher plant cell wall structure and degradation, there is justification to mention several factors which complicate the interpretation of relaxation time and ^{13}C CPMAS NMR studies.

In the cross-polarization process,[15] ^1H magnetization is at equilibrium with the surroundings (lattice) and is proportional to the number of ^1Hs in the sample and is inversely proportional to the lattice "spin temperature". In order to observe the dilute ("hot") ^{13}C species, these spins are induced to thermally interact with the abundant ("cold") ^1Hs by turning on a rf field of amplitude H_{1C} along the x axis of the ^{13}C rotating frame. The magnitude of H_{1C} is arranged to fulfill the Hartman-Hahn condition ($\gamma_H H_{1H} = \gamma_C H_{1C}$) so that the two spin systems come to equilibrium with respect to spin temperature or magnetization. The rate at which ^1H-induced ^{13}C magnetization increases is equal to T_{CH}^{-1}. At longer contact times, the ^{13}C signal diminishes due to ^1H rotating frame spin-lattice relaxation ($T_{1\rho H}$).

A. Dehydration

In Table 1,[16] CPMAS NMR ^{13}C $T_{1\rho}$ and T_{CH} data, calculated from the Na^+-polygalacturonate (NaPGA; Sigma Chemical Co.) carbonyl (C=O) signal, are presented as a function of degree of dryness of the powders as well as the presence of other, non-acid sugar, constituents (e.g., cellulose). As shown previously,[15,17,18] T_{CH}, the reciprocal ^1H-^{13}C polarization transfer rate,

Table 1
EFFECT OF DEHYDRATION AND CELLULOSE ON C=O RECIPROCAL ^1H–^{13}C POLARIZATION TRANSFER RATE (T_{CH}) AND ^{13}C $T_{1\rho}$[16]

Drying technique/treatment	T_{CH} ± SE/μs	13C=O $T_{1\rho}$ ± SE/ms
Freeze-dried NaPGA + cellulose	298.0 ± 22	18 ± 1.00
Freeze-dried NaPGA[a]	214.0 ± 13	23 ± 1.00
NaPGA-lab[b]	143.0 ± 10	7 ± 0.32

Note: Peak intensities were used for the calculation of ^{13}C $T1_\rho$ and T_{CH} relaxation parameters by standard literature least square methods.[31] SEs were calculated from the least squares analysis of all data.

[a] Dissolved in H_2O and freeze dried immediately before CPMAS NMR experiment.
[b] Same as above without freeze drying (e.g., from the bottle).

depends upon a correlation time (τ_c) as well as ^{13}C–^1H angular (with respect to the static magnetic field) and distance terms. C=O T_{CH} differences between the two NaPGA samples is probably related to differences in the degree of polymer hydration since the treatment labeled "freeze dried-NaPGA" (Table 1) was lyophilized immediately prior to T_{CH} measurement. The "NaPGA-lab" treatment was stored at room temperature without prior lyophilization or other special drying treatment. Other evidence for differences in polymer hydration is that the ^{13}C $T_{1\rho}$ data for these same samples were different by a factor of approximately three which argues for rather large differences in carboxylate group motion. These data indicate that improper drying techniques can cause significant differences in the rate of cross-polarization, via a increase in the average number of near neighbor ^{1}Hs (T_{CH}), or an increase in the molecular motion (^{13}C $T_{1\rho}$) and, thus, could be problematic with regard to quantitative comparisons between cell wall treatments as well as comparisons of relaxation parameters. In this work, when performing solid state magnetic resonance experiments with *M. pumila* cell wall material,[3-7] great care was taken to thoroughly dehydrate the samples via critical point drying procedures followed by storage *in vacuo* over dehydrated silica gel. This dehydration technique is preferred over freeze drying when working with intact native matrices since the specimens are not subjected to the surface tension forces, or the freezing and sublimation boundary conditions, associated with lyophilization which could alter the morphology of the sample.

Other evidence[6] for H_2O-induced perturbations in the cross polarization process and ^{1}H spin-lattice relaxation are presented in Figure 1. In solids, the rate of ^{1}H spin-lattice relaxation ($\sim T_{1H}^{-1}$) is mediated by ^{1}H spin diffusion such that, in homogeneous samples with rapidly moving ^{1}H-rich domains (e.g., bound water or methyl groups), polymer T_{1H} values will be inversely proportional to the molar concentration of these domains,[19] assuming all other complicating factors are equivalent (i.e., molecular size). These data (Figure 1) indicate that polygalacturonan Ca^{2+} salts carry a certain degree of bound H_2O which is not easily removed by ordinary drying techniques since polyuronide T_{1H} values declined as the carboxylate group binding sites filled with the divalent cation. If ionically associated bound H_2O were not present, one would expect either no change in T_{1H} or an increase due to the cross-linking effect of Ca^{2+} on PGA motion. Other evidence for the presence of a significant degree of bound H_2O in "dehydrated" polymer salts is demonstrated (Figure 1, insert) by a decrease in both ring carbon (COH) and C=O T_{CH}. The C=O T_{CH} values are about two times larger than those corresponding to ring carbons because PGA's acid functional group is unprotonated.[17,18] As the available divalent cation binding sites filled the C=O T_{CH}s diminished more (about 36%) than the COH values indicating that the bound water was associated with the cationic species. As these findings indicate, it is important that

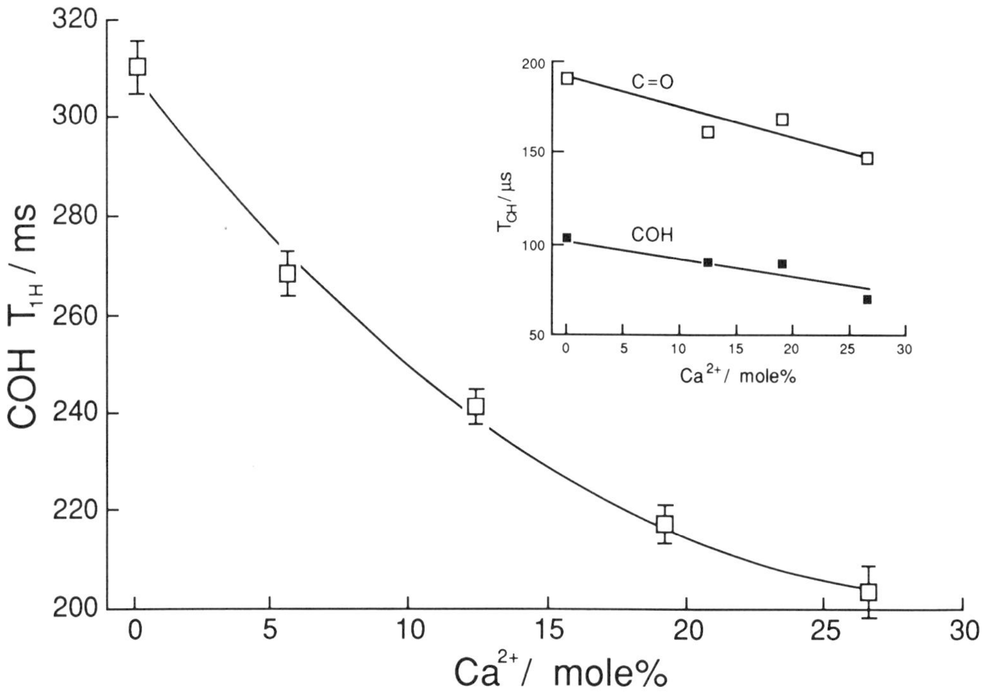

FIGURE 1. ^{1}H spin-lattice relaxation times (T_{1H}) and reciprocal ^{1}H to ^{13}C polarization transfer rates (T_{CH}) of dehydrated Ca^{2+} polygalacturonic acid salts. Vertical bars represent asymptotic SEs. About 500 mg polygalacturonic acid (H^+ form) was equilibrated in 25 ml of 800, 400, 100, 50, and 2 m*M* $CaCl_2$ in 75% (v/v) ethanol:H_2O to achieve 27 ± 2, 19 ± 1, 13 ± 4, 5 ± 1, and 0.15 ± 0.015 mol% bound Ca^{2+}, respectively. Samples were washed repeatedly with absolute ethanol, vacuum dried on scintered glass, and stored *in vacuo* at approximately 160 torr over silica gel for 2 to 3 weeks prior to CPMAS NMR experiments. Peak intensities were used for the calculation of ^{1}H T_1 and T_{CH} (the reciprocal ^{1}H–^{13}C polarization transfer rate) relaxation parameters by standard literature least square methods.[31] SEs were calculated from the least squares analysis of all data. (From Irwin, P. L., Sevilla, M. D., and Stoudt, C. L., *Biochim. Biophys. Acta*, 842, 76, 1985.)

relaxation time studies of native polymer materials have equivalent levels of bound divalent cations. In the *M. pumila* cell wall experiments reported herein, the concentration of bound nonparamagnetic divalent cations were equivalent (about $1 \times 10^{-5} \pm 5 \times 10^{-8}$ mol· g^{-1}).[15]

B. Paramagnetic Ion Contamination

Another problem related to solid matrices which contain anionic components is paramagnetic ion contamination which can induce extreme perturbations in various relaxation processes due to alterations in the local magnetic field (Figure 2, insert).[5] In this experiment, Cu^{2+}-PGA was utilized as the paramagnetic species. Even at low concentrations of bound Cu^{2+} (e.g., 0.05 mol%), the ratio of C=O to anomeric C signal (contact time = 0.8 ms[5]) is about 0.9; this ratio is about 1.0[16] without added Cu^{2+}. The difference between the contact time for maximal signal response ($S_{i,max}$ ^{1}H–^{13}C contact time;[16] Figure 2) of both anomeric and carbonyl carbons is equivalent (e.g., the slopes are equal), regardless of bound Cu^{2+} level, indicating that the differential signal response (Figure 2, insert) was due to a selective loss (about 10%) of the ^{13}C=O signal, possibly due to nonrandom Cu^{2+} binding as has been reported previously.[4-6] It is possible to eliminate most paramagnetic species from acid sugar-containing wall materials via titration of the carboxylate group. Unfortunately, this procedure is also likely to remove a portion of the matrix polysaccharides.[3] Bound Cu^{2+} is more problematic than species such as Mn^{2+} since it binds more tightly.[5,6] Indeed, the titration of the Cu^{2+}-PGA powders with acidified EtOH could

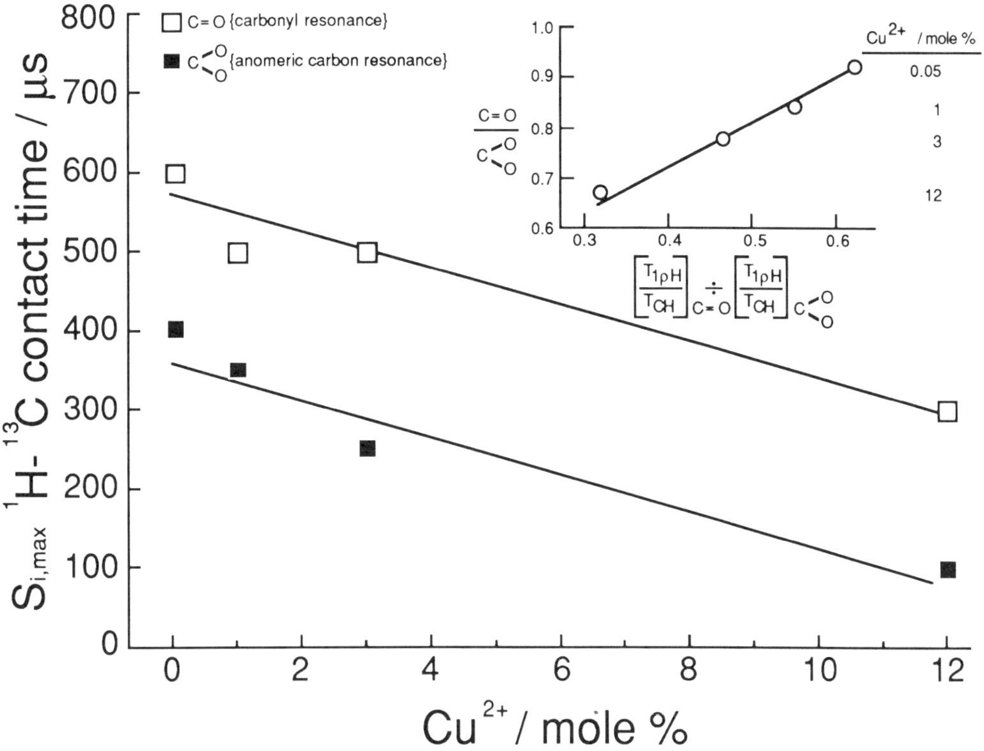

FIGURE 2. ^{1}H–^{13}C thermal contact time for the observed maximum carbonyl and anomeric carbon signal ($S_{i,max}$ ^{1}H-^{13}C contact time) as a function of mole percent bound Cu^{2+} on a polygalacturonic acid matrix.[16] Inset figure: dependency of the ratio of carbonyl and anomeric carbon resonance integrals on the ratio of $T_{1\rho H}/T_{CH}$ for carbonyl and anomeric carbon resonances at different levels of bound Cu^{2+}. For these experiments, about 500 mg dry PGA powder was added to 25 ml of 100, 10, 1, and 0.1 m*M* $CuCl_2$ solutions in 95% (v/v) ethanol:H_2O and vortexed. After 7 h at room temperature, the samples were washed three times with an equal volume of absolute ethanol. The samples were dehydrated as described in Figure 1. (From Irwin, P. L., Gerasimowicz, W. V., Pfeffer, P. E., and Fishman, M., *Agric. Food Chem.*, 1985, 1197, 1985.)

not completely eliminate the paramagnetic effect.[16] Cr contamination of laboratory glassware (from certain glass cleaning products) is a common source of undesired paramagnetic species.[16] Because of these factors, it is advisable to select cell wall matcrials to be utilized for solid state NMR experiments, via atomic absorption spectrophotometry or EPR, which have no observable paramagnetic ion species. One should also make sure all laboratory glassware is paramagnetic ion free. The native apple cell wall materials discussed in this manuscript were examined by atomic absorption spectrophotometry as well as electron paramagnetic resonance spectroscopy (EPR) for the presence of paramagnetic species; none were detected.

C. Differential ^{13}C Signal Response Based Upon Proximity to Near Neighbor ^{1}Hs

The last problem associated with CPMAS NMR of cell wall systems relates to quantitative comparisons of the ratio of carboxylate carbonyl resonances (C=O) or primary carbons ($-CH_3$) to total carbohydrate ring carbon resonances (COH + anomeric C). As shown before,[5] such comparisons are likely to be quantitatively spurious if these samples have extremely different rates of ^{1}H to ^{13}C polarization transfer (T_{CH}^{-1}) and/or rotating frame spin-lattice relaxation ($T_{1\rho H}^{-1}$) as exemplified in Figure 3.[16] In this graph, two calculated curves (COH T_{CH} = 150 µs, C=O T_{CH} = 400 µs; the $T_{1\rho H}$ for both is 4 ms) of relative signal vs. contact time are presented for directly protonated COH and remotely protonated C=O groups. The contact time for maximal signal response for each functional group is extremely different. Also, the relative signal

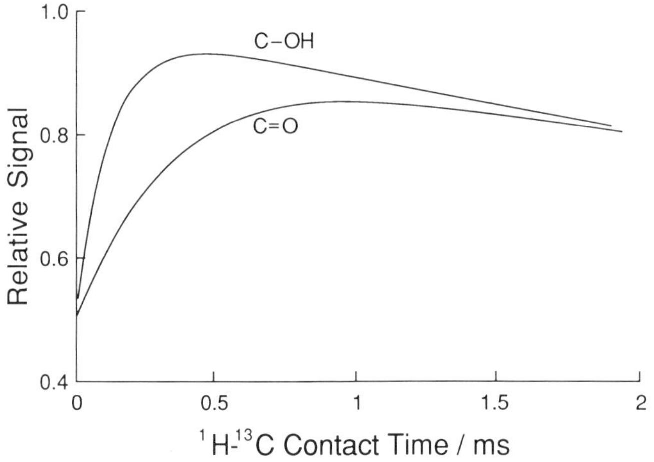

FIGURE 3. Calculated curves of ^1H to ^{13}C polarization transfer for nonspecific carbohydrate ring carbon (COH: $T_{CH} = 150\ \mu s$; $T_{1\rho H} = 4$ ms) and carbonyl (C=O: $T_{CH} = 400\ \mu s$; $T_{1\rho H} = 4$ ms) carbon resonances as a function of ^1H-^{13}C contact time. In this relationship:

$$\int_0^{M_{t=\infty}} \frac{dM}{[M - M_{max}]} + \int_{M_{t=0}}^{M_{t=\infty}} \frac{dM}{M} = \int_0^t -\frac{dt}{T_{CH}} + \int_0^t -\frac{dt}{T_{1\rho H}}$$

M_{max} is the maximum magnetization observable assuming the $T_{1\rho H}$ process is negligible; t is the ^1H-^{13}C thermal contact period.

amplitudes do not approach equivalence until relatively long contact times whereupon a 15 to 20% reduction in total potential signal was observed due to $T_{1\rho H}$-mediated relaxation. Another example of this effect[20] is shown (Figures 4 and 5) for CPMAS studies on *Nitella translucens*[21,22] cell walls. This algal species is interesting because the cell walls are quite large (up to several centimeters; can be treated as crystalline materials) and are made up almost entirely of primary wall constituents without the interconnecting middle lamella (the nodes were excised to facilitate the removal of the cytoplasmic components) between adjacent cell walls. In this example, one immediately notes the relatively large 19 ppm δ_{CH3} resonance (Figure 4) which is typical for the primary C-6 of rhamnose or the methyl resonance of acetylated sugars; this constituent is ordinarily unobservable due to low concentrations in other primary cell wall/middle lamellar matrices, such as the apple system. The relative mole percentage of rhamnose or acetylated sugars in these algal cell walls is valuable to know since rhamnose is an important constituent of pectic polysaccharides and because this particular sugar was unreported in previous primary structural studies.[21,22] Unfortunately, because of methyl group rotation,[19] the cross-polarization process for the 19 ppm δ_{CH3} is extremely different than the $\delta_{C=O}$ functionality or any other carbohydrate resonance (Figure 5) and, therefore, complicates the estimation of the relative molar composition of this sugar. In Figure 5, it is clear that the signal intensity, all of which were normalized such that the maximum signal was unity, for this CH_3 group diminished approximately 40% as the C=O magnetization approached its maximum at about 0.8 ms of ^1H–^{13}C thermal contact. Even if these functional groups had equivalent T_{CH} and $T_{1\rho H}$ values quantitation would, at best, be difficult since the 19-ppm resonance is part of the shoulder of the major ring carbon resonances.

In the experiments reported herein, the contact time (0.8 ms) utilized for apple cell wall/middle lamellar matrices, which were compared quantitatively, produced precise, but not necessarily accurate, quantitative information[5] even though C=O T_{CH}s between compared

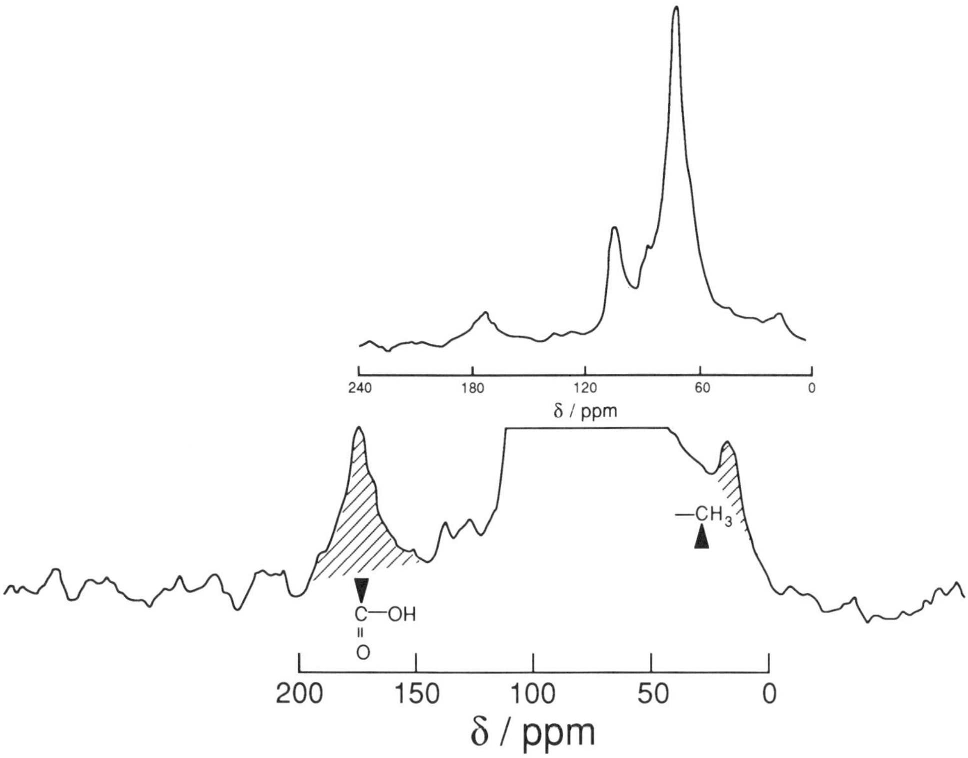

FIGURE 4. ^{13}C CPMAS NMR spectra of dehydrated *Nitella translucens* primary cell walls.[20] The 8-kHz wide NMR spectra were obtained at a ^{13}C frequency of 15 MHz; the ^{1}H decoupling field strength was 11 G. A total of 1024 (1k) data points were sampled and 0 filled to 4096 (4k) for data acquisition. All chemical shifts (δ) were assigned relative to the methyl resonance (δ = 17.36 ppm) of hexamethylbenzene. Samples were spun approximately 2.4 kHz at the "magic angle" in a Kel-F bullet type rotor. This spectrum represents 20,000 transients collected with a $^{1}H–^{13}C$ thermal contact period (contact time) of 0.8 ms. Samples were prepared by removing nodal lamellae and carefully squeezing out the cytoplasmic constituents in the presence of a 1% [w/v] solution of sodium dodecyl sulfate. Samples were dehydrated and critical point dried as described in Figure 8's caption.

treatments were different by a factor of about 50% (day 21 C=O T_{CH} = 381 ± 38 μs, day 0 C=O T_{CH} = 251 ± 14 μs; Figure 8).

III. THE PRIMARY STRUCTURE, MOLECULAR MOTION, AND DEGRADATION OF HIGHER PLANT CELL WALL CONSTITUENTS

A. A Comparative Study of the Molecular Motion of Cellulose, Ca PGA Gels, and Intact Cell Wall Complexes[23]

As mentioned previously, the major structural carbohydrate constituents of the primary cell wall and middle lamellar complex of higher plants are cellulose (β 1→4 glucans), hemicellulose (β 1→3 and 4 glucans, xyloglucans, xylans) and pectin (α 1→4 galacturonans with rhamnose, galactose, and other neutral sugars). McKay and co-workers[23] have utilized broadline (e.g., samples in a hydrated, but solid, state) ^{1}H magnetic resonance spectroscopic techniques to study the molecular motion of perdeuterated cellulose, Na and Ca PGA gels and isolated *Phaseolus vulgaris* hypocotyl cell walls (Figures 6 and 7) which have been fully "hydrated" in $^{2}H_2O$. Figure 6 shows typical ^{1}H time (spectra B and C, for cell wall and 8% Ca PGA gels, respectively) and frequency domain (spectrum A, Fourier transform of time domain spectrum C) spectra for the PGA gels and cell walls. Second moments ($\langle \nu^2 \rangle$, Equation 1) were calculated from all spectra; the $\langle \nu^2 \rangle$ is useful because, in solids, it is a function of the dipolar interactions between near

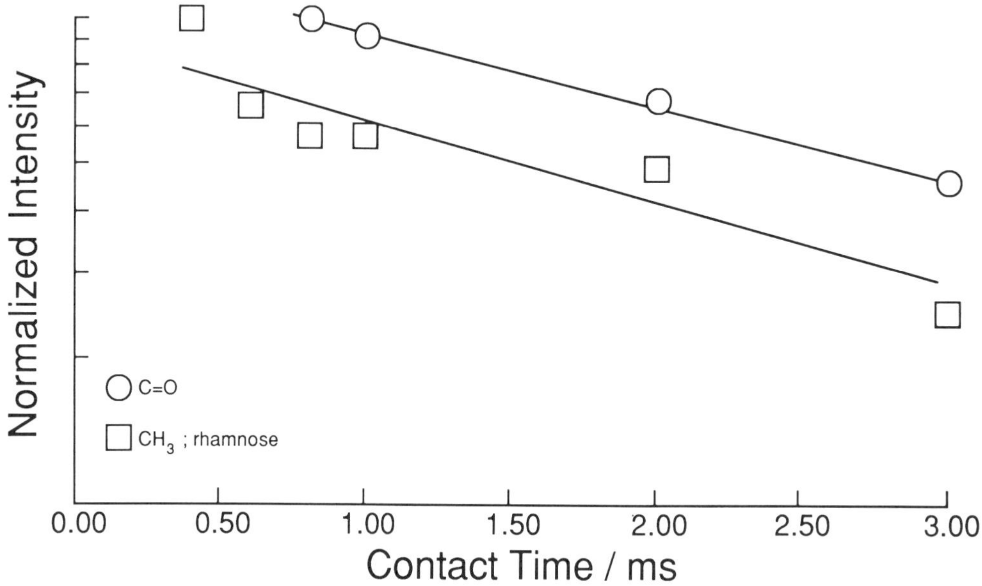

FIGURE 5. Contact time study of the carbonyl resonance (open circles) and the methyl group resonance (open squares) of dehydrated *Nitella translucens* cell walls.[20] Experimental conditions were as noted in Figure 4.

neighbor spins and is easily related to the spatial architecture of the ^{1}Hs in the various samples (Equation 2) as shown below:

$$<\nu^2> = \int_0^{\infty} \frac{(\omega - \omega_0)^2 f(\omega)}{2\pi} d\omega \qquad (1)$$

and

$$<\nu^2> = \frac{9 \gamma^4 h^2}{20 N} \sum_j \sum_{k \neq j} \frac{1}{r_{jk}^6} \qquad (2)$$

In these relations, f(ω) is the ^{1}H NMR lineshape, ω_0 is the angular Larmor frequency of the nuclei (ω = πν), N is equivalent to the number of ^{1}Hs in the sample, r is the average distance between neighboring ^{1}Hs; γ, as well as h, have their usual values. The early part of the time domain spectrum (the free induction decay or FID) represents the motionally rigid components while the later part of the total FID results from the more mobile components. For cellulose, the early part of the FID decayed approximately 40% faster (a larger $\langle \nu^2 \rangle$) than the equivalent part of the intact cell wall system, which, in turn, was approximately 20% faster than the early part of the Ca^{2+} PGA gel's FID (Figure 6, subspectra B and C). This finding indicates that the time domain spectrum of the cell wall is roughly equal to the summation of the gel/cellulose FIDs. This finding makes sense since these individual constituents are about equal to each other in the primary wall complex of higher plants. Figure 7 shows a plot of the ratio of the cell wall $\langle \nu^2 \rangle$ ($\langle \nu^2 \rangle^T$ = total $\langle \nu^2 \rangle$; $\langle \nu^2 \rangle^R$ = rigid, or broadline, component $\langle \nu^2 \rangle$) as a function of sample temperature. Clearly, as the hydrating species begins to melt at about –5 to –10°C, the mobile component undergoes a greater motional averaging of the dipolar interactions than the rigid component

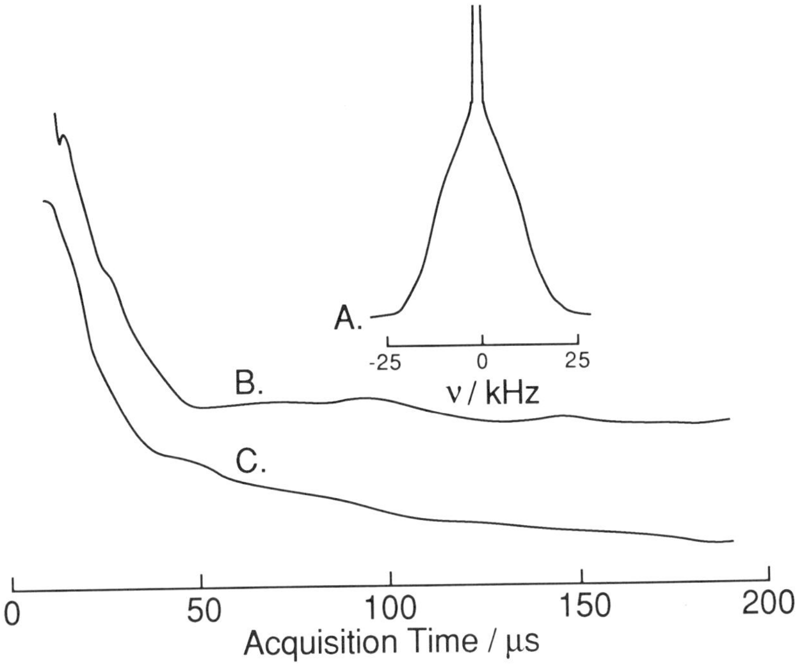

FIGURE 6. Frequency (A) and time domain (B and C) spectra of *Phaseolus vulgaris* (B) cell walls and CaPGA gels (A and C). ^{1}H-NMR results were acquired at 90 MHz. Na^+ Polygalacturonic acid (PGA) was obtained from Sigma Chemical Co. A 2.5% [w/v] solution of PGA was made with ^2H_2O whereupon it was lyophilized and redissolved in ^2H_2O to make an 8% [w/v] solution. This solution was induced to gel by dialysis against a 10 m*M* $CaCl_2$ (in ^2H_2O) solution for 18 h. *Phaseolus vulgaris* cell wall matrices were isolated from dark-grown 8-d-old hypocotyls. Hypocotyl sections, which had been stored at –20°C, were pulverized in a dry ice:ethanol bath, thawed, and ultrasonicated to remove bound cytoplasmic fragments. The resultant slurry was allowed to settle until two layers appeared; the upper layer (cell walls) were decanted and centrifuged at 200 xg for 15 min. The pelleted wall fragments were suspended in ^1H_2O and centrifuged twice to remove any residual cytoplasmic contamination whereupon the cell wall matrices were allowed to exchange with several changes of ^2H_2O. Excess ^2H_2O was removed by vacuum filtration and the sample was weighed and packed into a 10-mm NMR tube. The samples prepared in this fashion contained about 4.7 g ^2H_2O per gram cell wall.[23] For this figure, individual spectra were digitized from the original article[23] and recreated on a 32 bit microcomputer. (From MacKay, A. L., Bloom, M., Tepfer, M., and Taylor, I. E. P., *Biopolymers*, 21, 1521, 1982. With permission.)

since $\langle v^2 \rangle^R$ diminishes only 16% while $\langle v^2 \rangle^T$ drops more than 60 to 70% and argues that the mobile cell wall components resulted from the most hydrated portions of the wall matrix. Unfortunately, ^2H_2O, which was used in these studies as the hydrating species, has completely different hydration properties, at least with homopolygalacturonans,[20] than does ^1H_2O. In this laboratory, it has been noted with spin-labeled PGA that ^2H_2O is a worse solvent than ^1H_2O. In fact, EPR spin echo modulation studies on such samples hydrated in ^2H_2O show fewer bound ^2H_2O molecules than one would expect. Similar results[20] have been noted in ^2H_2O NMR spin echo studies of hydrated PGA.

B. Polygalacturonan Structural Similarity between Tissues of Different Physiological Stage

The major functionalities observed in ^{13}C CPMAS NMR spectra (Figure 8)[5] of apple fruit cell wall material are the anomeric ($\delta = 105$ ppm) and nonspecific carbohydrate ring carbon ($\delta = 74$ ppm) resonances. These signals are not specific to any particular cell wall or middle lamellar component but result from the combined resonances of all carbohydrate ring carbons. However,

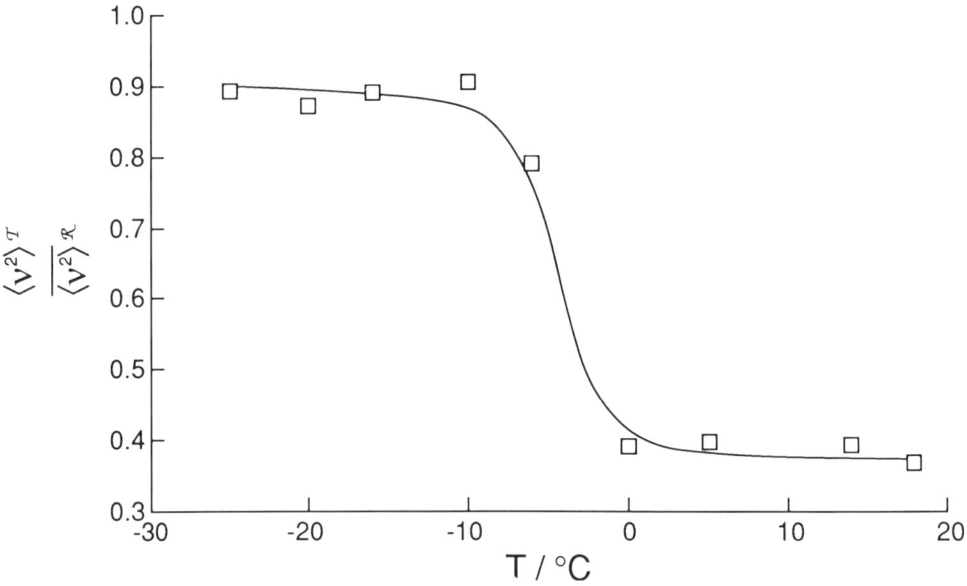

FIGURE 7. Dependency of $\langle v^2\rangle^T/\langle v^2\rangle^R$ on temperature in ^2H_2O-hydrated *Phaseolus vulgaris* cell walls. $\langle v^2\rangle^T$ and $\langle v^2\rangle^R$ represent the total components and the broadline component of the ^1H frequency domain spectral second moments, respectively. The authors[23] estimated that the accuracy of the $\langle v^2\rangle$ measurements were ±5%. For this figure, individual data points from the original were digitized, $\langle v^2\rangle^T$ divided by $\langle v^2\rangle^R$ and plotted on a 32-bit microcomputer.

the carbonyl resonance ($\delta = 172$ ppm) results from cell wall components that contain carboxylate groups, most of which are due to the acid sugar-containing polymers.[24] Spectra (Figure 8) for the day 0 and 21 apple cell wall complex show that the ratio of the C=O to anomeric carbon resonance areas was about 0.3.[5] Assuming a 5% (w/w) cell wall protein contribution to the C=O signal,[24] the day 0 and 21 treatments would have a total uronide content of approximately 24 and 25%, respectively. The degree of methyl esterification ($\delta = 54$ ppm) was also found to be similar between the two treatments (about 54 ± 5%). Assuming proper conditions for quantitation exist (T_{CH} << contact time << $T_{1\rho H}$ {i.e., $T_{1\rho H}/T_{CH} >> 1$}), these findings argue that the primary structure, or basic composition, of cell wall polyuronides does not change significantly during the course of the cell separation process.

Also arguing for structural similarity is the fact that the sequential[4] divalent cation binding behavior of these samples were virtually identical (Figure 9).[6,25] In Figure 9, day 0 and 21 (C=O $T_{1H} = 333 \pm 12$ and 123 ± 15 ms, respectively) cell wall-bound Mn^{2+} line widths (ΔH_{pp}) are plotted as a function of cell wall bound Mn^{2+} concentration. ΔH_{pp}s were determined by comparing an empirical line broadening factor to computer-generated first derivative spectra.[4] As the paramagnetic ion lattice filled, the ΔH_{pp}s, which have been shown to be a measure of Mn^{2+} near neighbor distances,[4,6] were equivalent at either physiological stage. These data indicate that the lattice constants (κ) and, therefore, the lattice site molecular sizes[25] were similar. A comparable result was found with the Cu^{2+} salts.[6,25]

C. Changes in Cell Wall Polygalacturonan Conformation and Flexibility during the Time Coincident with Dissolution of the Middle Lamella

Cell wall to wall separation in climacteric fruits, such as apple, has long been assumed to result from the direct action of various hydrolytic enzymes.[26,27] However, in apple, the wall separation process[8] is unusual since no random-cleaving polygalacturonase[9,10] (endoPG) has been found while, concurrently, solubilization of polygalacturonans, the major constituent of

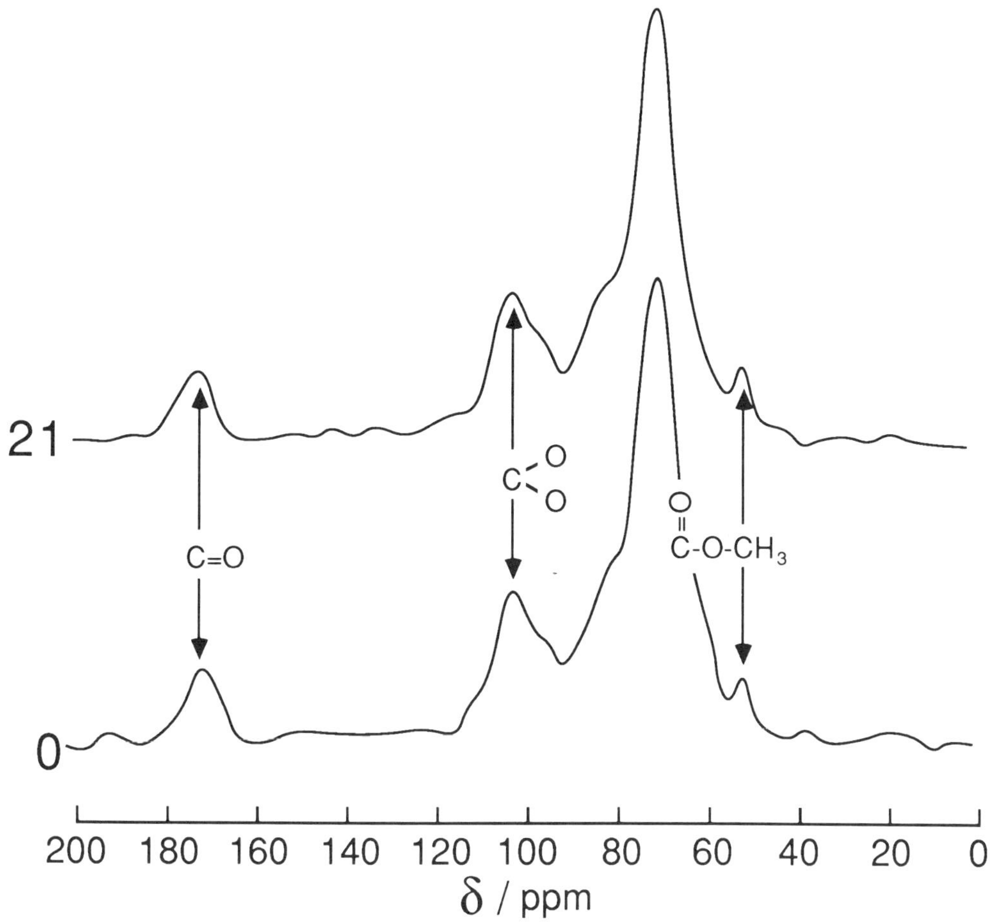

FIGURE 8. ^{13}C-CPMAS NMR spectra of dehydrated *Malus pumila* fruit cortical tissue ripened 0 and 21 d. Spectral conditions were as described in Figure 4. Peak areas were determined by triangulating to the baseline and taking three planimeter measurements per resonance. Spectra used for quantitative measurements were acquired by utilizing a 0.8-ms ^{1}H-^{13}C thermal contact time. Apple fruit (cv. Golden Delicious) were obtained with the cooperation of the Beltsville Agricultural Research Center. Fruit were picked, randomized, and immediately stored at approximately 0°C in 1% (v/v) O_2. After 1 month storage, fruit were removed and ripened for periods of 0, 4, 11, and 21 d at 20°C. Cortical tissues were cut into approximately 5 × 5 × 3 mm sections and dipped in chilled 0.01 *M* $CaCl_2$ to inhibit oxidation. Cortical tissues were vacuum infiltrated consecutively 2 h or more in each of the following solutions: 20, 40, 60, and 80% ethanol:H_2O. The ethanol fixed cell wall complex was then equilibrated in 95% ethanol for 1 d with repeated solution replacement. Samples were further reduced in size to 1-mm sections and equilibrated 1 h each in three changes of absolute ethanol (100 ml/500 mg dry weight equivalent) followed by critical point drying. (From Irwin, P. L., Gerasimowicz, W. V., Pfeffer, P. E., and Fishman, M., *J. Agric. Food Chem.*, 1985, 1197, 1985.)

pectic substances, does increase.[11] Because matrix and primary cell wall polygalacturonans are somehow involved in the changes of plant tissue structure, it is of biochemical interest to study the localized molecular motions of uronide-containing polymers in intact cell walls during the cell separation process by utilizing noninvasive techniques, such as ^{13}C CPMAS NMR and solid state EPR spectroscopy.

Assuming the number of directly bonded and H_2O of hydration ^{1}Hs remain constant, nonspecific carbohydrate ring carbon ^{1}H relaxation times (COH T_{1H}) are measures of polymer reorientation rates or motions averaged across all cell wall components. However, relaxation times measured from the C=O resonance are inversely proportional to the motion of either carboxylate groups specifically, as measured by ^{13}C=O $T_{1\rho}$, or to the motion or conformational flexibility of entire uronide-containing polymers, as measured by ^{1}H $T_{1\rho}$ and ^{1}H T_1.

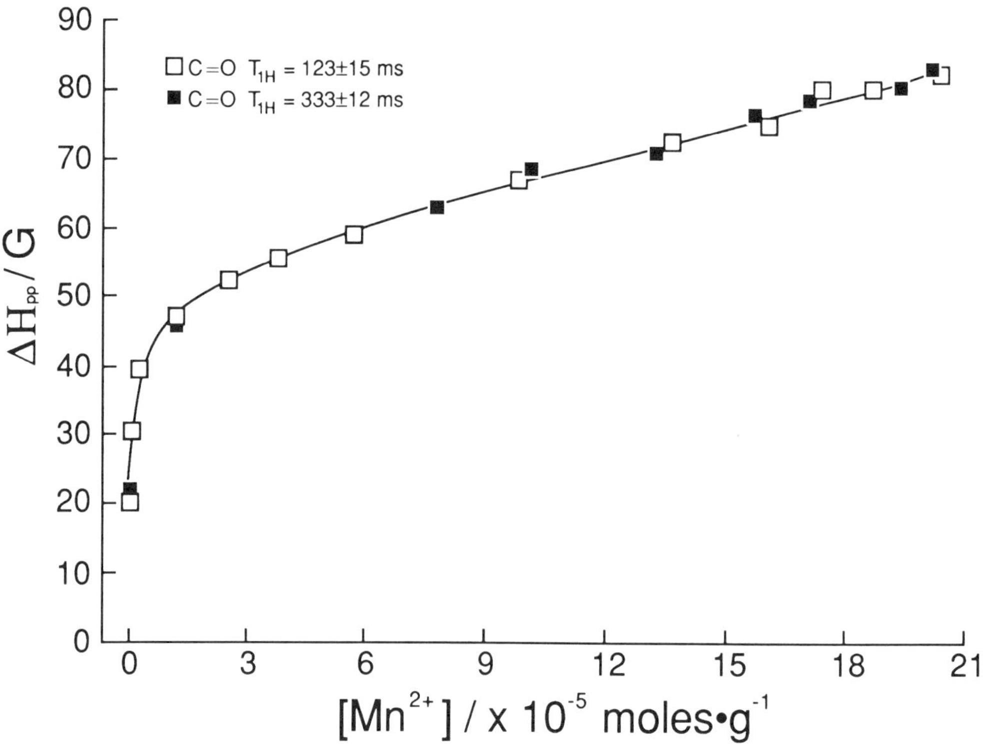

FIGURE 9. Cell wall bound Mn^{2+} concentration dependency of Mn^{2+} line width (ΔH_{pp}).[25] Open squares represent measurements taken from cell wall samples having a C=O T_{1H} of 123 ± 15 ms. Closed squares represent measurements taken from cell wall samples having a C=O T_{1H} of 333 ± 12 ms. Intact cortical tissues of apple fruit were prepared as detailed in Figure 8 and equilibrated at room temperature in 10 ml of various[6] $MnCl_2$ concentrations in order to achieve a wide range of bound paramagnetic ions; critical point drying followed thereafter. All EPR spectra were obtained on a Varian Series E-109B spectrometer* at various temperatures. Mn^{2+} spectral parameters were 1.6-mW microwave power; 4 to 32 min scan time; 1.6 to 2.0 kG scan range; 3224 G center field setting (H_0); about 9.1-GHz microwave frequency; 4-G modulation amplitude; and 100-kHz modulation frequency.

Rotating frame spin-lattice relaxation times of the C=O resonance (Figure 10),[3] which are measures of motions in the 10 to 100-kHz range,[13] were measured from intact, ethanol-dehydrated and critical point dried tissues obtained from apple fruit before and after 21 days of ripening at 20°C. ^{13}C=O $T_{1\rho}$ values decreased approximately 15% over this time period; the day 21 treatment $T_{1\rho C}$ was approximately 13% smaller than the NaPGA $T_{1\rho C}$ upon lyophilization within a cellulose matrix (33% [w/w]; Table 1).[16] These data argue that some alteration has occurred in the size or higher order structure of pectic substances within the intact cell wall matrix. Confirmation of differences in cell wall carboxylate group motion were also obtained by way of $^{23}Na^+$ T_1 measurements (Figure 10, insert) of the polygalacturonan Na^+ salts in intact cell wall matrices. Cell wall-bound Na^+ T_2s were estimated to be on the order of 0.22 ms ($T_1/T_2 >> 1$) from the line width of the $^{23}Na^+$ resonance. From this knowledge one can argue that the change in $^{23}Na^+$ T_1s with time was indicative of decreasing Na^+ binding site τ_c or increasing carboxylate group mobility. Therefore, the alteration of $^{23}Na^+$ T_1s tends to agree with the $T_{1\rho C}$ results.

Evidence for increasing cell wall complex polygalacturonan main chain mobility, as a

* Reference to brand or firm name does not constitute endorsement by the U.S. Department of Agriculture over others of a similar nature not mentioned.

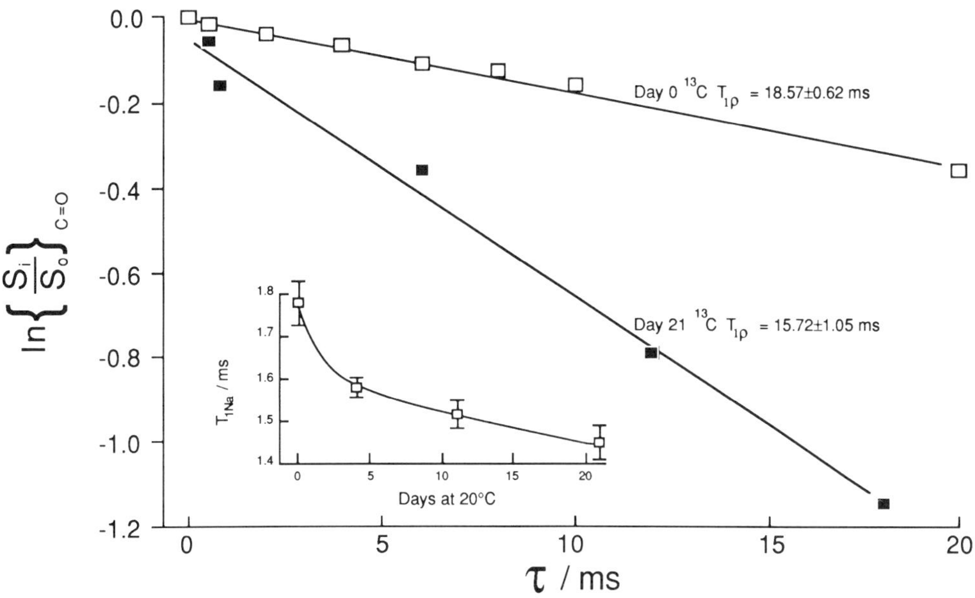

FIGURE 10. Natural log of normalized carbonyl signal intensities as a function of ^{13}C spin-locking time (τ) for cell wall samples ripened 0 and 21 d. Inset figure: plot of $^{23}Na^+$ T_1s as a function of ripening time at 20°C. The calculated relaxation times are reported ± S.E. Approximately 2 g of the fixed cell wall material from each treatment was washed three times in 180 ml H_2O and equilibrated 1 d in 1 *M* NaCl (pH 7) at 20°C. The samples were washed to remove unassociated Na^+, equilibrated in absolute ethanol and critical point dried. Three or more 30-mg dry cell wall samples were washed three times in 10 ml pH 2 H_2O in order to remove bound Na^+. Na^+ was analyzed by atomic absorption spectrophotometry using standard methods. ^{23}Na Spectra (±25-kHz sweep width) were obtained at 500 MHz (^1H frequency) by the University of Illinois School of Chemical Sciences NSF regional NMR facility (Dr. Eric Oldfield and Mr. Benjamin Montez). Samples were spun 3.7 kHz at the "magic angle". Peak intensities (100-Hz line broadening) were used for the calculation of T_1s. The pulse sequence for the Na^+ experiments consisted of a train of ten 90° pulses separated with delay time of 10 μs followed by a variable delay (0.05 to 50 ms) during which T_1 relaxation occured. Calculations were performed utilizing standard least squares analysis.[15] (From Irwin, P. L., Pfeffer, P. E., Gerasimowicz, W. V., Pressey, R., and Sams, C. E., *Phytochemistry*, 23, 2239, 1984.)

function of the cell separation process or specific enzymatic (PG II from tomato) degradation, are presented in Figures 11 and 12, respectively. In these experiments, spin-lattice relaxation times measured from the carbonyl resonance (C=O T_{1H}) were proportional to cell separation force (Figure 11)[3] or PG II-induced soluble polyuronide production (Figure 12).[3] Nonspecific carbohydrate ring carbon ^1H T_1s (δ_{COH} = 74 ppm; Table 2) displayed no such clear trend. The COH T_{1H} (Table 2) data might reflect an initial perturbation in total cell wall complex polysaccharide motion due to changes in the structure or architecture of the pectic polysaccharides, which make up nearly [1/3] of the total cell wall complex, followed by a gradual relaxation back to their initial state of mobility. The specific degradation of apple cell wall polygalacturonans by means of an endoPG induces a 42% (from about 160 ms to about 91 ms)[3] drop in nonspecific ring carbon T_{1H}s which tends to support the concept that a change in the organization or size of cell wall-bound polygalacturonans induces motional perturbations in the entire complex. Unfortunately, due to the diversity of polymers represented by the total ring carbon signal, one cannot attribute these observed T_{1H} alterations to any one component. Cell wall ^1H $T_{1\rho}$ values (Table 2) remained relatively constant during this time course study, indicating that lower frequency modes of reorientation[28] have not been altered.

Cell wall separation-associated perturbations in polygalacturonan main chain flexibility have also been confirmed utilizing paramagnetic ion spin-spin coupling differences as a

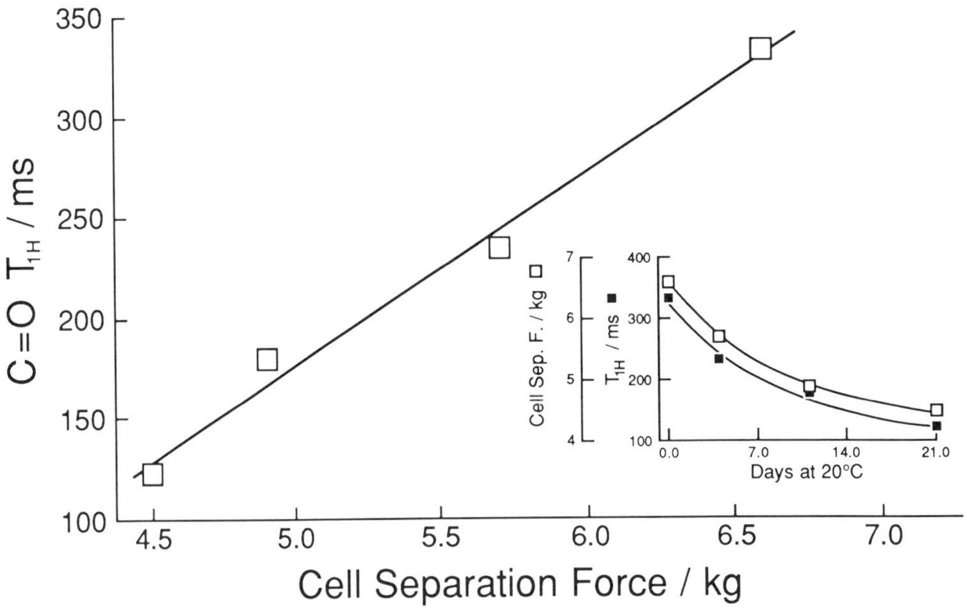

FIGURE 11. *Malus pumila* fruit ^{1}H spin-lattice relaxation time for the carbonyl resonance (C=O T_{1H}) and cell separation force as a function of ripening time at 20°C. At each time interval, stored fruit were randomly selected and cell separation force measured with an Instron Universal testing device equipped with a 0.79-cm (OD) plunger. Peak intensities were used for the calculation of ^{13}C $T_{1\rho}$, ^{1}H T_1, as well as ^{1}H $T_{1\rho}$ relaxation parameters by standard literature least square methods.[31] SEs were calculated from the least squares analysis of all data. (From Irwin, P. L., Pfeffer, P. E., Gerasimowicz, W. V., Pressey, R., and Sams, C. E., *Phytochemistry*, 23, 2239, 1984.)

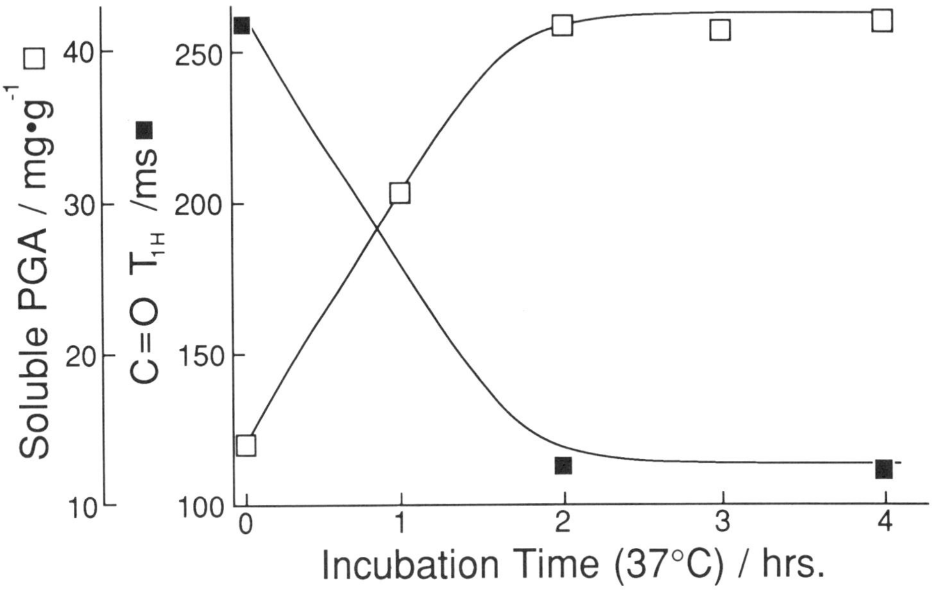

FIGURE 12. Changes in carbonyl ^{1}H spin-lattice relaxation times (C=O T_{1H}) and water soluble polyuronides (Soluble PGA) as a function of time at 37°C in the presence of tomato fruit PG II, an endo-cleaving polygalacturonase. Samples were dehyrated and critical point dried as noted previously. (From Irwin, P. L., Pfeffer, P. E., Gerasimowicz, W. V., Pressey, R., and Sams, C. E., *Phytochemistry*, 23, 2239, 1984.)

Table 2
^{1}H SPIN-LATTICE (T_{1H}) AND ROTATING FRAME SPIN-LATTICE ($T_{1\rho H}$) RELAXATION TIMES DETERMINED FROM CARBONYL ($\delta_{C=O}$ = 172 PPM) AND/OR NONSPECIFIC RING CARBON (δ_{COH} = 74 PPM) RESONANCES IN INTACT APPLE CELL WALL SAMPLES AT DIFFERENT TIMES AT 20°C[3]

	T_{1H} ± SE/ms	$T_{1\rho H}$ ± SE/ms	
Days at 20°C	δ = 74 ppm	δ = 172 ppm	δ = 74 ppm
0	245 ± 3	4.3 ± 0.4	4.2 ± 0.2
4	142 ± 15	—	—
11	160 ± 4	—	—
21	207 ± 4	4.6 ± 0.5	4.1 ± 0.2

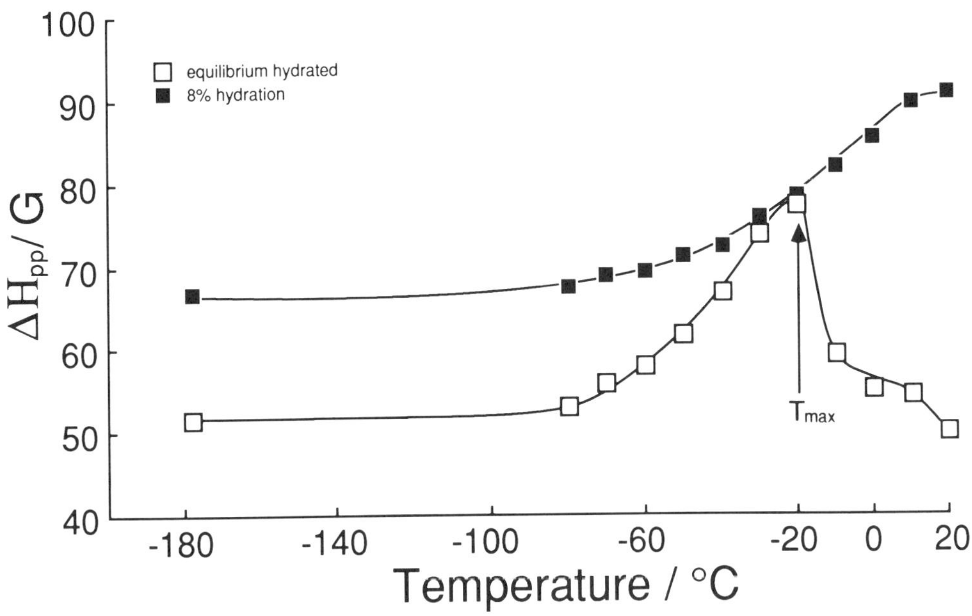

FIGURE 13. Temperature dependence of cell wall-bound Mn^{2+} line widths (ΔH_{pp}) for slightly hydrated (8% [w/w], closed squares) and equilibrium hydrated (about 70% [w/w], open squares) cell wall complex. Hydrated samples were prepared by equilibrating the critical point dried cell wall matrices various times (0.5 to 0.75, 1 to 1.5, 2 to 3, and 4 to 6 h for 21.7 ± 0.3, 32.9 ± 0.5, 42.0 ± 0.4, and 51.7 ± 0.5% [w/w] bound H_2O, respectively) in a saturated water vapor chamber at 160 torr and 22°C. (From Irwin, P. L., Sevilla, M. D., and Stroudt, C. L., *Biochim. Biophys. Acta*, 842, 76, 1985.)

function of sample temperature (Figures 13 and 14).[6] In these studies, Mn^{2+}-Mn^{2+} nearest neighbor distance parameters (d) were calculated as follows:

$$\Delta H_{pp} = 2\left(\frac{3}{5} \cdot g^4 \beta^4 h^{-2} \cdot S(S + 1) \cdot \frac{\kappa^2}{d^6}\right)^{1/2} \tag{3}$$

The constants g, β, and h have their usual values, and S represents the total electron spin of Mn^{2+} {5/2}. This calculation is based on the fact that the ΔH_{pp} for a Gaussian line is equal to two times

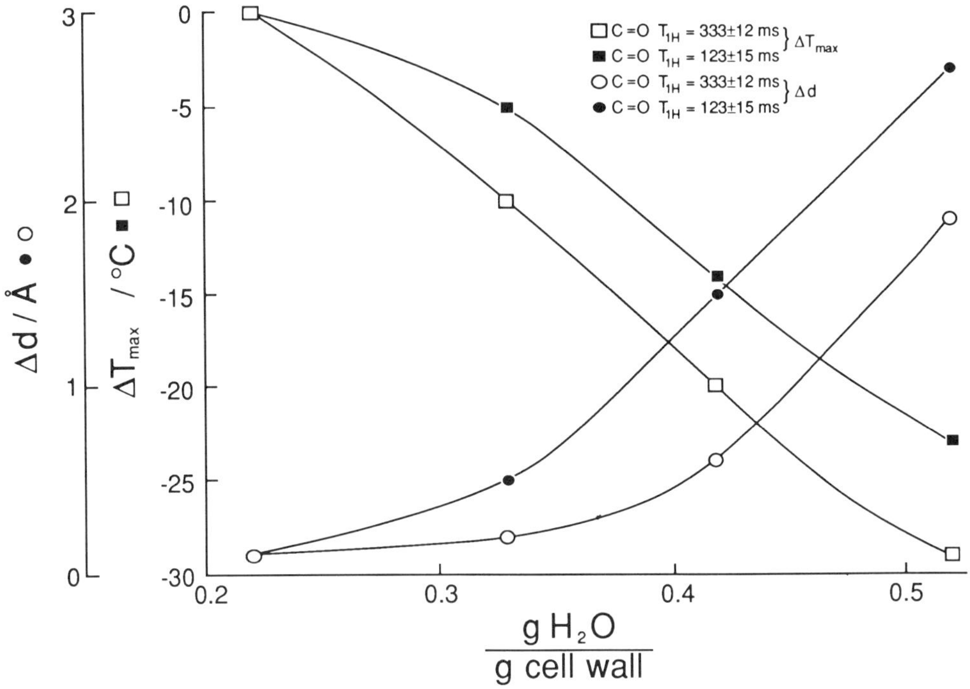

FIGURE 14. Dependence of Dd and DT_{max} on Mn^{2+} binding site-associated spin-lattice relaxation times (C=O T_{1H}) as a function of degree of cell wall hydration. Dd is the calculated change in d between 20°C and T_{max} assuming k = 2.5 throughout. This calculation assumes that all DH_{pp} perturbations were due to distance, as opposed to motional, effects. (From Irwin, P. L., Sevilla, M. D., and Stroudt, C. L., *Biochim. Biophys. Acta*, 842, 76, 1985.)

the square root of the second moment.[29,30] In Equation 3, d is the distance (in units of Å) between neighboring spins; κ ($κ^2$ is an approximation of the number of strong spin-spin interactions per point dipole), which is the lattice constant,[6] depends on the arrangement of ions in the lattice, and is approximately one for a two spin system, $\sqrt{3}$ for a linear sequential array, and from $\sqrt{8}$ up to $\sqrt{12}$ for a three-dimensional hexagonal close packing array.[4,6] It has been shown before that divalent cations bind to cell wall polyuronides in a spatially sequential fashion.[4,6] As a consequence of this special case of cooperative ion binding, bound paramagnetic ion ΔH_{pp}s are large, even at very low concentrations (Figure 9), due to their close spatial proximity to near neighbor paramagnetic species. In Figure 13 are shown cell wall bound Mn^{2+} (1.89×10^{-5} mol·g^{-1} = about 3 mol%) ΔH_{pp}s as a function of sample temperature and degree of hydration. As the paramagnetic ion cell wall lattice approached equilibrium hydration, an unusual temperature dependency became apparent. The 20 and –180°C ΔH_{pp}s were approximately equal; however, ΔH_{pp}s increased to a maximum at about –20°C (T_{max}) and fell normally thereafter. Initially, it was thought that this maxima was due to a conformational change in polygalacturonan blocks as a consequence of the freezing of bound H_2O. However, further experiments (Figure 14) indicated that, as the degree of hydration increased, the T_{max} diminished and argues strongly against the ΔH_{pp} maxima being associated with the freezing of bound H_2O. For these calculations[6] (ΔT_{max}, Figure 14), 320 ΔH_{pp} observations were made which encompassed several degrees of hydration, all temperatures (T), levels of bound Mn^{2+} ([Mn]) and 2 paramagnetic ion lattice structures (as represented by different C=O T_{1H}s). All the ΔH_{pp} data were fit to a statistical model quadratic with respect to both [Mn] and T. The partial first derivatives were taken with respect to T and [Mn] and set to zero such that

$$\frac{\delta \Delta H_{pp}}{\delta T} = 0$$

$$a_1 T + b_1[Mn] = c_1 \tag{4}$$

and

$$\frac{\delta \Delta H_{pp}}{\delta [Mn]} = 0$$

$$a_2T + b_2[Mn] = c_2 \tag{5}$$

T_{max} (T where $\partial \Delta H_{pp}/\partial T = 0$) values were then calculated by the method of simultaneous equations whereupon

$$T_{max} = \begin{vmatrix} c_1 & b_1 \\ c_2 & b_2 \end{vmatrix} \div \begin{vmatrix} a_1 & b_1 \\ a_2 & b_2 \end{vmatrix} \tag{6}$$

ΔT_{max} is the change in T_{max} relative to the lowest level of hydration. These data demonstrate that the cell wall anionic polymer ligand structure affects the T_{max} as well as Δd, the calculated change in the Mn^{2+} near neighbor distance parameter between 20°C and T_{max}. Thus, during the approach to T_{max}, Mn^{2+} binding sites with a short C=O T_{1H} conformationally flex about 0.7 ± 0.2 Å more than cell wall polygalacturonans of the corresponding treatment at equivalent levels of hydration.

IV. CONCLUSION

^{13}C CPMAS NMR, ^{1}H broadline NMR, and EPR are useful in studying certain chemical and physical properties of plant cell walls and matrix polysaccharides. With specific ^{13}C or other specific isotopic labeling of cell wall components, solid-state NMR could be of even more utility, albeit at a great cost. However, problems associated with incomplete dehydration, paramagnetic ion-induced perturbations of relaxation behavior as well as nonrepresentative functional group ^{1}H-^{13}C polarization transfer should be addressed before interpretation of the data.

Various NMR relaxation time and EPR data clearly indicate that plant cell wall polygalacturonans increase in mobility or flexibility during the cell separation process and dissolution of the middle lamella.[8] Such observations are important because of the noninvasive nature of these experiments. Most techniques for cell wall polysaccharide characterization provide little structural information about their behavior and arrangement within the matrix. Certain techniques could yield spurious results[2] due to the requisite extraction and solubilization processes possibly resulting in alterations of the polymer's primary and higher order structure.

The ^{13}C CPMAS NMR data reported herein could be interpreted as evidence for the specific action of an unidentified endo-cleaving polygalacturonan hydrolase.[3] An exoPG, which produces monomers upon hydrolysis, should have resulted in an overall loss of C=O signal since the uronic acid monomer is soluble in most of the ethanol:H_2O solutions used to prepare these samples; this was not observed. Perturbations in polygalacturonan main chain flexibiliy could also be due to alterations in the structure of either pectin's neutral sugar fraction or some other near neighbor polymer species. However, in light of newer information[25] about the structural similarity of cell wall polygalacturonans during the cell separation process, the latter interpretation appears to be more likely.

REFERENCES

1. **Hall, M. A.,** Cell wall structure in relation to texture, in *Quality in Stored and Processed Vegetables and Fruits,* Goodenough, P. W. and Atkin, R. A., Eds., Academic Press, London, 1981, 53.
2. **Darvill, A., McNeil, M., Albersheim, P., and Delmer, D.,** The primary cell walls of flowering plants, in *The Biochemistry of Plants,* Vol. 1, Academic Press, New York, 1980, 91.
3. **Irwin, P. L., Pfeffer, P. E., Gerasimowicz, W. V., Pressey, R., and Sams, C. E.,** Ripening-related perturbations in apple cell wall nuclear spin dynamics, *Phytochemistry,* 23, 2239, 1984.
4. **Irwin, P. L., Sevilla, M. D., and Shieh, J. J.,** ESR evidence for sequential divalent cation binding in higher plant cell walls, *Biochim. Biophys. Acta,* 805, 186, 1984.
5. **Irwin, P. L., Gerasimowicz, W. V., Pfeffer, P. E., and Fishman, M.,** ^1H-^{13}C polarization transfer studies of uronic acid polymer systems, *J. Agric. and Food Chem.,* 1197, 1985.
6. **Irwin, P. L., Sevilla, M. D., and Stoudt, C. L.,** ESR spectroscopic evidence for hydration- and temperature-dependent spatial perturbations of a higher plant cell wall paramagnetic ion lattice, *Biochim. Biophys. Acta,* 842, 76, 1985.
7. **Irwin, P. L., Sevilla, M. D., Shieh, J. J., and Stoudt, C. L.,** Paramagnetic ion spin-spin coupling as direct evidence for cooperative ion binding to higher plant cell walls, in *Chemistry and Function of Pectin,* Fishman, M. L. and Jen, J. J., Eds., American Chemical Society, Washington, D.C., 1986, chap. 14.
8. **Ben-Aire, R., Kisley, N., and Frenkel, C.,** Ultrastructural changes in the cell walls of ripening apple and pear fruit, *Plant Physiol.,* 64, 197, 1979.
9. **Knee, M.,** Fruit softening. III. Requirement for oxygen and pH effects, *J. Exp. Bot.,* 33, 1263, 1982.
10. **Morris, E. R.,** Physical probes of polysaccharides conformation and interactions, *Food Chem.,* 6, 15, 1980.
11. **Bartley, I. M.,** β-Galactosidase activity in ripening apples, *Phytochemistry,* 13, 2107.
12. **Havens, J. R. and Koenig, J. L.,** Applications of high-resolution carbon-13 nuclear magnetic resonance spectroscopy to solid polymers, *Appl. Spectrosc.,* 37, 226, 1983.
13. **Schaefer, J. and Stejskal, E. O.,** High resolution C-13 NMR of solid polymers, *Top. in C-13 NMR Spectrosc.,* 3, 283, 1979.
14. **Yannoni, C. S.,** High-resolution NMR in solids: the CPMAS experiment, *Acc. Chem. Res.,* 15, 201, 1982.
15. **Pines, A., Gibby, M. G., and Waugh, J. S.,** Proton-enhanced NMR of dilute spins in solids, *J. Chem. Phys.,* 59, 569, 1973.
16. **Irwin, P. L.,** unpublished data, 1985.
17. **Alemany, L. B., Grant, D. M., Pugmire, R. J., Alger, T. D., and Zilm, K. W.,** Cross-polarization and magic angle sample spinning NMR spectra of model organic compounds. I. Highly protonated molecules, *J. Am. Chem. Soc.,* 105, 2133, 1983.
18. **Alemany, L. B., Grant, D. M., Pugmire, R. J., Alger, T. D., and Zilm, K. W.,** Cross-polarization and magic angle sample spinning NMR spectra of model organic compounds. II. Molecules of low or remote protonation, *J. Am. Chem. Soc.,* 105, 2142, 1983.
19. **Andrew, E. R., Bryant, D. J., and Cashell, E. M.,** Proton magnetic relaxation of proteins in the solid state: molecular dynamics of ribonuclease, *Chem. Phys. Lett.,* 69, 551, 1980.
20. **Irwin, P. L. and Chamulitrat, W.,** unpublished data, 1987.
21. **Anderson, D. M. W. and King, N. J.,** Polysaccharides of the Characeae. II. The carbohydrate content of *Nitella translucens, Biochim. Biophys. Acta,* 52, 441, 1961.
22. **Anderson, D. M. W. and King, N. J.,** Polysaccharides of the Characeae. III. The carbohydrate content of *Chara australis, Biochim. Biophys. Acta,* 52, 449, 1961.
23. **MacKay, A. L., Bloom, M., Tepfer, M., and Taylor, I. E. P.,** Broadline proton magnetic resonance study of cellulose, pectin and bean cell wall, *Biopolymers,* 21, 1521, 1982.
24. **Knee, M. and Bartley, I. M.,** Composition and metabolism of cell wall polysaccharides in ripening fruits, in *Recent Advances in the Biochemistry of Fruits and Vegetables,* Friend, J. and Rhodes, M. J. C., Eds., Academic Press, London, 133, 1981.
25. **Irwin, P. L., Sevilla, M. D., and Chamulitrat, W.,** Homopolygalacturonan molecular size in plant cell wall matrices via paramagnetic ion and nitroxyl amide dipolar spin-spin interactions, *Biophys. J.,* 54, 337, 1988.
26. **Dilley, D. R.,** Enzymes, in *The Biochemistry of Fruits and their Products,* Hulme, A. C., Ed., Academic Press, London, 1970, 195.
27. **Knee, M.,** Changes in structural polysaccharides of apples ripening during storage, in *Facteurs et Regulation de la Maturation des Fruits,* No. 238, Centre del la Recherche Scientifique, Paris, 1974, 341.
28. **Cornell, B. A., Davenport, J. B., and Separovic, F.,** Low-frequency motion in membranes; the effect of cholesterol and proteins, *Biochim. Biophys. Acta,* 337, 689, 1982.
29. **Van Vleck, J. H.,** The dipolar broadening of magnetic resonance lines in crystals, *Phys. Rev.,* 74, 1168, 1948.
30. **Poole, C. P.,** *Electron Spin Resonance,* 2nd ed., John Wiley & Sons, New York, 1982, chap. 12.
31. **Draper, N. and Smith, H.,** *Applied Regression Analysis,* 2nd ed., John Wiley & Sons, New York, 1981, chap. 1.

Chapter 12

SOLID-STATE NMR STUDIES OF TREATED WOOD AND LIGNIN

James F. Haw

TABLE OF CONTENTS

I. INTRODUCTION

A. Lignin Utilization

After cellulose, lignin is the most abundant organic material on earth. It is a major constituent of all vascular plants and a major renewable source of reduced carbon. As such, lignin is potentially one of our most important raw materials. Most of the technology associated with lignin regards its elimination from wood in the production of paper. Approximately 200 million pounds/d of lignin are produced in the U.S. by the pulping of wood. Unfortunately, this potentially valuable resource is burned for its minimal heating value (2 to 3¢/lb) rather than using it as a starting material for more valuable products.

Possible useful products which could be made from lignin include fuels, fertilizers, adhesives, surfactants, and dispersants. Extensive reviews of the structure and chemistry of lignin have been published elsewhere.[1] Many processes used to treat wood or lignin involve selected reagents including various alkaline or acidic catalysts, oxidative or reductive reagents, or selected modifiers used to make copolymers. While these processes have diverse approaches, they either depolymerize lignin into smaller and more useful compounds or modify the lignin to form polymeric or copolymeric materials.

B. CP/MAS NMR

The development of new strategies for the utilization of lignin will depend, in part, on the availability of analytical techniques for characterization of the complex materials produced by such strategies. A number of investigators, especially Nimz and co-workers have demonstrated[2-4] the value of solution state ^{13}C-NMR for studies of soluble lignin fractions. Its applicability, however, is limited by the necessity of modifying or degrading the sample to obtain a soluble fraction. Fortunately, in the last decade, it has become possible to obtain high-resolution ^{13}C-NMR spectra of solid samples.[5-7] This is possible through a combination of technical advances designed to overcome the line-broadening effects and sensitivity problems that formerly precluded high-resolution studies of typical solids. The first technique is high-power proton decoupling which is used to eliminate a major line-broadening interaction due to dipole-dipole coupling between ^{13}C and the protons in the sample. A double-resonance experiment called cross-polarization (CP) is used to increase the intrinsic sensitivity of ^{13}C and to increase the rate at which the signal can be time averaged.

In addition to the above techniques, which involve high-power radio frequency pulses, the sample is usually rotated about a particular angle, called the magic angle, which is 54.7° from the direction of the magnetic field, B_0. This angle is important because many of the line-broadening interactions in NMR depend on the geometric term $3\cos^2\Theta-1$, and this term has a value of zero when Θ is selected to be the magic angle. When the sample is spun very rapidly (i.e., 4 kHz = 240,000 rpm) about an axis inclined at the magic angle, the line broadening due to chemical shift anisotropy (CSA) is removed. If high-power proton decoupling is also used, high-resolution ^{13}C spectra of solids may be observed. CP is normally used with the other techniques, and the complete experiment is called the CP/MAS experiment.

The chemical shifts observed in CP/MAS experiments are essentially identical to those observed in solution. Confirmation of structural assignments for CP/MAS spectra is facilitated by variants of the basic CP/MAS experiment that select particular structural units for observation. The most important example of such assignment aids is the interrupted-decoupling experiment[8] which provides spectra with signals from quaternary carbons and methyl carbons exclusively. With spectra such as these, detailed assignments are possible and the modification of lignin or wood samples by thermal and/or chemical means can be monitored without the complication and uncertainty of further modification for the purpose of solubility.

In the last several years, there have been a number of reports of the use of ^{13}C-CP/MAS NMR to characterize plant materials and the effects of various treatment processes on plant materials.

^{13}C-CP/MAS NMR has been used for the characterization of wood,[9-11] steam exploded wood,[10] steam hydrolyzed wood,[12] wood pulps,[11] herbaceous plants,[13] seeds,[14,15] pyrolyzed wood,[16] model sludge components,[17] lignin,[18,19] cellulose[20-22] and autohydrolyzed wood.[23] In the last-cited study it was possible to quantify the lignin and to estimate the hemicellulose remaining in the solid products of four biomass treatment processes, on the basis of the ^{13}C-CP/MAS NMR spectra of those products. It has been shown that quantitative, as well as qualitative information about lignin composition is obtained from ^{13}C-CP/MAS studies.[11] Further details of the CP/MAS experiment and its application to plant constituents may be found in other chapters of this volume.

^{13}C-CP/MAS NMR has recently proved to be a highly informative method for the study of lignins and other solids obtained from wood treatment processes. This is illustrated in the remainder of this chapter which summarizes some of the results of three studies: rapid steam hydrolysis (RASH) of red oak,[12] low temperature lignin pyrolysis,[16] and the kraft pulping process.[11] RASH treatment has been selected as a representative process for the partial depolymerization of lignin. Although pyrolysis is not likely to produce useful products from lignin, its study is important in that most treatment processes involve heat treatment in presence of selected reagents, and it is useful to understand the effects of heat alone in sorting out the roles of solvent, heat, and added reagents. The effects of kraft pulping on the lignin spectra have been studied because this process represents an important industrial source of lignin, and the properties of kraft lignins are, therefore, of great interest.

Rapid steam hydrolysis[24] (RASH) is similar to steam explosion in that wood chips are subjected to high-pressure steam at temperatures in the range of 200 to 280°C in both processes. The RASH process differs from the steam explosion process in that no sudden decompression is used in the former process. The RASH process also involves the continuous introduction of steam during the treatment period, with volatile and water-soluble products (principally depolymerized hemicellulose and depolymerized lignin) continuously removed from the reactor and recovered with the steam condensate. One advantage of the RASH process is that cellulose recovery is somewhat better than with steam explosion under similar conditions, presumably due to the continuous removal of acetic acid in the RASH process.

All of the ^{13}C-NMR spectra in this chapter were obtained at 25.27 MHz with a home-built spectrometer at Colorado State University. The pulse repetition time was 1 s and the CP contact time was 1 ms. The ^{1}H irradiation field was 12 G, and 1-K data points were zero-filled to 4-K points. Sleeve-type bullet spinners with sample volumes of 0.4 cm^3 were used and were spun at approximately 3.5 kHz. Chemical shifts were measured with respect to tetramethylsilane via hexamethylbenzene as a secondary substitution reference (aromatic peak at 132.3 ppm). A total of 12,000 to 40,000 scans were accumulated for each spectrum.

To a first approximation, wood is a mixture (or possibly a copolymer) of lignin, cellulose, and noncellulosic polysaccharides, termed hemicellulose. The principal repeat units of conifer lignin are derivatives of guaiacylpropane, I. Hardwood lignins also contain repeat units derived from syringylpropane (Structure I with a second methoxy group at position 5). The anhydroglucose repeat unit (Structure II) of cellulose is also shown.

Figure 1a is the ^{13}C-CP/MAS spectrum of ground, untreated red oak wood. This spectrum is qualitatively similar to spectra of pine wood[9,11] and ash wood[10] for which spectral assignments have recently been made. Drawing on the work in those previous studies, peak assignments can be made for the spectrum in Figure 1a as described below.

The peak at 172 ppm is principally due to the carbonyl carbons of acetyl groups and carboxylic acid groups in hemicellulose. The region between 160 ppm and 143 ppm consists primarily of signals due to the oxygen-substituted aromatic carbons of lignin. There are two partially resolved signals in this region, a more-intense signal at 153 ppm and a less-intense signal at 148 ppm. The 153-ppm signal is prominent in the spectrum of ash wood, but is of much

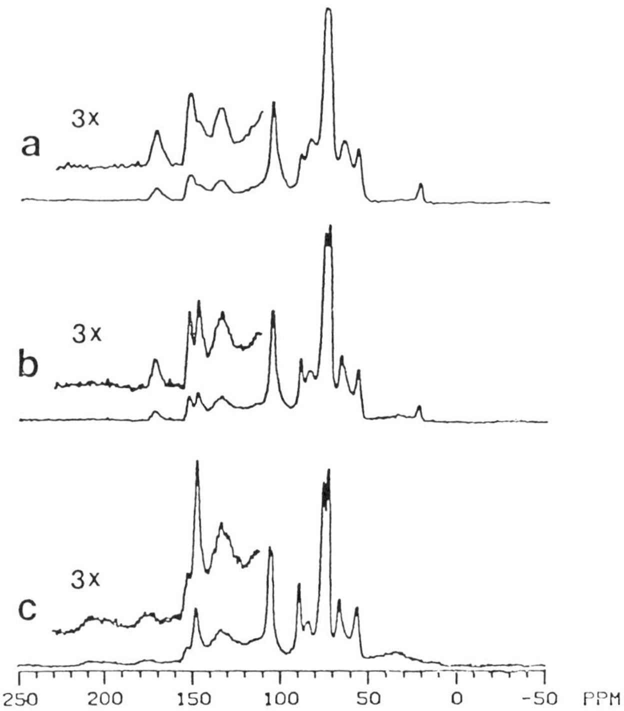

FIGURE 1. ^{13}C-CP/MAS NMR spectra of red oak wood: (a) untreated red oak, (b) red oak subjected to RASH treatment at 230°C for 1.5 min, and (c) red oak subjected to RASH treatment at 270°C for 1 min. (From Haw, J. F., Maciel, G. E., and Biermann, C. J., *Holzforschung*, 38, 327, 1984. With permission.)

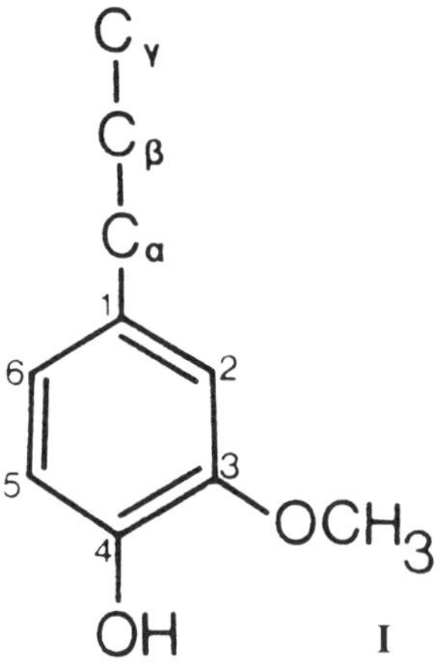

I

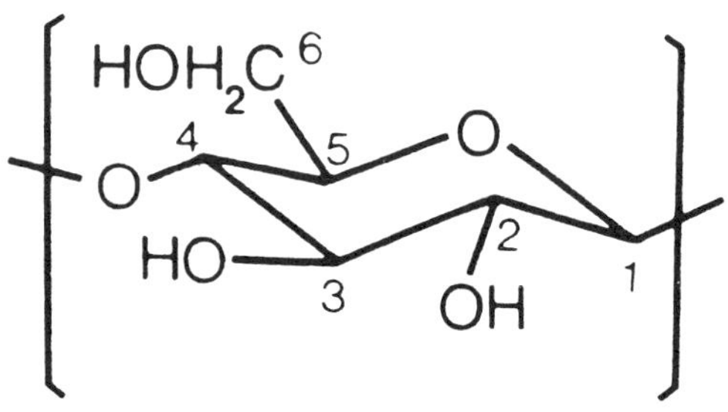

II

lower intensity in the spectra of pine wood samples. On that basis alone, it is reasonable to assign the 153-ppm peak to one or more of the carbons in syringyl units, since the presence of syringyl units in hardwood lignin but not in softwood lignin is a major compositional difference between these two lignins. The relative intensity of the 153-ppm signal is not dramatically reduced in an interrupted-decoupling experiment indicating that this signal is primarily due to nonprotonated carbons (interrupted-decoupling spectrum not shown). A more definitive assignment of the 153-ppm peak can be made on the basis of a previously reported liquid-solution ^{13}C-NMR study[3] of a large number of dimeric and monomeric lignin model compounds. The only ^{13}C-NMR signals observed in that previous study in the 151- to 157-ppm region that were assigned to nonpro-

Table 1
^{13}C-NMR SPECTRAL ASSIGNMENTS FOR HARDWOOD LIGNINS

Chemical shift (range) (ppm)	Assignment
210—190	Aldehyde and ketone carbonyls
180—166	Ester carbonyls
153	S3/5(4-0-R)
148	S3/5(4-0-H), G3/4
137	S1/4(4-0-R), G1(4-OH)
134	S1/4(4-0-H), G1(4-OH)
120	G6
115	G5
112	G2
106	S2/6
84	C-β in β-0-4
74	C-α-OH in β-0-4
60	C-γ-OH
52—15	Aliphatic C not bound to oxygen

Note: G, guaiacyl; S, syringyl.

tonated carbons were assigned to C-3 and C-5 of syringyl units that are O-alkylated at C-4 (154 ppm), and C-4 of guaiacyl units with a ketone group at the α position on the propyl chain (152 ppm). Ludemann and Nimz also studied[3] the liquid-solution ^{13}C-NMR spectrum of the milled-wood lignin obtained from beech and reported that for that material the signal assigned to C-3 and C-5 of O-alkylated syringyl units was much more intense than the signal assigned to C-4 of guaiacyl units with a ketone group at the α position. On that basis, the signal at 153 ppm in Figure 1a is assigned primarily to C-3 and C-5 of syringyl units O-alkylated at C-4. The abundance of O-alkylation at C-4 in red oak lignin implied by this assignment for the spectrum in Figure 1a is hardly surprising; aryl-alkyl ether linkages involving C-4 are firmly established as a major linkage in lignin.[1]

The spectral region in Figure 1a between 150 ppm and 142 ppm consists of signals due to the oxygen-substituted aromatic carbons of a number of guaiacyl and syringyl derivatives, including C-3 and C-5 of syringyl units which are not alkylated at C-4.[3] The signal intensity in Figure 1a between 143 and 130 ppm is primarily due to C-1 carbons of guaiacyl and syringyl groups, but several other carbon-types, including C-4 of nonalkylated syringyl groups, also contribute intensity to this region.[3] The intensity between ~130 and ~110 ppm in Figure 1a is primarily due to C-2, C-5, and C-6 of guaiacyl units. Signal intensity due to C-2 and C-6 of syringyl units will be assigned on the basis of the spectrum of HC1 lignin from red oak *(vide infra).*

The remaining features in Figure 1a are assigned as follows. The signal at 57 ppm is due to methoxy groups, principally those of lignin. The signal at 22 ppm is assigned to the methyl carbon of acetyl groups in hemicellulose. The low-intensity signals in the region from 50 ppm to 27 ppm are assigned to alkyl carbons not attached to oxygen, such as certain carbons found in the propyl chains in lignin and in extractives. Representative spectral assignments are summarized in Table 1.

Having presented a detailed assignment of the spectrum of untreated red oak, we now consider the effects of RASH treatment on red oak, as reflected in the spectra of the resulting solid products. Figure 1b is the ^{13}C-CP/MAS NMR spectrum of the solid residue obtained by RASH treatment of red oak at 230°C for 1.5 min. In the absence of an internal standard, it is not easy to directly compare the absolute intensities of signals in one spectrum to those in another spectrum. However, it is possible to make comparisons of relative intensities obtained from different spectra. The intensity assigned to the methyl carbon of the acetyl group at 22 ppm in Figure 1b relative to the other signals in that spectrum is only ~30% lower than the corresponding relative intensity in Figure 1a (untreated red oak). This result indicates that the generation of acetic acid by the action of steam on hemicellulose is slow for red oak chips at 230°C. Evidence for the partial removal of hemicellulose is provided by the reduction of the ratio of peak heights at 84 and 89 ppm from 1.3 in the spectrum of untreated red oak wood (Figure 1a) to 0.82 in the spectrum of the 230°C RASH solids (Figure 1b). The 84-ppm/89-ppm peak-height ratio has been previously applied as a convenient semi-quantitative indicator of the removal of hemicellulose in biomass treatment processes.[23] Further spectroscopic evidence for the reduced hemicellulose content in the 230°C RASH solids is afforded by the substantial increase in the splitting of the two signals at 75 ppm and 73 ppm. These signals are due to cellulose, and their resolution can be degraded by the presence of hemicellulose, which is known to contribute intensity between these signals.[9]

A very significant difference between the spectrum of the 230°C RASH solids (Figure 1b) and the spectrum of the untreated wood (Figure 1a) is the change in the relative intensities of the 153-ppm and 148-ppm peaks. RASH treatment at 230°C decreases the relative intensity of the 153-ppm peak while increasing the relative intensity of the 148-ppm peak. Based on the assignments of these peaks *(vide supra)*, these data clearly show that RASH treatment at 230°C partially depolymerizes red oak lignin by hydrolyzing the aryl-alkyl linkages involving C-4 of syringyl units. Other transformations of the lignin are also likely, but these cannot be identified conclusively on the basis of the spectra in Figures 1a and b. For example, dealkylation of a guaiacyl unit will shift its liquid-solution C-4 resonance from 150 ppm to 147 ppm,[3] but this change is not clearly evident in Figures 1a and 1b, possibly because of overlap with signals from other components.

Hydrolysis of aryl-alkyl ether linkages is even more apparent in the ^{13}C-CP/MAS NMR spectrum of the solids obtained by a 1-min RASH treatment at 270°C (Figure 1c). The relative intensity of the 153-ppm signal is substantially reduced in Figure 1c relative to Figures 1a and b, with a corresponding substantial increase in the 148-ppm signal. This result indicates that substantial hydrolysis of aryl-alkyl ether linkages is achieved by RASH treatment at 270°C. The increase in the relative intensities of the aromatic signals following RASH treatment at 270°C is at least in part due to the preferential removal of carbohydrates (principally hemicellulose) with the steam condensate, resulting in a concentration of the lignin in the RASH solids.

The 84-ppm/89-ppm peak-height ratio is only 0.75 in Figure 1c (relative to the ratio 1.3 in Figure 1a), a result consistent with significant removal of hemicellulose. Removal of hemicellulose with these conditions has been previously demonstrated on the basis of wet-chemical analysis.[24] The methyl carbon signal of the acetyl group (22 ppm) is completely absent in Figure 1c. The ^{13}C spectrum in the region between 112 ppm and 60 ppm in Figure 1c is very similar to the corresponding region of previously-reported spectra of cellulose I.[20,22]

Figure 1c also shows low-intensity carbonyl signals. The very broad signal between 220 ppm and 190 ppm is assigned to aldehyde and ketone groups, which are probably associated with lignin. The peak between 170 and 183 ppm is characteristic of esters and carboxylic acids. Since Figure 1c shows no evidence of hemicellulose, the esters and/or carboxylic acid groups are probably associated with lignin.

It has been demonstrated previously[24] that some lignin appears in the steam condensate during RASH treatment at temperatures above 210°C. A lignin fraction was obtained by filtration of the steam condensate from the 270°C run and subjected to NMR analysis. To serve as a basis of comparison, a ^{13}C-CP/MAS spectrum of the HCl lignin[25] obtained from red oak was acquired (Figure 2a). HCl lignin is believed to be much more similar to native lignin than is Klason lignin, which is highly condensed and unreactive. Figure 2a has features which are similar to the features assigned to lignin in the spectrum of untreated red oak wood (Figure 1a). In particular, the peak at 153 ppm is more intense than the peak at 148 ppm, indicating that the HCl lignin has a substantial amount of aryl-alkyl ether linkages. Figure 2a also shows

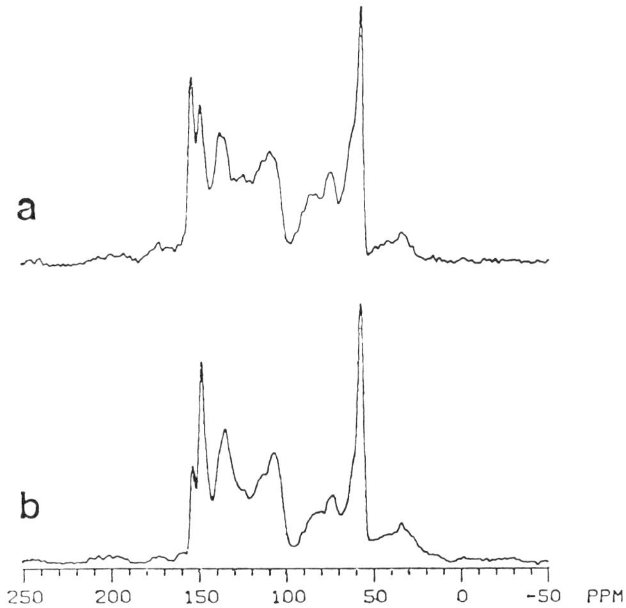

FIGURE 2. ^{13}C-CP/MAS NMR spectra of red oak lignins: (a) HCl lignin derived from red oak, and (b) solids obtained from steam condensate following RASH treatment of red oak at 270°C for 1 min. (from Haw, J. F., Maciel, G. E., and Biermann, C. J., *Holzforschung*, 38, 327, 1984. With permission.)

significant intensity in the 110 ppm to 103 ppm region, which can be attributed to C-2 and C-6 of syringyl units on the basis of the liquid-solution work of Ludemann and Nimz.[3] This intensity is masked in Figures 1a to c by overlap with the intense signal due to C-1 of cellulose (105 ppm).

The peaks at 83 and 74 ppm in Figure 2a are due to oxygen-substituted aliphatic carbons. It is not clear on the basis of Figure 2a whether these signals are due to lignin or to a small amount of carbohydrates. The peak at 56 ppm in Figure 2a is due to methoxy groups. The 61-ppm shoulder on this peak is significantly attenuated in an interrupted decoupling experiment. These data suggest that the 61-ppm shoulder peak is due to an oxymethylene group. Figure 2a also shows low-intensity aliphatic and carbonyl signals similar to those observed in the spectrum of the 270°C RASH solids (Figure 1c).

The ^{13}C-CP/MAS spectrum of the lignin obtained from the steam condensate of the 270°C RASH treatment is presented in Figure 2b. Comparison of Figure 2b with Figure 2a (HC1 lignin) reveals differences consistent with the steam condensate lignin being significantly more depolymerized than is HC1 lignin. In particular, compared with the patterns seen for the HC1 lignin (Figure 2a), the 153-ppm signal in Figure 2b is of reduced relative intensity, while the relative intensity of the 148-ppm signal is increased. These results show that the aryl-alkyl ether linkages have been extensively hydrolyzed in the formation of steam condensate lignin. This result was also observed for the lignin remaining in the RASH solids *(vide supra)*.

This study has shown that ^{13}C-CP/MAS NMR can characterize the extent of aryl-alkyl ether hydrolysis in hardwood lignins during biomass treatment by steam hydrolysis. This is possible because several of the carbons in syringyl and guaiacyl groups have chemical shifts which differ significantly (3- to 5-ppm differences) on the basis of whether or not C-4 is alkylated. The effect is most clearly seen for C-3 and C-5 of syringyl units, which give a single well-resolved peak at 153 ppm when C-4 is alkylated, while the corresponding resonance is at 148 ppm when the aryl-alkyl ether linkage is cleaved.

On the basis of interrupted decoupling experiments and the liquid-solution ^{13}C-NMR work of Ludemann and Nimz, it is possible to make a fairly detailed assignment of the ^{13}C-CP/MAS spectra of red oak lignins. The close agreement between the liquid-solution chemical shifts of lignin model compounds reported in that previous study and the solid-state chemical shifts observed for lignin in the present study is satisfying.

C. Low-Temperature Lignin Pyrolysis

Recently, the application of ^{13}C-CP/MAS NMR to the study of low-temperature pyrolysis of two lignins derived from sweetgum has been reported.[16] In that work, ^{13}C-CP/MAS NMR, FT-IR, elemental analysis, and thermogravimetric analysis (TGA) were used to study the low temperature pyrolysis of steam exploded and HC1 lignin from sweetgum. It was possible to deduce a variety of changes in the lignins as a result of pyrolysis. One-gram samples of lignin were pyrolyzed for 2 min in a pyrex tube immersed in a sand bath at various temperatures under a flowing nitrogen atmosphere. ^{13}C-CP/MAS spectra of the pyrolyzed HC1 sweetgum lignins are shown in Figure 3. Analogous spectra for pyrolyzed steam exploded sweetgum lignins are shown in Figure 4. Chemical shift assignments for these spectra are summarized in Table 1. The changes in these NMR spectra features upon pyrolysis provide insight into the pyrolysis process.

^{13}C-CP/MAS spectra of three representative HC1 lignin samples are presented in Figure 3. Figure 3a is the spectrum of the unpyrolyzed material. The ^{13}C-CP/MAS spectrum of HC1 red oak lignin was discussed previously. The changes of several of these spectra features upon pyrolysis provide important insight into the changes in the solid residues during lignin pyrolysis (Figures 3b and 3c). The intense peak at 153 ppm in Figure 3a is due to C-3 and C-5 of syringyl units involved in ether linkages at C-4 (principally β-0-4). The less intense peak at 148 ppm in Figure 3a is also due to C-3 and C-5 of syringyl units, but only for those with a free phenolic group at C-4. The relative intensities of these two signals can be used as a measure of the degree to which β-0-4 linkages are cleaved during the processing of wood, as was done in the steam hydrolysis study. The relative intensities of these peaks in Figure 3a indicate that the majority of syringyl units in unpyrolyzed HC1 lignin are involved in ether linkages. Comparison of Figure 3a with the spectrum of pyrolyzed HC1 lignin (Figures 3b, 250°C; Figure 3c, 280°C) reveals that these linkages are readily cleaved under the low-temperature pyrolysis conditions.

The peaks in Figure 3a at 84 ppm and 74 ppm and the shoulder at 60 ppm are due to aliphatic carbons bound to oxygen. For the alkyl chain of a β-0-4 syringylglycerol ether unit, these assignments correspond to C-β, C-α, and C-γ, respectively. It was possible to assign these signals to lignin, because a glucose analysis procedure revealed that the cellulose contents of these materials were low.[12] Comparison of Figure 3a with the pyrolyzed HC1 lignin spectra (Figure 3b, 3c) suggests that dehydration of the alkyl chains occurs readily in lignin pyrolysis. This is evidenced by a reduction in the intensities of the peaks at 84 ppm and 74 ppm as a result of pyrolysis. The 60-ppm shoulder, although not as well resolved as in Figure 3a, is still present in Figures 3b and 3c, suggesting that primary alcohols are still present in lignin chars after pyrolysis at 280°C.

Comparing the 52 ppm to 15 ppm regions in the three spectra of Figure 3, one notes an increase in the number of aliphatic carbons not adjacent to oxygen (Table 1) in the residue as a result of pyrolysis. This is consistent with alkyl chain dehydration, provided that the alkenes formed in such dehydration reactions react rapidly under pyrolysis conditions to give alkyl chains. The formation of alkenes and carbonyls in lignin pyrolysis has been previously proposed on the basis of model compound studies.[26,27]

The ^{13}C-NMR spectra in Figure 3 show that the accumulation of carbonyl functionality in the HC1 lignin pyrolysis residues is very limited, in contrast to the results obtained in the previously reported model compound studies.[26,27] The peak intensities between 210 ppm and 166 ppm in Figures 3b and 3c (pyrolyzed HC1 lignin) are only slightly higher than in the spectrum of the unpyrolyzed material (Figure 3a). This observation suggests that most carbonyl functionality, if formed in lignin pyrolysis, is efficiently lost as volatile species. The evolution of formaldehyde and carbon dioxide during the pyrolysis of kraft lignin has been previously noted in an FT-IR evolved gas study.[28] The constant relative intensity of the methoxy carbon signal (56 ppm) in Figure 3 indicates that this group is stable over the temperature range of our study.

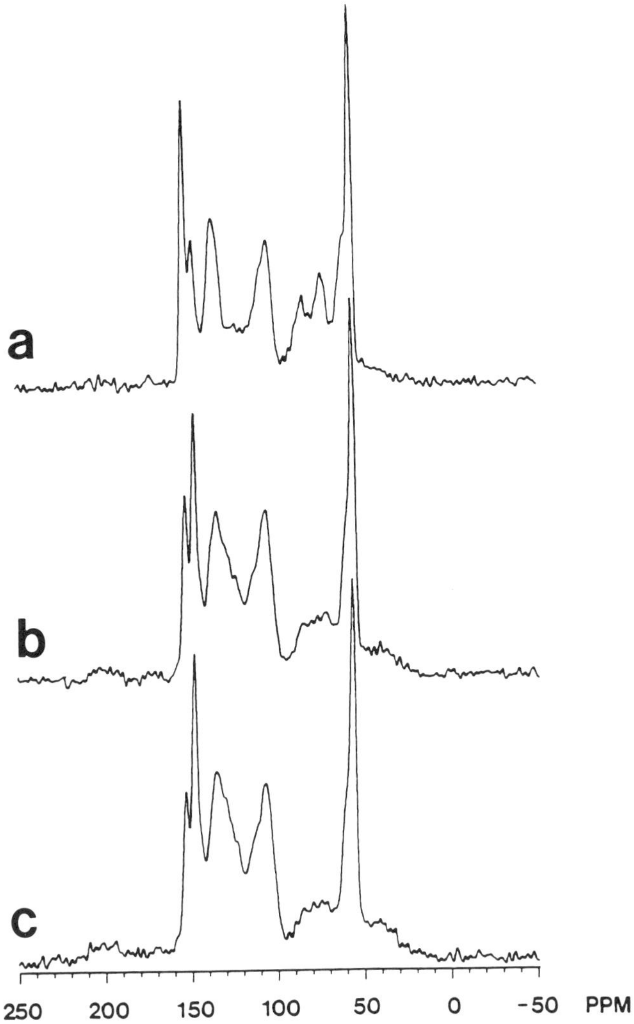

FIGURE 3. ^{13}C-CP/MAS NMR spectra of the HCl lignin obtained from sweetgum: (a) unpyrolyzed, (b) pyrolyzed at 250°C, and (c) pyrolyzed at 280°C. (From Haw, J. F. and Schultz, T. P., *Holzforschung*, 39, 289, 1985. With permission.)

The spectra in Figure 3 show that no major reorganization of the aromatic ring structures occurs at or below 280°C on the time scale of our pyrolysis experiment (2 min). The spectral changes in the aromatic regions (160 ppm to 100 ppm) of Figure 3 can be accounted for by ether cleavage alone, without recourse to cracking mechanisms or ring fusion.

^{13}C-CP/MAS NMR spectra of the samples obtained by pyrolyzing HC1 lignin at 300, 325, and 335°C, respectively (not shown), follow the trend shown for pyrolysis at 250 and 280°C (Figure 3). In particular, the 153-ppm peak is barely visible as a shoulder in the spectrum of the 335°C pyrolysis product.

The above observations for HC1 lignin pyrolysis largely apply to the results obtained for the steam-exploded lignin. Figure 4 shows the ^{13}C-CP/MAS NMR spectra of unpyrolyzed steam-exploded lignin (Figure 4a) and the residues of its pyrolysis at 250°C (Figure 4b) and 280°C (Figure 4c). The several differences between the spectra in Figure 4 (steam-exploded lignins) and Figure 3 (HC1 lignins) are more a matter of degree than of kind. For example, the spectrum of the unpyrolyzed steam-exploded lignin (Figure 4a) shows a greater extent of ether cleavage than does the spectrum of the unpyrolyzed HC1 lignin (Figure 3a). This effect is due to steam hydrolysis, as was observed in the ^{13}C-CP/MAS NMR study of the rapid steam hydrolysis of red

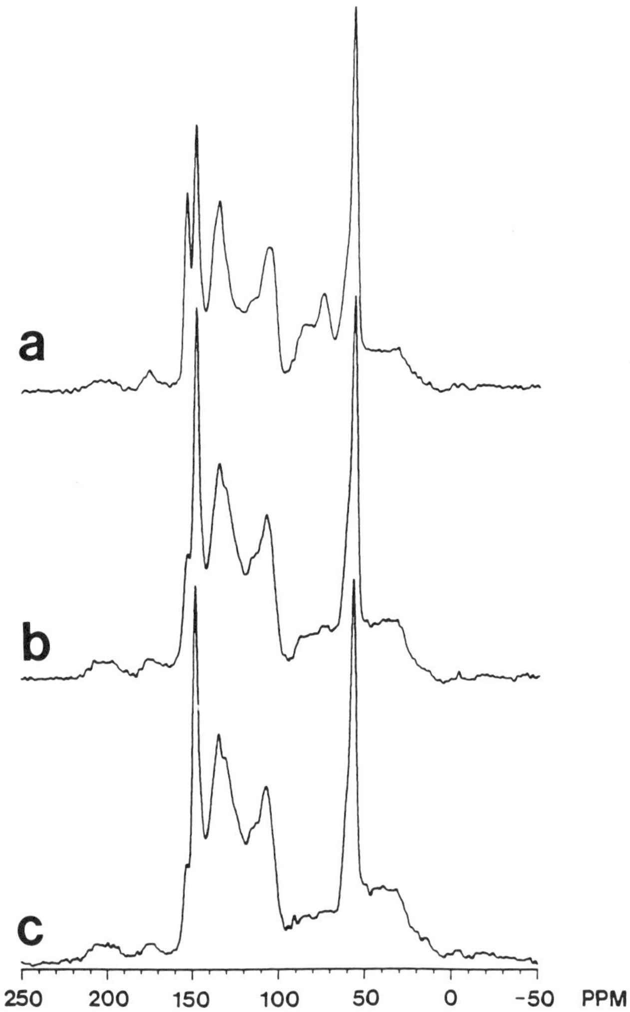

FIGURE 4. ^{13}C-CP/MAS NMR spectra of the lignin obtained from steam-exploded sweetgum: (a) unpyrolyzed, (b) pyrolyzed at 250°C, and (c) pyrolyzed at 280°C. (From Haw, J. F., and Schultz, T. P., *Holzforschung*, 39, 289, 1985. With permission.)

oak (*vide supra,* Section II.A). The trends observed in the pyrolysis of the steam-exploded lignin (Figure 4), however, are identical to those for HCl lignin (Figure 3). These same trends are also evident in the ^{13}C-CP/MAS NMR spectra of the samples obtained by pyrolyzing steam-exploded lignin at 220 and 235°C, respectively (not shown).

In conclusion, ^{13}C-CP/MAS NMR spectra provide fairly detailed insight into the structural changes of the residue (char) obtained in low-temperature lignin pyrolysis. These structural changes may be summarized by the following observations.

1. Aryl-alkyl ether linkages are extensively cleaved in the pyrolysis residue.
2. Dehydration of the alkyl chains occurs, approximately in parallel with ether cleavage.
3. Nonoxygen-substituted aliphatic material accumulates in the residue, possibly due to the condensation of alkenes initially formed in the dehydration step.
4. Carbonyl-containing species do not accumulate in the solid residue.
5. Methoxy groups are not cleaved below 335°C.

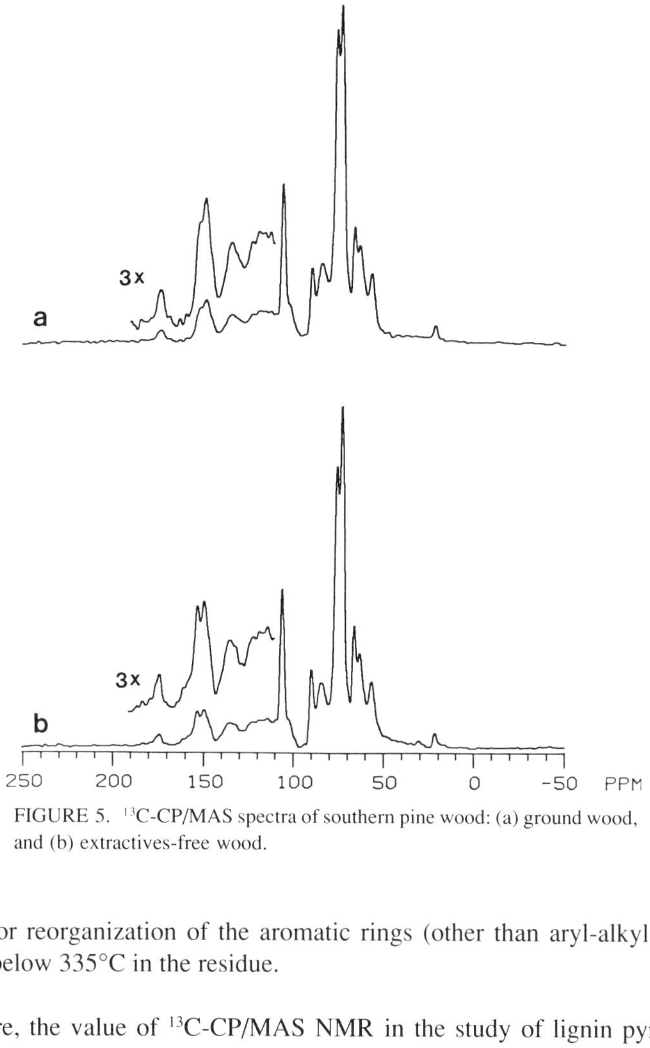

FIGURE 5. ^{13}C-CP/MAS spectra of southern pine wood: (a) ground wood, and (b) extractives-free wood.

6. No major reorganization of the aromatic rings (other than aryl-alkyl ether cleavage) occurs below 335°C in the residue.

Furthermore, the value of ^{13}C-CP/MAS NMR in the study of lignin pyrolysis has been demonstrated.

D. Kraft Pulping

The most important chemically modified wood product, wood pulp, has been produced commercially since the mid nineteenth century, when chemical pulping replaced mechanical pulping. Important processes in pulping include delignification and fiberization. Pulp production in the U.S. alone is on the order of 5×10^7 tons per year.

All pulp samples[12] were prepared in a 10-l electrically heated stainless steel rotating digester. All preparations used 1000 g of an oven-dried equivalent of air-dried loblolly pine *(Pinus taeda* L.*)* woodchips and a liquor-to-wood ratio of 4.5 (w/w). The digester was evacuated with a water aspirator for 20 min after it was filled (with the exception of the acid sulfite cook), but prior to starting the heating. The kraft cooks were carried out with 20% active alkali and 25% sulfidity (194 g of NaOH, 63 g of Na_2S). Three kraft pulps were prepared with various cook times. The lignin contents of these pulps (as determined by Kappa number or permanganate number) were: kraft 1, 14.5%; kraft 2, 4.3%; kraft 3, 2.2%.

Figure 5 shows the 25-MHz ^{13}C-CP/MAS spectra[15] of ground southern pine wood (Figure 5a) and extractives-free southern pine wood (hereafter referred to as extractives-free wood, Figure 5b). Wood extractives constitute a diverse collection of materials, including phenols, terpenes, and resin acids. Comparison of Figure 5a and Figure 5b shows that the extractives contribute some intensity to various regions of the pine wood spectrum. Since pulping removes most extractives, the extractives-free wood spectrum will be the basis of comparison used here for exploring modification of the lignin and carbohydrate fractions of wood in pulping.

^{13}C-CP/MAS spectra of pine wood were assigned by Kolodjeski and co-workers.[9] The assignments offered here are somewhat more detailed and form the basis for comparison with spectra of wood pulps. The intensity in the spectral region from 160 ppm to 110 ppm in Figure 5b is due to the aromatic ring carbons of lignin. The intensity between ~160 ppm and ~143 ppm is specifically due to oxygen-substituted aromatic ring carbons (e.g., carbons 3 and 4 in structure 1). The signals due to carbon types, α, β, and γ are overlapped with signals due to the carbohydrate components of wood and are not generally resolved. The signal at 56 ppm is due to methoxy carbons, principally of lignin, although approximately 10% are known from previous work[29] to be methoxy groups in some of the less common monomer units of hemicellulose, specifically the glucuronic acid in xylan.

The single peak at 105 ppm in Figure 5b is assigned to the C-1 carbon of the anhydroglucose repeat unit, 2. The sharper signal at 89 ppm and the broader high-shielding shoulder at 84 ppm are due to cellulose carbon C-4. The apparent doublet in the 70- to 80-ppm region is due to the C-2, C-3, and C-5 carbons. The peak at 66 ppm and shoulder at 63 ppm are due to the C-6 carbon of cellulose. These features are characteristic of cellulose I, the native (natural) form of cellulose. ^{13}C-CP/MAS spectra of hemicellulose preparations have also been reported.[9] Many of the spectral features are similar to those of cellulose, although the hemicellulose signals tend to be broader, reflecting the heteropolymeric nature of hemicellulose and the severity of the isolation procedures. Figure 5b shows three resolved signals that are readily assignable to hemicellulose. The broad, poorly resolved signal at 103 ppm (visible as a shoulder on the high-shielding side of the 105 ppm peak) is due to hemicellulose carbons that are structurally analogous to cellulose carbon C-1. The signals at 22 ppm and 174 ppm are due to acetate groups, which are known to be present in hemicellulose.[29] Some of the intensity of the 174 ppm peak could be due to other carbonyl species present in hemicellulose (e.g., CO_2H groups of uronic acids).

Spectra of the three kraft pulps are shown in Figure 6, together with the spectrum of extractives-free wood (Figure 6a) for comparison. The spectrum of the short-cook pulp (kraft 1, Figure 6b) shows substantial, but incomplete, loss of lignin relative to the lignin content of extractives-free wood. The signals assigned to the carbonyl and methyl carbons of acetate groups in hemicellulose appear to be completely absent in the spectrum of kraft 1, even when the time-domain signal is processed with a moderate exponential weighting function for sensitivity enhancement prior to Fourier transformation. However, the acetate carbon signals are not good indicators of hemicellulose content, since acetate groups can be lost selectively by hydrolysis. The high-shielding shoulder on the cellulose C-1 peak, assigned to hemicellulose, is reduced in the kraft 1 spectrum relative to its intensity in the spectrum of extractives-free wood. Further evidence of the selective depletion of hemicellulose by kraft pulping is provided by the enhanced splitting in Figure 6b of the two peaks in the 80 to 70 ppm region. Hemicellulose is known to contribute intensity in the region between these peaks.[9] The shoulders on the cellulose C-4 and C-6 signals are reduced (but not eliminated) in the spectrum of kraft cook 1, relative to their intensities in the extractives-free wood spectrum. These observations are consistent with substantial removal of lignin and preferential loss of hemicellulose at short cook times. Preferential, but incomplete, removal of hemicellulose is known to occur in the early stages of pulping.[30]

The spectrum of the kraft pulp obtained from an intermediate cook time (kraft 2) is shown in Figure 6c. It is readily apparent from the aromatic and methoxy regions of the spectrum that a significant further reduction in lignin content is achieved by pulping for a longer period of time. There are no apparent changes in the carbohydrate signals relative to those in Figure 6b, indicating that there are no obvious modifications in the structure or morphology of cellulose in the later stages of kraft pulping. The shoulders on the cellulose C-4 and C-6 signals in the

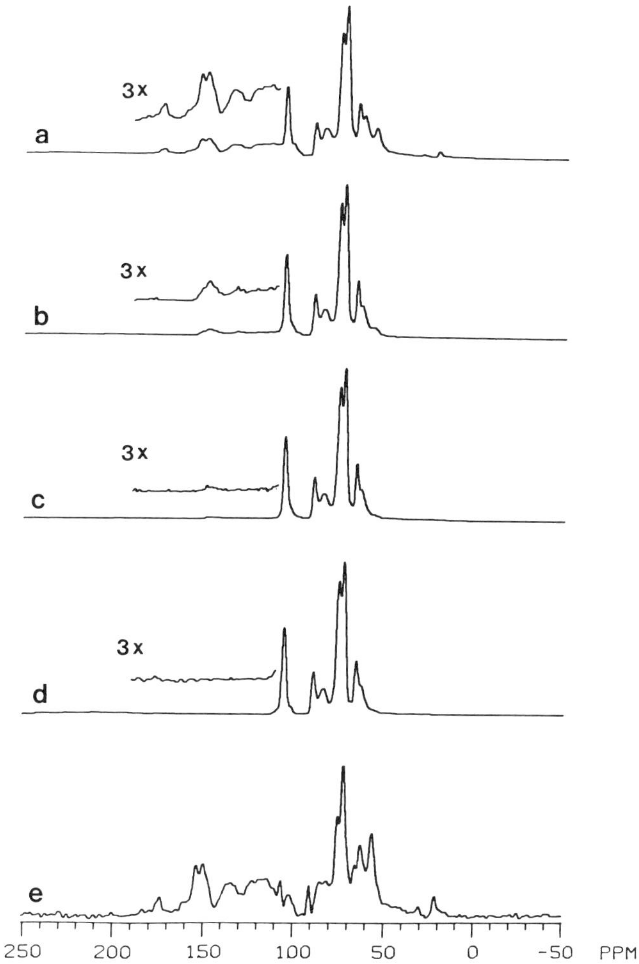

FIGURE 6. ^{13}C-CP/MAS NMR spectra of kraft pulps: (a) extractives-free wood, (b) kraft cook 1 (short cook time), (c) kraft cook 2 (intermediate cook time), (d) kraft cook 3 (long cook time), and (e) difference spectrum obtained by the subtraction of (d) from (e).

spectrum of kraft 2 pulp are identical in intensity and appearance with those in the spectrum of kraft 1 pulp. ^{13}C-CP/MAS NMR spectra of hemicellulose preparations have intensity in the regions of the C-4 and C-6 shoulders.[9] The failure to observe further decreases in shoulder intensity with longer cook times supports the idea that selective removal of hemicellulose occurs at short cook times but not at moderate cook times.[30] The residual shoulders are due, in part, to cellulose.

The NMR spectrum of a kraft pulp (kraft 3) that was cooked somewhat longer than the one discussed above is shown in Figure 6d. This spectrum bears a strong resemblance to that of cellulose I.[20] There is no evidence in the spectrum of this pulp (or of any other pulp which were examined) for the presence of cellulose polymorphs other than cellulose I. No changes in the spectrum of this pulp were observed following overnight drying *in vacuo* at 145°C (spectrum not shown). Lignin signals are not clearly visible in Figure 6d; the lignin content by Kappa number is 2.2% for this sample. In order to show more clearly the changes in wood during kraft pulping, the spectrum of kraft pulp 3 (Figure 6d) was digitally subtracted from the spectrum of extractives-free wood (Figure 6a). Such subtractions produce difference spectra which empha-

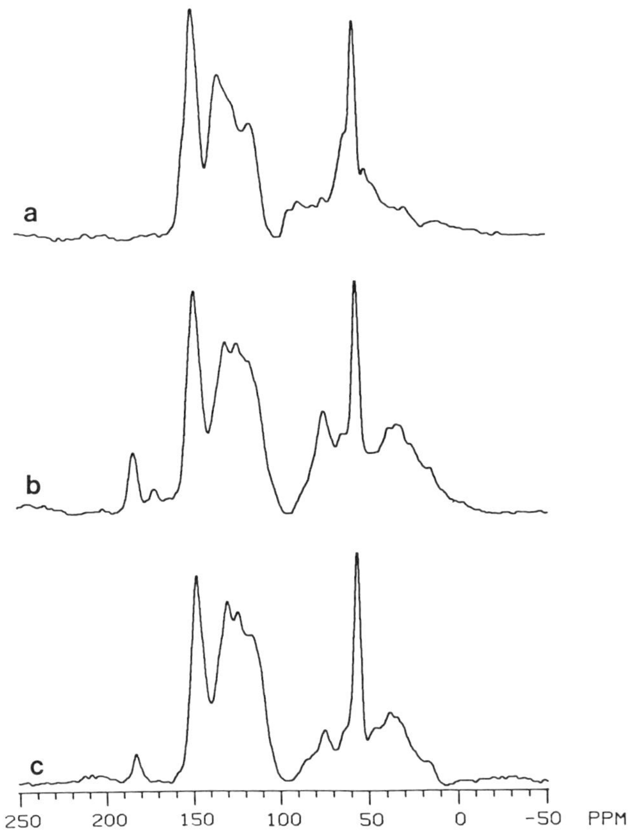

FIGURE 7. ^{13}C-CP/MAS NMR spectra of kraft lignins: (a) Klason lignin, (b) kraft cook 2 lignin, and (c) kraft cook 3 lignin.

size the features of one spectrum that are distinct from those of another spectrum. The weighting factor used in the subtraction was selected to minimize the resulting cellulose signals, without introducing negative peaks. The result of this subtraction is shown in Figure 6e. Signals from the lignin and hemicellulose components are clearly seen in this mode of presentation.

Kraft lignin preparations were obtained from kraft cooks 2 and 3 by neutralizing mother liquor samples with CO_2; precipitated lignin was collected by filtration. Many of the features of the kraft 2 and kraft 3 lignin spectra (Figure 7b and Figure 7c, respectively) are very similar to those in the spectrum of Klason lignin (Figure 7a). Comparisons of kraft lignins and Klason lignin must be done with caution; these two types of lignins are prepared under greatly differing conditions, and both have properties differing from those of native lignin. The spectra of the kraft lignins (Figures 7b,c) each have a signal at 183 ppm which is not present in the Klason lignin spectrum (Figure 7a) and which is consistent with carboxylate anions. This peak lies outside the range of chemical shifts of esters. Uronic acids of hemicellulose could contribute some intensity to this signal.

The spectra of the kraft lignins each have a signal at 74 ppm which is absent in the spectrum of Klason lignin. This signal is probably due to carbohydrate and resembles that of hemicellulose.[9] The signal at 74 ppm is much smaller in Figure 7c than in Figure 7b. This presumably reflects a greater extent of alkaline hydrolysis at longer reaction times. The spectra of the kraft lignins each have a broad peak, or complex of peaks, centered at 30 to 40 ppm. This feature is of much lower intensity in the spectrum of Klason lignin.

III. CONCLUSIONS

The above three studies have illustrated the use of ^{13}C-CP/MAS NMR in the study of processes for the treatment of wood or lignin. The utility of solid state NMR for probing the chemical changes induced by such processes has been demonstrated for the specific examples of steam hydrolysis, pyrolysis, and pulping.

The most serious limitation apparent from the above studies is spectral resolution. Lignin and polysaccharides are complex polymeric materials, and highly detailed studies of structure and reactivity in the solid state will require more resolution than is currently available with existing CP/MAS techniques. Future applications of solid-state NMR to plant materials would be greatly facilitated by the development of new methods and/or the improvement of existing techniques.

The present state of the art, however, is not too bad. Substantially better resolution is observed in CP/MAS spectra of lignin than with, for example, coal. Even without further refinements in NMR methods, there is much that can still be learned from solid-state NMR studies of the chemistry of wood and lignin. Such studies will, no doubt, be forthcoming.

REFERENCES

1. **Sarkanen, K. V. and Ludwig, C. H.,** Lignins: occurrence, formation, structure and reactions, Wiley-Interscience, New York, 1971, 695.
2. **Nimz, H. H. and Ludemann, H.-D.,** Kohlenstoff-13-NMR-Spektren von Ligninen. VI. Lignin-und DHP-Acetate, *Holzforschung*, 30, 33, 1976.
3. **Ludemann, H.-D. and Nimz, H.,** Carbon-13 nuclear magnetic resonance spectra of lignins, *Biochem. Biophys. Res. Commun.*, 52, 1162, 1973.
4. **Nimz, H. H., Robert, D., Faix, O., and Nemr, M.,** Carbon-13 NMR spectra of lignins, 8, *Holzforschung*, 35, 16, 1981.
5. **Yannoni, C. S.,** High resolution NMR in solids: the CPMAS experiment, *Acc. Chem. Res.*, 15, 201, 1982.
6. **Fyfe, C. A.,** *Solid State NMR for Chemists*, CFC Press, Guelph, Ontario, 1983.
7. **Maciel, G. E.,** High-resolution nuclear magnetic resonance of solids, *Science*, 226, 282, 1984.
8. **Opella, S. J. and Frey, M. H.,** Selection of nonprotonated carbon resonance in solid-state nuclear magnetic resonance, *J. Am. Chem. Soc.*, 101, 5854, 1979.
9. **Kolodziejski, W., Frye, J. S., and Maciel, G. E.,** Carbon-13 nuclear magnetic resonance spectrometry with cross polarization and magic-angle spinning for analysis of lodgepole pine wood, *Anal. Chem.*, 54, 1419, 1982.
10. **Taylor, M. G., Deslandes, Y., Bluhm, T., Marchessault, R. H., Vincendon, M., and Saint-Germain, J.,** Solid-state ^{13}C NMR characterization of wood, *Tappi*, 66(6), 92, 1983.
11. **Haw, J. F., Maciel, G. E., and Schroeder, H. A.,** Carbon-13 nuclear magnetic resonance spectrometric study of wood and wood pulping with cross polarization and magic-angle spinning, *Anal. Chem.*, 56, 1323, 1984.
12. **Haw, J. F., Maciel, G. E., and Biermann, C. J.,** Carbon-13 nuclear magnetic resonance study of rapid steam hydrolysis of red oak, *Holzforschung*, 38, 327, 1984.
13. **Maciel, G. E., Haw, J. F., Smith, D. H., Gabrielson, B. C., and Hatfield, G. R.,** Carbon-13 nuclear magnetic resonance of herbaceous plants and their components using cross polarization and magic angle spinning, *J. Agric. Food Chem.*, 33, 185, 1985.
14. **Haw, J. F. and Maciel, G. E.,** Carbon-13 nuclear magnetic resonance spectrometry of oil seeds with cross polarization and magic-angle spinning, *Anal. Chem.*, 55, 1262, 1983.
15. **O'Donnell, D. J., Ackerman, J. J. H., and Maciel, G. E.,** Comparitive study of whole seed protein and starch content via cross polarization-magic angle spinning carbon-13 nuclear magnetic resonance spectroscopy, *J. Agric. Food Chem.*, 29, 514, 1981.
16. **Haw, J. F. and Schultz, T. P.,** Carbon-13 CP/MAS NMR and FT-IR study of low-temperature lignin pyrolysis, *Holzforschung*, 39, 289, 1985.

17. **Pfeffer, P. E., Gerasimowicz, W. V., and Piotrowski, E. G.,** Effect of paramagnetic iron on quantitation in carbon-13 cross polarization magic angle spinning nuclear magnetic resonance spectrometry of heterogeneous environmental matrices, *Anal. Chem.*, 56, 734, 1984.
18. **Maciel, G. E., Bartuska, V. J., Ackerman, J. J. H., O'Donnell, D. J., and Hawkins, B. L.,** A ^{13}C NMR study of four lignins in the solid and solution states, *Makromol. Chem.*, 182, 2297, 1981.
19. **Bartuska, V. J., Maciel, G. E., Bolker, H. I., and Fleming, B. I.,** Structural studies of lignin isolation procedures ^{13}C NMR, *Holzforschung*, 34, 214, 1980.
20. **Atalla, R. H., Gast, J. C., Sindorf, D. W., Bartuska, V. J., and Maciel, G. E.,** ^{13}C NMR Spectra of cellulose polymorphs, *J. Am. Chem. Soc.*, 102, 3249, 1980.
21. **Earl, W. L. and VanderHart, D. L.,** High-resolution magic angle sample spinning ^{13}C NMR of solid cellulose I, *J. Am. Chem. Soc.*, 102, 3251, 1980.
22. **Earl, W. L. and VanderHart, D. L.,** Observations by high-resolution carbon-13 nuclear magnetic resonance of cellulose related to morphology and crystal structure, *Macromolecules*, 14, 570, 1981.
23. **Haw, J. F., Maciel, G. E., Kinden, J. C., and Murphy, V. G.,** Carbon-13 nuclear magnetic resonance spectrometric study of autohydrolyzed and organosolv-treated lodgepole pinewood using cross polarization and magic-angle spinning, *Holzforschung*, 39, 99, 1985.
24. **Biermann, C. J., Schultz, T. P., and McGinnis, G. D.,** Rapid steam hydrolysis/extraction of mixed hardwoods as a biomass pretreatment, *J. Wood Chem. Technol.*, 4, 111, 1984.
25. **Papadopoulos, J., Chem, C.-L., and Goldstein, I. S.,** The behavior of lignin during hydrolysis of sweetgum wood with concentrated hydrochloric acid at moderate temperatures, *Holzforschung*, 35, 283, 1981.
26. **Brezny, R., Mihalow, V., and Kovacik, V.,** Low temperature thermolysis of lignins I. Reaction of β-0-4 model compounds, *Holzforschung*, 37, 199, 1983.
27. **Brezny, R., Surina, I., and Kosik, M.,** Low temperature thermolysis of lignins II. Thermofractography and thermal analysis of β-0-4 model compounds, *Holzforschung*, 38, 19, 1984.
28. **Fenner, R. A. and Lephart, J. O.,** Examination of the thermal decomposition of kraft pine lignin by fourier transform evolved gas analysis, *J. Agric. Food Chem.*, 29, 846, 1981.
29. **Browning, B. L.,** *The Chemistry of Wood*, R. E. Brieger, Huntington, NY, 1975.
30. **Rydholm, S. A.,** *Pulping Processes*, Interscience, New York, 1965.

Chapter 13

PROBING METABOLISM WITH DOUBLE-CROSS POLARIZATION SOLID-STATE NMR

George T. Coker, III and Jacob Schaefer

TABLE OF CONTENTS

I. INTRODUCTION

In many studies of metabolism, investigators are interested in determining how a particular compound is either synthesized or metabolized. What they are really asking is how are chemical bonds being synthesized or broken. The typical way that this is studied is to introduce a molecule within which there is an atom which is either radioactive or an unusual stable isotope.[1-3] The fate of the molecule is then deduced by monitoring the tagged atom. For the last several years, there has been an alternate and more direct way to monitor the fate of ^{13}C-^{15}N bonds. This procedure, double cross-polarization magic angle spinning (DCPMAS) NMR, allows the investigator to determine the percentage of ^{15}Ns (or ^{13}C) bound to ^{13}C (or ^{15}N).

NMR has several advantages over the use of radiochemistry to monitor tagged molecules. NMR allows the investigator to determine easily the chemical nature of the ^{15}N or ^{13}C. When radioisotopes are used, samples must be treated with either specific enzymes or chemicals to detect specifically labeled atoms.[4,5] This frequently entails extensive handling of the sample with associated losses. NMR uses stable isotopes as tracers. Compounds labeled with ^{15}N are commercially available whereas compounds labeled with ^{13}N, the radioisotope of nitrogen, must be synthesized and used within 2 h[3] because of the short half-life of ^{13}N (10 min). Stable isotopes are nontoxic whereas radioisotopes may have deleterious effects on tissues.[6] NMR is a nondestructive technique, thus permitting the exact same sample to be analyzed in a number of ways. NMR may obtain the same information as more convential techniques, but at a reduced cost in time and effort (e.g., Table 1).

DCPMAS NMR is a solid-state NMR technique. There are several other advantages to using solid-state NMR vs. solution NMR. Solution NMR can only measure compounds that are rotating rapidly in solution. Solid NMR can examine cellular constituents *in vivo* that are not freely soluble, such as nucleic acids, proteins, or insoluble material, such as cell walls. In dried samples the low molecular weight compounds that are visible to solution NMR are also detected by solid NMR. The combination of high-power resonant decoupling and high-speed mechanical sample spinning has allowed liquid-like spectra to be obtained from solid material. The spectra thus obtained from solid NMR can be used to determine quantitatively the amount of a particular isotope that is present in a particular compound whereas the same is not always true of solution NMR. In succeeding sections, examples are presented in which DCPMAS NMR has been used to examine different areas of metabolism of agricultural importance.

II. PLANT METABOLISM

A. Amino Acid Metabolism in Developing Legumes

1. Asparagine

In developing cotyledons of legumes, the proper ratios of amino acids necessary for the synthesis of storage protein must be synthesized from a carbon input of sucrose and a nitrogen input of asparagine and glutamine.[7] The mechanism by which this is regulated is not known. For example, the asparaginyl residues of protein may either be derived directly from asparagine or they may arise from *de novo* synthesis. Schaefer et al.[8] used DCPMAS NMR to determine the extent to which these two pathways form the asparaginyl residues of seed storage protein.

Immature soybean cotyledons were grown in culture with [4-^{13}C; amide-^{15}N]asparagine, a mixture of [4-^{13}C]asparagine plus [amide-^{15}N]asparagine, or a mixture of [4-^{13}C; amide-^{15}N]asparagine plus natural abundance asparagine. After 14 d, the cotyledons were harvested, lyophilized, and extracted with 80% ethanol to remove soluble amino acids. The cotyledons were then analyzed by DCPMAS NMR (Figure 1). Two spectra from each sample were obtained. The ^{15}N CPMAS spectra (Figure 1, top) account for all the ^{15}N present in the sample. Double-cross difference spectra (bottom) were obtained by subtracting a ^{15}N CPMAS spectrum from a spectrum that arises from those ^{15}N that were not bound to ^{13}C. The difference spectra represent those ^{15}N that are bonded to ^{13}C (see Section V.A for further details).

Table 1
COMPARISON OF METHODS TO DETERMINE THE EXTENT TO WHICH BACTERIAL PEPTIDOGLYCAN IS CROSS-LINKED

Conventional[a]
- Grow cells
- Harvest cells
- Isolate cell walls
- Digest cell walls with two enzymes that are not commercially available
- Gel filtration
- Assay column fractions
- Gel filtration of pooled fractions
- Assay column fractions
- Preparative electrophoresis
- Acid hydrolysis
- Amino acid analysis
- Total time ~7 d

NMR[a]
- Grow cells with [ε-^{15}N]lysine
- Harvest cells
- Lyophilize cells
- Analyze dried cells with NMR (machine time ~1 d)
- Total time 2 d

[a] The conventional procedure was summarized from Nakel et al.[5] and the NMR procedure was summarized from Jacob et al.[24] We estimated the time necessary to do both procedures.

The ^{13}C-^{15}N amide bond of the asparaginyl residues of cotyledons fed double-labeled asparagine may arise via the direct incorporation of asparagine into protein. Alternatively, asparagine entering the cotyledons may be hydrolyzed to produce [^{15}N]ammonium and [4-^{13}C]aspartate.[9] A ^{13}C-^{15}N bond may be introduced into protein indirectly by the resynthesis of double-labeled asparagine which is subsequently used for protein synthesis. The latter pathway was shown to exist by the formation of a ^{13}C-^{15}N bond when cotyledons were fed [4-^{13}C]asparagine and [amide-^{15}N]asparagine (Table 2). The two pathways were distinguished by measuring the extent to which natural abundance perturbs the introduction of ^{13}C-^{15}N into protein. Direct incorporation of a ^{13}C-^{15}N bond from double-labeled asparagine would not be affected by the presence of natural abundance asparagine, but the indirect pathway would be affected. The flux along these two pathways was represented by r:

$$r = \frac{f(\text{undiluted}) - f(\text{diluted})}{f(\text{undiluted}) - f(\text{diluted, calculated})}$$

where f represents the relative number of ^{15}Ns that is bonded to ^{13}C under various labeling conditions: f(undiluted) is 17, f(diluted) is 12, and f(diluted,calculated) is 8.5 (17/2) (Table 2). Thus, r = 0.6 and about 40% of the asparagine residues were formed by the direct incorporation of exogenous asparagine. These results imply that there is some mechanism by which asparagine entering the cotyledons is shunted into one pathway or the other. The basic techniques and approaches used in this experiment can easily be extended to examining the incorporation of other amino acids into protein, and purines or pyrimidines into nucleic acids.

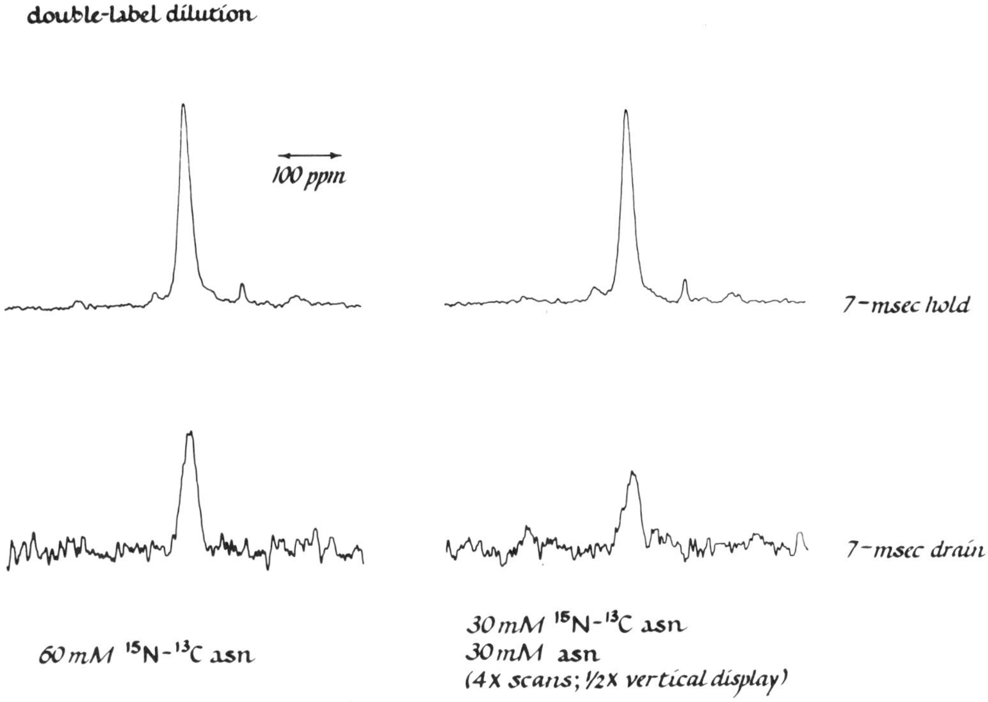

FIGURE 1. ^{15}N-CPMAS and DCPMAS NMR spectra of soybean cotyledons cultured on L-[4-^{13}C; ^{15}N]asparagine. Immature cotyledons were cultured for 14 d on either 60 m*M* double-labeled asparagine (left) or 30 m*M* double-labeled asparagine plus 30 m*M* natural abundance asparagine (right). The cotyledons were extracted with ethanol and the insoluble material was lyophilized. The top spectra represent the total ^{15}N present in the sample. The major peak represents amide nitrogen. The bottom spectra represent those ^{15}Ns that are bound to ^{13}C. The bottom spectra were obtained at 6× the data accumulation as the top spectra and are displayed at 2× the vertical scale.

Table 2
INCORPORATION OF ^{15}N-^{13}C INTO COTYLEDON PROTEIN

Asparagine	^{15}N as % of total protein nitrogen incorporated	% amide ^{15}N adjacent to ^{13}C
30 m*M* [4-^{13}C] plus 30 m*M* [amide-^{15}N]	34	5
60 m*M* [4-^{13}C; amide-^{15}N]	38	17
30 m*M* [4-^{13}C; amide-^{15}N] plus 30 m*M* natural abundance	18	12

Note: Immature soybean cotyledons were grown in media containing asparagine at a final concentration of 60 m*M*.[8] After 14 d, the cotyledons were harvested and lyophilized. The free amino acids were extracted with 80% ethanol. The remaining insoluble material was analyzed by ^{15}N-CPMAS NMR and ^{15}N-DCPMAS NMR.

2. *Allantoin*

Coker et al.[10] used ^{15}N DCPMAS NMR to examine ureide metabolism in developing soybean cotyledons. The ureides, allantoin and allantoic acid, are the major forms of nitrogen transported to the shoot from the root in N_2 fixing soybeans.[11] To distinguish between several possible pathways for the degradation of allantoin and the use of the nitrogen for protein synthesis, immature cotyledons were grown in media containing allantoin labeled in specific atoms with ^{13}C and ^{15}N. By analyzing the labeled cotyledons with a combination of CPMAS and DCPMAS and DCPMAS NMR, it was determined that all of the carbon and nitrogen bonds were broken during its metabolism and the subsequent nitrogen was reassimilated via glutamine synthetase and GOGAT.

3. Methionine

The addition of methionine to cotyledons grown *in vitro* stimulated the growth of the cotyledons by 17%.[12] This was attributed to the specific enhancement of protein synthesis by additional methionine. Using ^{15}N-CPMAS NMR it was possible to quantitate how much ^{15}N in cotyledons grown with [^{15}N]methionine was present in amino acids.[13] By examining cotyledons grown with [2-^{13}C; ^{15}N]methionine with DCPMAS NMR, it was determined how much of the ^{15}N-NMR signal consisted of methionine that had not been metabolized. From these and other results, it was concluded that the high levels of exogenous methionine allow the slow step of methionine syntheses, i.e., the transfer of methyl group to homocysteine, to be bypassed and thus increase the rate at which a number of compounds can be synthesized.

4. Proline

The control of transformation of undifferentiated callus to differentiated tissue is not well understood. Certain hormone treatments and reduced sources of nitrogen are known to increase this event. Skokut et al.[14] using CPMAS and DCPMAS determined the utilization of several amino acids during regeneration of alfalfa callus to embryos. Alanine and proline increased the number of embyros obtained by up to sevenfold when the proper hormones were added. When these amino acids were compared to amino acids, such as glutamate and glycine which did not enhance embryo formation, there were no differences in the uptake or utilization of nitrogen between tissues from alfalfa lines that could or could not regenerate. In tissues where regeneration was suppressed, the level of intracellular free amino acids was elevated.

When regeneration was suppressed, free proline was formed from nitrate or ammonium, and less nitrogen from proline was used in the synthesis of protein. When [^{15}N]proline was fed to the tissues, a large percentage of the proline was directly incorporated into protein as evidenced by a prominent assymmetry of the amide peak.[14] The nitrogen of free proline results in a peak at 30 ppm; when present as a part of an amide bond, it results in a peak around 110 ppm. To check whether the direct incorporation of proline was a result of the synthesis of cell-wall protein which is rich in hydroxyproline, alfalfa tissue was grown with [2-^{13}C; ^{15}N]proline. As measured by DCPMAS NMR, there was no difference in the metabolism of proline in regenerating and nonregenerating tissues and no evidence for enhanced cell-wall protein synthesis in nonregenerating tissues.

B. Protein Turnover

During senescence, the protein of leaves of legumes is degraded and the resulting nitrogen and carbon is shipped to the developing fruits.[15] One of the questions that arises is to what extent does protein turnover prior to senescence provide nitrogen and carbon for new growth. Schaefer et al.[16] used DCPMAS NMR to address this question.

Soybean plants were massively labeled with ^{15}N from $^{15}NH_4{}^{15}NO_3$ over a period of 7 weeks. The plants were then placed in a chamber and allowed to fix $^{13}CO_2$ for 7 d. Several leaflets of different developmental ages were harvested and lyophilized. The plants continued to grow and developmentally similar leaflets were harvested 10 to 20 d later. The exposure to $^{13}CO_2$ effectively tagged the ^{13}C-^{15}N bonds of the protein synthesized. To the extent that this labeled protein turned over, the ^{13}C-^{15}N bonds were broken. If a particular bond was broken, the ^{13}C will be diluted with $^{12}CO_2$ or lost to the environment via photorespiration.

In a manner similar to that of the asparagine experiments (Section II.A.1 and V.A), the

Table 3
PROTEIN TURNOVER AS MEASURED BY TURNOVER OF ^{13}C-^{15}N BONDS

^{13}C main chain concentration			
C1	C2	ΔT (d)	$T_{1/2}$ (d)
6.9	5.6	5	20
8.5	4.2	15	17
6.1	4.3	20	45
9.1	2.2	40	20
8.2	5.9	15	38
			Avg 28

Note: Soybean plants were massively labeled with ^{15}N and then labeled with $^{13}CO_2$ for 7 d. After 1 d of labeling with $^{13}CO_2$, one trifolate leaf was harvested. A number of days later (ΔT) a developmentally similar trifolate was harvested. The concentration of ^{13}C in the amide bond of protein for the leaf harvested first (C1) and second (C2) was determined by ^{15}N-DCPMAS NMR as described in the text. From these values the half-life ($T_{1/2}$) of the protein was determined as described in Section II.B.

fraction of amide ^{15}N that have adjacent ^{13}C was determined. An estimate of the rate at which protein was turning over was made (Table 3). The turnover rate was determined by assuming a completely ^{15}N-labeled protein pool and a simple first-order ^{13}C scrambling of the protein, $dc/dt = -kc$ where c is the isotopic concentration of the protein carbon pool and k is the turnover rate. The time constant of the turnover was between 20 and 40 d.

III. BACTERIAL METABOLISM

A. Glyphosate

Roundup is one of the biggest selling herbicides in the world. One of the reasons for this popularity is that glyphosate, the active ingredient, is rapidly degraded in the soil. Moore et al.[17] isolated a soil bacterium, PG 2982, that could grow on glyphosate as the sole source of phosphate indicating that this bacteria could break the bond between carbon and phosphorous. Jacob et al.[18,19] using a combination of solid-state ^{13}C and ^{15}N-CPMAS and DCPMAS, deduced the pathways by which glyphosate is degraded and the regulation of these pathways in PG 2982. This is one of the few examples of NMR being used to determine metabolism which was previously unknown.

One of the initial questions was, on which side of the nitrogen is the carbon nitrogen bond broken? To answer this, cultures of PG 2982 grown on either [2-^{13}C; ^{15}N]glyphosate or [3-^{13}C; ^{15}N]glyphosate were analyzed by DCPMAS NMR. There are two major ^{15}N peaks in CPMAS NMR spectra of ethanol-extracted bacteria grown on either of the [^{15}N]glyphosates (Figure 2). The peak at 85 ppm represents nitrogen in the amide bond of protein; the peak at 215 ppm represents the N-7 found in purines. There is no double-cross difference signal in the bacteria grown on [3-^{13}C; ^{15}N]glyphosate demonstrating that this ^{13}C-^{15}N bond was completely broken. In contrast, there are two prominent peaks in the difference spectrum of bacteria grown on [2-^{13}C; ^{15}N]glyphosate. Jacob et al.[18] concluded that [2-^{13}C; ^{15}N]glycine formed from labeled glyphosate was used to synthesize purines and protein. They were also able to determine that the ^{13}C enrichment next to these ^{15}N was about 90% (see Section V.A). Thus glycine was incorporated intact into both purine and protein.

Jacob et al.[19] extended these observations using ^{13}C-DCPMAS NMR. The same basic

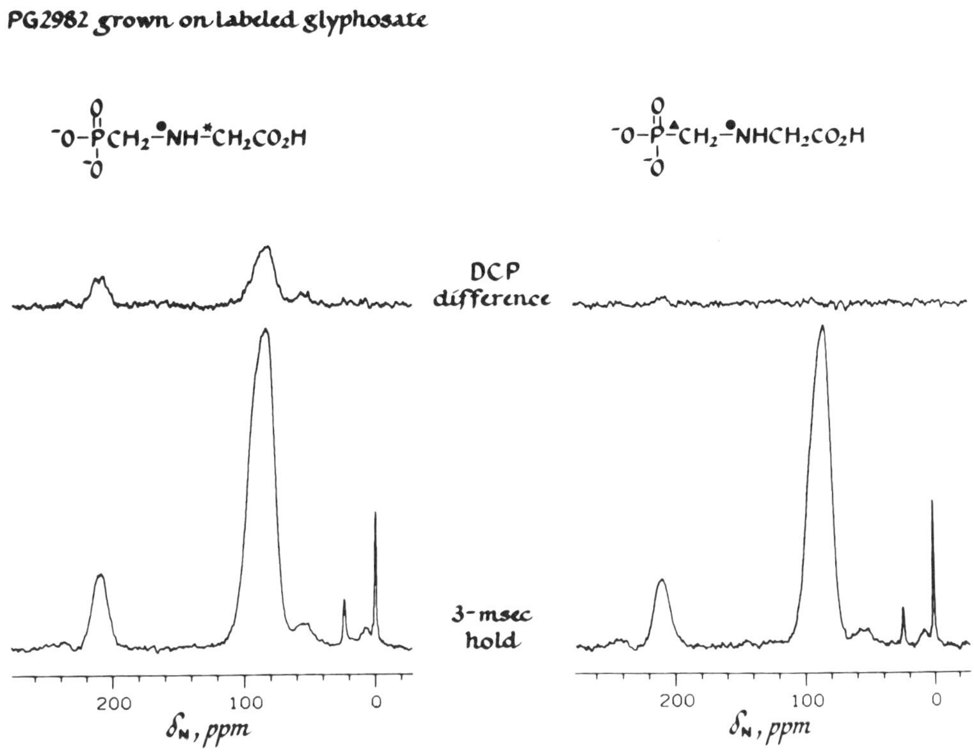

FIGURE 2. ^{15}N-DCPMAS (top) and CPMAS (bottom) NMR spectra of *Pseudomonas* PG2982 grown on [2-^{13}C,^{15}N]glyphosate (left) and [3-^{13}C,^{15}N]glyphosate (right). The peaks at 210, 85, 60, 10, and 0 ppm represent aromatic (purine), amide, guanidino, amino, and NH_4^+ nitrogen, respectively. The peak at 22 ppm is an unknown metabolite. (From Jacob, G. S., Schaefer, J., Stejskal, E. O., and McKay, R. A., *J. Biol. Chem.*, 260, 5899, 1985. With permission.)

procedures used for ^{15}N-DCPMAS NMR were employed except that the ^{13}C were polarized from the ^{1}Hs and then the polarization of the ^{13}C was drained by adjacent ^{15}N. The ^{15}N enrichment next to ^{13}C was determined. Generally this is a more difficult experiment to analyze because of the background level ^{13}C, but this can be overcome with massive ^{13}C enrichments in the carbons of interest. Using ^{15}N-DCPMAS NMR no ^{13}C-^{15}N bonds were found in intact cells when [3-^{13}C; ^{15}N]glyphosate was fed to bacteria, but with the enhanced sensitivity of ^{13}C NMR, a bond between the C-8 and N-7 of purines was found (Table 4). This bond arises from the synthesis of purines from the C-3 of glyphosate entering the tetrahydrofolate pool of C-1 intermediates and the ^{15}N of glyphosate entering via glycine. The C-2 of glyphosate also enters the same bond (Table 4). As shown by ^{13}C DCPMAS NMR, the glycine from glyphosate was incorporated intact into purines, but this depends on the cellular demand for glycine. If exogenous glycine was provided, the carbon and nitrogen were not incorporated specifically as glycine, but were integrated via general cell metabolites (Figure 3). Thus, by analyzing the end products using ^{15}N and ^{13}C solid-state NMR, Jacob et al.[18,19] were able to map quantitatively the fluxes of glyphosate metabolism in PG 2982 without having to detect any of the intermediates.

B. Nitrogen Fixation

The synthesis of sufficient nitrogenase by *Klebsiella pneumoniae* to grow on N_2 upon exhaustion of fixed nitrogen typically takes 15 to 20 h.[20] This lag period can be shortened to 6 to 8 h with the addition of serine, but cannot be reduced by the addition of a number of other amino acids including glycine. Jacob et al [21] were able to show, using DCPMAS and CPMAS NMR, that the advantage of serine over other amino acids was that it supplied both carbon and nitrogen in the forms needed for the synthesis of ribosomal RNA which precedes the induction

Table 4
^{15}N ENRICHMENTS OF NITROGENS OF METABOLITES OF GLYPHOSATE IN BACTERIA

	Bond	^{15}N enrichment
[2-^{13}C; ^{15}N]glyphosate	Amide	3
	[8-C; 7-N]purine	25
	[5-C; 7-N]purine	90
	[2-C,N]seryl residue	90
	[2-C;N]glycyl residue	90
[3-^{13}C; ^{15}N]glyphosate	Amide	5
	[8-C; 7-N]purine	25

Note: Bacteria (Strain PG2982) were grown in media containing 30 m*M* ammonium and 1 m*M* of the double-labeled glyphosate as the sole sources of nitrogen. Labeled cells were harvested and lyophilized. The dried material was analyzed by DCPMAS ^{13}C NMR. ^{15}N enrichments were estimated to be ±10%.

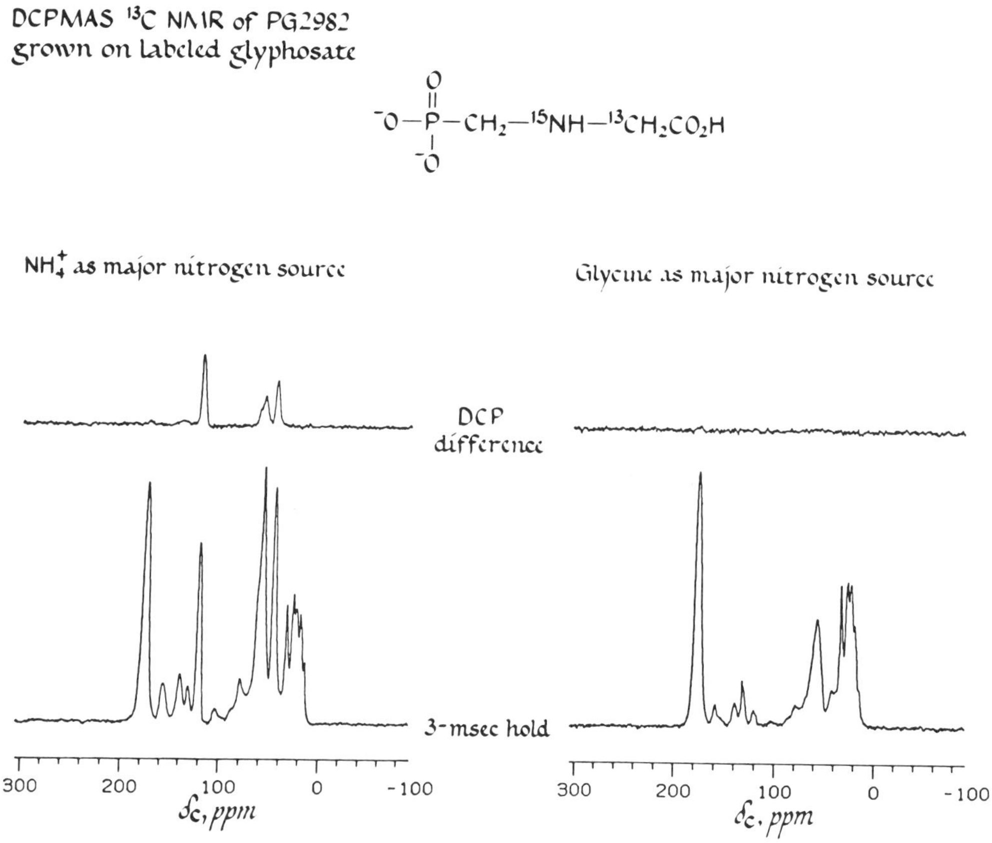

FIGURE 3. ^{13}C-DCPMAS (top) and CPMAS (bottom) NMR spectra of *Pseudomonas* PG2982 grown on [2-^{13}C,^{15}N]glyphosate. The DCPMAS signals at 119, 45, and 56 ppm arise from the formation of [5-^{13}C,7-^{15}N]purine, [1-^{13}C,^{15}N]glycyl, and [2-^{13}C,^{15}N]seryl residues, respectively, (From Jacob, G. S., Garbow, J. R., Schaefer, J., and Kishore, G. M., *J. Biol. Chem.*, 262, 1522, 1987. With permission.)

of nitrogenase. They were able to establish that although glycine could be used for the synthesis of the N-7 of purine, as could serine, the nitrogen of glycine was not readily used in the synthesis of the other purine nitrogens (N-1, 3, and 9).

IV. INSOLUBLE BIOLOGICAL MATRICES

A. Insect Cuticle

One of the major advantages to CPMAS AND DCPMAS NMR is that intact samples can be examined. Thus solid-state NMR can be used to examine materials that are not easily extracted. The structure of the cross-links in the cuticle of the tobacco hornworm was examined by Schaefer et al.[22] using CPMAS and DCPMAS. Because of the difficulties in extracting material, previous workers were at best able to determine the composition of the cuticle and guess as to the nature of the bonds holding the cuticle together. Previous workers[23] proposed that the main structural components, chitin and protein, were linked directly together via cross-links formed between the epsilon groups of lysine groups and the chitin. Schaefer et al.[22] examined whether lysine formed the cross-links by injecting tobacco hornworm larvae with [ε-^{15}N]lysine. The lyophilized cuticles of the larvae were analyzed by ^{15}N-CPMAS NMR. Most of the ^{15}N labeled amino groups of lysine were scrambled to amide nitrogen and no cross-links were detected. However only 1% of the cuticular amino acids are lysine. In contrast, about 10% of the amino acids are histidine, which may also participate in cross-links. CPMAS NMR spectra of larvae fed [1,3-$^{15}N_2$]histidine show that a substantial percentage of the histidyl-nitrogen was not scrambled and became part of a cross-link (Figure 4). By feeding larvae both [1,3-$^{15}N_2$]histidine and dopamine labeled in the ring carbons with ^{13}C, Schaefer et al.[22] were able to show the formation of a ^{15}N-^{13}C bond using DCPMAS NMR (Figure 5). From this as well as additional information, they proposed a structure for the aromatic cross-link of the cuticle of tobacco hornworm (Figure 6).

B. Bacterial Cell Wall

The cell walls of bacteria consist of a peptidoglycan (Figure 7) which is held together by peptide bonds between short peptides of adjacent strands. Jacob et al.[24] used DCPMAS to analyze the cross-linked cell walls of bacteria. Bacteria grown with [ε-^{15}N]lysine and D-[1-^{13}C]alanine were harvested and subjected to ^{15}N-DCPMAS NMR (Figure 8). Jacob et al.[24] found that the ^{13}C-^{15}N signal is enriched ~20%. Having previously shown that D-[1-^{13}C]alanine is not scrambled they concluded that ~80% of the D-alanine for peptidoglycan arises from *de novo* synthesis. Using similar techniques, Schaefer et al.[24] were able to establish that there are no terminal D-alanine or D-ala-D-ala units in uncross-linked chains of the peptidoglycan.

C. Future Directions

In a similar manner, DCPMAS might be used to examine the structure of other biological matrices, such as the cell walls of higher plants. Plant cell walls consist of up to 20% protein, but the nature of the bonds holding the protein and carbohydrates together is only beginning to be unraveled.[26,27] One protein, extensin, appears to be fixed in the cell wall by linkages between tyrosines.[28] Extensin also contains 7% lysine[29] which may be used in cross-linking the protein to other proteins, carbohydrates, or lignin. Other proteins in cell walls which contain a higher percentage of histidine[30] might used for cross-linking in a manner similar to that of the insect cuticle.

V. TECHNICAL DETAILS

A. Experimental Procedures

DCPMAS NMR is an extension of CPMAS NMR. Protons are used to polarize either carbons

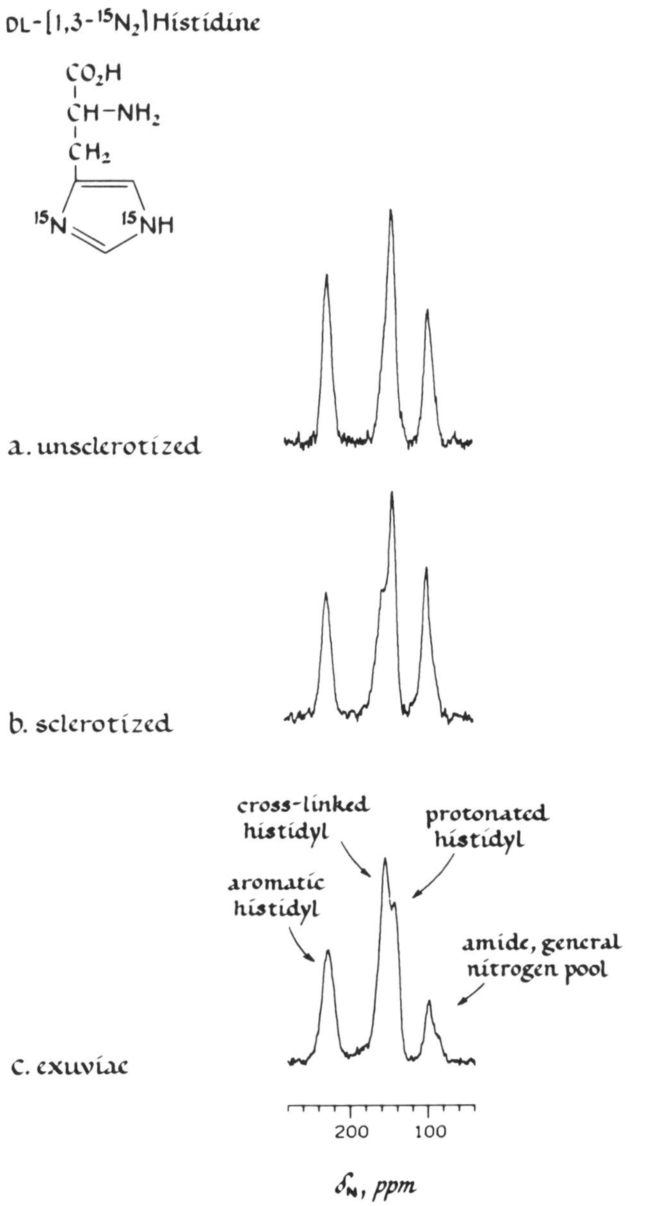

FIGURE 4. ^{15}N-NMR of tobacco hornworm cuticle. Immature larvae were injected with L-[1,3-$^{15}N_2$]histidine and sacrificed before the cuticle formed (unsclerotized), as the cuticle was forming (sclerotized), or when the cuticle was completely formed (exuviae). The material was analyzed by ^{15}N-CPMAS NMR. The horizontal scale is in parts per million downfield from solid ammonium sulfate as an external reference. (From Schaefer, J., Kramer, K. J., Garbow, J. R., Jacob, G. S., Stejskal, E. O., Hopkins, T. L., and Speirs, R. O., *Science*, 235, 1200, 1987. With permission.)

or nitrogens (Figure 9). Then for DCPMAS NMR, the proton radio frequency is turned off and the spins of the ^{15}N and ^{13}C are allowed to interact. Because this interaction falls off as the inverse of distance to the sixth power, only ^{13}C and ^{15}N bonded together will interact. After a period of time, the spins of either ^{15}N or ^{13}C can be observed. In practice, two spectra are obtained concurrently. For example, the first pulse sequence with ^{13}C radio frequency off resonance results in a CPMAS signal arising from all of the ^{15}N present. The next pulse sequence with the

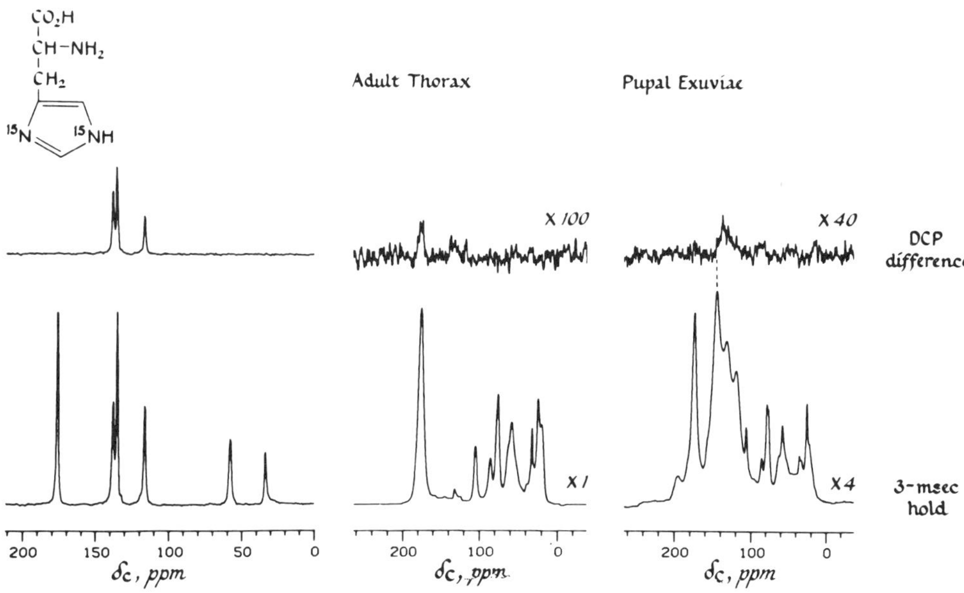

FIGURE 5. CPMAS and DCPMAS NMR of tobacco hornworm cuticle. Single (bottom) and double (top) CPMAS ^{13}C-NMR spectra of [1,3-$^{15}N_2$]histidine (left), adult thoracic cuticle labeled with L-[1,3-$^{15}N_2$]histidine and D-[UL-^{13}C]glucose (middle), and mature pupal cuticle labeled with [1,3-$^{15}N_2$]histidine and [ring-$^{13}C_6$]dopamine (right) are presented. The top spectrum represents only those ^{13}Cs bonded to ^{15}N (DCP difference); the bottom spectrum represents all ^{13}Cs. The ^{13}C label from glucose appears predominantly in chitin peaks, such as those near 80 ppm, while that from dopamine enhances the intensity of the aromatic region of the spectrum from 120 to 160 ppm. (From Schaefer, J., Kramer, K. J., Garbow, J. R., Jacob, G. S., Stejskal, E. O., Hopkins, T. L., and Speirs, R. O., *Science*, 235, 1200, 1987. With permission.)

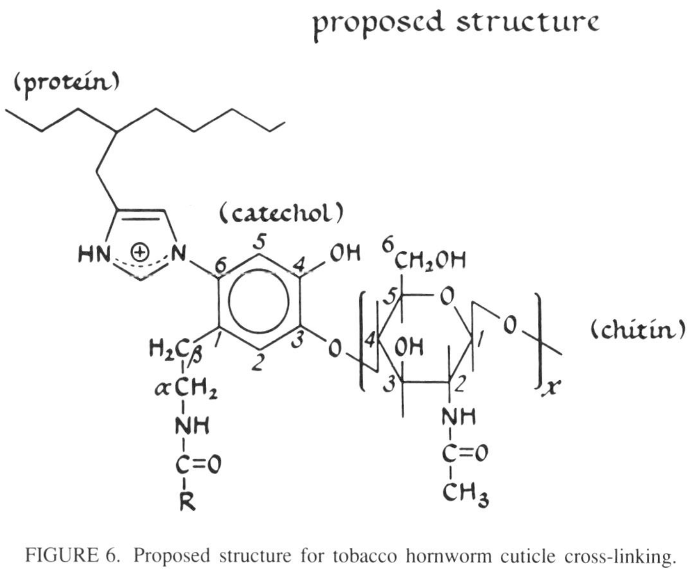

FIGURE 6. Proposed structure for tobacco hornworm cuticle cross-linking. (From Schaefer, J., Kramer, K. J., Garbow, J. R., Jacob, G. S., Stejskal, E. O., Hopkins, T. L., and Speirs, R. O., *Science*, 235, 1200, 1987. With permission.)

^{13}C radio frequency on resonance results in a DCPMAS signal representing the ^{15}N not bonded to the ^{13}C. The two spectra are subtracted from one another resulting in the DCPMAS difference spectrum. This spectrum represents the ^{15}N that is bound to ^{13}C. In a similar manner, ^{13}C DCPMAS signals can also be obtained.

When the ^{13}C is on resonance, an additional drain on the ^{15}N is present such that for the ^{15}N

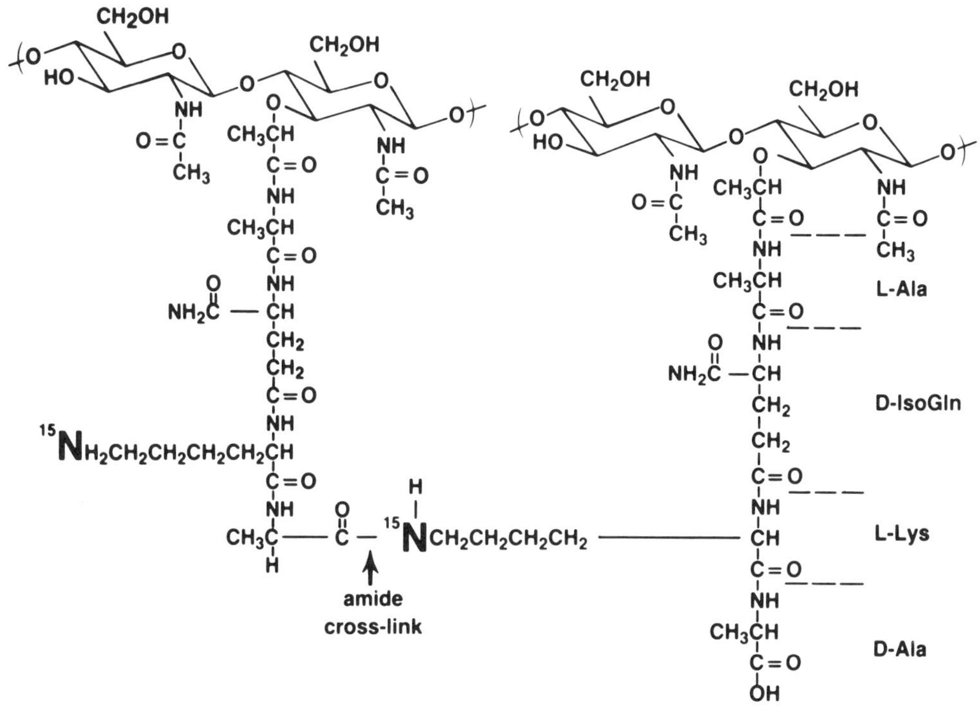

FIGURE 7. Structure of peptidoglycan.

signal $S = S_0 \exp(-t/T_d)$ where S_0 represents the signal obtained without the ^{13}C drain, t is the time that the ^{13}C and ^{15}N are allowed to interact, and T_d is the time constant for the drain. The time constant represents the efficiency by which the ^{15}N polarizes the ^{13}C. This value depends on several factors including the length of the ^{13}C-^{15}N bond. The two spectra are subtracted from one another to obtain the relative difference spectrum, delta S. The fraction of ^{13}C next to ^{15}N, f, can be read from Figure 10, provided the T_d for the particular bond is known. For example, in Section III, Jacob et al.[18] determined the extent to which doubly labeled glycine was being incorporated intact into purines and into proteins. The ratio of the difference spectrum to that of the CPMAS spectrum for the purine signal was 0.43 and for the amide signal was 0.18. The T_d for the purine nitrogen was 5 ms and for the amide signal was 14 ms. By combining these values and reading off Figure 10, it was determined that the ^{13}C enrichment next to these ^{15}N was 90%.

The analysis is somewhat more complicated when protein turnover was being analyzed (Section II.B).[16] In this case, the ^{15}N can be found in three types of linkages with respect to ^{13}C, all with different T_ds. The ^{15}N may be bound to a ^{13}C carbonyl carbon, a ^{13}C aliphatic carbon, or it may have both a ^{13}C carbonyl and a ^{13}C aliphatic carbon. Thus instead of only a single rate by which polarization is drained from ^{15}N, there are three possible rates. Schaefer et al.[16] showed that these simplify to an equation with a form similar to that shown above.

B. Equipment

Only one DCPMAS spectrometer has been reported in routine use.[31] This machine is built around an Oxford 89-mm vertical bore superconducting solenoid operating at 200 MHz for protons. The spectrometer's three radio frequency channels (^1H, ^{13}C, and ^{15}N) are each equipped with 1.5-kW transmitters. The single-coil probe is triple tuned using remote tuning elements.[32]

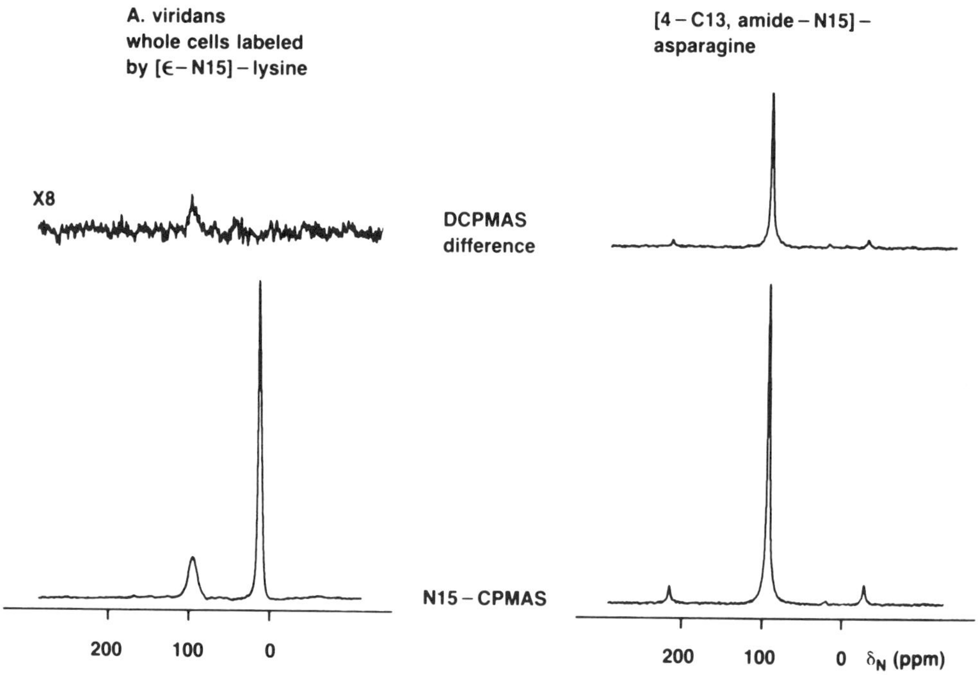

FIGURE 8. ^{15}N-CPMAS and DCPMAS NMR of lyophilized whole cells of *Aerococcus viridans* labeled with L-[ε-^{15}N]lysine and DL-[1-^{13}C]alanine. The spectrum at the top (left) represents ^{15}N bonded to ^{13}C; the bottom (left) spectrum represents all ^{15}N. The peak at 95 ppm represents amide nitrogen and the peak at 14 ppm represents amino nitrogen. The spectra at the right was obtained from solid [4-^{13}C; amide-^{15}N]asparagine.

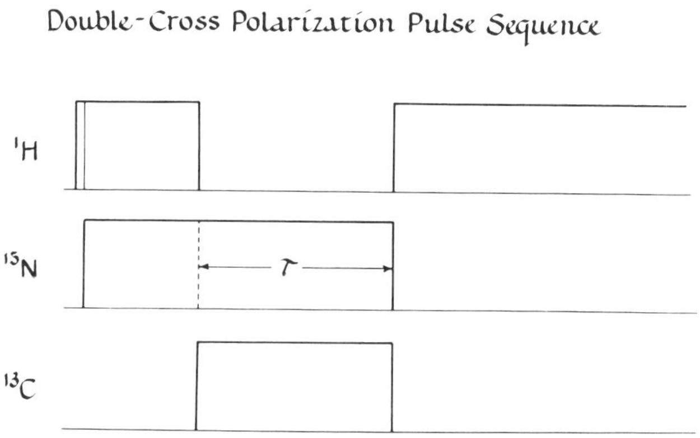

FIGURE 9. DCPMAS NMR pulse sequence. A ^{15}N direct difference signal is accumulated by shifting the ^{13}C radio frequency field off resonance for one scan and on resonance for the next. The second NMR signal is subtracted from the first. The time, t, during which the ^1H is off is called the ^{15}N hold time when the ^{13}C radio frequency field is off resonance, and the drain time when the ^{13}C is on resonance. Data acquisition occurs with the ^1H radio frequency field turned back on for dipolar decoupling.

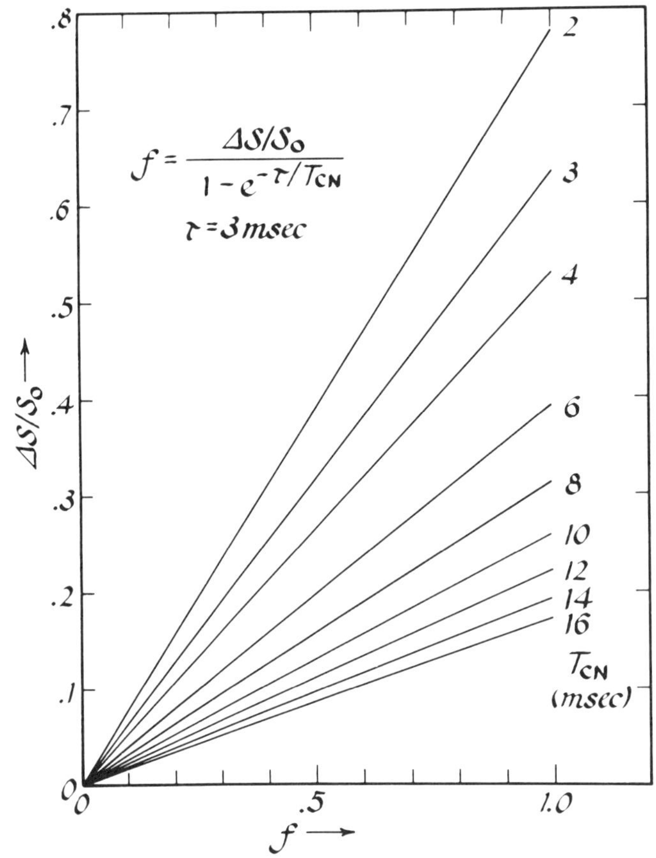

FIGURE 10. The fraction, f, of the ^{15}N atoms of a chemical bond with ^{13}C neighbors as a function of the ^{15}N DCP difference signal, ΔS, observed for a carbon-nitrogen contact time, t, of 3 ms for typical values of the carbon-nitrogen cross-polarization transfer time T_{CN} The difference signal is scaled by the CPMAS signal after the ^{15}N spin lock, S_0. The calculated lines correspond to $f = (\Delta S/S_0)/(1 - e^{-t/T_{CN}})$.

The coil is 13 mm in diameter and permits the use of rotors with a 1-g payload. In operation, both the radio frequency pulse amplitudes and the magic angle spinning speed are controlled by active feedback circuits.[33] This is necessary because of the sensitivity of the ^{13}C-^{15}N cross-polarization transfer to slight Hartmann-Hahn mismatches. The operation of the spectrometer is monitored twice a minute by a dedicated microcomputer to ensure that no crucial parameter has drifted outside of established limits. In practice, this has meant successful 2-week long double-cross data accumulations involving 1,000,000 scans with no operator intervention.[22]

C. Cost

The cost of the isotopically labeled material can vary widely. [^{13}C]sodium bicarbonate and [^{15}N]nitrate are available for as little as \$70/g. Glycine labeled either with ^{13}C or ^{15}N is \$125/g. The costs will go up dramatically for compounds that are labeled with both ^{13}C and ^{15}N in specific atoms. For example [2-^{13}C, ^{15}N]glyphosate costs \$4000/g. However, these costs must also be balanced against the amount of time necessary to obtain similar results with other methods. A number of groups had been trying to identify the pathway of glyphosate metabolism for several years using [^{14}C]glyphosate. Jacob et al.[18] using labeled glyphosate and DCPMAS NMR with one experiment were able to establish the major pathways by which glyphosate was metabolized. It took about 4 man-months and 2 d of NMR time. Although radioactive isotope was cheaper per experiment, the answer was obtained more rapidly using NMR.

VI. FUTURE DEVELOPMENTS

A. Detection of ^{13}C-^{13}C Labeled Chemical Bonds by Solids NMR

One approach to the detection of a ^{13}C-^{13}C labeled spin pair in a biological solid is by the solid-state analog of a liquid-state 2D COSY experiment.[34] In this experiment, homonuclear chemical shifts are connected by the effect of scalar coupling between the labeled pair during the two-dimensional (2D) mixing time.[35] The coupling generates so-called "off-diagonal" peaks in a 2D contour plot. For solids, direct ^{13}C-^{13}C dipolar coupling is weak enough that searching for minor peaks from a label, just off a dense and cluttered natural-abundance diagonal, is impractical.[34]

Multiple-quantum coherence has also been used in liquid-state experiments to identify homonuclear labeled pairs.[36] Unfortunately, the solid-state version of this experiment requires choosing a pulse spacing *comparable* to the inverse of the ^{13}C-^{13}C dipolar coupling of 2 to 4 kHz, but *short* compared to the inverse of the magic-angle spinning speed of 3 to 6 kHz. These are mutually exclusive demands. More involved multiple-quantum experiments using synchronized pulsing and spinning[37] overcomes this limitation, at least in part, but overall sensitivity of the multiple-quantum experiment for spinning solids is still questionable.

A more straightforward approach uses ^{13}C-^{13}C spin diffusion to establish connectivity and so to characterize ^{13}C-labeled chemical bonds. This experiment begins with a cross-polarization transfer from protons to carbons. The resulting ^{13}C magnetization is then restored to the static magnetic field direction by a 90° pulse. The third step is selective inversion by one of the lines arising from the label by a string of weak pulses, the so-called "DANTE" irradiation.[38] If this magnetization is sampled immediately, inverted, and the spectrum without DANTE irradiation subtracted, the resulting difference spectrum consists only of the inverted line from the labeled carbon, together with its spinning sidebands.[39] The complicated natural-abundance signal has been removed. If we allow, say, 10 ms to elapse after the DANTE irradiation and before the sampling, ^{13}C-^{13}C spin diffusion will connect carbons which are directly coupled (i.e., ^{13}C which are part of a labeled chemical bond). Spin diffusion is slow enough that more distant ^{13}Cs are not connected. Now the difference spectrum will contain both the original labeled peak, and a new peak due to spin diffusion. Characteristic isotropic shifts as well as chemical shift tensors (from spinning sidebands) can be used to identify the ^{13}C-labeled bond. This method has been used successfully to identify the binding sites of [15,16-$^{13}C_2$] labeled acetylenic steroid to estradiol dehydrogenase.[40]

B. Detection of Labeled Chemical Bonds Involving One Quadrupolar Nucleus (^{13}C-^{17}O, ^{13}C-2D)

Oldfield[41] has shown that CPMAS ^{17}O-NMR is possible in some inorganic solids using only the central ^{17}O transition of this spin 5/2 quadrupolar nucleus. While even the central transition may be too broad (300 kHz) to observe in biological solids (excluding phosphates) because of unfortuitous ^{17}O relaxation parameters, indirect detection of ^{13}C-NMR in a double-cross polarization experiment may be practical.

The key to practicality is sensitivity. Compared to ^{13}C-^{15}N double-cross, the ^{13}C-^{17}O experiment is at a disadvantage because the central transition represents only about 15% of the total intensity and because commercial isotopic enrichment of ^{17}O is less than 60%. However, the ^{13}C-^{17}O experiment is at an advantage because the ^{13}C-^{17}O dipolar coupling is three times greater than ^{13}C-^{15}N coupling and magic-angle spinning modulation of the transfer process may not be an interferring factor.[42]

Possibly a simple approach to the detection of ^{13}C-^{17}O labeled bonds is to avoid a genuine cross-polarization step altogether. Instead, by inserting ^{17}O 180° pulses synchronized with the magic-angle rotor (one pulse every half rotor period), ^{17}O dipolar dephasing of the carbon magnetization will not refocus with each rotor period. The dephasing due to ^{13}C-^{17}O dipolar coupling will then appear as a "drain" on the carbon magnetization. Appropriate difference spectra will reveal only those carbons coupled to ^{17}O. This is an approach similar to that suggested by Lippmaa[43] for the observation of carbon chemical shift anisotropies by synchronized spinning and carbon 180° pulses.

Both this approach and the standard double cross-polarization experiment should be appropriate for the observation of ^{13}C-2D and ^{15}N-2D labeled bonds. The 2D NMR spectrum is well characterized and about 150 kHz in width.[44] Nevertheless, the motivation for developing a ^{13}C-2D double cross-polarization has been less compelling than that for ^{13}C-^{17}O, possibly because of the popularity of 2D NMR direct detection of labels.

REFERENCES

1. **Mudd, S. H. and Datko, A. H.,** Methionine methyl group metabolism in *Lemna*, *Plant Physiol.*, 81, 103, 1986.
2. **Dungey, N. O. and Davies, D. D.,** Protein turnover in the attached leaves of non-stressed and stressed barley seedlings, *Planta*, 154, 435, 1982.
3. **Schubert, K. R. and Coker, G. I., III,** Nitrogen and carbon assimilation in N_2-fixing plants. Short-term studies using ^{13}N and ^{11}C, in *Short-Lived Radionuclides in Chemistry and Biology*, Root, J. W. and Krohn, K. A., Eds., American Chemical Society, Washington, D.C., 1981.
4. **McClure, P. R., Coker, G. T., III, and Schubert, K. R.,** Carbon dioxide fixation in roots and nodules of *Alnus glutinosa*. I. Role of phosphoenolpyruvate carboxylase and carbamyl phosphate synthetase in dark CO_2 fixation, citrulline synthesis, and N_2 fixation, *Plant Physiol.*, 71, 652, 1983.
5. **Nakel, M., Ghuysen, J.-M., and Kandler, O.,** Wall peptidolycan in *Aerococcus viridans* strains 201 Evans and ATCC 11563 and in *Gaffkya homari* strain ATCC 10400, *Biochemistry*, 10, 2170, 1971.
6. **Cooper, P. C. and Burgess, A. W.,** Biosynthetic labeling with ^{32}P: radiation damage to mammalian cells, *Anal. Biochem.*, 144, 32, 1985.
7. **Rainbird, R. M., Thorne, J. H., and Hardy, R. W. F.,** The role of amides, amino acids and ureides in the nutrition of developing soybean seeds, *Plant Physiol.*, 74, 329, 1984.
8. **Schaefer, J., Skokut, T. A., Stejskal, E. O., McKay, R. A., and Varner, J. E.,** Asparagine amide metabolism in developing cotyledons, *Proc. Natl. Acad. Sci. U.S.A.*, 78, 5978, 1981.
9. **Skokut, T. A., Varner, J. E., Schaefer, J., Stejskal, E. O., and McKay, R. A.,** NMR determination of asparagine and glutamine nitrogen utilization for synthesis of storage protein in developing cotyledons of soybean in culture, *Plant Physiol.*, 69, 308, 1982.
10. **Coker, G. T., III and Schaefer, J.,** ^{15}N and ^{13}C NMR determination of allantoin metabolism in developing soybean cotyledons, *Plant Physiol.*, 77, 129, 1985.
11. **McClure, P. R., Israel D. W., and Volk, R. J.,** Evaluation of the relative ureide content of xylem sap as an indicator of N_2 fixation in soybeans, *Plant Physiol.*, 66, 720, 1980.
12. **Thompson, J. F., Madison, J. T., Waterman, M. A., and Muenster, A-M. E.,** Effect of methionine on growth and protein composition of cultured soybean cotyledons, *Phytochemistry*, 20, 941, 1981.
13. **Coker, G. T., III, Garbow, J. R., and Schaefer, J.,** ^{15}N and ^{13}C NMR determination of methionine metabolism in developing soybean cotyledons, *Plant Physiol.*, 83, 698, 1987.
14. **Skokut, T. A., Manchester, J., and Schaefer, J.,** Regeneration in alfalfa tissue culture. Stimulation of somatic embryo production by amino acids and N-15 NMR determination of nitrogen utilization, *Plant Physiol.*, 79, 579, 1985.
15. **Pate, J. S.,** Transport and partitioning of nitrogenous solutes, *Annu. Rev. Plant Physiol.*, 31, 313, 1980.
16. **Schaefer, J., Skokut, T. A., Stejskal, E. O., McKay, R. A., and Varner, J. E.,** Estimation of protein turnover in soybean leaves using magic angle double cross-polarization nitrogen-15 nuclear magnetic resonance, *J. Biol. Chem.*, 256, 11574, 1981.
17. **Moore, J. K., Braymer, H. D., and Larson, A. D.,** Isolation of a *Pseudomonas* sp. which utilizes the phosphonate herbicide glyphosate, *Appl. Environ. Microbiol.*, 46, 316, 1983.
18. **Jacob, G. S., Schaefer, J., Stejskal, E. O., and McKay, R. A.,** Solid-state NMR determination of glyphosate metabolism in a *Pseudomonas* sp., *J. Biol. Chem.*, 260, 5899, 1985.

19. **Jacob, G. S., Garbow, J. R., Schaefer, J., and Kishore, G. M.,** Solid-state NMR studies of regulation of *N*-(phosphonomethyl)glycine and glycine metabolism in *Pseudomonas* sp. Strain PG2982, *J. Biol. Chem.*, 262, 1552, 1987.
20. **Lindsay, H. L.,** Physiological Studies of Nitrogen Fixation by Cells and Cell-Free Extracts of *Aerobacter aerogenes,* Ph.D. thesis, University of Wisconsin, Madison, 1963.
21. **Jacob, G. S., Schaefer, J., Garbow, J. R., and Stejskal, E. O.,** Solid-state NMR studies of *Klebsiella pneumoniae* grown under nitrogen-fixing conditions, *J. Biol. Chem.*, 262, 254, 1987.
22. **Schaefer, J., Kramer, K. J., Garbow, J. R., Jacob, G. S., Stejskal, E. O., Hopkins, T. L., and Speirs, R. D.,** Direct measurement of insect cuticle cross-linking by solid-state ^{13}C and ^{15}N NMR, *Science,* 235, 1200, 1987.
23. **Lipke, H., Sugumaran, M., and Henzel, W.,** in *Advances in Insect Physiology,* Vol. 17, Berridge, M. H., Treherne, J. E., and Wigglesworth, V. B., Eds., Academic Press, New York, 1984, 1.
24. **Jacob, G. S., Schaefer, J., and Wilson, G. E., Jr.,** Solid-state ^{13}C and ^{15}N nuclear magnetic resonance studies of alanine metabolism in *Aerococcus viridans (Gaffkya homari)*, *J. Biol. Chem.*, 260, 2777, 1985.
25. **Schaefer, J., Garbow, J. R., Jacob, G. S., Forrest, T. M., and Wilson, G. E., Jr.,** Characterization of peptidoglycan stem lengths by solid-state ^{13}C and ^{15}N NMR, *Biochem. Biophys. Res. Commun.*, 137, 736, 1986.
26. **Lamport, D. T. A. and Epstein, L.,** in *Current Topics in Plant Biochemistry and Physiology,* Vol. 2, Randall, D. D., Blevins, D. G., Larson, R. L., and Rapp, B. J., Eds., University of Missouri, Columbia, 1983, 73.
27. **Cooper, J. B., Chen, J. A., and Varner, J. E.,** in *Structure, Function, and Biosynthesis of Plant Cell Walls,* Dugger, W. M. and Bartnicki-Garcia, S., Eds., American Society of Plant Physiologists, Rockville, MD, 1984, 75.
28. **Fry, S. C.,** Isodityrosine, a new cross-linking amino acid from plant cell wall glycoprotein, *Biochem. J.*, 204, 449, 1982.
29. **Van Holst, G. J. and Varner, J. E.,** Reinforced polyproline II conformation in a hydroxyproline rich cell wall glycoprotein from carrot root, *Plant Physiol.*, 74, 247, 1984.
30. **Stevenson, T. T., McNeil, M., Darvill, A. G., and Albersheim, P.,** Structure of plant cell walls. XVIII. An analysis of the extracellular polysaccharides of suspension cultured sycamore cells, *Plant Physiol.*, 80, 1012, 1986.
31. **Schaefer, J., McKay, R. A., and Stejskal, E. O.,** Double cross-polarization NMR of solids, *J. Magn. Reson.*, 34, 443, 1979.
32. **McKay, R. A.,** U.S. Patent 4,446,431 (May 1, 1984).
33. **McKay, R. A.,** 10th Annual Meeting of the Federation of Analytical Chemistry and Spectroscopy Societies, Philadelphia, September 1983.
34. **Menger, E. M., Vega, S., and Griffin, R. G.,** Double-quantum observation of carbon couplings in solids, *J. Magn. Reson.*, 56, 338, 1984.
35. **Freeman, R. and Morris, G. A.,** Experimental chemical-shift correlation maps in nuclear magnetic resonance spectroscopy, *J. Chem. Soc. Chem. Commun.*, 684, 1978.
36. **Bax, A., Freeman, R., and Frenkiel, T. A.,** An NMR technique for tracing out the carbon skeleton of an organic molecule, *J. Am. Chem. Soc.*, 103, 2102, 1981.
37. **Meier, B. H. and Earl, W. L.,** Excitation of multiple quantum transitions under magic angle spinning, *J. Chem. Phys.*, 85, 4905, 1986.
38. **Morris, G. A. and Freeman, R.,** Selective excitation in Fourier-transform nuclear magnetic resonance, *J. Magn. Reson.*, 29, 433, 1978.
39. **Caravatti, P., Bodenhausen, G., and Ernst, R. R.,** Selective pulse experiments in high-resolution solid-state NMR, *J. Magn. Reson.*, 55, 88, 1983.
40. **Auchus, R. J., Covey, D. F., Bork, V., and Schaefer, J.,** Solid-state NMR observation of cysteine and lysine Michael adducts of inactivated estradiol dehydrogenase, *J. Biol. Chem.*, 263, 11640, 1988.
41. **Oldfield, E.,** 13th Annual Meeting of the Federation of Analytical Chemistry and Spectroscopy Societies, St. Louis, September 28 to October 3, 1986; **Waltar, T. H., Turner, G. L., and Oldfield, E.,** Oxygen-17. Cross-polarization NMR spectroscopy of inorganic solids, *J. Magn. Reson.*, 76, 106, 1988.
42. **Schaefer, J., Stejskal, E. O., Garbow, J. R., and McKay, R. A.,** Quantitative determination of the concentrations of ^{13}C-^{15}N chemical bonds by double cross-polarization NMR, *J. Magn. Reson.*, 59, 150, 1984.
43. **Lippmaa, E., Alla, M., and Tuherm, T.,** in *Proc. XIX Cong. Ampere,* Heidelberg, West Germany, 1977, 113.
44. **Spiess, H. W.,** Molecular dynamics of solid polymers as revealed by deuteron NMR, *Colloid Polym. Sci.*, 261, 193, 1983.

Chapter 14

SOLID-STATE NMR APPROACHES FOR RESOLVING COMPONENTS OF PLANT MATERIALS

James S. Frye and Gary E. Maciel

TABLE OF CONTENTS

I. INTRODUCTION

Two basic approaches to chemical component analysis are (1) isolation of components by chemical or physical fractionation, or (2) use of an analytical technique that can simultaneously observe and discriminate all components or capable of suppression of contributing signals from all components but one. This chapter will discuss NMR methods for analyzing complex mixtures using the second approach; such mixtures may be the product of an attempted physical separation, or the sample may be a relatively intact specimen of plant material. An analytical technique capable of characterizing components in complex mixtures may be used to address several fundamental questions:

1. Can components be analyzed reliably and efficiently without separation steps?
2. Do the fractionation procedures alter the structures of the components?
3. How do the components interact in the intact material?

The remaining sections are in approximately historical order, starting with the simplest, well-characterized techniques, going on to novel experiments, and concluding with a brief treatment of anticipated "next generation" NMR experiments that have been demonstrated but not yet extensively applied to plant materials. Most examples involve ^{13}C, which is the most studied nuclide in organic solid-state NMR, but the techniques can also be applied to other spin-$^{1}/_{2}$ cases, such as ^{31}P and ^{15}N.

II. CHEMICAL SHIFT

From a chemist's vantage point, differences in nuclear shielding by electrons, the chemical shifts, usually constitute the most useful features of an NMR spectrum. For samples studied in solution, this effect normally determines the frequency range of a spectrum, with other interactions providing fine structure or influencing intensities and peak widths. Fast Brownian motion averages the orientation dependence of the shielding tensor to its isotropic value, δ_{iso}, and this parameter has proven extremely useful in applications of NMR to chemical structure determinations.

In the solid state, the orientation dependence of the chemical shift (chemical shift anisotropy, CSA) is generally not averaged to δ_{iso} because large-amplitude, rapid motion is absent.[1] In a single crystal, the chemical shift varies with sample orientation in the magnetic field. In an amorphous or polycrystalline sample (e.g., most plant materials), all orientations are present in a random, virtually continuous distribution, and the resultant signal is spread over a powder pattern that may cover a frequency range comparable to the entire range of δ_{iso} values in the spectrum. For the most commonly encountered example, ^{13}C, sp^2-hybridized carbons have CSAs on the order of 200 ppm, and sp^3 CSA are often on the order of 40 ppm.[2,3] This type of line-broadening influence is termed "inhomogeneous" because it arises from the superposition of a multitude of relatively narrow component peaks with different chemical shifts and in some respects behaves in an NMR experiment like the consequences of an inhomogeneous magnetic field.

A different inhomogeneous broadening mechanism is "chemical shift dispersion," in which the measured peak width arises from a multitude of overlapping isotropic chemical shifts (or CSA patterns for static solids). This effect is commonly seen in both solid- and solution-state NMR spectra of polymers and complex mixtures and it often represents a major source of signal broadening in solid-state NMR.[4]

The geometric relationship in which the orientation dependence of the chemical shift tensor is embodied contains the factor $(1 - 3\cos^2\theta)$, where θ represents the angle between the magnetic field direction and an axis of the atom-based principal axis system of the chemical shift tensor.

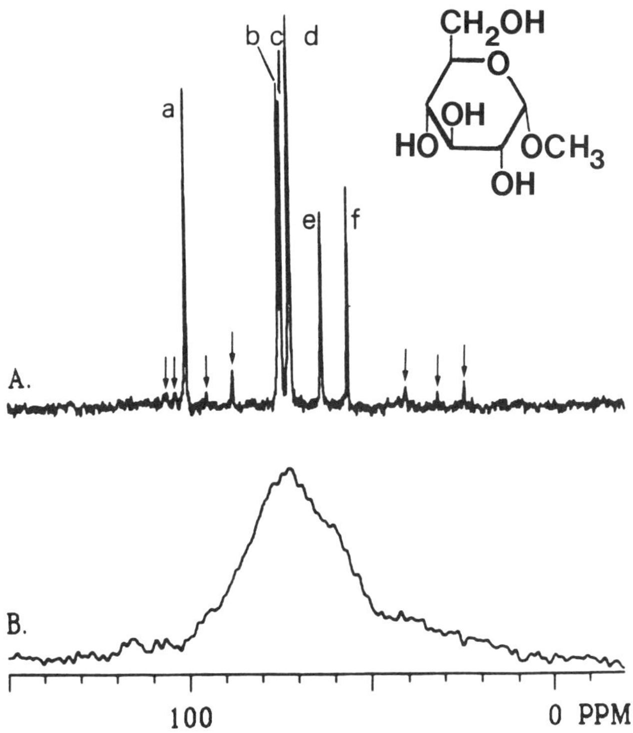

FIGURE 1. Comparison of ^1H-decoupled ^{13}C spectra of α-methylglucoside with (A) and without (B) MAS, taken at 50.1-MHz ^{13}C frequency. MAS rate is 1600 Hz. Isotropic peaks are labeled a to f; spinning sidebands are denoted by arrows. (From Zeigler, R. C., Wind, R. A., and Maciel, G. E., *J. Magn. Reson.*, 79, 299, 1988. With permission.)

This geometric factor offers an alternative averaging scheme to the random Brownian motion of solutions. If the sample is rotated rapidly about an axis inclined at an angle, β, to the magnetic field direction, the CSA will be scaled by the factor $(1 - 3\cos^2\beta)$. At $\beta = 54.7°$, this factor becomes zero and the broadening effect of the CSA vanishes, hence the terms "magic angle" and "magic-angle spinning" (MAS).[1] In this context "rapidly" means a rotor frequency larger than the CSA width in Hertz. A typical ^{13}C MAS spectrum is presented in Figure 1.

Most solid-state ^{13}C NMR utilizes magnetic field strengths of 1.4 to 7.0 T, corresponding to proton frequencies of 60 to 300 MHz and ^{13}C frequencies of 15 to 75 MHz. A typical CSA range of 200 ppm thus corresponds to a frequency width of 3 to 15 kHz. Recent progress in MAS designs have advanced spinning rates beyond 20 kHz, but most commercial spectrometers operate routinely at 2 to 8 kHz, and this is clearly not rapid with respect to CSA except at the lower magnetic field strengths. MAS is nonetheless useful because the consequence of moderate spinning rate is not broad peaks but the appearance of spinning sidebands spaced at the rotor frequency and approximately mapping out the CSA pattern.[5] MAS at a rate less than the frequency corresponding to 100 ppm (for ^{13}C) generally yields spinning sidebands that seriously interfere with the isotropic peaks. There are more sophisticated NMR techniques that can suppress the spinning sidebands and partially compensate for the signal intensity distributed over the spinning sideband manifold; these are used routinely at the higher field strengths.[6-10] However, the resulting "sideband-free" spectrum will not in general accurately reflect relative concentrations of carbon types because each signal can relax at different rates during the various waiting times in the experiment. The obvious solution to the spinning sideband problem is to spin more rapidly, thereby decreasing the sideband intensities and removing them from the normal region of δ_{iso}. This can also compromise quantitation by partially averaging the ^{1}H-^{13}C dipolar coupling responsible for cross-polarization (see Section III).

The usual goal of a MAS experiment is to generate a solid-state NMR spectrum that approaches the resolution achieved in a liquid-state experiment. For ^{1}H-decoupled (see below) ^{13}C MAS spectra of plant materials, the peak widths are seldom less than 0.5 ppm, but this is sufficient to permit a separation of components according to differences in δ_{iso}, exactly as is done in solution NMR. Most sp^3-hybridized carbon peaks have $\delta_{iso} < 90$ ppm from tetramethylsilane. For plant materials, an important exception is the carbohydrate anomeric carbon, e.g., C_1 of glucose, which is generally in the range of 95 to 105 ppm. Other carbohydrate peaks appear in the range 60 to 90 ppm. The other major structural component, lignin, is often the main source of aromatic peaks (120 to 160 ppm), but lignin also contains such aliphatic groups as $-CH_2-$, -CHOH-, and $-O-CH_3$. Hemicellulose contributes acetyl peaks around 175 ppm (carboxyl) and 25 ppm (methyl), in addition to peaks in the carbohydrate region of 60 to 110 ppm. Proteins contain carbon types that appear all across the ^{13}C spectrum, but since every amino acid residue contributes to the amide region (about 170 ppm) and α-C region (about 55 ppm), these envelopes are particularly prominant in relatively protein-rich materials such as leaves and seeds. Oils and other extractives are largely aliphatic. Figure 2 presents several ^{13}C spectra of untreated herbage samples, illustrating the full chemical shift range of common plant components.

Although there is a good deal of overlap in the spectral regions of these components, differences in δ_{iso} values have provided the basis for most of the literature on NMR studies of plant materials in both liquid- and solid-states. The δ_{iso} values almost always provide a starting point for the more complicated experiments that yield additional discriminations.

III. DIPOLAR COUPLING

Even without CSA, the ^{13}C line width in an organic solid would be broadened by ^{13}C-^{1}H dipolar coupling to several tens of kHz in the absence of ^{1}H decoupling.[1] Rapid tumbling in solution averages the coupling to zero, and this interaction then generally determines the ^{13}C longitudinal relaxation time, T_{1C}.[11] MAS at routine speeds is usually ineffective at removing the effects of ^{13}C-^{1}H dipolar couplings because strong ^{1}H-^{1}H dipolar coupling causes the ^{1}H-^{13}C broadening to be homogeneous, i.e., each contributing signal is broad, and the MAS rate would have to be much faster than the ^{1}H-^{1}H dipolar coupling frequency to achieve useful narrowing. Fortunately the protons can be decoupled from each other and from the carbons with a strong radio-frequency (rf) field, a high-power version of the usual ^{1}H decoupling in solution spectra aimed at removing J_{CH} splittings. The major limitation is sample and hardware heating from the deposition of 100 W or so of ^{1}H rf power during data aquisition; for this reason the decoupler is always gated off during the relaxation delay between repetitions. The high-power decoupling also eliminates the potentially useful J_{CH} splittings from the spectrum.

The strong ^{13}C-^{1}H dipolar coupling is turned to advantage in the cross polarization (CP) experiment.[1,12] Simultaneous irradiation of the ^{13}C and ^{1}H spins at power levels adjusted to the "Hartmann-Hahn match" condition, $\gamma_C H_{1C} = \gamma_H H_{1H}$,[13] allows the transfer of ^{1}H polarization to the ^{13}C spins, resulting in up to a four-fold signal enhancement and, more important, allowing signal averaging to be carried out with repetition rates governed by ^{1}H relaxation times (T_{1H}) rather than by the ^{13}C relaxation times (T_{1C}).[12] The CP experiment permits a distinction between liquid and solid components in a mixture in that a nonviscous liquid does not have static components of the C-H dipolar coupling to provide a CP mechanism and therefore should not generate a signal in the CP spectrum.[10] Conversely, a single-pulse excitation with fairly rapid repetition and low-power decoupling should reveal mainly liquid-state components. Intermediate situations such as rapid anisotropic motion (e.g., biological membranes) can compromise the distinction.[14,15]

A sample often contains carbons with various strengths of ^{13}C-^{1}H dipolar coupling. The

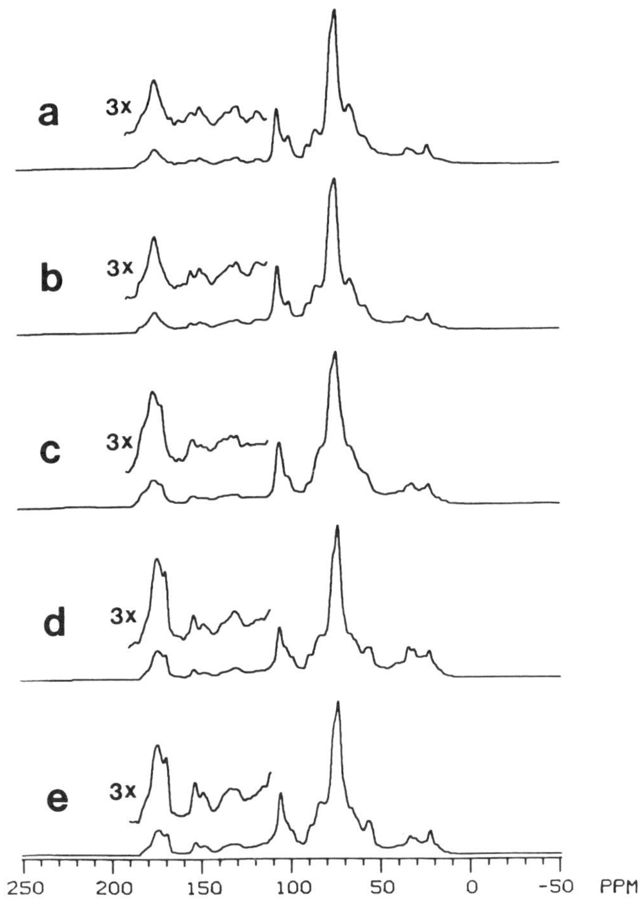

FIGURE 2. ^{13}C CP/MAS spectra of herbage samples: (a) forage sorghum; (b) sorghum X Sudan grass; (c) Jerusalem artichoke; (d) kochia; and (e) Russian thistle. ^{13}C frequency is 25.27 MHz; MAS rate is 3.5 kHz. (From Maciel, G. E., Haw, J. F., Smith, D. H., Gabrielsen, B. C., and Hatfield, G. R., *J. Agric. Food Chem.*, 33, 185, 1985. With permission.)

dipole-dipole energy falls off with the cube of the through-space distance, so any ^{13}C nucleus that is far from the nearest ^{1}H will be weakly coupled. The cleanest distinction is between carbon atoms with or without directly bonded hydrogens. However, this distinction is compromised by rapid, large-amplitude motion that can partially average dipolar coupling even for carbons with directly attached protons. The most common example of this case is the methyl group, which undergoes rapid rotation at room temperature in typical solid samples.

Within the CP experiment, there are several approaches for distinguishing strong from weak C-H dipolar coupling. One approach utilizes the CP dynamics to make distinctions. The ^{13}C CP intensity grows during the CP period with an exponential time constant, T_{CH}, toward a maximum value, and T_{CH} is dependent on the ^{13}C-^{1}H dipolar coupling. Short CP (or "contact") time favors strongly coupled subsystems, whereas a longer contact time tends to blur the distinction.[10,16,17] This has the advantage of being inherently quantitative, but the corresponding disadvantage is that the kinetic measurements required for quantitation are quite time consuming.[18]

The most commonly applied dipolar "filter" is the interrupted decoupling (or "dipolar dephasing") experiment,[19] in which the CP period is followed by a short time during which the decoupler is switched off and the ^{13}C spins evolve under the ^{13}C-^{1}H dipolar interaction

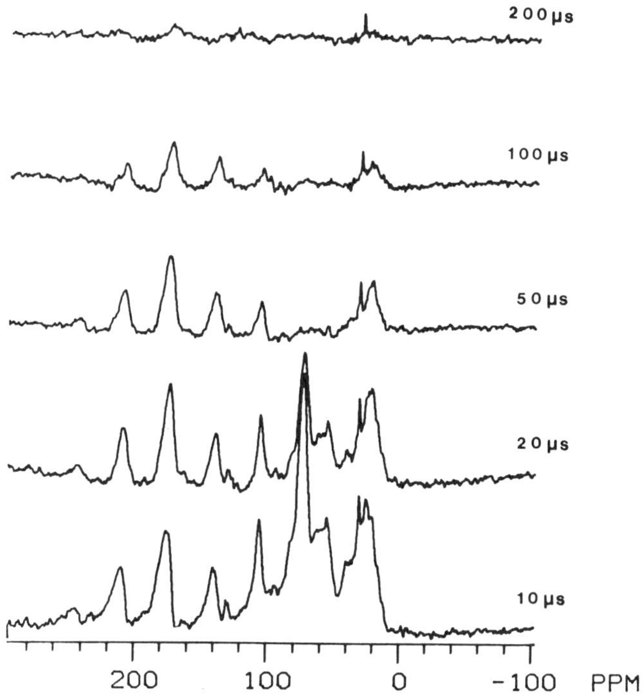

FIGURE 3. ^{13}C CP/MAS spectra of soybeans with interrupted decoupling for durations noted. ^{13}C frequency is 50.3 kHz; MAS rate is 1.7 kHz. (From Haw, J. F. and Maciel, G. E., *Anal. Chem.*, 55, 1262, 1983. With permission.)

(somewhat averaged by MAS). The decoupler is switched on again for data aquisition, so the observed spectrum consists of peaks that are at least as narrow as those of an ordinary ^{13}C MAS experiment with ^{1}H decoupling. Magnetization from strongly coupled carbons has time to dephase, and the peaks are much attenuated or absent entirely from the spectrum. With the usual interrupt time of 40 to 70 μs, this experiment suppresses carbons with directly bonded hydrogens with little attenuation of other carbons. The usual effect of interrupted decoupling is, therefore, a "subspectrum" of unprotonated and methyl carbons.[19,20] A simple example is the suppression of starch signals in the interrupted decoupling spectrum of red clover seeds, resulting in a spectrum of only protein carbons.[21] Another case is shown in Figure 3 for the CP/MAS spectrum of whole soybeans taken with various interrupt times.[10] The common exception for methyl groups arises from rapid rotation which almost always averages the ^{13}C-^{1}H dipolar coupling enough to let the carbon resonances survive. A second, less prevalent, exception can occur for rapidly rotating solvent molecules trapped in a solid lattice.[22] In principle, the dipolar dephasing dynamics can be measured and fitted to structural models, and a two-dimensional version has been demonstrated,[23] but theoretical treatments of the spin dynamics are exceedingly complex, and the true utility of the method lies in qualitatively distinguishing CH and CH_2 groups from C and CH_3 moieties in a single experiment.

The Hartmann-Hahn match requirement can be used as a dipolar filter because highly mobile systems with relatively weak ^{13}C-^{1}H and ^{1}H-^{1}H dipolar couplings provide CP signals whose intensities are much more sensitive to error in the match than are signals from tightly coupled systems.[24] This effect is a consequence of the relatively narrow ^{1}H line width for weakly coupled systems; overlap between the ^{13}C and ^{1}H peaks in the rotating frame decreases rapidly with increasing mismatch, causing the CP time constant T_{CH} to be longer and the ^{13}C intensity lower for a given contact time. Intentional mismatch has been utilized to suppress the CP subspectrum of triglycerides in intact soybeans, leaving the starch and protein subspectra virtually unaffected.[10] Sensitivity to Hartmann-Hahn mismatch can also be a liability in that random or systematic fluctuations in rf transmitter power levels can lead to fluctuations in peak intensities that are much larger than would be expected from the signal-to-noise ratio of a single scan. This effect may preclude the use of CP in relaxation studies of highly mobile species. Such mobile systems generally have sufficiently short T_1 values and narrow signals that the usual liquid-state experiments (in conjunction with MAS) will yield reliable relaxation times.[25]

The CP concept can be extended to a third nucleus in a "double CP" experiment.[26] The major application has been CP from ^{1}H to ^{13}C to ^{15}N in doubly enriched samples, providing an ^{15}N subspectrum of nuclei bonded to specifically enriched carbon sites and thereby following a C-N bond diad through a biosynthetic sequence.[27-29] With sufficient care, the ^{13}C-^{15}N group can be analyzed quantitatively.[30] This experiment puts stringent constraints on instrument stability but can offer very detailed structural information about biological systems.

IV. ^{1}H SPIN DYNAMICS

In the absence of state-of-the-art NMR trickery, rapid spin "flip-flops" between protons in the solid state average potential chemical distinctions in the ^{1}H spectrum so that all protons in a physical domain relax with a single value of T_{1H} and exhibit a large homogeneous line width that totally obscures δ_{iso} and CSA. The normal single-pulse ^{1}H spectrum of a proton-rich solid, with or without MAS, is, therefore, not particularly informative about chemical structure. This uniformity in the proton spin properties can be turned to advantage, however, in that the high-resolution ^{13}C CP/MAS spectrum can be used to probe domain structure in a sample. If ^{13}C signals arising from different molecules or structural moieties (e.g., lignin and cellulose) reveal different T_{1H} values, this would provide strong evidence that the molecules lie in separate spatial domains, with typical separations of at least several tens of Å.[17,31] The appearance of equal T_{1H} values within a given ^{13}C CP/MAS experiment is consistent with intimate mixing at the molecular level, though distinct domains can *coincidentally* have the same ^{1}H relaxation times.[10] The ^{13}C spectra for molecules in the two domains do not have to be totally resolved; all that is needed is a single resolved peak from each species.

An inversion-recovery T_{1H} segment in a CP experiment can be used to suppress all carbon peaks arising from one domain component, if the T_{1H} values are sufficiently different. In the traditional inversion-recovery mode, the ^{1}H magnetization decays from $-M_0$ to $+M_0$, crossing zero for a relaxation period of 0.693 T_{1H}. At this point the "nulled" protons will not provide ^{13}C intensity in an immediately following CP period, and that component will be suppressed in the ^{13}C spectrum.

As in the liquid state, T_1 is determined by fluctuating local magnetic fields with a frequency component at the Larmor frequency, which is normally tens of MHz to a few hundred MHz.[11] In the CP experiment, the spin-locked ^{1}H magnetization decays toward zero during the contact time; this decay follows a time constant $T_{1\rho H}$, "proton T_1 in the rotating frame." The observed ^{13}C intensity, therefore, grows with a time contant T_{CH} and, at sufficiently long contact time, decays with the time constant $T_{1\rho H}$, which is sensitive to fluctuations with characteristic frequencies on the order of the precession rate about the rf field, typically on the order of 50 kHz. As for the case of T_{1H}, efficient ^{1}H-^{1}H coupling forces all hydrogens to follow the same $T_{1\rho H}$ in an individual physical domain, so measurement of $T_{1\rho H}$ provides a second test for domain heterogeneity in a sample.[31] If two components have very different values of $T_{1\rho H}$, then a long contact time should allow the magnetization of the component with the faster $T_{1\rho H}$ to decay nearly to zero, leaving a subspectrum of the other component. This "filter" has allowed the detection of only carbohydrate and protein components (short T_{CH} and $T_{1\rho H}$) with a contact time of 100 μs and only the triglyceride fraction (long T_{CH} and $T_{1\rho H}$) at 20 ms contact time in whole soybeans.[10]

For quantitative assays based on ^{13}C CP/MAS, one must at least determine T_{CH}, $T_{1\rho H}$, and T_{1H} for each peak of interest. These parameters may, if one is fortunate, determine a contact time and repetition time that provide a quantitative spectrum or, more generally, provide correction factors for compensating the measured relative intensities for differences in spin dynamics. T_{CH} and $T_{1\rho H}$ can be determined from experiments in which the ^{13}C NMR intensity is measured as a function of contact time (see Chapter 9). When T_{CH} and $T_{1\rho H}$ have been determined, the following equation can be used to calculate the "corrected" ^{13}C intensity, $M_{C\infty}$, (the signal intensity which would be obtained if CP were infinitely rapid and $T_{1\rho H}$ decay were infinitely slow) from the observed magnetization at contact time, τ.[18]

$$M(\tau) = [(T_{1\rho H})/(T_{1\rho H} - T_{CH})][\exp(-\tau/T_{1\rho H}) - \exp(-\tau/T_{CH})]M_{C\infty} \tag{1}$$

V. ^{13}C SPIN DYNAMICS

In contrast to the case of ^1H, in which relaxation behavior is averaged out by ^1H-^1H flip-flops, the ^{13}C spins offer a view of *local* atomic motions. T_{1C} reports on motion at the ^{13}C resonance frequency (typically 15 to 75 MHz), and the low natural isotopic abundance and MAS ensure that ^{13}C-^{13}C spin diffusion is not effective at averaging relaxation behavior even across one molecule or major structural unit of a macromolecule. T_{1C} can be measured in the CP mode,[32] so signal averaging can proceed with a fairly rapid repetition rate, but the measurement of relatively long T_{1C} values still requires long relaxation periods. If a particular region of the spectrum has severely overlapping signals from carbons with different T_{1C} values, the usual inversion-recovery experiment can be used to provide a null signal for one component. Since each carbon site typically has a unique T_{1C}, this type of filter has to be applied to one peak at a time rather than to an entire sub-spectrum of a given molecule.

If T_{1C} is obtained in the inversion-recovery mode that results in an exponential decay to zero,[32,33] and if one component has much shorter T_{1C} values for all carbons, possibly as a result of increased mobility, then the subspectrum of that component could be suppressed by using an intermediate relaxation time that allows all fast-T_{1C} peaks to decay to zero, but leaves significant intensity in slow-T_{1C} peaks. This scheme has been used to suppress the contribution from the amorphous component in a combination of amorphous and crystalline cellulose.[34]

The appearance of biexponential T_{1C} decay of a single peak can be interpreted as evidence for the presence of two unresolved peaks with different T_{1C}. This interpretation is compromised in principle by the practical necessity of leaving the ^1H decoupler turned off during the relaxation period between the 180 and 90° pulses. The ^{13}C spins are then relaxing as a C-H coupled system, and the relaxation behavior can be biexponential even for a peak arising from a single structural site;[35] thus T_{1C} experiments in liquids are normally conducted with decoupling during relaxation periods.[11] This is not practical for solid-state ^{13}C experiments because of the larger T_{1C} values typical of solids and the much higher ^1H rf power levels employed for decoupling. In most cases, the ^{13}C magnetization does relax with a simple exponential time constant even in the absence of continuous ^1H decoupling, but biexponential behavior alone should not be considered conclusive evidence for a composite peak.

VI. CHEMICAL SHIFT ANISOTROPY

CSA can be measured directly for a single crystal with goniometer equipment,[36] or from the powder pattern in a static-sample experiment on a simple molecule,[2] or from the dominant powder pattern from a specifically enriched sample. An indirect method relies on an analysis of spinning sideband intensities;[5] these approaches are not generally applicable to plant materials because large single crystals are not available, the static ^{13}C spectrum is a complex superposition of many individual powder patterns (Figure 1), and the ^{13}C MAS spectra are sufficiently complex that spinning sidebands are not clearly distinguishable from the isotropic peaks. A relatively new experiment utilizes a two-dimensional FT approach to extract the CSA for each peak that can be resolved in the MAS spectrum.[37] The spinning axis is "flipped" between 90° and the magic angle during the time between the evolution and detection periods. During the evolution time period the ^{13}C spins evolve at $\beta = 90°$, yielding a CSA behavior scaled by $-1/2$. The spinning axis is then rotated to $\beta = 54.7°$ for acquiring the high-resolution MAS spectrum. The two-dimensional plot consists of the MAS spectrum along one axis and the powder patterns (scaled by $1/2$, i.e., reversed left-to-right and half as broad) along the other axis. The primary benefit of this experiment is that CSA provides three chemical shift parameters for each MAS peak instead of just one. A second advantage bears on the study of mobility, since molecular motion that averages CSA can be identified directly if one knows the unaveraged CSA pattern.

A third benefit is the opportunity to detect gross orientation of an atom-based coordinate system (the chemical shift tensor). If a bulk sample consists of small crystals that are partially oriented by the spinning, the resulting CSA "powder patterns" exhibit intensity distortions because not all orientations are present in random amounts; typically the CSA pattern has "dips" corresponding to the disfavored orientations with compensating "bumps" at the favored ones (Figure 4).[38] If, for example, the lignin or cellulose molecules in wood are significantly aligned with respect to the grain, the derived CSA patterns should show intensity distortions that depend on the orientation of the grain with respect to the spinning axis of the sample holder. The single publication of this application revealed no evidence of net molecular ordering over a 0.9 cm^3 volume of wood, as illustrated in Figure 5.[39] This does not rule out molecular ordering on the scale of individual cell walls, however.

A new variant on this approach has recently been demonstrated in which the spinning is interrupted for an evolution period under the full CSA, followed by data acquisition under MAS.[40] The product of this "stop-and-go" experiment is again a two-dimensional plot that provides a CSA pattern for each resolved peak in the MAS spectrum.

VII. CRAMPS

High-resolution solid-state ^1H spectra are best acquired with a multiple-pulse experiment that averages the ^1H-^1H dipolar coupling with a train of intense, short, carefully controled rf pulses during data aquisition. This multiple-pulse sequence is combined with MAS (to average ^1H CSA) in the combined rotation and multiple-pulse spectroscopy (CRAMPS) experiment[41] to provide ^1H spectra with peak widths as small as 0.2 ppm, similar to the limiting resolution in a ^{13}C MAS spectrum.[42] The ^1H δ_{iso} range is only about 15 ppm for typical organic samples, so there is greater opportunity for peak overlap, and adsorbed water is always a potential interference. The optimum sample size is only a few mg, compared with hundreds of mg for ^{13}C samples. A multiple-pulse version of the interrupted decoupling experiment[43] suppresses components with strong H-H dipolar coupling. Considerable work has been published on fossil fuels,[41,44,45] and a preliminary report on CRAMPS of humic materials has appeared,[46] but the spectra in Figure 6 are, to our knowledge, the first of plant materials.

The spectra exhibit differences that are reasonable in terms of established properties of the materials. The lodgepole pine cellulose spectrum provides the best resolution, with at least four peaks or shoulders in the normal carbohydrate range of 3.5 to 5.5 ppm. The lignin spectrum has significant intensity in the ^1H aromatic region of 6 to 7.5 ppm, along with decreased resolution in the aliphatic region. The spectrum of extractives-free lodgepole pine wood appears as expected for a reasonable combination of the two major components. A multiple-pulse interrupted decoupling experiment on the whole wood shows that all of the protons are tightly coupled, with all intensity almost completely dephased in a 15-μs interrupt time. The most surprising aspect of these spectra is that the samples dried for 12 h at room temperature under

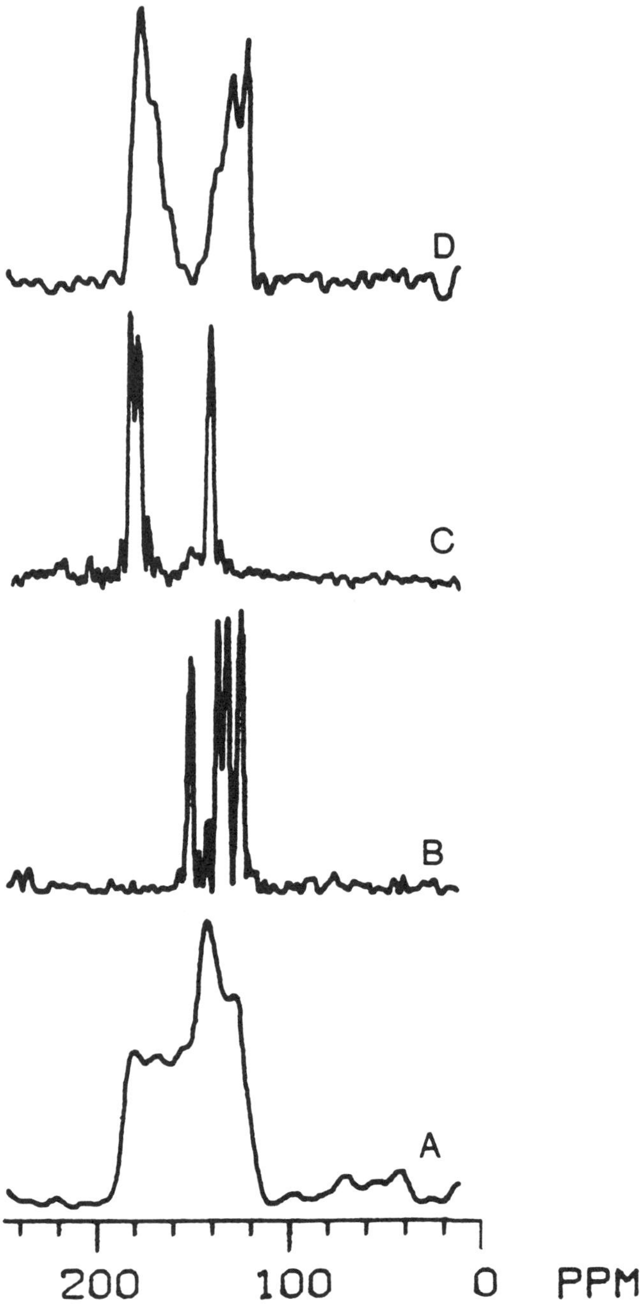

FIGURE 4. ^{13}C CSA patterns for the 149-ppm isotropic chemical shift of homovanillic alcohol, obtained with the "angle-flipping" experiment: (A) powdered sample; (B) single crystal at one orientation relative to the spinning axis; (C) the same single crystal at another orientation; and (D) collection of a few crystals. ^{13}C frequency is 25.3 MHz; spinning rate is 2.4 kHz. (From Hatfield, G. R., Sardashti, M., and Maciel, G. E., *Anal. Chem.*, 59, 1659, 1987. With permission.)

dynamic vacuum yield spectra virtually identical to those obtained from samples at equilibrium with water vapor in the atmosphere. Either the undried samples already have a very low water content, which is not likely, or the spin dynamics fortuitously suppress the water signal.

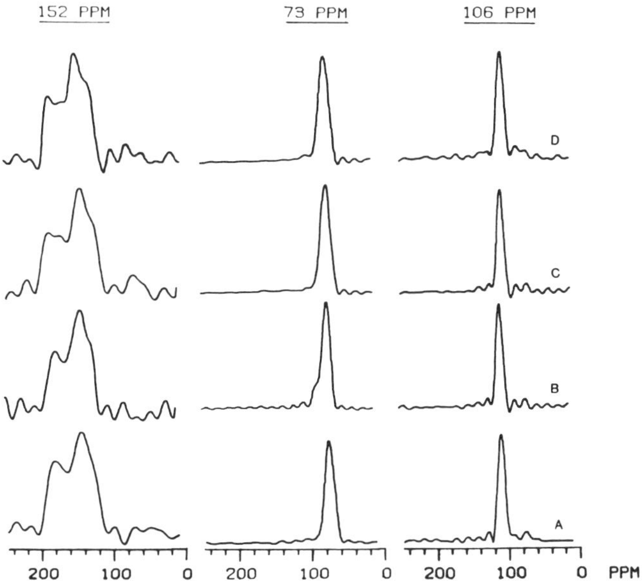

FIGURE 5. ^{13}C CSA patterns for three signals in the spectrum of Eucalyptus polyanthemus wood; conditions as in Figure 5: (A) grain collinear with spinning axis; (B) grain at 45°; (C) grain at 90°; and (D) powdered wood. (From Hatfield, G. R., Sardashti, M., and Maciel, G. E., *Anal. Chem.*, 59, 1659, 1987. With permission.)

VIII. IMAGING

NMR imaging techniques[47,48] are currently most extensively applied to medical research and diagnosis, but since the early development stages, plant materials have been favored subjects.[48,49] ^{1}H imaging normally detects fluid components such as water[48,50] or oil,[48] with solid components visualized by the absence of signal. Such studies of structure or water flow have included seeds,[48,51] fruits,[49,52] roots,[50,53,54] wood,[48,55,56] and stems.[51,57] Spatial resolution has recently been extended well below the the 1-mm level.[51,54,57]

Methods for direct imaging of the solid components are desirable but much more difficult, because the large NMR line widths from solids tend to obscure the effects of the magnetic field gradients on the spectrum, thereby significantly decreasing the spatial resolution. The typical short T_{2H} of a rigid material is an advantage in studies directed toward the fluid components of plants because the solid "background" signals from cellulose and lignin are easily suppressed. Some solid imaging has been done on simple compounds,[58] and more sophisticated narrowing techniques such as multiple-pulse decoupling and MAS with coherently rotating gradient fields[59,60] offer possibilities for direct solid images and possibly even spatially selective spectra. It is conceivable that all of the solid-state NMR techniques described in this chapter will be combined to effect resolution of components in space, time, and frequency domains to probe structure and dynamics nondestructively over the entire range from the atomic to the macroscopic scale.

ACKNOWLEDGMENT

The ^{1}H CRAMPS spectra were obtained at the Colorado State University Regional NMR Center, funded by National Science Foundation Grant No. CHE-8616437.

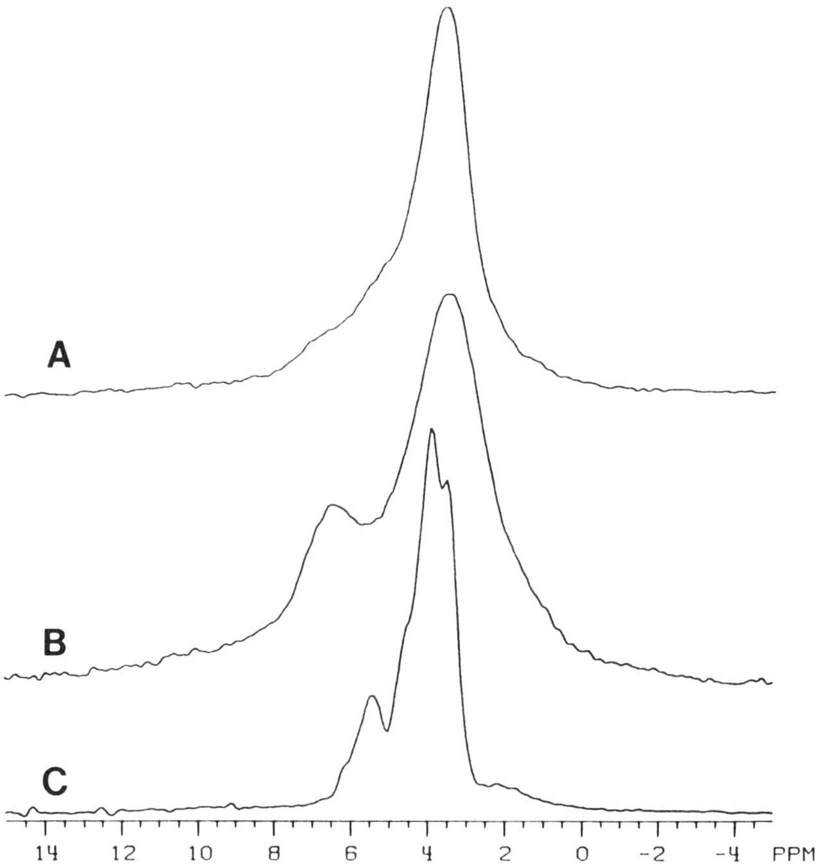

FIGURE 6. ^{1}H CRAMPS spectra of wood and its major constituents; ^{1}H frequency is 187 MHz, using the "BR-24" multiple-pulse sequence with 2 kHz MAS rate.[42] (A) lodgepole pine extractives-free wood; (B) hypochlorite lignin derived from (A); and (C) cellulose derived from (A).

REFERENCES

1. **Schaefer, J. and Stejskal, E. O.,** High-resolution ^{13}C NMR of solid polymers, in *Topics in C-13 NMR Spectroscopy*, Vol. 3, Levy, G. C., Ed., Wiley-Interscience, New York, 1979, 283.
2. **Facelli, J. C., Grant, D. M., and Michl, J.,** Carbon-13 shielding tensors: experimental and theoretical determination, *Acc. Chem. Res.*, 20, 152, 1987.
3. **Duncan, T. M.,** ^{13}C chemical shieldings in solids, *J. Phys. Chem. Ref. Data,* 16, 125, 1987.
4. **VanderHart, D. L., Earl, W. L., and Garroway, A. N.,** Resolution in ^{13}C NMR of organic solids using high power proton decoupling and magic-angle spinning, *J. Magn. Reson.*, 44, 361, 1981.
5. **Herzfeld, J. and Berger, A. E.,** Sideband intensities in NMR spectra of samples spinning at the magic angle, *J. Chem. Phys.*, 73, 6021, 1980.
6. **Dixon, W. T.,** Spinning-sideband-free NMR spectra, *J. Magn. Reson.*, 44, 220, 1981.
7. **Dixon, W. T., Schaefer, J., Sefcik, M. D., Stejskal, E. O., and McKay, R. A.,** Quantitative chemical composition of materials such as humic soils, lignins, and coals by high-resolution carbon-13 NMR, *J. Magn. Reson.*, 45, 173, 1981.
8. **Dixon, W. T.,** Spinning-sideband-free and spinning sideband-only NMR spectra in spinning samples, *J. Chem. Phys.*, 77, 1800, 1982.

9. **Dixon, W. T., Schaefer, J., Sefcik, M. D., Stejskal, E. O., and McKay, R. A.,** Total suppression of sidebands in CPMAS C-13 NMR, *J. Magn. Reson.*, 49, 341, 1982.
10. **Haw, J. F. and Maciel, G. E.,** Carbon-13 nuclear magnetic resonance spectrometry of oil seeds with cross polarization and magic-angle spinning, *Anal. Chem.*, 55, 1262, 1983.
11. **Lyerla, J. R., Jr. and Levy, G. C.,** Carbon-13 nuclear spin relaxation, in Topic in *Carbon-13 NMR Spectroscopy*, Vol. 1, Levy, G. C., Ed., John Wiley & Sons, New York, 1974, 79.
12. **Pines, A., Gibby, M. G., and Waugh, J. S.,** Proton-enhanced NMR of dilute spins in solids, *J. Chem. Phys.*, 59, 569, 1973.
13. **Hartmann, S. R. and Hahn, E. L.,** Nuclear double resonance in the rotating frame, *Phys. Rev.*, 128, 2042, 1962.
14. **Akutsu, H. and Kyogoku, Y.,** Phase transitions of phosphatidylcholine bilayers as revealed by cross-polarization efficiency of P-31 NMR, *Biochim. Biophys. Acta*, 774, 293, 1984.
15. **Frye, J., Albert, A. D., Selinsky, B. S., and Yeagle, P. L.,** Cross polarization P-31 nuclear magnetic resonance of phospholipids, *Biophys. J.*, 48, 547, 1985.
16. **Sullivan, M. J. and Maciel, G. E.,** Structural resolution in the carbon-13 nuclear magnetic resonance spectrometric analysis of coal by cross polarization and magic-angle spinning, *Anal. Chem.*, 54, 1606, 1982.
17. **Sullivan, M. J. and Maciel, G. E.,** Spin dynamics in the carbon-13 nuclear magnetic resonance spectrometric analysis of coal by cross polarization and magic-angle spinning, *Anal. Chem.*, 54, 1615, 1982.
18. **Hatfield, G. R., Maciel, G. E., Erbatur, O., and Erbatur, G.,** Qualitative and quantitative analysis of solid lignin samples by carbon-13 nuclear magnetic resonance spectrometry, *Anal. Chem.*, 59, 172, 1987.
19. **Opella, S. J. and Frey, M. H.,** Selection of nonprotonated carbon resonances in solid-state nuclear magnetic resonance, *J. Am. Chem. Soc.*, 101, 5854, 1979.
20. **Alemany, L. B., Grant, D. M., Alger, T. G., and Pugmire, R. J.,** Cross polarization and magic angle spining NMR spectra of model organic compounds. III. Effect of the ^{13}C-^{1}H dipolar interaction on cross polarization and carbon-proton dephasing, *J. Am. Chem. Soc.*, 105, 6697, 1983.
21. **O'Donnell, D. J., Ackerman, J. J. H., and Maciel, G. E.,** Comparative study of whole seed protein and starch content via cross polarization-magic angle spinning carbon-13 nuclear magnetic resonance spectroscopy, *J. Agric. Food Chem.*, 29, 514, 1981.
22. **Early, T. A., Haw, J. F., Bax, A., Maciel, G. E., and Puar, M. S.,** Carbon-13 nuclear magnetic resonance studies of a solid inclusion complex, *J. Phys. Chem.*, 88, 324, 1984.
23. **Wilson, M. A.,** Soil organic matter maps by nuclear magnetic resonance, *J. Soil Sci.*, 35, 209, 1984.
24. **Mehring, M.,** *High Resolution NMR in Solids*, 2nd ed., Springer-Verlag, Berlin, 1983,. chap. 4.
25. **Forbes, J., Husted, C., and Oldfield, E.,** High-field, high-resolution proton "magic-angle" sample-spinning nuclear magnetic resonance spectroscopic studies of gel and liquid crystalline lipid bilayers and the effects of cholesterol, *J. Am. Chem. Soc.*, 110, 1059, 1988.
26. **Schaefer, J., McKay, R. A., and Stejskal, E. O.,** Double-cross-polarization NMR of solids, *J. Magn. Reson.*, 34, 443, 1979.
27. **Schaefer, J., Stejskal, E. O., and McKay, R. A.,** Cross-polarization NMR of N-15 labeled soybeans, *Biochem. Biophys. Res. Commun.*, 88, 274, 1979.
28. **Schaefer, J., Skokut, T. A., Stejskal, E. O., McKay, R. A., and Varner, J. E.,** Asparagine amide metabolism in developing cotyledons of soybean, *Proc. Natl. Acad. Sci. U.S.A.*, 78, 5978, 1981.
29. **Schaefer, J., Skokut, T. A., Stejskal, E. O., McKay, R. A., and Varner, J. E.,** Estimation of protein turnover in soybean leaves using magic angle double cross-polarization nitrogen-15 nuclear magnetic resonance, *J. Biol. Chem.*, 256, 11574, 1981.
30. **Schaefer, J., Stejskal, E. O., Garbow, J. R., and McKay, R. A.,** Quantitative determination of the concentrations of ^{13}C-^{15}N chemical bonds by double cross-polarization NMR, *J. Magn. Reson.*, 59, 150, 1984.
31. **Gerasimowicz, W., Hicks, K. B., and Pfeffer, P. E.,** Evidence for the existence of associated lignin-carbohydrate polymers as revealed by carbon-13 solid-state NMR spectroscopy, *Macromolecules*, 17, 2957, 1984.
32. **Torchia, D. A.,** The measurement of proton-enhanced carbon-13 T_1 values by a method which suppresses artifacts, *J. Magn. Reson.*, 30, 613, 1978.
33. **Freeman, R. and Hill, H. D. W.,** Fourier transform study of NMR spin-lattice relaxation by "progressive saturation," *J. Chem. Phys.*, 54, 3367, 1971.
34. **Horii, F.,** The structure of cellulose as studied by CP/MAS ^{13}C NMR spectroscopy, in *Nuclear Magnetic Resonance in Agriculture*, Pfeffer, P. E. and Gerasimowicz, W. V., Eds., CRC Press, Boca Raton, FL, 1989, chap. 10.
35. **Alger, T. D., Freeman, R., and Grant, D. M.,** Carbon-13 T_1 measurements under proton coupled and decoupled conditions, *J. Chem. Phys.*, 57, 2168, 1972.
36. **Veeman, W. S.,** ^{13}C chemical shift tensors in organic single crystals, *Philos. Trans. R. Soc., London Ser. A*, 299, 629, 1981.

37. **Bax, A., Szeverenyi, N. M., and Maciel, G. E.,** Chemical shift anisotropy in powdered solids studied by 2D FT NMR with flipping of the spinning axis, *J. Magn. Reson.*, 55, 494, 1983.
38. **Maciel, G. E., Szeverenyi, N. M., and Sardashti, M.,** Chemical-shift-anisotropy powder patterns by the two-dimensional angle-flipping approach. Effects of crystallite packing, *J. Magn. Reson.*, 64, 365, 1985.
39. **Hatfield, G. R., Sardashti, M., and Maciel, G. E.,** Analysis of molecular orientational order in solid samples by nuclear magnetic resonance: application to lignin and cellulose in wood, *Anal. Chem.*, 59, 1659, 1987.
40. **Zeigler, R. C., Wind, R. A., and Maciel, G. E.,** The stop-and-go spinning technique, *J. Magn. Reson.*, 79, 299, 1988.
41. **Gerstein, B. C.,** High-resolution n.m.r. in solids with strong homonuclear dipolar broadening: combined multiple-pulse decoupling and magic angle spinning, *Philos. Trans. R. Soc. London Ser. A*, 299, 521, 1981.
42. **Bronnimann, C. E., Hawkins, B. L., Zhang, M., and Maciel, G. E.,** Combined rotation and multiple pulse spectroscopy as an analytical proton nuclear magnetic resonance technique for solids, *Anal. Chem.*, 60, 1743, 1988.
43. **Bronnimann, C. E., Zeigler, R. C., and Maciel, G. E.,** Proton NMR study of the dehydration of the silica gel surface, *J. Am. Chem. Soc.*, 110, 2023, 1988.
44. **Rosenberger, H. and Scheler, G.,** Charakterisierung einiger Braunkohlen der DDR mit Hilfe der hochauflösenden magnetischen Protonenresonanzspektroskopie in Festkörper, *Z. Chem.*, 23, 34, 1983.
45. **Bronnimann, C. E. and Maciel, G. E.,** ^{1}H NMR spectra of coals of various ranks, obtained by the CRAMPS technique, *Org. Geochem.*, in press.
46. **Frye, J. S., Bronnimann, C. E., and Maciel, G. E.,** Solid-state NMR of humic materials, in *NMR of Humic Substances and Coal,* Wershaw, R. L. and Mikita, M. A., Eds., Lewis Publishers, Chelsea, MI, 1987.
47. **Lauterbur, P. C.,** Image formation by induced local interactions: examples employing nuclear magnetic resonance, *Nature*, 242, 190, 1973.
48. **Lauterbur, P. C.,** Magnetic resonance zeugmatography, *Pure Appl. Chem.*, 40, 149, 1974.
49. **Hinshaw, W. S., Bottomley, P. A., and Holland, G. N.,** A demonstration of the resolution of NMR imaging in biological systems, *Experientia*, 35, 1268, 1979.
50. **Rogers, H. H. and Bottomley, P. A.,** *In situ* nuclear magnetic resonance imaging of roots: influence of soil type, ferromagnetic particle content, and soil water, *Agron. J.*, 79, 957, 1987.
51. **Connelly, A., Lohman, J. A. B., Loughman, B. C., Quiquampoix, H., and Ratcliffe, R. G.,** High resolution imaging of plant tissues by NMR, *J. Exp. Bot.*, 38, 1713, 1987.
52. **van As, H. and Schaafsma, T. J.,** Noninvasive measurement of plant water flow by nuclear magnetic resonance, *Biophys. J.*, 45, 469, 1984.
53. **Bottomley, P. A., Rogers, H. H., and Foster, T. H.,** NMR imaging shows water distribution and transport in plant root systems *in situ*, *Proc. Natl. Acad. Sci. U.S.A.*, 83, 87, 1986.
54. **Brown, J. M., Johnson, G. A., and Kramer, P. J.,** *In vivo* magnetic resonance microscopy of changing water content in *Pelargonium hortorum* roots, *Plant Physiol.*, 82, 1158, 1986.
55. **Wang, P. C. and Chang, S. J.,** Nuclear magnetic resonance imaging of wood, *Wood Fiber Sci.*, 18, 308, 1986.
56. **Hall, L. D. and Rajanayagam, V.,** Evaluation of the distribution of water in wood by use of three dimensional proton NMR volume imaging, *Wood Sci. Technol.*, 20, 329, 1986.
57. **Johnson, G. A., Brown, J., and Kramer, P. J.,** Magnetic resonance microscopy of changes in water content in stems of transpiring plants, *Proc. Natl. Acad. Sci. U.S.A.*, 84, 2752, 1987.
58. **Szeverenyi, N. M. and Maciel, G. E.,** NMR spin imaging of magnetically dilute nuclei in the solid state, *J. Magn. Reson.*, 60, 460, 1984.
59. **Wind, R.A. and Yannoni, C. S.,** U.S. Patent 4,301,410, November 17, 1981.
60. **Cory, D. G., van Os, J. W. M., and Veeman, W. S.,** NMR images of rotating solids, *J. Magn. Reson.*, 76, 543, 1988.

NMR Imaging

Chapter 15

MAGNETIC RESONANCE IMAGING IN AGRICULTURAL RESEARCH

George Gassner

TABLE OF CONTENTS

I. INTRODUCTION

Magnetic resonance imaging (MRI) offers the biologists working in agriculture a technique that is non-invasive and non-destructive. Images obtained from magnetic resonance properties of nuclei provide the biologist with a tool for the *in vivo* analysis of the dynamics of growth, flow, and movement in developing and modulating systems. MRI is useful in evaluating normal events, experimental perturbations, diseases, symbiosis, and parasitism. When combined with localized spectroscopy,[1] MRI provides a unique experimental approach because volumes of 125 nl can be analyzed spectroscopically. These volumes can be selected by superimposing a square cursor representing a cubic volume onto a cathode ray tube (CRT) video screen projection of an MRI. At the end of the experiment, the organism subjected to magnetic resonance (MR) and MRI remains intact in contrast to *in vitro* analysis.

Knowledge of the fundamentals of magnetic resonance is critical to the productive use of MRI technology for biological research. The novice MR biologist accustomed to microscopic images produced by light and electron scattering and absorbance characteristics needs to understand that the MRI "morphology" relates to relaxation times.

The utility of MR as a tool for studying biological systems is well founded.[2,3] A general introduction to high resolution MR methods is presented in Chapter 1. Beyond this text, explanations of the magnetic resonance phenomenon are available to (1) satisfy an introductory understanding,[4-8] (2) fulfill the need for a more complex grasp of the concepts that describe MRI and spectroscopy,[9-12] and (3) prepare for hands-on-operation of MR equipment.[13,14]

An understanding of the following features of MR should give the biologist sufficient information to appreciate studies cited in this chapter. Chemical nuclei (^{1}H, ^{10}B, ^{13}C, ^{14}N, ^{19}F, ^{31}P, etc.) possess magnetic moments and exhibit alignment when placed in a magnetic field. The most abundant nuclei in biological tissue with a magnetic moment are protons. These nuclei provide strong signals while naturally occurring nuclei of oxygen and carbon give relatively weak resonances. Several other nuclei of biological interest require stronger magnetic fields which could be perilous to health, damaging to tissue through radio frequency (RF), heating, and distortion of signal by coil loading from the tissue.

The MR process starts after short bursts of RF displace the nuclei in a magnetic field to higher energy states. Following the cessation of the RF, the nuclei realign while emitting characteristic energies which are detected by a RF receiver. The nature and site of the nuclei and their chemical and physical dynamics can be determined after the reduction of the RF data is received. These features are central to the production of an MR image.

Lauterbur demonstrated the capacity of an MR spectrometer to provide an image by directing a linear field gradient along a coordinate of the spectrometer magnetic field.[10] Any two or more volumes of relevant nuclei along the altered magnetic field gradient experience different net magnetic field strengths. The consequence of nuclei realignment following displacement by an RF in the presence of a linear field gradient is the production of two or more frequencies; the frequencies are proportional to the magnetic field. The relative spatial position of two volumes is visualized after the superimposition of the back projections of the frequency profiles that are generated following three independent linear magnetic gradients.

The above brief description of an MR imaging process, projection-reconstruction technique, is intended to give a rudimentary concept to the reader. The projection-reconstruction technique is relatively simple and algorithms for image reconstructions are readily available. However, as an *in vivo* imaging technique, it has the disadvantages of motion artifacts and field inhomogeneities that result in image distortion. A pulse and gradient scheme for a more sophisticated MRI experiment is shown in Figure 1.[11] The reader is referred to the literature on Fourier transformation imaging technique to learn about many modifications leading to data acquisition for the production of MR images.[9,12,15-17]

The primary objective of this chapter is to give an overview of MRI applications that relate

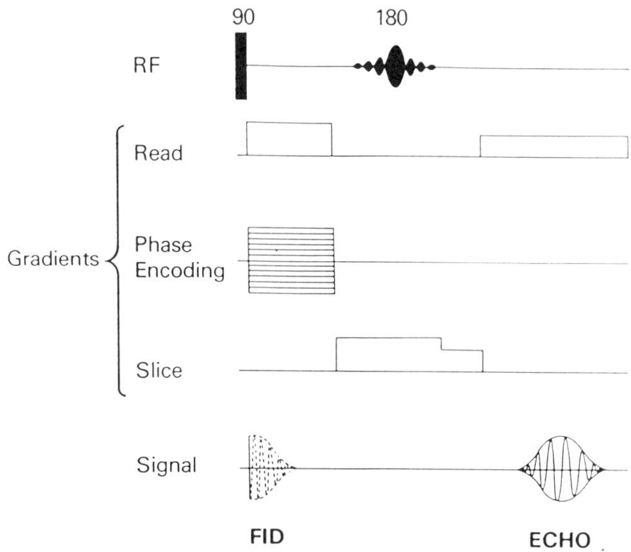

FIGURE 1. Pulse and gradient scheme used to acquire images similar to those seen in Figure 6. A 90° pulse excites a signal, followed by a 180° pulse to refocus the free induction decay (FID) and subsequent detection of a spin echo encoded using selective RF pulses and three orthogonal field gradients. The spin echo data matrix measured as a function of incremented phase encoding gradients, was Fourier transformed to produce an MRI.[11]

to agricultural research. Three categories of MRI *in vivo* studies are reviewed: animals, plants, and insects. Now, that MRI technology has matured to a level where proven instrumentation is available, a wealth of MRI experimentation has been published giving agricultural scientists unique opportunities to use MRI in the pursuit of elusive research problems. Animal scientists can benefit from techniques and data accumulated in a wide variety of vertebrates. Published MRI data cover topics including: tissue injury, edema, necrosis, neoplasms, and blood flow. Plant scientists can utilize the same technology to study normal, experimental, and diseased plants. Because of recent developments, entomologists can now use MRI of small scale subjects. The following review will consider the use of MRI to study morphology and physiology in animals, plants, and insects.

II. ANIMAL MRI

MRI studies of animals (vertebrates) have been directed for the most part toward advancing medicine. Numerous references to and several reviews about human MRI are available. For the period from January 1985 to September 1987, the National Library of Medicine Literature Search on MRI (U.S. Government Document #HE20.314/2:87-11) gave 959 citations. Much of this data can provide important information for agricultural researchers. Building on this foundation, scientists can use MRI to investigate problems of nutrition, tissue and whole body composition, genetic selection, naturally occurring organ abnormalities, embryonic development, glandular function, blood flow, and morphological and biochemical changes related to xenobiotics and as they relate to the advancement of agriculture.

This review will focus primarily on nonhuman MRI. Aside from humans, sheep, rats, rabbits, dogs, cats, mice, monkeys, and chick embryos have served as subjects for MRI. These animals and the MRI data derived from them demonstrated the utility of MRI as a noninvasive tool for use by medical practitioners and researchers. The animal studies clearly show that agricultural researchers, too, can study their subjects alive and intact.

MRI offers a unique perspective for researching animal science problems dealing with metabolism, nutrition, physiology, development, and genetics. Multinuclear techniques, image enhancement, and new imaging operational procedures in the following examples demonstrate that MRI is well established as a technology which can be readily adapted to agricultural research.

One approach for MRI studies of metabolism involves image enhancement. When image enhancement reagents are injected into the organism, then the consequences of changing from a noninvasive to an invasive technique must be considered. Toxic effects may alter the physiology or prevent repetition of an experiment on the same subject. Chemical shifts may distort the MRI so that it does not correspond to the MRI of a noninjected specimen. These disadvantages do not preclude the usefulness of enhancement techniques as indicated by the following examples. A chemical shift reagent (dysprosium) is useful for distinguishing between intracellular and extracellular components. This procedure was applied in conjunction with ^{23}Na MRI to provide evidence for a pumplike system and a sodium gradient within the bovine lens.[18] The artificial blood substrate, perflourocarbon (PFC), enhances visualization of the vascular systems of the rat in imaging both flourine and proton signals.[19,20] Signals from ^{19}F MR provides a means of monitoring the rate of blood pool and extramuscular space dilution.[21] The interpretations of *in vivo* MRI of PFC in heart, lung, liver, and spleen of the rat are corroborated by post-mortem studies.

Three additional features of ^{19}F MRI are noteworthy. A diffuse ^{19}F MRI signal indicates tissue perfusion.[22] Nontoxic levels of ^{19}F also appear useful in MRI visualization of the rat liver, large bowel, and stomach.[23] The lungs of rodents show a signal strength enhancement of 90% when PFC compounds are breathed as neat liquids using pure oxygen.[24]

Several contrast agents have been examined with respect to pharmocology, toxicity, and pharmakinetic parameters.[25] Gadolinium-diethylenetriamine-pentaacetic acid [(Gd)DTPA] enhances the whole body MRI of rats and organ systems, such as dog heart MRI, at levels that are nontoxic (Gd)DTPA.[26] Both (Gd)DTPA and iron(III)-ethylenebis-(2-hydroxyphenyl-glycine)$^-$ [Fe (FHPG)$^-$] enhances hepatobiliary system MRI.[27,28] Although Fe (FHPG)$^-$ shows some promise, it requires more work. Macrocyclic compounds containing copper and manganese are effective at one tenth the dose used for (Gd)DTPA.[29,30] Nitroxide stable free radical (NSFR) appears useful as a contrast agent for studying the dog central nervous system.[31] The low toxicity, readily available supply, and sufficient sensitivity combined with an efficient quadrapole relaxation mechanism give D_2O potential as an exogenous MR spin label.[32] Manipulating the physical parameters of the data collected also improve MR images. Studies on the normal and ischemic myocardium show that T_2 weighted precontrast images provide equivalent or better images than T_1 weighted postcontrast images.[33-35] Several references to paramagnetic ions and their complexes as contrast agents in MR imaging have been reviewed.[36]

Animal metabolism studies using MRI benefit from multinuclear techniques which use a nucleus (e.g., ^{1}H) to visualize the structure of interest and record spectra of other nuclei (e.g., ^{13}P, ^{13}C, ^{23}Na) associated with a particular aspect of metabolism. Multinuclear techniques provide comparisons among images of protons, sodium, and phosphorous in the same animal.[37,38] Simultaneous signals generated from various phosphorous groups displayed images from the spectral contributions of each group. Time required for sodium imaging compares favorably with the time needed to generate a proton image.[39] Metabolites labeled with ^{13}C and subjected to a decoupling process, i.e., irradiation with an RF magnetic field matching proton frequency and receiving ^{13}C signals, give signals similar in strength to ^{31}P. Natural abundance ^{13}C signals from magnetic resonance spectroscopy (MRS) clearly detect *in vivo* fats and glycogen tissue excised from living rabbits.[40] Future combination of ^{13}C MRS with MRI will permit more effective studies of animal body fat than have been achieved previously using proton MR only.[41] The combined MRS/MRI method will make it possible to visualize body fat deposits and simultaneously determine the size of the total unsaturated pool, the degree of polyunsaturation,

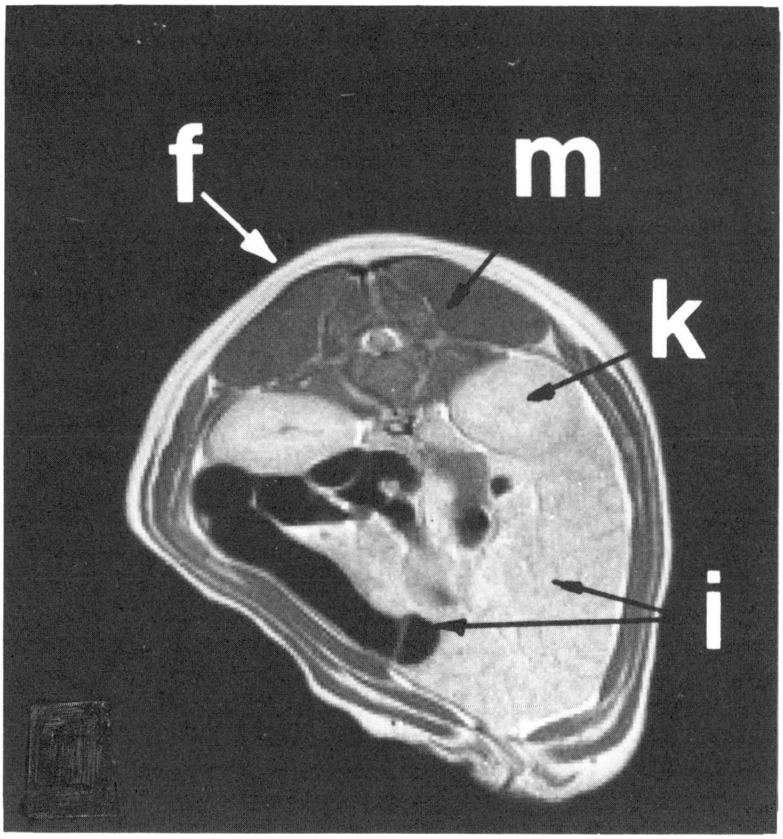

FIGURE 2. MRI of a pig abdomen from a 10-mm thick cross-sectional slice using a Seimens Magnetom whole body imager operating at 0.5 T (21 MHz). Note details of soft tissues: f, fat; k, kidney; i, intestine; m, muscle. (Courtesy of A. D. Mitchell and P. Wang.)

and measure the amount of body fat present. MRI of events in the storage and metabolism of glycogen and other ^{13}C metabolites may also be combined with MRS in future biochemical studies.

Anatomical investigation using MRI makes it possible to see tissues that would be transparent to X-rays and opaque to the emissions of other wavelengths of the electromagnetic spectrum. A whole body MR imaging instrument can be used to produce images of a pig carcass from which the amount and distribution of fat in the body could be determined (Figure 2).[42] By using different pulse excitation schemes, images can be selectively produced from water or lipid protons.[43] Alternatively, overall proton spectra of the pig body can be collected. From these spectra, the relative water and lipid proton peak heights provide a means for estimating the water or lipid content of pork tissue (Figures 3 and 4). Similar data on water and lipid distribution can be collected using an iron core magnet modified for small animal imaging.[44]

Anatomical features of the human finger,[45-47] whole body cross-section, and sections of the coronal plane of the rabbit head, although of low resolution at first,[48] demonstrated that MR had practical potential for anatomical studies. Subsequently, high-resolution MR microscopy images of the developing chick embryo were correlated with light microscopy histological studies (Figure 5).[49] Pathological conditions, such as tumors,[50] liver disease,[51] renal abscess,[52] hyperthyroid conditions,[53-55] and lung edema are amenable to MRI diagnosis.

MRI of paramagnetic ion labels provides a means for evaluating pathological conditions the destruction of the blood-brain barrier,[56-58] and the effects of therapeutic drugs on tumors.[59] MR

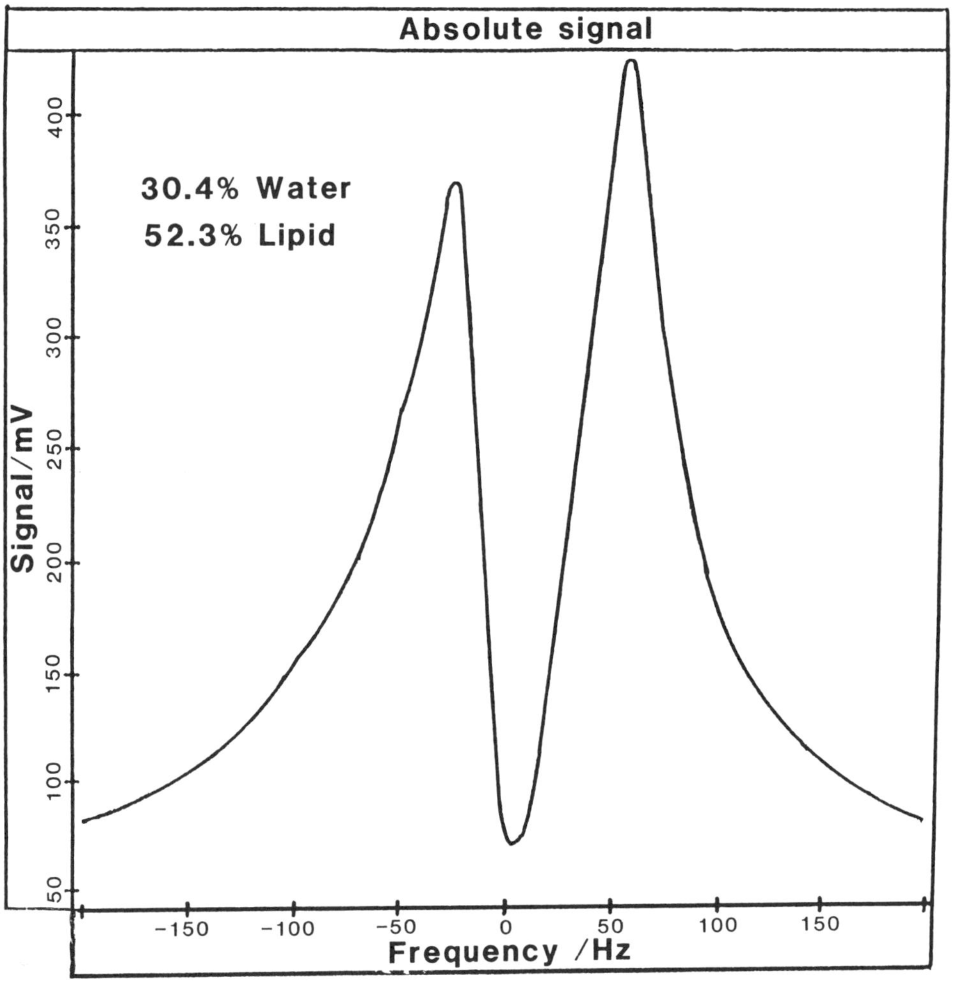

FIGURE 3. Proton spectra from same spectrometer used for Figure 2 MRI. Note the water proton peak has a greater magnitude than the lipid proton peak.

images of radiation induced cerebral injury correlate with cerebral gross and microscopic changes.[60] These examples clearly show that diagnosis and response to treatments can be visualized noninvasively by the pathologist and veterinarian.

Advances in the technology make it possible to utilize MRI instrumentation as a serious investigative tool. Modification of a 1.5-T MRI unit gave 1-mm sliced proton images in several small animals with 50×50 μm picture elements (pixels) in an image.[61] The image data allow anatomical evaluation of brain ventricals, adrenal glands, and microvascular structure in the kidneys and liver in small animals. The determination of dog left ventricular myocardial mass and volume demonstrates the utility of MRI to examine the substructure of vital organs.[62-64]

Monoclonal antibodies for tagging specific targets within an organism combined with appropriate chemical shift reagents for MRI enhancement would be an invaluable advancement for MRI technology. Some applications using antibodies to rat synthetic polymers, such as peptides, proteins, and carbohydrates linked to DTPA have been considered.[65] However, (Gd)DTPA linked to carcinoembryonic antigen gave no enhancement of human colon carcinoma implants in hamsters.[66] Other monoclonal antibody (Gd)DTPA experiments have met with limited success. Possibly, the gamma-camera or positron imaging would be more useful

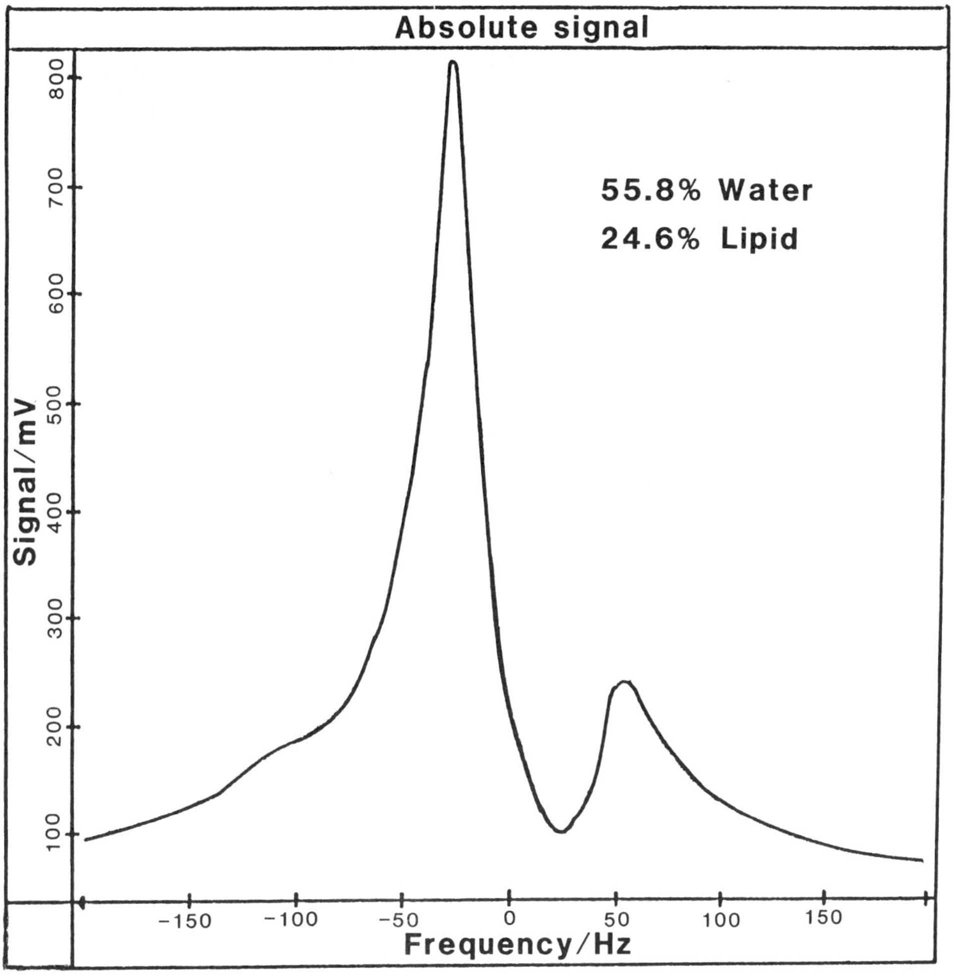

FIGURE 4. Proton spectra from same spectrometer used for Figure 2 MRI. Note the water proton peak has a lesser magnitude than the lipid proton peak. (Courtesy of A. D. Mitchell and P. Wang.)

than MRI in the detection of monoclonal antibodies target cells and tissues *in vivo*.[67] However, antimyosin monoclonal antibody phosphorous MR images show promise in the diagnosis of tissue transplant rejections.[68] The need to stop action for visualization of cardiovascular phenomena and the need to follow sequential events of biological growth and function led to the development of "gated" and "movie" MRI data collection procedures.[69] Gating procedures for MRI generally use the input from the electrocardiographic (EKG) "QRS complex" events, with a delay that coincides with either the systolic or diastolic responses, to activate MR pulse sequences.[70-73] The gating technique, combined with velocity dependent phase contrast and an image subtraction procedure, enhances all vessels that carry pulsatile flow and eliminates static structures. This method provides MR images of the major arterial vessels but not a measure of flow. However, MRI can be used to measure blood volume crossing each voxel based on using small phase shifts and the analysis of the complete image data as complex numbers.[70] Gated MRI can be useful also in proton imaging of the heart and accurately determining the left ventricular mass or ^{23}Na imaging of an isolated perfused working heart.[74-76] Combined proton MRI and ^{31}P MR spectroscopy gives a visualization and measure of phosphate fluxes related to sugar phosphates, creatine phosphates, α- and β-ATP, and inorganic phosphate in heart metabolism.[72]

MR "movies" are more significant than the term movie might imply. As a MR procedure, it

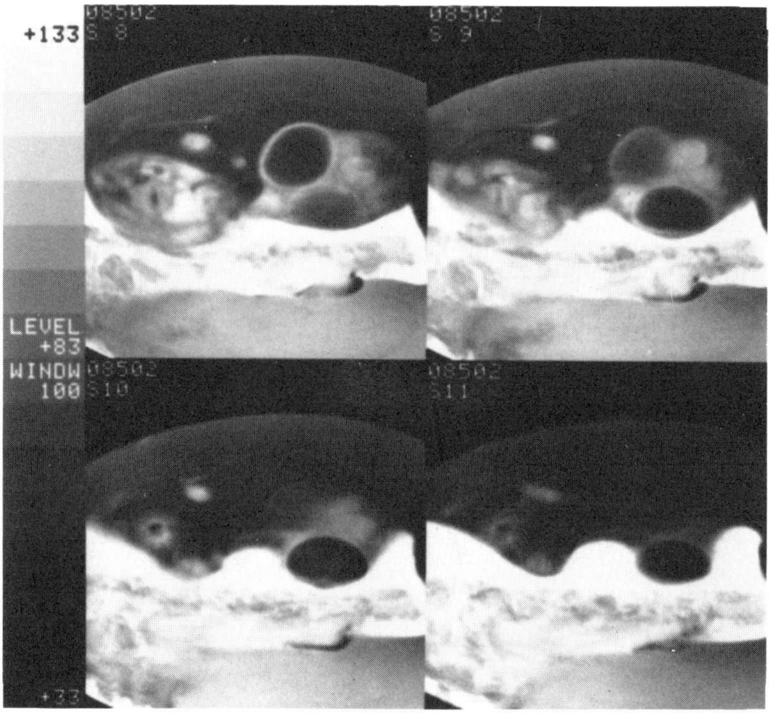

FIGURE 5. MRI was performed on live chick embryos. The signal-to-noise ratio was sufficient to allow unequivocal indentification of the individual chambers of the heart, spinal cord, ventricles in the brain, and vascular structures in the liver of a live 11 d embryo. The 4 slices were selected from a set of 16 1.25-mm thick contiguous slices with in-plane resolution of 50×50 μm.[49]

permits real time imaging of the dynamic configuration of a functioning organ or other anatomic structures *in vivo*. The technique is based on the echo-planar method which permits data collection over milliseconds instead of seconds.[46,50] Ordidge and his colleagues used echo planar MR gated to EKG events to produce the first MR movies, a six-frame loop demonstrating heart motion in arbitrary phases.[69] The MR movie procedure continues to be improved to the point where chemical shifts,[77] diffusion, and flow are accessible using MRI. It is now possible to measure the time course of brain and kidney physiological functions with a time frame of 1 s. Nongated MRI of brain tissue, thorax, and abdomen at end of diastole and end of expiratory rest are possible using a fast low-angle shot (FLASH) technique.[78-80] The FLASH procedure does not refocus dephased transverse magnetizations, and motions with longer time constants are averaged out. FLASH and FAST (fast angle shot tomography) should prove useful for many biochemical MRI experiments where the duration of a physiological event is brief.[81,82]

Interpretation of MRI data calls for extreme caution, especially following invasive techniques such as surgery and medication. High signal abnormality may relate to a postoperative condition.[83] In the case of allograft imaging, T_2 intensity differences may give an indication of cardiac rejection.[84,85] On the other hand, differences in T_2 intensities may not reliably indicate a renal allograft rejection.[86,87]

Exposure to MRI attenuates fentanyl-induced analgesia in mice.[88] During morphine-induced anesthesia, MRI abolishes nocturnal induced analgesia and attenuates the morphine response in the day time.[89,90] Also, when animals are anesthetized and tracheotimized, MRI spectrographs show reduced glycogen while storage hepatic glycogen signals are seen after the administration of α- and β- adenergic blockers.[40] The effects of diet and anatomical structure can also mislead. The relaxation times of skeletal muscle are dependent on diet and muscle fiber type and the T_1

and T_2 of liver and gallbladder bile varies with fasting conditions.[91-93] Additional research is required to understand these perturbations.

Presently, MRI appears to be a safe procedure for animal diagnostics and research. The biological basis of limits on exposure to static, time-varying (excluding RF fields), and RF magnetic field effects have been examined under the auspices of the National Radiological Protection Board (Britain, U.K.) and guidelines for MRI have been documented.[94] Also, in 1982, the Food and Drug Administration Bureau of Radiological Health (now called Center of Devices and Radiology Health) prepared a document entitled, Guidelines for Evaluation Electromagnetic Exposure Risks for Trials of Clinical NMR Systems. These guidelines set the operational limitations to 2.0 T or less. Presently, guidelines for a premarket notification document number 510 (k) for magnetic resonance diagnostic device application review are being prepared for medical applications.[95,96] As noted above, certain anesthesia procedures complicate MR interpretations, therefore, the MR spectroscopist must be cautioned that MRI may be mechanically noninvasive, but it can physically and chemically interact with the tissues being studied. Minute amounts of ferromagnetic material and some nonferromagnetic metals may interfere with MRI and displacement of these materials in the magnetic field are potential biological hazards.[97]

Heat is another potential biological hazard, especially when small metal implants in the specimen interact with changing magnetic and RF fields.[98] Heating caused by RF deposition at 1.9 T using a 80-mm diameter surface coil decoupler tuned to 80.9 MHz, results from 82% of the applied power being deposited as heat as indicated when calorimetric measurements are made on a phantom sample that approximates a human limb.[99] The heat generated when pregnant mice are exposed to 3.5 to 12 kT/s with pulse periods ranging over 0.33 to 0.56 ms at various times during gestation gave no indication of any adverse effect on the litter numbers and growth rates;[100] nor did MRI have any effect on the blood chemistry of sheep although slight elevations in body temperatures occurred.[101]

It appears that rats do not suffer debilitating effects from MRI. No significant changes were determined for survival, body growth, organ weights, stress related to male hormone levels,[102] and effects on memory;[103] nor were significant short or long-term changes found. Rats exposed to MRI equivalent magnetic and RF fields for nearly 1/2 h display no changes in their behavior when tested immediately and 3 months after the exposure.[104] No pathology was indicated 2 years after rat eye exposure to a 2.7-T field and a 29-MHz RF pulse at 800-ms intervals.[105]

III. PLANT MRI

The relatively slow mobile proton distribution in plants made their early MR images a possibility.[45] These images and proton images of fruit, nuts, and stems gave a clear indication that plant specimens surrounded by an ambient atmosphere make good subjects for MRI.[106,107] Changing water content in roots, stems, and seed, and the degree of water bound to them are readily observed using MRI techniques (Figure 6).[108-111] For large plants, like trees, MR images clearly shows growth rings and knots. Kiln-dried wood soaked in water show similar features. Other characteristics associated with tree anatomy require more research.[112,113]

The observation that plant parts buried in various types of soils and artificial media were also candidates for MRI was an unexpected surprise (Figures 7 and 8).[111,114-116] Subsequently, several naturally occurring soil media were identified as being suitable for root MRI investigations.[117] It was anticipated that the magnetic qualities of the various plant root support matrices would distort the MRI, rendering the data useless. However, the data recorded is of a useful quality, although the magnetic properties of the soils and soil substitutes yield varying MR images. These studies showed that patterns in root extent and distribution and detection of decay could be visualized. This finding led to the suggestion that MRI could be useful in studying plant injury involving disease, herbicides, pollutants, and physical stress. The ability to image fungi in plant vascular systems could be of considerable importance to the plant pathologists.[118]

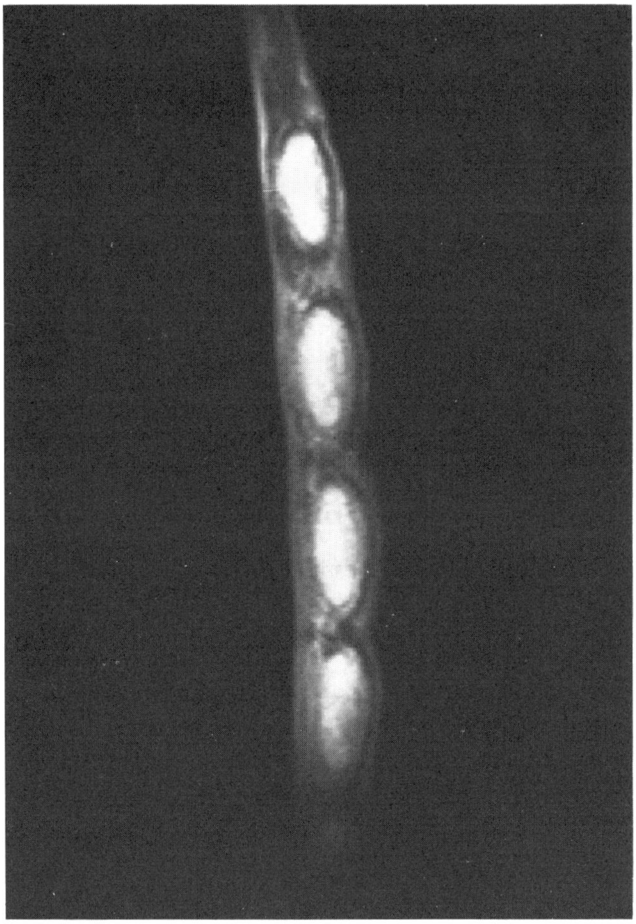

FIGURE 6. MRI of developing seeds within a pod from a green bean. A puzzling problem challenging plant physiologists is determining changes in seed water content as they dehydrate within the fleshy environment of vegetables and fruits. NMR imaging can determine changes in water content of developing fruits and vegetables while they remain attached to the plant. (Courtesy of J. Brown.)

The power and versatility of using *in vivo* MRS for studying plant metabolism is the subject of an earlier review.[119] The earlier review discussed the utility of observing and quantifying mobile metabolites, studying compartmentalization, tracing metabolites, and measuring rates of ATP synthesis. Many of these *in vivo* MRS studies could be enhanced when combined with MRI. A more recent report shows the influence of spin parameters on plant MRI contrast and evaluates the usefulness of determining the physical and chemical states of plants from MR images.[120] Advances in plant MRI suggest that the combined visual, chemical, and physical analysis has the potential to solve agricultural problems.

The first step in visualizing mobile metabolites is to characterize the medium which supports them. In plants water is the medium most often associated in moving metabolites. MRI can show changes in root cortices water content during transpiration (Figure 9). The changes in water content in roots of seedlings has been imaged with a resolution of 2 mm and when doped with Cu_2SO_4 a resolution of 0.6 mm.[114,116] Introduction of an MRI slice definition of 1.25 mm achieved spatial resolution of 0.05 mm.[121,122] The replacement of tissue water by D_2O from data collected from a matrix of 64×64 points and can be visualized with a pixel resolution of about 30×30 μm, (Figure 10).[123] In another plant study, an improvement of in-plane resolution (50

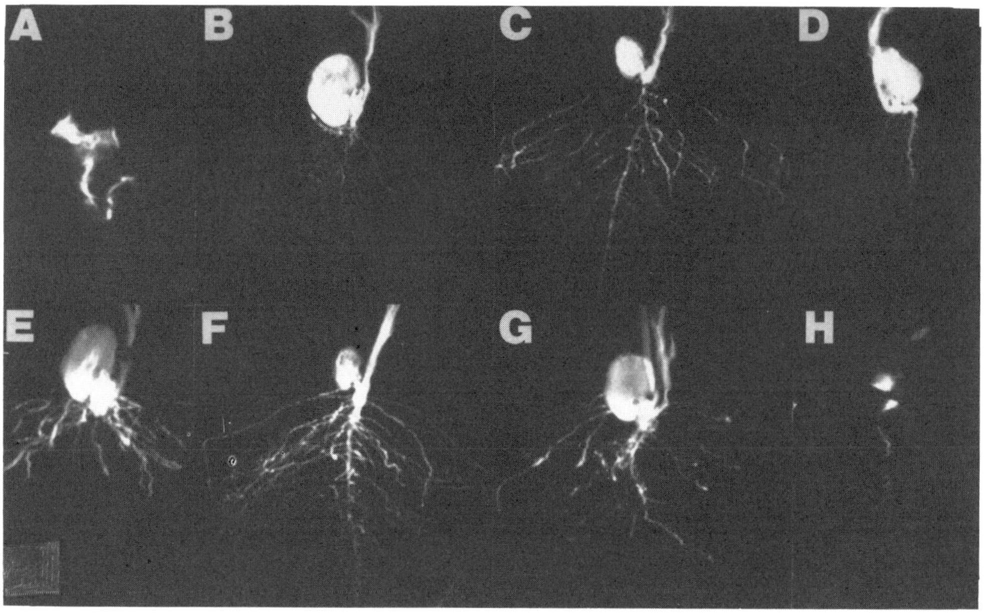

FIGURE 7. Proton MR images of *Vicia faba* L. in various soil types in plastic pots 30 d after germination using a General Electric spectrometer at 1.5 T (64 MHz), 1-m bore magnet with a 140-mm diameter MR receiver coil. Cahaba sandy loam (A), Wynnville fine sandy loam (B), Lucy loamy sand (C), Lakeland sand (E), Kinston loamy sand (F), Blanton loamy sand (G), and Houston clay (H). Soil water contents are 9, 24, 80, 14, 21, 33, 43, and 193 g/Kg, respectively. Hypogeal cotyledons appear as bright ovals in B through G; stems ascend and root systems descend. Soil-air interfaces occur just above the cotyledons. (Courtesy of H. H. Rogers and P. A. Bottomley.)

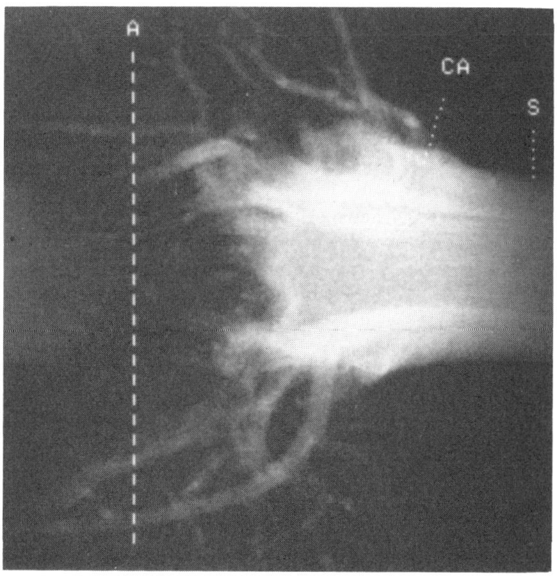

FIGURE 8. MRI of a geranium stem cutting rooted in a synthetic phenolic foam substrate growing undisturbed within a container. Tissues within the stem (S) are clearly seen. Adventitious roots originate from callus tissue (CA) and are nondestructively imaged growing in the foam. (Courtesy of J. Brown.)

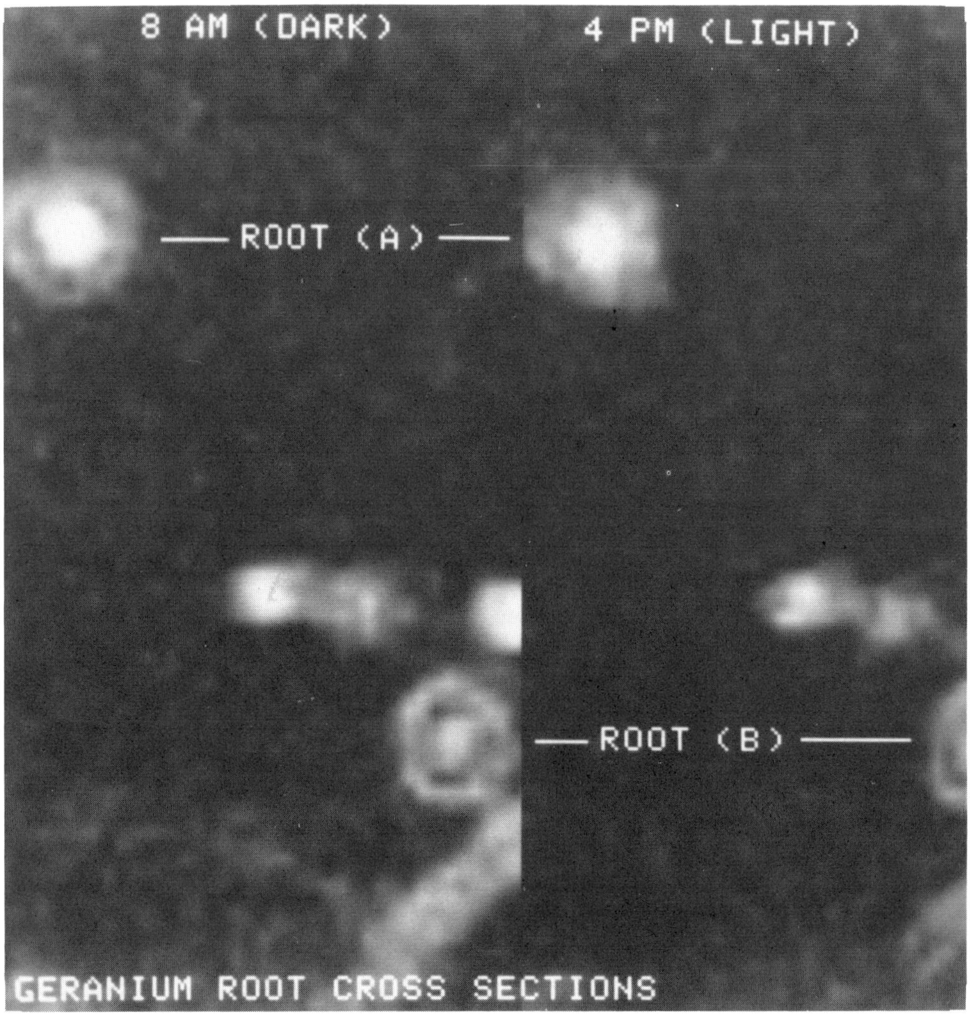

FIGURE 9. A transverse section was acquired through the geranium root system. Root sections are shown at many angles to the acquisition plane. Most sections show a bright central vascular system, dark cortex, and bright epidermal region. Roots A and B from before and during active transpiration. In the dark, root A shows bright regions across the cortex appearing like spokes on a wheel and their function is puzzling, but may be involved with radial transport across the cortex. After 8 h of active transpiration, both root sections show apparent changes in the cortex. Regions similar to these have been observed in roots of several plant species as well as in roots growing in different soil types. (Courtesy of J. Brown.)

μm × 50 μm or better) was recorded from MRI of germinating seeds, roots of seedlings, and stems of young plants.[11] The usefulness of MRI to support plant physiology research is established. In addition to the H_2O-D_2O exchange, the demonstration of Mn^{2+} uptake in maize roots and changes in the distribution of water and lipid in germinating soybeans demonstrate the feasibility of characterizing and visually localizing chemical dynamics within an intact plant.[11]

Crops dependent on irrigation are subject to sodium toxicity. Noninvasive, *in vivo,* real time measurements of sodium transport parameters in millet cells and roots of salt grass (Figure 11) using ^{23}Na MR demonstrated that the rates and amounts of sodium efflux correlate with the ability of these plants to adapt to saline stress.[124] Sodium-23 MRI (Figure 12) of salt grass indicated that the concentration of the salt in the root vascular system approximated the amount of salt in the growth media. Proton MRI (Figure 13) of these same roots, giving 25% better

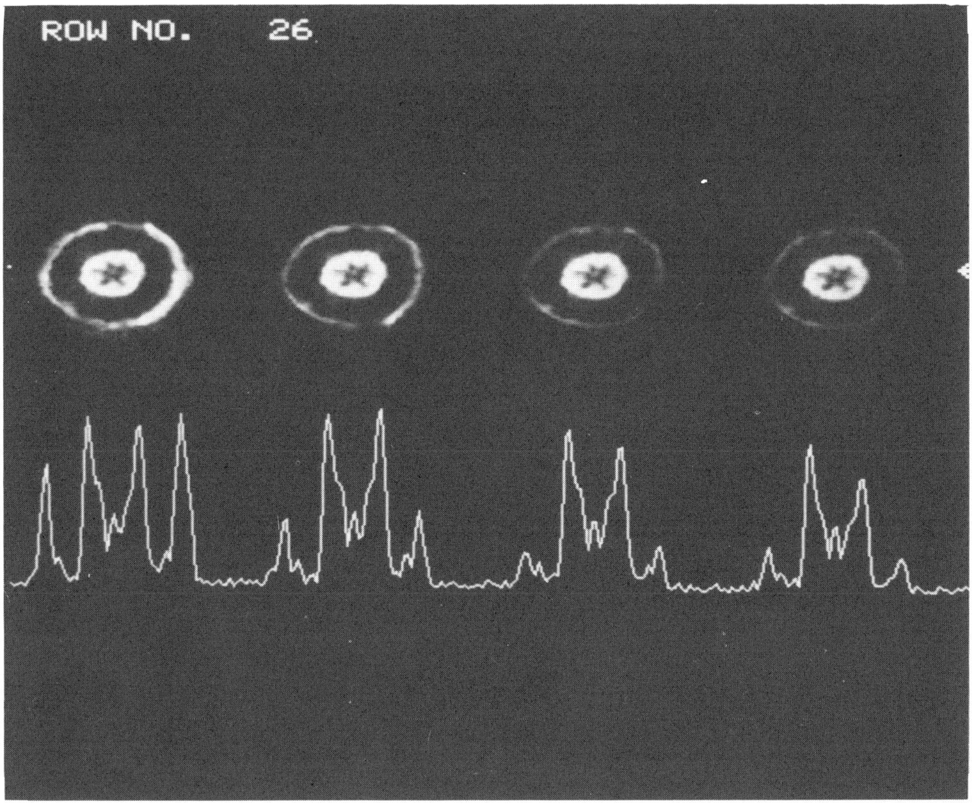

FIGURE 10. Proton MR images of roots of freshly germinated maize seed in a 1.5-mm internal diameter glass tube. Images were obtained using a 3-mm solenoid imaging coil in a Bruker AM200SY spectrometer with a 54-mm bore, 4.7-T magnet. From left to right the images show progressive replacement of water by D_2O. The first image is just before the introduction of D_2O, the second image at the start of D_2O flow, the third image after 4 min and the fourth image after 14 min of D_2O flow. The spectral profiles are from resonance data collected across the root plane indicated by an arrow positioned to the right of the image series. (Courtesy of J. A. B. Lohmann and R. G. Ratcliffe.)

resolution than ^{23}Na, showed that the vascular system was the most reasonable source of the ^{23}Na signal. Concentrations of ^{23}Na as low as 1 m*M* can be detected in 1 min by MR. The 100% natural abundance of ^{23}Na and its magnetogyric ratio of 9.25% of proton makes ^{23}Na a candidate nucleus for plant MR images with a spatial resolution of 2 to 3 mm and a time of image production of about 5 min.[125]

IV. INSECT MRI

The first MRI studies of whole insects *in vivo* could be considered as one-dimensional.[126,127] These studies produce proton spectra that allowed the researchers to discriminate between insect infested and noninfested seeds. Changes in proton peak intensity reflected the rate of insect development. Movement of the insect resulted in chemical shift perturbations over 10 min measurement periods.[127] Image artifacts produced when an MRI specimen moves can provide useful information. Motion distortions of a *Manduca sexta* pupa during an MRI experiment gave evidence for active ventilation of air sacs (Figure 14).[128]

Now that technology has advanced, MR images can clearly depict insects in relation to plants, environment, and infestation. Just as plant roots can be examined undisturbed in their natural soil matrices (see above) using MRI, so can the insects that inhabit the soil be investigated. Thus an avenue of study is opened that will allow understanding of the complexities of the biological

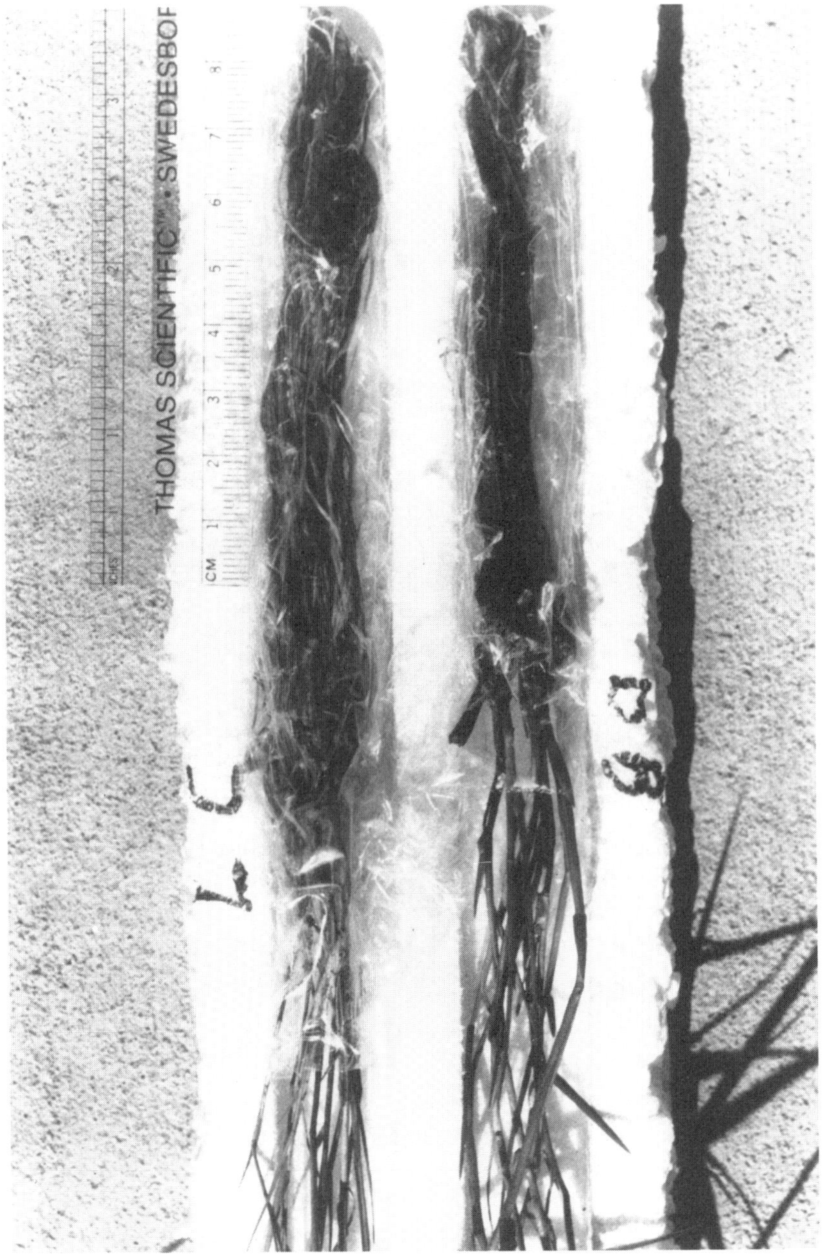

FIGURE 11. Two specimens of hydroponically grown *Distichlis spicata* used in the NMR imaging experiments. The plants were mature individuals that had been grown in Linsmaier-Skoog medium supplemented with 250 m*M* NaCl. The plants were removed from the growth medium, the roots were blotted dry of clinging medium, and then wrapped in plastic film and put into a shaped, styrofoam holder. (Courtesy of L. O. Sillerud, J. W. Heyser, C, H. Han, and M. W. Bitensky.)

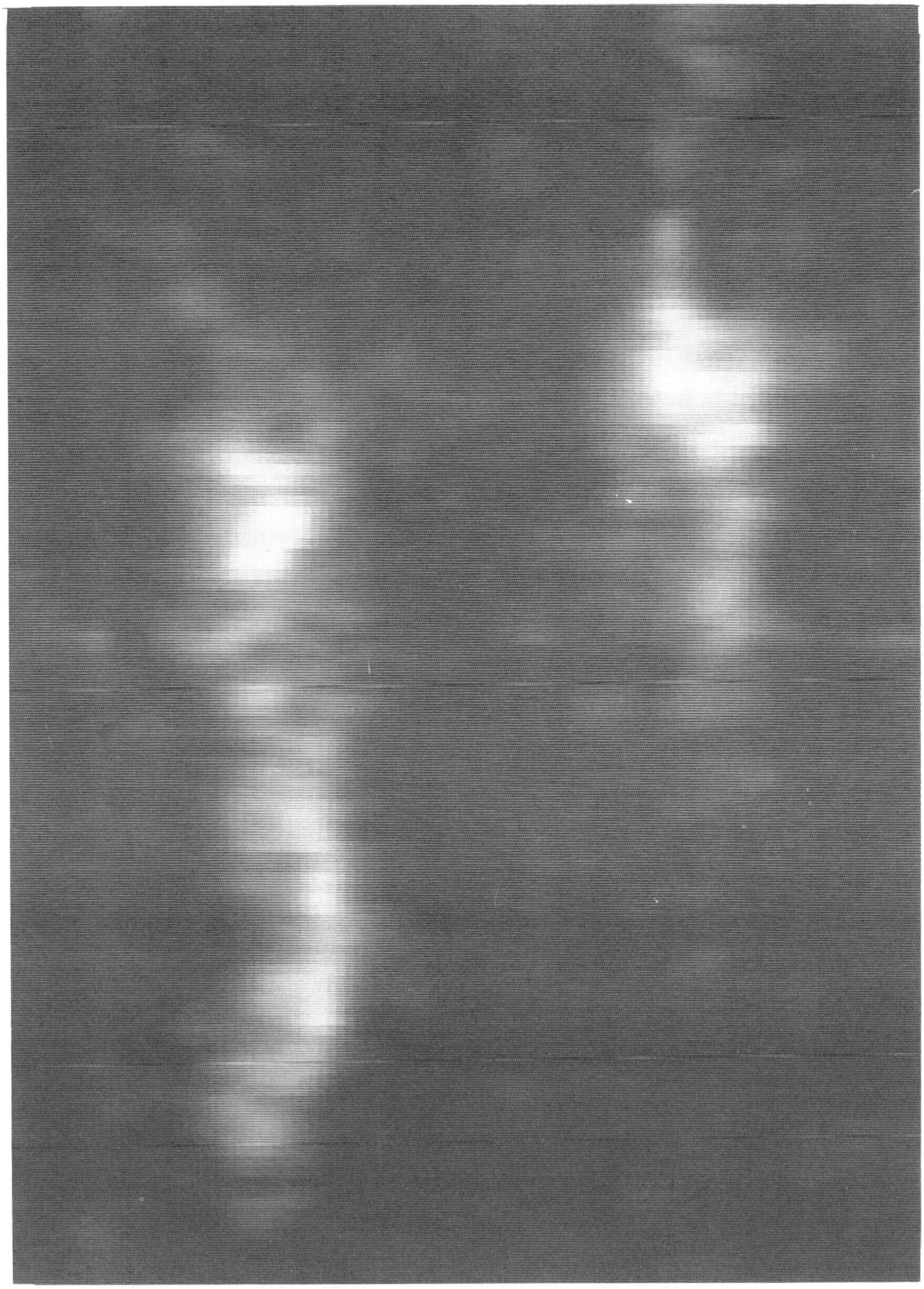

FIGURE 12. Sodium-23 (52.9 MHz) MR images of the root portion of the *D. spicata* plant shown in Figure 7. The image resolution is about 2.5 mm with 128 × 128 pixels. This spin-echo image is the result of 100 averages taken at a rate of 20 per second, for a total imaging time of about 8 min. (Courtesy of L. O. Sillerud, J. W. Heyser, C. H. Han, and M. W. Bitensky.)

processes involving insects, soils, and plant root systems. MRI data can be interpreted in conjuction with X-ray computed tomography (CT). MR imaging, which requires several minutes for data acquisition, can provide information related to metastable physiological

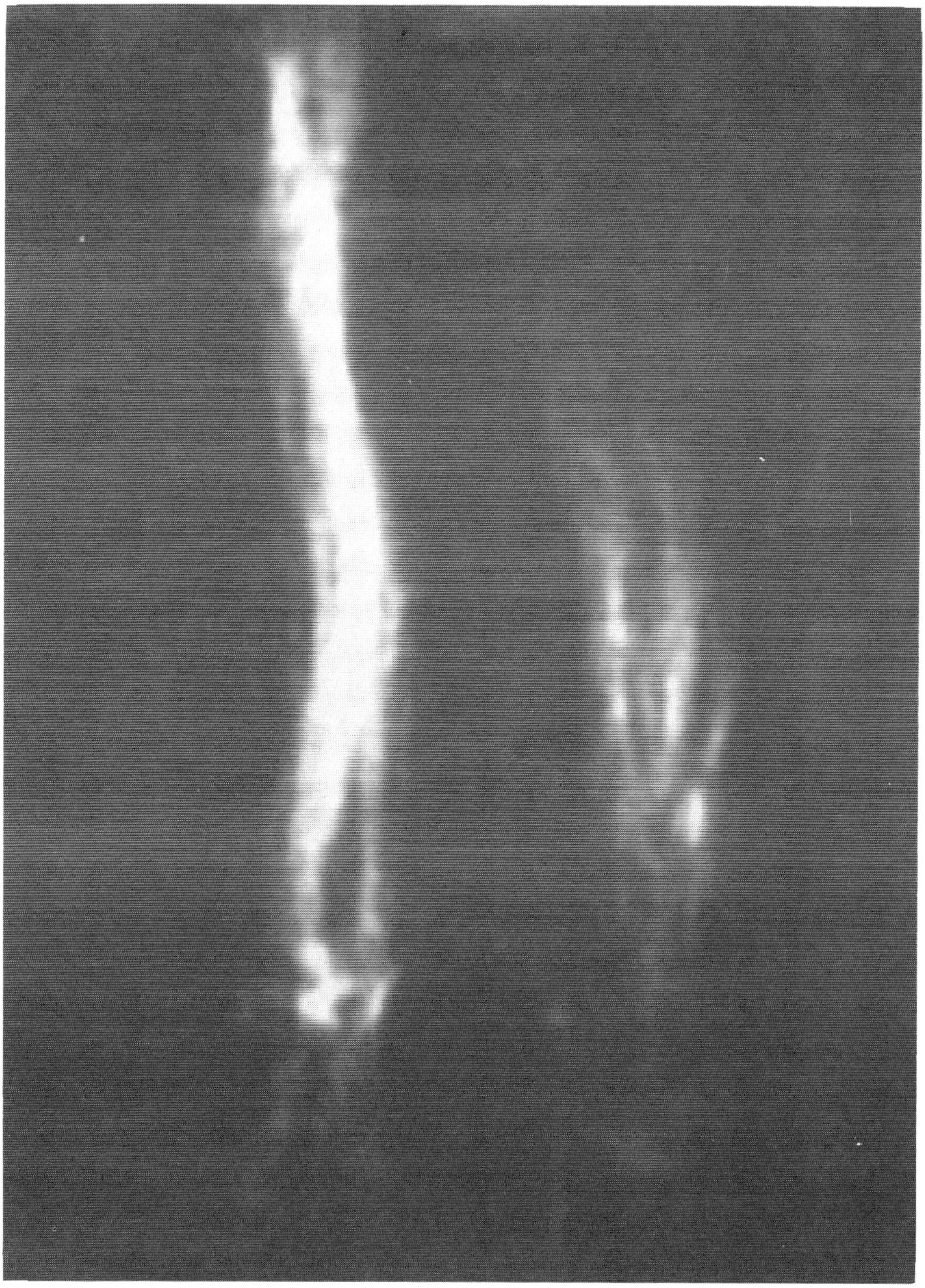

FIGURE 13. A 200-MHz proton NMR image of the same plant sample as shown in Figures 7 and 8. The resolution is about 670 μm/pixel with 256 ¥ 256 pixels comprising the image. The imaging time was about 4 min with a data accumulation rate of about 1 per second. (Courtesy of L. O. Sillerud, J. W. Heyser, C. H. Han, and M. W. Bitensky.)

conditions. By incorporating CT in an MRI study, behavioral characteristics, such as movement, can be evaluated, since CT data can be obtained in microseconds (Figure 15).[129] A 200-MHz spectrometer connected to a 54-mm bore 4.7 T super-conducting magnet combined MRI with localized MR spectroscopy using a 2-D technique.[130-132] This system, made acquisition of

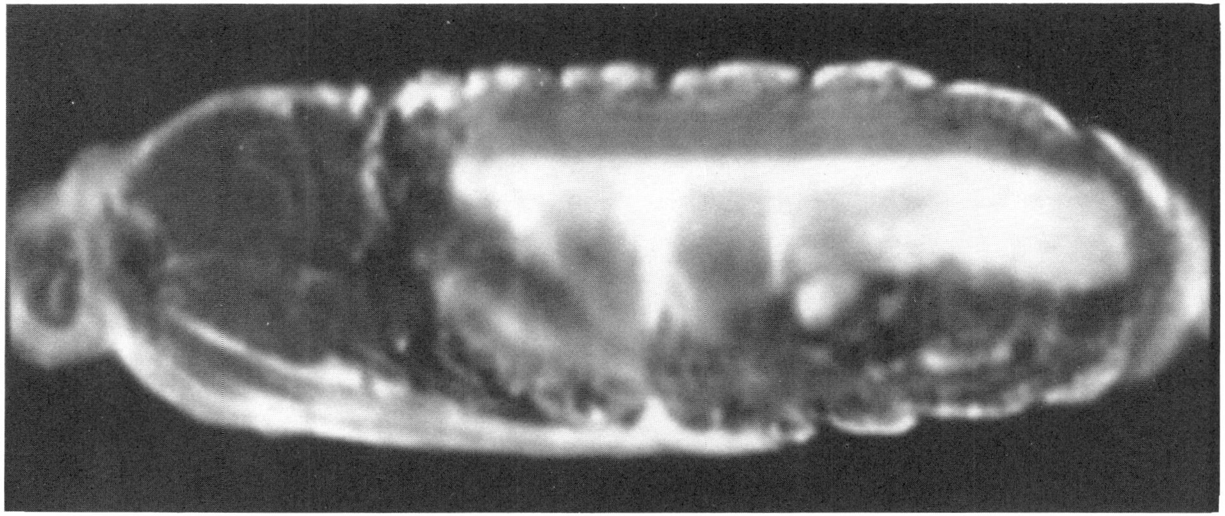

FIGURE 14. The development of a pupal tobacco hornworm, *Manduca sexta* (L.) was nondestructively observed with NMR microscopy. This image was selected from a set of 16 contiguous slices each with an effective thickness of 1.25 mm and in-plane resolution of 100 × 100 μm. Structures including ganglia and testes (moth balls) are clearly seen. Advantages of NMR imaging over standard histological techniques are (1) NMR is noninvasive, (2) NMR distinguishes among tissues which vary in proton content and proton relaxation "environment", (3) it allows the detection of motion within an organism, and (4) it avoids fixation artifacts.[128]

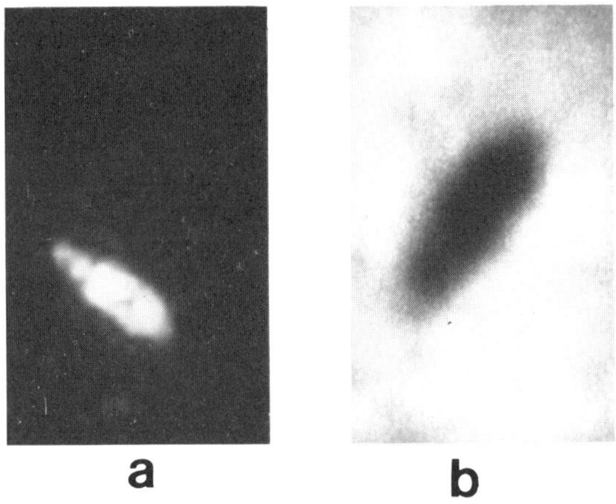

FIGURE 15. (a) Proton image of an adult carabid beetle, *Harpalus pennsylvanicus* in Lakeland sand. Image was obtained using Philips Gyroscan imaging system (0.5 T, 21.3 MHz) and a 30-cm diameter Helmholtz pair of saddle shaped RF coils. (Courtesy of J. Cheshire.) (b) CT image of the same beetle shown in Figure 10(a). CT image was produced using a Siemens Somatom DR3 medical scanner applying a 125 KV, 450 mA-s X-ray beam and detecting with a circular array of 512 fixed scintillation sensors. (Courtesy of J. Cheshire.)

reasonably good images of a developing locust embryo possible (Figure 16). The locust was placed in a 3-mm diameter solenoid coil designed to produce an intense and homogeneous RF field. Separate images of the water and lipid distribution in three to four adjacent sagittal slices through the egg were obtained; the in-plane pixel resolution was approximately 100 μm and the slice thickness 500 μm. Combined with the imaging, spectroscopy of a selected volume within the developing locust was performed (Figure 11 A, B, and C). Proton spectra of selected cubic volumes were examined within the developing locust egg by placing a square cursor over the CRT projected image and using the Image Selected *In Vivo* Spectroscopy technique (ISIS).[1] The MRI volume selection studies permitted measurements of "water" and "lipid" proton changes in discrete body compartments throughout the course of locust embryonic development.

V. CONCLUSION

MRI methods can measure both dynamic processes and static conditions of biological systems. Documented achievements using MRI provide a powerful base for the benefit of agriculture research on animals, plants, and insects. MRI referenced work on animals exceeds the combined plant and insect citations by a power of three. The preponderance of animal MRI work is understandable, since the greatest emphasis in MRI research has come from medical researchers. Many opportunities for using MRI in agricultural research are reported in the preceding pages. Opportunities for MRI of solids, beyond these pages,[133,134] will give researchers spatial data that will enhance their understanding of the chemical structure of agricultural materials and products.

Given these possibilities, the researcher must be aware of the limitations of the MR analytical method. The availability and sensitivity of the biologically interesting nuclei that are detectable by MR vary. Protons, for example, have a high sensitivity and natural abundance in relation to that of ^{13}C. Since high resolution signals are characteristic of the most mobile nuclei in the cellular and extracellular milieu, the researcher must consider their mobility as well as their

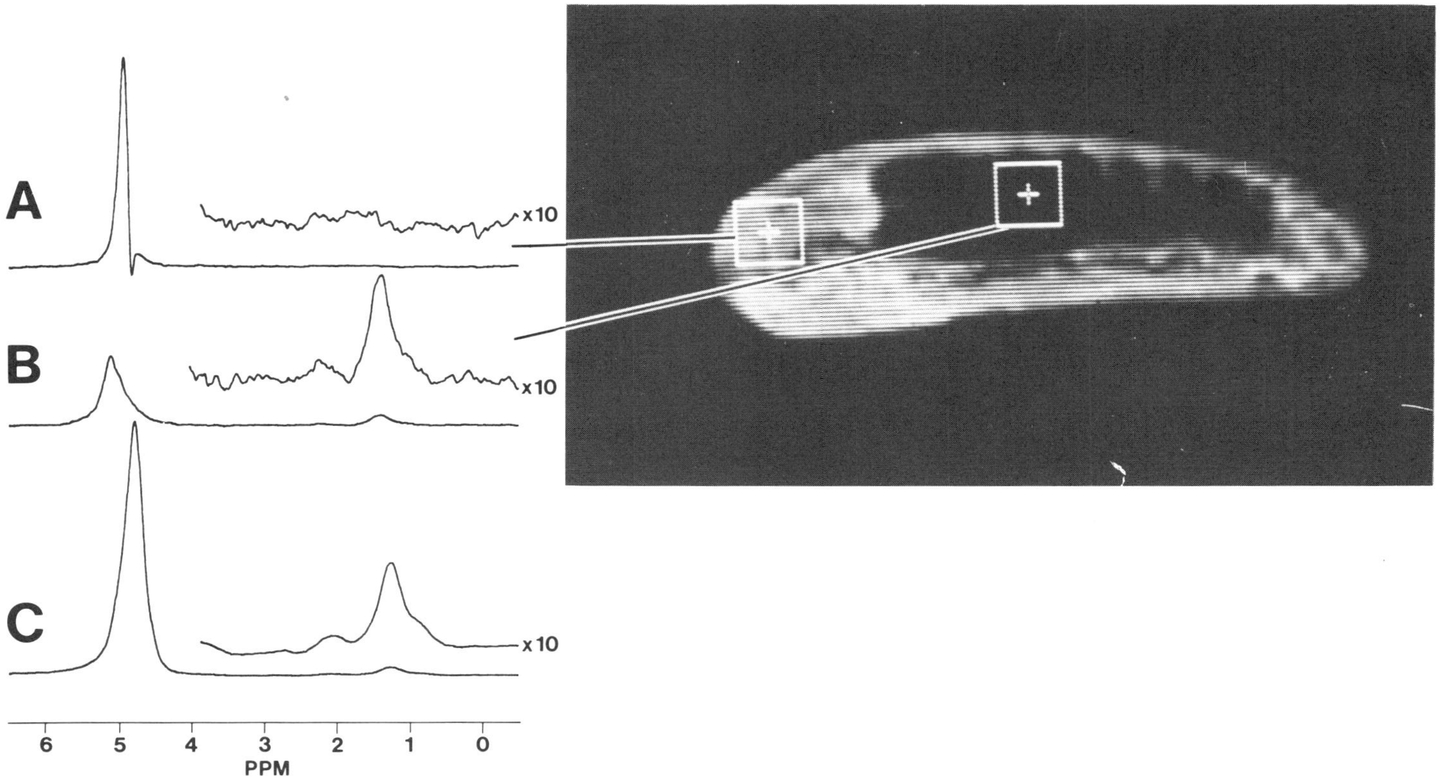

FIGURE 16. Image selected *in vivo* spectrocopy (ISIS) of proton spectra from the image shown indicates that the volume dorsal to the stomatodeum (A) at 66% development is essentially void of lipid while the hemocel (B) contains lipid. Also, the larger line width of the water resonance in (B) as compared with (A) indicates a difference in transverse relaxation time between the two types of tissue. This partly explains the lower intensity of signals from the hemocel as apparent in the image. Note the overall spectrum (C). The volumes of interest were selected with reference to the water images. Consequently, the square cursor accurately delineates the region that contributes to the water intensity in the localized spectrum. Inherent to the chemical shift difference between the water and the lipid signals, a chemical shift artifact exists in the spectra. Due to this artifact, the volume that contributes to the lipid intensity is displaced with respect to the square cursor by a small amount. The size and the position of the selected volume were chosen such as to ensure that the water and lipid signals originate from the same tissue sample. (From Gassner, G. and Lohman, J. A. B., *Proc. Natl. Acad. Sci. U.S.A.*, 84, 5297, 1987. With permission.)

sensitivity and natural abundance. The most critical MRI limitations are the physical processes involved in specimen excitation, data acquisition, and interpretation of results. Perseverance and careful selections of appropriate parameters for the MRI experiment can provide data about real relationships among the many complex chemistries in biology and invaluable advances for agriculture.

ACKNOWLEDGMENTS

The author thanks the figure contributors for this courtesy, his colleagues for their critical reading and criticisms, and a special thanks to Michelle A. Baumgartner for constant diligence and attention to detail in assembling the literature and entering the text into the word processor.

REFERENCES

1. **Ordidge, R. J., Connelly, A., and Lohman, J. A. B.,** Image-selected *in vivo* spectroscopy (ISIS). A new technique for spacially selective NMR spectroscopy, *J. Magn. Reson.*, 66, 283, 1986.
2. **Gadian, D. G.,** *Nuclear Magnetic Resonance and its Applications to Living Systems*, Oxford University Press, New York, 1982.
3. **Campbell, I. D. and Dwek, R. A.,** *Biological Spectroscopy*, Benjamin/Cummings, Menlo Park, CA, 1984.
4. **Dumoulin, C. L.,** Nuclear magnetic resonance imaging, *Spectroscopy*, 2, 32, 1987.
5. **Morris, P. G.,** *Nuclear Magnetic Resonance Imaging in Medicine and Biology*, Clarendon, Oxford, 1986.
6. **Kuchel, P. W.,** NMR of biological samples, *Crit. Rev. Anal. Chem.*, 12, 155, 1981.
7. **Smith, S. L.,** Nuclear magnetic resonance imaging, *Anal. Chem.*, 57, 595A, 1985.
8. **Traficante, D. D.,** NMR concepts, in *NMR Spectroscopy Techniques*, Dybowski, C., Ed., Marcel Dekker, New York, 1987.
9. **Hoult, D. I.,** NMR imaging techniques, *Br. Med. Bull.*, 40, 2, 1984.
10. **Lauterbur, P. C.,** Magnetic resonance zeugmatography, *Pure Appl. Chem.*, 40, 149, 1974.
11. **Connelly, A., Lohman, J. A. B., Loughman, B. C., Quiquampoix, H., and Ratcliffe, R. G.,** High resolution imaging of plant tissues by NMR, *J. Exp. Bot.*, 38, 1713, 1987.
12. **Jones, J. P., Partain, C. L., Mitchell, M. R., Patton, J. A., Stephens, W. H., Price, P. R., Kulkarni, M. V., and James, H. E. J.,** Principles of magnetic resonance, in *Magnetic Resonance Annual*, Kressel, H. Y., Ed., Raven Press, New York, 1985.
13. **Fukushima, E. and Roeder, S. B. W.,** *Experimental Pulse NMR — A Nuts and Bolts Approach*, Addison-Wesley, Reading, MA, 1981.
14. **Derome, A. E.,** Modern NMR techniques for chemistry research, in *Organic Chemistry Series*, Vol. 5, Baldwin, J. E., Ed., 1987.
15. **Hoult, D. I., Hedges, L. K., and Phil, D.,** Spatial localization by rotating frame techniques, *Ann. N.Y. Acad. Sci.*, 508, 366, 1987.
16. General Electric Co., Medical Systems Group, NMR: A Perspective on Imaging, Publication #5484, 30, 1984.
17. **Morris, P. G.,** *Nuclear Magnetic Resonance Imaging in Medicine and Biology*, Clarendon Press, Oxford, 1986.
18. **Garner, W. H., Hilal, S. K., Lee, S.-W., and Spector, A.,** Sodium-23 magnetic resonance imaging of the eye and lens, *Proc. Natl. Acad. Sci. U.S.A.*, 83, 1901, 1986.
19. **Joseph, P. M., Fishman, J. E., Mukherji, B., and Sloviter, H. A.,** *In vivo* 5194F NMR imaging of the cardiovascular system, *J. Comput. Assisted Tomography*, 9, 1012, 1985.
20. **Joseph, P. M., Fishman, J. E., Mukherji, B., and Sloviter, H. A.,** Imaging the vascular system using flourine-19 magnetic resonance imaging, *Magn. Reson. Imaging*, 4, 114, 1986.
21. **Nunnally, R. L., Babcock, E. E., Horner, S. D., and Peshock, R. M.,** Fluorine-19 NMR spectroscopy and imaging investigations of myocardial perfusion and cardiac function, *Magn. Reson. Imaging*, 3, 399, 1985.
22. **Joseph, P. M., Yuasa, Y., Kundel, H. L., Mukherji, B., and Sloviter, H. A.,** Magnetic resonance imaging of flourine in rats infused with artifical blood, *Invest. Radiol.*, 20, 504, 1985.
23. **McFarland, E., Koutcher, J. A., Rosen, B. R., Teicher, B., and Brady, T. J.,** *In vivo* 5194F NMR imaging, *J. Comput. Assisted Tomography*, 9, 8, 1985.

24. **Thomas, S. R., Clark, L. C., Jr., Ackerman, J. L., Pratt, R. G., Hoffmann, R. E., Busse, L. J., Kinsey, R. A., and Samaratunga, R. C.,** Magnetic resonance imaging of the lung liquid perflourocarbons, *J. Comput. Assisted Tomogr.*, 10, 1, 1986.
25. **Bonnemain, B., Lautrow, J., Meyer, D., and Doucet, D.,** Nonradiological contrast media, *Rech Imagirie Med.*, 160, 61, 1987.
26. **Runge, V. M., Clanton, J. A., Wehr, C. J., Partain, C. L., and James, A. E.,** Gated magnetic resonance of acute myocardial ischemia in dogs: applications of multiecho techniques and contrast enhancement with GD DTPA, *Magn. Reson. Imag.*, 3, 255, 1985.
27. **Saha, G. B., Bateson, B., Meaney, T., Go, R. T., MacIntyre, W. J., O'Donnell, W. J., and Feiglin, D. H. I.,** Gadolinium diisopropyliminodiacetic acid a potential contrast agent for magnetic resonance imaging of hepatobiliary system, *J. Nucl. Med.*, 27, 882, 1986.
28. **Greif, W. L., Buxton, R. B., Lauffer, R. B., Saini, S., Stark, D. D., Wedeen, V. J., Rosen, B. R., and Brady, T. J.,** Pulse sequence optimization for MR imaging using a paramagnetic hepatobiliary contrast agent, *Radiology*, 157, 461, 1985.
29. **Jackels, S. C., Kroos, B. R., Hinson, W. H., Karstaedt, N., and Moran, P. R.,** Paramagnetic macrocyclic complexes as contrast agents for magnetic resonance imaging proton nuclear relaxation rate enhancement in aqueous solution and in rat tissues, *Radiology*, 159, 525, 1986.
30. **Taeuber, U., Weinmann, H.-J., Panzer, M., Acksteiner, B., Vollert, B., and Schulze, P. E.,** Whole-body autoradiographic studies in rats with gadolinium-diethylenetriaminepentaacetic acid a new contrast agent for magnetic resonance imaging, *Arzneim. Forsch.*, 36, 1081, 1986.
31. **Brasch, R. C., Nitecki, E. D., Brant-Zawadzki, M. N., Enzmann, D., and Tozer, T. N.,** NSFR: NMR contrast agent for the CNS, *Radiology*, 144, 4, 1982.
32. **Ewy, C. S., Babcock, E. E., and Ackerman, J. J. H.,** Deuterium NMR spin-imaging of deuterium oxide a potential exogenous magnetic resonance imaging label, *Magn. Reson. Imag.*, 4, 407, 1986.
33. **Tscholakoff, D., Higgins, C. B., Sechtem, U., and McNamara, M. T.,** Occlusive and reperfused myocardial infarcts: effect of gadolinium diethylenetriaminepentaacetic-acid on electrocardiographic-gated magnetic resonance imaging, *Radiology*, 160, 515, 1986.
34. **Manabe, Y., Longley, C., and Furmanski, P.,** High-level conjugation of chelating agents into immunoglobins use of an intermediary poly-l-lysine-diethylenetriaminepentaacetic-acid carrier, *Biochim. Biophys. Acta*, 883, 460, 1986.
35. **Marincek, B. and Friedburg, H.,** Calculated T-1 and T-2 magnetic resonance images and phosphorus-31 magnetic resonance spectra in rabbit thigh V-2 carcinoma to monitor the effect of steroid administration, *Invest. Radiol.*, 2, 701, 1985.
36. **Ratkovic, S.,** Paramagnetic ions and their complexes as contrast agents in NMR imaging, *Farm. Vestn. (Ljubljana)*, 38, 195, 1987.
37. **Hilal, S. K., Maudsley, A. A., Perman, W., Silver, A. J., Mawad, M. E., Ganti, S. R., and Sane, P.,** Multi nuclear imaging with NMR, *Radiology*, 144, 945, 1982.
38. **Bartels, G.,** Magnetic resonance spectroscopy for *in vivo* demonstration of biochemical processes within the body, *Electromedica*, 53, 104, 1985.
39. **Maudsley, A. A. and Hilal, S. K.,** Biological aspects of sodium-23 imaging, *Br. Med. Bull.*, 40, 165, 1984.
40. **Alger J. R. and Schulman, R. G.,** High resolution 5134C spectroscopy, *Br. Med. Bull.*, 40, 160, 1984.
41. **Fuller, M. F., Foster, M. A., and Hutchinson, J. M. S.,** Estimation of body fat by NMR imaging, *Proc. Nutr. Soc.*, 44, 108A, 1985.
42. **Mitchell, A. D. and Wang, P.,** personal communication, 1988.
43. **Braun, M., Jung, W. I., Lutz, O., and Oeschey, R.,** Selective non-excitation of water on fat protons in magnetic resonance imaging, *Z. Naturforsch, Teil A*, 42, 1391, 1987.
44. **Blatter, D. D., Morris, A. H., Ailion, D. C., Cutillo, A. G., and Case, T. A.,** Asymmetric spin echo sequences a simple new method for obtaining NMR ^{1}H spectral images, *Invest. Radiol.*, 20, 845, 1985.
45. **Mansfield, P. and Maudsley, A. A.,** Line scan proton spin imaging in biological structures by NMR, *Phys. Med. Biol.*, 21, 847, 1976.
46. **Mansfield, P. and Maudsley, A. A.,** Planar spin imaging by NMR, *J. Magn. Reson.*, 27, 101, 1977.
47. **Damadian, R.,** Tumor detection by nuclear magnetic resonance, *Science*, 171, 1151, 1971.
48. **Hinshaw, W. S., Andrew, E. R., Bottomley, P. A., Holland, G. N., Moore, W. S., and Worthington, B. S.,** Display of cross sectional anatomy by NMR imaging, *Br. J. Radiol.*, 51, 273, 1978.
49. **Bone, S. N., Johnson, G. A., and Thompson, D. V. M.,** Three-dimensional magnetic resonance microscopy of the developing chick embryo, *Invest. Radiol.*, 21, 782, 1986.
50. **Pykett, I. L. and Mansfield, P.,** A line scan image study of a tumorous rat leg by NMR, *Phys. Med. Biol.*, 23, 961, 1978.
51. **Stark, D. D., Bass, N. M., Moss, A. A., Bacon, B. R., McKerrow, J. H., Cann, C. E., Goldberg, H. I., and Brito, A.,** NMR imaging of experimentally induced liver disease, *Radiology*, 148, 743, 1983.
52. **Hamlin, D. J., Ackerman, N., Kaude, J. V., Fitzsimmons, J. R., and Gaskin, J. M.,** Magnetic resonance imaging of renal abscess in an experimental animal model, *Acta Radiol. Diagn.*, 26, 315, 1985.

53. **Cahill, P. T., Peterson, M., Knowles, J. R., and Kneeland, B.,** MR imaging used to follow the course of I-131 therapy for feline hyperthyroidism, *Radiology*, 153, 208, 1984.
54. **Cutillo, A. G., Morris, A. H., Ailion, D. C., Durney, C. H., Case, T. A., Akhtari, M., and Watanabe, F.,** Determination of lung water distribution by NMR imaging in pulmonary edema, *Am. Rev. Respir. Dis.*, 133, A19, 1986.
55. **Hayes, C. E., Case, T. A., Ailion, D. C., Morris, A. H., Cutillo, A., Blackburn, C. W., Durney, C. H., and Johnson, S. A.,** Lung water quantitation by NMR imaging, *Science*, 216, 1313, 1982.
56. **Runge, V. M., Foster, M. A., Clanton, J. A., Smith, F. W., Lukehart, C. M., Hutchinson, J. M. S., Mallard, J. R., Partain, C. L., and James, A. E., Jr.,** Particulate oral NMR contrast agents, *Int. J. Nucl. Med. Biol.*, 5, 419, 1985.
57. **Saini, S., Stark, D. D., Hahn, P. F., Bousquet, J.-C., Introcasso, J., Wittenberg, J., Brady, T. J., and Ferrucci, J. T., Jr.,** Ferrite particles a superparamagnetic MR contrast agent for enhanced detection of liver carcinoma, *Radiology*, 162, 217, 1987.
58. **Saini, S., Stark, D. D., Hahn, P. F., Wittenberg, J., Brady, T. J., and Ferrucci, J. T., Jr.,** Ferrite particles a superparamagnetic MR contrast agent for the reticuloendothelial system, *Radiology*, 162, 211, 1987.
59. **Naruse, S., Hoarikawa, Y., Tanaka, C., Hirakawa, K., Nishikawa, H., and Yoshizaki, K.,** Significance of proton relaxation time measurement in brain edema cerebral infarction and brain tumors, *Magn. Reson. Imaging*, 4, 293, 1986.
60. **Daniels, D. L., Pech, P., Moulder, J., Pojunas, K., Williams, A. L., and Haughton, V. H.,** Characterization of cerebral radiation changes with magnetic resonance imaging, *Radiology*, 153, 142, 1984.
61. **Johnson, G. A., Gewalt, S., and Thompson, M. B.,** Ultra high magnetic resonance imaging for small animal models, *Magn. Reson. Imag.*, 3, 178, 1985.
62. **Keller, A. M., Peshock, R. M., Malloy, C. R., Fitzgerald, C. P., Buja, L. M., Nunnally, R. L., Parkey, R. W., and Willerson, J. T.,** In vivo measurement of myocardial mass using magnetic resonance imaging. *Am. Heart Assoc. Circ. Suppl.*, 114, 489, 1985.
63. **Keller, A. M., Peshock, R. M., Malloy, C. R., Buja, I. M., Nunnally, R., Parkey, R. W., and Willerson, J. T.,** *In vivo* measurement of myocardial mass using NMR imaging, *J. Am. Coll. Cardiol.*, 8, 113, 1986.
64. **Lui, P., Leavitt, M., Southern, J., Thompson, R. C., Johnston, D. L., Brady, T. J., and Okada, R. D.,** Validation of left ventricular mass and volumes calculated from computerized 3-D reconstruction of magnetic resonance surface coil tomograms, *Am. Heart Assoc. Circ. Suppl.*, 114, 490, 1985.
65. **Khaw, B. A., Torchilin, V. P., Klibanov, A. L., Gold, H. K., Yasuda, T., Smirnov, V. N., and Haber, E.,** Modification of monoclonal antibody with diethylenetriaminepentaactic-acid linked synthetic polymers possible application in magnetic resonance imaging, *Am. Heart Assoc. Monogr.*, 114, 111, 1985.
66. **Unger, E. C., Totty, W. G., Neufeld, D. M., Otsuka, F. L., Murphy, W. A., Welch, M. S., Connett, J. M., and Philpott, G. W.,** Magnetic resonance imaging using gadolinium labeled monoclonal antibody, *Invest. Radiol.*, 20, 693, 1985.
67. **Anderson-Berg, W. T., Strand, M., Lempert, T. E., Rosenbaum, A. E., and Joseph, P. M.,** NMR and gamma camera tumor imaging using gadolinium-labeled monoclonal antibodies, *J. Nucl. Med.*, 27, 829, 1986.
68. **Hall, T. S., Baumgartner, W. A., Borkon, A. M., Lafrance, N. D., Traill, T. A., Norris, S., Hutchins, G. M., Brawn, J., and Reitz, B. A.,** Diagnosis of acute cardiac rejection with antimyosin monoclonal antibody phosphorous NMR imaging two-dimensional echocardiography and endocardial biopsy, *J. Heart Transplant*, 5, 419, 1986.
69. **Ordidge, R. J., Mansfield, P., Doyle, M., and Coupland, R. E.,** Real time movie images by NMR, *Br. J. Radiol.*, 55, 729, 1982.
70. **Wedeen, V. J., Meuli, R. A., Edelman, R. R., Geller, S. C., Frank, L. R., Brady, T. J., and Rosen, B. R.,** Projective imaging of pulsatile flow with magnetic resonance, *Science*, 230, 946, 1985.
71. **Florentine, M., Grosskreutz, C., Chang, W., Ehrhardt, J., Dunn, V., Skorton, D. J., and Marcus, M.,** Gated magnetic resonance imaging of the *in vivo* heart: determination of left ventricular mass in dogs, *Pediatr. Res.*, 19, 127A, 1985.
72. **Toyo-oka, T., Umeda, M., Eguchi, K., Nagayama, K., and Hosoda, S.,** Combined in vivo study of 514H-NMR imaging and EKG-gated 5314P-NMR spectroscopy in rat cardiac cycle, *Am. Heart Assoc. Circ. Suppl.*, 114, 492, 1985.
73. **Herfkens, R. J., Hedlund, L. W., Dietz, J., and Coleman, R. E.,** The pulsatile nature of blood flow by NMR imaging, *Invest. Radiol.*, 20, S48, 1985.
74. **Florentine, M. S., Grosskreutz, C. L., Chang, W., Hartnett, J. A., Dunn, V. D., Ehrhardt, J. C., Fleagle, S. R., Collins, S. M., Marcus, M. L., and Skorton, D. J.,** Measurement of left ventricular mass *in vivo* using gated NMR imaging, *J. Am. Coll. Cardiol.*, 8, 107, 1986.
75. **Maddahi, J., Crues, J., Berman, D., Mericle, J., Becerra, A., Garcia, E., Henderson, R., and Bradley, W.,** Validation of magnetic resonance imaging for noninvasive quantification of left ventricular myocardial mass, *Am. Heart Assoc. Circ. Suppl.*, 114, 491, 1985.
76. **Delayre, J. L., Ingwall, J. S., Malloy, C., and Fossel, E. T.,** Gated sodium-23 NMR image of an isolated perfused working rat heart, *Science*, 212, 935, 1981.

77. **Doyle, M., Rzedzian, R., Mansfield, P., and Coupland, R. E.,** Dynamic NMR cardiac imaging in a piglet, *Br. J. Radiol.*, 56, 925, 1983.
78. **Haase, A., Matthaei, D., Haenicke, W., and Frahm, J.,** Dynamic digital subtraction imaging using fast low-angle shot magnetic resonance movie sequences, *Radiology*, 160, 537, 1986.
79. **Frahm, J., Haase, A., Matthaei, D., Haenicke, W., and Merboldt, K. D.,** NMR imaging today: images, specificity, and movies, *Biol. Chem., Hoppe-Seylers Z. Physiol. Chem.*, 366, 787, 1985.
80. **Matthaei, D., Frahm, J., Haase, A., Haenicke, W., and Merboldt, K.-D.,** Three-dimensional flash magnetic resonance imaging of thorax and abdomen without triggering or gating, *Magn. Reson. Imag.*, 4, 381, 1986.
81. **Frahm, J.,** Rapid flash NMR imaging, *Naturwissenschaften*, 74, 415, 1987.
82. **Haase, A., and Matthaei, D.,** Biophysical medical aspects of FAST NMR-imaging, *Med. Prog. Technol.*, 13, 57, 1987.
83. **Latack, J. T., Gabrielson, T. O., Knake, J. E., Gebarski, S. S., Whitaker, J. B., and Yang, P. J.,** Differentiation of malignancy from normal postoperative brain changes by MR imaging, *Radiology*, 153, 142, 1984.
84. **Aherne, T., Yee, E. S., Tscholakoff, D., Gollin, G., Higgins, C. B., and Ebert, P. A.,** A noninvasive in vivo study of cardiac rejection using magnetic resonance imaging (MRI), *J. Cardiovasc. Surgery*, 26, 77, 1985.
85. **Aherne, T., Tscholakoff, D., Finkbeiner, W., Sechtem, U., Derugin, N., Yee, E., and Higgins, C. B.,** Magnetic resonance imaging of cardiac transplants: the evaluation of rejection of cardiac allografts with and without immunosuppression, *Circulation*, 74, 145, 1986.
86. **Terrier, F., Hricak, H., Revel, D., Alpers, C., Levine, J., and Higgins, C. B.,** *In vitro* T-1 and T-2 measurements of the spleen in immune reaction: experimental study in the rat, *Radiology*, 153, 208, 1984.
87. **Terrier, F., Hricak, H., Revel, D., Alpers, C., Bretan, P., Ehrman, R. L., and Feduska, N. J.,** Magnetic resonance imaging in the diagnosis of acute renal allograft rejection and its differentiation from acute tubular necrosis. Experimental study in the dog, *Invest. Radiol.*, 20, 617, 1985.
88. **Teskey, G. C., Ossenkopp, K.-P., Prato, F. S., and Kavaliers, M.,** Exposure to magnetic resonance imaging attenuates fentanyl-induced analgesia in mice, *Soc. Neurosci. Abstr.*, 12, 909, 1986.
89. **Ossenkopp, K. P., Kavaliers, M., Prato, F. S., Teskey, G. C., Sestini, E., and Hirst, M.,** Exposure to nuclear magnetic resonance imaging procedure attenuates morphine-induced analgesia in mice, *Life Sci.*, 37, 1507, 1985.
90. **Ossenkopp, K. P., Kavaliers, M., Prato, F. S., and Hirst, M.,** Effects of magnetic fields from NMR imaging and other sources on opioid analgesia in mice, *Fed. Proc., Fed. Am. Soc. Exp. Biol.*, 43, 748, 1985.
91. **Mardini, I. A., McCarter, R. J. M., and Fullerton, G. D.,** NMR relaxation times of skeletal muscle dependence on fiber type and diet, *Magn. Reson. Imag.*, 4, 393, 1986.
92. **Kang, Y. S., Pope, C. F., and Gore, J. C.,** Alterations in magnetic resonance relaxation of normal canine gallbladder bile during fasting, *Magn. Reson. Imag.*, 4, 399, 1986.
93. **Demas, B. E., Hricak, H., Moseley, M., Wall, S. D., Moon, K., Goldberg, H. I., and Margulis, A. R.,** Gallbladder bile: An experimental study in dogs using MR imaging and proton MR spectroscopy, *Radiology*, 157, 453, 1985.
94. **Saunders, R. D. and Smith, H.,** Safety aspects of NMR clinical imaging, *Br. Med. Bull.*, 40, 148, 1984.
95. **Athey, T. W.,** personal communication, 1988.
96. **Czerski, P. and Athey, T. W.,** Safety of magnetic resonance in *in vivo* diagnostic examinations: theoretical and clinical considerations, Bureau of Radiological Health, U.S. Food and Drug Administration, September 24, 1987.
97. **New, P. F. J., Brady, T. J., Pykett, I. L., Buonanno, F. S., Kistler, J. P., Burt, C. T., Hinshaw, W. S., Davis, K. R., Taveras, J. M., Newhouse, J. H., Goldman, M. R., and Pohost, G. M.,** NMR imaging artifacts and potential hazards of ferromagnetic and nonferromagnetic surgical and dental materials and devices, *Radiology*, 144, 945, 1982.
98. **Davis, P. L., Crooks, S. L., Arakawa, M., McRee, R., Kaufmann, L., and Margulis, A. R.,** Potential hazards in NMR imaging: heating effects of changing magnetic fields and RF fields on small metallic implants, *Am. J. Roentgenol.*, 137, 857, 1981.
99. **Alger, J. R.,** Decoupler heating with surface coils at 1.9 T, *Ann N.Y. Acad. Sci.*, 508, 414, 1987.
100. **McRobbie, D. and Foster, M. A.,** Pulsed magnetic field exposure during pregnancy and implications for NMR foetal imaging: a study with mice, *Magn. Reson. Imag.*, 3, 231, 1985.
101. **Hilbert, D. M., Zawieja, D., Schaefer, D. J., and Barber, B. J.,** Effect of magnetic resonance imaging on body temperatures of sheep, *Fed. Proc., Fed. Am. Soc. Exp. Biol.*, 44, 1194, 1985.
102. **Prato, F. S., Teskey, G. C., Ossenkopp, K.-P., Innis, N. K., and Harris, P.,** Long-term effects of exposure to magneitc resonance imaging no significant changes in survivability, body growth, organ weights, and stress related hormone levels in male rats, *Soc. Neurosci. Abstr.*, 12, 1484, 1986.
103. **Innis, N. K., Ossenkopp, K.-P., Prato, F. S., and Sestini, E.,** Behavioral effects of exposure to NMR imaging. II. Spatial memory tests, *Magn. Reson. Imag.*, 4, 281, 1986.

104. **Ossenkopp, K.-P., Innis, N. K., Prato, F. S., and Sestini, E.,** Behavioral effects of exposure to NMR imaging. I. Open-field behavior and passive avoidance learning in rats, *Magn. Reson. Imag.*, 4, 275, 1986.
105. **Sacks, E., Worgul, B. V., Merriam, G. R., Jr., and Hilal, S.,** The effects of NMR imaging on ocular tissues, *Arch. Opthalmol.*, 104, 890, 1986.
106. **Hinshaw, W. S., Bottomley, P. A., and Holland, G. N.,** A demonstration of the resolution of NMR imaging in biological systems, *Experientia*, 35, 1268, 1979.
107. **Eccles, C. D., Callaghan, P. R., and Jenner, C. F.,** Measurement of the self-diffusion coefficient of water as a function of position in wheat grain using nuclear magnetic resonance imaging, *Biophys. J.*, 53, 77, 1988.
108. **Brown, J. V., Johnson, G. A., and Kramer, P. J.,** *In vivo* magnetic resonance microscopy of changing water content in *Pelargonium hortorum* roots, *Plant Physiol.*, 82, 1158, 1986.
109. **Johnson, G. A., Brown, J., and Kramer, P. J.,** Magnetic resonance microscopy of changes in water content in stems of transpiring plants. *Proc. Natl. Acad. Sci. U.S.A.*, 84, 2752, 1987.
110. **Eccles, C. D. and Callaghan, P. T.,** High resolution imaging. The NMR microscope, *J. Magn. Reson.*, 68, 393, 1986.
111. **Brown, J.,** personal communication, 1988.
112. **Hall, L. D., Rajanayagam, V., Stewart, W. A., and Steiner, P. R.,** Magnetic resonance imaging of wood, *Can. J. For. Res.*, 16, 2, 423, 1986.
113. **Wang, P. C. and Chang, S. J.,** Nuclear magnetic resonance imaging of wood, *Wood Fiber*, 18, 308, 1986.
114. **Bottomley, P. A., Rogers, H. H., and Foster, T. H.,** NMR imaging shows water distribution and transport in plant root systems *in situ*, *Proc. Natl. Acad. Sci. U.S.A.*, 83, 87, 1986.
115. **Rogers, H. H., Bottomley, P. A., and Foster, T. H.,** Application of nuclear magnetic resonance imaging to plant root studies, *Proc. Intl. Conf. Soil Dynamics*, 5, 1152, 1985.
116. **Omasa, K., Onoe, M., and Yamada, H.,** NMR Imaging for measuring root system and soil water content, *Environ. Control Biol.*, 23, 99, 1985.
117. **Rogers, H. H. and Bottomley, P. A.,** *In situ* nuclear magnetic resonance imaging of roots: influence of soil type, ferromagnetic particle content, and soil water, *Agron. J.*, 79, 957, 1987.
118. **Ikeda, T., Fukaya, T., Nomura, Y., and Yoshikawa, K.,** NMR imaging of the paranasal sinuses, pharynx, and oral cavity, *Pract. Otol. Kyoto*, 78, 513, 1985.
119. **Roberts, J. K. M.,** Study of plant metabolism *in vivo* using NMR spectroscopy, *Annu. Rev. Plant Physiol.*, 35, 375, 1984.
120. **Schneider, W., Dreher, W., Boernert, P., and Gaessler, A.,** NMR tomography of plants, *Stud. Biophys.*, 121, 153, 1987.
121. **Brown, J. M., Fonteno, W. C., Cassel, D. K., and Johnson, G. A.,** Computer tomography and magnetic resonance imaging of water and root distribution for container-grown plants, *HortScience*, 21, 738, 1986.
122. **Brown, J. M., Johnson, G. A., and Kramer, P. J.,** *In vivo* magnetic resonance microscopy of changing water content in *Pelargonium hortorum* roots, *Plant Physiol.*, 82, 1158, 1986.
123. **Lohman, J. A. B. and Ratcliff, R. G.,** personal communication, 1986.
124. **Sillerud, L. O., Heyser, J. W., Han, C. H., and Bitensky, M. W.,** Application of carbon-13 and sodium-23 NMR in the study of plants, animal and human cells, in *NMR in Living Systems*, Axenrod, T. and Ceccarelli, G., Eds., D. Reidel, Dordrecht, The Netherlands, 1986.
125. **Sillerud, L. O.,** personal communication, 1987.
126. **Street, M. W.,** Nuclear magnetic resonance for determining hidden insect infestation in stored grains, *J. GA Entomol. Soc.*, 6, 249, 1971.
127. **Chambers, J., McKevitt, N. J., and Stubbs, M. R.,** Nuclear magnetic resonance spectroscopy for studying the development and detection of the grain weevil, *Sitophilus granarius*, (L.) (*Coleoptera:* Curculionidae), within wheat kernels, *Bull. Entomol. Res.*, 74, 707, 1984.
128. **Conner, W. E., Johnson, G. A., Cofer, G. P., and Dittrich, K.,** Magnetic resonance microscopy: *in vivo* sectioning of a developing insect, *Experientia*, 44, 11, 1988.
129. **Cheshire, J.,** personal communication, 1987.
130. **Lohman, J. A. B. and Gassner, G.,** NMR micro-imaging at high magnetic fields, Recent Developments in Medical NMR Symposium, Antwerp, Belgium, December 13 to 16, 1985.
131. **Lohman, J. A. B. and Gassner, G.,** Proton NMR *in vivo* imaging: a tool for studying embryo development of small organisms, *Ann. N.Y. Acad Sci.*, 508, 435, 1987
132. **Gassner G. and Lohman, J. A. B.,** Combined proton NMR imaging and spectral analysis of locust embryonic development, *Proc. Natl. Acad. Sci. U.S.A.*, 84, 5297, 1987.
133. **Chingas, G. C., Miller, J. B., and Garroway, A. N.,** NMR images of solids, *J. Magn. Reson.*, 66, 530, 1986.
134. **Cory, D. G., van Os, J. W. M., and Veeman, W. S.,** NMR images of rotating solids, *J. Magn. Reson.*, 76, 543, 1988.

Index

INDEX

A

B

C

D

E

F

N

O

P

Q

R

S

T

U

V

W